A FIRST COURSE IN NUMERICAL ANALYSIS

INTERNATIONAL SERIES IN PURE AND APPLIED MATHEMATICS

E. H. Spanier, G. Springer, and P. J. Davis, Consulting Editors

AHLFORS: Complex Analysis
BENDER AND ORSZAG: Advanced Mathematical Methods for Scientists and Engineers
BUCK: Advanced Calculus
BUSACKER AND SAATY: Finite Graphs and Networks
CHENEY: Introduction to Approximation Theory
CHESTER: Techniques in Partial Differential Equations
CODDINGTON AND LEVINSON: Theory of Ordinary Differential Equations
CONTE AND DE BOOR: Elementary Numerical Analysis: An Algorithmic Approach
DENNEMEYER: Introduction to Partial Differential Equations and Boundary Value Problems
DETTMAN: Mathematical Methods in Physics and Engineering
GOLOMB AND SHANKS: Elements of Ordinary Differential Equations
HAMMING: Numerical Methods for Scientists and Engineers
HILDEBRAND: Introduction to Numerical Analysis
HOUSEHOLDER: The Numerical Treatment of a Single Nonlinear Equation
KALMAN, FALB, AND ARBIB:Topics in Mathematical Systems Theory
LASS: Vector and Tensor Analysis
MCCARTY: Topology: An Introduction with Applications to Topological Groups
MONK: Introduction to Set Theory
MOORE: Elements of Linear Algebra and Matrix Theory
MOURSUND AND DURIS: Elementary Theory and Application of Numerical Analysis
PEARL: Matrix Theory and Finite Mathematics
PIPES AND HARVILL: Applied Mathematics for Engineers and Physicists
RALSTON AND RABINOWITZ: A First Course in Numerical Analysis
RITGER AND ROSE: Differential Equations with Applications
RITT: Fourier Series
RUDIN: Principles of Mathematical Analysis
SHAPIRO: Introduction to Abstract Algebra
SIMMONS: Differential Equations with Applications and Historical Notes
SIMMONS: Introduction to Topology and Modern Analysis
SNEDDON: Elements of Partial Differential Equations
STRUBLE: Nonlinear Differential Equations

SECOND EDITION

A FIRST COURSE IN NUMERICAL ANALYSIS

ANTHONY RALSTON
Department of Computer Science
State University of New York at Buffalo

PHILIP RABINOWITZ
Division of Applied Mathematics
Weizmann Institute, Rehovot, Israel

McGRAW-HILL BOOK COMPANY

New York St. Louis San Francisco Auckland Bogotá Düsseldorf
Johannesburg London Madrid Mexico Montreal New Delhi
Panama Paris São Paulo Singapore Sydney Tokyo Toronto

To our children

Jonathan, Geoffrey, Steven, Elizabeth

Yehoshua, Yaffa, Yehuda, Yona, Yoel

A FIRST COURSE IN NUMERICAL ANALYSIS

1234567890 DODO 78321098

This book was set in Times Roman. The editors were A. Anthony Arthur, Carol Napier, and Shelly Levine Langman; the production supervisor was Milton J. Heiberg. The drawings were done by ANCO/Boston.
R. R. Donnelley & Sons Company was printer and binder.

Library of Congress Cataloging in Publication Data

Ralston, Anthony.
A first course in numerical analysis.

(International series in pure and applied mathematics)
Includes bibliographies and index.
1. Numerical analysis. I. Rabinowitz, Philip, joint author.
II. Title.
QA297.R3 1978 519'.4 77-10643
ISBN 0-07-051158-6

CONTENTS

Preface xiii
Notation xvii

CHAPTER ONE INTRODUCTION AND PRELIMINARIES 1

1.1 What Is Numerical Analysis? 1
1.2 Sources of Error 2
1.3 Error Definitions and Related Matters 4

1.3-1 Significant Digits **1.3-2** Error in Functional Evaluation
1.3-3 Norms

1.4 Roundoff Error 9

1.4-1 The Probabilistic Approach to Roundoff; A Particular Example

1.5 Computer Arithmetic 12

1.5-1 Fixed-Point Arithmetic **1.5-2** Floating-Point Numbers
1.5-3 Floating-Point Arithmetic **1.5-4** Overflow and Underflow
1.5-5 Single- and Double-Precision Arithmetic

1.6 Error Analysis 20

1.6-1 Backward Error Analysis

1.7 Condition and Stability 22

Bibliographic Notes 24
Bibliography 24
Problems 24

CHAPTER TWO APPROXIMATION AND ALGORITHMS 31

2.1 Approximation 31

2.1-1 Classes of Approximating Functions **2.1-2** Types of Approximations **2.1-3** The Case for Polynomial Approximation

2.2 Numerical Algorithms 39
2.3 Functionals and Error Analysis 42
2.4 The Method of Undetermined Coefficients 44

Bibliographic Notes 46
Bibliography 46
Problems 47

CHAPTER THREE INTERPOLATION 52

3.1 Introduction 52
3.2 Lagrangian Interpolation 54
3.3 Interpolation at Equal Intervals 56

3.3-1 Lagrangian Interpolation at Equal Intervals **3.3-2** Finite Differences

3.4 The Use of Interpolation Formulas 63
3.5 Iterated Interpolation 66
3.6 Inverse Interpolation 68
3.7 Hermite Interpolation 70
3.8 Spline Interpolation 73
3.9 Other Methods of Interpolation; Extrapolation 78

Bibliographic Notes 79
Bibliography 80
Problems 81

CHAPTER FOUR NUMERICAL DIFFERENTIATION, NUMERICAL QUADRATURE, AND SUMMATION 89

4.1 Numerical Differentiation of Data 89
4.2 Numerical Differentiation of Functions 93
4.3 Numerical Quadrature: The General Problem 96

4.3-1 Numerical Integration of Data

4.4 Gaussian Quadrature 98
4.5 Weight Functions 102
4.6 Orthogonal Polynomials and Gaussian Quadrature 104
4.7 Gaussian Quadrature over Infinite Intervals 105
4.8 Particular Gaussian Quadrature Formulas 108

4.8-1 Gauss-Jacobi Quadrature **4.8-2** Gauss-Chebyshev Quadrature **4.8-3** Singular Integrals

4.9 Composite Quadrature Formulas 113
4.10 Newton-Cotes Quadrature Formulas 118

4.10-1 Composite Newton-Cotes Formulas **4.10-2** Romberg Integration

4.11 Adaptive Integration 126
4.12 Choosing a Quadrature Formula 130
4.13 Summation 136

4.13-1 The Euler-Maclaurin Sum Formula **4.13-2** Summation of Rational Functions; Factorial Functions **4.13-3** The Euler Transformation

Bibliographic Notes 145
Bibliography 146
Problems 148

CHAPTER FIVE THE NUMERICAL SOLUTION OF ORDINARY DIFFERENTIAL EQUATIONS 164

5.1 Statement of the Problem 164
5.2 Numerical Integration Methods 166

5.2-1 The Method of Undetermined Coefficients

5.3 Truncation Error in Numerical Integration Methods 171
5.4 Stability of Numerical Integration Methods 173

5.4-1 Convergence and Stability **5.4-2** Propagated-Error Bounds and Estimates

5.5 Predictor-Corrector Methods 183

5.5-1 Convergence of the Iterations **5.5-2** Predictors and Correctors **5.5-3** Error Estimation **5.5-4** Stability

5.6 Starting the Solution and Changing the Interval 195

5.6-1 Analytic Methods **5.6-2** A Numerical Method **5.6-3** Changing the Interval

5.7 Using Predictor-Corrector Methods 198

5.7-1 Variable-Order–Variable-Step Methods **5.7-2** Some Illustrative Examples

5.8 Runge-Kutta Methods 208

5.8-1 Errors in Runge-Kutta Methods **5.8-2** Second-Order Methods **5.8-3** Third-Order Methods **5.8-4** Fourth-Order Methods **5.8-5** Higher-Order Methods **5.8-6** Practical Error Estimation **5.8-7** Step-Size Strategy **5.8-8** Stability **5.8-9** Comparison of Runge-Kutta and Predictor-Corrector Methods

5.9 Other Numerical Integration Methods 224

5.9-1 Methods Based on Higher Derivatives **5.9-2** Extrapolation Methods

5.10 Stiff Equations 228

Bibliographic Notes 233
Bibliography 234
Problems 236

CHAPTER SIX FUNCTIONAL APPROXIMATION: LEAST-SQUARES TECHNIQUES 247

6.1 Introduction 247
6.2 The Principle of Least Squares 248
6.3 Polynomial Least-Squares Approximations 251

6.3-1 Solution of the Normal Equations **6.3-2** Choosing the Degree of the Polynomial

6.4 Orthogonal-Polynomial Approximations 254
6.5 An Example of the Generation of Least-Squares Approximations 260
6.6 The Fourier Approximation 263

6.6-1 The Fast Fourier Transform **6.6-2** Least-Squares Approximations and Trigonometric Interpolation

Bibliographic Notes 274
Bibliography 275
Problems 276

CHAPTER SEVEN FUNCTIONAL APPROXIMATION: MINIMUM MAXIMUM ERROR TECHNIQUES 285

7.1 General Remarks 285
7.2 Rational Functions, Polynomials, and Continued Fractions 287
7.3 Padé Approximations 293
7.4 An Example 295
7.5 Chebyshev Polynomials 299
7.6 Chebyshev Expansions 301
7.7 Economization of Rational Functions 307

7.7-1 Economization of Power Series **7.7-2** Generalization to Rational Functions

7.8 Chebyshev's Theorem on Minimax Approximations 311
7.9 Constructing Minimax Approximations 315

7.9-1 The Second Algorithm of Remes **7.9-2** The Differential Correction Algorithm

Bibliographic Notes 320
Bibliography 320
Problems 322

CHAPTER EIGHT THE SOLUTION OF NONLINEAR EQUATIONS 332

8.1 Introduction 332
8.2 Functional Iteration 334

8.2-1 Computational Efficiency

8.3 The Secant Method 338
8.4 One-Point Iteration Formulas 344
8.5 Multipoint Iteration Formulas 347

8.5-1 Iteration Formulas Using General Inverse Interpolation
8.5-2 Derivative Estimated Iteration Formulas

8.6 Functional Iteration at a Multiple Root 353
8.7 Some Computational Aspects of Functional Iteration 356

8.7-1 The δ^2 Process

8.8 Systems of Nonlinear Equations 359
8.9 The Zeros of Polynomials: The Problem 367

8.9-1 Sturm Sequences

8.10 Classical Methods 371

8.10-1 Bairstow's Method **8.10-2** Graeffe's Root-squaring Method **8.10-3** Bernoulli's Method **8.10-4** Laguerre's Method

8.11 The Jenkins-Traub Method 383
8.12 A Newton-based Method 392
8.13 The Effect of Coefficient Errors on the Roots; Ill-conditioned Polynomials 395

Bibliographic Notes 397
Bibliography 399
Problems 400

CHAPTER NINE THE SOLUTION OF SIMULTANEOUS LINEAR EQUATIONS 410

9.1 The Basic Theorem and the Problem 410
9.2 General Remarks 412
9.3 Direct Methods 414

9.3-1 Gaussian Elimination **9.3-2** Compact Forms of Gaussian Elimination **9.3-3** The Doolittle, Crout, and Cholesky Algorithms **9.3-4** Pivoting and Equilibration

9.4 Error Analysis 430

9.4-1 Roundoff-Error Analysis

9.5 Iterative Refinement 437
9.6 Matrix Iterative Methods 440
9.7 Stationary Iterative Processes and Related Matters 443

9.7-1 The Jacobi Iteration **9.7-2** The Gauss-Seidel Method **9.7-3** Roundoff Error in Iterative Methods **9.7-4** Acceleration of Stationary Iterative Processes

9.8 Matrix Inversion 450
9.9 Overdetermined Systems of Linear Equations 451
9.10 The Simplex Method for Solving Linear Programming Problems 457
9.11 Miscellaneous Topics 465

Bibliographic Notes 468
Bibliography 470
Problems 472

CHAPTER TEN THE CALCULATION OF EIGENVALUES AND EIGENVECTORS OF MATRICES 483

10.1 Basic Relationships 483

10.1-1 Basic Theorems **10.1-2** The Characteristic Equation
10.1-3 The Location of, and Bounds on, the Eigenvalues
10.1-4 Canonical Forms

10.2 The Largest Eigenvalue in Magnitude by the Power Method 492

10.2-1 Acceleration of Convergence **10.2-2** The Inverse Power Method

10.3 The Eigenvalues and Eigenvectors of Symmetric Matrices 501

10.3-1 The Jacobi Method **10.3-2** Givens' Method
10.3-3 Householder's Method

10.4 Methods for Nonsymmetric Matrices 513

10.4-1 Lanczos' Method **10.4-2** Supertriangularization
10.4-3 Jacobi-Type Methods

10.5 The *LR* and *QR* Algorithms 521

10.5-1 The Simple *QR* Algorithm **10.5-2** The Double *QR* Algorithm

10.6 Errors in Computed Eigenvalues and Eigenvectors 536

Bibliographic Notes 538
Bibliography 539
Problems 541

Index 549

PREFACE

The 12 years since the publication of the first edition have been a time of great progress in numerical analysis. This progress is epitomized by the development in recent years of the first complete subroutine packages for digital computers—so-called *mathematical software*—programs which can accept as input the parameters of a problem in a particular area of numerical analysis and which will generally produce as output the solution to the problem within the accuracy desired (or, rarely, a statement that the problem is insoluble or not solvable to the desired accuracy) without the necessity for the user to choose the method of solution. This has been made possible by the development of methods—or classes of methods—which are responsive to all or nearly all the difficulties or sensitivities which can occur in numerical analytic problems.

It is hardly surprising, therefore, that the first edition is now badly out of date in many respects. In this edition we have brought all chapters in the book up to date as of 1977 and have deleted material no longer of interest because it has been superseded by more modern techniques. Some additional material has been deleted solely because of space limitations.

The scope of the book itself may be easily gleaned from the Table of Contents. Here we note only the major additions to and changes from the first edition.

Chapter 1. New sections on Norms, Error Analysis, and Condition and Stability; rewritten section on Computer Arithmetic.

Chapter 2. New sections on Numerical Algorithms, Functionals, and the Method of Undetermined Coefficients.

Chapter 3. New section on Splines.

Chapter 4. New sections on Adaptive Integration and the Euler Transformation; rewritten sections on Numerical Differentiation.

Chapter 5. New sections on Variable Order, Variable Step Methods, Extrapolation Methods, and Stiff Equations; revised section on Runge-Kutta Methods.

Chapter 6. New section on the Fast Fourier Transform.

Chapter 7. New section on the Differential Correction Algorithm.

Chapter 8. New sections on the Jenkins-Traub Method and a Newton-based Method for the zeros of polynomials; rewritten sections on Systems of Nonlinear Equations and on the general problem of the Zeros of Polynomials.

Chapter 9. New sections on Overdetermined Systems and the Simplex Method; rewritten sections on Direct Methods and Error Analysis.

Chapter 10. New sections on the Inverse Power Method and Jacobi-type Methods for nonsymmetric matrices; rewritten section on the *QR* Algorithm.

In addition, there are many changes in other sections to improve clarity and to reflect advances since the publication of the first edition.

It will ordinarily not be possible to cover all the material in this book in a full-year course in numerical analysis. Rather than suggest topics for inclusion or exclusion, we would leave this to the instructor's own taste and experience. The fact that the subjects of each of Chapters 3 through 10 have themselves been the subjects of at least one book apiece should serve to emphasize to the student that a course taught from this book is indeed a first course in numerical analysis. Moreover, we have not covered such topics as the numerical solution of partial differential equations, integral equations, or boundary-value problems. These topics properly fall in the domain of advanced numerical analysis. Since the basis of much of advanced numerical analysis is the solution of systems of linear equations and the calculation of eigenvalues, these topics have been purposely placed at the end of this volume.

In each of Chapters 3 through 10 there are a number of illustrative examples whose purpose is to enhance the student's understanding of the relevant numerical method. Since a morass of numbers is more likely to impede this aim than otherwise, the numbers in these examples have, where possible, been kept simple.

In this edition we have added problems at the end of each chapter corresponding to the new and rewritten sections. The problems fall generally into four categories:

1. Simple proofs of topics considered in the text.
2. Algebraic manipulations and derivations, which would not add materially to the understanding of the student if included in the text, but which may nevertheless be instructive.
3. Computational problems.

4. Proofs and derivations of results which are an extension of the subject matter in the text.

Although the especially difficult problems have been starred (*), the student will find few really easy problems. One of the major purposes of the separately published *Hints and Answers* is to help the student solve those problems found particularly difficult. The student should be prepared to find minor discrepancies between calculated numerical answers and those in this manual. These will generally be the result of the idiosyncrasies of roundoff error on the computer on which the calculations have been performed.

The few bibliographic references in the text itself are to topics outside the scope of numerical analysis or to topics not suitable for problems. The Bibliographic Notes and Bibliographies at the end of each chapter have been brought up to date for this edition. They are meant to guide the student to basic sources from which a deeper understanding of the subject matter of this book may be obtained. For this reason no attempt has been made to make the bibliography exhaustive, and comparatively few foreign-language references have been included.

Anthony Ralston
Philip Rabinowitz

NOTATION

Below is a list of symbols and notation used in this book. Amplified explanations are given when necessary at the first use of the symbol or notation. The reader is cautioned that some symbols may have more than one meaning but it is hoped that they are unambiguous; for example, $P_n(x)$ is used as the notation for the Legendre polynomial of degree n in Chap. 4 and thereafter, but in Chap. 2 this notation is also used for a general polynomial of degree n.

A. Problems and references

Meaning	Symbol or example	Page first defined or used
1. References in text to problems at end of each chapter: numbers in braces	{2}	5
2. References to bibliography at end of each chapter: name followed by date in parentheses	Feller (1950)	10
3. Problems of more than ordinary difficulty: asterisk next to problem number	***10**	26

B. General mathematical notation

Meaning	Symbol or example	Page first defined or used
1. Approximately equal	$\approx$	6
2. Binomial coefficient	$\binom{n}{k}$	36
	$(m)_k$	59

Meaning	Symbol or example	Page first defined or used
3. Closed interval	$[a,b]$	1
4. Conjugate transpose of vector or matrix: superscript asterisk*	$\mathbf{v}^*$	445
5. Continued fraction	$\frac{C_1\vert}{\vert x + D_1} + \frac{C_2\vert}{\vert x + D_2} + \cdots$	292
6. Derivatives:		
(a) Single, double, or triple prime	$f''(\xi_1)$	1
(b) Lowercase roman superscript	$f^{\text{iv}}(\xi_2)$	1
(c) Letter or number in parentheses	$f^{(n)}(x)$	43
7. Determinant of a matrix	$\vert A\vert$	412
	$\det(L_{n-1})$	423
8. Evaluation of a quantity at a point	$\vert_n$	210
9. Factorial function	$x^{(n)}$	141
10. Functional	$F(f)$	42
11. Inner product		
(a) T for transpose	$\mathbf{v}^T\mathbf{v}$	6
(b) * for conjugate transpose	$\mathbf{x}^*\mathbf{x}$	487
12. Integer functions		
(a) Ceiling	$\lceil x\rceil$	15
(b) Floor	$\lfloor x\rfloor$	15
13. Norm		
(a) function: L_p	$\Vert f\Vert_p$	7
(b) matrix: general	$\Vert A\Vert$	7
(c) matrix: Euclidean	$\Vert A\Vert_E$	8
(d) vector: L_p	$\Vert \mathbf{v}\Vert_p$	7
(e) vector: Euclidean	$\Vert \mathbf{v}\Vert$	6
14. Open interval	(a_i, m_i)	40
15. Order of magnitude	$O(h^2)$	174
16. Sequences of functions or numbers: indexed quantity in braces (cf. A1 above)	$\{x^n\}$	33
17. Spectral radius of a matrix	$\rho(A)$	8
18. Transform pair	$G_j \leftrightarrow g_k$	264
19. Vector: boldface English or Greek letters		
(a) Column	$\mathbf{v}$	6
(b) Row (T for transpose)	$\mathbf{v}^T$	6
20. Vector function	$\mathbf{f} = [f_1 f_2 \cdots f_n]$	359

C. Specific mathematical symbols

Symbol	Meaning	Page first defined or used
1. $B_n(x)$	Bernstein polynomial	3
2. $B_n(x)$	Bernoulli polynomial	136
3. B_n	Bernoulli number	136

Symbol	Meaning	Page first defined or used
4. EI	Efficiency index	337
5. erf(x)	Error function	26
6. $H_n(x)$	Hermite polynomial	107
7. I	Identity matrix	433
8. $I_a^b R(x)$	Cauchy index	368
9. $J_n(x; \alpha, \beta)$	Jacobi polynomial	108
10. $l_j(x)$	Lagrangian interpolation polynomial	53
11. $L_n(x)$	Laguerre polynomial	106
12. $P_n(x)$	Legendre polynomial	99
13. $S_n(x)$	Chebyshev polynomial of second kind	111
14. T	Transpose of matrix or column vector	6
15. $T_n(x)$	Chebyshev polynomial	109
16. $T_r^*(x)$	Shifted Chebyshev polynomial	327
17. tr (A)	Trace of a matrix	26
18. $\mathbf{x}_c$	Computed solution of linear system	413
19. $\mathbf{x}_t$	True solution of linear system	413
20. ∇	Backward-difference operator	58
21. ∇	Gradient operator	362
22. Δ	Forward-difference operator	58
23. δ	Central-difference operator	58
24. δ_{jk}	Kronecker delta	54
25. $\Gamma(x)$	Gamma function	108
26. μ	Mean central-difference operator	83
27. $\cup$	Set union	486

CHAPTER

ONE

INTRODUCTION AND PRELIMINARIES

1.1 WHAT IS NUMERICAL ANALYSIS?

That numerical analysis is both a science and an art is a cliché to specialists in the field but is often misunderstood by nonspecialists. Is calling it an art and a science only a euphemism to hide the fact that numerical analysis is not a sufficiently precise discipline to merit being called a science? Is it true that "numerical analysis" is something of a misnomer because the classical meaning of analysis in mathematics is not applicable to numerical work? In fact, the answer to both these questions is no. The juxtaposition of science and art is due instead to an uncertainty principle which often occurs in solving problems, namely that to determine the best way to solve a problem may require the solution of the problem itself. In other cases, the best way to solve a problem may depend upon a knowledge of the properties of the functions involved which is unobtainable either theoretically or practically. A simple example will illustrate this. Two common methods for estimating

$$\int_a^b f(x)\,dx$$

are the trapezoidal rule and the parabolic rule. The error incurred, i.e., the difference between the true value of the integral and the approximation, in the former is $-(b-a)^3 f''(\xi_1)/12n^2$, where $n+1$ is the number of points at which we evaluate $f(x)$ in $[a, b]$ and ξ_1 is some (unknown) point in $[a, b]$. For the parabolic rule the error is $-(b-a)^5 f^{\text{iv}}(\xi_2)/180n^4$, where again $n+1$ is the number of points at which $f(x)$ is evaluated and ξ_2 is an unknown point in $[a, b]$. Which do we use, especially when $f(x)$ is such that its derivatives are not reasonably calculable? As numerical analysis is a science, it has provided us with these two methods and the errors incurred when we use them, but as it is an art, it requires us to use our intuition, experience, and knowledge of functions "like" $f(x)$ to choose that method best suited to our particular problem.

As a science, then, numerical analysis is concerned with the processes by which mathematical problems can be solved by the operations of arithmetic. Sometimes this will involve the development of algorithms to solve a problem already in a form in which the solution can be found by arithmetic means, e.g., simultaneous linear equations. Often it will involve replacing quantities which cannot be calculated arithmetically, e.g., derivatives or integrals, by approximations which permit an approximate solution to be found. In this case, we shall naturally be interested in the errors incurred in our approximation. But in any case, the tools we shall use in developing the processes of numerical analysis will be the tools of exact mathematical analysis as classically understood.

As an art, numerical analysis is concerned with choosing that procedure (and suitably applying it) which is "best" suited to the solution of a particular problem. This implies the need for anyone who wishes to practice numerical analysis to develop experience and with it—it is hoped—intuition. We have therefore provided numerous examples to illustrate the numerical methods discussed. The purpose of these examples is to help the reader understand principles and develop an insight into computational processes. To further these aims, the numbers, where permissible, have been kept simple.

Numerical analysis is a very different discipline today from what it was 25 years ago at the advent of the high-speed digital computer. High-speed computation has revolutionized numerical analysis as an art and given enormous impetus to its development as a science. Our orientation in this book will be entirely toward methods which are particularly useful on digital computers. This does not mean ignoring either traditional desk calculators or modern hand-held electronic calculators, including those which are programmable. Most of the methods to be discussed are applicable on hand-held and desk calculators as well as on digital computers, but we recognize that despite the rapidly increasing popularity of hand-held calculators, the overwhelming majority of significant numerical computations are performed on digital computers.

1.2 SOURCES OF ERROR

Numerical answers to problems generally contain errors which arise in two areas: those inherent in the mathematical formulation of the problem and those incurred in finding the solution numerically. The former category includes the error incurred when the mathematical statement of a problem is only an *approximation to the physical situation.* Such errors are often negligible, as in the case of neglecting relativistic effects in problems in classical mechanics. If they are not negligible, then no matter how accurate the numerical computations, there will be a significant error in the result. Another source of inherent error is the *inaccuracies in the physical data.* Such errors are also generally negligible when they are caused by inaccuracies in physical constants, e.g., the gravitational constant. But when they are the result of errors in empirical data, the worth of a computed solution must be carefully weighed against these errors. Moreover, because such errors are usually random, treating them analytically may be quite difficult. Such errors will play significant roles in Chaps. 6 and 9.

There are three main sources of computational error. The first, familiar to all users of desk calculators or pencil and paper and to all programmers, is the gross error, or *blunder*. Digital computers have enormously reduced the probability that calculational blunders will occur, but of course the possibility of a programming blunder which results in the correct calculation of the wrong result is always present. In addition, undetected bugs in the compiler or other system software are not very unusual. When the correctness or reasonableness of a computed solution cannot readily be verified, the possibility of a blunder should not be ignored. It is the two other sources of computational error which will chiefly interest us here, however.

The first of these sources is that caused by solving not the problem as formulated but rather some approximation to it. This is usually caused by the replacement of an infinite, i.e., summation or integration, or infinitesimal, i.e., differentiation, process by a finite approximation. Some examples of this are:

1. Calculation of an elementary function, e.g., $\sin x$, by using the first n terms of its infinite Taylor-series expansion
2. Approximation of the integral of a function by a finite summation of functional values, as in the trapezoidal rule
3. Solution of a differential equation by replacing the derivatives by approximations to them, e.g., difference quotients
4. Solution of the equation $f(x) = 0$ by the Newton-Raphson method, an iterative process which in general converges only in the limit as the number of iterations goes to infinity

We shall denote this type of error in all its various forms as *truncation error*, since it often is the result of truncating an infinite process to get a finite process. In all the numerical procedures considered in this book, we shall be interested in estimating, or at least bounding, this error (to know it would, of course, be to eliminate it!).

The other source of error of importance to us is that caused by the fact that arithmetic calculations can almost never be carried out with complete accuracy. Most numbers have infinite decimal representations which must be rounded. But even if the data of a problem can be expressed exactly by finite decimal representations, division may introduce numbers which must be rounded and multiplication may produce more digits than can reasonably be retained. The error we introduce by rounding a number is called *roundoff error*. Like the errors in empirical data, roundoff error has a random character which makes it difficult to deal with. We shall consider these difficulties in more detail in Sec. 1.4.

1.3 ERROR DEFINITIONS AND RELATED MATTERS

In the previous section, we relied upon the fact that the reader undoubtedly has a good intuitive notion of error. Here we shall formalize the concept of error. The two basic definitions are

Definition 1.1

$$\text{True value} = \text{approximate value} + \text{error}$$

Definition 1.2

$$\text{Relative error} = \frac{\text{error}}{\text{true value}}$$

Thus, denoting error by E and relative error by RE, if we approximate $\frac{1}{3}$ by .333, we have

$$E = \tfrac{1}{3} \times 10^{-3} \qquad \text{and} \qquad RE = 10^{-3}$$

Generally, we shall be interested in E (which is sometimes called absolute error) rather than RE, but when the true value of a quantity is very small or very large, relative errors are more meaningful. For example, if the true value of a quantity is 10^{15}, an error of 10^6 is probably not serious, but this is more meaningfully expressed by saying that $RE = 10^{-9}$. In actual computation of the relative error, we shall often replace the unknown true value by the computed approximate value.

1.3-1 Significant Digits

Let x be a real number which in general has an infinite decimal representation. We shall say that x has been correctly rounded to a d-decimal-place number, which we denote by $x^{(d)}$, if the *roundoff error* ϵ is such that†

$$|\epsilon| = |x - x^{(d)}| \leq \tfrac{1}{2} \times 10^{-d} \tag{1.3-1}$$

† When $|\epsilon| = \frac{1}{2} \times 10^{-d}$, we have the choice of rounding "up" or "down." That is, the number .2775500 ... rounded to four decimal places may be either .2776 or .2775. Commonly, such numbers are always rounded "up" or always "down," but, since it is usually desirable to avoid any bias in roundoff, a rule such as rounding so that the last digit is always even (or always odd) is desirable in lengthy computations.

Thus, if $x = 6.74399666 \ldots$, $x^{(3)} = 6.744$ and $x^{(7)} = 6.7439967$.

If y is any approximation to a true value x, then the kth decimal place of y is said to be *significant* if

$$|x - y| \leq \tfrac{1}{2} \times 10^{-k} \tag{1.3-2}$$

Therefore, every digit of a correctly rounded number is significant.†

It might seem natural now to define the number of significant digits in y to be all the digits of y satisfying the definition of significant digit. Thus, if $y_1 = .9863$ is an approximation to $x_1 = .98632$ and if $y_2 = .0028$ is an approximation to $x_2 = .00278$, we would say y_1 and y_2 both have four significant digits. But are y_1 and y_2 equally "significant"? If they are the final answers to a computation and we are interested in absolute error, then they certainly are. But if they are intermediate numbers in a calculation which are to be used later as divisors, then they most certainly are not: for the magnitude of the error in $1/y_1$ is much less than that in $1/y_2$ {2}. We shall therefore avoid the use of the notion of the number of significant digits in a number.

The example above indicates that numbers with leading zeros can cause substantial magnification of error when used as divisors. In numerical calculations subtraction (or addition of numbers of opposite sign) is the most common source of numbers with leading zeros. Suppose all the digits in $y_1 = 2.78493$ and $y_2 = 2.78469$ are significant. Then the error in $y_1 - y_2 = .00024$ is in magnitude at most 10^{-5} (the sum of the maximum magnitudes of the errors in y_1 and y_2). But, if $y_1 - y_2$ is later used as a divisor, the magnitude of the error will be greatly increased. Note that the relative error in y_1 is less than $\frac{1}{2} \times 10^{-5}/2.784925 \approx 1.8 \times 10^{-6}$, while that in $y_1 - y_2$ may be as much as $10^{-5}/.00023 \approx 4.3 \times 10^{-2}$. Thus we can say that if the sum or difference of two numbers causes a large increase in relative error, then if this result is later used as a divisor, substantial magnification of error may occur. The phenomenon, often called *subtractive cancellation*, is one of which the numerical analyst must always be aware.

1.3-2 Error in Functional Evaluation

An important general problem is that of estimating the error in the evaluation of $f(y_1, \ldots, y_n)$, where

$$y_i = x_i - \epsilon_i \tag{1.3-3}$$

† It is not satisfactory to define the kth digit of y as significant only if x and y are identical when rounded to k digits. For suppose $x = 3.76512$ and $y = 3.7648$. Then the 6 would not be significant, but the 4 would be. By our definition, both are significant.

with x_i the true value of the ith variable and ϵ_i the error. Then

$$f(x_1, \ldots, x_n) - f(y_1, \ldots, y_n) = \sum_{i=1}^{n} \epsilon_i \frac{\partial f}{\partial x_i} + \frac{1}{2}\left(\sum_{i=1}^{n} \epsilon_i \frac{\partial}{\partial x_i}\right)^2 f + \cdots \tag{1.3-4}$$

where the partial derivatives are to be evaluated at $(y_1, \ldots, y_n)$. Since we shall usually be able to assume that terms containing products of the errors are small compared with the first term on the right-hand side of (1.3-4), we can write

$$f(x_1, \ldots, x_n) - f(y_1, \ldots, y_n) \approx \sum_{i=1}^{n} \epsilon_i \frac{\partial f}{\partial x_i} \tag{1.3-5}$$

Equation (1.3-5) can serve as a convenient tool in estimating errors in functional evaluation, although in special cases more terms in (1.3-4) may need to be retained. In Secs. 1.5 and 1.6 error analyses for some special cases of $f(y_1, \ldots, y_n)$ are considered.

1.3-3 Norms

In dealing with vectors, matrices, and functions, the problem arises of measuring their size in some convenient form. This is usually done by means of a *norm*, a real-valued function with properties which generalize the usual Euclidean concept of length.

If we consider a vector $\mathbf{v} = (v_1, \ldots, v_n)^T$,† then the Euclidean length of $\mathbf{v}$, which we shall for the moment denote by $\|\mathbf{v}\|$ (read the norm of $\mathbf{v}$), is given by

$$\|\mathbf{v}\| = (\mathbf{v}^T\mathbf{v})^{1/2} = \left(\sum_{i=1}^{n} v_i^2\right)^{1/2} \tag{1.3-6}$$

This norm has the following properties:

Property 1 $\|\mathbf{v}\| \geq 0$ and $\|\mathbf{v}\| = 0$ if and only if $\mathbf{v} = \mathbf{0}$; that is, $v_i = 0$, $i = 1, \ldots, n$.

Property 2 $\|\alpha\mathbf{v}\| = |\alpha| \cdot \|\mathbf{v}\|$ for any real (or complex) number α.

Property 3 $\|\mathbf{u} + \mathbf{v}\| \leq \|\mathbf{u}\| + \|\mathbf{v}\|$, the triangle inequality.

† The superscript T applied to any vector or matrix denotes the transpose. All vectors are assumed to be column vectors so that a row vector is represented as the transpose of a (column) vector.

Properties 1 and 2 are immediately obvious, while Property 3 is a consequence of the Cauchy-Schwarz inequality {6}.

The norm defined by (1.3-6) is not the only function which has the above three properties. In fact, any function which has these properties is called a *vector norm*. Thus, a generalization of (1.3-6) is the L_p norm

$$\|\mathbf{v}\|_p = \left(\sum_{i=1}^{n} |v_i|^p\right)^{1/p} \tag{1.3-7}$$

defined for any $p \geq 1$. If we let p tend to infinity, we get {6} the *maximum*, or L_∞, norm

$$\|\mathbf{v}\|_\infty = \max_{1 \leq i \leq n} |v_i| \tag{1.3-8}$$

A further generalization yields the *weighted* L_p or $L_{p,\mathbf{w}}$ norm

$$\|\mathbf{v}\|_{p,\mathbf{w}} = \left(\sum_{i=1}^{n} w_i |v_i|^p\right)^{1/p} \tag{1.3-9}$$

where $\mathbf{w} = (w_1, \ldots, w_n)^T$ is a vector of *positive* weights w_i, $i = 1, \ldots, n$. The proof that these are indeed norms is left to the reader {6}. The most frequently used norms in practice are the L_1, L_2, and L_∞ norms and their weighted counterparts.

For functions $f(x)$ continuous on a finite interval $[a, b]$, any real functional (see Sec. 2.3) which satisfies properties 1 to 3 is a norm. Thus, corresponding to the L_p vector norm, we have the L_p norm for functions given by

$$\|f\|_p = \left[\int_a^b |f(x)|^p \, dx\right]^{1/p} \tag{1.3-10}$$

and corresponding to the maximum norm, we have the L_∞, or *uniform*, norm

$$\|f\|_\infty = \max_{a \leq x \leq b} |f(x)| \tag{1.3-11}$$

The counterpart to the weighted L_p norm for functions is

$$\|f\|_{p,w} = \left[\int_a^b w(x)|f(x)|^p \, dx\right]^{1/p} \tag{1.3-12}$$

where $w(x)$ is a nonnegative weight function which does not vanish on some subinterval of $[a, b]$.

Returning now to vector norms, we can use them to define norms on a matrix A as follows:

$$\|A\| = \max_{\|\mathbf{x}\| = 1} \|A\mathbf{x}\| \tag{1.3-13}$$

Such a norm is called a *subordinate* or *induced norm*. That this norm has properties 1 to 3 is immediate {7}. In addition, for square matrices, we have from (1.3-13):

Property 4

$$\|AB\| \leq \|A\| \cdot \|B\| \tag{1.3-14}$$

In view of this property of subordinate norms, we shall define a *matrix* norm as a norm which satisfies Property 4. For the L_1 and L_∞ vector norms, the corresponding subordinate matrix norms are given respectively by {7}

$$\|A\|_1 = \max_{1 \leq j \leq n} \sum_{i=1}^{n} |a_{ij}| \tag{1.3-15}$$

and

$$\|A\|_\infty = \max_{1 \leq i \leq n} \sum_{j=1}^{n} |a_{ij}| \tag{1.3-16}$$

so that $\|A\|_1 = \|A^T\|_\infty$. The computation of the matrix norm induced by other vector norms, including the L_2, or Euclidean, norm (1.3-6), is quite difficult.

From the definition (1.3-13), it follows {8} that for any vector norm and the corresponding subordinate matrix norm,

$$\|A\mathbf{x}\| \leq \|A\| \cdot \|\mathbf{x}\| \tag{1.3-17}$$

for all matrices A and vectors $\mathbf{x}$. Any matrix norm for which (1.3-17) holds with respect to a given vector norm is said to be *consistent* or compatible with that vector norm.

An example of a matrix norm which is not a subordinate norm is given by the *Euclidean matrix norm* (sometimes called the Frobenius norm)

$$\|A\|_E = \left(\sum_{i=1}^{n} \sum_{j=1}^{n} a_{ij}^2 \right)^{1/2} \tag{1.3-18}$$

That this is indeed a matrix norm is quite simple to prove {7}. And that it is not a subordinate norm follows from the fact that for all subordinate norms, $\|I\| = 1$, where I is the unit matrix, whereas $\|I\|_E = \sqrt{n}$. Thus the Euclidean matrix norm is a generalization of the Euclidean vector norm but is not subordinate to it. However, it is consistent with the Euclidean vector norm {8}. In addition, we have that {8}

$$\|A\|_2 \leq \|A\|_E \leq n^{1/2} \|A\|_2 \tag{1.3-19}$$

Closely related to the notion of matrix norm is that of the *spectral radius* $\rho(A)$ of a matrix, defined by

$$\rho(A) = \max_{1 \leq i \leq n} |\lambda_i| \tag{1.3-20}$$

where λ_i are the eigenvalues of A. It is easy to show {9} that for any consistent matrix norm

$$\rho(A) \leq \|A\| \tag{1.3-21}$$

Further, we have, using the definition of matrix norm (1.3-13) and Theorem 10.6, that

$$\|A\|_2^2 = \rho(A^T A) \tag{1.3-22}$$

From this it follows that if A is a symmetric matrix, then

$$\|A\|_2 = \rho(A) \tag{1.3-23}$$

For this reason, the L_2 matrix norm is also called the *spectral norm.*

The use of norms is very important in measuring the quality of a computed approximation $\mathbf{v}_c$ to a desired true result $\mathbf{v}_t$. For any norm, $\|\mathbf{v}_c - \mathbf{v}_t\|$ is a measure of the deviation of the approximation. We shall use this approach in Chap. 9, where some additional results on norms are given.

1.4 ROUNDOFF ERROR

If the solution of a problem requires many thousands or even millions of arithmetic operations, each of which is performed using rounded numbers, it is intuitively clear that the accumulated roundoff error may significantly affect the result. It is true that, given the computation to be performed and the roundoff rules to be applied, the roundoff error at each step is determined. Thus roundoff is not a random process. But a priori it is essentially impossible to determine what the roundoff error will be in the millionth or even the hundredth step of the computation. Therefore, a probabilistic approach to roundoff error is helpful in understanding how this error accumulates.

1.4-1 The Probabilistic Approach to Roundoff: A Particular Example

Consider the addition of n numbers $\{a_i\}$ each correctly rounded to d decimal places. Since the error in each number is no greater than $\frac{1}{2} \times 10^{-d}$, the accumulated error in the sum is no greater than $n/2 \times 10^{-d}$. But since most errors will be less in magnitude than $\frac{1}{2} \times 10^{-d}$, and since the differing signs in the errors will cause some cancellation of error, we expect the accumulated error to be substantially less than $n/2 \times 10^{-d}$ in general. Our object here is to consider just what the distribution of this error is.

Denote by ϵ_i the roundoff error in a_i, and for convenience, let us, without loss of generality, take $d = 0$ in what follows. Then ϵ_i takes on values in the interval $[-\frac{1}{2}, \frac{1}{2}]$, and if we assume each value in this range is equally likely, then the probability density of ϵ_i is shown in Fig. 1.1, which indicates

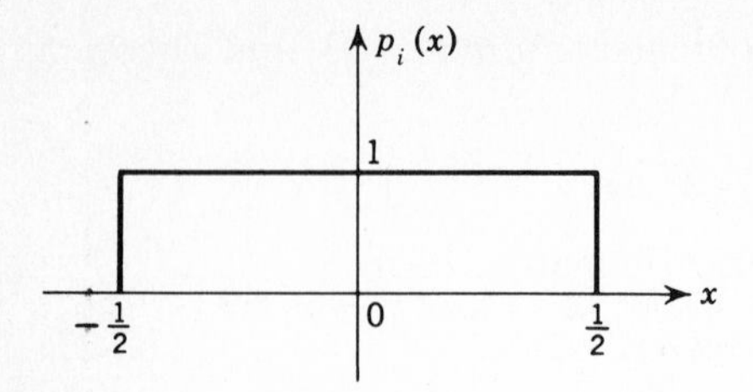

Figure 1.1 Graph of $p_i(x)$, the probability density of ϵ_i.

that

$$\Pr(x_1 \le \epsilon_i \le x_2) = \begin{cases} x_2 - x_1 & -\frac{1}{2} \le x_1;\ x_2 \le \frac{1}{2} \\ x_2 + \frac{1}{2} & x_1 < -\frac{1}{2};\ -\frac{1}{2} \le x_2 \le \frac{1}{2} \\ \frac{1}{2} - x_1 & -\frac{1}{2} \le x_1 \le \frac{1}{2};\ x_2 > \frac{1}{2} \\ 1 & x_1 < -\frac{1}{2};\ x_2 > \frac{1}{2} \end{cases} \tag{1.4-1}$$

Now let us consider what happens when a_i is added to a_j, assuming that the roundoff errors in the two numbers are independent. Then {11} the probability density of the error $\epsilon_{ij} = \epsilon_i + \epsilon_j$ in $a_i + a_j$ is shown in Fig. 1.2.

From Fig. 1.1 we can calculate the variance σ_i^2 of $p_i(x)$ as

$$\sigma_i^2 = \int_{-1/2}^{1/2} x^2\, dx = \tfrac{1}{12} \tag{1.4.2}$$

and from Fig. 1.2, the variance of $p_{ij}(x)$ is†

$$\sigma_{ij}^2 = \int_{-1}^{0} x^2(1 + x)\, dx + \int_{0}^{1} x^2(1 - x)\, dx = \tfrac{1}{6} \tag{1.4-3}$$

Corresponding results for the sum of three {11} and four numbers would suggest a result which is only hinted at by the results for one and two numbers, namely that in the limit as $n \to \infty$, the density function for the sum of n numbers approaches the normal density function with zero mean and variance $n\sigma_i^2$ with σ_i given by (1.4-2). Thus, if ϵ is the accumulated roundoff error for n additions,

$$\Pr(x \le \epsilon \le x + dx) \to \frac{1}{(\pi n/6)^{1/2}} e^{-6x^2/n}\, dx \tag{1.4-4}$$

as $n \to \infty$, where the function on the right-hand side of (1.4-4) is the normal density function with zero mean and standard deviation $\sigma = (n/12)^{1/2}$. This result can be derived rigorously using the central-limit theorem of probability theory [see Feller (1950), pp. 191ff.], which states that the distribution of a sum of n mutually independent random variables with a common distribution approaches a normal distribution as $n \to \infty$ with mean equal to that of the common distribution and variance equal to n times the variance of the

† Or we can calculate σ_{ij} using the result that the variance of the sum of independent distributions with zero mean is the sum of the variances.

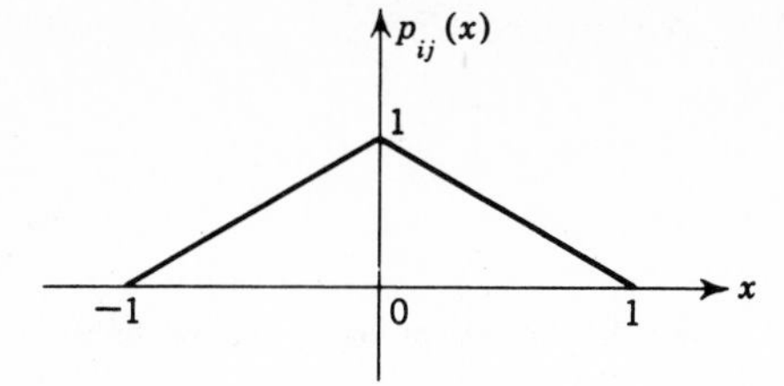

Figure 1.2 Graph of $p_{ij}(x)$, the probability density of ϵ_{ij}.

common distribution. [When $d \neq 0$, the only changes that must be made are to multiply all variances by 10^{-2d} and to multiply x and dx on the right-hand side of (1.4-4) by 10^d.]

In actual fact the normal density function gives a very good approximation to the distribution of $\epsilon_1 + \cdots + \epsilon_n$ for quite small n; so let us suppose that we can use the right-hand side of (1.4-4) as the probability density for the error in our sum of n numbers. The *probable error*, which is defined as that positive value of x for which the probability is $\frac{1}{2}$ that the magnitude of ϵ will exceed x, is given by

$$.6745\sigma = \frac{.6745}{\sqrt{12}}\sqrt{n} \tag{1.4-5}$$

Thus, while the error cannot exceed $n/2$ in magnitude, the probable error is proportional to the square root of n. For the sum of n numbers, then, the probable error is substantially less than the maximum error, for lengthy computations, i.e., large n, by an order of magnitude or more. This is a general result which we would intuitively expect to hold also for more complicated computations where the analysis of roundoff is too complex to carry out.

This result leads to what seems like a paradoxical situation in error analysis. On the one hand, in numerical computations we are usually interested in bounding the error incurred. That is, when we finish a computation, we like to be able to say that the computed result differs by *no more than* the error bound from the true result. In deriving such error bounds, we shall then, for the roundoff component of the error, choose its maximum value. On the other hand, we now realize that by so doing we are generally being unduly conservative. This paradox can be resolved only in the context of a particular situation. If roundoff error is small compared with truncation error, then using its maximum value will not make the error bound unrealistic. But if roundoff is the dominant error, the unrealistic error bound may have to be replaced by an estimate—generally a conservatively high estimate—of the expected error in order to get a usable result. Because $\sqrt{n}$ appears in the probable error while n appears in the maximum error in the above example, a common way to estimate the roundoff error in a long computation is to find a bound on the error and then to replace n (where n is a measure of the number of computations) by $\sqrt{n}$.

A problem closely related to that considered above is forming the weighted sum of the n numbers

$$\sum_{i=1}^{n} \alpha_i a_i \tag{1.4-6}$$

If the roundoff errors in the a_i's all have the density function shown in Fig. 1.1, the result, analogous to the one above, is that the density function of the error in the sum approaches a normal density function with zero mean and variance

$$\sigma^2 = \frac{1}{12} \sum_{i=1}^{n} \alpha_i^2 \tag{1.4-7}$$

Linear combinations of the form (1.4-6) occur often in numerical analysis, as we shall see. The coefficients α_i may often be chosen arbitrarily except for a constraint of the form

$$\sum_{i=1}^{n} \alpha_i = \alpha > 0 \tag{1.4-8}$$

An obvious question then is: What values of the α_i lead to the best roundoff behavior in the sum? The answer depends on what we mean by "best." If we only wish to minimize the bound on the roundoff error, then all sets of α_i which are all positive are equally good. But if $\alpha_1 = \alpha$ and all the other α_i are zero, then the worst case will often be realized. A more reasonable definition of "best" would be that set of α_i which minimizes σ^2 in (1.4-7), since this will lead to good roundoff behavior on the average (why?). Minimizing σ^2 subject to the constraint (1.4-8) is a problem easily solved using the Lagrangian-multiplier technique {15}. The answer is that the α_i are all equal to α/n. Therefore, the roundoff properties of any sum of the form (1.4-6) can be judged by comparing the sum of the squares of the α_i with α^2/n.

1.5 COMPUTER ARITHMETIC

The example in the previous section was idealized in the sense that it dealt with numbers without regard to the number of digits in each number. On digital computers, of course, there are restrictions on the number of digits in each number. Therefore in order to perform roundoff-error analyses of numerical alogorithms implemented on digital computers, it is first necessary to understand how computers perform arithmetic. This is the subject of this section.

1.5-1 Fixed-Point Arithmetic

One mode of computer arithmetic, *fixed-point arithmetic*, is quite similar to ordinary pencil-and-paper arithmetic. It deals with numbers expressed in the familiar way as a sequence of digits (although these digits are almost always *binary* digits rather than decimal ones). The crucial difference from ordinary arithmetic is the assumption that the (binary or decimal) *point* is *fixed*, usually at the left-hand end of the number, i.e., all fixed-point numbers less than 1 in magnitude, but on some computers at the right-hand end, i.e., all numbers integers. The problems with using fixed-point arithmetic in nontrivial calculations are twofold:

1. Since the actual numbers ones deals with are seldom so well-behaved as to allow one to deal entirely with numbers less than 1 in magnitude (or with only integral numbers), the user of the computer, i.e., the programmer, must arrange to keep track of an *imaginary* point which represents the correct position of the point. Particularly troublesome, for example, is the need for lining these imaginary points up correctly when adding two numbers.
2. Equally annoying is the necessity for assuring that the result of any arithmetic operation is, from the point of view of the computer, a valid number in the fixed-point system. For example, if all numbers are less than 1 in magnitude, the sum $.5 + .6$ cannot be performed and is said to *overflow*. Protection against this phenomenon must be achieved by appropriately *shifting* the two arguments, a process called *scaling*, which can be extremely tedious in large calculations.

For these reasons fixed-point arithmetic is only very rarely used in nontrivial calculations on computers. We shall consider it no further in this book but shall instead assume that all calculations are performed in floating-point arithmetic.

1.5-2 Floating-Point Numbers

Numbers in a computer are typically of a standard length l, which includes the sign of the number and its digits (but see also Sec. 1.5-5). In floating-point arithmetic l consists of three parts:

One bit† s for the sign of the number
p bits for the fractional part, or *mantissa*, m
q $(= l - p - 1)$ bits for the *exponent* e

as shown in Fig. 1.3. In this representation the number is interpreted by the computer as

$$\pm m \times 2^{e-d} = \pm m \times 2^f \tag{1.5-1}$$

† Hereafter we shall assume a binary representation of numbers in a computer and shall use the standard contraction *bit* for binary digit.

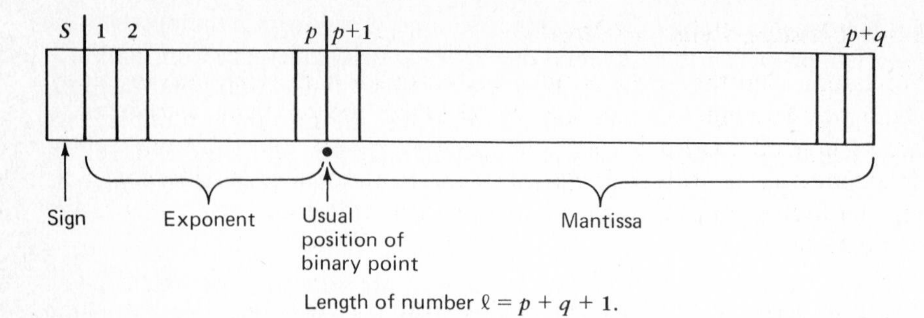

Figure 1.3 Floating-point number representation.

where d, the displacement, is chosen so that the exponents have a range from $-d$ ($e = 0$) to $2^q - d - 1$ ($e = 2^q - 1$). To make this range approximately symmetric about zero, d would normally be chosen equal to 2^{q-1}, so that the range of exponents would be from -2^{q-1} to $2^{q-1} - 1$. In what follows we shall replace $e - d$ with f, which we shall assume has a set of possible integer values approximately symmetric about 0, and, for convenience, we shall ignore the adjustments to the exponents which must be made because of d.

In most floating-point calculations the mantissas are *normalized*, which means that the most significant, i.e., the first, bit is a 1.† Thus, if the binary point of the mantissa is assumed to be at its left end (in a few computers it is assumed to be at the right end),

$$.5 \leq |m| < 1 \tag{1.5-2}$$

In the analysis which follows we shall assume that floating-point numbers have the form (1.5-1) and satisfy (1.5-2).

Before considering arithmetic in floating-point numbers we note one property of such numbers which is often significant in numerical calculations (see, for example, Example 2.2 in Sec. 2.2). This property is the density of these numbers on the real line. Suppose p, the number of mantissa bits, is 24. Then on the interval $[0, 1)$ (exponent part f equals 0), there are 2^{24} floating-point numbers equally spaced $1/2^{24}$ apart. Similarly on any interval $[2^f, 2^{f+1})$ there are 2^{24} equally spaced numbers, but their density is $2^f/2^{24}$. Thus, for example, between $2^{20} = 1{,}048{,}576$ and $2^{21} = 2{,}097{,}152$ there are $2^{24} = 16{,}777{,}216$ floating-point numbers, but the spacing between successive numbers is $\frac{1}{16}$. The particular lesson to be learned from this is that when comparing two nearby large floating-point numbers, e.g., in testing the convergence of an iterative process, it is almost always advisable to compare the difference *relative* to the magnitude of the numbers.

† On many computers which are essentially binary, floating-point numbers are nevertheless interpreted in hexadecimal (base 16). In this form (1.5-1) and (1.5-2) become

$$\pm m \times 16^f \qquad \tfrac{1}{16} \leq |m| < 1$$

1.5-3 Floating-Point Arithmetic

We assume that the reader is sufficiently familiar with computers to understand why floating-point representation avoids the problems considered in the section on fixed-point numbers. In this section we shall analyze the errors inherent in arithmetic using floating-point numbers, first by considering the four arithmetic operations and then with an example of a larger computation.

The basic operations. We consider the result of operating on two l-digit normalized floating-point numbers x and y to produce an l-digit result, which we denote by

$$fl(x \text{ op } y)$$

where op is $+$, $-$, $*$, or $/$. We assume in each case that the mantissa of the result is first normalized and then rounded.† The rounded value m_r to p bits of a mantissa m of more than p bits is defined as

$$m_r = \begin{cases} 2^{-p}\lfloor 2^p m + \frac{1}{2} \rfloor & m > 0 \\ 2^{-p}\lceil 2^p m - \frac{1}{2} \rceil & m < 0 \end{cases} \tag{1.5-3}$$

where the *floor function* $\lfloor x \rfloor$ is the largest integer less than or equal to x and the *ceiling function* $\lceil x \rceil$ is the smallest integer greater than or equal to x. For positive numbers (1.5-3) expresses the familiar rule of adding 1 in the $(p+1)$st position. Considering the mantissa only, rounding therefore results in an absolute error bounded by

$$|\epsilon| \le 2^{-p-1} \tag{1.5-4}$$

and a relative error bounded by $2^{-p-1}/\frac{1}{2} = 2^{-p}$.

Now we consider briefly algorithms for the four arithmetic operations and the resultant relative errors. We begin with multiplication and division because in floating point they are much easier to understand than addition and subtraction. We assume the existence of a $2p$-bit accumulator although in fact $p + 2$ bits are sufficient {19}.

Multiplication The exponents are added, and the mantissas are multiplied. If the resulting mantissa is unnormalized, it is normalized and the exponent is adjusted. Then the mantissa is rounded to p bits.

To analyze the error let

$$x = m_x 2^{f_x} \qquad y = m_y 2^{f_y} \tag{1.5-5}$$

Then

$$xy = m_x m_y \times 2^{f_x + f_y} \tag{1.5-6}$$

where

$$\tfrac{1}{4} \le |m_x m_y| < 1 \tag{1.5-7}$$

† The rounding may cause overflow, which then requires a renormalization.

since m_x and m_y are both assumed to satisfy (1.5-2). Therefore, normalization of $m_x m_y$ involves a left shift of at most 1. The rounded mantissa of $fl(x * y)$ is then either

$$m_x m_y + \epsilon \qquad \text{or} \qquad 2m_x m_y + \epsilon \tag{1.5-8}$$

with ϵ satisfying (1.5-4). Thus

$$\begin{aligned} fl(x * y) &= \begin{cases} (m_x m_y + \epsilon)2^{f_x + f_y} & |m_x m_y| \geq \frac{1}{2} \\ (2m_x m_y + \epsilon)2^{f_x + f_y - 1} & \frac{1}{2} > |m_x m_y| \geq \frac{1}{4} \end{cases} \\ &= m_x m_y \times 2^{f_x + f_y} \times \begin{cases} 1 + \dfrac{\epsilon}{m_x m_y} & |m_x m_y| \geq \frac{1}{2} \\ 1 + \dfrac{\epsilon}{2m_x m_y} & \frac{1}{2} > |m_x m_y| \geq \frac{1}{4} \end{cases} \\ &= x * y(1 + \epsilon_m) \end{aligned} \tag{1.5-9}$$

where

$$|\epsilon_m| \leq 2|\epsilon| \leq 2^{-p} \tag{1.5-10}$$

Our result is that the computed product is the exact product times a factor $1 + \epsilon_m$ with ϵ_m bounded as in (1.5-10). Thus the bound on the relative error in multiplication is the same as the bound on the relative error in rounding a number.

Division Here the exponents are subtracted; one-half the numerator mantissa is divided by the denominator mantissa (to avoid quotients greater than 1), and the result is normalized and then rounded. Proceeding as above and assuming $y \neq 0$, we have

$$\frac{x}{y} = \frac{m_x/2}{m_y} 2^{f_x - f_y + 1} \tag{1.5-11}$$

with

$$\frac{1}{4} \leq \left| \frac{m_x}{2m_y} \right| < 1 \tag{1.5-12}$$

Then by an analysis precisely similar to that above

$$fl\left(\frac{x}{y}\right) = \frac{x}{y}(1 + \epsilon_d) \tag{1.5-13}$$

where

$$|\epsilon_d| \leq 2^{-p} \tag{1.5-14}$$

Addition and subtraction Here the mantissa of the operand of smaller magnitude, say y, is shifted $f_x - f_y$ places to the right, the resulting mantissas are added (or subtracted), the result is normalized (by shifting right if the result overflowed), f_x is adjusted accordingly, and then the result is rounded. Thus

$$x \pm y = (m_x \pm m_y \times 2^{f_y - f_x})2^{f_x} \tag{1.5-15}$$

where if the true sum or difference is nonzero,

$$2^{-p} \le |m_x \pm m_y \times 2^{f_y - f_x}| < 2 \tag{1.5-16}$$

Normalization consists of a left shift of t places, where

$$-1 \le t \le p - 1 \tag{1.5-17}$$

with -1 corresponding to a right shift of 1. In general t is such that

$$2^{-t-1} \le |m_x \pm m_y \times 2^{f_y - f_x}| < 2^{-t} \tag{1.5-18}$$

Then

$$\begin{aligned} fl(x \pm y) &= [(m_x \pm m_y \times 2^{f_y - f_x})2^t + \epsilon]2^{f_x - t} \\ &= (x \pm y)\left(1 + \frac{2^{-t}\epsilon}{m_x \pm m_y \times 2^{f_y - f_x}}\right) \\ &= (x \pm y)(1 + \epsilon_a) \end{aligned} \tag{1.5-19}$$

where, using (1.5-18),

$$|\epsilon_a| \le 2|\epsilon| \le 2^{-p} \tag{1.5-20}$$

In all the floating-point arithmetic operations, therefore, the relative error in the result has a magnitude no greater than 1 in the least significant position of the mantissa.

It is important to note that if, instead of our assumption of the existence of a $2p$-bit accumulator, there were only a p-bit accumulator, our results would be quite different. In particular, instead of (1.5-19) and (1.5-20) we would have {20}

$$fl(x \pm y) = (x \pm y)(1 + \epsilon_a) \tag{1.5-21}$$

with

$$|\epsilon_a| \le \tfrac{3}{2} \times 2^{-p} \tag{1.5-22}$$

Error analysis of a floating-point computation. Consider the sum

$$\sum_{i=1}^{n} x_i = x_1 + x_2 + \cdots + x_n \tag{1.5-23}$$

Suppose first that we add successive terms and use the results of the previous section. Define

$$\begin{aligned} s_1 &= x_1 \\ s_r &= fl(s_{r-1} + x_r) = (s_{r-1} + x_r)(1 + \epsilon_r) \qquad r = 2, \ldots, n \end{aligned} \tag{1.5-24}$$

with $|\epsilon_r| \le 2^{-p}$. Then

$$\begin{aligned} s_n &= (s_{n-1} + x_n)(1 + \epsilon_n) \\ &= [(s_{n-2} + x_{n-1})(1 + \epsilon_{n-1}) + x_n](1 + \epsilon_n) \\ &\cdots\cdots\cdots\cdots\cdots\cdots\cdots\cdots \\ &= x_1(1 + \eta_1) + x_2(1 + \eta_2) + \cdots + x_n(1 + \eta_n) \end{aligned} \tag{1.5-25}$$

where

$$1 + \eta_r = (1 + \epsilon_r)(1 + \epsilon_{r+1}) \cdots (1 + \epsilon_n) \qquad r = 2, \ldots, n$$
$$\eta_1 = \eta_2 \tag{1.5-26}$$

Therefore

$$(1 - 2^{-p})^{n-r+1} \leq 1 + \eta_r \leq (1 + 2^{-p})^{n-r+1} \qquad r = 2, \ldots, n \tag{1.5-27}$$

and finally

$$fl\left(\sum_{r=1}^{n} x_i\right) = \left(\sum_{i=1}^{n} x_i\right)\left(1 + \frac{\sum_{i=1}^{n} x_i \eta_i}{\sum_{i=1}^{n} x_i}\right) \tag{1.5-28}$$

From this result we note the following:

1. If the true sum is small relative to the x_i's, the relative error can be large. This is as we would expect since this case corresponds to *subtractive cancellation* as the sum is developed.
2. Since the bounds of η_i decrease with i, the smallest upper bound on the relative error is obtained if the numbers are added in increasing order of magnitude (although in practice this does not necessarily give the smallest error {21}).

Now suppose in computing (1.5-23) that the double-precision ($2p$-bit) mantissas obtained at each stage are not rounded after normalization but instead are used directly in the next stage. Assuming that the accumulator can accommodate only $2p$ bits, the error in each addition using (1.5-22) with p replaced by $2p$ is such that

$$s_n = x_1(1 + \bar{\eta}_1) + x_2(1 + \bar{\eta}_2) + \cdots + x_n(1 + \bar{\eta}_n) \tag{1.5-29}$$

with

$$(1 - \tfrac{3}{2} \times 2^{-2p})^{n-r+1} \leq 1 + \bar{\eta}_r \leq (1 + \tfrac{3}{2} \times 2^{-2p})^{n-r+1} \qquad r = 2, \ldots, n$$
$$\bar{\eta}_1 = \bar{\eta}_2 \tag{1.5-30}$$

When we compare (1.5-27) and (1.5-30), the advantage of retaining the double-precision mantissa is clear since the bound on $\bar{\eta}_r$ is far smaller than that on η_r. Even if the double-precision s_n in (1.5-29) is then rounded to single precision, the result is only to multiply the right-hand side of (1.5-29) by $1 + \eta$ with η bounded by 2^{-p}.

Whereas the calculation of sums like (1.5-23) is not very common, the calculation of *inner products*

$$\sum_{i=1}^{n} x_i y_i = x_1 y_1 + \cdots + x_n y_n \tag{1.5-31}$$

is very common, particularly, in numerical linear algebra. Reasoning similar to that above implies the great desirability of using double-precision products $x_i y_i$ in computing the sum (1.5-31) {22}.

More generally readers will have been well served by this section if they recognize that for significant floating-point computations a careful error analysis is both worthwhile and nontrivial.

1.5-4 Overflow and Underflow

The discussion above ignored the possibility that the result of any floating-point operation will not be representable in the floating-point representation scheme of the computer. But this can occur as the result of any arithmetic operation in floating point.

The magnitude of the largest number which can be expressed in the form (1.5-1) is

$$M \times 2^F \tag{1.5-32}$$

where F is the largest positive exponent (usually $2^{q-1} - 1$) and $M = 1 - 2^{-p}$ is a mantissa all of whose bits are 1. Floating-point *overflow* results when the result of a floating-point operation has a magnitude greater than that in (1.5-32). This can happen with any arithmetic operation. For example, with $q = 8$ and therefore $F = 128 - 1 = 127$, multiplying $\frac{1}{2} \times 2^{70}$ and $\frac{1}{2} \times 2^{80}$ gives a result greater than (1.5-32). Similarly the difference of $\frac{1}{2} \times 2^{127}$ and $-\frac{1}{2} \times 2^{127}$ also overflows.

Underflow results when the result of a floating-point operation is nonzero but too small to be expressed in the form (1.5-1). If numbers are required to be normalized, the smallest number expressible in the form (1.5-1) is $\frac{1}{2} \times 2^{-F}$, where $-F$ is the most negative exponent allowed, usually -2^{q-1}. With $q = 8$ this is -128. Then, for example, $\frac{1}{2} \times 2^{-80}$ divided by $\frac{1}{2} \times 2^{50}$ is a number too small to be expressed. Similar examples could be given for the other three operations. If nonnormalized numbers are allowed, the smallest nonzero number in magnitude is smaller than in the normalized case but underflow is still possible with any arithmetic operation {23}.

Overflow is almost always the result of an error in the computation. For underflow, however, it is usually sufficient to replace the result by 0 although there are exceptions to this.

1.5-5 Single- and Double-Precision Arithmetic

Arithmetic performed on either fixed- or floating-point numbers of length l is called *single-precision* arithmetic. The example in Sec. 1.5-3 indicated an instance where it is desirable to retain intermediate values of length greater than l. In some calculations it is desirable to be able to use numbers of length $2l$ throughout or in some significant part of the calculation. Such arithmetic is called *double-precision* arithmetic, and the numbers are called double-precision numbers.

Many computers on which the standard length of numbers for arithmetic is l have machine-language instructions for doing double-precision arithmetic at least on floating-point numbers. On computers without this facility the effect of double-precision arithmetic can be obtained by suitable programming. Occasionally triple- or even higher-precision arithmetic is necessary to achieve a needed level of accuracy. Such higher precision is almost always accomplished by programming.

The relationship of double-precision floating-point numbers to single-precision floating-point numbers varies somewhat from computer to computer. But, using our previous notation, the most usual arrangement is to have a $(p + l)$-bit mantissa and, as with single precision, a q-bit exponent.

1.6 ERROR ANALYSIS

The intuitive approach to error analysis is to start from the errors in the initial data and take into account the successive roundoff errors as the calculation proceeds in order to compute estimates of, or bounds on, the error in the result. For example, if x_1, x_2, and x_3 may each have an error, e_1, e_2, or e_3 bounded, respectively, by ϵ_1, ϵ_2, and ϵ_3, then, using overbars to denote true values,

$$\begin{aligned} x_1 + x_2x_3 &= (\bar{x}_1 + e_1) + (\bar{x}_2 + e_2)(\bar{x}_3 + e_3) \\ &= \bar{x}_1 + \bar{x}_2\bar{x}_3 + e_1 + \bar{x}_2e_3 + \bar{x}_3e_2 + e_2e_3 \end{aligned} \tag{1.6-1}$$

and the error in the exact computed value is bounded by

$$\epsilon_1 + |\bar{x}_2|\epsilon_2 + |\bar{x}_3|\epsilon_3 + \epsilon_2\epsilon_3 \tag{1.6-2}$$

Using this technique, as in the previous section, we can compute absolute and relative error bounds for each arithmetic operation, and using them, we can (theoretically at least) compute error bounds for long sequences of calculations. There are two main drawbacks to this *forward error-analysis* procedure:

1. The resulting bounds are often very *conservative;* i.e., the true error is unlikely to be near the bound. We saw one example of this in Sec. 1.4-1. The result is that we may be forced to use estimates of the probable error rather than bounds on the error, thereby incurring an obvious risk.
2. The analysis itself is often extremely difficult and/or tedious for complex calculations.

Notwithstanding these drawbacks, forward error analysis can lead to useful results and has been responsible for some notable results in numerical analysis. For many calculations, however, *backward error analysis*, which we now consider, is to be preferred.

1.6-1 Backward Error Analysis

The essence of backward error analysis is to take the result of a calculation and determine from it the range of initial data which could have given rise to it. Why is this a useful thing to do? One reason is that it is not unusual to determine that the range of initial data which could give rise to the result is comparable to the *inherent* errors in these data caused by observational errors or by the roundoff committed when exact (decimal) numbers are converted to inexact binary numbers on input to the computer.

The analyses in Sec. 1.5-3 of floating-point operations were backward error analyses. For example:

1. Equation (1.5-9) indicates that the floating-point product of two numbers is the true product of the true x and a y whose relative difference from the y actually used is ϵ_m. Since the relative error committed by rounding an exact y to p mantissa bits has a magnitude bounded by 2^{-p}, we note that the y which would give the calculated floating-point product as a true product differs from the y used by no more than this roundoff-error bound.
2. Equation (1.5-19) indicates that the floating-point sum of two numbers is the true sum of two summands each of which has a relative difference from the summands actually used of ϵ_a, where, as above, the relative difference is no more than what could be expected from rounding each summand to p mantissa bits.
3. Equation (1.5-25) indicates that the floating-point sum of n numbers is the true sum of n numbers each of which differs from the true summand x_i by a relative difference of less than η_i. In this case, use of (1.5-27) indicates that the relative difference between the summands which would give the true results and those actually used may be substantially greater than a single rounding error.

The reader might easily conclude that the last example, the only one of the three of a nontrivial calculation, is the most typical and that in general backward error analysis does not lead to results which show that the result obtained could have come from initial data differing from those actually used by only the input roundoff error in the data. But a look at (1.5-30) indicates that if the double-precision mantissas are retained, the factors which could make the calculated sum the true sum differ very little from the actual data used. That is, even after s_n in (1.5-29) is rounded to single precision, the error associated with x_r is

$$(1 + \bar{\eta}_r)(1 + \epsilon) \tag{1.6-3}$$

with $\bar{\eta}_r$ given by (1.5-30) and ϵ bounded by 2^{-p}. The 2^{-2p} term in $\bar{\eta}_r$ means that the product in (1.6-3) is very close to $1 + \epsilon$ unless n is very large. And, of course, the bound on ϵ is just that of the maximum magnitude of a single roundoff error.

The use of backward error analysis together with such techniques as the use of double-precision arithmetic in certain portions of a calculation can lead to some very striking results which attest to the value of this type of error analysis. One such result will be developed in detail in Sec. 9.4.

1.7 CONDITION AND STABILITY

Instability is a phenomenon which results when individual roundoff errors propagate through a calculation with increasing effect. In this section we shall discuss this phenomenon and the related subject of the *condition* of a problem.

The system of differential equations

$$\begin{aligned} y_1' &= y_2 \\ y_2' &= y_1 \end{aligned} \tag{1.7-1}$$

has the general solution

$$\begin{aligned} y_1 &= a_1 e^x + a_2 e^{-x} \\ y_2 &= a_1 e^x - a_2 e^{-x} \end{aligned} \tag{1.7-2}$$

With the initial conditions

$$y_1(0) = -y_2(0) = 1 \tag{1.7-3}$$

the constants a_1 and a_2 are determined to be

$$a_1 = 0 \qquad \text{and} \qquad a_2 = 1 \tag{1.7-4}$$

Now suppose that Eqs. (1.7-1) with initial conditions (1.7-3) are solved numerically by any method whatsoever whose aim is to compute y_1 and y_2 at a sequence of points $x_1, x_2 \ldots$. The effect of roundoff error is to compute a numerical solution of (1.7-1) with initial conditions perturbed from those of (1.7-3). But even the slightest perturbation of (1.7-3) will result in a nonzero a_1. Since the a_1 terms in (1.7-2) are rising exponentials, any nonzero a_1, no matter how small, will result in an e^x term which dominates the e^{-x} term for sufficiently large x. Therefore, it is not possible to compute a solution to (1.7-1) with initial conditions (1.7-3) which, for sufficiently large x, will not result in an arbitrarily large error relative to the true solution. This problem is therefore inherently unstable or, to use the more common term in numerical analysis, *ill-conditioned.*

A second example of ill condition is one discussed at greater length in

Sec. 8.13. The polynomial

$$f(z) = (z + 1)(z + 2) \ldots (z + 20) \tag{1.7-5}$$

has zeros $-1, -2, \ldots, -20$. However, the polynomial

$$f(z) + 2^{-23}z^{19} \tag{1.7-6}$$

has five pairs of complex zeros (one of which is $-19.502 \pm 1.940i$), and among its 15 real zeros one is -20.847. Thus, a change in one coefficient of $f(z)$ (which, by the way, is equal to 210) by $2^{-23} \approx 10^{-7}$ results in a large change in the zeros. The roundoff in any numerical method for computing the zeros of (1.7-5) (given, of course, in the form $z^{20} + a_1 z^{19} + \cdots$) has the effect that the zeros computed are the true zeros of a polynomial with perturbed coefficients. Thus, computing the zeros of (1.7-5) is an ill-conditioned problem (although with multiple-precision arithmetic accurate zeros can nevertheless be calculated). Although, of course, we know the zeros of (1.7-5) without computing, it turns out (see Sec. 8.13) that high-degree polynomials generally tend to be ill-conditioned with respect to computation of their zeros.

A well-conditioned problem, therefore, is one in which small changes in the data of a problem result, in some sense, in small changes in the solution, and an ill-conditioned problem has the opposite property. These two examples should illustrate how important it is to know or to be able to estimate the *condition* of a problem before attempting a numerical solution of it.

Condition is related to how sensitive the solution of a problem is to changes in its data. Stability, on the other hand, is concerned with the numerical method used to solve a problem, in particular with the behavior of the roundoff errors introduced in the numerical solution. For example, the problem

$$y' = -y \qquad y(0) = 1 \tag{1.7-7}$$

has the solution $y = e^{-x}$ and is very well-conditioned in that a small change in the initial condition $[y(0) = 1 + \epsilon]$ results in a small change in the solution $[y = (1 + \epsilon)e^{-x}]$. But there are numerical methods for solving this equation (one such, Milne's method, is discussed in Sec. 5.5-2) which give useless results for medium to large values of x because the roundoff errors have the effect of introducing a spurious rising-exponential solution which soon overwhelms the true, falling-exponential solution. Or, in other words, the roundoff errors introduced at one stage of the computation propagate with increasing magnitude in later stages. Such a numerical method is said to be *unstable* although, as is the case with Milne's method, it may be unstable for some problems but stable for others.

The distinction between the condition of a problem and the stability of a method is a most important one to understand. An ill-conditioned problem can be solved accurately, if this is possible at all, only by very careful calculation, e.g., multiple-precision arithmetic, quite aside from the method used. A well-conditioned problem can be solved accurately by any numerical method which is stable for that problem. A method which is unstable for a particular (say, well-conditioned) problem may give accurate results in the calculation—before the propagation of roundoff error has overwhelmed the true solution—but inevitably will give bad results if used long enough.

BIBLIOGRAPHIC NOTES

Most texts on numerical analysis contain an introductory chapter (or chapters) with material in some degree similar to this chapter. In particular we refer the reader to Dahlquist, Björck, and Anderson (1974) and Young and Gregory (1972).

The best book available on floating-point arithmetic, roundoff-error analysis, and error analysis more generally is Wilkinson (1964). Knuth (1969) also has an extensive and excellent discussion of many of these matters. A recent general textbook with a good chapter on floating-point arithmetic is Shampine and Allen (1973). Hamming (1973) devotes a chapter to roundoff error. For a deeper understanding of the probabilistic theory underlying roundoff-error analysis, see Feller (1950). For more details on the algorithms by which computers perform arithmetic see Knuth (1969), Flores (1960, 1963), and Richards (1955).

BIBLIOGRAPHY

Dahlquist, G., and Å. Björck (1974): *Numerical Methods* (trans. N. Anderson), Prentice-Hall, Inc., Englewood Cliffs, N.J.

Feller, W. (1950): *Probability Theory and Its Applications*, vol. 1, John Wiley & Sons, Inc., New York.

Flores, I. (1960): *Computer Logic*, Prentice-Hall, Inc., Englewood Cliffs, N.J.

——— (1963): *The Logic of Computer Arithmetic*, Prentice-Hall, Inc., Englewood Cliffs, N.J.

Hamming, R. W. (1973): *Numerical Methods for Scientists and Engineers*, 2d ed., McGraw-Hill Book Company, New York.

Henrici, P. (1964): *Elements of Numerical Analysis*, John Wiley & Sons, Inc., New York.

Knuth, D. E. (1969): *The Art of Computer Programming*, vol. 2, Addison-Wesley Publishing Company, Inc., Reading, Mass.

Newman, M. (1962): "Matrix Computations," in J. Todd (ed.), *Survey of Numerical Analysis*, pp. 225–254, McGraw-Hill Book Company, New York.

Richards, R. K. (1955): *Arithmetic Operations in Digital Computers*, D. Van Nostrand Company, Inc., Princeton, N.J.

Shampine, L. F., and R. C. Allen, Jr. (1973): *Numerical Computing: An Introduction*, W. B. Saunders Company, Philadelphia.

Wilkinson, J. H. (1964): *Rounding Errors in Algebraic Processes*, Prentice-Hall, Inc., Englewood Cliffs, N.J.

Young, D. M., and R. T. Gregory (1972): *A Survey of Numerical Mathematics*, vol. 1, Addison-Wesley Publishing Company, Inc., Reading, Mass.

PROBLEMS

Section 1.3

1 Let all the numbers in the following calculations be correctly rounded to the number of digits shown: (*a*) 1.1062 + .947; (*b*) 23.46 − 12.753; (*c*) (2.747)(6.83); (*d*) 8.473/.064. For each calculation, determine the smallest interval in which the result, using true instead of rounded values of the quantities, must lie.

2 Let $y_1 = .9863$ and $y_2 = .0028$ be correctly rounded approximations to x_1 and x_2, respectively. Find the maximum magnitude of the difference between the calculated values of $1/y_1$ and $1/y_2$ and the true values.

3 Use (1.3-5) to approximate the error incurred when correctly rounded numbers are used to compute (*a*) the product of n numbers; (*b*) the quotient of two numbers; (*c*) a power of a number where the power is known exactly; (*d*) a power of a number where the power is also in error.

4 (*a*) Use the results of parts (*a*) and (*b*) of Prob. 3 to get estimates of the errors in the computations of Prob. 1*c* and *d*. Where there are discrepancies between these bounds and those calculated in Probs. 1 and 2, what causes them?

(*b*) Repeat part (*a*) on the computations of Prob. 2.

5 (*a*) Use the results of Prob. 3*c* and *d* to get estimates of the errors in (i) $(6.45)^{1/32}$ and (ii) $(8.47)^{.643}$ if $\frac{1}{32}$ is exact and all other numbers are correctly rounded as shown.

(*b*) Suppose you wish the values of (i) cos 1.473, (ii) $\tan^{-1}$ 2.621, (iii) ln 1.471, (iv) $e^{2.653}$ but in each case have a table at an interval of .01 in the argument. Use (1.3-5) to get an estimate of the error made by using the nearest values in the tables to the given arguments.

(*c*) Suppose each of the values in part (*b*) is a correctly rounded value. Estimate the maximum error incurred using the nearest tabular values.

6 (*a*) Derive the Cauchy-Schwarz inequality

$$\left(\sum_{i=1}^{n} u_i v_i\right)^2 \le \left(\sum_{i=1}^{n} u_i^2\right)\left(\sum_{i=1}^{n} v_i^2\right)$$

(*b*) Use this to show that the Euclidean vector norm (1.3-6) satisfies property 3 for norms.
(*c*) Show that the L_1 and L_∞ norms satisfy properties 1 to 3.
(*d*) Derive the Hölder inequality

$$|\mathbf{u}^T\mathbf{v}| \le \|\mathbf{u}\|_p \|\mathbf{v}\|_q$$

where p, q are numbers greater than 1 such that $1/p + 1/q = 1$.
(*e*) Use this to show that the L_p norm (1.3-7) satisfies property 3 for norms for $1 < p < \infty$.
(*f*) Show that the weighted L_p norm (1.3-9) satisfies properties 1 to 3 for norms.

(*g*) Show that $\lim_{p\to\infty} \left(\sum_{i=1}^{n} |v_i|^p\right)^{1/p} = \max_{1\le i\le n} |v_i|$.

7 (*a*) Show that (1.3-13) defines a matrix norm.
(*b*) Given a matrix A and vector $\mathbf{x}$ show that

$$\|A\mathbf{x}\|_\infty = \max_i \left|\sum_{j=1}^{n} a_{ij}x_j\right|$$

(*c*) Thus derive that

$$\|A\mathbf{x}\|_\infty \le \sum_{j=1}^{n} |a_{rj}| \cdot |x_j|$$

where r is that value of i which maximizes the sum in part (*b*).
(*d*) By choosing appropriate values of x_j, show that

$$\|A\|_\infty = \max_i \left(\sum_{j=1}^{n} |a_{ij}|\right)$$

(*e*) Similarly show that

$$\|A\|_1 = \max_j \left(\sum_{i=1}^{n} |a_{ij}| \right)$$

(*f*) Show that the matrix norm (1.3-18) is a vector norm in n^2 space for the vector $(a_{11}, \ldots, a_{1n}, a_{21}, \ldots, a_{nn})$ so that properties 1 to 3 for norms hold.

(*g*) Show that $\|AB\|_E^2 \leq \|A\|_E^2 \|B\|_E^2$ so that (1.3-18) is indeed a norm [Ref.: Newman (1962), p. 228.]

8 (*a*) By considering the vector $\mathbf{y} = \mathbf{x}/\|\mathbf{x}\|$, show that (1.3-17) holds for any vector $\mathbf{x}$.

(*b*) Show that for any given vector norm, the value of the norm of any matrix A given by a consistent norm is not less than the value given by the subordinate matrix norm.

(*c*) Use the Cauchy-Schwarz inequality to show that $\|A\mathbf{x}\|_2^2 \leq \|A\|_E^2 \|\mathbf{x}\|_2^2$, so that the Frobenius norm is consistent with the L_2 vector norm.

(*d*) The trace of a matrix, tr (A), is defined to be $\sum_{i=1}^{n} a_{ii}$ and can be shown to be equal to $\sum_{i=1}^{n} \lambda_i$, where the λ_i are the eigenvalues of A. Show that $\|A\|_E^2 = \text{tr}\,(A^T A)$.

(*e*) Use this together with (1.3-22) to show that $\|A\|_E \leq n^{1/2} \|A\|_2$.

9 Let λ be an eigenvalue of A and $\mathbf{x} \neq \mathbf{0}$ the corresponding eigenvector, so that $A\mathbf{x} = \lambda\mathbf{x}$. By taking norms of both sides of this equation and using (1.3-17), prove (1.3-21).

Section 1.4

***10** (*a*) Let ϵ have a probability density $p(x)$ on $(-\infty, \infty)$. Show that the probability that ϵ does not exceed x is given by

$$P(x) = \int_{-\infty}^{x} p(x)\, dx$$

$P(x)$ is called the probability distribution of ϵ.

(*b*) Use this result to show that if ϵ_1 and ϵ_2 are independently distributed variables with probability densities $p_1(x)$ and $p_2(x)$, the probability distribution of $\epsilon_1 + \epsilon_2$ is given by

$$\iint_{s+t<x} p_1(s) p_2(t)\, ds\, dt$$

(*c*) Manipulate this integral to show that the probability density of $\epsilon_1 + \epsilon_2$ is given by

$$\int_{-\infty}^{\infty} p_1(x - t) p_2(t)\, dt$$

***11** (*a*) Use the result of the previous problem to verify Fig. 1.2.

(*b*) Use this result to find the probability density of $\epsilon_1 + \epsilon_2 + \epsilon_3$ when each of these variables is independently distributed as shown in Fig. 1.1.

(*c*) Compare the results of (*a*) and (*b*) with the corresponding normal density functions given by (1.4-4) with $n = 2$ and 3 by plotting the graphs of the functions.

12 The error function is defined by

$$\text{erf}\,(x) = \frac{2}{\sqrt{\pi}} \int_0^x e^{-t^2}\, dt$$

(*a*) Show that

$$\lim_{x \to \infty} \text{erf}\,(x) = 1$$

Hint: Consider erf (x) erf (y) and use polar coordinates.

(*b*) Use this result and (1.4-4) to show that the normal distribution function corresponding to the normal density function with zero mean and variance $n/12$ is given by $\frac{1}{2} + \frac{1}{2}$ erf $[(6/n)^{1/2}x]$.

(*c*) Finally deduce that the probability that ϵ in (1.4-4) is less than x in magnitude is given by erf $[(6/n)^{1/2}|x|]$.

13 (*a*) Use a table of the error function to verify Eq. (1.4-5).

(*b*) How large must n be so that the probable error is less than one-tenth the maximum error $n/2$? For this n, use part (*c*) of Prob. 12 to estimate the probability that the error is greater in magnitude than three-quarters of the maximum error.

***14** Consider the iteration

$$x_{i+1} = \alpha x_i + \beta \qquad x_0 = a \qquad \beta \neq 0 \text{ or } a(1-\alpha)$$

(*a*) Show that $|\alpha| < 1$ in order for the iteration to converge, i.e., in order for $\lim_{i\to\infty} x_i$ to exist.

(*b*) Let ϵ_i be the accumulated roundoff error in x_i, and let δ_i be the roundoff error introduced in the calculation of x_i from x_{i-1} (with δ_0 the roundoff error in x_0). Assume α and β are known exactly and that all x_i are correctly rounded to d decimals. Find a bound δ on δ_i assuming the arithmetic is performed as efficiently as possible.

(*c*) Use the iteration equation to derive a difference equation relating ϵ_i, ϵ_{i-1}, and δ_i.

(*d*) Use this result to show by induction that

$$\epsilon_i = \sum_{j=0}^{i} d_{ij}\delta_j \qquad \text{where } d_{ii} = 1$$

$$d_{ij} = \alpha d_{i-1,j};\ i > j \geq 0$$

(*e*) Thus, deduce that

$$d_{ij} = \alpha^{i-j}$$

and therefore that

$$|\epsilon_i| \leq \frac{1}{1-|\alpha|}\delta$$

(*f*) Use the results of parts (*b*), (*d*), and (*e*) and Eq. (1.4-7) to show that the variance of the probability density of ϵ_i is given by

$$\sigma^2 = \frac{10^{-2d}}{2}\frac{1-\alpha^{2i+2}}{1-\alpha^2} \leq \frac{10^{-2d}}{12}\frac{1}{1-\alpha^2}$$

[Ref.: Henrici (1964), pp. 309–314.]

15 Use a Lagrangian multiplier to show that σ^2 in (1.4-7) is minimized subject to the constraint (1.4-8) when all the α_i's are equal.

Section 1.5

16 For each of the four arithmetic operations derive bounds on rounded fixed-point calculation of the form

$$fi(x \text{ op } y) = x \text{ op } y + \epsilon$$

with ϵ suitably bounded. Assume p-digit fixed-point numbers in the range $[-1, 1]$ and no overflow in the result.

17 Consider a floating-point representation (1.5-1) with $p = 27$, $q = 8$. Which of the 10 million numbers 9,000,000.0, 9,000,000.1, ..., 9,999,999.9 have the same representations in this format? Why must it be true in general that in any floating-point system some real numbers must have the same representation?

18 Prove the following results about the floor and ceiling functions:
(*a*) $\lfloor x \rfloor = \lceil x \rceil$ if and only if x is an integer.
(*b*) $\lfloor x \rfloor + 1 = \lceil x \rceil$ if and only if x is not an integer.
(*c*) $-\lfloor x \rfloor = \lceil -x \rceil$ for all x.

***19** (*a*) Let the floating-point mantissa length $p = 4$. By considering the addition of $.1000 \times 2^7$ and $-.1001 \times 2^2$ show that a $(2p + 1)$-bit accumulator must be available in general to get the corrected rounded result when adding two floating-point numbers.

(*b*) After m_y has been shifted $f_x - f_y$ places to the right in a $(p + 2)$-bit accumulator, replace m_y by

$$\operatorname{sgn}(m_y)2^{-p-2}\lfloor 2^{p+2}|m_y| \rfloor \qquad \text{if } \operatorname{sgn}(m_y) = \operatorname{sgn}(m_x)$$

$$\operatorname{sgn}(m_y)2^{-p-2}\lceil 2^{p+2}|m_y| \rceil \qquad \text{if } \operatorname{sgn}(m_y) \neq \operatorname{sgn}(m_x)$$

Show that this has no effect if $f_x < f_y + 3$ $(f_x \geq f_y)$.

(c) If $f_x \geq f_y + 3$, show that the replacement of part (*b*) results in $|m_x + m_y| > \frac{1}{4}$ and thus infer that the bit which governs the rounding is now correctly in the $p + 1$ or $p + 2$ position in the accumulator.

(*d*) Finally deduce that an appropriate application of the transformation in part (*b*) will allow a $(p + 2)$-bit accumulator to be used to get the correctly rounded result in floating-point addition.

(*e*) Show that even with a transformation like that in part (*b*), a $(p + 1)$-bit accumulator cannot give a correctly rounded result in all cases. [Ref.: Knuth (1969).]

20 Consider floating-point addition with a p-bit accumulator on numbers with p-bit mantissas and let $\epsilon = 2^{-p}$. Assume that when the sum of two mantissas overflows, the overflow bit is retained and may be shifted right.

(*a*) Show that if no overflow occurs when the two mantissas (one of which may have been shifted) are added, then

$$fl(x + y) = x(1 + \epsilon_x) + y(1 + \epsilon_y)$$

where $|\epsilon_x|$ and $|\epsilon_y|$ are both bounded by ϵ.

(*b*) Show that if overflow occurs, the addition may incur two rounding errors bounded by, respectively, $2^{b-1}2^{-1-p}$ and $2^{b-1}2^{-p}$, where b is the exponent of the computed sum.

(*c*) Thus deduce that in the case described in part (*b*)

$$fl(x + y) = (x + y)(1 + \epsilon_{xy}) \qquad \text{where } |\epsilon_{xy}| \leq \tfrac{3}{2} \times 2^{-p}$$

[Ref.: Wilkinson (1964).]

21 (*a*) Explain why, in practice, adding a sequence of floating-point numbers in increasing order of magnitude may not give the smallest error.

(*b*) Suppose using an 8-digit accumulator and the algorithm (1.5-24) you wish to add

$$.1025 \times 10^4 + (-.9123) \times 10^3 + (-.9663) \times 10^2 + (-.9315) \times 10^1$$

Show that these numbers can be added without incurring any rounding errors but that adding them in order of increasing magnitude results in some rounding errors. [Ref.: Wilkinson (1964).]

22 Consider the calculation of the inner product

$$s_n = fl(x_1 y_1 + x_2 y_2 + \cdots + x_n y_n)$$

(*a*) Show that

$$t_r = fl(x_r y_r) \qquad r = 1, \ldots, n$$

$$s_1 = t_1$$

$$s_r = fl(s_{r-1} + t_r) \qquad r = 2, \ldots, n$$

is an algorithm to calculate s_n.

(*b*) From this algorithm deduce relations analogous to (1.5-24).

(*c*) Finally deduce relations analogous to Eqs. (1.5-25) to (1.5-26) and (1.5-29) to (1.5-30). [Ref.: Wilkinson (1964).]

23(*a*) Using the notation of (1.5-1), what are the largest and smallest unnormalized floating-point numbers in magnitude?

(*b*) Show that underflow and overflow are possible with unnormalized numbers for all arithmetic operations.

***24** Consider a double-precision arithmetic scheme where a $2d$-digit number is held in two d-digit words. Suppose that the computer can add two d-digit numbers at a time and can recognize when overflow occurs.

(*a*) Devise an algorithm for adding two such double-precision numbers assuming all numbers are positive.

(*b*) If both halves of the double-precision number have a sign associated with them, and if the signs are required to be the same, modify this algorithm so that it can handle addition of positive and negative numbers so that the sum will also have the same sign in both halves.

In both parts assume there is no overflow in the overall sum.

Section 1.6

25 Find a bound on the relative error in computing the quantity on the left-hand side of (1.6-1) if the calculation is performed using floating-point arithmetic.

26 Perform a backward error analysis on

$$p_n = fl(x_1 x_2 \cdots x_n)$$

to show that the relative error in the product is of the same order of magnitude as the individual roundoff errors made when the true values of x_i are rounded to obtain the x_i's inside the computer.

Section 1.7

27 (*a*) Consider the linear system

$$2x + 6y = 8$$

$$2x + 6.00001y = 8.00001$$

Is this problem well-conditioned or ill-conditioned? Why?

(*b*) Consider

$$2x + 6y = 8$$

$$2x + 5.99999y = 8.00002$$

Solve this system exactly. What does the solution tell you about your answer to part (*a*)?

(*c*) Give a geometric interpretation of your answers to parts (*a*) and (*b*).

28 Consider the differential equation

$$y'' - y = 0$$

(*a*) For what initial conditions

$$y(0) = a \qquad y'(0) = b$$

is the problem of solving this equation stable?

(*b*) What is the relation of part (*a*) to (1.7-1)?

29 For the equation $f(x) = 0$ consider solving this equation by the algorithm

$$\text{Guess } x_0$$

$$x_{i+1} = F(x_i)$$

where $F(x)$ is somehow derived from $f(x)$. Suppose it can be shown that

$$\bar{x} = \lim_{i \to \infty} x_i$$

exists and that $f(\bar{x}) = 0$. What can you say about the stability of such a method of solving $f(x) = 0$?

CHAPTER

TWO

APPROXIMATION AND ALGORITHMS

2.1 APPROXIMATION

We have said that numerical analysis is concerned with the solution of mathematical problems by arithmetic processes. Clearly, then, the need to approximate nonarithmetic quantities by arithmetic quantities—and to ascertain the errors associated with such approximations—lies at the heart of much of numerical analysis. In a given situation there will usually be several possible methods of obtaining the desired approximation. Which of these to choose depends upon which of various possible criteria are used to judge how efficacious a given approximation is. The following simple example will illustrate what these criteria are and how they affect the choice of an approximation.

Suppose that we are given a value of x and we wish to calculate $\sqrt{x}$. The following are among the possibilities open to us:

1. Use the classic method learned by most people in grammar school which begins by pairing off digits on either side of the decimal point.
2. Look up x in a table of square roots and if x lies between two arguments in the table, interpolate (see Chap. 3) to find $\sqrt{x}$.
3. Use any one of a number of iterative techniques (see Chap. 8) to compute $\sqrt{x}$.

Our object here is not to decide which of these methods should be used but to discuss the considerations that must precede any such decision. We naturally assume that all the methods "work," i.e., that they all lead to a result which can reasonably be considered an approximation to $\sqrt{x}$. The basic question we must answer is: What error can we tolerate in the result?

This question not only recognizes the importance of the errors, both truncation and roundoff, incurred in an approximation but, more subtly, implies the importance of being able to estimate or bound these errors. The latter is a consideration of first importance in choosing a method for the solution of a problem. Only when we have methods where it is possible to estimate or bound the error can we then try to compare these methods on the basis of the magnitudes of the errors to which they lead. Each of the above methods for computing the square root has an error which can be estimated or bounded and for which the truncation component, at least, can be made arbitrarily small by carrying the computation far enough, e.g., by computing as many decimal places as desired in the first method. Therefore, on the basis of error considerations, each is a reasonable candidate for computing the square root.

The reader may have noted that the question above is ambiguous. What do we mean by error in the result? Absolute error or relative error? Do we wish to bound the error for all x in some interval, or shall we be satisfied with a small average error (where now "average" is ambiguous)? These queries need to be answered in practice, but our purpose here has been to point out *that the primary aim of any approximation is to achieve some desired degree of accuracy* and implicit in this aim is the assumption that the accuracy can indeed be estimated.

In one important sense our approach to numerical analysis in this book will be basically pragmatic; i.e., except for techniques of special theoretical interest, we shall concentrate on methods which are usable in practice. Thus, for a given method, we shall usually also wish to answer the question: How fast can a solution be computed using a given method?

In the case of the first method above for the calculation of $\sqrt{x}$, this question is easily answered since given the number of decimal places desired in the square root, this is a finite process with the number and type of calculations strictly determined. Using the second method, interpolation, we must first choose an interpolation formula which will achieve the desired accuracy. Having done this, however, the amount of calculation is again strictly determined.† But using iterative processes, the situation is different. Our assumption that the method would "work" is equivalent to assuming that the iteration converges. But the amount of computation required to achieve the desired degree of accuracy depends on the rate of convergence. Therefore, determining rates of convergence in iterative processes will always be of importance to us.

Generalizing then from our example of $\sqrt{x}$, we conclude that the pri-

† This is really a simplification of the truth; see Sec. 3.5.

mary aim of an approximation is to achieve some desired degree of accuracy and to be such that the accuracy can be estimated. Second, we are also interested in the amount of computation required to achieve the approximation. With these heuristic notions behind us, we now proceed to consider the general problem of approximation.

2.1-1 Classes of Approximating Functions

Much of the approximation done in numerical analysis consists of approximating a function $f(x)$ by some combination—most often a linear combination—of functions drawn from some particular class of functions. The most familiar example of this is the approximation of $f(x)$ by the first N terms of its Taylor-series expansion. Another familiar example occurs in the trapezoidal rule, in which $f(x)$ is approximated by a sequence of straight lines. In the former example and for each straight line in the trapezoidal rule, the approximation is a linear combination of functions from the class $\{x^n\}$, $n = 0, 1, \ldots$. More generally, we may consider the class $\{p_n(x)\}$, where $p_n(x)$ is a polynomial of degree n. Another class which is suggested by the importance of periodic functions is the class of Fourier functions $\{\sin nx, \cos nx\}$, $n = 0, 1, \ldots$. There are, of course, a number of other classes of functions which would lead to useful approximations in particular cases. Especially worth mentioning are:

Rational functions, which will play an important role in Chap. 7 and which can be used, whereas polynomials cannot, to approximate functions with poles.
Piecewise polynomial functions, i.e., functions which are different polynomials on different subintervals, which will be used in Sec. 3.8.
Exponential functions.

But for general application, polynomials and the Fourier functions are by far the most important, with the former predominating. Since this assertion about polynomial approximations is basic to our study of numerical analysis, in the next few pages we shall attempt to justify it.

2.1-2 Types of Approximations

Let $f(x)$ be a function which we wish to approximate using the class of functions $\{g_n(x)\}$, $n = 0, 1, \ldots$. Suppose we approximate $f(x)$ by the linear combination

$$f(x) \approx a_0 g_0(x) + a_1 g_1(x) + \cdots + a_m g_m(x) \tag{2.1-1}$$

where the a_i, $i = 0, 1, \ldots, m$ are constants. We shall call (2.1-1) an approximation of *linear* type to $f(x)$. Because the analysis of approximations involving nonlinear combinations of the approximating functions, like most nonlinear analysis, is very difficult, we shall be concerned almost entirely with approximations of linear type. In Chap. 7, however, we shall make extensive use of approximations of *rational* type which have the form

$$f(x) \approx \frac{a_0 g_0(x) + a_1 g_1(x) + \cdots + a_m g_m(x)}{b_0 g_0(x) + b_1 g_1(x) + \cdots + b_k g_k(x)} \tag{2.1-2}$$

The crux of the approximation problem is the criterion to use in choosing the constants in (2.1-1) and (2.1-2). Three methods of doing this lead to three types of approximations of major importance:

1. *Exact* or *interpolatory* approximations, in which the constants are chosen so that on some fixed set of points x_i, $i = 1, \ldots, p$, the approximation and its first r_i derivatives (where r_i is a nonnegative integer) agree with $f(x)$ (except for roundoff).
2. *Least-squares* approximations, in which the object is to minimize the integral of the square of the difference between $f(x)$ and its approximation (perhaps multiplied by a suitable weighting function) over an interval $[a, b]$ or, more commonly, to minimize the weighted sum of the squares of the error over a discrete set of points of $[a, b]$. In the language of Sec. 1.3-3 we wish to minimize the L_2 norm.
3. *Minimum maximum error* approximations, where the aim is to minimize the maximum magnitude of the difference between $f(x)$ and its approximation (again perhaps suitably weighted) on an interval $[a, b]$. In the terminology of Sec. 1.3-3, we wish to minimize the L_∞, or uniform norm.

These heuristic definitions have been given here in order to orient the reader to what follows. They will be made properly precise in later chapters. Of the three types, exact approximations are generally easier to derive and analyze (i.e., errors can be more easily estimated) than the other two. Chapters 3 to 5 will be exclusively concerned with exact approximations. Because of their particular advantages in certain applications, we shall discuss least-squares and minimum maximum error approximations in Chaps. 6 and 7, respectively.

2.1-3 The Case for Polynomial Approximation

By a polynomial approximation we mean one of the form (2.1-1), where each $g_i(x)$ is a polynomial. The computational case for the use of polynomials as the approximating functions follows directly from the fact that a digital computer can perform only the computational operations of arithmetic. A piecewise rational function is then the *most general kind of function that can be evaluated directly on a digital computer.*† Thus, in approximating a function $f(x)$ using any other class of functions $\{g_n(x)\}$, we must first evaluate the functions $g_n(x)$ using some approximation of each $g_n(x)$ as a polynomial in x or rational function of x. As a word of caution, though, it is easy to overemphasize this advantage of the class $\{x^n\}$. Thus, for example, in Chap. 6 we shall indicate how Fourier approximations can be calculated with only a single evaluation of a sine and a cosine.

One property that the powers of x and the trigonometric functions (as well as exponential functions) have in common is that an approximation using either of these classes changes its coefficients but not its form if the origin of the coordinate system is changed. Thus, if $P(x)$ is a polynomial or rational function, so is $P(x + \alpha)$, and if $T(x)$ is a linear or rational approximation using sines and cosines, so is $T(x + \alpha)$. Approximations using powers of x have the further advantage that if the scale of the variable is changed, again the coefficients but not the form of the approximation are changed. Thus $P(kx)$ is still a polynomial in x. But this property does not hold for approximations using sines and cosines. That is, in general, for noninteger k, $\sin nkx$ is not a member of the class $\{\sin nx\}$. Finally we note that if $P(x)$ and $Q(x)$ are polynomials, so is $P(Q(x))$ and if $R(x)$ and $S(x)$ are rational functions, so is $R(S(x))$.

One further obvious analytic advantage of polynomials is the ease with which they can be manipulated in general and differentiated and integrated in particular. In classical analysis, this enables us, for example, to express the remainder term in a Taylor series in closed form, and, as we shall see, there are analogous advantages in numerical analysis. A similar analytic advantage is also possessed by the Fourier functions.

All the advantages of the class $\{x^n\}$ that we have mentioned would be for naught if there were no analytic basis for our hope that we can achieve arbitrarily high accuracy with this class. We assume the reader is familiar with the result [Courant and Hilbert (1953), p. 65] that the set of functions $\{x^n\}$ is complete over any interval $[a, b]$; that is, for any piecewise continuous function $f(x)$, given any $\epsilon > 0$, there exists an n and coefficients $a_0, \ldots, a_n$ such that

$$\int_a^b \left[f(x) - \sum_{i=0}^{n} a_i x^i \right]^2 dx < \epsilon \tag{2.1-3}$$

Since sines and cosines also form a complete set, there is a result analogous to (2.1-3) for them. This result assures us that we can achieve arbitrarily good least-squares approximations using linear combinations of polynomials. To show that we can achieve arbitrarily small minimum maximum

† This is a slight exaggeration. More general functions can be evaluated using the logical operations or operations on the magnitude of a number which are available on all computers. But for all practical purposes, the statement above is true.

error with linear combinations of polynomials, we shall now prove the classical theorem of Weierstrass on polynomial approximation. As a corollary of this theorem, we shall then obtain a similar result for Fourier approximations.

The Weierstrass Approximation Theorem

Theorem 2.1 (Weierstrass) If $f(x)$ is continuous on a finite interval $[a, b]$, then, given any $\epsilon > 0$, there exists an n $[= n(\epsilon)]$ and a polynomial $P_n(x)$ of degree n such that

$$|f(x) - P_n(x)| < \epsilon \tag{2.1-4}$$

for all x in $[a, b]$.

Thus, by requiring continuity instead of piecewise continuity, we achieve uniform approximation instead of approximation in the mean as in (2.1-3).

PROOF Our proof of this theorem is due to Bernstein (1912). It is achieved by constructing a sequence of polynomials which converges uniformly to $f(x)$. Without loss of generality, we let $a = 0$, $b = 1$ since any other interval can be reduced to $[0, 1]$ by a simple change of variable {11}. On this interval, the Bernstein polynomial of degree n is defined by

$$B_n(x) = \sum_{k=0}^{n} \binom{n}{k} x^k (1-x)^{n-k} f\left(\frac{k}{n}\right) \tag{2.1-5}$$

We shall show that

$$\lim_{n \to \infty} B_n(x) = f(x) \tag{2.1-6}$$

uniformly in $[0, 1]$. To prove this theorem, we require the following lemma.

Lemma 2.1 The following identities are true:

$$\begin{aligned} \sum_{k=0}^{n} \binom{n}{k} x^k (1-x)^{n-k} &= 1 \\ \sum_{k=0}^{n} \frac{k}{n} \binom{n}{k} x^k (1-x)^{n-k} &= x \\ \sum_{k=0}^{n} \frac{k^2}{n^2} \binom{n}{k} x^k (1-x)^{n-k} &= \left(1 - \frac{1}{n}\right) x^2 + \frac{1}{n} x \end{aligned} \tag{2.1-7}$$

PROOF To derive all three identities, we use the binomial expansion

$$(p+q)^n = \sum_{k=0}^{n} \binom{n}{k} p^k q^{n-k} \tag{2.1-8}$$

The first identity follows immediately when $p + q = 1$. The other two follow by differentiating (2.1-8), respectively, once and twice with respect to p and then setting $p + q = 1$ {12}.

Combining the identities in (2.1-7), we have

$$\sum_{k=0}^{n} \left(\frac{k}{n} - x\right)^2 \binom{n}{k} x^k (1-x)^{n-k} = \frac{x(1-x)}{n} \tag{2.1-9}$$

Multiplying the first identity in (2.1-7) by $f(x)$ and subtracting $B_n(x)$, we have

$$f(x) - B_n(x) = \sum_{k=0}^{n} \left[f(x) - f\left(\frac{k}{n}\right)\right] \binom{n}{k} x^k (1-x)^{n-k} \tag{2.1-10}$$

Since $f(x)$ is continuous on $[0, 1]$, it is both uniformly continuous and bounded. Therefore, there exist δ and M such that

$$\begin{aligned} |f(x_1) - f(x_2)| &< \frac{\epsilon}{2} \qquad \text{if } |x_1 - x_2| < \delta;\ x_1, x_2 \in [0, 1] \\ |f(x)| &< M \qquad x \in [0, 1] \end{aligned} \tag{2.1-11}$$

where ϵ is as given in the statement of the theorem.

For any x, we divide the points k/n, $k = 0, \ldots, n$, into two sets A and B such that

$$\frac{k}{n} \text{ is} \begin{cases} \text{in } A & \text{if } \left|\dfrac{k}{n} - x\right| < \delta \\ \text{in } B & \text{otherwise} \end{cases}$$

Then, using the first of the relations (2.1-11) and the first of the identities (2.1-7), we have

$$\left|\sum_A \left[f(x) - f\left(\frac{k}{n}\right)\right] \binom{n}{k} x^k (1-x)^{n-k}\right| < \frac{\epsilon}{2} \sum_A \binom{n}{k} x^k (1-x)^{n-k} \le \frac{\epsilon}{2} \tag{2.1-12}$$

Using the second of the relations (2.1-11) and (2.1-9), we have

$$\begin{aligned} \left|\sum_B \left[f(x) - f\left(\frac{k}{n}\right)\right] \binom{n}{k} x^k (1-x)^{n-k}\right| &\le 2M \sum_B \binom{n}{k} x^k (1-x)^{n-k} \\ &= 2M \sum_B \frac{(k/n - x)^2}{(k/n - x)^2} \binom{n}{k} x^k (1-x)^{n-k} \\ &\le \frac{2M}{\delta^2} \frac{x(1-x)}{n} \le \frac{M}{2n\,\delta^2} \end{aligned} \tag{2.1-13}$$

since $0 \le x(1-x) \le \frac{1}{4}$ on $[0, 1]$. Now if we choose n so that

$$\frac{M}{2n\delta^2} < \frac{\epsilon}{2} \tag{2.1-14}$$

and combine (2.1-12) and (2.1-13) in (2.1-10), the theorem is proved by identifying $P_n(x)$ with $B_n(x)$.

For periodic functions, we have an analogous theorem.

Theorem 2.2 (Weierstrass) If $F(t)$ is a periodic continuous function of period 2π, then, given any $\epsilon > 0$, there exists an n [$= n(\epsilon)$] and a trigonometric sum

$$S_n(t) = a_0 + \sum_{k=1}^{n} (a_k \cos kt + b_k \sin kt) \tag{2.1-15}$$

such that

$$|F(t) - S_n(t)| < \epsilon \tag{2.1-16}$$

for all t.

Using Theorem 2.1, this theorem is not hard to prove; we leave the proof to a problem {13}.

In a sense, then, Theorem 2.1 justifies the use of polynomial approximations since it guarantees that a polynomial *can* be found with an arbitrarily small maximum deviation from $f(x)$ on $[a, b]$. In fact, since the proof is constructive, the reader may well think that we have solved the problem of minimum maximum error approximations. Furthermore, our reasons for preferring exact approximations—ease of derivation and error analysis—may seem to be weak in the light of Theorem 2.1. But unfortunately we have neither solved the minimum maximum error problem nor really weakened the case for exact approximation. The reasons for this are as follows:

1. In deriving minimum maximum error approximations, we shall be concerned with finding, for example, that polynomial of degree n which has the minimum maximum error as an approximation to $f(x)$. The Bernstein polynomial of degree n is by no means this polynomial.
2. The usual situation with exact approximations is that we are given only a sequence of points x_i and the corresponding $f(x_i)$, that is, a table. But, using Bernstein polynomials, we are forced to use particular values of $f(x)$ which may not be available. Moreover, the Bernstein polynomial of degree n is no help in deriving the polynomial of degree $n + 1$. This is a serious drawback, as will be made clearer in Chap. 3.
3. The ease with which it appears possible to bound the error in a Bernstein polynomial is a mirage, because the error bound tends to be extremely conservative. The following example will illustrate this.

Example 2.1 What degree Bernstein polynomial will guarantee an error of less than .2 in approximating e^x on [0, 1]?

We have the inequality on [0, 1]

$$|e^{x_1} - e^{x_2}| < e|x_1 - x_2|$$

which follows from the mean-value theorem. Therefore, with $\epsilon = .2$, taking $\delta < .2/2e$ satisfies the first relation of (2.1-11). Since $M = e$, in order to satisfy (2.1-14) we need

$$n > \frac{M}{\epsilon\delta^2} > \frac{e^3}{2 \times 10^{-3}} > 10{,}000$$

However, if we compute the Bernstein polynomial of degree 2, we find

$$B_2(x) = (1 - x)^2 + 2x(1 - x)e^{1/2} + x^2 e$$

and by direct calculation

$$|e^x - B_2(x)| < .11$$

on [0, 1]. Therefore; the estimate resulting from the proof of Theorem 2.1 is extremely conservative. Moreover, using the techniques of Chap. 7, we can achieve a substantially smaller maximum error than .11 with a quadratic approximation to e^x on [0, 1].

Our conclusion, then, is that while the Bernstein polynomials lead to an elegant constructive proof of the Weierstrass theorem, they do not in themselves generally give useful polynomial approximations.

The assurance that Theorem 2.1 gives us that we can find polynomial approximations which have arbitrarily small error, combined with the computational and other analytic advantages mentioned previously in this section, make a strong case for polynomial approximations. Their dominance in the next four chapters will not surprise the reader.

The case for polynomial approximation, however, is not so good as to rule out all other types. In Chap. 7 our approximating functions will always be polynomials, but we shall make extensive use of rational approximations of the form (2.1-2). Approximations using functions other than polynomials are of considerable importance also. We have already noted the importance of periodic functions and the fact that the Fourier functions satisfy results analogous to those discussed in this section for polynomials.

The importance of *band-limited functions*, functions whose constituent periodicities are known to be bounded, in many servomechanism and related problems is another indication of the importance of approximations using the Fourier functions. We shall touch on such approximations in Sec. 6.6, but for a more extensive treatment we refer the reader to Hamming (1973).

2.2 NUMERICAL ALGORITHMS

If numerical analysis is concerned with the processes by which mathematical problems can be solved by the operations of arithmetic, then the practice of numerical analysis requires that a problem statement be turned into a sequence of arithmetic operations which convert the *data* of the problem into the *results*. The sequence of arithmetic operations and the set of decisions which indicate which operation in the sequence to perform next constitute a rule, a recipe for solving the given problem. Such a rule in mathematics and computer science is called an *algorithm*.

One of the most active areas of research in computer-related mathematics is the analysis of algorithms. This subject treats such matters as the speed of convergence of algorithms, the accuracy of algorithms, and the relationship of one algorithm for solving a particular problem to another for the same problem. Strictly speaking, as a mathematical science, numerical analysis is concerned with

The development of algorithms
The analysis of algorithms
The computer implementation of algorithms

In this section we shall illustrate the basic characteristics of algorithms by an example of a simple but effective algorithm.

Example 2.2

Bisection

Given: $f(x)$ continuous on $[a, b]$ and such that

$$f(a)f(b) < 0$$

Problem: Find a point α such that

$$f(\alpha) = 0$$

We define an iterative algorithm as follows. Let $I_i = (a_i, b_i)$, $i = 0, 1, \ldots, I_0 = (a, b)$ define a sequence of intervals and let m_i be the midpoint of I_i. Let

$$I_{i+1} = \begin{cases} (a_i, m_i) & \text{if } f(a_i)f(m_i) < 0 \\ (m_i, b_i) & \text{if } f(m_i)f(b_i) < 0 \end{cases} \tag{2.2-1}$$

Continue until $f(m_i) = 0$ or the length of I_i divided by the maximum of 1 and $|m_i|$ (see below) is less than some given $\epsilon > 0$, in which case approximate α by m_i.

Using a rather formal notation we could describe this algorithm as follows:

Input

$a, b, \epsilon, f(x)$

Algorithm

$I_0 \leftarrow (a, b)$; $m_0 \leftarrow (a + b)/2$

for $i = 0, 1, 2, \ldots$ **do**

if $f(m_i) = 0$ **then** $\alpha \leftarrow m_i$; **stop**

```
            if f(a_i)f(m_i) < 0    then I_i ← (a_i, m_i)
                                   else I_i ← (m_i, b_i)
            endif
            m_i ← (a_i + b_i)/2
            if length (I_i)/max(|m_i|, 1) < ε then α ← m_i; stop
        endfor
    Output
        α
```

The strictly mathematical analysis of this algorithm, like that of any algorithm, consists of showing that

1. If it terminates, the result is the desired one; i.e., the algorithm is *effective*.
2. The algorithm does terminate.

First we show that

$$\lim_{i \to \infty} m_i = \alpha_0 \tag{2.2-2}$$

where α_0 is some point at which $f(x) = 0$.
The proof is quite straightforward:

1. By our hypotheses, at least one α such that $f(\alpha) = 0$ exists.
2. It also follows from (2.2-1) that each interval I_i contains a point α such that $f(\alpha) = 0$.
3. Since m_i is the midpoint of I_i, it follows from (2.2-1) that the length of I_{i+1} is one-half the length of I_i. Therefore

$$\lim_{i \to \infty} \text{length } (I_i) = 0 \tag{2.2-3}$$

   Thus, eventually each I_i contains just one point α_0 such that $f(\alpha_0) = 0$ and this α_0 must be the point to which the m_i converge.

The proof that the algorithm must terminate is even simpler. If $f(m_i) = 0$ at any stage, it terminates. If not, then (2.2-3) indicates that the condition length $(I_i) < \epsilon$ must be satisfied at some point.

The computational analysis, however, is not quite so trivial.

1. Roundoff may cause a result with a value of m_i not within ϵ of the true value. For suppose at some stage of the iteration roundoff causes the evaluation of $f(m_i)$ to have a sign different from its true sign. The effect of this will be that the final value of the root cannot be closer to the true root α than $|m_i - \alpha|$ (why?). While it is true that this problem is very unlikely to arise except when m_i is already quite close to α, nevertheless it is quite possible for it to occur well before $|m_i - \alpha| < \epsilon$. It is worth noting that this phenomenon is likely to occur only when $f(x)$ is quite "flat" in a neighborhood of α, in which case an error in m_i larger than ϵ may be tolerable.
2. The algorithm may not terminate. For if ϵ is too small, roundoff in the calculation of the length of I_i divided by m_i, even though it involves only a single subtraction and division, could prevent this length from ever being less than ϵ. Note here the importance of using length $(I_i)/\max(|m_i|, 1)$ rather than just length (I_i). If m_i is large, two successive floating-point numbers in the neighborhood of m_i may not be very close (cf. Sec. 1.5-2).

Therefore, length $(I_i) = b_i - a_i$ may never get very small, and if ϵ is small, a test of length $(I_i) < \epsilon$ might never be satisfied. When $|m_i| > 1$, the effect of dividing by m_i is to normalize the length of the interval, thus making it much less likely (but not impossible) that the test will never be satisfied. The reason for using max $(|m_i|, 1)$ as a divisor is to protect against a root very near to 0.

Both these effects are remote but not impossible. A good *computational algorithm*, i.e., the implementation of the algorithm on a computer, should be prepared for either eventuality. Our purpose here is not to consider how this might be done but merely to introduce the potential difficulties in order to alert the reader to computational problems not apparent in the mathematical statement of an algorithm.

2.3 FUNCTIONALS AND ERROR ANALYSIS

The analysis of the truncation error in an approximation, particularly in the case of polynomial approximation, is significantly facilitated using the concept of a *functional*. Given a family $\mathscr{F}$ of functions defined on an interval $[a, b]$, we define a functional $F(f)$ to be a mapping from $\mathscr{F}$ into the set of real (or complex) numbers which assigns a unique number to each function $f \in \mathscr{F}$. For example, $I(f) = \int_a^b f(x)\, dx$ is a functional, as is the point functional $f(x)$, where x is a fixed point in $[a, b]$. So also are the point derivatives $f^{(k)}(x)$ when they exist. A functional F is linear if

$$F(\alpha f + \beta g) = \alpha F(f) + \beta F(g) \tag{2.3-1}$$

All the functionals mentioned above are linear. However, the functional $\|f\|_2 = [\int_a^b f^2(x)\, dx]^{1/2}$ is not.

In many branches of numerical analysis, e.g., in evaluating $I(f)$ defined above, we are concerned with evaluating a particular (usually linear) functional. Since this may not always be possible, we must, as noted in Sec. 2.1, find an approximation to this functional. One standard method of approximating $F(f)$ is to approximate f by some other function g, usually a polynomial p_n, and then use $F(g)$ or $F(p_n)$ as an approximation to $F(f)$. We do this by choosing a g for which we can evaluate $F(g)$ exactly. For example, to approximate $I(f)$, we evaluate $I(p_n)$, where $p_n(x)$ is a good approximation to $f(x)$ in $[a, b]$. In these cases, it is important to determine an expression for the error in the approximation. Now since g, the approximation of f, depends upon f, $F(g)$ is itself a functional of f, say $G(f)$. We define the error functional as

$$E(f) = F(f) - G(f) = F(f) - F(g) \tag{2.3-2}$$

If F is linear, we have $E(f) = F(f - g)$.

Example 2.3 Let $F(f) = I(f) = \int_a^b f(x)\, dx$, and let us approximate $f(x)$ by the linear polynomial

$$p_1(x) = f(a) + \frac{x - a}{b - a}[f(b) - f(a)]$$

Then $$F(p_1) = \int_a^b p_1(x)\,dx = \frac{b-a}{2}[f(a) + f(b)] = G(f)$$

and [see Eq. (4.10-8)] $E(f) = F(f) - G(f) = -[(b-a)^3]f''(\xi)/12$ for some ξ such that $a < \xi < b$.

In general we shall be considering approximations to integrals and derivatives and functional values at particular points of a function $f(x)$ by linear combinations of integrals and derivatives and functional values at other points. Thus we can write

$$E(f) = \int_a^b [\alpha_0(x)f(x) + \alpha_1(x)f'(x) + \cdots + \alpha_n(x)f^{(n)}(x)]\,dx$$

$$- \sum_{i=1}^{j_0} \beta_{i0}\, f(x_{i0}) - \sum_{i=1}^{j_1} \beta_{i1} f'(x_{i1}) - \cdots - \sum_{i=1}^{j_n} \beta_{in}\, f^{(n)}(x_{in}) \quad (2.3\text{-}3)$$

where the $\alpha_k(x)$ are piecewise continuous over $[a, b]$. Usually this error functional will vanish when f is a polynomial of a specific degree n or less.

Theorem 2.3 (Peano) Let $f(x)$ have a continuous $(n+1)$st derivative in $[a, b]$, and let a linear functional $F(f)$ of f be approximated by a linear functional $G(f)$ such that $E(f)$ as given by (2.3-3) vanishes when f is any polynomial of degree n or less. Then

$$E(f) = \int_a^b f^{(n+1)}(t)K(t)\,dt \quad (2.3\text{-}4)$$

where $$K(t) = \frac{1}{n!} E_x[(x-t)_+^n] \quad (2.3\text{-}5)$$

and $$(x-t)_+^n = \begin{cases} (x-t)^n & x \geq t \\ 0 & x < t \end{cases} \quad (2.3\text{-}6)$$

The notation E_x means the linear functional E is applied to the x variable in its argument [$(x-t)_+^n$ in (2.3-5)]. The function $K(t)$ is called the *Peano kernel* for the linear functional E. It is also called an *influence function* for E.

PROOF By Taylor's theorem with exact remainder,

$$f(x) = f(a) + f'(a)(x-a) + \cdots + \frac{f^{(n)}(a)(x-a)^n}{n!}$$

$$+ \frac{1}{n!}\int_a^x f^{(n+1)}(t)(x-t)^n\,dt \quad (2.3\text{-}7)$$

The integral remainder can be written as

$$\frac{1}{n!}\int_a^b f^{(n+1)}(t)(x-t)_+^n\,dt$$

We apply E to both sides of Eq. (2.3-7) and use the fact that E vanishes for polynomials of degree n or less:

$$E(f(x)) = \frac{1}{n!} E_x \int_a^b f^{(n+1)}(t)(x-t)_+^n \, dt \tag{2.3-8}$$

Now since E has the form (2.3-3), we can interchange E_x and the integral; hence,

$$E(f) = \frac{1}{n!} \int_a^b f^{(n+1)}(t) E_x(x-t)_+^n \, dt \tag{2.3-9}$$

which completes the proof.

Corollary 2.1

$$|E(f)| \le \max_{a \le x \le b} (|f^{(n+1)}(x)|) \int_a^b |K(t)| \, dt \tag{2.3-10}$$

Corollary 2.2 If, in addition, $K(t)$ does not change its sign on $[a, b]$, then

$$E(f) = \frac{f^{(n+1)}(\xi)}{(n+1)!} E(x^{n+1}) \qquad a < \xi < b \tag{2.3-11}$$

PROOF Under the additional hypothesis we can use the mean-value theorem for integrals and obtain

$$E(f) = f^{(n+1)}(\xi) \int_a^b K(t) \, dt \tag{2.3-12}$$

Now we insert $f(x) = x^{n+1}$ in (2.3-12) to get

$$E(x^{n+1}) = (n+1)! \int_a^b K(t) \, dt \tag{2.3-13}$$

This yields (2.3-11).

2.4 THE METHOD OF UNDETERMINED COEFFICIENTS

A common situation in numerical analysis is to be given the form of the linear functional by which f is to be approximated and the points at which $f(x)$ and/or its derivatives can be evaluated but not the weighting coefficients in the linear functional. Thus we are given

$$F(f) = \sum_{j=0}^{n} \sum_{i=1}^{m_j} w_{ij} f^{(j)}(x_{ij}) + E(f) \tag{2.4-1}$$

where $f^{(j)}(x)$ is given at the points $x_{ij}, i = 1, \ldots, m_j, j = 0, \ldots, n$, and we wish to determine the w_{ij} so that $E(f)$ has certain properties. The method of undetermined coefficients, when it can be applied, is a simple and straightforward way of doing this.

Suppose our aim is that $E(f) = 0$ when $f(x)$ is a polynomial of degree N or less, where $N = (\sum_{j=0}^{n} m_j) - 1$. To this end, we replace f in (2.4-1) successively by $1, x, \ldots, x^N$. Then the requirement that $E(f) = 0$ for these functions yields a system of $N + 1$ linear equations in the $N + 1$ unknowns w_{ij}

$$\sum_{j=0}^{n} \sum_{i=1}^{m_j} w_{ij}(x^k)^{(j)}_{x=x_{ij}} = F(x^k) \qquad k = 0, \ldots, N \tag{2.4-2}$$

We assume here that $F(x^k)$ can be evaluated exactly. The linear system (2.4-2) will have a unique solution yielding the coefficients w_{ij} provided the matrix of coefficients is nonsingular. In many situations we are assured that this is so; e.g., when only functional values are given ($n = 0$), or when a particular derivative $f^{(j)}(x)$ is prescribed at x_{ij} only if $f^{(j-1)}(x)$ is also given, it can be shown that the matrix is nonsingular.

Sometimes we may want to keep a few parameters free to achieve some other end than accuracy as measured by the highest-degree polynomial for which $E(f)$ vanishes. This can be done by reducing the number of equations in (2.4-2).

Example 2.4 Given $f(a)$, $f'(a)$, and $f(b)$, compute approximations to

$$f\left(\frac{a+b}{2}\right) \qquad \text{and} \qquad \int_a^b f(x)\,dx$$

Here $N = 2$, $x_{10} = a$, $x_{11} = a$, and $x_{20} = b$. In the first case, we get the following system of linear equations

$$\begin{aligned} 1w_{10} + 0w_{11} + 1w_{20} &= 1 \\ aw_{10} + 1w_{11} + bw_{20} &= \frac{a+b}{2} \\ a^2w_{10} + 2aw_{11} + b^2w_{20} &= \left(\frac{a+b}{2}\right)^2 \end{aligned} \tag{2.4-3}$$

and the solution is $w_{10} = \frac{3}{4}$, $w_{11} = (b-a)/4$, $w_{20} = \frac{1}{4}$, so that we have the approximation

$$f\left(\frac{a+b}{2}\right) \approx \tfrac{3}{4}f(a) + \frac{b-a}{4}f'(a) + \tfrac{1}{4}f(b) \tag{2.4-4}$$

In the second case, we need only replace the right-hand side of (2.4-3) by $\int_a^b x^k\,dx$, $k = 0, 1, 2$. The solution of the resulting system is $w_{10} = \frac{2}{3}(b-a)$, $w_{11} = \frac{1}{6}(b-a)^2$, $w_{20} = \frac{1}{3}(b-a)$, so that we have the approximation

$$\int_a^b f(x)\,dx \approx \frac{b-a}{6}[4f(a) + 2f(b) + (b-a)f'(a)] \tag{2.4-5}$$

To get the error in these approximations, we can apply Peano's theorem. In the first case, the Peano kernel does not change sign, and so we can use (2.3-13) to derive {24}

$$E(f) = \frac{(a-b)^3}{48}f'''(\xi) \qquad a < \xi < b$$

Therefore, we can rewrite (2.4-4) as

$$f\left(\frac{a+b}{2}\right) = \tfrac{3}{4}f(a) + \frac{b-a}{4}f'(a) + \tfrac{1}{4}f(b) + \frac{(a+b)^3}{48}f'''(\xi) \qquad a < \xi < b$$

Similarly, in the second case, we can apply Peano's theorem and rewrite (2.4-5) as {24}

$$\int_a^b f(x)\,dx = \frac{b-a}{6}[4f(a) + 2f(b) + (b-a)f'(a)] + \frac{b-a}{72}[(a-b)^3 + 6ab^2]f'''(\xi)$$

BIBLIOGRAPHIC NOTES

Section 2.1 A number of works treat the general problem of approximation in much greater depth than we have done here. In particular, we recommend Achieser (1956), Natanson (1964), Cheney (1966), Davis (1963), Rice (1964, 1965), and Rivlin (1969). Some of the more specific works on approximation will be referred to in later chapters.

The cases for and against polynomial approximation and the case for trigonometric approximation are well presented in Hamming (1973). Our proof of the Weierstrass theorem is due to Bernstein (1912) and can also be found in Achieser (1956) and Rivlin (1969). For a quite different proof, see Courant and Hilbert (1953).

Section 2.2 The best and most comprehensive references on the development and analysis of computer algorithms is the series of books by Knuth (1968, 1969, 1973).

Section 2.3 For more on Peano's theorem see Sard (1963) and Davis (1963).

BIBLIOGRAPHY

Achieser, N. I. (1956): *Theory of Approximation* (trans. C. J. Hyman), Frederick Ungar Publishing Co., New York.

Bernstein, S. N. (1912): Démonstration du théorème de Weierstrass fondée sur le calcul de probabilité, *Proc. Kharkow Math. Soc.*, vol. 13.

Cheney, E. W. (1966): *Introduction to Approximation Theory*, McGraw-Hill Book Company, New York.

Courant, R., and D. Hilbert (1953): *Methods of Mathematical Physics*, vol. 1 (translated), Interscience Publishers, Inc., New York.

Davis, P. J. (1963): *Interpolation and Approximation*, Blaisdell Publishing Company, New York.

Hamming, R. W. (1973): *Numerical Methods for Scientists and Engineers*, 2d ed., McGraw-Hill Book Company, New York.

Knuth, D. E. (1968, 1969, 1973): *The Art of Computer Programming*, vols. 1–3, Addison-Wesley Publishing Company, Inc., Reading, Mass.

Macon, N. (1963): *Numerical Analysis*, John Wiley & Sons, Inc., New York.

Natanson, I. P. (1964): *Constructive Function Theory*, vol. 1, Frederick Ungar Publishing Co., New York.

Rice, J. R. (1964, 1965): *The Approximation of Functions*, vols. 1 and 2, Addison-Wesley Publishing Company, Inc., Reading, Mass.

Rivlin, T. J. (1969): *An Introduction to the Approximation of Functions*, Blaisdell Publishing Company, Waltham, Mass.

Sard, A. (1963): Linear Approximation, *Math. Surv.* 9, American Mathematical Society, Providence, R.I.

PROBLEMS

Section 2.1

1 Find a first-degree polynomial approximation $P(x) = ax + b$ to $\sin x$ on $[0, \pi/2]$:
(*a*) Using the first nonzero term of the Maclaurin expansion of $\sin x$.
(*b*) Which minimizes $\int_0^{\pi/2} [P(x) - \sin x]^2 \, dx$.
(*c*) Which minimizes $\int_0^{\pi/2} x[P(x) - \sin x]^2 \, dx$.
For all three approximations, draw a graph of the error $E(x) = \sin x - P(x)$. In what sense is the approximation (*a*) an exact approximation in the terminology of Sec. 2.1? What are the significant characteristics of the error in (*a*) compared with those in the least-squares approximations (*b*) and (*c*)?

***2** Now consider the problem of finding an approximation to $\sin x$ of the form $P(x) = ax + b$ which minimizes

$$\max_{x \in [0, \pi/2]} |P(x) - \sin x|$$

(*a*) Prove that any such $P(x)$ must be such that $E(x) = \sin x - P(x)$ has at least two zeros in $[0, \pi/2]$.

(*b*) Starting from the result of Prob. 1*c*, derive an approximation to $\sin x$ of the form $ax + b$ with a smaller maximum error.

3 For parts (*b*) and (*c*) of Prob. 1, derive the equations for a and b when $\sin x$ is replaced by $f(x)$.

4 Repeat Probs. 1 and 3 when $P(x)$ is replaced by the quadratic approximation $Q(x) = ax^2 + bx + c$.

5 Consider the continued-fraction expansion for the inverse tangent

$$\tan^{-1} t \approx \cfrac{t}{1 + \cfrac{t^2}{3 + \cfrac{t^2}{5 + \cfrac{\ddots}{\qquad \cfrac{t^2}{2n+1}}}}}$$

(*a*) Show that for any n this is equivalent to approximating $\tan^{-1} t$ by a rational function.

(*b*) For $n = 1$ and 2 compute the rational approximations and draw the graph of the error in the approximations on the interval $[0, 1]$.

(*c*) For each of the approximations of part (*b*), consider the analogous approximation derived by truncating the Maclaurin-series expansion for $\tan^{-1} t$ after the term of degree equal to the sum of the degrees of numerator and denominator in part (*b*). Draw the graphs of the errors and compare with those in part (*b*).

6 Derive an approximation to e^x of the form

$$e^x \approx \frac{ax + b}{x + c}$$

which at $x = 0$ has the same value and the same first two derivatives as e^x. Draw the graph of the error for $-1 \le x \le 1$ (cf. Sec. 7.4).

***7** Let $\sin x$ be approximated by the first three nonzero terms of its Maclaurin expansion.
(*a*) For any x find a bound on the magnitude of the truncation error.

(*b*) Convert the coefficients to correctly rounded 10-decimal-digit numbers.

(*c*) If $|x| < 1$ is correctly rounded to 10 decimal digits, find a bound on the error in the correctly rounded value of x^2.

(*d*) Use this bound to derive an approximate bound on the roundoff error incurred in evaluating the approximation in part (*b*) if all intermediate quantities are rounded to 10 decimals. Again assume $|x| < 1$ and neglect all quantities with order of magnitude less than 10^{-d}.

(*e*) For $x = \pi/4$ perform the calculation carrying 10 decimals at every stage. Compare the result with the correct value. Is the error caused by truncation and roundoff within expected limits?

8 Neglecting roundoff error, how many terms of the Maclaurin expansion for $\sin x$ are required to obtain a maximum error of less than 10^{-7} in the range $[0, \pi/2]$? In the range $[0, \pi]$?

9 Let $f(x)$ be approximated by

$$f(x) \approx P_1(x)f(a_1) + P_2(x)f(a_2)$$

where $P_1(x)$ and $P_2(x)$ are linear polynomials.

(*a*) Find $P_1(x)$ and $P_2(x)$ if the approximation is to have no error at $x = a_1$ and a_2 independent of the function $f(x)$. What is the advantage of having $P_1(x)$ and $P_2(x)$ independent of $f(x)$?

(*b*) Find an expression for the error in the approximation when $f(x) = x^2 + ax + b$. How do you explain the dependence of the result on a and b?

***10** A common iterative method for calculating the square root of a number a uses the formula

$$x_{n+1} = \frac{1}{2}\left(x_n + \frac{a}{x_n}\right) \qquad n = 1, 2, \ldots$$

where x_1 is an arbitrary positive number.

(*a*) Prove that if $\lim_{n\to\infty} x_n$ exists, this limit is $\sqrt{a}$.

(*b*) Prove that if $a \le x_n \le 1$, then $a \le x_{n+1} \le 1$.

(*c*) Prove that if $0 < a < 1$, then the iteration does converge, that is $\lim_{n\to\infty} x_n$ does exist.

11 (*a*) Derive the change of variable which transforms any finite interval $[a, b]$ to $[0, 1]$.

(*b*) By making the appropriate change of variable in (2.1-5), calculate the Bernstein polynomials of degree 1, 2, and 3 for $f(x) = \sin x$ on the interval $[0, \pi/2]$. Draw the graph of $\sin x - B_n(x)$ in each case and for $n = 1$ and 2 compare these errors with the corresponding errors in Probs. 1 and 4.

(*c*) Using the inequality (2.1-14), what value of n *guarantees* a smaller maximum error than that found in part (*b*) for $n = 1$?

12 (*a*) Derive the identities (2.1-7). (*b*) Using these identities, derive (2.1-9).

***13** Let $F(t)$ be a periodic continuous function of period 2π and define

$$\phi(t) = \frac{F(t) + F(-t)}{2} \qquad \psi(t) = \frac{F(t) - F(-t)}{2} \sin t$$

(*a*) Use Theorem 2.1 to show that, given any $\epsilon > 0$, there exist polynomials $P(x)$ and $Q(x)$ such that

$$|\phi(t) - P(\cos t)| \le \frac{\epsilon}{4} \qquad |\psi(t) - Q(\cos t)| \le \frac{\epsilon}{4}$$

for *all* t.

(*b*) Thus deduce that

$$|F(t) \sin t - U(t)| \le \frac{\epsilon}{2} \qquad \text{where} \qquad U(t) = Q(\cos t) + P(\cos t) \sin t$$

is a trigonometric sum. Similarly, show that there is another trigonometric sum $V(t)$ such that

$$\left| F\left(\frac{\pi}{2} - t\right) \sin t - V(t) \right| \leq \frac{\epsilon}{2}$$

(*c*) Use the two inequalities of part (*b*) to deduce that

$$\left| F(t) - U(t) \sin t - V\left(\frac{\pi}{2} - t\right) \cos t \right| \leq \epsilon$$

and from this complete the proof of Theorem 2.2. [Ref.: Achieser (1956), pp. 32–33.]

***14** (*a*) If $F(t)$ in Prob. 13 is also even, show how the proof can be considerably simplified.

(*b*) For the function

$$F(t) = \begin{cases} t - 2m\pi & 2m\pi \leq t < (2m+1)\pi \\ 2m\pi - t & (2m-1)\pi \leq t < 2m\pi \end{cases}$$

where m takes on all positive and negative integral values, use this simplification to derive an approximation of the form (2.1-15) to $F(t)$ using first- and second-degree Bernstein polynomials. How do you explain the relation between the two results? Draw the graph of the error.

15 For $f(x) = 1/(1 + x^2)$ calculate the Bernstein polynomials for $n = 2, 3, 4$ on the interval $[-1, 1]$ and draw the graph of the error in each case. (This function is an example of a case where *exact* polynomial approximation leads to certain difficulties; see p. 65 *n*.)

Section 2.2

16 Let $X_1, X_2, \ldots, X_n$ be distinct real numbers. We wish to find m and j such that

$$m = \max_{1 \leq k \leq n} X_k = X_j$$

for which j is as large as possible.

(*a*) Prove that the following is an algorithm for this problem:

$j \leftarrow n$; $k \leftarrow n - 1$; $m \leftarrow X_n$

L: **if** $k = 0$ **stop**

if $X_k > m$ **then** $j \leftarrow k$; $m \leftarrow X_k$

$k \leftarrow k - 1$

go to L

(*b*) Let A be the number of times the "then" part of the "if ... then" statement is executed. If p_{nk} represents the probability that $A = k$, show that

$$p_{nk} = \frac{1}{n} p_{n-1,k-1} + \frac{n-1}{n} p_{n-1,k}$$

(*c*) Use this to deduce that the *generating function*

$$G_n(z) = \sum_{k=1}^{n} p_{nk} z^k$$

is given by

$$\frac{1}{z+n} \binom{z+n}{n}$$

(*d*) Show that mean$(A) = G_n'(1)$ and thus deduce that

$$\text{mean }(A) = H_n - 1$$

where H_n is the nth harmonic number $\left(\sum_{i=1}^{n} \frac{1}{i}\right)$.

[Ref.: Knuth (1968), pp. 95–99.]

***17** Suppose instead of the algorithm of the previous problem we wish to find the two largest among distinct $X_1, \ldots, X_n$, $n \geq 2$.

(*a*) Prove that the following is an algorithm for this problem.

$k \leftarrow n - 2$; $m' \leftarrow X_{n-1}$; $m \leftarrow X_n$

L: **if** $m' > m$ **then** $t \leftarrow m'$; $m' \leftarrow m$; $m \leftarrow t$

if $k = 0$ **stop**

if $X_k > m'$ **then** $m' \leftarrow X_k$

$k \leftarrow k - 1$

go to L

(*b*) As a function of n, find the average number of times the exchange in the line labeled L is performed.

(*c*) Find the average number of times the step $m' \leftarrow X_k$ is performed.

18 The Euclidean algorithm to find the greatest common divisor d of two integers m and n is

$c \leftarrow m$; $d \leftarrow n$

L: $q \leftarrow c/d$; $r \leftarrow c - qd$ (integer division)

if $r = 0$ **then stop**

$c \leftarrow d$; $d \leftarrow r$

go to L

(*a*) Under what conditions on m and n does this algorithm terminate?

(*b*) When it terminates, prove that it is correct.

(*c*) Design a modification of this algorithm which will also produce integers a and b such that $am + bn = d$.

19 (*a*) Consider the bisection algorithm for a function $f(x)$ which is only piecewise continuous in $[a, b]$ but for which $f(a)f(b) < 0$. What can be said about its convergence?

(*b*) Apply bisection to $f(x) = x^3 - 2x - 5$ with $a = 0$, $b = 3$.

20 Let $f(x)$ be continuous and differentiable in a neighborhood of α such that $f(\alpha) = 0$. Consider the iteration

$$x_{i+1} = x_i - f(x_i) \qquad i = 1, 2, \ldots$$

where x_0 is an initial approximation to α.

(*a*) Derive a necessary condition on $f(x)$ in a neighborhood of α for the convergence of the sequence $\{x_i\}$ to α.

(*b*) Apply this method to the function of Prob. 19*b* with $x_0 = 3$.

(*c*) Apply this method to $f(x)/x^3$, where $f(x)$ is the function of Prob. 19*b*, again with $x_0 = 3$. Explain the difference in the results.

Section 2.3

21 If $\mathscr{F}$ is the family of all functions defined and continuous on $[a, b]$, which of the following are linear functionals where f is in $\mathscr{F}$?

(i) $\int_a^b x^2 f(x)\, dx$ (ii) $f'(a) + f'(b) - f\left(\frac{a+b}{2}\right)$ (iii) $\int_a^b [f(x)]^2\, dx$

22 (*a*) What is the Peano kernel for the functional

$$E(f) = -f(x_0) + 3f(x_1) - 3f(x_2) + f(x_3)$$

where $x_{i+1} - x_i = h$, $i = 0, 1, 2$, and if f is continuous and infinitely differentiable?

(*b*) Show that this kernel does not change sign on $[x_0, x_3]$ and thus apply (2.3-13) to compute $E(f)$.

23 Use Peano's theorem to find the error in Simpson's rule:

$$\int_{-1}^{1} f(x)\, dx \approx \tfrac{1}{3}[f(-1) + 4f(0) + f(1)]$$

24 Derive the two error terms in Example 2.4.

Section 2.4

25 Given $f(a)$, $f(b)$, $f'(a)$, and $f'(b)$, compute approximations to $f((a + b)/2)$ and $\int_a^b f(x)\, dx$ and compare the results with those of Example 2.4.

26 (*a*) Given $f(a)$, $f(b)$, $f'(a)$, and $f'(b)$, compute an approximation to $\int_a^b f(x)\, dx$ using $N = 2$.

(*b*) Determine the free parameter in the solution so that the sum of the squares of the coefficients w_{ij} is minimized. Why might you wish to do this?

CHAPTER

THREE

INTERPOLATION

3.1 INTRODUCTION

Interpolation lies at the heart of classical numerical analysis. There are two main reasons for this. The first is that in hand computation there is continual need to look up the value of a function in a table. In order to find the value of the function at nontabulated arguments, it is necessary to interpolate. Moreover, the highly accurate tables at small increments of the argument that we take for granted today are mostly of comparatively recent origin. Therefore, classical numerical analysts developed an extremely sophisticated group of interpolation methods. Today the need to interpolate arises comparatively seldom; e.g., on digital computers we almost always generate the value of a function directly rather than interpolate in a table of values (see Chap. 7). And when the need to interpolate in a table does arise, the small increments in the arguments in most tables mean that quite simple techniques, e.g., linear or quadratic interpolation, will usually suffice. Thus, while every numerical analyst must know how to interpolate, he will seldom, if ever, have use for the more sophisticated interpolation techniques.

Why then start the main body of this book with a chapter on interpolation? The answer to this question is provided by the second of the reasons mentioned at the beginning of this section, namely that interpolation formulas are the starting points in the derivations of many methods in other areas of numerical analysis. Almost all the classical methods of numerical differentiation, numerical quadrature, and numerical integration of ordinary differential equations are directly derivable from interpolation formulas.

While modern numerical analysis does not rely so heavily on interpolation formulas in these areas, their importance and usefulness are still great, as we shall see in Chaps. 4 and 5. This, then, is ample motivation for treating interpolation at the outset of this book.

Because we are especially interested in digital-computer applications, our approach to interpolation will not emphasize interpolation formulas based on difference techniques since they are seldom used on computers. Nevertheless, we shall not ignore finite differences because of their great usefulness in hand computation and, even on digital computers, for certain applications (see, for example, Sec. 4.13-1).

Suppose we have a function $f(x)$ which is known (perhaps along with certain of its derivatives) at a set of points. These points will hereafter be called the *tabular points* because interpolation so often takes place in a table of functional values. The object of interpolation is to *estimate* values of the function at nontabular points and—at least—to *bound* the error between the estimated and true values. Our approach will be to approximate $f(x)$ by a function $y(x)$ which, at the tabular points, has the same values as $f(x)$ (and perhaps the given derivative values, if any). Thus, in the language of the previous chapter, we shall be using exact approximations. In this chapter we shall consider only the case where $y(x)$ is a polynomial or a function which is a piecewise polynomial. In the last section of Chap. 6 we shall consider the case in which $y(x)$ is a linear combination of trigonometric functions.

We shall usually be concerned with interpolation using only values of the function at the tabular points. Thus our interpolation formula has the form

$$f(x) = \sum_{j=1}^{n} l_j(x) f(a_j) + E(x) = y(x) + E(x) \tag{3.1-1}$$

although the more general formula

$$f(x) = \sum_{j=1}^{n} \sum_{i=0}^{m_j} A_{ij}(x) f^{(i)}(a_j) + E(x) \tag{3.1-2}$$

is also of interest, particularly some special cases of it. Our object is to determine the $l_j(x)$ so that

$$E(a_j) = 0 \qquad j = 1, \ldots, n \tag{3.1-3}$$

independent of the function $f(x)$. In general, however, for nontabular points

$$E(x) \neq 0 \tag{3.1-4}$$

Our two aims then, are to determine the $l_j(x)$ so that (3.1-3) is satisfied and to find a representation for $E(x)$ which will enable us to estimate or at least bound the error for values of $x \neq a_j$, $j = 1, \ldots, n$.

3.2 LAGRANGIAN INTERPOLATION

In this section we consider the case where there are no restrictions on the spacing of the tabular points. In Sec. 3.3 we shall consider the case of equally spaced abscissas. Even in the general situation we consider here, however, the determination of the polynomials $l_j(x)$ is straightforward. Since we wish the error at the tabular points to be zero independent of $f(x)$, it follows using (3.1-1) and (3.1-3) that

$$l_j(a_k) = \delta_{jk} \qquad j, k = 1, \ldots, n \tag{3.2-1}$$

where δ_{jk} is the *Kronecker delta*.† Since $l_j(x)$ is to be a polynomial, this requires that it have a factor

$$(x - a_1)(x - a_2) \cdots (x - a_{j-1})(x - a_{j+1}) \cdots (x - a_n) \tag{3.2-2}$$

and since $l_j(a_j) = 1$, we can write

$$l_j(x) = \frac{(x - a_1) \cdots (x - a_{j-1})(x - a_{j+1}) \cdots (x - a_n)}{(a_j - a_1) \cdots (a_j - a_{j-1})(a_j - a_{j+1}) \cdots (a_j - a_n)} \tag{3.2-3}$$

Note that there are other possible polynomial representations of $l_j(x)$ but (3.2-3) is the only possible polynomial of degree $n - 1$ and no polynomial of lesser degree is possible (why?). It is notationally convenient to write $l_j(x)$ as

$$l_j(x) = \frac{p_n(x)}{(x - a_j)p_n'(a_j)} \qquad p_n'(a_j) = \left.\frac{dp_n}{dx}\right|_{x=a_j} \tag{3.2-4}$$

where

$$p_n(x) = \prod_{i=1}^{n} (x - a_i) \tag{3.2-5}$$

To find an expression for $E(x)$, we consider the function

$$F(z) = f(z) - y(z) - [f(x) - y(x)]\frac{p_n(z)}{p_n(x)} \tag{3.2-6}$$

with $y(x)$ as in (3.1-1). The function $F(z)$ *as a function of* z has $n + 1$ zeros at the points $a_1, \ldots, a_n$ and x [assume for now that x in (3.2-6) is not one of the tabular points]. Therefore, by applying Rolle's theorem n times‡

$$F^{(n)}(z) = f^{(n)}(z) - y^{(n)}(z) - [f(x) - y(x)]\frac{n!}{p_n(x)} \tag{3.2-7}$$

has at least one zero in the interval spanned by $a_1, \ldots, a_n$ and x. Calling this zero $z = \xi$ and noting that $y^{(n)}(z) = 0$ since $l_j(z)$ is a polynomial of degree $n - 1$, we have

† $\delta_{jk} = 0$ unless $j = k$, in which case $\delta_{jk} = 1$.

‡ Here and throughout the book, we shall assume that the functions involved are differentiable as many times as necessary for the discussion.

$$0 = F^{(n)}(\xi) = f^{(n)}(\xi) - [f(x) - y(x)]\frac{n!}{p_n(x)} \tag{3.2-8}$$

from which, using (3.1-1), it follows that

$$E(x) = \frac{p_n(x)}{n!} f^{(n)}(\xi) \tag{3.2-9}$$

where ξ, which is an unknown function of x, lies in the interval spanned by $a_1, \ldots, a_n$ and x. Although x in (3.2-6) was restricted to be a nontabular point, $E(x)$, as given by (3.2-9), holds for both tabular and nontabular points (why?).

Equation (3.1-1) with the $l_j(x)$ given by (3.2-4) and $E(x)$ by (3.2-9) is called the *Lagrangian interpolation formula.* When $n = 2$, $y(x)$ is the familiar formula for linear interpolation

$$y(x) = \frac{x - a_2}{a_1 - a_2} f(a_1) + \frac{x - a_1}{a_2 - a_1} f(a_2) \tag{3.2-10}$$

The polynomials $l_j(x)$ are called *Lagrangian interpolation polynomials.* Our derivation of the Lagrangian formula has been equivalent to finding that polynomial of degree $n - 1$ which passes through the points $[a_j, f(a_j)]$, $j = 1, \ldots, n$ {3}. Therefore, as we would expect, (3.2-9) indicates that this formula is *exact*, that is, $E(x) = 0$ for all x, for polynomials of degree $n - 1$ or less. In general, an interpolation formula which is exact for polynomials of degree r is said to have an *order of accuracy* r or to be of *order* r.

The use of the Lagrangian interpolation formula is straightforward. To estimate $f(x)$ at a nontabular point, we merely compute $y(x)$ as given by (3.1-1) using (3.2-4) and (3.2-5) to compute the polynomials $l_j(x)$. If we can estimate or bound the nth derivative of $f(x)$, then the error can be estimated or bounded using (3.2-9).

Example 3.1 Let $f(x) = \ln x$. Given the table of values

x	.40	.50	.70	.80
$\ln x$	−.916291	−.693147	−.356675	−.223144

estimate the value of ln .60.

With $a_1 = .40$, $a_2 = .50$, $a_3 = .70$, and $a_4 = .80$, we calculate from (3.2-4)

$$l_1(.60) = -\tfrac{1}{6} \qquad l_2(.60) = \tfrac{2}{3} \qquad l_3(.60) = \tfrac{2}{3} \qquad l_4(.60) = -\tfrac{1}{6}$$

and from (3.1-1) we get the approximation

$$\ln .60 \approx -.509975$$

The true value is $\ln .60 = -.510826$. From (3.2-9) we get

$$E(.60) = \frac{p_4(.60)}{4!} \frac{-6}{\xi^4} = \frac{-.0004}{4} \frac{1}{\xi^4}$$

In the interval (.4, .8), $10^4/4096 < 1/\xi^4 < 10^4/256$, so that

$$-\tfrac{1}{4096} > E(.60) > -\tfrac{1}{256}$$

and indeed the difference between the approximate and true values lies within this error.

An alternate approach to that in this section which must lead to the same polynomial (why?) and which is more convenient in certain applications is given by the Newton interpolation formula using divided differences {22}.

3.3 INTERPOLATION AT EQUAL INTERVALS

In many applications of interpolation, the tabular points are equally spaced. For this reason it is worthwhile to consider the simplifications of the Lagrangian formula that can be made in this case.

3.3-1 Lagrangian Interpolation at Equal Intervals

Let the equal spacing be h, so that

$$a_{j+1} - a_j = h \qquad j = 1, \ldots, n-1 \tag{3.3-1}$$

For reasons of symmetry and computational convenience, it is common to take n odd and let

$$x = a_r + hm \tag{3.3-2}$$

where $r = (n+1)/2$. Thus $m = 0$ corresponds to the center of the interval spanned by the tabular points. Using (3.3-2), $p_n(x)$ and $l_j(x)$ can be expressed as functions of m. In particular, from (3.2-3) it follows that $l_j(m)$ is independent of h and can thus be tabulated as a function of m. When we use (3.3-2) and write $f(a_r + hm)$ as $f(m)$, the Lagrangian interpolation formula becomes

$$f(m) = \sum_{j=1}^{n} l_j(m) f(a_j) + \frac{h^n p_n(m)}{n!} f^{(n)}(\xi) \tag{3.3-3}$$

where

$$p_n(m) = (m - r + 1)(m - r + 2) \cdots m(m+1) \cdots (m + r - 1) \tag{3.3-4}$$

Table 3.1 is a short tabulation of the Lagrangian interpolation polynomials $l_j(m)$ for $n = 5$. Clearly, when m and n are such that the $l_j(m)$ are tabulated, the use of (3.3-3) is quite straightforward on a hand calculator. On a digital computer, it will seldom be convenient to store such a table; rather it will be easier to generate the values of $l_j(m)$ using (3.2-4).

Table 3.1 Values of the Lagrangian interpolation polynomials for $n = 5$ ($x = a_3 + hm$)

m	$l_1(m)$	$l_2(m)$	$l_3(m)$	$l_4(m)$	$l_5(m)$	
0	.0000	.0000	1.0000	.0000	.0000	0
.2	.0144	−.1056	.9504	.1584	−.0176	−.2
.4	.0224	−.1536	.8064	.3584	−.0336	−.4
.6	.0224	−.1456	.5824	.5824	−.0416	−.6
.8	.0144	−.0896	.3024	.8064	−.0336	−.8
1.0	.0000	.0000	.0000	1.0000	.0000	−1.0
1.2	−.0176	.1024	−.2816	1.1264	.0704	−1.2
1.4	−.0336	.1904	−.4896	1.1424	.1904	−1.4
1.6	−.0416	.2304	−.5616	.9984	.3744	−1.6
1.8	−.0336	.1824	−.4256	.6384	.6384	−1.8
2.0	.0000	.0000	.0000	.0000	1.0000	−2.0
	$l_5(m)$	$l_4(m)$	$l_3(m)$	$l_2(m)$	$l_1(m)$	m

Example 3.2 Using the same data as in Example 3.1 plus the true value of ln .60, estimate the value of ln .54.

We have $h = .1$; using Table 3.1 with $m = -.6$, we get from (3.3-3)

$$\ln .54 \approx -.0416 \ln .40 + .5824 \ln .50 + .5824 \ln .60 - .1456 \ln .70$$

$$+ .0224 \ln .80 = -.616143$$

whereas the true value is $-.616186$.

When the values of $l_j(m)$ are not tabulated, for hand computation, instead of (3.3-3) it is preferable to use the finite-difference interpolation formulas, which we shall discuss in Sec. 3.4. Before proceeding to discuss finite differences, however, we emphasize that *there is one and only one polynomial of degree $n - 1$ that takes on the values of $f(x)$ at the n tabular points* (why?). In what follows, we shall write interpolation formulas in a form very different from (3.1-1) or (3.3-3). But as long as these formulas involve polynomials passing through the same n tabular points, they will be identical to the Lagrangian interpolation formula.

3.3-2 Finite Differences

In textbooks on classical numerical analysis, the calculus of finite differences and the interpolation, differentiation, and integration formulas based on it were always of central importance. This is because, for work on desk calculators, finite differences are a wonderfully convenient tool. Aside from their advantages for hand computation, there are certain special applications for which finite differences are invaluable (see, for example, Sec. 4.13-2). Also they are used extensively—although generally in a quite simple form—in the numerical solution of partial differential equations and boundary-value problems of ordinary differential equations on digital computers.

Definitions As in Sec. 3.3-1, let the interval between successive tabular points be h. Then we define:

1. The kth *forward difference* of $f(x)$ as

$$\Delta^k f(x) = \Delta^{k-1} f(x+h) - \Delta^{k-1} f(x) \qquad k = 1, 2, \ldots$$
$$\Delta^0 f(x) = f(x) \tag{3.3-5}$$

Thus, for example,

$$\Delta^1 f(x) \equiv \Delta f(x) = f(x+h) - f(x) \tag{3.3-6}$$
$$\Delta^2 f(x) = \Delta f(x+h) - \Delta f(x) = f(x+2h) - 2f(x+h) + f(x) \tag{3.3-7}$$

In fact, it should be clear from this definition that any order difference can be written as a linear combination of functional values as in (3.3-6) and (3.3-7). The general form of this linear combination, whose derivation we leave to a problem {7}, is

$$\Delta^j f(x) = \sum_{k=0}^{j} (-1)^{j-k} \binom{j}{k} f(x+kh) \tag{3.3-8}$$

where the binomial coefficient $\binom{j}{k} = \dfrac{j!}{k!\,(j-k)!}$.

2. The kth *backward difference* as

$$\nabla^k f(x) = \nabla^{k-1} f(x) - \nabla^{k-1} f(x-h) \qquad k = 1, 2, \ldots$$
$$\nabla^0 f(x) = f(x) \tag{3.3-9}$$

3. The kth *central difference* as

$$\delta^k f(x) = \delta^{k-1} f(x+\tfrac{1}{2}h) - \delta^{k-1} f(x-\tfrac{1}{2}h) \qquad k = 1, 2, \ldots$$
$$\delta^0 f(x) = f(x) \tag{3.3-10}$$

Note that if x is a tabular point, only even central differences involve tabular points (why?).

A property of differences that we shall have use for later is that the first difference of a polynomial of degree n is a polynomial of degree $n-1$ {8}. Therefore, *the nth difference of a polynomial of degree n is a constant, and the* $(n+1)$*st difference is identically zero*. The properties of finite differences and the formulas based upon them can be derived by operational calculus using the difference operators Δ, ∇, and δ; we leave a consideration of this approach to a problem {9}.

The lozenge diagram In the remainder of this section, we shall denote $\Delta^j f(a_k)$ by $\Delta^j f_k$ with a corresponding notation for backward and central differences. Furthermore, we shall change our previous notation slightly and let the tabular points have both positive and negative subscripts. When we calculate differences, it is convenient to set up a *difference table*, as in Fig. 3.1, in which each entry after the second column is the difference of the two immediately to its left. The use of forward differences in the table is arbitrary; backward differences could just as easily have been used (but not central differences—why?).

a_{-4}	f_{-4}		$\Delta^2 f_{-5}$		$\Delta^4 f_{-6}$	
		Δf_{-4}		$\Delta^3 f_{-5}$		$\Delta^5 f_{-6}$
a_{-3}	f_{-3}		$\Delta^2 f_{-4}$		$\Delta^4 f_{-5}$	
		Δf_{-3}		$\Delta^3 f_{-4}$		$\Delta^5 f_{-5}$
a_{-2}	f_{-2}		$\Delta^2 f_{-3}$		$\Delta^4 f_{-4}$	
		Δf_{-2}		$\Delta^3 f_{-3}$		$\Delta^5 f_{-4}$
a_{-1}	f_{-1}		$\Delta^2 f_{-2}$		$\Delta^4 f_{-3}$	
		Δf_{-1}		$\Delta^3 f_{-2}$		$\Delta^5 f_{-3}$
a_0	f_0		$\Delta^2 f_{-1}$		$\Delta^4 f_{-2}$	
		Δf_0		$\Delta^3 f_{-1}$		$\Delta^5 f_{-2}$
a_1	f_1		$\Delta^2 f_0$		$\Delta^4 f_{-1}$	
		Δf_1		$\Delta^3 f_0$		$\Delta^5 f_{-1}$
a_2	f_2		$\Delta^2 f_1$		$\Delta^4 f_0$	
		Δf_2		$\Delta^3 f_1$		$\Delta^5 f_0$
a_3	f_3		$\Delta^2 f_2$		$\Delta^4 f_1$	
		Δf_3		$\Delta^3 f_2$		$\Delta^5 f_1$
a_4	f_4		$\Delta^2 f_3$		$\Delta^4 f_2$	

Figure 3.1 Forward difference table.

Example 3.3 Using the data of Example 3.2 with one point added at either end, compute the difference table.

The result is

x	$\ln x$	Δ	Δ^2	Δ^3	Δ^4	Δ^5	Δ^6
.30	−1.203973						
		.287682					
.40	−.916291		−.064538				
		.223144		.023715			
.50	−.693147		−.040823		−.011062		
		.182321		.012653		.005959	
.60	−.510826		−.028170		−.005103		−.003534
		.154151		.007550		.002425	
.70	−.356675		−.020620		−.002678		
		.133531		.004872			
.80	−.223144		−.015748				
		.117783					
.90	−.105361						

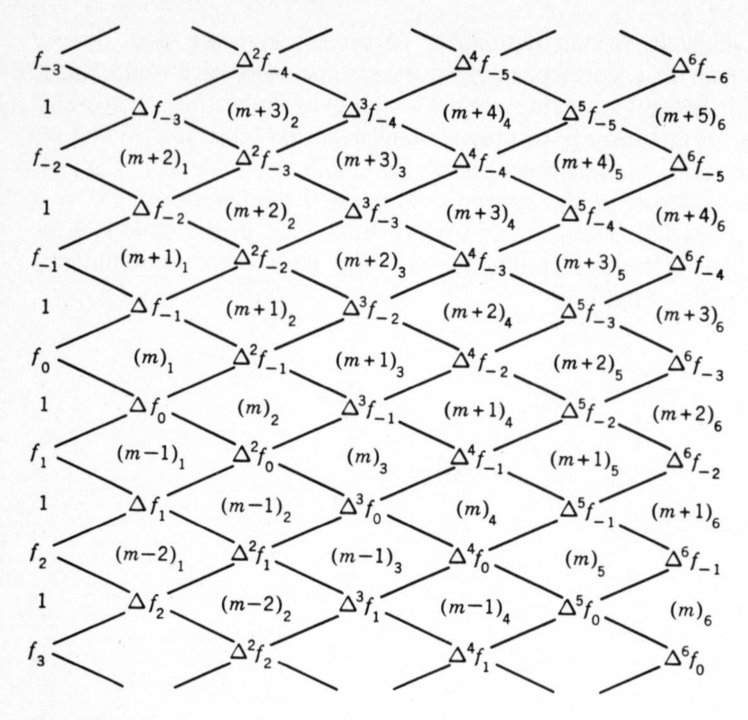

Figure 3.2 The lozenge diagram.

If to Fig. 3.1 we add connecting lines and binomial coefficients† as in Fig. 3.2, we can use this modified difference table, called a *lozenge* or *Fraser diagram*, to generate most of the interesting finite-difference interpolation formulas. To generate such an interpolation formula, we proceed as follows:

I. Start at an entry in the first (functional value) column and proceed along *any* path in the lozenge diagram; i.e., if a segment terminates on a difference, the path may be continued along any of the other three paths leading from the difference. End the path at any difference.

II. Then construct the formula by:

   A. Writing down the functional value at which the path started.

   B.1. For every left-to-right segment in the path *add* a term consisting of the difference on which the segment *terminates* multiplied by the binomial coefficient directly *below* this difference if the slope of the segment is positive and directly *above* if the slope of the segment is negative, and

---

† $$(m+k)_n = \frac{(m+k)(m+k-1)\cdots(m+k-n+1)}{n!} = \binom{m+k}{n}$$

In this section we let m be such that $x = a_0 + hm$ [cf. (3.3-2)].

2. For every right to left segment subtract a term consisting of the difference at which the segment *originates* multiplied by the binomial coefficient directly *below* this difference if the slope of the segment is positive, i.e., if the segment goes downward and to the left, and directly *above* if the slope is negative.

These rules imply that if at a given difference we change direction from right to left to left to right, this difference does not appear in the interpolation formula. As an example of the opposite situation, the path

$$\begin{matrix} & (m)_1 \\ \nwarrow & \\ & \Delta f_0 \\ \nearrow & \\ & (m-1)_1 \end{matrix}$$

gives rise to the terms

$$(m-1)_1 \, \Delta f_0 - (m)_1 \, \Delta f_0$$

For example, starting at f_0, proceeding along lines sloping downward to the right and terminating with the nth difference, we get, writing $y(a_0 + hm)$ as $y(m)$,

$$y(m) = f_0 + (m)_1 \, \Delta f_0 + (m)_2 \, \Delta^2 f_0 + \cdots + (m)_n \, \Delta^n f_0 = \sum_{j=0}^{n} (m)_j \, \Delta^j f_0 \tag{3.3-11}$$

This formula, called *Newton's forward formula*, will be discussed in more detail below.

The value of the procedure outlined above is contained in the statement that any formula derived by this procedure which terminates with an nth difference is *algebraically equivalent* to an equal-interval Lagrangian formula which uses the tabular points involved in the terminating difference. [For example, the nth difference in (3.3-11) involves the points $a_0, \ldots, a_n$; see (3.3-8).] The proof of this assertion requires that we show that

1. At least one formula has this property. Below we shall prove that Newton's forward formula has the desired property.
2. All formulas which terminate with the same difference no matter by what path they reach that difference are algebraically equivalent. We leave the proof of this to a problem {10}.

Finite-difference interpolation formulas We prove first that Eq. (3.3-11), Newton's forward formula, is algebraically equivalent to the Lagrangian interpolation formula at equal intervals for the $n + 1$ points $a_0, \ldots, a_n$. Since $(m)_n$ is a polynomial of degree n in m, it is sufficient to prove that $y(i)$ in (3.3-11) equals f_i, $i = 0, \ldots, n$, for then $y(m)$ would be the unique polynomial of degree n passing through the $n + 1$ points f_i. Using (3.3-8) in (3.3-11), we get

$$y(i) = \sum_{j=0}^{n} (i)_j \, \Delta^j f_0 = \sum_{j=0}^{n} \sum_{k=0}^{j} (-1)^{j-k} (i)_j \binom{j}{k} f_k$$

$$= \sum_{k=0}^{n} \sum_{j=k}^{n} (-1)^{j-k} (i)_j \binom{j}{k} f_k \qquad i = 0, \ldots, n \tag{3.3-12}$$

The coefficient of f_r in $y(i)$ is then given by

$$\sum_{j=r}^{n} (-1)^{j-r} (i)_j \binom{j}{r} \tag{3.3-13}$$

For $r > i$ this coefficient is zero since $(i)_j = 0$ if $i < j$. When $r = i$, the only nonzero term in (3.3-13) is that for $j = r$ and equals 1. When $r < i$, (3.3-13) can be written

$$\sum_{j=r}^{i} (-1)^{j-r} \binom{i}{j} \binom{j}{r} \tag{3.3-14}$$

which, by suitable manipulation {12}, can be shown to vanish. Thus, the right-hand side of (3.3-12) is just f_i, which completes the proof.

Using the lozenge diagram, we can generate the following interpolation formulas.

Newton's backward formula Starting at f_0 and proceeding along lines sloping upward and to the right, we get

$$y(m) = f_0 + (m)_1 \, \Delta f_{-1} + (m + 1)_2 \, \Delta^2 f_{-2} + \cdots + (m + n - 1)_n \, \Delta^n f_{-n} \tag{3.3-15}$$

which is equivalent to a Lagrangian formula using the points $a_0, a_{-1}, \ldots, a_{-n}$. In fact, this formula is more conveniently expressed in terms of backward differences {13}.

Gauss' forward formula Here we proceed in a zigzag, downward and to the right, then upward and to the right, then downward and to the right, etc. The result is

$$y(m) = f_0 + (m)_1 \, \Delta f_0 + (m)_2 \, \Delta^2 f_{-1} + (m + 1)_3 \, \Delta^3 f_{-1} + (m + 1)_4 \, \Delta^4 f_{-2} + \cdots \tag{3.3-16}$$

Gauss' backward formula Here we proceed as in Gauss' forward formula except that the first step is upward and to the right. The formula is

$$y(m) = f_0 + (m)_1 \Delta f_{-1} + (m+1)_2 \Delta^2 f_{-1}$$
$$+ (m+1)_3 \Delta^3 f_{-2} + (m+2)_4 \Delta^4 f_{-2} + \cdots \qquad (3.3\text{-}17)$$

Both Gaussian formulas are conveniently expressed in terms of central differences {13}.

Because of the result stated in Sec. 3.3-2, each of these formulas is algebraically equivalent to the Lagrangian formula which uses the same tabular points. The errors in these formulas therefore are given by (3.2-9). In the next section we shall indicate why it is useful to be able to express the same interpolation formula in a number of different forms.

If we take the mean of Gauss' forward and backward formulas as given by (3.3-16) and (3.3-17), we get *Stirling's interpolation formula*

$$y(m) = f_0 + \frac{m}{2}(\Delta f_0 + \Delta f_{-1}) + \tfrac{1}{2}[(m)_2 \Delta^2 f_{-1}$$
$$+ (m+1)_2 \Delta^2 f_{-2}] + \cdots \qquad (3.3\text{-}18)$$

Stirling's formula can be conveniently expressed in terms of central differences {13}.

Bessel's interpolation formula is the mean of the Gaussian forward formula given by (3.3-16) and a Gaussian backward formula launched not from f_0 but from f_1. It has the form

$$y(m) = \tfrac{1}{2}(f_0 + f_1) + (m - \tfrac{1}{2}) \Delta f_0 + \tfrac{1}{2}(m)_2(\Delta^2 f_0 + \Delta^2 f_1) + \cdots \qquad (3.3\text{-}19)$$

Note that when (3.3-17) is modified to consider launching from f_1, m must be replaced by $m - 1$ so that the origin of m is still at a_0. Bessel's formula is also conveniently written using central differences {13}.

Some other interpolation formulas which can be obtained by manipulating the ones derived in this section are considered in a problem {14}.

3.4 THE USE OF INTERPOLATION FORMULAS

With the exception of Stirling's formula terminated with an odd difference and Bessel's formula terminated with an even difference, all the interpolation formulas we have derived are algebraically equivalent over the same set of tabular points. For equally spaced data, the ease with which difference tables can be generated makes the finite-difference interpolation formulas more convenient than the Lagrangian formula for hand computation. To get some insight into which of the finite-difference formulas to use in a given application (why does it matter if they are all equivalent?), let us consider interpolation in a table of values.

One of the great advantages of the finite-difference interpolation formulas is the ease with which added terms of the formula can be used merely by calculating higher differences in the table of Fig. 3.1. For example, if we add the value of ln x at 1.0 to the table of Example 3.3, we can calculate a new row of differences and thereby get a difference of order 7 in the table. Commonly, we do not know a priori how many terms in a given interpolation formula will be sufficient to achieve the accuracy we desire. Therefore, we generally add terms to the formula by computing higher differences until the contribution of the added terms is so small that the number of decimal places of interest to us has stabilized. [If by use of (3.2-9) we can bound the error, all well and good; often, however, it will be difficult to estimate, much less bound, the derivative term in (3.2-9).] It is desirable then to use that interpolation formula which gives the best results at every stage of the computation.

Consider the problem of estimating ln .65 using the data in the difference table of Example 3.3. Suppose that a priori we do not know how many differences will be required to obtain the accuracy we need.† If all the data in the table will be required to achieve the desired accuracy, it makes no appreciable difference which finite-difference interpolation formula we use because all will be algebraically equivalent. But if it is possible that a sufficiently accurate result can be obtained using fewer than six differences, we should choose our interpolation formula with some care.

Let us compare the use of Newton's backward formula and Gauss' forward formula. If we may need to use all the data in the table, the Newtonian formula must use $x_0 = .9$, that is, $m = -2.5$ and the Gaussian formula must use $x_0 = .6$ ($m = .5$). But while these two formulas will be algebraically equivalent if terms through the sixth difference are used, for smaller numbers of terms, they will not be equivalent {17}. Therefore, which should we choose?

This question is most easily answered by considering the error term (3.2-9). The only term in the error that we can control is the $p_n(x)$ term. To minimize the magnitude of $p_n(x)$, we should choose the tabular points so that the value of x at which we wish to interpolate is as near as possible to the center of the interval spanned by the tabular points (why?). Therefore, the answer to our question is that the Gaussian formula is to be preferred in the above example because when the number of differences used is small, it more nearly satisfies the condition above than the Newtonian formula.

From the above it follows that Newton's backward formula has its chief value when we wish to interpolate near the end of a table, for in this case there would not be a sufficient number of differences available for the Gaussian formula. For example, to estimate ln .85 using the data of Example 3.3, if we used Gauss's forward formula with $x_0 = .8$, we could only use the terms through the second difference {18}. Similarly, Newton's forward formula is chiefly valuable near the beginning of a table. But when there are a substantial number of tabular points available on either side of the interpolation

† For such a simple function as this, we could, of course, estimate the error using (3.2-9).

point, a Gaussian formula is more desirable than either Newtonian formula. In particular, Stirling's formula (which is just the average of two Gaussian formulas) terminated with an even difference (so that it is equivalent to a Lagrangian formula) is useful when m can be chosen near zero; similarly, Bessel's formula terminated with an odd difference is useful when m is near $\frac{1}{2}$. The justification for these conclusions is considered in {18}.

Example 3.4 Use Newton's backward formula with $x_0 = .9$ and Gauss' forward formula with $x_0 = .6$ to find an estimate of ln .65.

Using (3.3-15), (3.3-16), and the data of Example 3.3, we can construct the following table:

Number of differences used	Newton's backward formula ($m = -2.5$)	Gauss' forward formula ($m = .5$)
0	$-.105361$	$-.510826$
1	$-.399819$	$-.433751$
2	$-.429346$	$-.430229$
3	$-.430869$	$-.430701$
4	$-.430762$	$-.430821$
5	$-.430791$	$-.430792$
6	$-.430774$	$-.430775$

The true value is ln. 65 $\approx -.430783$. As we expect, when the number of differences is small, the Gaussian formula is more accurate than the Newtonian formula, although both give the same value, except for roundoff, when all six differences are used. (Why is the Newtonian formula more accurate when four differences are used?) Using (3.2-9) we can verify that the error at every stage is within expected bounds {17}.

The reader may well think that any desired degree of accuracy could be achieved merely by increasing the number of terms used in any interpolation formula (finite difference or Lagrangian).† In fact, interpolation series formed by letting the number of tabular points go to infinity are generally only asymptotically convergent; i.e., as we add more points the error first decreases and then at some point starts to increase and grow without bound.‡ One reason for this eventual divergence of interpolation series is connected with the fact that the nth derivative of all but some entire functions (functions with no singularities in the complex plane) eventually grows without bound as n increases (see Sec. 4.9). Even for entire functions, however, the interpolation series may fail to converge {19}. We note that in practice the desired degree of accuracy in interpolation can almost always be achieved; i.e., the asymptotic convergence is generally very good indeed.

† We ignore here the fact that the growth of roundoff error with higher differences limits the accuracy attainable with finite-difference interpolation formulas.

‡ The classic example of this behavior is the very well-behaved function $f(x) = 1/(1 + x^2)$, which is considered by Steffensen (1950, pp. 35–38).

The discussion above deals with the case in which the interval spanned by these points increases without bound as more tabular points are added. Suppose now instead that we have a fixed finite interval $[a, b]$ and a sequence $\{S_n\}$ of sets of interpolation points

$$S_n = \{x_j^{(n)}, j = 1, \ldots, n\}$$

such that each S_n is contained in $[a, b]$. If $p_n(x)$ is then an interpolation polynomial based on S_n, we would expect that if the maximum distance between adjacent points $x_j^{(n)}$ goes to zero as n goes to ∞, the $p_n(x)$ would converge uniformly to $f(x)$ as n goes to ∞. However, this is not necessarily the case, as indicated by the classical example of Runge, in which S_n consists of equidistantly spaced points in $[-1, 1]$ and $f(x) = 1/(x^2 + 25)$. This function is quite smooth and is differentiable infinitely often, but the $p_n(x)$ do not converge uniformly to $f(x)$. This nonconvergence is called the *Runge effect*. On the other hand, there are sequences $\{S_n\}$ for which the $p_n(x)$ do converge uniformly to $f(x)$ provided only that $f(x)$ satisfies some quite mild conditions. One such is the sequence of *Chebyshev nodes*

$$S_n = \left\{-\cos\frac{(2j-1)\pi}{2n}, j = 1, \ldots, n\right\}$$

for which the $p_n(x)$ converge uniformly to $f(x)$ on $[-1, 1]$ provided that $f(x)$ has a bounded derivative on this interval.

3.5 ITERATED INTERPOLATION

An important advantage of finite-difference interpolation formulas over the Lagrangian formula would seem to be the property of the former that enables a term to be added to them merely by adding one tabular point and computing an additional row of differences. As we demonstrated in Example 3.4, this enables us to generate a sequence of *interpolants* each one involving one more tabular point than the previous one. Therefore, the convergence of the interpolation procedure can be tested easily. But suppose, given the Lagrangian interpolation formula using n points, we wish to add one point to get higher accuracy. A look at (3.2-4) indicates that even if we have saved the values of $p_n'(a_j), j = 1, \ldots, n$, each $l_j(x), j = 1, \ldots, n$, requires some recalculation and we must also calculate $l_{n+1}(x)$. Our purpose in this section is to show how this seeming disadvantage of the Lagrangian formula can be overcome. We shall do this by using *iterated interpolation*, in which a sequence of interpolants in the Lagrangian context is generated without the need for substantial recalculation of coefficients when going from n to $n + 1$ points.

Denote by $y_{n_1, \ldots, n_k}(x)$ the Lagrangian interpolation formula using the points $a_{n_1}, \ldots, a_{n_k}$, which we do not require to be equally spaced. Then in particular we can write

$$y_{1,2,\ldots,n}(x) = \frac{1}{a_n - a_{n-1}} \begin{vmatrix} y_{1,2,\ldots,n-1}(x) & a_{n-1} - x \\ y_{1,2,\ldots,n-2,n}(x) & a_n - x \end{vmatrix} \tag{3.5-1}$$

This equation can be verified by noting that the right-hand side, which is a polynomial of degree $n - 1$, takes on the values $f(a_i)$ at the points a_i, $i = 1, \ldots, n$. Equation (3.5-1) then indicates how a Lagrangian formula of order n can be generated from lower-order formulas. By use of the following table, we can generalize the result of (3.5-1) to achieve our object [note that $y_i(x) = f(a_i)$]:

$$\begin{array}{llllll}
a_1 & a_1 - x & y_1(x) & & & \\
a_2 & a_2 - x & y_2(x) & y_{1,2}(x) & & \\
a_3 & a_3 - x & y_3(x) & y_{1,3}(x) & y_{1,2,3}(x) & \\
\cdots & \cdots & \cdots & \cdots & \cdots & \\
a_n & a_n - x & y_n(x) & y_{1,n}(x) & y_{1,2,n}(x) \cdots & y_{1,2,3,\ldots,n}(x)
\end{array} \tag{3.5-2}$$

The entries in each column of the table can be generated from the entries in the previous column by analogy with (3.5-1). For example

$$y_{1,2,n}(x) = \frac{1}{a_n - a_2} \begin{vmatrix} y_{1,2}(x) & a_2 - x \\ y_{1,n}(x) & a_n - x \end{vmatrix} \tag{3.5-3}$$

The entries on the diagonal in (3.5-2) are just what we were seeking. They form a sequence of Lagrangian interpolants each of which incorporates one more tabular point than the previous one. Further, since each entry in (3.5-2) is calculated using a formula analogous to (3.5-1), the process is easily mechanized. Iterated interpolation is thus well suited to digital-computer application and, for points not equally spaced, is also convenient for hand computation.

Example 3.5 Use iterated interpolation to calculate ln .54 using the data of Example 3.2. Corresponding to the table (3.5-2), we get

a_i	$a_i - x$	$y_i(x)$				
.40	−.14	−.916291				
.50	−.04	−.693147	−.603889			
.60	.06	−.510826	−.632466	−.615320		
.70	.16	−.356675	−.655137	−.614139	−.616029	
.80	.26	−.223144	−.673690	−.613196	−.615957	−.616144

We are not bound to use the natural order of the points as above. Consider instead interated interpolation using the ordering below:

a_i	$a_i - x$	$y_i(x)$				
.50	−.04	−.693147				
.60	.06	−.510826	−.620219			
.40	−.14	−.916291	−.603889	−.615320		
.70	.16	−.356675	−.625853	−.616839	−.616029	
.80	.26	−.223144	−.630480	−.617141	−.615957	−.616144

The difference in the two final values from the result of Example 3.2 is the result of roundoff since the same five points were used in all computations. Note, however, that the first interpolant (−.620219) in the second calculation is substantially more accurate than that in the first calculation (−.603889). This occurs because in the second computation we arranged the data so that the magnitudes in the $a_i - x$ column would be increasing. In this way the value of $p_n(x)$ in the error term is minimized at every stage (cf. the discussion of Sec. 3.4). Therefore, if we order the tabular points so that the magnitudes in the $a_i - x$ column are increasing, each interpolant tends to be the best possible ["tends" because the value of the derivative in the error *may* be greater when $p_n(x)$ is smaller]. In this way the convergence of the interpolation (as judged by the difference in two successive interpolants or by the stabilization of a certain number of decimal places) will tend to be most rapid. In this example only the first interpolant is improved because only the tabular point at .40 is out of the best possible order.

3.6 INVERSE INTERPOLATION

In Chap. 8 we shall be concerned with the solution of the general nonlinear equation $f(x) = 0$. One of our basic tools in the solution of this equation will be inverse interpolation, which we now consider briefly. The solution of $f(x) = 0$ is one example of the common numerical problem of finding the zero of a function. Another case where this occurs is in the numerical integration of an ordinary differential equation (see Chap. 5), when we would like to know that value of the independent variable for which the dependent variable, i.e., the solution of the differential equation, is zero. Inverse interpolation provides us with a straightforward and powerful way to find such zeros of functions.†

Let the function whose zero (or zeros) we wish to find be $y = f(x)$ and suppose it is tabulated at a series of points (which need not necessarily be equally spaced), so that we have

x	x_1	x_2	$\cdots$	x_n
$y = f(x)$	$f(x_1)$	$f(x_2)$	$\cdots$	$f(x_n)$

(3.6-1)

† We note in passing that even the Newton-Raphson method for the solution of $f(x) = 0$ can be considered to be an application of inverse interpolation; see Chap. 8.

Now let us suppose that on the interval $[x_1, x_n]$, $f(x)$ satisfies the conditions of the inverse-function theorem, i.e., in particular that $f'(x) \neq 0$, so that we can write $x = g(y)$, where g is the function inverse to f. Therefore, finding the value of $g(0)$ is equivalent to finding a zero of $f(x)$. To estimate $g(0)$ we first write the table (3.6-1) as

y	$f(x_1)$	$f(x_2)$	$\cdots$	$f(x_n)$
$x = g(y)$	x_1	x_2	$\cdots$	x_n

(3.6-2)

Now in the context of interpolation let $f(x_1), \ldots, f(x_n)$ be the tabular points of the independent variable y (not equally spaced in general) and let $x_1, \ldots, x_n$ be the functional values at these points. Then, if we use a Lagrangian interpolation formula to approximate $g(y)$ by a polynomial and then interpolate at the point $y = 0$, we get the desired approximation to $\alpha = g(0)$.

Example 3.6 Given the data

x	.1	.2	.3	.4	.5
$f(x)$	.70010	.40160	.10810	$-.17440$	$-.43750$

find an approximate value of the zero of $f(x)$ between .3 and .4.

Our approach will be to use iterated interpolation. We therefore first arrange the data in order of increasing magnitude of $f(x)$ (cf. Example 3.5) and then use the technique of the previous section to generate the table

$f(x_i)$	$f(x_i) - 0$	x_i				
.10810	.10810	.3				
$-.17440$	$-.17440$	.4	.33827			
.40160	.40160	.2	.33683	.33783		
$-.43750$	$-.43750$	.5	.33963	.33737	.33761	
.70010	.70010	.1	.33652	.33792	.33771	.33765

The data for $f(x)$ are in fact values of the function $x^4 - 3x + 1$, which has a zero .33767 correctly rounded to five places.

Expressed in terms of $g(y)$ the error in inverse Lagrangian interpolation is

$$E(y) = \frac{p_n(y)}{n!} g^{(n)}(\xi) \tag{3.6-3}$$

Derivatives of g can be expressed in terms of f, and although this relation is not simple (see {1}, Chap. 8), there is a power of $f'(x)$ in the denominator of each derivative of $g(y)$, for example,

$$g'(y) = \frac{1}{f'(x)} \qquad g''(y) = \frac{-f''(x)}{[f'(x)]^3}$$

Therefore, although we can carry through the process of inverse interpolation even if $f'(x)$ vanishes in $[x_1, x_n]$, we would expect the accuracy to be very poor in this case. Indeed, at a point at which $f'(x)$ vanishes, the inverse function may not exist. When $f'(x)$ vanishes near the zero, however, we can often find the zero by an iterative process involving linear inverse interpolation {27}. In general, linear inverse interpolation may even be preferable to higher-order inverse interpolation, which can be very problematical, since the quality of the interpolation depends on how well the inverse function can be approximated by a polynomial, but we usually have this information only for the original function and not for its inverse.

3.7 HERMITE INTERPOLATION

In this section we consider the case $m_j = 1$ in (3.1-2) for $j = 1, \ldots, r$, that is, we suppose that the first derivative as well as the function is known at r of the n tabular points. In place of (3.1-1) we have then

$$f(x) = \sum_{j=1}^{n} h_j(x) f(a_j) + \sum_{j=1}^{r} \bar{h}_j(x) f'(a_j) + E(x) = y(x) + E(x) \quad (3.7\text{-}1)$$

where now the approximation $y(x)$ is given by

$$y(x) = \sum_{j=1}^{n} h_j(x) f(a_j) + \sum_{j=1}^{r} \bar{h}_j(x) f'(a_j) \quad (3.7\text{-}2)$$

and $h_j(x)$ and $\bar{h}_j(x)$ are both polynomials. Again using the criterion of exact approximation, we require that the error term $E(x)$ be such that

$$\begin{aligned} E(a_j) &= 0 \qquad j = 1, \ldots, n \\ E'(a_j) &= 0 \qquad j = 1, \ldots, r \end{aligned} \quad (3.7\text{-}3)$$

In analogy with Eq. (3.2-1) in the Lagrangian case, this leads to the following conditions that must be satisfied by the $h_j(x)$ and $\bar{h}_j(x)$:

$$\begin{aligned} h_j(a_k) &= \delta_{jk} \qquad j, k = 1, \ldots, n \\ \bar{h}_j(a_k) &= 0 \qquad j = 1, \ldots, r;\ k = 1, \ldots, n \\ h'_j(a_k) &= 0 \qquad j = 1, \ldots, n;\ k = 1, \ldots, r \\ \bar{h}'_j(a_k) &= \delta_{jk} \qquad j, k = 1, \ldots, r \end{aligned} \quad (3.7\text{-}4)$$

Since there are $n + r$ conditions to satisfy in (3.7-3), we expect that $y(x)$ will have to be a polynomial of degree $n + r - 1$; that is, we shall approximate $f(x)$ by a polynomial of degree $n + r - 1$ passing through $f(a_j)$, $j = 1, \ldots, n$, and having derivatives $f'(a_j)$, $j = 1, \ldots, r$. In deriving the $h_j(x)$ and $\bar{h}_j(x)$, we shall use the notation

$$p_n(x) = (x - a_1) \cdots (x - a_n)$$
$$p_r(x) = (x - a_1) \cdots (x - a_r)$$
$$l_{jn}(x) = \frac{p_n(x)}{(x - a_j)p_n'(a_j)} \qquad j = 1, \ldots, n$$
$$l_{jr}(x) = \frac{p_r(x)}{(x - a_j)p_r'(a_j)} \qquad j = 1, \ldots, r \tag{3.7-5}$$

To satisfy the conditions on $h_j(x)$ we set

$$h_j(x) = \begin{cases} t_j(x) l_{jn}(x) l_{jr}(x) & j = 1, \ldots, r \\ l_{jn}(x) \dfrac{p_r(x)}{p_r(a_j)} & j = r + 1, \ldots, n \end{cases} \tag{3.7-6}$$

where $t_j(x)$ is a linear polynomial so that $h_j(x)$ is of degree $n + r - 1$. As given by (3.7-6), $h_j(x)$ satisfies all the conditions of (3.7-4) except $h_j(a_j) = 1$, $j = 1, \ldots, r$, and $h_j'(a_j) = 0$, $j = 1, \ldots, r$. To satisfy these we must have

$$\left.\begin{aligned} t_j(a_j) &= 1 \\ t_j'(a_j) + l_{jn}'(a_j) + l_{jr}'(a_j) &= 0 \end{aligned}\right\} \quad j = 1, \ldots, r \tag{3.7-7}$$

Similarly, if we set

$$\bar{h}_j(x) = s_j(x) l_{jr}(x) l_{jn}(x) \qquad j = 1, \ldots, r \tag{3.7-8}$$

with $s_j(x)$ a linear polynomial, we must have

$$\left.\begin{aligned} s_j(a_j) &= 0 \\ s_j'(a_j) &= 1 \end{aligned}\right\} \quad j = 1, \ldots, r \tag{3.7-9}$$

in order to satisfy (3.7-4). Linear functions satisfying (3.7-7) and (3.7-9) are easily found to be {28}

$$t_j(x) = 1 - (x - a_j)[l_{jn}'(a_j) + l_{jr}'(a_j)] \qquad s_j(x) = x - a_j \tag{3.7-10}$$

This completes the determination of $h_j(x)$ and $\bar{h}_j(x)$.

To find $E(x)$ we proceed in a manner similar to the Lagrangian case. Let

$$F(z) = f(z) - y(z) - [f(x) - y(x)] \frac{p_n(z)p_r(z)}{p_n(x)p_r(x)} \tag{3.7-11}$$

with x not one of the tabular points. This function has $n + r + 1$ zeros (double zeros at $a_1, \ldots, a_r$, single zeros at $a_{r+1}, \ldots, a_n$ and x) so that by a generalization of Rolle's theorem, there exists a ξ in the interval spanned by $a_1, \ldots, a_n$ and x such that

$$0 = F^{(n+r)}(\xi) = f^{(n+r)}(\xi) - [f(x) - y(x)] \frac{(n + r)!}{p_n(x)p_r(x)} \tag{3.7-12}$$

Thus
$$E(x) = \frac{p_n(x)p_r(x)}{(n+r)!} f^{(n+r)}(\xi) \tag{3.7-13}$$

This relation also is correct if x is one of the tabular points (why?).

The interpolation formula (3.7-1) then becomes

$$f(x) = \sum_{j=1}^{n} h_j(x)f(a_j) + \sum_{j=1}^{r} \bar{h}_j(x)f'(a_j) + \frac{p_n(x)p_r(x)}{(n+r)!} f^{(n+r)}(\xi) \tag{3.7-14}$$

with

$$h_j(x) = \begin{cases} \{1 - (x - a_j)[l'_{jn}(a_j) + l'_{jr}(a_j)]\}l_{jn}(x)l_{jr}(x) & j = 1, \ldots, r \\ l_{jn}(x)\dfrac{p_r(x)}{p_r(a_j)} & j = r+1, \ldots, n \end{cases} \tag{3.7-15}$$

$$\bar{h}_j(x) = (x - a_j)l_{jr}(x)l_{jn}(x) \qquad j = 1, \ldots, r \tag{3.7-16}$$

and is called the *modified Hermite interpolation formula.* When $r = n$, the formula is

$$f(x) = \sum_{j=1}^{n} h_j(x)f(a_j) + \sum_{j=1}^{n} \bar{h}_j(x)f'(a_j) + \frac{p_n^2(x)}{(2n)!} f^{(2n)}(\xi) \tag{3.7-17}$$

with
$$h_j(x) = [1 - 2(x - a_j)l'_j(a_j)]l_j^2(x) \qquad \bar{h}_j(x) = (x - a_j)l_j^2(x) \qquad j = 1, \ldots, n \tag{3.7-18}$$

where we have replaced $l_{jn}(x)$ by $l_j(x)$. Equation (3.7-17) is the *Hermite interpolation formula,* also called the formula for osculatory interpolation.

Both the Hermite and modified Hermite formulas can be useful interpolation formulas. They also serve as useful theoretical tools in other areas of numerical analysis, as we shall see in Chaps. 4, 5, and 8.

Example 3.7 Given the table below of the natural logarithm and its derivative

x	$\ln x$	$1/x$
.40	−.916291	2.50
.50	−.693147	2.00
.70	−.356675	1.43
.80	−.223144	1.25

estimate the value of ln .60 using the Hermite interpolation formula.

From (3.7-18) we get

$$h_1(.60) = \tfrac{11}{54} \qquad h_2(.60) = \tfrac{8}{27} \qquad h_3(.60) = \tfrac{8}{27} \qquad h_4(.60) = \tfrac{11}{54}$$

$$\bar{h}_1(.60) = \tfrac{1}{180} \qquad \bar{h}_2(.60) = \tfrac{2}{45} \qquad \bar{h}_3(.60) = -\tfrac{2}{45} \qquad \bar{h}_4(.60) = -\tfrac{1}{180}$$

and from (3.7-17)

$$\ln .60 \approx -.510824$$

whereas the true value is $-.510826$. Using (3.7-13), the error is bounded by

$$-.000031 \approx \frac{-1}{32768} < E(.60) < -\frac{1}{2^{23}} \approx -.0000001$$

so that the excellent agreement between the interpolated and true values is to be expected.

3.8 SPLINE INTERPOLATION

As mentioned in Sec. 3.4, it is possible that a sequence of interpolation polynomials $\{p_n(x)\}$ over a fixed finite interval need not converge even to a smooth function. And if we investigate the behavior of these polynomials between the interpolation points we find that in many cases the polynomials oscillate quite violently while the function varies smoothly. And the higher the degree of the polynomial, the worse the situation becomes. A way to overcome this problem is by using piecewise low-order interpolating polynomials on subintervals of the given interval. With them, the oscillation between points is not significant, so that they can imitate the behavior of the function. However, the resulting function pieced together from the individual low-order polynomials may not be smooth. Since we wish to imitate the behavior of smooth functions, a requirement for these piecewise functions is that the resulting function pieced together be smooth. Such functions, with the maximal degree of smoothness, are called *splines* and we shall now describe them formally.

Let the interval $I = [a, b]$ be divided into $n - 1$ subintervals $a = a_1 < a_2 < \cdots < a_{n-1} < a_n = b$ not necessarily of equal length. A *spline* $S(x)$ of degree m is a function defined on I which:

1. Coincides with a polynomial of degree m on each subinterval $I_i = [a_{i-1}, a_i]$, $i = 2, \ldots, n$.
2. Has continuous derivatives up to order $m - 1$.

The abscissas $\{a_i\}$ are called the *nodes* or *knots* of the spline. A spline $S(x)$ is said to interpolate to the data points (a_i, y_i) if $S(a_i) = y_i$, $i = 1, \ldots, n$.

The word spline derives from the instrument often used by draftsmen in fairing a curve through data points. The simplest spline, that of degree 1, is a piecewise linear function which is not very smooth but very useful if the spacing between nodes is small. In fact, every table of functional values in which linear interpolation is used leads to an approximation of the underlying function by a linear spline. Splines of degree 2 can be defined, but since there is only one degree of freedom in their definition, there is a lack of symmetry in their determination with relation to the endpoints of the interval. Furthermore, the resulting functions are not sufficiently smooth. Thus the most prevalent spline in use is the cubic spline, which involves two parameters chosen to reflect the behavior at the endpoints of the interval. One type of end condition is that $S''(a) = S''(b) = 0$. The cubic spline which satisfies these conditions is called the *natural cubic spline*. A second condition, which generally yields better results, is

$$S'(a) = f'(a) \qquad S'(b) = f'(b)$$

This, of course, requires knowledge of the derivative at the endpoints.

Since we shall subsequently be dealing exclusively with cubic interpolatory splines, we shall drop the adjectives and simply write spline. One of several representations of splines can be derived as follows. Set

$$h_i = a_i - a_{i-1} \qquad i = 2, 3, \ldots, n \tag{3.8-1}$$

Since $S(x)$ is piecewise cubic, $S'(x)$ is piecewise quadratic and $S''(x)$ is piecewise linear and continuous. Hence, we can write

$$S''(x) = M_{i-1}\frac{a_i - x}{h_i} + M_i\frac{x - a_{i-1}}{h_i} \qquad \text{on } I_i \tag{3.8-2}$$

for certain constants M_i, where in fact

$$S''(a_i) = M_i \qquad i = 1, 2, \ldots, n \tag{3.8-3}$$

Integrating (3.8-2) twice and writing the arbitrary linear function in the form indicated, we obtain

$$S(x) = M_{i-1}\frac{(a_i - x)^3}{6h_i} + M_i\frac{(x - a_{i-1})^3}{6h_i} + c_i(a_i - x) + d_i(x - a_{i-1}) \tag{3.8-4}$$

on I_i. Since we wish the spline to interpolate at the knots, we have that $S(a_{i-1}) = y_{i-1}$ and $S(a_i) = y_i$. This determines the c_i and d_i, yielding

$$\begin{aligned} S(x) = {} & M_{i-1}\frac{(a_i - x)^3}{6h_i} + M_i\frac{(x - a_{i-1})^3}{6h_i} \\ & + \left(y_{i-1} - \frac{M_{i-1}h_i^2}{6}\right)\frac{a_i - x}{h_i} \\ & + \left(y_i - \frac{M_i h_i^2}{6}\right)\frac{x - a_{i-1}}{h_i} \end{aligned} \tag{3.8-5}$$

on I_i. Differentiating (3.8-5), we obtain

$$S'(x) = -M_{i-1}\frac{(a_i - x)^2}{2h_i} + M_i\frac{(x - a_{i-1})^2}{2h_i} + \frac{y_i - y_{i-1}}{h_i} - \frac{M_i - M_{i-1}}{6}h_i \tag{3.8-6}$$

on I_i. We note the particular values

$$S'(a_i^-) = \frac{h_i}{6} M_{i-1} + \frac{h_i}{3} M_i + \frac{y_i - y_{i-1}}{h_i}$$
$$S'(a_i^+) = -\frac{h_{i+1}}{3} M_i - \frac{h_{i+1}}{6} M_{i+1} + \frac{y_{i+1} - y_i}{h_{i+1}} \tag{3.8-7}$$

Since $S'(x)$ is required to be continuous, these two values must be equal; this yields the equations

$$\frac{h_i}{6} M_{i-1} + \frac{h_i + h_{i+1}}{3} M_i + \frac{h_{i+1}}{6} M_{i+1} = \frac{y_{i+1} - y_i}{h_{i+1}} - \frac{y_i - y_{i-1}}{h_i} \qquad i = 2, 3, \ldots, n-1 \tag{3.8-8}$$

These form a set of $n - 2$ linear equations in $M_1, \ldots, M_n$, so that two more conditions must be added. Once the M's are determined, the interpolation spline is completely determined through (3.8-5). We shall abbreviate Eqs. (3.8-8) by setting

$$\sigma_i = \frac{y_i - y_{i-1}}{h_i} \qquad i = 2, 3, \ldots, n$$
$$\lambda_i = \frac{h_{i+1}}{h_i + h_{i+1}} \qquad \mu_i = 1 - \lambda_i \qquad d_i = \frac{6(\sigma_{i+1} - \sigma_i)}{h_i + h_{i+1}} \qquad i = 2, 3, \ldots, n-1 \tag{3.8-9}$$

We then get the set of equations

$$\mu_i M_{i-1} + 2M_i + \lambda_i M_{i+1} = d_i \qquad i = 2, 3, \ldots, n-1 \tag{3.8-10}$$

We shall write the two additional conditions in the form

$$2M_1 + \lambda_1 M_2 = d_1 \qquad \mu_n M_{n-1} + 2M_n = d_n \tag{3.8-11}$$

and indicate below several possible choices of the constants $\lambda_1, d_1, \mu_n, d_n$. The combined system now becomes

$$\begin{bmatrix} 2 & \lambda_1 & 0 & & & & \\ \mu_2 & 2 & \lambda_2 & & & \bigcirc & \\ 0 & \mu_3 & 2 & & & & \\ & & & \ddots & & & \\ & & & & 2 & \lambda_{n-2} & 0 \\ & \bigcirc & & & \mu_{n-1} & 2 & \lambda_{n-1} \\ & & & & 0 & \mu_n & 2 \end{bmatrix} \begin{bmatrix} M_1 \\ M_2 \\ M_3 \\ \vdots \\ M_{n-2} \\ M_{n-1} \\ M_n \end{bmatrix} = \begin{bmatrix} d_1 \\ d_2 \\ d_3 \\ \vdots \\ d_{n-2} \\ d_{n-1} \\ d_n \end{bmatrix} \tag{3.8-12}$$

with a tridiagonal matrix.

1. If one selects $\lambda_1 = d_1 = \mu_n = d_n = 0$, then $M_1 = M_n = 0$. This yields the natural spline as defined above.
2. Using the second endpoint condition proposed above

$$S'(a) = y'_1 \qquad S'(b) = y'_n \tag{3.8-13}$$

we obtain
$$\lambda_1 = 1 = \mu_n \qquad d_1 = \frac{6}{h_2}\left(\frac{y_2 - y_1}{h_2} - y'_1\right)$$

$$d_n = \frac{6}{h_n}\left(y'_n - \frac{y_n - y_{n-1}}{h_n}\right) \tag{3.8-14}$$

3. A third possibility is to choose $a_i, i = 2, \ldots, n-1$, as the spline nodes, i.e., not require the endpoints to be nodes, and to impose the conditions $S(a_1) = y_1$ and $S(a_n) = y_n$.

In this case, the index i in (3.8-10) runs from 3 to $n - 2$, and the two additional conditions are similar in form to those of (3.8-11). The values of λ_2 and d_2 can readily be evaluated by letting $x = a_1$ in (3.8-5) and setting $S(a_1) = y_1$. Similarly, by setting $x = a_n$, we can find the values of λ_{n-1} and d_{n-1} {30}. This scheme, which does not require any additional information, appears to be better than choice 1 for the average function that comes up in practice since the second derivative does not generally vanish at the endpoints of the interval.

From (3.8-9) we see that $0 < \lambda_i < 1$ and $0 < \mu_i < 1$ for $i = 2, \ldots, n-1$. If, therefore, $|\lambda_1| < 2$, $|\mu_n| < 2$, the matrix in (3.8-12) will be diagonally dominant. In this case, it can be shown (see Sec. 9.1) that unique solutions to (3.8-12) will exist for arbitrary $d_1, \ldots, d_n$. Thus, in cases 1 and 2, we are assured of the existence of the spline. Similarly, in case 3, a solution always exists {30}.

Splines have the following important properties which can be readily proved {31, 32}.

1. Given data points (a_i, y_i), $i = 1, \ldots, n$. Of all the functions $f(x)$ with continuous second derivatives which interpolate to these data the spline $S(x)$ which also satisfies $S''(a) = S''(b) = 0$ uniquely minimizes the integral

$$E(g) = \int_a^b [g''(x)]^2 \, dx \tag{3.8-15}$$

Similarly, of all the functions $f(x)$ with continuous second derivatives which interpolate to these data and satisfy $f'(a) = y'_1, f'(b) = y'_n$ the spline $S(x)$ which also satisfies (3.8-13) uniquely minimizes (3.8-15).

2. If we define $h = \max_i h_i$ and let $S(x)$ be the natural spline interpolating $f(x)$ at a_i, $i = 1, \ldots, n$, where $f(x)$ has a continuous second derivative, then

$$\max_{a \le x \le b} |f(x) - S(x)| \le h[hE(f)]^{1/2} \qquad (3.8\text{-}16)$$

$$\max_{a \le x \le b} |f'(x) - S'(x)| \le [hE(f)]^{1/2} \qquad (3.8\text{-}17)$$

A similar theorem holds for the spline $S(x)$ satisfying $S'(a) = f'(a)$, $S'(b) = f'(b)$.

A stronger, albeit only asymptotic, result states that if $f(x)$ has a continuous fourth derivative, and if $\max_i (h/h_i) \le \beta < \infty$ as $h \to 0$ for a fixed β, then

$$\max_{a \le x \le b} |f^{(k)}(x) - S^{(k)}(x)| = O(h^{4-k}) \qquad k = 0, 1, 2 \qquad (3.8\text{-}18)$$

Example 3.8 Determine the natural cubic spline $S(x)$ which interpolates to the values of y_i at the points a_i, $i = 1, \ldots, 5$, where

$a_i = .25$	.30	.39	.45	.53
$y_i = .5000$	.5477	.6245	.6708	.7280

We have that $h_2 = .05$, $h_3 = .09$, $h_4 = .06$, $h_5 = .08$, so that using (3.8-9) leads to

$$\lambda_2 = \tfrac{9}{14} \qquad \mu_2 = \tfrac{5}{14} \qquad \lambda_3 = \tfrac{2}{5} \qquad \mu_3 = \tfrac{3}{5} \qquad \lambda_4 = \tfrac{4}{7} \qquad \mu_4 = \tfrac{3}{7}$$

$$\sigma_2 = .9540 \qquad \sigma_3 = .8533 \qquad \sigma_4 = .7717 \qquad \sigma_5 = .7150$$

$$d_2 = -4.3157 \qquad d_3 = -3.2640 \qquad d_4 = 2.4300$$

Inserting these values into (3.8-10), we get

$$\begin{aligned}
\tfrac{5}{14}M_1 + 2M_2 + \tfrac{9}{14}M_3 \qquad\qquad &= -4.3157 \\
\tfrac{3}{5}M_2 + 2M_3 + \tfrac{2}{5}M_4 \qquad &= -3.2640 \\
\tfrac{3}{7}M_3 + 2M_4 + \tfrac{4}{7}M_5 &= -2.4300
\end{aligned}$$

Since we wish to find a natural spline, we have that $M_1 = M_5 = 0$, so that we are left with a tridiagonal system of three equations. Solving this system by the algorithm given in Sec. 9.11 [cf. (9.11-8) to (9.11-10)], we find that

$$M_2 = -1.8806 \qquad M_3 = -.8226 \qquad M_4 = -1.0261$$

Inserting the values of the M_i's, h_i's, x_i's, and y_i's into (3.8-5), we get for $S(x)$ the following representations:

$$S(x) = \begin{cases} -6.2687(x - .25)^3 + 10(.30 - x) + 10.9697(x - .25) & x \in [.25, .30] \\ -3.4826(.39 - x)^3 - 1.5974(x - .30)^3 & \\ \qquad + 6.1138(.39 - x) + 6.9518(x - .30) & x \in [.30, .39] \\ -2.3961(.45 - x)^3 - 2.8503(x - .39)^3 & \\ \qquad + 10.4170(.45 - x) + 11.1903(x - .39) & x \in [.39, .45] \\ -2.1377(.53 - x)^3 + 8.3987(.53 - x) + 9.1000(x - .45) & x \in [.45, .53] \end{cases}$$

3.9 OTHER METHODS OF INTERPOLATION; EXTRAPOLATION

In interpolation, as in all branches of numerical analysis, there will be special cases in which methods superior to the general ones derived in this chapter can be derived and used without an unreasonable expenditure of effort. One example of this is the case of periodic functions in which methods based on Fourier-series approximations may be preferable to the polynomial approximations of this chapter; for more on this, see Sec. 6.6-1. Another example is the use of rational functions instead of polynomials. While there are some theoretical and practical problems with rational interpolation {37}, which, of course, includes polynomial interpolation as a special case, it warrants more serious attention than is normally given to it. In fact, whenever the function to be interpolated is known to have a special functional character, an approximation based on this known functional character may be desirable {34}.

Although we are restricting ourselves in this book to functions of a single variable, interpolation of functions of two or more variables can often be effected by a sequence of interpolations using the formulas of this chapter {36}.

This chapter is entitled Interpolation, but it has been equally about extrapolation. Interpolation and extrapolation are, in fact, two aspects of the same type of procedure. Of the two, interpolation is much more common than extrapolation. The reason for this is straightforward and practical. We argued in Sec. 3.4 that $p_n(x)$ is minimized when x is as nearly as possible in the center of the interval spanned by the tabular points. Conversely, as x moves *outside* the interval spanned by the tabular points, as is the case in extrapolation, the factors $x - a_i$ in $p_n(x)$ grow, and therefore, the error tends to grow unless x is very close to one of the endpoints of the interval I spanned by the tabular points. Furthermore, from (3.2-9) we note that the best a priori bound for $E(x)$ based on (3.2-9) is given by

$$|E(x)| \le \frac{M_n(x)|p_n(x)|}{n!} \tag{3.9-1}$$

where
$$M_n(x) = \max_{\xi \in I_x} |f^{(n)}(\xi)| \tag{3.9-2}$$

and I_x is the interval spanned by $a_1, \ldots, a_n$ and x. Now as x moves further from I so that the length of I_x increases, $M_n(x)$ can only increase and this also contributes to the growth of the error bound. Of course, for a particular value of x, $f^{(n)}(\xi)$ and hence also $E(x)$ may be very small in magnitude, but we cannot know a priori when this occurs and must rely only on the error bound (3.9-1).

Thus extrapolation is inherently a more inaccurate process than interpolation and must always be used with extreme caution. When extrapolation must be used in some form (see, for example, Chap. 5), the value of x should be restricted to be as near the interval spanned by the tabular points as possible.

BIBLIOGRAPHIC NOTES

Sections 3.1 to 3.4 The topics covered in these sections will be found in virtually any textbook on numerical analysis. In particular, excellent discussions of interpolation can be found in Hildebrand (1974), Kopal (1961), and Kuntzmann (1959). The orientation in these books as well as in most other numerical analysis texts is much more toward difference and divided-difference techniques {22} than in this book. An excellent though somewhat older reference to classical interpolation techniques is Steffensen (1950). A modern treatment of the subject is contained in Davis (1963). Hartree (1958) and Whittaker and Robinson (1948) contain a number of practical hints for special situations.

The coefficients of both the Lagrangian and finite-difference interpolation formulas have been extensively tabulated. A bibliography of these tables will be found in Fletcher, Miller, Rosenhead, and Comrie (1962). A convenient collection of the more useful formulas is given by Davis and Polonsky (1964).

The error term in the Lagrangian formula is discussed in a more general context in Milne (1949). The derivation of the error term used here can also be found in Scarborough (1962). For another approach, see Hildebrand (1974) or Kopal (1961).

Our discussion of the lozenge diagram follows closely that of Hamming (1973); see also Kopal (1961). The use of difference methods in the construction of mathematical tables is considered by Fox (1957*b*); see also Fox (1957*a*).

The operational techniques introduced in Prob. 9 are further considered in the problems after the next chapter. A thorough discussion of these techniques will be found in Hildebrand (1974). The convergence of interpolation series {19} is considered by Erdos and Turan (1937) and by Davis (1963).

Section 3.5 The basic references on iterated interpolation are the papers by Aitken (1932) and Neville (1934).

Section 3.6 Inverse interpolation will be considered in much greater detail in Chap. 8; for tables of coefficients for particular cases of inverse interpolation using differences see Salzer (1943, 1944, 1945).

Section 3.7 Hermite interpolation is discussed in many texts [e.g., Hildebrand (1974), Kopal (1961)].

Section 3.8 The treatment of splines in this section follows that of Ahlberg, Nilson, and Walsh (1967). The minimal property of natural cubic splines was discovered by Holladay (1957). The result (3.8-18) appears in Birkhoff and de Boor (1964). Further applications of splines in numerical analysis will be found in Greville (1967, 1969).

Section 3.9 A detailed discussion of interpolation in several variables is given by Steffensen (1950); see also Pearson (1920). The theory of rational interpolation is treated by Meinguet (1970), while algorithms for implementing rational interpolation appear in Bulirsch and Rutishauser (1968).

BIBLIOGRAPHY

Ahlberg, J. H., E. N. Nilson, and J. L. Walsh (1967): *The Theory of Splines and Their Applications*, Academic Press, Inc., New York.

Aitken, A. C. (1932): On Interpolation by Iteration of Proportional Parts, without the Use of Differences, *Proc. Edinb. Math. Soc.*, vol. 3, ser. 2, pp. 56–76.

Birkhoff, G., and C. de Boor (1964): Error Bounds for Spline Interpolation, *J. Math. Mech.*, vol. 13, pp. 827–835.

Bulirsch, R., and H. Rutishauser (1968): Interpolation und genäherte Quadratur, pp. 232–319 in *Mathematische Hilfsmittel des Ingenieurs*, pt. III (R. Sauer and I. Szabo, eds.), Springer-Verlag, Berlin.

Davis, P. J. (1963): *Interpolation and Approximation*, Blaisdell Publishing Company, Waltham, Mass.

——— and I. Polonsky (1964): Numerical Interpolation, Differentiation, and Integration, pp. 875–924 in Handbook of Mathematical Functions (M. Abramowitz and I. A. Stegun, eds.), *Nat. Bur. Stad. Appl. Math. Ser.* 55.

Erdos, P., and P. Turan (1937): On Interpolation, I, *Ann. Math.*, vol. 38, pp. 142–155.

Fletcher, A., J. C. P. Miller, L. Rosenhead, and L. J. Comrie (1962): *An Index of Mathematical Tables*, 2d ed., Addison-Wesley Publishing Company, Inc., Reading, Mass.

Fox, L. (1957*a*): Minimax Methods in Table Construction, pp. 233–244 in *On Numerical Approximation* (R. E. Langer, ed.), The University of Wisconsin Press, Madison, Wis.

——— (1957*b*): *The Use and Construction of Mathematical Tables*, vol. I, National Physical Laboratory Mathematical Table Series, London.

Greville, T. N. E. (1967): Spline Functions, Interpolation and Numerical Quadrature, pp. 156–168 in *Mathematical Methods for Digital Computers*, vol. 2 (A. Ralston and H. S. Wilf, eds.), John Wiley & Sons, Inc., New York.

——— (ed.) (1969): *Theory and Application of Spline Functions*, Academic Press, Inc., New York.

Hamming, R. W. (1973): *Numerical Methods for Scientists and Engineers*, 2d ed., McGraw-Hill Book Company, New York.

Hartree, D. R. (1958): *Numerical Analysis*, 2d ed., Oxford University Press, Fair Lawn, N.J.

Hildebrand, F. B. (1974): *Introduction to Numerical Analysis*, 2d ed., McGraw-Hill Book Company, New York.

Holladay, J. C. (1957): A Smoothest Curve Approximation, *MTAC*, vol. 11, pp. 233–243.

Kopal, Z. (1961): *Numerical Analysis*, 2d ed., John Wiley & Sons, Inc., New York.

Kuntzmann, J. (1959): *Méthodes numériques, interpolation—dérivées*, Dunod, Paris.

Meinguet, J. (1970): On the Solubility of the Cauchy Interpolation Problem, pp. 137–163 in *Approximation Theory* (A. Talbot, ed.), Academic Press, Inc., London.

Milne, W. E. (1949): The Remainder in Linear Methods of Approximation, *J. Res. Natl. Bur. Std*, vol. 43, pp. 501–511.

Neville, E. H. (1934): Iterative Interpolation, *J. Indian Math. Soc.*, vol. 20, pp. 87–120.

Pearson, K. (1920): *On the Construction of Tables and on Interpolation*, II, *Bivariate Interpolation*, University of London, Tracts for Computers III, Cambridge University Press, New York.
Salzer, H. E. (1943): Tables of Coefficients for Inverse Interpolation with Central Differences, *J. Math. and Phys.*, vol. 22, 210–224.
——— (1944): Tables of Coefficients for Inverse ınterpolation with Advancing Differences, *J. Math. and Phys.*, vol. 23, pp. 75–102.
——— (1945): Inverse lnterpolation for Eight, Nine, Ten and Eleven Point Direct Interpolation, *J. Math. and Phys.*, vol. 24, pp. 106–108.
Scarborough, J. B. (1962): *Numerical Mathematical Analysis*, 5th ed., The Johns Hopkins Press, Baltimore.
Steffensen, J. F. (1950): *Interpolation*, Chelsea Publishing Company, New York.
Whittaker, E. T., and W. Robinson (1948): *The Calculus of Observations*, 4th ed., Blackie & Son, Ltd., Glasgow.

PROBLEMS

Section 3.2

1. (*a*) If n is the order of a Lagrangian interpolation formula, show that

$$\sum_{j=1}^{n} a_j^k l_j(x) = x^k \qquad k = 0, \ldots, n-1$$

where the a_j are the tabular points.

(*b*) For $n = 3$ and equally spaced tabular points, compute $\max_{[a_1, a_3]} |l_j(x)|$ for $j = 1, 2, 3$. Use Table 3.1 to estimate the bounds on $l_j(x)$ for $n = 5$. Use these results to make an inference on the importance of roundoff error in interpolation using equally spaced data.

Section 3.3

2 (*a*) Using equally spaced data and a three-point Lagrangian formula, find a bound on $h^3 f'''(x)$ which, on the interval spanned by the three points, assures a truncation error of less than 10^{-d}, where d is an integer.

(*b*) Similarly, find a bound on $h^5 f^{\text{v}}(x)$ when using a five-point Lagrangian formula.

(*c*) Use these results to estimate the maximum value of h, for both the three- and five-point cases, that can be used to interpolate (i) $\sin x$ on $[-\pi, \pi]$, (ii) e^x on $[-4, 4]$, and (iii) $\sin 100x$ on $[-\pi, \pi]$, with a truncation error of less than 10^{-10}.

3 (*a*) Show that $y(x)$ in the Lagrangian interpolation formula is the unique polynomial of degree $n - 1$ passing through the points $[a_j, f(a_j)]$.

(*b*) Use the Lagrangian interpolation formula to find the cubic passing through the points $(-3, -1)$, $(0, 2)$, $(3, -2)$, $(6, 10)$.

4 (*a*) Do the computation of Examples 3.1 and 3.2 with the same tabular points when $f(x) = \sin x$.

(*b*) Repeat part (*a*) using $\tan^{-1} x$.

***5** Consider the following table for the Bessel functions $J_p(x)$, $p = 0, 1, 2, 3, 4, 5$ correctly rounded to four decimal places

x	$J_0(x)$	$J_1(x)$	$J_2(x)$	$J_3(x)$	$J_4(x)$	$J_5(x)$
2.0	.2239	.5767	.3528	.1289	.0340	.0070
2.1	.1666	.5683	.3746	.1453	.0405	.0088
2.2	.1104	.5560	.3951	.1623	.0476	.0109
2.3	.0555	.5399	.4139	.1800	.0556	.0134
2.4	.0025	.5202	.4310	.1981	.0643	.0162
2.5	−.0484	.4971	.4461	.2166	.0738	.0195
2.6	−.0968	.4708	.4590	.2353	.0840	.0232
2.7	−.1424	.4416	.4696	.2540	.0950	.0274
2.8	−.1850	.4097	.4777	.2727	.1067	.0321
2.9	−.2243	.3754	.4832	.2911	.1190	.0373
3.0	−.2601	.3391	.4861	.3091	.1320	.0430

(*a*) Suppose you wished to interpolate to find values of $J_0(x)$ at $x = 2.05 + .1j, j = 0, \ldots, 9$. Use the relation

$$J'_p(x) = -J_{p+1}(x) + \frac{p}{x} J_p(x)$$

to find a bound on the truncation error in the worst case using (i) linear interpolation; (ii) a Lagrangian three-point formula. Which of these methods would you use if you wished to *guarantee* a *total* error in the result for every j of less than 5×10^{-4} in magnitude?

(*b*) Carry out the interpolation using this method.

(*c*) Repeat parts (*a*) and (*b*) to find values of $J_1(x)$ at $x = 2.05 + .1j, j = 1, \ldots, 9$.

(*d*) How many correctly rounded decimal places for $J_0(x)$ would have to be given for the use of a five-point Lagrangian formula to give significantly higher accuracy than the three-point formula?

6 Use the data of Prob. 5 and a three-point Lagrangian formula to approximate (*a*) $J_p(2.07)$, (*b*) $J_p(2.405)$, (*c*) $J_p(2.64)$, (*d*) $J_p(2.91)$, with $p = 0, 1, 2$.

7 Derive (3.3-8).

8 Show that the first difference (forward, backward, or central) of a polynomial of degree n is a polynomial of degree $n - 1$. Thus deduce that the nth difference of a polynomial of degree n is a constant and the $(n + 1)$st is zero.

9 Difference operators. Define the *shifting* operator E to be such that $Ef(x) = f(x + h)$. Using this and the definitions of Δ, ∇, and δ, establish the following identities: (*a*) $\Delta = E - 1$; (*b*) $\nabla = 1 - E^{-1}$; (*c*) $\delta = E^{1/2} - E^{-1/2}$. Then use these relations to derive relations between Δ and ∇ and between Δ and δ.

***10** (*a*) Using the rules of Sec. 3.3-2, show that any closed path of the form

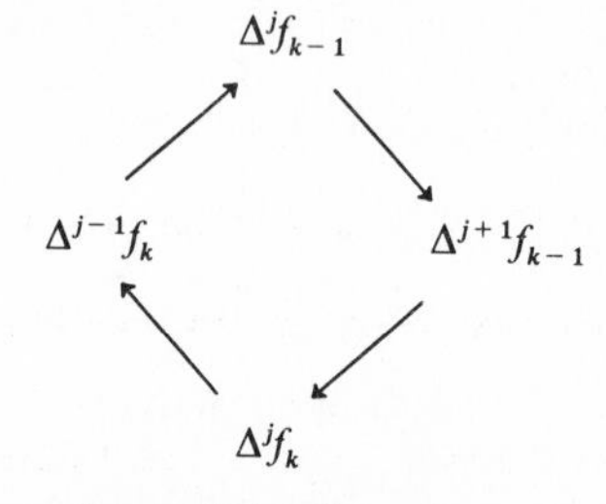

results in no contribution to any interpolation formula.

(*b*) Thus deduce that the path from $\Delta^{j-1}f_k$ to $\Delta^j f_{k-1}$ to $\Delta^{j+1}f_{k-1}$ results in the same contribution as the path from $\Delta^{j-1}f_k$ to $\Delta^j f_k$ to $\Delta^{j+1}f_{k-1}$. Similarly, show that the path from $\Delta^{j-1}f_k$ to $\Delta^j f_{k-1}$ to $\Delta^{j+1}f_{k-1}$ to $\Delta^j f_k$ and the path from $\Delta^{j-1}f_k$ to $\Delta^j f_k$ result in the same contribution. From these results deduce that *any closed path* contributes nothing.

(*c*) Show also that the path from f_{j+1} to Δf_j to f_j contributes nothing.

(*d*) Use the results of parts (*a*), (*b*), and (*c*) to deduce that all formulas which terminate on a given difference and start anywhere in the functional-value column are algebraically equivalent.

***11** (*a*) Given a table of values at an interval h, discuss how you would generate a new table ("subtabulate") at an interval ρh, $0 < \rho < 1$, by using an appropriate interpolation formula.

(*b*) Show that as $n \to \infty$, the left-hand side of (3.3-11) approaches $f(a_0 + hm)$ if the series on the right-hand side converges (see Prob. 19).

(*c*) Let the forward-difference operator with respect to the interval ρh be represented by Δ_1. Using (3.3-8) and the result of part (*b*), show that

$$\Delta_1^j f_0 = \sum_{i=0}^{\infty} \sum_{k=0}^{j} (-1)^{j-k} \binom{j}{k} \binom{k\rho}{i} \Delta^i f_0 \qquad j = 1, 2, \ldots$$

(*d*) Use the results of Prob. 9 to show that in operational form

$$\Delta_1^j f_0 = [(1 + \Delta)^\rho - 1]^j f_0$$

Use this to calculate Δ_1^j, $j = 1, 2, 3, 4$ in terms of Δ^k and ρ, retaining terms through Δ^4.

(*e*) Use the results of part (*c*) to subtabulate the data of Prob. 5 for $J_0(x)$ with (i) $\rho = \frac{1}{2}$; (ii) $\rho = \pm\frac{1}{3}$. Compare the results for $\rho = \frac{1}{2}$ with those of Prob. 5. How could you overcome the problems that arise near the end of the tabulation?

12 (*a*) Derive the identities

$$\text{(i)} \sum_{k=0}^{m} (-1)^k \binom{m}{k} = 0 \qquad \text{(ii)} \binom{r}{m}\binom{m}{k} = \binom{r-k}{m-k}\binom{r}{k}$$

(*b*) Use these results to show that $\sum_{j=r}^{i} (-1)^{j-r} \binom{i}{j}\binom{j}{r}$ vanishes and thus deduce that the right-hand side of (3.3-12) is f_i.

13 (*a*) Use the results of Prob. 9 to express Newton's backward formula in terms of backward differences at a_0.

(*b*) Similarly, express Gauss' forward and backward formulas in terms of central differences at a_0.

(*c*) Using the notation $\mu\delta^{2m+1}f_0 = \frac{1}{2}(\delta^{2m+1}f_{1/2} + \delta^{2m+1}f_{-1/2})$, express Stirling's and Bessel's formulas in terms of central differences.

14 (*a*) Show that in any finite-difference interpolation formula a difference of any order can be eliminated by using the relation $\delta^m f_{k+1} - \delta^m f_k = \delta^{m+1} f_k$ or a similar relation for forward and backward differences.

(*b*) Use this result and the result of Prob. 13*b* to eliminate the odd differences from Gauss' forward formula and thus derive *Everett's interpolation formula*

$$y(m) = (1 - m)f_0 + mf_1 - \frac{m(m-1)(m-2)}{3!}\delta^2 f_0 + \frac{(m+1)(m)(m-1)}{6}\delta^2 f_1 + \cdots$$

(This formula is useful in interpolating in tables which provide auxiliary tables of *even* central differences.)

(*c*) Similarly, eliminate the even differences in Gauss' forward formula to get *Steffensen's interpolation formula*

$$y(m) = f_0 + \frac{(m+1)m}{2!}\delta f_{1/2} - \frac{m(m-1)}{2}\delta f_{-1/2} + \frac{(m+2)(m+1)m(m-1)}{4!}\delta^3 f_{1/2} - \cdots$$

[Ref.: Hildebrand (1974), pp. 143–144, or Kopal (1961), pp. 50–54.]

***15** Throwback. (*a*) Use the result of Prob. 13*c* to show that the ratio of the coefficient B_4 of the fourth central difference in Bessel's formula to the coefficient B_2 of the second difference is $(m+1)(m-2)/12$ and that for $0 \le m \le 1$ this ratio varies between $-\frac{1}{6}$ and $-\frac{3}{16}$.

(*b*) Because this ratio varies very little on this interval, consider replacing B_4 by cB_2. Show that $B_4 - cB_2$ as a function of m has a maximum independent of c and two minima dependent on c on [0, 1]. Find the two values of c which equalize the minimum and maximum values of $B_4 - cB_2$ on this interval. Show that one of these values c_1 is very nearly equal to the average value of B_4/B_2 over [0, 1].

(*c*) Thus rewrite Bessel's formula as $y(m) = \frac{1}{2}(f_0 + f_1) + (m - \frac{1}{2})\delta f_{1/2} + B_2[\delta^2 f_0 + \delta^2 f_1 + c_1(\delta^4 f_0 + \delta^4 f_1)]$. This procedure is called *throwback*; i.e., we have thrown back the effect of the fourth difference onto the second difference. [Ref.: Kopal (1961), pp. 54–56.]

16 (*a*) Display the error terms for the Newtonian and Gaussian interpolation formulas terminated with the difference of order k in terms of h and m.

(*b*) Use these to derive the error terms for Stirling's and Bessel's formulas terminated with an odd or even difference.

Section 3.4

17 (*a*) What abscissas are involved in the calculation of each entry in the table in Example 3.4 for both the Gaussian and Newtonian formulas?

(*b*) Verify that the actual error using six differences is consistent with that calculated using the result of Prob. 16*a*.

18 (*a*) How many terms in Gauss's forward formula can be used if x_0 is (i) the next to last entry in a table; (ii) the fourth entry?

(*b*) Use (3.3-18) and (3.3-19) to show that when m is near zero, Stirling's formula is a desirable one to use and that when m is near one-half, Bessel's formula is desirable.

***19** (*a*) If h is fixed, show that in the limit as $n \to \infty$ Newton's forward formula, if it converges, becomes with $a_0 = 0$

$$f(x) = f_0 + \sum_{j=1}^{\infty} \frac{\Delta^j f_0}{j!h^j} x(x-h)\cdots[x-(j-1)h]$$

(*b*) For $f(x) = e^{ax}$ and $a_0 = 0$, show that $\Delta^r f_0 = (e^{ah} - 1)^r$.

(*c*) By using the result of part (*b*) in part (*a*), show that the ratio of the $k+1$ and k terms of the series is given by

$$\frac{e^{ah} - 1}{h(k+1)}(x - kh)$$

(*d*) By considering this ratio as $k \to \infty$, deduce that the series in part (*a*) converges if $e^{ah} < 2$ and diverges if $e^{ah} > 2$ unless x is a positive integral multiple of h, in which case it converges. (A more difficult result is that for $e^{ah} = 2$ the series converges if and only if $x > -h$.)

(*e*) Thus deduce that Newton's forward formula is an asymptotic series for e^{ax} when $e^{ah} > 2$. Contrast this with the convergence of the Taylor series for e^{ax} for all ax. In practice, why would we expect Newton's formula to be asymptotic even when $e^{ah} < 2$? [Ref.: Hildebrand (1974), pp. 154–156.]

20 Suppose you have a table of sin x at an interval $h = .1$. How many tabular points would have to be used in interpolating in this table to assure a truncation error of less than (*a*) 10^{-3}; (*b*) 10^{-4}; (*c*) 10^{-5}; independent of a_0 and m?

21 Use the data of Prob. 5 and a finite-difference interpolation formula to approximate (*a*) $J_p(2.07)$, (*b*) $J_p(2.405)$, (*c*) $J_p(2.64)$, (*d*) $J_p(2.91)$, with $p = 0, 1, 2$. In each case motivate your choice of a particular interpolation formula and compare the results with those of Prob. 6.

***22** Divided differences. The divided difference of order $k > 1$ of $f(x)$ is defined by

$$f[a_1, \ldots, a_k] = \frac{f[a_2, \ldots, a_k] - f[a_1, \ldots, a_{k-1}]}{a_k - a_1} \qquad \text{with } f[a_1] = f(a_1)$$

(*a*) Prove that

$$f[a_1, \ldots, a_k] = \sum_{i=1}^{k} \frac{f(a_i)}{(a_i - a_1) \cdots (a_i - a_{i-1})(a_i - a_{i+1}) \cdots (a_i - a_k)}$$

and thus deduce that the order of the arguments in a divided difference is immaterial.

(*b*) Use the data of Example 3.1 to generate a divided-difference table analogous to the difference table of Fig. 3.1.

(*c*) Show that

$$f[a_1, \ldots, a_{k-1}, x] = f[a_1, \ldots, a_k] + (x - a_k) f[a_1, \ldots, a_k, x]$$

and use this result to derive the formula

$$f(x) = f[a_1] + (x - a_1) f[a_1, a_2] + (x - a_1)(x - a_2) f[a_1, a_2, a_3] + \cdots + (x - a_1)(x - a_2) \cdots (x - a_{n-1}) f[a_1, \ldots, a_n] + E(x)$$

where

$$E(x) = p_n(x) f[a_1, \ldots, a_n, x]$$

This formula is called *Newton's divided-difference interpolation formula.*

(*d*) Deduce from part (*c*) that $(x - a_k) f[a_1, \ldots, a_n, x] \to 0$ as $x \to a_k$, $k = 1, \ldots, n$.

(*e*) Use this result to show that this formula must be algebraically equivalent to the Lagrangian interpolation formula which uses the tabular points $a_1, \ldots, a_n$. Thus deduce that

$$f[a_1, \ldots, a_n, x] = \frac{1}{n!} f^{(n)}(\xi)$$

where ξ is in the interval spanned by $a_1, \ldots, a_n$ and x.

(*f*) Use the results of parts (*c*) and (*e*) to show that when $a_i \to a$, $i = 1, \ldots, n$, the Newton divided-difference formula and therefore the Lagrangian formula are both equivalent to a Taylor series with remainder.

(*g*) Use Newton's divided-difference formula and the table of part (*b*) to approximate ln .60. Compare the result with Example 3.1.

(*h*) When the tabular points are equally spaced, show that

$$f[a_1, \ldots, a_k] = \frac{1}{(k-1)!\,h^{k-1}} \Delta^{k-1} f_1$$

and thus use part (*c*) to derive Newton's forward formula.

Section 3.5

23 (*a*) Show that the table of (3.5-2) can be replaced by the symmetrical arrangement

$$\begin{array}{cccccccc}
a_1 & a_1 - x & y_1(x) & & & & & \\
\vdots & \vdots & \vdots & y_{12}(x) & & & & \\
\vdots & \vdots & \vdots & y_{23}(x) & & y_{123}(x) & \vdots & \\
\vdots & \vdots & \vdots & & & \vdots & \vdots & y_{1,2,\ldots,n}(x) \\
\vdots & \vdots & \vdots & \vdots & & y_{n-2,n-1,n}(x) & \vdots & \\
\vdots & \vdots & \vdots & y_{n-1,n}(x) & & & & \\
a_n & a_n - x & y_n(x) & & & & &
\end{array}$$

What would the additional entries to the table be if the point a_{n+1} were used?

(*b*) Use the technique of part (*a*) to do the computation of Example 3.5. [Ref.: Neville (1934).]

24 Use iterated interpolation to do the calculations of parts (*a*) and (*b*) of Prob. 6. Compare the results with those of Probs. 6 and 21.

25 Interpolation near a singularity. Suppose you are given a tabulation of sine, cosine, and tangent as follows:

x	$\sin x$	$\cos x$	$\tan x$
1.566	.9999885	.0047963	208.49128
1.567	.9999928	.0037963	263.41125
1.568	.9999961	.0027963	357.61106
1.569	.9999984	.0017963	556.69098
1.570	.9999997	.0007963	1255.76559

Using these data and any interpolation formula, calculate tan 1.5695 by (*a*) using the tan x tabulation directly; (*b*) calculating sin 1.5695 and cos 1.5695. Discuss the reasons for the varying errors in the two results. [Ref.: Kopal (1961), p. 84.]

Section 3.6

26 Use the data of Prob. 5 and inverse interpolation to approximate the zero of $J_0(x)$ between 2 and 3 (cf. Prob. 6*b*).

27 Inverse interpolation near a singularity. Suppose we wished to calculate that value of x for which $\sin x = .9999950$ using the data of Prob. 25. Why will the procedure of Sec. 3.6 not work here? To solve this problem:

(*a*) Obtain an initial approximation $\bar{x}$ to x by linear inverse interpolation between $x_1 = 1.567$ and $x_2 = 1.568$.

(*b*) By direct interpolation, compute $\sin \bar{x}$.

If $\sin \bar{x} < .9999950$, replace x_1 by $\bar{x}$ and repeat this procedure. Otherwise, replace x_2 by $\bar{x}$. Continue until the process converges. What condition must $\sin x$ satisfy on $[x_1, x_2]$ in order for the process to converge?

Section 3.7

28 Derive Eqs. (3.7-10).

29 Use the data of Prob. 5 and Hermite interpolation to do the computation of parts (*a*) and (*b*) of Prob. 6 for $p = 0$. Compare with the previous results. Can the use of the Hermite interpolation formula be simplified for equally spaced data in a fashion analogous to that for the Lagrangian formula in Sec. 3.3-1?

Section 3.8

30 Determine the explicit form of λ_2, d_2, μ_{n-1}, and d_{n-1} in case 3 on p. 76 and show that $|\lambda_2| < 2$, $|\mu_{n-1}| < 2$.

31 (*a*) Verify that if $S(x)$ is a cubic spline, then

$$\int_a^b [f''(x)]^2\,dx - \int_a^b [S''(x)]^2\,dx = \int_a^b [f''(x) - S''(x)]^2\,dx + 2\int_a^b S''(x)[f''(x) - S''(x)]\,dx$$

for any twice-differentiable function $f(x)$.

(*b*) Integrating by parts, show that if $f(x_i) = S(x_i)$, $i = 1, \ldots, n$, where the x_i are the nodes of $S(x)$, then

$$\int_a^b S''(x)[f''(x) - S''(x)]\,dx = S''(b)[f'(b) - S'(b)] - S''(a)[f'(a) - S'(a)]$$

(*c*) Hence prove that the natural spline uniquely minimizes (3.8-15) among all functions with continuous second derivatives which interpolate the data at the points x_i, $i = 1, \ldots, n$.

(*d*) Similarly prove that the spline which satisfies (3.8-13) uniquely minimizes (3.8-15) among all such functions which in addition satisfy $f'(a) = y_1'$, $f'(b) = y_n'$.

32 (*a*) Using Rolle's theorem, show that for every $x \in [a_i, a_{i+1}]$ there exists a point $z \in [a_i, a_{i+1}]$ such that

$$f'(x) - S'(x) = \int_z^x [f''(t) - S''(t)]\,dt$$

(*b*) Using the Cauchy-Schwarz inequality, show that

$$|f'(x) - S'(x)| \le \left| \int_z^x [f''(t) - S''(t)]^2\,dt \right|^{1/2} |x - z|^{1/2}$$

(*c*) Prove (3.8-16) and (3.8-17).

33 (*a*) Determine the cubic spline $S(x)$ which interpolates to the data of Example 3.8 and which also satisfies the conditions

$$S'(.25) = 1.0000 \qquad S'(.53) = .6868$$

(*b*) Determine the cubic spline $S(x)$ with nodes a_2, a_3, a_4 which interpolates to the values of y_2, y_3, y_4 of Example 3.8 and which also satisfies the conditions $S(.25) = .5000$, $S(.53) = .7280$.

(*c*) The values of y_i given in Example 3.8 are equal to $a_i^{1/2}$, $i = 1, \ldots, 5$. Evaluate the integral $\int_{.25}^{.53} S(x)\,dx$ for the three splines given in Example 3.8 and parts (*a*) and (*b*) and compare the answers with the true value of $\int_{.25}^{.53} x^{1/2}\,dx = .1739$. Similarly, evaluate $S(.35)$ and $S'(.35)$ and compare the answers with $\sqrt{.35} = .5916$ and $1/2\sqrt{.35} = .8452$, respectively.

Section 3.9

34 Prony's method. Given $2n$ pairs (x_i, y_i), $i = 0, \ldots, 2n - 1$, such that $x_i = x_0 + ih$, to find $2n$ values a_j, b_j, $j = 1, \ldots, n$, such that $f(x_i) = y_i$, $i = 0, \ldots, 2n - 1$, where $f(x) = \sum_{j=1}^n a_j e^{b_j x}$.

(*a*) Show that this problem is equivalent to that of finding $2n$ values c_j, μ_j, $j = 1, \ldots, n$, such that $g(i) = y_i$, $i = 0, \ldots, 2n - 1$, where $g(x) = \sum_{j=1}^n c_j(\mu_j)^x$.

(*b*) From the equations $g(i) = y_i$, $i = 0, \ldots, 2n - 1$, taken $n + 1$ at a time, derive a system of linear equations for the coefficients d_k, $k = 0, \ldots, n - 1$, of the polynomial

$$p_n(x) = \sum_{k=0}^n d_k x^k \qquad d_n = 1$$

whose roots are μ_j, $j = 1, \ldots, n$.

(*c*) Having found the d_k and the μ_j, find the c_j by solving another system of n linear equations. Finally express the required a_j, b_j in terms of the c_j, μ_j, $j = 1, \ldots, n$.

(*d*) What happens if one of the roots of $p_n(x)$ is complex? [Ref.: Hildebrand (1974), pp. 457–462.]

35 Apply Prony's method to the following two sets of data

x_i	0	1	2	3
y_i	−.35	.11	.07	.03
y_i	−.35	.11	.08	.03

36 Interpolation of functions of two variables. Suppose we are given a function of two variables $f(x, y)$ tabulated at points (a_i, b_j), $i = 1, \ldots, n$, $j = 1, \ldots, m$.

(*a*) If we wish to approximate $f(x, y)$ at a nontabular point, show that we can do this by first interpolating to find $f(x, b_j)$ for a sequence of values of j and then using these values to interpolate to find $f(x, y)$ or vice versa.

(*b*) Given the table of values of the elliptic integral

$$E(x, y) = \int_0^y (1 - \sin^2 x \sin^2 t)^{1/2}\, dt$$

	x			
y	50°	54°	58°	62°
50°	0.8134	0.8060	0.7988	0.7920
52°	0.8414	0.8332	0.8251	0.8174
54°	0.8690	0.8598	0.8508	0.8422
56°	0.8962	0.8859	0.8759	0.8663

find an approximation to $E(55.4°, 53.1°)$ by (i) interpolating horizontally to find $E(55.4°, y)$ for $y = 50°, 52°, 54°, 56°$ and then interpolating vertically; (ii) interpolating vertically and then horizontally. If the desired point lies on a diagonal, for example, (52°, 51°), how could the interpolation procedure be simplified? [Ref.: Hildebrand (1974), p. 167.]

37 Rational interpolation. Consider the problem of finding a rational function $R_{mn}(x) = P_m(x)/Q_n(x)$, where $P_m(x) = \sum_{i=0}^m a_i x^i$ and $Q_n(x) = \sum_{i=0}^n b_i x^i$, such that $R_{mn}(x_j) = y_j$, $j = 1, \ldots, s$.

(*a*) Show that the number of free parameters in $R_{mn}(x)$ is $m + n + 1$ so that we must have $s = m + n + 1$.

(*b*) Show that if a rational interpolation function $R_{mn}(x)$ exists, it is unique.

(*c*) By considering the case $x_j = 0, 1, 2$, $y_j = 1, 3, 3$, $j = 1, \ldots, 3$, $m = n = 1$, show that a rational interpolation function does not always exist.

(*d*) By considering the case $x_j = 0, 2, 3$, $y_j = -1, 1, \frac{1}{2}$, $j = 1, \ldots, 3$, $m = n = 1$, show that even when a rational interpolation function exists, it need not be continuous on the interval spanned by the x_j. [Ref.: Bulirsch and Rutishauser (1968) p. 278.]

CHAPTER

FOUR

NUMERICAL DIFFERENTIATION, NUMERICAL QUADRATURE, AND SUMMATION

4.1 NUMERICAL DIFFERENTIATION OF DATA

Whereas in interpolation we are only concerned with the case where the function is represented by a table of values, in numerical differentiation we may be given either a table of values or a closed expression for the function. In the latter case, the given form will usually be a single function statement, and it will normally be possible to differentiate the expression analytically using one of the many available formula-manipulation languages. Indeed, analytic differentiation has proved useful in various areas of numerical analysis and should be applied more frequently. When analytic differentiation can be performed on a computer, there is little more that need be said. Numerical differentiation of functions which cannot (reasonably) be differentiated analytically is considered in the next section. In this section we shall be concerned with the differentiation of numerical data.

When the function is represented by a table of values, the most obvious approach is to differentiate the Lagrangian interpolation formula (3.1-1). This gives us

$$f^{(k)}(x) = \sum_{j=1}^{n} l_j^{(k)}(x) f(a_j) + \frac{d^k}{dx^k}\left[\frac{p_n(x)}{n!} f^{(n)}(\xi)\right] = y^{(k)}(x) + \frac{d^k}{dx^k}[E(x)] \quad (4.1\text{-}1)$$

In particular, for $k = 1$ we have

$$f'(x) = \sum_{j=1}^{n} l'_j(x) f(a_j) + \frac{d}{dx}\left[\frac{p_n(x)}{n!} f^{(n)}(\xi)\right] \tag{4.1-2}$$

where the derivative of $l_j(x)$ is easily calculated using (3.2-4). Determination of the error term in (4.1-1) or (4.1-2) presents a problem because ξ is an unknown function of x (see Sec. 3.2). However, it can be shown that

$$\frac{d^k}{dx^k}[E(x)] = \frac{f^{(n)}(\xi)}{(n-k)!} \prod_{j=1}^{n-k} (x - \eta_j) \tag{4.1-3}$$

where the $n - k$ distinct points η_j are independent of x and are known to lie in the intervals

$$a_j < \eta_j < a_{j+k} \qquad j = 1, \ldots, n-k$$

ξ depends on x and is in the smallest interval I containing x and the η_j.

If, in addition to function values at the data points, we are given values of some of the derivatives at several points, we can derive the coefficients $w_{ij}^{(k)}(x)$ in the formula

$$f^{(k)}(x) \approx \sum_{j=1}^{n} \sum_{i=0}^{m_j} w_{ij}^{(k)}(x) f^{(i)}(a_j) \tag{4.1-4}$$

using the method of undetermined coefficients {1}.

An alternative approach to numerical differentiation of data for approximating the first and second derivatives is to compute a cubic interpolating spline (see Sec. 3.8) and differentiate it. For any particular value x, we determine the two nodes a_j and a_{j+1} such that $a_j \le x < a_{j+1}$. Then $y'(x) \approx S'_j(x)$ and $y''(x) \approx S''_j(x)$, where $S_j(x)$ is that piece of the spline which is a cubic polynomial in $[a_j, a_{j+1}]$. By the result quoted in Sec. 3.8, we are assured that, asymptotically, as h, the maximum distance between nodes, goes to zero,

$$\max_{a \le x \le b} |f^{(k)}(x) - S^k(x)| = O(h^{4-k}) \qquad k = 1, 2$$

This indicates that the first and second derivatives of the interpolating spline approximate the behavior of the corresponding derivatives of the given function. By contrast, such a situation may not hold in the case of (4.1-1). We can see this more clearly if we consider the case of equally spaced tabular points. We then have that

$$y^{(k)}(x) = \frac{1}{h^k} \sum_{j=1}^{n} l_j^{(k)}(m) f(a_i) \qquad x = a_0 + hm \tag{4.1-5}$$

where the differentiation of $l_j(m)$ is with respect to m and we have used the fact that

$$\frac{dy}{dx} = \frac{dy}{dm}\frac{dm}{dx} = \frac{1}{h}\frac{dy}{dm} \tag{4.1-6}$$

In a similar fashion, we can differentiate the interpolation formulas expressed in difference form. For example, if we differentiate Newton's forward formula (3.3-11), we get

$$\frac{d^k}{dx^k} y(a_0 + hm) = \frac{1}{h^k} \frac{d^k}{dm^k} y(a_0 + hm)$$

$$= \frac{1}{h^k} \sum_{j=0}^{n} \frac{d^k}{dm^k} \binom{m}{j} \Delta^j f_0 = \frac{1}{h^k} \sum_{j=k}^{n} \frac{d^k}{dm^k} \binom{m}{j} \Delta^j f_0 \quad (4.1\text{-}7)$$

The appearance of the factor $1/h^k$ in (4.1-5) makes explicit a fact which was hidden in (4.1-1). In numerical differentiation formulas, we divide by small quantities. This in itself would not be critical, but since our final result is not large, the numerator must also be small. This can occur only if we subtract quantities of the same order of magnitude. This causes a large relative error in the result, especially if the original data are of low accuracy. In this situation, a better approach to numerical differentiation may be to first "smooth" the data and then to differentiate.

Example 4.1 Suppose we are given the empirical data (cf. Sec. 6.5)

x_i	$\bar{f}_i$	x_i	$\bar{f}_i$	x_i	$\bar{f}_i$
.1	5.1234	.4	5.9378	.7	7.9493
.2	5.3057	.5	6.4370	.8	9.0253
.3	5.5687	.6	7.0978	.9	10.3627

and we wish to find an approximation to the value of $f'(.5)$. Since the data points are equally spaced, we can use formulas of type (4.1-5). Inasmuch as it is desirable to choose points symmetrically about the given evaluation point, we shall approximate $f'(.5)$ by the three-point and five-point formulas

$$f'_3(x) = a_{-1} f(x - h) + a_0 f(x) + a_1 f(x + h)$$

$$f'_5(x) = b_{-2} f(x - 2h) + b_{-1} f(x - h) + b_0 f(x) + b_1 f(x + h) + b_2 f(x + 2h)$$

By the method of undetermined coefficients {1} or directly from (4.1-5) we find first that $-a_{-1} = a_1 = 1/2h$, $a_0 = 0$, so that [cf. (4.2-3)]

$$f'_3(x) = \frac{f(x + h) - f(x - h)}{2h}$$

and second that $b_{-2} = -b_2 = 1/12h$, $-b_{-1} = b_1 = 8/12h$, $b_0 = 0$, so that

$$f'_5(x) = \frac{1}{12h} [f(x - 2h) - 8f(x - h) + 8f(x + h) - f(x + 2h)]$$

Substituting the values in the table above, we find that

$$f'_3(.5) = 5.8000 \qquad \text{and} \qquad f'_5(.5) = 5.7495$$

The values given in the table are perturbations of the values of

$$f(x) = x^4 + 3x^3 + 2x^2 + x + 5$$

which can be thought to arise either from observational error or roundoff or both. The exact value is $f'(.5) = 5.7500$, and we see that $f'_5(.5)$ is indeed almost exactly equal to $f'(.5)$ while $f'_3(.5)$ is not too bad an approximation. If we approximate $f'(.5)$ using the least-squares polynomial (cf. Sec. 6.5)

$$y_4(x) = .9988x^4 + 2.9898x^3 + 2.0172x^2 + .9920x + 5.0010$$

we find that $y'_4(.5) = 5.7510$, which is also very good.

The small error in $f'(.5)$ was due to a fortunate cancellation of roundoff error. Since $f(x)$ is a fourth-degree polynomial, $f'_5(x)$ would be exact in the absence of roundoff; here we have achieved such accuracy because of cancellation of roundoff since the true values of $f(.4)$ and $f(.6)$ are respectively 5.9376 and 7.0976, so that

$$f(.6) - f(.4) = \bar{f}_6 - \bar{f}_4$$

That we are not always so fortunate can be seen from the following problem using the same data: compute approximations to $f'(.1)$ and $f''(.1)$. In this case it is more convenient to use differences based on formula (4.1-7) with $m = 0$. The formulas for the first and second derivatives are then, up to fourth differences, as follows:

$$hf'_0 \approx \Delta f_0 - \tfrac{1}{2}\,\Delta^2 f_0 + \tfrac{1}{3}\,\Delta^3 f_0 - \tfrac{1}{4}\,\Delta^4 f_0 \tag{4.1-8}$$

$$h^2 f''_0 \approx \Delta^2 f_0 - \Delta^3 f_0 + \tfrac{11}{12}\,\Delta^4 f_0 \tag{4.1-9}$$

We set up the following difference table:

$\bar{f}_i$	$\Delta\bar{f}_i$	$\Delta^2\bar{f}_i$	$\Delta^3\bar{f}_i$	$\Delta^4\bar{f}_i$
5.1234				
	.1823			
5.3057		.0807		
	.2630		.0254	
5.5687		.1061		−.0014
	.3691		.0240	
5.9378		.1301		
	.4992			
6.4370				

Using formula (4.1-8) starting with the approximation $hf'_0 \approx \Delta f_0$ and adding an additional difference each time, we find in succession the following approximation to $f'(.1)$:

1.823, 1.420, 1.504, 1.507

The exact value is 1.494, and the least-squares values is 1.489. For $f''(.1)$ the values found using (4.1-9) are in succession

8.07, 5.53, 5.40

The true value is 5.92, and the least-squares value is 5.95.

The influence of errors in the data is evident here, since, in the absence of errors, the formula using fourth differences should give the true answer for a fourth-degree polynomial. Thus, the initial error in the data of less than .01 percent becomes magnified tenfold in the first derivative and a hundredfold in the second in accordance with the general result that the roundoff error in the kth derivative is of the order of $1/h^k$ times the roundoff error in the data when the data are given at equally spaced points with spacing h.

4.2 NUMERICAL DIFFERENTIATION OF FUNCTIONS

When we must compute the derivative of a function which can be evaluated anywhere in a given interval I, the following procedure is preferable to those of the previous section. If $x + h$ and $x - h$ are in I, we begin with the Taylor expansions

$$f(x+h) = f(x) + hf'(x) + \frac{h^2}{2} f''(x) + \frac{h^3}{3!} f'''(x) + \cdots$$

$$f(x-h) = f(x) - hf'(x) + \frac{h^2}{2} f''(x) - \frac{h^3}{3!} f'''(x) + \cdots$$

Subtracting, dividing by $2h$, and rearranging terms, we find that

$$f'(x) = \frac{f(x+h) - f(x-h)}{2h} + \sum_{i=1}^{\infty} \frac{f^{(2i+1)}(x)}{(2i+1)!} h^{2i} \tag{4.2-1}$$

Similarly, we find that

$$f''(x) = \frac{f(x+h) - 2f(x) + f(x-h)}{h^2} + \sum_{i=1}^{\infty} \frac{2f^{(2i+2)}(x)}{(2i+2)!} h^{2i} \tag{4.2-2}$$

Thus, we can expect that by taking h small enough we can approximate $f'(x)$ and $f''(x)$, respectively, by

$$f'(x) \approx \frac{f(x+h) - f(x-h)}{2h} \tag{4.2-3}$$

$$f''(x) \approx \frac{f(x+h) - 2f(x) + f(x-h)}{h^2} \tag{4.2-4}$$

Our strategy would then be to evaluate (4.2-3) [or (4.2-4)] for a sequence of values of h tending to zero, stopping when we get agreement to the desired accuracy.

This procedure, however, is fraught with danger since eventually round-off will dominate the calculation. As h tends to zero, both $f(x+h)$ and $f(x-h)$ tend to $f(x)$, so that their difference tends to the difference of two almost equal quantities and thus contains fewer and fewer significant digits. Thus, it is meaningless to carry out this computation beyond a certain threshold value of h which is dependent on the accuracy with which $f(x)$ can be computed. If a bound on the relative error in the computation of $f(x)$ is given by R, that is, the computed value $\bar{f}(x) = f(x)(1 + r)$ with $|r| \le R$, then a bound on the relative error in (4.2-3) is given by R/h. If we wish a relative error of ϵ or less in the first derivative, we cannot allow h to go below R/ϵ. For the second derivative the threshold is $h^2 > 4R/\epsilon$.

If while computing (4.2-3) or (4.2-4) with successively smaller values of h

we find that we reach the threshold value of h before we have achieved agreement to the desired accuracy between several successive values, we can resort to the technique of Richardson extrapolation, based on expansions (4.2-1) and (4.2-2). This is a general technique which we shall now explain.

In the general case of Richardson extrapolation, we have a procedure for approximating a functional $\phi = \phi(f)$ of a given function f by a sequence, depending on a step length h, in which as $h \to 0$ the error has the asymptotic form

$$E = \sum_{j=1}^{\infty} a_j h^{\gamma_j} \qquad \gamma_1 < \gamma_2 < \cdots \tag{4.2-5}$$

where the a_j are constants which may depend on the function f but not on h. The a_j are unknown to us, but we assume knowledge of the γ_j. If the computation is carried out for two different values of h, h_1, and h_2, $h_2 < h_1$, resulting in two approximations ϕ_1 and ϕ_2, then the true value ϕ is given by

$$\phi = \phi_i + \sum_{j=1}^{\infty} a_j h_i^{\gamma_j} \qquad i = 1, 2 \tag{4.2-6}$$

Multiplying (4.2-6) with $i = 1$ by $h_2^{\gamma_1}$ and (4.2-6) with $i = 2$ by $h_1^{\gamma_1}$, subtracting, and solving for ϕ, we get

$$\phi = \frac{1}{h_1^{\gamma_1} - h_2^{\gamma_1}} (h_1^{\gamma_1}\phi_2 - h_2^{\gamma_1}\phi_1) + \sum_{j=2}^{\infty} a_j \frac{h_1^{\gamma_1} h_2^{\gamma_j} - h_2^{\gamma_1} h_1^{\gamma_j}}{h_1^{\gamma_1} - h_2^{\gamma_1}} \tag{4.2-7}$$

When we set $h_2 = \rho h_1$, $\rho < 1$, (4.2-7) becomes

$$\phi = \frac{\phi_2 - \rho^{\gamma_1}\phi_1}{1 - \rho^{\gamma_1}} + \sum_{j=2}^{\infty} \frac{\rho^{\gamma_j} - \rho^{\gamma_1}}{1 - \rho^{\gamma_1}} a_j h_1^{\gamma_j} = \phi_{12} + \sum_{j=2}^{\infty} b_j h_1^{\gamma_j} \tag{4.2-8}$$

which is of the form (4.2-6). If we wish to eliminate the term with $j = 2$ in (4.2-8), we must compute ϕ_3 with $h_3 = \rho h_2$ and compute ϕ_{23} in a similar fashion and then compute ϕ_{123} from ϕ_{12} and ϕ_{23}.

Let us formalize this procedure. If we denote ϕ_i by T_0^i, we can generate a triangular array of approximants T_m^i by the formula

$$T_m^i = \frac{T_{m-1}^{i+1} - \rho^{\gamma_m} T_{m-1}^i}{1 - \rho^{\gamma_m}} = T_{m-1}^{i+1} + \frac{T_{m-1}^{i+1} - T_{m-1}^i}{\rho^{-\gamma_m} - 1} \qquad i = 1, 2, \ldots \tag{4.2-9}$$

where the element T_{m-1}^1 corresponds to $\phi_{12\cdots m}$.

In some cases, a geometric growth of $1/h$ by a factor $1/\rho$ is not desirable or possible, and we are interested in a general sequence $\{h_i\}$. In this case, we can generate a table T_m^i by a simple algorithm only if the γ_j have a special structure, $\gamma_j = j\gamma + \delta$. For the usual case $\delta = 0$, we have that {6}

$$T_m^i = \frac{h_i^{\gamma} T_{m-1}^{i+1} - h_{i+m}^{\gamma} T_{m-1}^i}{h_i^{\gamma} - h_{i+m}^{\gamma}} = T_{m-1}^{i+1} + \frac{T_{m-1}^{i+1} - T_{m-1}^i}{(h_i/h_{i+m})^{\gamma} - 1} \tag{4.2-10}$$

The general case $\delta \neq 0$ is considered in a problem {7}. When both $h_i = \rho^{i-1} h_1$ and $\gamma_j = j\gamma$, we have a special case of both general cases, and we see that both (4.2-9) and (4.2-10) reduce to the same formula

$$T_m^i = T_{m-1}^{i+1} + \frac{T_{m-1}^{i+1} - T_{m-1}^i}{\rho^{-m\gamma} - 1} \tag{4.2-11}$$

Example 4.2 Compute the first derivative of $f(x) = -\cot(x)$ for $x = 0.04$. Take $h_1 = .0128$ and $\rho = \frac{1}{2}$. We get the following table using (4.2-11) with $\gamma = 2$ from (4.2-1), where $T_0(h) = [f(x+h) - f(x-h)]/2h$.

h	T_0	T_1	T_2	T_3
0.0128	696.6346914			
		623.4601726		
0.0064	641.7538023		625.3455055	
		625.2276722		625.3334226
0.0032	629.3592047		625.3336114	
		625.3269902		625.3334448
0.0016	626.3350438		625.3334474	
		625.3330438		625.3334257
0.0008	625.5835438		625.3334260	
		625.3334021		
0.0004	625.3959375			

The oscillating behavior of the values in the T_3 column precludes continuing, but we have achieved agreement to seven figures with the exact value 625.33344002.

If we allow ourselves more freedom in the choice of the h_i, we get the following result, using (4.2-10):

h	T_0	T_1	T_2	T_3	T_4	T_5	T_6
0.0256	1058.9377906						
		495.5225783					
0.0192	812.4436352		640.1425224				
		603.9875364		624.4282997			
0.0128	696.6346914		626.6381123		625.3572216		
		620.9754683		625.2991640		625.3330932	
0.0096	663.5337813		625.4479360		625.3339415		625.3334398
		624.3298191		625.3317679		625.3334344	
0.0064	641.7538023		625.3481040		625.3334485		
		625.0935328		625.3333435			
0.0048	634.4649344		625.3349836				
		625.2746209					
0.0032	629.3592047						

Here we have achieved nine-figure accuracy even though our first two approximations in the T_0 column are way off.

4.3 NUMERICAL QUADRATURE: THE GENERAL PROBLEM

We now consider the problem of numerical quadrature, namely the approximation of the linear functional

$$I(f) = \int_a^b f(x)\,dx \qquad -\infty \le a \le b \le \infty \tag{4.3-1}$$

Since $I(f)$ is linear, it makes sense to approximate it by the linear functional

$$Q(f) = \sum_{i=0}^{m} \sum_{j=1}^{n} A_{ij} f^{(i)}(a_{ij}) \tag{4.3-2}$$

so that we have the quadrature equation†

$$I(f) = Q(f) + E \tag{4.3-3}$$

or

$$\int_a^b f(x)\,dx = \sum_{j=1}^{n} \sum_{i=0}^{m} A_{ij} f^{(i)}(a_{ij}) + E \tag{4.3-4}$$

Setting $E = 0$ in (4.3-4) then gives us an approximation to the definite integral of $f(x)$ as a linear combination of values of $f(x)$ and its derivatives. The numerical-quadrature problem is to specify the A_{ij}'s and a_{ij}'s so that this approximation has desirable properties, i.e., achieves some desired accuracy.

Once again our approach will be that of exact polynomial approximation. That is, we shall attempt to choose the A_{ij}'s and a_{ij}'s so that E in (4.3-4) is zero when $f(x)$ is a polynomial of sufficiently low degree. We shall again restrict ourselves mainly to the case $m = 0$; that is, we shall try to express the integral as a linear combination of functional values alone, as is done, for example, in the trapezoidal rule. This is by far the most important case both theoretically and practically. With the restriction $m = 0$, we can rewrite (4.3-4) after some obvious changes in notation as

$$\int_a^b f(x)\,dx = \sum_{j=1}^{n} H_j f(a_j) + E \tag{4.3-5}$$

One equation of the form (4.3-5) can clearly be derived by integrating the Lagrangian interpolation formula (3.1-1). Without considering the details of this now, we can nevertheless see that since the Lagrangian formula is exact for polynomials of degree $n - 1$ or less, then so will the formula resulting from its integration. This suggests the question: With no a priori restrictions on the "abscissas" a_j (such as that they be equally spaced) and the "weights" H_j, what is the highest-degree polynomial for which E in (4.3-5) can be made zero? We call the degree of this polynomial the *order of*

† Here and in the remainder of this chapter, we shall generally denote the error by E instead of $E(x)$ because the variable x will not appear explicitly in the error term as it has previously.

accuracy of the formula. Since we have $2n$ constants at our disposal (n a_j's and n H_j's), we suspect that the answer is a polynomial of degree $2n - 1$. In the next section, we shall show that this is indeed the case.

We shall not explicitly consider the problem of evaluating the indefinite integral

$$y(x) = \int_{x_0}^{x} f(t)\,dt \tag{4.3-6}$$

in this chapter. This problem is equivalent to solving the differential equation

$$\frac{dy}{dx} = f(x) \qquad y(x_0) = 0 \tag{4.3-7}$$

and as such can be solved by the techniques of Chap. 5. For any specific value of x the methods of this chapter can, of course, be used to evaluate $y(x)$.

Another approach to indefinite integration which can also be used for definite integrals is to determine a linear approximation to $f(t)$

$$f(t) \approx \sum_{i=1}^{m} a_i \phi_i(t) \tag{4.3-8}$$

where the indefinite integrals $\psi_i(x)$

$$\psi_i(x) = \int_{x_0}^{x} \phi_i(t)\,dt \tag{4.3-9}$$

are available analytically. Then

$$\int_{x_0}^{x} f(t)\,dt \approx \sum_{i=1}^{m} a_i \psi_i(x) \tag{4.3-10}$$

gives a series approximation of the indefinite integral. Particular choices of $\phi_i(x)$ in the normalized case where $x_0 = -1$ and $-1 \le x \le 1$ are the Chebyshev polynomials of the first kind $T_i(x)$, the Chebyshev polynomials of the second kind $S_i(x)$, and the Legendre polynomials $P_i(x)$, described later in this chapter.

4.3-1 Numerical Integration of Data

In the succeeding sections, we shall be concerned with the numerical quadrature of functions that can be evaluated at any given point or that are tabulated at equally spaced points. A different problem is that of the integration of data. Here, we are given a set of pairs (a_i, f_i), $i = 1, \ldots, n$, and we wish to find an approximation to the integral of the function $f(x)$ which is represented more or less accurately by the given values f_i. We mention three approaches to this problem. The first is to compute an interpolating cubic spline $S(x)$, as in Sec. 3.8, and integrate $S(x)$ exactly. This yields the indefinite integral and hence also the definite integral between any two points in the integration interval. The resulting approximation is quite good, and, asymptotically, as h, the maximum spacing between the nodes, goes to zero, the error in the integral is of the order of h^4. A second approach which also yields the indefinite integral is to approximate $f(x)$ by a least-square polynomial $p_m(x)$ (Chap. 6) and approximate the integral of $f(x)$ by the integral of $p_m(x)$. This is especially to be recommended when the f_i are empirical data subject to experimental error.

The third approach is to use piecewise polynomial interpolation using cubic polynomials. If we define $p_j(x)$ to be the cubic interpolating to $f(x)$ at the four points $a_{j-1}, a_j, a_{j+1}, a_{j+2}$, then the approximation to the integral of $f(x)$ takes the form

$$\int_{a_k}^{a_l} f(x)\,dx = \sum_{j=k}^{l-1} \int_{a_j}^{a_{j+1}} \hat{p}_j(x)\,dx$$

where

$$\hat{p}_j(x) = p_j(x) \qquad 2 \le j \le n-1$$

$$\hat{p}_1(x) = p_2(x) \qquad \hat{p}_n(x) = p_{n-1}(x)$$

4.4 GAUSSIAN QUADRATURE

For now let us assume that a and b in (4.3-5) are finite. Then, if (4.3-5) is to be exact for polynomials of degree $2n-1$ or less, we can get a set of $2n$ equations for the $2n$ unknown constants by substituting $f(x) = x^k$, $k = 0, 1, \ldots, 2n-1$, into (4.3-5) and setting $E = 0$. We get

$$\alpha_k = \sum_{j=1}^{n} H_j a_j^k \qquad k = 0, \ldots, 2n-1 \tag{4.4-1}$$

where

$$\alpha_k = \int_a^b x^k\,dx = \frac{b^{k+1} - a^{k+1}}{k+1} \tag{4.4-2}$$

These nonlinear equations, if we can solve them and if the solution is real, will give us the abscissas and weights we desire. This algebraic approach to our problem is considered further in {10}, but we abandon it here in favor of an analytic approach which (1) will tell us without actually calculating the weights and abscissas whether or not they are real; (2) will enable us to determine E when $f(x)$ is not a polynomial of degree $2n-1$ or less; and (3) will enable us to show that the abscissas are the zeros of well-known polynomials. As we shall see, once the abscissas are known, the weights are easily calculable.

The starting point of our analytical approach is the Hermite interpolation formula (3.7-17)

$$f(x) = \sum_{j=1}^{n} h_j(x) f(a_j) + \sum_{j=1}^{n} \bar{h}_j(x) f'(a_j) + \frac{p_n^2(x)}{(2n)!} f^{(2n)}(\xi) \tag{4.4-3}$$

which is exact for polynomials of degree $2n - 1$ or less. Integrating (4.4-3) between a and b, we get

$$\int_a^b f(x)\,dx = \sum_{j=1}^{n} H_j f(a_j) + \sum_{j=1}^{n} \bar{H}_j f'(a_j) + E \tag{4.4-4}$$

where

$$H_j = \int_a^b h_j(x)\,dx \qquad \bar{H}_j = \int_a^b \bar{h}_j(x)\,dx \tag{4.4-5}$$

and

$$E = \int_a^b \frac{p_n^2(x)}{(2n)!} f^{(2n)}(\xi)\,dx \tag{4.4-6}$$

Since E is zero if $f(x)$ is a polynomial of degree $2n - 1$ or less, if we can choose the abscissas so that $\bar{H}_j = 0$, $j = 1, \ldots, n$, then (4.4-4) will have the form (4.3-5) with the desired properties. Thus $\bar{H}_j = 0$, $j = 1, \ldots, n$, is a sufficient condition to achieve our desired accuracy of order $2n - 1$. It is also necessary. To see this let $f(x) = \bar{h}_j(x)$ in (4.3-5). Since $\bar{h}_j(x)$ is a polynomial of degree $2n - 1$, $E = 0$. From (4.4-5) it follows that the left-hand side of (4.3-5) is $\bar{H}_j$, and from (3.7-4) it follows that the right-hand side is zero. Therefore, since we have put no restriction on the a_j's in (4.4-4), if we cannot find abscissas for which the $\bar{H}_j$'s are zero, then no formula of the type (4.3-5) with order of accuracy $2n - 1$ is possible.

Using (3.7-18) and (3.2-4), we have

$$\bar{H}_j = \int_a^b (x - a_j) l_j^2(x)\,dx = \int_a^b p_n(x) \frac{l_j(x)}{p_n'(a_j)}\,dx \tag{4.4-7}$$

Since $p_n(x)$ is a polynomial of degree n and $l_j(x)$ is a polynomial of degree $n - 1$, a sufficient condition for $\bar{H}_j = 0$, $j = 1, \ldots, n$, is for $p_n(x)$ to be *orthogonal* to all polynomials of degree $n - 1$ or less over $[a, b]$. This condition is also necessary; we leave the proof to a problem {11}. Without loss of generality, we may assume $[a, b] = [-1, 1]$,† in which case the orthogonal polynomial $p_n(x)$ is a multiple of $P_n(x)$, the Legendre polynomial of degree n. Since by definition $p_n(x)$ has leading coefficient 1, we have, using the standard definition of the Legendre polynomials {19},

$$p_n(x) = \frac{2^n (n!)^2}{(2n)!} P_n(x) \tag{4.4-8}$$

In the next section we shall prove that the zeros of the Legendre polynomial of any degree are real, so that this settles the question of the existence of real abscissas. The zeros of the Legendre polynomials have been tabulated for all values of n of practical interest; a short table of these zeros is given in Table 4.1.

† By the change of variable $y = [1/(b - a)](2x - a - b)$, the interval $[a, b]$ in x is replaced by the interval $[-1, 1]$ in y.

Table 4.1 Zeros of Legendre polynomials and corresponding weights

n	Abscissas a_j	Weights H_j
2	$\pm 0.577350 = \pm \dfrac{1}{\sqrt{3}}$	1
3	0	$\frac{8}{9}$
	± 0.774597	$\frac{5}{9}$
4	± 0.339981	0.652145
	± 0.861136	0.347855
5	0	0.568889
	± 0.538469	0.478629
	± 0.906180	0.236927

To find the weights we again use Eqs. (3.7-18) and (3.2-4) to get

$$\begin{aligned} H_j &= \int_{-1}^{1} h_j(x)\,dx = \int_{-1}^{1} [1 - 2l_j'(a_j)(x - a_j)]l_j^2(x)\,dx \\ &= \int_{-1}^{1} l_j^2(x)\,dx - 2l_j'(a_j)\int_{-1}^{1} (x - a_j)l_j^2(x)\,dx \\ &= \int_{-1}^{1} l_j^2(x)\,dx \end{aligned} \tag{4.4-9}$$

since the second integral, which by definition is $\bar{H}_j$, is zero. From (4.4-9) it is obvious that the weights are all positive; see Table 4.1. A simpler expression for the weights can be found by considering (4.3-5) with $f(x) = l_k(x)$ and using the weights and abscissas found above. Since $l_k(x)$ is a polynomial of degree $n - 1$, E is zero and we have

$$\int_{-1}^{1} l_k(x)\,dx = \sum_{j=1}^{n} H_j l_k(a_j) = H_k \tag{4.4-10}$$

since $l_k(a_j) = \delta_{kj}$. Together (4.4-9) and (4.4-10) imply

$$\int_{-1}^{1} l_j^2(x)\,dx = \int_{-1}^{1} l_j(x)\,dx \tag{4.4-11}$$

The error term as given by (4.4-6) can be simplified using the mean-value theorem for integrals since $p_n^2(x)$ is always positive. Using $[-1, 1]$ in place of $[a, b]$, we have

$$E = \frac{f^{(2n)}(\eta)}{(2n)!}\int_{-1}^{1} p_n^2(x)\,dx \tag{4.4-12}$$

where η lies in $(-1, 1)$.

Any quadrature formula whose abscissas and weights are subject to no constraints and which are determined so as to achieve a maximum order of accuracy is called a *Gaussian quadrature formula*. In particular, (4.3-5) with the abscissas given as the zeros of the Legendre polynomial of degree n and the weights by (4.4-10) is called a *Gauss-Legendre quadrature formula*, or simply a Gauss formula.

Example 4.3 Evaluate

$$\int_1^3 \frac{dx}{x}$$

using Gauss-Legendre quadrature with $n = 3$.

With the change of variable $y = x - 2$, the integral becomes

$$\int_{-1}^1 \frac{dy}{y+2}$$

Using Table 4.1 with $n = 3$, we have

$$\int_{-1}^1 \frac{dy}{y+2} \approx \frac{5}{9}\frac{1}{1.225403} + \frac{8}{9}\frac{1}{2} + \frac{5}{9}\frac{1}{2.774597} \approx 1.098039$$

whereas the true value of the integral is ln 3 = 1.098612. From Eq. (4.4-12) we have

$$E = \frac{f^{\mathrm{vi}}(\eta)}{6!}\int_{-1}^1 p_3^2(x)\,dx$$

Using (4.4-8) and anticipating some results from Sec. 4.6 [see (4.6-10)], we have

$$\int_{-1}^1 p_3^2(x)\,dx = \frac{4}{25}\int_{-1}^1 P_3^2(x)\,dx = \frac{8}{175}$$

so that
$$E = \frac{1}{15{,}750}\frac{6!}{(\eta+2)^7} = \frac{8}{175}\frac{1}{(\eta+2)^7}$$

Thus
$$.000021 \approx \frac{8}{175}\frac{1}{(3)^7} < E < \frac{8}{175} \approx .045714$$

and the actual error is indeed within these bounds.

Gaussian quadrature formulas were seldom used in practice before the advent of digital computers. This was because the use of "simple" numbers, i.e., integers and rational numbers, is much more convenient on desk calculators than the nonsimple numbers, which generally must be used in Gaussian quadrature calculations (cf. Table 4.1). For example, when using desk calculators, functions are often evaluated by table lookup, in which case simple values of the abscissas may mean that no interpolation is required. But on digital computers, functions are almost always evaluated using a rational or polynomial approximation of the type to be discussed in Chap. 7. In this case whether the numbers are simple or not generally makes no difference. On digital computers, therefore, the Gaussian quadrature formulas discussed in this section and the next four are practical for certain problems. In Sec. 4.11 we shall discuss the general problem of choosing a quadrature formula for a specific application.

4.5 WEIGHT FUNCTIONS

In this section we shall generalize the ideas of the previous section by considering in place of (4.3-5)

$$\int_a^b w(x)f(x)\,dx = \sum_{j=1}^{n} H_j f(a_j) + E \tag{4.5-1}$$

where $w(x)$, the *weight function*, does not appear on the right-hand side of (4.5-1). That it is not artificial to separate the integrand into two functions $w(x)$ and $f(x)$ is borne out by (1) the not uncommon need to evaluate the coefficients in an orthogonal polynomial expansion and (2) the frequency with which some functions appear in integrands, particularly when dealing with integrals over infinite intervals (see Sec. 4.7).

The advantages of the formulation (4.5-1) are also twofold: (1) computationally it will generally be easier to evaluate $f(a_j)$ than $w(a_j)f(a_j)$; and (2) it is often convenient to express the error term in terms of a derivative of $f(x)$ only, especially when the weight function or one of its derivatives is unbounded in the interval.

It is, of course, possible to treat *any* numerical quadrature problem in the form (4.5-1). That is, we can always consider splitting the integrand up into the product of two functions. But since, as we shall see below, the abscissas and weights are functions of $w(x)$, this would necessitate the evaluation of the weights and abscissas for each problem. Thus we shall consider only those weight functions which are practically or mathematically significant. Furthermore, as we shall also see below, it is important that $w(x)$ be of constant sign in $[a, b]$.

The evaluation of the weights and abscissas in the more general Gaussian quadrature formula (4.5-1) is again quite straightforward if we make use of the Hermite interpolation formula. This time we merely multiply (3.7-17) by $w(x)$ before integrating and obtain in this way

$$\int_a^b w(x)f(x)\,dx = \sum_{j=1}^{n} H_j f(a_j) + \sum_{j=1}^{n} \bar{H}_j f'(a_j) + E \tag{4.5-2}$$

where now

$$H_j = \int_a^b w(x)h_j(x)\,dx \qquad \bar{H}_j = \int_a^b w(x)\bar{h}_j(x)\,dx \tag{4.5-3}$$

Now, proceeding as in the previous section, we set $\bar{H}_j = 0$, $j = 1, \ldots, n$, and get as a necessary and sufficient condition on the abscissas that they be the zeros of a polynomial orthogonal with respect to $w(x)$ to all polynomials of lesser degree over $[a, b]$ {11}. That is,

$$\int_a^b w(x)u_{n-1}(x)p_n(x)\,dx = 0 \tag{4.5-4}$$

where $u_{n-1}(x)$ is any polynomial of degree $n-1$ or less. Then, corresponding to (4.4-9) and (4.4.10), we have

$$H_j = \int_a^b w(x)l_j^2(x)\,dx = \int_a^b w(x)l_j(x)\,dx \tag{4.5-5}$$

The error term becomes

$$E = \frac{1}{(2n)!}\int_a^b w(x)p_n^2(x)f^{(2n)}(\xi)\,dx \tag{4.5-6}$$

which, if $w(x)$ does not change sign in (a, b), can be written

$$E = \frac{f^{(2n)}(\eta)}{(2n)!}\int_a^b w(x)p_n^2(x)\,dx \tag{4.5-7}$$

where η lies in (a, b).

If, in fact, $w(x) \geq 0$ in $[a, b]$, then from (4.5-5) it follows that all the weights are positive, as we found for the Gauss-Legendre quadrature formula. Assuming $w(x)$ to satisfy this condition, we can now prove the following theorem.

Theorem 4.1 The abscissas defined in (4.5-4) are all real and distinct and lie within the interval (a, b).

The importance of having the abscissas real is clear. The importance of their being within the interval (a, b) is also clear if one considers the error term (4.5-7), whose magnitude depends upon the magnitude of $p_n(x)$. Furthermore, the integrand may not be defined outside the interval of integration.

PROOF Let $a_1, \ldots, a_m$ be the points of (a, b) where $p_n(x)$ changes sign. Then

$$(x-a_1)\cdots(x-a_m)p_n(x)$$

does not change sign in $[a, b]$. Since $p_n(x)$ is orthogonal to all polynomials of degree less than n with respect to $w(x)$ over $[a, b]$, we have

$$\int_a^b w(x)(x-a_1)\cdots(x-a_m)p_n(x)\,dx = 0 \tag{4.5-8}$$

unless $m = n$. But the integrand does not change sign. Therefore, the integral cannot be zero, and so $m = n$, which proves the zeros are real, distinct, and lie within (a, b).

Equation (4.5-1) is the general form of a Gaussian quadrature formula with the weights defined by (4.5-5), the abscissas by (4.5-4), and the error by (4.5-7). In the next section we consider in more detail the properties of the orthogonal polynomials defined by (4.5-4).

4.6 ORTHOGONAL POLYNOMIALS AND GAUSSIAN QUADRATURE

Let $\{\phi_n(x)\}$ be a sequence of polynomials, the degree of $\phi_n(x)$ being n, orthogonal with respect to a weight function $w(x)$ over an interval $[a, b]$. Let the coefficient of x^n in $\phi_n(x)$ be A_n, so that $\phi_n(x) = A_n p_n(x)$, where $p_n(x)$ is the orthogonal polynomial of the previous section. We introduce the coefficient A_n so that the orthogonal polynomials to be considered in this and later sections can be put in their standard form, which generally has leading coefficients not equal to 1. Note that the choice of the leading coefficient has no effect whatever on the abscissas or weights. The details of the derivation of such sequences of orthogonal polynomials are considered in the problems {13 to 16}. We have

$$\int_a^b w(x)\phi_i(x)\phi_j(x)\,dx = 0 \qquad i \neq j \tag{4.6-1}$$

Let

$$\alpha_k = \frac{A_{k+1}}{A_k} \qquad \gamma_k = \int_a^b w(x)\phi_k^2(x)\,dx \tag{4.6-2}$$

The basis of the results of this section is the *Christoffel-Darboux identity*

$$\sum_{k=0}^{n} \frac{\phi_k(x)\phi_k(y)}{\gamma_k} = \frac{\phi_{n+1}(x)\phi_n(y) - \phi_n(x)\phi_{n+1}(y)}{\alpha_n\gamma_n(x-y)} \tag{4.6-3}$$

the derivation of which is left to a problem {17}.

If we set $y = a_j$ in (4.6-3), where a_j is a zero of $\phi_n(x)$, we get

$$\sum_{k=0}^{n-1} \frac{\phi_k(x)\phi_k(a_j)}{\gamma_k} = -\frac{\phi_n(x)\phi_{n+1}(a_j)}{\alpha_n\gamma_n(x-a_j)} \tag{4.6-4}$$

Now, multiplying both sides of (4.6-4) by $w(x)\phi_0(x)$, integrating over $[a, b]$, and using (4.6-1), we get

$$\frac{\phi_0(a_j)}{\gamma_0}\gamma_0 = -\frac{\phi_{n+1}(a_j)}{\alpha_n\gamma_n}\int_a^b w(x)\frac{\phi_0(x)\phi_n(x)}{x-a_j}\,dx \tag{4.6-5}$$

From the definition of the Lagrangian interpolation polynomial we have

$$l_j(x) = \frac{p_n(x)}{(x-a_j)p_n'(a_j)} = \frac{\phi_n(x)}{(x-a_j)\phi_n'(a_j)} \tag{4.6-6}$$

Using this, (4.5-5), and the fact that $\phi_0(x)$ is a constant, we can rewrite (4.6-5) as

$$\begin{aligned} 1 &= -\frac{\phi_{n+1}(a_j)}{\alpha_n\gamma_n}\int_a^b w(x)\frac{\phi_n(x)}{x-a_j}\,dx \\ &= -\frac{\phi_{n+1}(a_j)\phi_n'(a_j)}{\alpha_n\gamma_n}\int_a^b w(x)l_j(x)\,dx = \frac{-\phi_{n+1}(a_j)\phi_n'(a_j)}{\alpha_n\gamma_n}H_j \end{aligned} \tag{4.6-7}$$

Thus
$$H_j = \frac{-A_{n+1}\gamma_n}{A_n\phi_{n+1}(a_j)\phi_n'(a_j)} \qquad j = 1, \ldots, n \tag{4.6-8}$$

which, given the orthogonal polynomials, is a much simpler way to calculate H_j than (4.5-5). The definition of γ_n allows an obvious simplification of (4.5-7) to give

$$E = \frac{\gamma_n}{A_n^2(2n)!} f^{(2n)}(\eta) \tag{4.6-9}$$

These results can be used to simplify the formulas of Sec. 4.4 on Gauss-Legendre quadrature. For the Legendre polynomials {19}

$$\gamma_n = \int_{-1}^{1} P_n^2(x)\, dx = \frac{2}{2n+1} \qquad A_n = \frac{(2n)!}{2^n(n!)^2} \tag{4.6-10}$$

so that H_j in terms of Legendre polynomials is

$$H_j = \frac{-2}{(n+1)P_{n+1}(a_j)P_n'(a_j)} \tag{4.6-11}$$

Some other similar forms of H_j are considered in {22}. For the error term we get in a corresponding manner

$$E = \frac{2^{2n+1}(n!)^4}{(2n+1)[(2n)!]^3} f^{(2n)}(\eta) \tag{4.6-12}$$

The results of this section and the next two are summarized in Table 4.4 on p. 114. In the next two sections, we shall use the results of this and the previous section to derive Gaussian quadrature formulas for particular weight functions.

4.7 GAUSSIAN QUADRATURE OVER INFINITE INTERVALS

For integrals over infinite intervals (which we shall assume are convergent), there are many possible approaches. We give two here: (1) use a knowledge of the integrand to bound the magnitude of the integral from some finite value to infinity by a positive constant $\epsilon > 0$ and then use a quadrature formula for the remaining finite interval; (2) use a quadrature formula especially developed for the infinite interval. The first of these two approaches requires no further discussion. The quadrature over the finite interval can be performed using Gauss-Legendre quadrature or one of the many methods to be presented later in this chapter. In this section we treat the latter case.

For numerical integration over infinite and semi-infinite intervals, it is convenient to use a weight function $w(x)$ which assures the convergence of

Table 4.2 Zeros of Laguerre polynomials and corresponding weights

n	Abscissas a_j	Weights H_j	n	Abscissas a_j	Weights H_j
2	0.585786	0.853553	4	4.536620	0.038888
	3.414214	0.146447		9.395071	0.000539
3	0.415775	0.711093	5	0.263560	0.521756
	2.294280	0.278518		1.413403	0.398667
	6.289945	0.010389		3.596426	0.075942
4	0.322548	0.603154		7.085810	0.003612
	1.745761	0.357419		12.640801	0.000023

the integral of $w(x)f(x)$ when $f(x)$ is a polynomial of arbitrary degree. For the semi-infinite interval (a, ∞) (for convenience we set $a = 0$), such a weight function is $w(x) = e^{-x}$. Therefore, the sequence of polynomials we require must be orthogonal over $(0, \infty)$ with respect to e^{-x}. Such a sequence of polynomials is the Laguerre polynomials [see Jackson (1941) and {20}]. The polynomial of degree n, $L_n(x)$, has leading coefficient $A_n = (-1)^n$. For the Laguerre polynomials

$$\gamma_n = \int_0^\infty e^{-x} L_n^2(x)\, dx = (n!)^2 \tag{4.7-1}$$

Then from (4.6-8) and (4.6-9)

$$H_j = \frac{(n!)^2}{L_n'(a_j) L_{n+1}(a_j)} \tag{4.7-2}$$

$$E = \frac{(n!)^2}{(2n)!} f^{(2n)}(\eta) \tag{4.7-3}$$

The *Gauss-Laguerre* quadrature formula then has the form

$$\int_0^\infty e^{-x} f(x)\, dx = \sum_{j=1}^{n} H_j f(a_j) + E \tag{4.7-4}$$

where the a_j's are the zeros of $L_n(x)$ and the H_j's are given by (4.7-2) and E by (4.7-3). Table 4.2 is a short listing of weights and abscissas for (4.7-4).

Example 4.4 Evaluate

$$\int_0^\infty x^7 e^{-x}\, dx$$

using $n = 3$ in (4.7-4).

Using Table 4.2, we have

$$\int_0^\infty x^7 e^{-x}\,dx \approx (.711093)(.415775)^7 + (.278518)(2.294280)^7$$

$$+ (.010389)(6.289945)^7 = 4139.9$$

whereas the true value of the integral is $7! = 5040$. The error given by (4.7-3) is

$$E = \frac{(3!)^2}{6!} 7!\,\eta = 252\eta$$

with η in $(0, \infty)$ so it cannot be bounded. Thus the substantial error between the approximate and true values is not surprising. This illustrates the fact that the Gauss-Laguerre quadrature formula and the Gauss-Hermite quadrature formula to be discussed below should be avoided if the derivative in the error term cannot be bounded on $(0, \infty)$. In this case the first technique mentioned in this section will usually be preferable. Note that in this particular example, $n = 4$ would have led to an exact result (except for roundoff) since the eighth derivative of $f(x)$ is zero.

A generalization of the weight function e^{-x} to $x^\beta e^{-\alpha x}$ is considered in {20}.

For the infinite interval $(-\infty, \infty)$ we choose, for the same reason as above, the weight function e^{-x^2}. The sequence of polynomials we require must be orthogonal over $(-\infty, \infty)$ with respect to this weight function. This sequence is the Hermite polynomials [see Jackson (1941) and {21}]. The polynomial $H_n(x)$ of degree n has leading coefficient $A_n = 2^n$.

For these polynomials

$$\gamma_n = \int_{-\infty}^{\infty} e^{-x^2} H_n^2(x)\,dx = \sqrt{\pi}\,2^n n! \tag{4.7-5}$$

Then from (4.6-8) and (4.6-9)

$$H_j = -\frac{2^{n+1} n! \sqrt{\pi}}{H_n'(a_j) H_{n+1}(a_j)} \tag{4.7-6}$$

$$E = \frac{n! \sqrt{\pi}}{2^n (2n)!} f^{(2n)}(\eta) \tag{4.7-7}$$

The *Gauss-Hermite quadrature formula* then has the form

$$\int_{-\infty}^{\infty} e^{-x^2} f(x)\,dx = \sum_{j=1}^{n} H_j f(a_j) + E \tag{4.7-8}$$

where the a_j's are the zeros of $H_n(x)$, the H_j's are given by (4.7-6), and E by (4.7-7). Table 4.3 is a short listing of weights and abscissas for (4.7-8).

Table 4.3 Zeros of Hermite polynomials and corresponding weights

n	Abscissas a_j	Weights H_j
2	± 0.707107	0.886227
3	0	1.181636
	± 1.224745	0.295409
4	± 0.524648	0.804914
	± 1.650680	0.081313
5	0	0.945309
	± 0.958572	0.393619
	± 2.020183	0.019953

4.8 PARTICULAR GAUSSIAN QUADRATURE FORMULAS

In this section we consider Gaussian quadrature over finite intervals using weight functions of considerable theoretical and practical importance.

4.8-1 Gauss-Jacobi Quadrature

We consider here the weight function $w(x) = (1 - x)^\alpha(1 + x)^\beta$, $\alpha, \beta > -1$. The polynomials orthogonal to this weight function over $[-1, 1]$ are the Jacobi polynomials $J_n(x; \alpha, \beta)$. They are generally defined so that the coefficient A_n of x^n in the polynomial of degree n is given by

$$A_n = \frac{1}{2^n n!} \frac{\Gamma(2n + \alpha + \beta + 1)}{\Gamma(n + \alpha + \beta + 1)} \tag{4.8-1}$$

and we can calculate {24}

$$\gamma_n = \int_{-1}^{1} (1 - x)^\alpha(1 + x)^\beta J_n^2(x; \alpha, \beta)\, dx$$

$$= \frac{2^{\alpha+\beta+1}}{n!\,(2n + \alpha + \beta + 1)} \frac{\Gamma(n + \alpha + 1)\Gamma(n + \beta + 1)}{\Gamma(n + \alpha + \beta + 1)} \tag{4.8-2}$$

Then proceeding as in the previous section, we get the *Gauss-Jacobi quadrature formula* (sometimes called the Mehler quadrature formula)

$$\int_{-1}^{1} (1 - x)^\alpha(1 + x)^\beta f(x)\, dx = \sum_{j=1}^{n} H_j f(a_j) + E \tag{4.8-3}$$

where {24}

$$H_j = -\frac{2n+\alpha+\beta+2}{n+\alpha+\beta+1}\frac{\Gamma(n+\alpha+1)\Gamma(n+\beta+1)}{\Gamma(n+\alpha+\beta+1)(n+1)!}$$

$$\times \frac{2^{\alpha+\beta}}{J_n'(a_j;\alpha,\beta)J_{n+1}(a_j;\alpha,\beta)} \tag{4.8-4}$$

and

$$E = \frac{\Gamma(n+\alpha+1)\Gamma(n+\beta+1)\Gamma(n+\alpha+\beta+1)}{(2n+\alpha+\beta+1)[\Gamma(2n+\alpha+\beta+1)]^2}$$

$$\times \frac{n!\,2^{2n+\alpha+\beta+1}}{(2n)!} f^{(2n)}(\eta) \tag{4.8-5}$$

The Gauss-Legendre quadrature formula of Sec. 4.4 is just a special case of the Gauss-Jacobi formula with $\alpha = \beta = 0$. In the next section, we consider another special case $\alpha = \beta = -\frac{1}{2}$ because of the importance of the weight function in this case and because it serves to introduce the Chebyshev polynomials, which will play an important role in Chap. 7.

4.8-2 Gauss-Chebyshev Quadrature

When $\alpha = \beta = -\frac{1}{2}$ so that $w(x) = 1/(1-x^2)^{1/2}$, (4.8-1) yields

$$A_n = \frac{(2n-1)!}{2^n n!\,(n-1)!} \tag{4.8-6}$$

but in this case it is customary to choose

$$A_0 = 1 \qquad A_n = 2^{n-1} \qquad n \geq 1 \tag{4.8-7}$$

Using (4.8-7), we can calculate {26}

$$\gamma_n = \frac{\pi}{2} \tag{4.8-8}$$

The orthogonal polynomials $J_n(x; -\frac{1}{2}, -\frac{1}{2})$ are generally denoted by $T_n(x)$ and are called Chebyshev polynomials of the first kind or usually just Chebyshev polynomials. The *Gauss-Chebyshev quadrature formula* has the form

$$\int_{-1}^{1} \frac{1}{(1-x^2)^{1/2}} f(x)\,dx = \sum_{j=1}^{n} H_j f(a_j) + E \tag{4.8-9}$$

where the a_j's are the zeros of $T_n(x)$,

$$H_j = -\frac{\pi}{T_n'(a_j)T_{n+1}(a_j)} \tag{4.8-10}$$

and
$$E = \frac{2\pi}{2^{2n}(2n)!} f^{(2n)}(\eta) \tag{4.8-11}$$

Equation (4.8-10) can be remarkably simplified by using the result {26} that

$$T_n(x) = \cos(n \cos^{-1} x) \tag{4.8-12}$$

Therefore, at a zero of $T_n(x)$ we have $\cos(n \cos^{-1} x) = 0$ and $\sin(n \cos^{-1} x) = \pm 1$. Thus

$$\begin{aligned} T_{n+1}(a_j) &= \cos[(n+1)\cos^{-1} a_j] \\ &= \cos(n \cos^{-1} a_j)a_j - \sin(n \cos^{-1} a_j)\sin(\cos^{-1} a_j) \\ &= \mp(1 - a_j^2)^{1/2} \end{aligned} \tag{4.8-13}$$

since a_j is a zero of $T_n(x)$. Further

$$T_n'(a_j) = \sin(n \cos^{-1} a_j)\frac{n}{(1 - a_j^2)^{1/2}} = \frac{\pm n}{(1 - a_j^2)^{1/2}} \tag{4.8-14}$$

Using (4.8-13) and (4.8-14) in (4.8-10), it follows that

$$H_j = \frac{\pi}{n} \tag{4.8-15}$$

From (4.8-12) it also follows that

$$a_j = \cos\frac{(2j-1)\pi}{2n} \qquad j = 1, \ldots, n \tag{4.8-16}$$

Thus all the weights are equal, and (4.8-9) can be written

$$\int_{-1}^{1} \frac{1}{(1 - x^2)^{1/2}} f(x)\,dx = \frac{\pi}{n}\sum_{j=1}^{n} f(a_j) + E \tag{4.8-17}$$

Example 4.5 Evaluate

$$I = \int_{-1}^{1} \frac{e^x\,dx}{(1 - x^2)^{1/2}}$$

correct to six decimal places. Applying (4.8-11), we find that the error E_n in using the n-point rule (4.8-17) is given by

$$E_n = \frac{2\pi}{2^{2n}(2n)!}e^{\eta} \qquad -1 < \eta < 1$$

so that
$$|E_n| \le \frac{2\pi e}{2^{2n}(2n)!} \equiv B_n$$

For $n = 4$, $B_n = 1.66 \times 10^{-6}$, and for $n = 5$, $B_n = 4.6 \times 10^{-9}$. Hence, we apply (4.8-17) with $n = 5$ to get

$$I \approx \frac{\pi}{5}\sum_{j=1}^{5} \exp\left[\cos\frac{(2j-1)\pi}{10}\right] = 3.977463$$

correct to six decimal places.

4.8-3 Singular Integrals

A common problem in numerical analysis is the need to evaluate integrals in which the integrand has a singularity. If the singularity results in an improper integral, as in the case of $1/(1-x)^{1/2}$ on $[0, 1]$, we assume the integral is convergent. But we must also consider singularities of the form $(1-x)^{1/2}$ on $[0, 1]$. For in both cases, a Gaussian quadrature approach without weight functions would lead to trouble because of the derivative of the integrand which appears in the error.

The Gauss-Chebyshev quadrature formula is a good example of the value of the weight-function approach to singular integrals. For if $f(x)$ in (4.8-9) is analytic, the effect of making the singular term $1/(1-x^2)^{1/2}$ the weight function is to remove this term from the summation and the error term on the right-hand side of (4.8-9). In this section we shall consider some other applications of this technique to common types of singularities. We shall restrict our discussion to integrands with singularities at the endpoints of the interval. However, by splitting the integral into two integrals, singularities in the interior of the interval can also be handled.

The general problem then is to find quadrature formulas of the form

$$\int_a^b s(x)f(x)\,dx = \sum_{j=1}^{n} H_j f(a_j) + E \tag{4.8-18}$$

where the weight function $s(x)$ is singular at one endpoint or both. Without loss of generality, we shall restrict ourselves to $[a, b] = [-1, 1]$ or $[0, 1]$, whichever is most convenient.

Case 1: $s(x) = (1-x^2)^{1/2}$ on $[-1, 1]$. This is the Jacobi weight function with $\alpha = \beta = \frac{1}{2}$. The resulting orthogonal polynomials are called Chebyshev polynomials of the second kind. Let the polynomial of degree n be $S_n(x)$ [sometimes denoted in the literature by $U_n(x)$]. Then, analogous to (4.8-12), we have the relation {30}

$$S_n(x) = \frac{\sin\,[(n+1)\cos^{-1} x]}{\sin\,(\cos^{-1} x)} \tag{4.8-19}$$

The abscissas in (4.8-18) are therefore {30}

$$a_j = \cos\frac{j\pi}{n+1} \qquad j = 1, \ldots, n \tag{4.8-20}$$

and, using (4.6-8), we can calculate {30}

$$H_j = \frac{\pi}{n+1}\sin^2\frac{j\pi}{n+1} \tag{4.8-21}$$

Using (4.6-9), we find {30}

$$E = \frac{\pi}{2^{2n+1}(2n)!}f^{(2n)}(\eta) \tag{4.8-22}$$

Case 2: $s(x) = 1/\sqrt{x}$ on $[0, 1]$. Here we have a singularity at one endpoint. By manipulation of the orthogonality integral (4.6-1), we can show that the orthogonal polynomial of degree n, $p_n(x)$ is given by {31}

$$p_n(x) = P_{2n}(\sqrt{x}) \tag{4.8-23}$$

where $P_{2n}(x)$ is the Legendre polynomial of degree $2n$. Corresponding to each positive zero α_j of $P_{2n}(x)$, there is then an abscissa of (4.8-18) given by

$$a_j = \alpha_j^2 \tag{4.8-24}$$

Using (4.6-8) and (4.8-23), we can also show that {31}

$$H_j = 2h_j \tag{4.8-25}$$

where h_j is the weight corresponding to α_j in the Gauss-Legendre formula of order $2n$. For the error we get {31}

$$E = \frac{2^{4n+1}[(2n)!]^3}{(4n+1)[(4n)!]^2} f^{(2n)}(\eta) \tag{4.8-26}$$

Example 4.6 Evaluate

$$\int_0^1 \frac{1+x}{\sqrt{x}}\, dx$$

using $n = 2$ in (4.8-18). From Table 4.1 and (4.8-24), we have

$$a_1 = (.339981)^2 = .115587 \qquad a_2 = (.861136)^2 = .741555$$

and

$$H_1 = 1.304290 \qquad H_2 = .695710$$

Therefore, our approximation to the integral is

$$\int_0^1 \frac{1+x}{\sqrt{x}}\, dx \approx (1.304290)(1.115587) + (.695710)(1.741555) \approx 2.666666$$

whereas the true value is $\frac{8}{3}$. Since $f(x) = 1 + x$, the derivative in (4.8-26) is zero and so, except for roundoff, the result is exact, as it should be.

Case 3: $s(x) = \sqrt{x}$ on $[0, 1]$. Again the singularity is at one endpoint, but this time the singularity is in the derivatives of the function. In a manner similar to case 2, we can show that {31}

$$p_n(x) = \frac{1}{\sqrt{x}} P_{2n+1}(\sqrt{x}) \tag{4.8-27}$$

Therefore, if α_j is a positive zero of $P_{2n+1}(x)$, the abscissa in (4.8-18) is given by

$$a_j = \alpha_j^2 \tag{4.8-28}$$

Again using (4.6-8), we find that the corresponding weight is given by {31}

$$H_j = 2h_j\alpha_j^2 \tag{4.8-29}$$

where h_j is the weight corresponding to α_j in the Gauss-Legendre formula of order $2n + 1$. Finally, for the error we get {31}

$$E = \frac{2^{4n+3}[(2n+1)!]^4}{(4n+3)[(4n+2)!]^2(2n)!} f^{(2n)}(\eta) \tag{4.8-30}$$

Case 4: $s(x) = [x/(1-x)]^{1/2}$ on $[0, 1]$. This time the weight function has a singularity at one endpoint, and its derivatives have a singularity at the other endpoint. In this case we get {32}

$$p_n(x) = \frac{1}{\sqrt{x}} T_{2n+1}(\sqrt{x}) \tag{4.8-31}$$

where $T_{2n+1}(x)$ is the Chebyshev polynomial of degree $2n + 1$. From this it follows that

$$a_j = \cos^2 \frac{(2j-1)\pi}{4n+2} \tag{4.8-32}$$

From (4.6-8) we get {32}

$$H_j = \frac{2\pi}{2n+1} a_j \tag{4.8-33}$$

and from (4.6-9)

$$E = \frac{\pi}{2^{4n+1}(2n)!} f^{(2n)}(\eta) \tag{4.8-34}$$

The results of Secs. 4.6 to 4.8 are summarized in Table 4.4.

4.9 COMPOSITE QUADRATURE FORMULAS

We must face the same problem in choosing n, the number of abscissas in a quadrature formula, that we faced in choosing the number of points in an interpolation formula. As n gets larger, the constant in the error term decreases but the order of the derivative increases. One problem in using large values of n is the difficulty of estimating high-order derivatives. Another is the fact we mentioned in Sec. 3.4 that ultimately the derivatives of all but certain entire functions increase without bound. This result is most easily derived by showing that any function $f(z)$ of a complex variable with bounded derivatives of all orders at a point z_0 must be entire. The Taylor-series expansion of $f(z)$ about z_0 is given by

$$f(z) = \sum_{j=0}^{\infty} A_j(z - z_0)^j \qquad A_j = \frac{f^{(j)}(z_0)}{j!} \tag{4.9-1}$$

Table 4.4 Summary of Gaussian quadrature formulas of the form $\int_a^b w(x)f(x)\,dx \approx \sum_{j=1}^{n} H_i f(a_j)$

Weight function $w(x)$	Interval $[a, b]$	Abscissas a_j are zeros of	Weights H_j given by Eq.	Error given by Eq.
1	$[-1, 1]$	$P_n(x)$	(4.6-11)	(4.6-12)
e^{-x}	$(0, \infty)$	$L_n(x)$	(4.7-2)	(4.7-3)
e^{-x^2}	$(-\infty, \infty)$	$H_n(x)$	(4.7-6)	(4.7-7)
$(1-x)^\alpha(1+x)^\beta$	$[-1, 1]$	$J_n(x; \alpha, \beta)$	(4.8-4)	(4.8-5)
$\dfrac{1}{(1-x^2)^{1/2}}$	$[-1, 1]$	$T_n(x) = \cos\dfrac{(2j-1)\pi}{2n}$	(4.8-15)	(4.8-11)
$(1-x^2)^{1/2}$	$[-1, 1]$	$S_n(x) = \cos\dfrac{j\pi}{n+1}$	(4.8-21)	(4.8-22)
$\dfrac{1}{\sqrt{x}}$	$[0, 1]$	$P_{2n}(\sqrt{x})$	(4.8-25)	(4.8-26)
$\sqrt{x}$	$[0, 1]$	$\dfrac{1}{\sqrt{x}}P_{2n+1}(\sqrt{x})$	(4.8-29)	(4.8-30)
$\left(\dfrac{x}{1-x}\right)^{1/2}$	$[0, 1]$	$\dfrac{1}{\sqrt{x}}T_{2n+1}(\sqrt{x})$	(4.8-33)	(4.8-34)

If all the derivatives of $f(z)$ at z_0 are bounded so that $|f^{(j)}(z_0)| < M$ for some M and all j, then from (4.9-1)

$$|f(z)| \le \sum_{j=0}^{\infty} |A_j|\,|z - z_0|^j < M \sum_{j=0}^{\infty} \frac{|z - z_0|^j}{j!} = Me^{|z - z_0|} \tag{4.9-2}$$

Since this implies the Taylor series converges for all finite z, $f(z)$ is entire.

When $f(z)$ is not entire, we can estimate the rate of growth of the derivatives by writing the coefficients in (4.9-1) in the form

$$A_j = \frac{1}{2\pi i}\oint_C \frac{f(z)}{(z - z_0)^{j+1}}\,dz \tag{4.9-3}$$

where C is any circle centered at z_0 of radius less than the radius of convergence R of the Taylor series. If L is a bound on the magnitude of $f(z)$ in the circle $|z - z_0| < R$, then it follows from (4.9-3) that

$$|f^{(j)}(z_0)| = j!\,|A_j| \le \frac{Lj!}{R^j}$$

Therefore, the derivatives may grow eventually as fast as $j!/R^j$. Since in fact rapid growth of derivatives may start for quite low derivatives (see, for example, Example 4.5) and because of the difficulty in estimating high derivatives, there is a tendency not to use high-order quadrature formulas. But low-order ones may not be sufficiently accurate. Consider, for example, the error term in the Gauss-Legendre quadrature formula (4.6-12). Suppose $f(x)$ is such that we can estimate its derivatives only up to order 4 but that for $n = 1$ and 2 the bound on E does not assure us of the accuracy we desire. To avoid this predicament, we can proceed as follows: (1) break up the interval $[a, b]$ into a number, say m, of subintervals; (2) on each subinterval apply a quadrature formula and sum the results.

The effectiveness of this technique depends upon the fact (which may have been obscured by our use of the interval $[-1, 1]$) that the error in all the finite-interval quadrature formulas we have developed is proportional to some power of the length of the interval. To see this, consider the problem of evaluating

$$\int_a^b f(x)\,dx \tag{4.9-4}$$

by Gauss-Legendre quadrature. We first make the change of variable

$$y = \frac{1}{b-a}(2x - a - b) \qquad x = \frac{1}{2}[(b-a)y + a + b] \tag{4.9-5}$$

which changes (4.9-4) to

$$\int_a^b f(x)\,dx = \frac{b-a}{2}\int_{-1}^{1} g(y)\,dy \tag{4.9-6}$$

with
$$g(y) = f[\tfrac{1}{2}(b-a)y + \tfrac{1}{2}(a+b)]$$

When Gauss-Legendre quadrature is used, the error in $\int_{-1}^{1} g(y)\,dy$ is given by (4.6-12) as

$$E = \frac{2^{2n+1}(n!)^4}{(2n+1)[(2n)!]^3} g^{(2n)}(\bar{\eta}) \qquad \bar{\eta} \in (-1, 1) \tag{4.9-7}$$

where the derivative is with respect to y. But using (4.9-5), we have

$$\frac{d}{dy} g(y) = \frac{dx}{dy}\frac{d}{dx} f(x) = \frac{b-a}{2} f'(x) \tag{4.9-8}$$

Thus the error incurred in $\int_a^b f(x)\,dx$ is

$$E = \left(\frac{b-a}{2}\right)^{2n+1} \frac{2^{2n+1}(n!)^4}{(2n+1)[(2n)!]^3} f^{(2n)}(\eta) \qquad \eta \in (a, b) \tag{4.9-9}$$

If now we write

$$\int_a^b f(x)\,dx = \int_a^{(a+b)/2} f(x)\,dx + \int_{(a+b)/2}^{b} f(x)\,dx \tag{4.9-10}$$

and apply to each of the integrals on the right-hand side of (4.9-10) a Gauss-Legendre formula with n points, it follows as above that the error is

$$E = \frac{1}{2^{2n+1}}\left(\frac{b-a}{2}\right)^{2n+1} \frac{2^{2n+1}(n!)^4}{(2n+1)[(2n)!]^3}[f^{(2n)}(\eta_1) + f^{(2n)}(\eta_2)]$$

$$\eta_1 \in \left(a, \frac{a+b}{2}\right);\ \eta_2 \in \left(\frac{a+b}{2}, b\right) \qquad (4.9\text{-}11)$$

which, assuming continuity of the $2n$th derivative of $f(x)$, can be written

$$E = \frac{1}{2^{2n}}\left(\frac{b-a}{2}\right)^{2n+1} \frac{2^{2n+1}(n!)^4}{(2n+1)[(2n)!]^3} f^{(2n)}(\eta_3) \qquad \eta_3 \in (a, b) \qquad (4.9\text{-}12)$$

Thus dividing the interval in two and using an n-point formula in both intervals has brought an extra factor of $1/2^{2n}$ into the error term, leaving everything else the same [except that the $2n$th derivative is evaluated at different points in (4.9-9) and (4.9-12)]. If, instead of two intervals, we had divided $[a, b]$ into m intervals, the added factor would have been $1/m^{2n}$ {36}. Thus the procedure outlined above is indeed an effective way of performing numerical quadrature more accurately by adding abscissas to the interval $[a, b]$ but without increasing the order of the quadrature formula used.

Performing the integral over $[a, b]$ of $f(x)$ by the method outlined here is familiar to all users of the trapezoidal and parabolic rules, which we shall consider in Sec. 4.10-1. As an example in the Gaussian context let us consider evaluating (4.9-4) by dividing $[a, b]$ into m intervals and using a Gauss-Legendre quadrature formula with $n = 2$ over each interval. Using Table 4.1, the integral is approximated as

$$\int_a^b f(x)\,dx \approx \frac{h}{2}\sum_{j=0}^{m-1} f\left\{a + \frac{h}{2}\left[1 + \frac{1}{\sqrt{3}} + 2j\right]\right\}$$
$$+ f\left\{a + \frac{h}{2}\left[1 - \frac{1}{\sqrt{3}} + 2j\right]\right\} \qquad (4.9\text{-}13)$$

where $h = (b-a)/m$. The error in (4.9-13) is given, using (4.9-12) with 2^{2n} replaced by m^{2n}, by

$$E = \frac{(b-a)^5}{4320m^4} f^{\text{iv}}(\eta) = \frac{mh^5}{4320} f^{\text{iv}}(\eta) \qquad (4.9\text{-}14)$$

A quadrature formula of the type (4.9-13), which is the sum of a number of quadrature formulas over subintervals, is called a *composite quadrature formula*, a *composite rule*, or sometimes a *compound rule*. From (4.9-14) we see that if $f^{\text{iv}}(\eta)$ is bounded in $[a, b]$, by increasing the number of subintervals we can make the error arbitrarily small. In general, if the derivative in the error term is bounded on the interval of integration, then, in the limit as the number of subintervals $m \to \infty$, the approximation must converge to the true value of the integral as $1/m^{2n}$. Indeed even if this derivative is not bounded, any composite rule will converge to the integral of a Riemann-integrable function provided only that the sum of the weights of the underlying rule equals the length of the basic integration interval and that the abscissas lie in that interval. This is true since any composite rule can be written as follows as a weighted sum of Riemann sums. Let

$$Q(f) = \sum_{i=1}^{n} H_i f(x_i) \qquad \sum_{i=1}^{n} H_i = 1 \qquad 0 \le x_i \le 1 \tag{4.9-15}$$

be a quadrature rule approximating $I(f) = \int_0^1 f(x)\,dx$. Then the composite rule using m subintervals $Q_m(f)$ has the form

$$Q_m(f) = \sum_{j=0}^{m-1} \sum_{i=1}^{n} \frac{H_i}{m} f\left(\frac{1}{m}(x_i + j)\right) = \sum_{i=1}^{n} H_i \sum_{j=0}^{m-1} \frac{1}{m} f\left(\frac{1}{m}(x_i + j)\right) \tag{4.9-16}$$

The right-hand formula is a weighted sum of the Riemann sums

$$S_i = \sum_{j=0}^{m-1} \frac{1}{m} f\left(\frac{1}{m}(x_i + j)\right) \qquad i = 0, \ldots, m-1 \tag{4.9-17}$$

Hence, if S_i converges to $I(f)$ as $m \to \infty$ for all j, then $Q_m(f)$ will also converge to $I(f)$ since

$$\lim_{m\to\infty} Q_m(f) = \lim_{m\to\infty} \sum_{i=1}^{n} H_i S_i = \sum_{i=1}^{n} H_i\left(\lim_{m\to\infty} S_i\right) = \sum_{i=1}^{n} H_i I(f) = I(f) \tag{4.9-18}$$

For this reason composite quadrature formulas are used in most numerical quadrature work.

A further advantage of composite formulas will be indicated in Secs. 4.10-1 and 4.10-2, where we shall develop a method to get a more accurate result than that given by two or more composite quadratures each using a different number of subintervals.

Example 4.7 Evaluate the integral of Example 4.3 using (4.9-13) with $m = 2$.

We have $h = 1$; therefore,

$$\int_1^3 \frac{dx}{x} \approx \frac{1}{2} \sum_{j=0}^{1} \left\{ \frac{1}{1 + \frac{1}{2}\left[\left(1 + \frac{1}{\sqrt{3}}\right) + 2j\right]} + \frac{1}{1 + \frac{1}{2}\left[\left(1 - \frac{1}{\sqrt{3}}\right) + 2j\right]} \right\}$$

$$\approx 1.097713$$

Using (4.9-14), the error is bounded by

$$.000046 \approx \frac{1}{90}\frac{1}{(3)^5} < E < \frac{1}{90} \approx .011111$$

and, in fact, we have achieved a substantially more accurate result than in Example 4.3 at the cost of using four abscissas instead of three.

4.10 NEWTON-COTES QUADRATURE FORMULAS

We noted in Sec. 4.4 the advantage of having abscissas and weights which are simple numbers when using a desk calculator. To the reasons for this adduced previously we might add that simplicity of numbers also serves to reduce blunders. In this section we shall develop a class of quadrature formulas that are ideal in this sense for hand-calculator work. Most readers are undoubtedly familiar with the most common members of this class—the trapezoidal formula and Simpson's rule. It should not be thought, however, that this class of formulas is applicable mainly for hand rather than automatic computation. Frequently a member of this class is the best method to use on a digital computer.

A quadrature formula in which the abscissas are constrained to be equally spaced is called a *Newton-Cotes quadrature formula*. They are thus, in particular, ideally suited for the numerical quadrature of tabulated functions. Newton-Cotes formulas of practical value fall into one of two classes: (1) closed formulas, where the endpoints of the interval are abscissas, and (2) open formulas, where the endpoints are not abscissas and the other abscissas are symmetrically placed with respect to the endpoints. Other types, such as half-open formulas, are possible but are seldom used in practice for quadrature.

Closed Newton-Cotes quadrature formulas have the form

$$\int_a^b w(x)f(x)\,dx = \sum_{j=0}^{n} H_j\, f(a+hj) + E \tag{4.10-1}$$

where $h = (b-a)/n$ and $w(x)$ is a weight function. Thus (4.10-1) involves $n+1$ abscissas; we have summed from 0 to n (in contrast to our previous notation) for notational simplicity in what follows in this section. Also for convenience of notation let $a_j = a + hj$ so that $a = a_0$ and $b = a_n$. Then (4.10-1) becomes

$$\int_{a_0}^{a_n} w(x)f(x)\,dx = \sum_{j=0}^{n} H_j\, f(a_j) + E \tag{4.10-2}$$

We have the $n+1$ weights H_j at our disposal, so that we expect to be able to make (4.10-2) exact for polynomials of degree n or less. In fact we shall see that when n is even, we get exactness for polynomials of degree $n+1$ also {38} provided the weight function $w(x)$ is symmetric on the interval $[a_0, a_n]$.

To determine the weights, we use the Lagrangian interpolation formula with the abscissas chosen as above, multiply both sides by $w(x)$, and integrate between a_0 and a_n:

$$\int_{a_0}^{a_n} w(x)f(x)\,dx = \sum_{j=0}^{n} H_j\, f(a_j) + \frac{1}{(n+1)!}\int_{a_0}^{a_n} w(x)p_{n+1}(x)f^{(n+1)}(\xi)\,dx \tag{4.10-3}$$

where $$H_j = \int_{a_0}^{a_n} w(x)l_j(x)\,dx \tag{4.10-4}$$

and $p_{n+1}(x) = (x - a_0)\cdots(x - a_n)$. Because of the $(n + 1)$st derivative in the error term, the H_j's given by (4.10-4) make (4.10-2) exact for polynomials of degree n or less. These H_j's are called *Cotes numbers.*

The simplification of the error term in (4.10-3) is considerably more difficult than before because even if $w(x)$ does not change sign in $[a_0, a_n]$, $p_{n+1}(x)$ does. Thus the second law of the mean cannot be applied. In order to simplify the error, we assume that $w(x)$ is an even function with respect to the center of the interval, that is, $w[x - (a_0 + a_n)/2] = w[(a_0 + a_n)/2 - x]$, and consider first the case where n is even. Integrating by parts, we have

$$\begin{aligned} E &= \frac{1}{(n+1)!}\int_{a_0}^{a_n} w(x)p_{n+1}(x)f^{(n+1)}(\xi)\,dx \\ &= -\frac{1}{(n+1)!}\int_{a_0}^{a_n} q(x)\frac{d}{dx}[f^{(n+1)}(\xi)]\,dx \end{aligned} \tag{4.10-5}$$

where $$q(x) = \int_{a_0}^{x} w(t)p_{n+1}(t)\,dt \tag{4.10-6}$$

The part integrated out in (4.10-5) is zero because $q(a_0) = 0$ and $q(a_n)$ is also zero. The latter follows because $p_{n+1}(x)$ is an odd function of $x - (a_0 + a_n)/2$ (why?), and we have assumed $w(x)$ is an even function of the same argument.

In the remainder of this section we shall consider the particular case $w(x) = 1$. We leave to a problem {39} the result that in this case $q(x)$ is of constant sign in $[a_0, a_n]$. Then since it can be shown that $(d/dx)[f^{(n+1)}(\xi)] = [1/(n + 2)]f^{(n+2)}(\eta)$, with η in the same interval as ξ, [Ralston (1963)] we get the result that

$$E = -\frac{f^{(n+2)}(\eta)}{(n+2)!}\int_{a_0}^{a_n} q(x)\,dx = \frac{f^{(n+2)}(\eta)}{(n+2)!}\int_{a_0}^{a_n} xp_{n+1}(x)\,dx \tag{4.10-7}$$

the latter result following from an integration by parts {39}. Therefore, with n even the Newton-Cotes closed formula with $w(x) = 1$ is exact for polynomials of degree $n + 1$ or less. When n is odd, the derivation is somewhat more difficult {40}; the result is that

$$E = \frac{f^{(n+1)}(\eta)}{(n+1)!}\int_{a_0}^{a_n} p_{n+1}(x)\,dx \tag{4.10-8}$$

which is the result that would be obtained if the second law of the mean could in fact be applied to the error term in (4.10-3).

For the case $w(x) = 1$ the weights are given in Table 4.5 for values of n from 1 to 8 (where $n + 1$ is the number of abscissas). The value of H_j in (4.10-2) is given by hAW_j, where h is the spacing of the abscissas. Also given are the coefficients of $h^{k+1}f^{(k)}(\eta)$ in the error term, where $k = n + 1$ if n is odd

Table 4.5 Weights and error-term coefficients for Newton-Cotes closed formulas

n	A	W_0	W_1	W_2	W_3	W_4	Error coefficient
1	1/2	1	1				−1/12
2	1/3	1	4	1			−1/90
3	3/8	1	3	3	1		−3/80
4	2/45	7	32	12	32	7	−8/945
5	5/288	19	75	50	50	75	−275/12,096
6	1/140	41	216	27	272	27	−9/1400
7	7/17,280	751	3577	1323	2,989	2989	−8183/518,400
8	4/14,175	989	5888	−928	10,496	−4540	−2368/467,775

and $n + 2$ if n is even. Because of the symmetry of the weights, only a portion of the complete table need be given. The case $n = 1$ is the trapezoidal formula, and $n = 2$ is Simpson's rule.

For the same reasons discussed in the previous section, high-order Newton-Cotes formulas are seldom used. Note that some of the weights for $n = 8$ are negative. In fact, it can be shown {41} that only for $n \leq 7$ and $n = 9$ are all the weights positive. The sum of the weights is always the length of the interval (why?); therefore if some of the weights are negative, this adversely affects roundoff error [cf. (1.4-7)].

Open Newton-Cotes formulas have the form

$$\int_{a_0}^{a_n} w(x) f(x)\,dx = \sum_{j=1}^{n-1} H_j f(a_j) + E \tag{4.10-9}$$

In this case we have $n - 1$ weights. For n odd (4.10-9) is exact for polynomials of degree $n - 2$ or less, but for n even, as in the case of the closed formulas, we get a bonus in that (4.10-9) is exact for polynomials of degree $n - 1$ or less {42}. The derivation of the error term is similar to that for the closed formulas {42}.

Table 4.6 is the analog of Table 4.5 for open formulas. The error coefficients are the coefficients of $h^{k+1} f^{(k)}(\eta)$, where $k = n$ for n even and $n - 1$ for n odd. Note that the sign of the error coefficient in the open rules is the opposite of the sign in the closed rules.

For $n = 2$, we have the *midpoint rule*, which is better known in the form it takes when the interval has length h instead of $2h$

$$\int_a^{a+h} f(x)\,dx = hf\left(a + \frac{h}{2}\right) + \frac{h^3}{24} f''(\xi) \qquad a < \xi < a + h \tag{4.10-10}$$

The composite midpoint rule is competitive with the composite trapezoidal rule both with respect to efficiency and error term. It is preferable to the

Table 4.6 Weights and error-term coefficients for Newton-Cotes open formulas

n	A	W_1	W_2	W_3	Error coefficient
2	2	1			$\frac{1}{3}$
3	$\frac{3}{2}$	1	1		$\frac{3}{4}$
4	$\frac{4}{3}$	2	-1	2	$\frac{14}{45}$
5	$\frac{5}{24}$	11	1	1	$\frac{95}{144}$
6	$\frac{3}{10}$	11	-14	26	$\frac{41}{140}$

latter when the integrand has singularities at the endpoints of the integration interval. Aside from the midpoint rule, there is seldom any advantage in using a Newton-Cotes open-type formula in preference to the closed formula using the same number of abscissas (see Sec. 4.12), but as we shall see in Chap. 5, open-type formulas do have application to the numerical integration of ordinary differential equations. The weights in the other open-type formulas are all positive only for $n = 3$ and 5.

Newton-Cotes formulas with weight functions other than $w(x) = 1$ are sometimes of use also, particularly for singular integrands of the type considered in Sec. 4.8-3. We shall not consider such weight functions here but refer the reader to the references in the bibliographic notes.

4.10-1 Composite Newton-Cotes Formulas

The same reasons for wishing to use Gaussian quadrature formulas in composite rules apply to Newton-Cotes formulas. Moreover, there is one further advantage to composite rules involving closed Newton-Cotes formulas. Since the endpoints of each subinterval (except for the endpoints of the whole interval) are abscissas for two subintervals, the total number of abscissas used is not equal to the number m of subintervals times the number $n + 1$ of points in each subinterval but is $m - 1$ less than this number. This can be best illustrated using $n = 1$ in Table 4.5. Suppose the interval $[a, b]$ is divided into m subintervals of length h and we use (4.10-2) with $n = 1$ on each subinterval. Then our approximation is

$$\int_a^b f(x)\,dx \approx h(\tfrac{1}{2}f_0 + f_1 + f_2 + \cdots + f_{m-2} + f_{m-1} + \tfrac{1}{2}f_m) \qquad (4.10\text{-}11)$$

where $f_j = f(a + hj)$ and $h = (b - a)/m$. Although there are m subintervals and each application of (4.10-2) uses two abscissas, there are only $m + 1$ abscissas in (4.10-11). Equation (4.10-11) is, of course, the trapezoidal rule.

The error is given by

$$E = -\frac{(b-a)^3}{12m^2} f''(\eta) = -\frac{mh^3}{12} f''(\eta) = -\frac{(b-a)h^2}{12} f''(\eta) \quad (4.10\text{-}12)$$

where η is in (a, b). As in the case of (4.9-13), the error can be made arbitrarily small by making m sufficiently large [assuming $f''(\eta)$ is bounded in (a, b)†].

As another example, we divide the interval $[a, b]$ into $m/2$ intervals of length $2h$ (m even) and use (4.10-2) with $n = 2$ over each subinterval. Then we get

$$\int_a^b f(x)\,dx \approx \frac{h}{3}(f_0 + 4f_1 + 2f_2 + 4f_3 + \cdots + 4f_{m-3} + 2f_{m-2} + 4f_{m-1} + f_m) \quad (4.10\text{-}13)$$

with $h = (b-a)/m$. The error is given by

$$E = -\frac{(b-a)^5}{180m^4} f^{\text{iv}}(\eta) = -\frac{mh^5}{180} f^{\text{iv}}(\eta) = -\frac{(b-a)h^4}{180} f^{\text{iv}}(\eta) \quad (4.10\text{-}14)$$

Equation (4.10-13) is the parabolic rule.

Similar composite rules could be derived using other Newton-Cotes closed formulas or using Newton-Cotes open formulas, although in the latter case, because the endpoints are not abscissas, we do not get a reduction in the total number of abscissas. One advantage of the low-order composite rules, particularly the trapezoidal rule, is the lack of fluctuation in the coefficients which results in good roundoff properties [cf. (1.4-7)]. When an integral is approximated using a large value of m in a composite rule, roundoff error may become a significant factor. Since the expected roundoff error will be minimized when the coefficients are most nearly equal, the trapezoidal rule has almost ideal roundoff properties. Higher-order Newton-Cotes formulas in which some of the weights may even be negative have quite bad roundoff properties. Note that the Gaussian composite rule (4.9-13) has ideal roundoff properties.

Suppose we use (4.10-11) to estimate the $\int_a^b f(x)\,dx$ for two separate values of m, m_1 and m_2. Let I be the true value of the integral and I_1 and I_2 the respective approximations. Then, using (4.10-12),

$$I = I_1 - \frac{(b-a)^3}{12m_1^2} f''(\eta_1) \qquad I = I_2 - \frac{(b-a)^3}{12m_2^2} f''(\eta_2) \quad (4.10\text{-}15)$$

where η_1 and η_2 are both in (a, b). Now suppose we assume that the two second derivatives are equal and eliminate these derivatives from (4.10-15),

† In fact, since the trapezoidal rule is a Riemann sum, it converges as $m \to \infty$ for any function $f(x)$ continuous on $[a, b]$.

obtaining

$$I \approx \frac{m_2^2 I_2 - m_1^2 I_1}{m_2^2 - m_1^2} = I_2 + \frac{m_1^2}{m_2^2 - m_1^2}(I_2 - I_1) \tag{4.10-16}$$

The value of this approximation clearly depends on how good the assumption is that the two second derivatives are equal. Suppose that we have a sequence of approximations to I, corresponding to an increasing number of subintervals, which appear to be converging monotonically. Then this assumption is probably good since the differences in the errors are probably being caused mainly by the m^2 term in the denominator of (4.10-12). Equation (4.10-16) is just another example of Richardson extrapolation, discussed in Sec. 4.2.

From a computational point of view, it is desirable to choose $m_2 = 2m_1$, for in this case all the abscissas used in the computation with m_1 subintervals are also abscissas for the m_2 subintervals calculation (why?), thus reducing the number of evaluations of $f(x)$. This consideration will be important in the next section.

Example 4.8 Evaluate the integral of Example 4.3 using (4.10-11) with $m = 2$ and 4 and then use (4.10-16) to obtain a third approximation.

When $m = 2$, we have

$$\int_1^3 \frac{dx}{x} \approx 1[\tfrac{1}{2}(\tfrac{1}{1}) + \tfrac{1}{2} + \tfrac{1}{2}(\tfrac{1}{3})] = \tfrac{7}{6}$$

When $m = 4$

$$\int_1^3 \frac{dx}{x} \approx \tfrac{1}{2}[\tfrac{1}{2} + \tfrac{2}{3} + \tfrac{1}{2} + \tfrac{2}{5} + \tfrac{1}{2}(\tfrac{1}{3})] = \tfrac{67}{60}$$

Then from (4.10-16)

$$\int_1^3 \frac{dx}{x} \approx \frac{67}{60} + \frac{4}{16-4}(-.05) = \tfrac{11}{10}$$

with the true result $\ln 3 = 1.098612$. Note that the Richardson extrapolation gives an improved value, but even this value is not so good as that using (4.9-13), where the error term has the fourth power of the number of subintervals in the denominator.

4.10-2 Romberg Integration

Any computation based on a fixed spacing h is then a candidate for Richardson extrapolation. To extend the procedure outlined here, we may consider performing M computations each using a different value of h, that is, a different number of subintervals in the quadrature context, and then eliminate the first $M - 1$ terms in (4.2-5). An intuitive objection to carrying this procedure too far would be that because of our discussion of high-order derivatives in Sec. 4.9, we would expect the higher powers of h in the error term to be accompanied by increasingly large coefficients. Therefore, we should not be surprised if this procedure did not converge. However, in the case of the trapezoidal rule, it does converge and leads to the following elegant and useful method of Romberg.

It is perhaps not obvious that the error in the trapezoidal rule can be expressed in the form (4.2-5). But, as we shall derive in Sec. 4.13-1 [cf. (4.13-17)], the trapezoidal rule can indeed be written as

$$\int_a^b f(x)\,dx = h(\tfrac{1}{2}f_0 + f_1 + f_2 + \cdots + f_{m-1} + \tfrac{1}{2}f_m) + \sum_{j=1}^{\infty} a_j h^{2j} \tag{4.10-17}$$

with $h = (b-a)/m$, where the a_j's depend only on a, b, and $f(x)$ [cf. (4.10-12)] and where the series on the right-hand side of (4.10-17) is asymptotic [cf. Example 4.11]. Now let

$$J = \int_a^b f(x)\,dx \tag{4.10-18}$$

and let

$$T_{0,k} = \frac{b-a}{2^k}(\tfrac{1}{2}f_0 + f_1 + f_2 + \cdots + f_{2^k-1} + \tfrac{1}{2}f_{2^k}) \tag{4.10-19}$$

be the trapezoidal-rule approximation for 2^k subintervals. Then

$$T_{0,k} = J - \sum_{j=1}^{\infty} a_j \left(\frac{b-a}{2^k}\right)^{2j} \tag{4.10-20}$$

We define

$$T_{1,k} = \tfrac{1}{3}(4T_{0,k+1} - T_{0,k}) \qquad k = 0, 1, \ldots \tag{4.10-21}$$

Using (4.10-20), we have

$$\begin{aligned} T_{1,k} &= J - \sum_{j=1}^{\infty} \frac{1}{3}\left(\frac{4}{2^{2j}} - 1\right) a_j \left(\frac{b-a}{2^k}\right)^{2j} \\ &= J - \sum_{j=2}^{\infty} \frac{1}{3}\left(\frac{4}{2^{2j}} - 1\right) a_j \left(\frac{b-a}{2^k}\right)^{2j} \qquad k = 0, 1, \ldots \end{aligned} \tag{4.10-22}$$

Equation (4.10-22) states that if we perform the trapezoidal rule to approximate J using a spacing $h_1 = (b-a)/2^{k+1}$ and $h_2 = (b-a)/2^k$, the resulting approximation has a leading term in the error of the order of h_2^4. This equation is therefore directly analogous to (4.2-7). The approximation $T_{1,k}$ is, in fact, precisely the parabolic rule for 2^k subintervals {47}.

Now, in general, we define

$$T_{m,k} = \frac{1}{4^m - 1}(4^m T_{m-1,k+1} - T_{m-1,k}) \qquad \begin{matrix} k = 0, 1, \ldots \\ m = 1, 2, \ldots \end{matrix} \tag{4.10-23}$$

The approximation $T_{2,k}$ is the composite rule found using the Newton-Cotes closed formula with $n = 4$ and 2^k subintervals, but for $m > 2$, there is no direct relation between $T_{m,k}$ and a Newton-Cotes composite rule {47}.

Using (4.10-24), we can write

$$T_{m,k} = \sum_{j=0}^{m} c_{m,m-j} T_{0,k+j} \tag{4.10-24}$$

That is, each $T_{m,k}$ is a linear combination of trapezoidal rules using $2^k, 2^{k+1}, \ldots, 2^{k+m}$ subintervals. Moreover, analogously to (4.10-22), we can show that the leading term in the error of $T_{m,k}$ as an approximation to J is of the order of $[(b-a)/2^k]^{2(m+1)}$ {47}. We would arrange the calculations as indicated in the following table:

$$\begin{array}{llllll}
T_{0,0} & & & & & \\
T_{0,1} & T_{1,0} & & & & \\
T_{0,2} & T_{1,1} & T_{2,0} & \ddots & & \\
\vdots & \vdots & & & \ddots & \\
T_{0,m} & T_{1,m-1} & \cdots & \cdots & \cdots & T_{m,0}
\end{array} \tag{4.10-25}$$

That is, using the $m+1$ evaluations of the trapezoidal rule in the first column, we can then, using (4.10-23), calculate all the remaining entries in (4.10-25). We know that if $f(x)$ is Riemann-integrable in $[a, b]$, then as $m \to \infty$, $T_{0,m}$ converges to J. But here we are interested in proving that as $m \to \infty$, $T_{m,k}$ converges to J. Note that since $T_{m,k}$ has a leading term in its error of the order of $[(b-a)/2^k]^{2(m+1)}$ while that of $T_{0,k}$ is $[(b-a)/2^k]^2$, we would hope that if $T_{m,k}$ converges, it would converge much more rapidly than $T_{0,k}$.

Using (4.10-24), we can write

$$\begin{bmatrix} T_{0,0} \\ T_{1,0} \\ \vdots \\ T_{m,0} \end{bmatrix} = \begin{bmatrix} c_{00} & & & \bigcirc \\ c_{11} & c_{10} & & \\ \vdots & & \ddots & \\ c_{mm} & \cdots & \cdots & c_{m0} \end{bmatrix} \begin{bmatrix} T_{0,0} \\ \vdots \\ \vdots \\ T_{0,m} \end{bmatrix} \tag{4.10-26}$$

We need first a means for generating the coefficients c_{mj}. We leave to a problem {48} the proof of the result that the c_{mj} are the coefficients of

$$t_m(z) = \sum_{j=0}^{m} c_{mj} z^j = \frac{(4-z)(4^2-z)\cdots(4^m-z)}{(4-1)(4^2-1)\cdots(4^m-1)} \tag{4.10-27}$$

In particular

$$\sum_{j=0}^{m} |c_{mj}| = t_m(-1) = \prod_{j=1}^{m} \frac{4^j+1}{4^j-1} < \prod_{j=1}^{\infty} \frac{4^j+1}{4^j-1} \tag{4.10-28}$$

Since this infinite product converges {48}, the sum of magnitudes of the elements in any row of (4.10-26) is bounded. It also follows from (4.10-27) that $c_{m,m-j} \to 0$ as $m \to \infty$ {48}; that is, each column in (4.10-26) converges to zero. Finally, $\sum_{j=0}^{m} c_{mj} = t_m(1) = 1$. The coefficient matrix in (4.10-26) therefore satisfies the conditions for Toeplitz convergence [see Zygmund (1952)], which means that since $T_{0,m}$ converges to J, so does $T_{m,0}$. In Example 4.10 in Sec. 4.12, we shall consider a problem in which $T_{m,0}$ not only converges but converges much more rapidly than $T_{0,m}$.

Equation (4.10-24) can also be written in terms of ordinates as

$$T_{m,k} = h \sum_{j=0}^{2^{m+k}} d_{jm} f(a + jh) \qquad h = \frac{b-a}{2^{m+k}} \tag{4.10-29}$$

We leave to a problem {49} the proof of the result that d_{jm} is positive for all j and m. Therefore, in contrast to the higher-order Newton-Cotes formulas, the coefficients in Romberg integration do not change sign.

We shall not derive the error term in Romberg integration here, but it can be shown [see Bauer, Rutishauser, and Stiefel (1963)] that this error can be expressed in the standard form of a constant times h^{2m+2} times a $2m+2$ derivative of $f(x)$ with h as in (4.10-29). The convergence of the sequence $\{T_{m,0}\}$ implies that this constant becomes small very rapidly (why?).

The Romberg integration scheme is of great theoretical interest and has also been used in some quadrature routines and even in some adaptive schemes; however, it turns out that the number of functional evaluations needed grows too rapidly with m. However, in the light of (4.2-10) we can choose a sequence $\{h_n\}$ other than the sequence $\{h/2^n\}$ used above and still have the benefits of Richardson extrapolation. The formulas for generating the T table (4.10-25) are only slightly more complicated than (4.10-23). The particular choice of $\{h_n\} = \{(b-a)/p_n\}$, where $\{p_n\} = \{1, 2, 3, 4, 6, 8, 12, \ldots\}$, has been recommended by Oliver (1971) as optimal in the sense of giving the best accuracy with the least amount of roundoff for a fixed amount of computation. Here, the p_n are all the integers of the form 2^k and $3(2^k)$, so that all evaluations of the integrand are used in the computation of succeeding trapezoidal sums.

4.11 ADAPTIVE INTEGRATION

Whenever we compute an integral numerically by some quadrature formula, we cannot know how good the approximation is unless we know something about the error. There are various ways to estimate the error. We can try to bound the derivative in the error term, but, as we noted earlier, this is usually unsatisfactory because, on the one hand, computing the derivatives of a complicated function or otherwise estimating them can become very tedious and, on the other hand, due to the variation in the higher derivatives, a bound on the derivative usually yields an error estimate far in excess of the true error, so that we end up evaluating the integrand at many more points than are actually needed. Another possibility is to compute a sequence of approximations to the integral and stop when two or three of them agree to the number of significant figures desired. This sequence may consist of composite rules of the same underlying primitive rule, Romberg approximations (as discussed in the previous section), Gauss rules of increasing order, etc., with an attempt made to try to use information generated previously in subsequent approximations. A variation of this idea is to compute sequences of pairs of comparable rules using different evaluation points and stop when there is agreement in one or several successive pairs. In all these schemes the integral is treated in a global manner; i.e., we always generate approximations over the original integration interval. Now, while it is relatively easy to implement these schemes, they may involve a considerable amount of unnecessary computation. By treating the entire interval in a uniform manner, they do not distinguish between the places where the integrand is well-behaved and those where it is not. Bad behavior in a small part of the interval can cause the numerical approximation over the entire interval to have a substantial error. A remedy for this situation would be to isolate those sections of the interval where the function is ill-behaved and work hard there while approximating the integral where the integrand is well-behaved with a relatively small amount of computation. The implementation of this idea leads to an *adaptive* integration scheme.

Many such schemes exist; we shall describe one in some detail and make some comments on others.

The scheme which we shall describe is an adaptive version of Simpson's rule which yields an approximation $I_{as}(f)$ to

$$I(f) = I_a^b(f) = \int_a^b f(x)\, dx \tag{4.11-1}$$

such that, it is hoped, either

$$|I(f) - I_{as}(f)| < \epsilon \tag{4.11-2}$$

or

$$|I(f) - I_{as}(f)| < \epsilon I(|f|) \tag{4.11-3}$$

for some preassigned tolerance $\epsilon > 0$. Here we shall use the absolute error estimate (4.11-2). The estimate (4.11-3) can be treated similarly with successively better approximations to $I(|f|)$ as more functional values are computed.

Our approximation $I_{as}(f)$ will have the form

$$I_{as}(f) = \sum_{i=0}^{n-1} R_2[a_i, a_{i+1}]f \tag{4.11-4}$$

where $R_2 f$ is the Simpson five-point rule, i.e., the parabolic rule with $m = 4$ [cf. (4.10-13)]:

$$R_2[a_i, a_{i+1}]f = \frac{h_i}{12}\left\{f(a_i) + 4f\left(a_i + \frac{h_i}{4}\right) + 2f\left(a_i + \frac{h_i}{2}\right) + 4f\left(a_i + \frac{3h_i}{4}\right) + f(a_{i+1})\right\} \tag{4.11-5}$$

where $h_i = a_{i+1} - a_i$ and the a_i's are a subdivision of $[a, b]$:

$$a = a_0 < a_1 < \cdots < a_{n-1} < a_n = b \tag{4.11-6}$$

The subintervals $[a_i, a_{i+1}]$ are, in general, of different length, but each has length $H/2^r$ for some r, where $H = b - a$. A subinterval of length $H/2^r$ is said to have *level* r. The various subintervals are determined by taking the level 0 interval $[a, b]$, subdividing it into two level 1 subintervals $[a, a + H/2]$, $[a + H/2, b]$, then perhaps subdividing either or both of these into two level 2 subintervals and continuing this way until (4.11-6) is obtained. Whether or not a given level r subinterval is divided into two level $r + 1$ subintervals is determined by a comparison of the result of (4.11-5) on this subinterval with the result of Simpson's rule:

$$R_1[a_i, a_{i+1}]f = \frac{h_i}{6}\left\{f(a_i) + 4\left(a_i + \frac{h_i}{2}\right) + f(a_{i+1})\right\} \tag{4.11-7}$$

as we shall now describe.

At a particular stage in the computation, suppose we have a subinterval $[a_i, a_i + h]$ at level r, so that the numbers $a_0, a_1, \ldots, a_i$ have already been determined and the approximation

$$\sum_{j=1}^{i} R_2[a_{j-1}, a_j]f \approx \int_a^{a_i} f(x)\,dx \tag{4.11-8}$$

has been accepted; i.e., the estimated error is less than $(a_i - a)\epsilon/H$. The rest of the values of the final subdivision are not yet available [although a subset of these numbers together with the corresponding function values is available and can be used, e.g., to estimate $I(|f|)$]. Now $R_1[a_i, a_i + h]$ is usually available from a calculation at a lower level, together with the values $f(a_i)$, $f(a_i + h/2)$, and $f(a_i + h)$. To get $R_2[a_i, a_i + h]f$ and compare it with $R_1[a_i, a_i + h]f$ we thus need to compute only two new values $f(a_i + h/4)$ and $f(a_i + 3h/4)$. We then check to see if

$$|R_2[a_i, a_i + h]f - R_1[a_i, a_i + h]f| < \epsilon(r) \tag{4.11-9}$$

where a choice for $\epsilon(r)$, described below, is given by (4.11-15). If so, we say that this interval has converged, we add $a_{i+1} = a_i + h$ to the list of accepted points, we add $R_2[a_i, a_{i+1}]f$ to the left-hand side of (4.11-8) to get an approximation to $I_a^{a_{i+1}}(f)$, and we go on to the next subinterval to the right of a_{i+1}.

If (4.11-9) does not hold, we subdivide $[a_i, a_{i+1}]$ into two subintervals $[a_i, a_i + h/2]$, $[a_i + h/2, a_i + h]$, on level $r + 1$, and start working on the interval $[a_i, a_i + h/2]$, saving all previously computed values for future use. Thus we proceed from left to right and store for subsequent attention various intervals to the right of $[a_i, a_i + h]$ at various levels between 1 and r. We finish when a subinterval $[a_{n-1}, b]$ converges.

Now we consider the choice of $\epsilon(r)$. If we can choose $\epsilon(r)$ such that

$$|I_{a_i}^{a_{i+1}}(f) - R_2[a_i, a_{i+1}]f| < \frac{\epsilon}{2^r} \tag{4.11-10}$$

where the subinterval $[a_i, a_{i+1}]$ is at level r, then it is easy to show {54} that $I_{as}(f)$ as given by (4.11-4) satisfies (4.11-2). To determine an appropriate $\epsilon(r)$, we consider the error in Simpson's rule on $[\alpha, \alpha + h]$:

$$I_\alpha^{\alpha+h}(f) - R_1[\alpha, \alpha + h]f = -\frac{h^5}{90}f^{(4)}(\xi_1) = -\frac{h^5}{90}f^{(4)}\left(\alpha + \frac{h}{2}\right) + O(h^6) \tag{4.11-11}$$

provided $f(x)$ has a continuous fifth derivative on $[\alpha, \alpha + h]$ and where $\alpha < \xi_1 < \alpha + h$. The error using the five-point rule is

$$\begin{aligned} I_\alpha^{\alpha+h}(f) - R_2[\alpha, \alpha + h]f &= -\frac{h^5}{1440}f^{(4)}(\xi_2) \\ &= -\frac{h^5}{1440}f^{(4)}\left(\alpha + \frac{h}{2}\right) + O(h^6) \\ \alpha < \xi_2 < \alpha + h \end{aligned} \tag{4.11-12}$$

If we neglect the terms in h^6, then

$$\begin{aligned} R_2[\alpha, \alpha + h]f - R_1[\alpha, \alpha + h]f &= \left(\frac{h^5}{1440} - \frac{h^5}{90}\right)f^{(4)}\left(\alpha + \frac{h}{2}\right) \\ &= -\frac{15h^5}{1440}f^{(4)}\left(\alpha + \frac{h}{2}\right) \end{aligned} \tag{4.11-13}$$

so that

$$|I - R_2| = \tfrac{1}{15}|R_2 - R_1| \tag{4.11-14}$$

Therefore, if $|R_2 - R_1| < 15\epsilon/2^r$, then $|I - R_2| < \epsilon/2^r$ and so we choose

$$\epsilon(r) = \frac{15\epsilon}{2^r} \tag{4.11-15}$$

The neglect of the $O(h^6)$ terms is compensated for by the fact that we are working with magnitudes of errors [cf. (4.11-10)], whereas the errors on the various subintervals are usually of both signs. Therefore there tends to be some cancellation so that the actual error neglecting $O(h^6)$ terms is usually much less than the tolerance ϵ. For further refinements of this particular scheme, see the references in the Bibliographic Notes.

If we analyze this scheme, we see that its components are a pair of quadrature rules with error estimates containing the same power of h and a way of choosing which subinterval to tackle next, both in case of success, i.e., convergence, and failure, i.e., no convergence. In the various adaptive quadrature schemes in existence, other pairs of rules or, more generally, sequences of rules are used, other methods of choosing the next subinterval are considered, and subdivisions other than bisection are proposed. For example, another choice for the next subinterval is that one on which the estimated error is maximal. It has been estimated that millions of reasonable adaptive-integration schemes can be constructed, each one having some claim to superior performance. However, any one of the routines currently available should do the trick for most of the problems arising in practice.

Of course, the scheme described in this section is not foolproof. Thus, if one knows how the routine works, one can cook up a function which will "beat" the routine by having it vanish at a sufficient number of appropriate points. This in turn can be countered by making a random division of the initial interval into two subintervals and applying the adaptive scheme separately on each of these subintervals. In general, as in so many problems in numerical analysis, there is a tradeoff between efficiency and *robustness*, where the latter is a quality of an algorithm which in almost every case either gives the correct answer within the accuracy desired or exits with an indication that it cannot achieve this accuracy. Only in very rare cases, therefore, will a robust algorithm give an incorrect answer and claim it to be correct. For one-shot problems, robustness is almost always more important than efficiency. But when a series of integrals is to be computed, efficiency becomes important. The user must then decide whether a saving of time is worth the risk of possibly incorrect results. If the integrands are well-behaved, this risk will be very slight. Only in the case of difficult integrands does it become difficult to decide.

Example 4.9 Use adaptive integration to calculate

$$\int_0^{10} e^x \, dx = e^{10} - 1 = 22{,}025.46579$$

with an error of less than .001.

Using the error criterion (4.11-15) and using R_2 and R_1 as defined in (4.11-5) and (4.11-7), we require 47 abscissas to calculate the integral. These are

0 to $3\frac{1}{8}$ at a spacing of $\frac{5}{8}$
$3\frac{1}{8}$ to $6\frac{1}{4}$ at a spacing of $\frac{5}{16}$
$6\frac{1}{4}$ to $8\frac{29}{32}$ at a spacing of $\frac{5}{32}$
$8\frac{29}{32}$ to 10 at a spacing of $\frac{5}{64}$

The calculated integral is 22,025.46607

which differs from the desired value by .00028. If, using (4.11-14), the error is estimated at each stage and these errors are accumulated, the estimated error in the result is $-.00028$ which, added to the calculated result, would give full accuracy to the 10 significant digits carried in the calculation.

4.12 CHOOSING A QUADRATURE FORMULA

If we have no more than a few integrals to compute *and* the integrand can be evaluated at any point, the use of an adaptive-quadrature routine is indicated and any fully tested one will do. If the above situation does not hold, we must make some decisions about choosing a quadrature rule.

Table 4.7 Error-term coefficients in Newton-Cotes closed and open formulas

Number of abscissas	Error coefficient	
	Closed	Open
2	$-1/12$	$1/36$
3	$-1/2880$	$7/23{,}040$
4	$-1/6480$	$19/90{,}000$
5	$-1/1{,}935{,}360$	$41/39{,}191{,}040$

Consider first the problem of tabulating the integral of a tabulated function. Because the data are available at equal intervals, we would almost certainly choose a Newton-Cotes formula in order to avoid interpolation. If derivatives of the tabulated function can be evaluated, the Euler-Maclaurin formula of the following section can be used with good results. It gives very good accuracy with only a small additional computational effort, namely the computation of derivatives of odd order at the endpoints of the integration interval. If the derivatives can only be bounded, and if the derivatives do not grow too rapidly, we can choose n, the order of the Newton-Cotes formula, so that the error bound assures us of the accuracy we desire.

We can, of course, use open or closed Newton-Cotes formulas. Tables 4.5 and 4.6 might lead us to believe that, for the same number of abscissas, the closed-type formula is clearly superior. Consider the case where the number of abscissas is 3. Using $n = 2$ in Table 4.5, the error is $-\frac{1}{90}h^5 f^{\text{iv}}(\eta)$ and using $n = 4$ in Table 4.6, the error is $\frac{14}{45}h^5 f^{\text{iv}}(\eta)$. But note that h is $(b - a)/2$ in the first case and h is $(b - a)/4$ in the second. So, in fact, the errors are, respectively, $-[(b - a)^5/2880]f^{\text{iv}}(\eta)$ and $[7(b - a)^5/23{,}040]f^{\text{iv}}(\eta)$, and the open formula is slightly better. In Table 4.7 the coefficients of the corresponding error terms, with h replaced by $(b - a)/n$, are given for n equals 2 to 5. For 2 and 3, the open formula is slightly better, but for 4 and 5 the closed formula is better; for $n > 5$ the advantage of the closed formula becomes more marked.

If we cannot bound the derivatives, then no matter what order Newton-Cotes formula we use, we have no assurance that we shall achieve the desired accuracy. Here the safest technique is to use a composite formula and increase the number of subintervals until the number of decimal places of accuracy that we desire has stabilized, i.e., remains unchanged, in two successive approximations.

Example 4.10 Use the trapezoidal rule to evaluate the integral of Example 4.3 with an error of less than $\frac{1}{2}$ in the fourth decimal place.

In Example 4.8 we used the trapezoidal rule with $m = 2$ and 4 and got the results

$m = 2$: 1.166667
$m = 4$: 1.116667

In order to use those abscissas at which $f(x)$ has already been calculated, we continue to double the number of subintervals. The results are

$m = 8$: 1.103211
$m = 16$: 1.099768
$m = 32$: 1.098902
$m = 64$: 1.098685
$m = 128$: 1.098630

At this point the third decimal place has surely stabilized and the fourth appears to be in error by no more than $\frac{1}{2}$ unit. Applying (4.10-15) to the results for $m = 64$ and $m = 128$, we get $I \approx 1.098612$ which is, in fact, accurate to six decimal places. Applying Romberg integration to this problem, we get for the table (4.10-25)

1.333333			
1.166667	1.111111		
1.116667	1.100000	1.099259	
1.103211	1.098726	1.098641	1.098631
1.099768	1.098620	1.098613	1.098613
1.098902	1.098613	1.098613	1.098613
1.098685	1.098613	1.098613	1.098613
1.098630	1.098612	1.098612	1.098612

so that, except for round-off error, convergence is achieved after the use of only 16 subintervals (fifth line of table).

Of course, higher-order composite rules than the trapezoidal rule can also be used. The parabolic rule, for example, ultimately converges more rapidly than the trapezoidal rule because of the $1/m^4$ term in (4.10-14) in contrast to the $1/m^2$ term in (4.10-12). Because of the high-order derivative in the error term, higher-order composite rules, i.e., composite rules derived from higher-order Newton-Cotes formulas, are seldom used. Indeed our general conclusion is that the simplicity and rapid convergence of Romberg integration make the trapezoidal rule the best Newton-Cotes formula to use for most problems.

It is important to realize that because of the results of our discussion in Sec. 4.9 on the growth of higher-order derivatives, it is not true in general that any desired degree of accuracy can be achieved by an arbitrarily high-order formula. To illustrate this, following Hildebrand (1974), let us consider

$$\int_{-4}^{4} \frac{dx}{1 + x^2} = 2 \tan^{-1} 4 \approx 2.6516353$$

In Table 4.8 the results of evaluating this integral using varying numbers of abscissas and various closed Newton-Cotes formulas are shown. It is reasonably clear that the Newton-Cotes formulas of high order are not converging. In contrast, the parabolic rule is oscillating about the true solution with

Table 4.8 Evaluation of $\int_{-4}^{4} \frac{dx}{1+x^2}$ by Newton-Cotes techniques

Number of abscissas n	Newton-Cotes formula of order n	Parabolic rule	Trapezoidal rule	Romberg integration
3	5.490	5.490	4.235	5.490
5	2.278	2.478	2.918	2.278
7	3.329	2.908	2.701	
9	1.941	2.573	2.659	2.584
11	3.596	2.695	2.6511	
17		2.6477	2.6505	2.6542
33		2.651627	2.65135	2.65186
65		2.6516353	2.65156	2.651631
129		2.6516353	2.651617	2.6516353

decreasing amplitude, and the trapezoidal rule is converging monotonically to the true result. Both the parabolic and trapezoidal rules must converge to the true solution with increasing n, and eventually, the parabolic rule must, as it does, give better results than the trapezoidal rule (why?). For purposes of comparison, Table 4.8 also contains the results using Romberg integration. For 17 or more abscissas, Romberg gives better results than the trapezoidal rule, and for 129 or more abscissas, it would give better results than the parabolic rule except for roundoff.

The singularity of the integrand at $x = \pm i$, which limits the radius of convergence of the Taylor-series expansion, is the cause of the difficulty with the higher-order Newton-Cotes formulas. But the reader will agree that the example chosen is not so contrived that we would expect occurrence of this phenomenon to be rare.

If the function whose integral we wish to calculate is not tabulated but is given analytically and we must calculate many integrals of the same type, i.e., over the same interval and with similar integrands, then we must consider Gaussian as well as Newton-Cotes formulas for use on an automatic computer. If a single quadrature formula—perhaps of high order—is to be used, the case for using some Gaussian formula is very good. The reason for this is that the Gaussian formulas achieve higher accuracy with the use of fewer abscissas than Newton-Cotes methods. This is illustrated in Table 4.9 for the Gauss-Legendre formula. The integral in all cases is considered to be over the interval $[-1, 1]$. The order of accuracy n is the highest-degree polynomial integrated exactly by the method, and the error coefficient is the coefficient of $f^{(n+1)}(\eta)$ in the error term. Thus not only do the Gaussian formulas require calculation of fewer values of $f(x)$ but also they have slightly more favorable error terms. In this case then, the general superiority of the Gaussian formulas is clear.

Table 4.9 Comparison of accuracy of Gaussian and Newton-Cotes (closed) formulas

Order of accuracy n	Gauss-Legendre formula: Number of abscissas	Gauss-Legendre formula: Error coefficient	Newton-Cotes formula: Number of abscissas	Newton-Cotes formula: Error coefficient
3	2	1/135	3	$-1/90$
5	3	1/15,750	5	$-1/15{,}120$
7	4	1/3,472,875	7	$-1/3{,}061{,}800$
9	5	8.08×10^{-10}	9	-1.21×10^{-9}

Furthermore, in contrast to the Newton-Cotes case, if $f(x)$ is continuous and $w(x)$ is any integrable, nonnegative weight function, then any desired degree of accuracy is obtainable by using a sufficiently high-order Gaussian quadrature formula. To see this let $\epsilon > 0$ be given and let $p(x)$ be a polynomial, say of degree N, such that $|f(x) - p(x)| < \epsilon$ on $[a, b]$. The existence of such a polynomial is guaranteed by the Weierstrass approximation theorem (page 36). Now, using the notation of (4.5-1),

$$\begin{aligned} |E| &= \left| \int_a^b w(x)f(x)\,dx - \sum_{j=1}^n H_j f(a_j) \right| \\ &\le \left| \int_a^b w(x)[f(x) - p(x)]\,dx \right| \\ &\quad + \left| \int_a^b w(x)p(x)\,dx - \sum_{j=1}^n H_j p(a_j) \right| \\ &\quad + \left| \sum_{j=1}^n H_j[p(a_j) - f(a_j)] \right| \end{aligned} \tag{4.12-1}$$

If $2n - 1 \ge N$, the second term on the right is zero (why?). Then, since the sum of the weights in any Gaussian quadrature formula is the integral of the weight function over the interval [to see this let $f(x) = 1$ in (4.5-1)], and since all the Gaussian weights are positive, we have

$$|E| \le \epsilon \int_a^b w(x)\,dx + \epsilon \int_a^b w(x)\,dx = 2\epsilon \int_a^b w(x)\,dx \tag{4.12-2}$$

which proves the assertion that the error can be made arbitrarily small. Therefore, using a sequence of Gaussian formulas with increasing n will lead to a convergent sequence of approximations. However, because it is inconvenient to have to store a sequence of Gaussian weights and abscissas in a computer,† because it is difficult to estimate the magnitude of the error term

† It is also possible to generate them internally at a cost of $O(n^2)$ operations.

for all but quite small n, and because of the efficiency of using composite rules, this approach is less used in practice than it should be. Note that this proof of convergence is valid for any sequence of quadrature formulas of increasing order provided only that the weights are all positive, a condition which is not satisfied by Newton-Cotes formulas. But this condition is satisfied by Romberg integration [cf. (4.10-29)], and therefore the above argument also provides a proof of the convergence of the Romberg technique.

Thus, when there are many similar integrals to calculate, our strategy would be to do first an exploratory computation on a few integrals to determine the order of the Gaussian formula which gives the desired accuracy and then to use this formula for the entire series of integrals. If one wants to check the accuracy, one can do so, for example, with a second Gaussian formula of higher order.

We should also mention the possibility of transforming the integrand by a change of the independent variable. This can sometimes be used to eliminate the singularity in an improper integral. For example, if $f(x)$ is continuous in $[0, 1]$, the change of variable $t^n = x$ transforms the integral

$$\int_0^1 x^{-1/n} f(x)\, dx \qquad n \geq 2 \tag{4.12-3}$$

into

$$n \int_0^1 f(t^n) t^{n-2}\, dt \tag{4.12-4}$$

which is a proper integral. A more generally applicable transformation is

$$\int_a^b f(x)\, dx = \int_{g(a)}^{g(b)} f(g(t)) g'(t)\, dt \tag{4.12-5}$$

where $g(t)$ is a continuously differentiable function whose derivative does not vanish on $[a, b]$. By a proper choice of the function $g(t)$ it may be possible to evaluate the transformed integral more efficiently than the original one.

In the present chapter we have covered only the basic topics in numerical quadrature. Thus we have not mentioned the difficult problem of integration of highly oscillatory integrands and have only briefly touched upon infinite intervals and integrands with singularities. Nor have we mentioned multiple integrals. Finally we have not gone deeply into the important question of error analysis. Let us only point out that the Peano theorem (Sec. 2.3) has proved very useful for estimating the errors of quadrature formulas. Details of the above topics and of others can be found in the literature cited in the Bibliographic Notes.

4.13 SUMMATION

Our interest in this section is in approximating the sum

$$\sum_{j=0}^{n} f(x_0 + jh) \tag{4.13-1}$$

where n may be infinite. As a by-product of the development of methods to do this, we shall arrive at another and useful technique for numerical quadrature.

4.13-1 The Euler-Maclaurin Sum Formula

We begin by introducing the *Bernoulli polynomials* $B_k(x)$, defined to be the polynomials of degree k which are the coefficients of $t^k/k!$ in the expansion

$$\frac{t(e^{xt} - 1)}{e^t - 1} = \sum_{k=0}^{\infty} B_k(x) \frac{t^k}{k!} \tag{4.13-2}$$

from which it follows that

$$\begin{aligned} B_k(0) &= 0 \qquad k \geq 0 \\ B_k(1) &= 0 \qquad k \neq 1 \end{aligned} \tag{4.13-3}$$

By expanding the left-hand side of (4.13-2) in a power series we can compute the first few Bernoulli polynomials as

$$\begin{aligned} B_0(x) &= 0 \qquad & B_1(x) &= x \\ B_2(x) &= x^2 - x \qquad & B_3(x) &= x^3 - \frac{3x^2}{2} + \tfrac{1}{2}x \end{aligned} \tag{4.13-4}$$

The expansion of the left-hand side of (4.13-2) with the $e^{xt} - 1$ term deleted can be written

$$\frac{t}{e^t - 1} = \sum_{k=0}^{\infty} B_k \frac{t^k}{k!} \tag{4.13-5}$$

where the constants B_k are the *Bernoulli numbers*.† Of the many identities involving Bernoulli polynomials and numbers, the following three are of particular interest to us:

$$B_{2k+1} = 0 \qquad k > 0 \tag{4.13-6}$$

$$B'_{2k}(x) = 2kB_{2k-1}(x) \qquad k > 1 \tag{4.13-7}$$

$$B'_{2k+1}(x) = (2k + 1)[B_{2k}(x) + B_{2k}] \qquad k \geq 0 \tag{4.13-8}$$

† The Bernoulli polynomials and numbers are defined somewhat differently by different authors. Our notation follows generally that of Bromwich (1947). See Steffensen (1950) for a somewhat different notation. These polynomials should not be confused with the Bernstein polynomials (p. 36).

We leave the derivations to a problem {57}. The first few nonzero Bernoulli numbers are $B_0 = 1$, $B_1 = -\frac{1}{2}$, $B_2 = \frac{1}{6}$, $B_4 = -\frac{1}{30}$, $B_6 = \frac{1}{42}$, $B_8 = -\frac{1}{30}$, $B_{10} = \frac{5}{66}$, $B_{12} = -\frac{691}{2730}$, $B_{14} = \frac{7}{6}$, $B_{16} = -\frac{3617}{510}$, $B_{18} = 43{,}867/798$, $B_{20} = -174{,}611/330$. Now we define

$$X_k = \frac{1}{(2k)!}\int_0^h B_{2k}\left(\frac{y}{h}\right) f^{(2k)}(x+y)\,dy \tag{4.13-9}$$

where $f(x)$ is the function of (4.13-1). Integrating by parts twice and using (4.13-3), (4.13-7), and (4.13-8), we get

$$h^2 X_k - X_{k-1} = \frac{B_{2k-2}}{(2k-2)!}[f^{(2k-3)}(x+h) - f^{(2k-3)}(x)] \qquad k > 1 \tag{4.13-10}$$

When $k = 1$ we have, also by integration by parts,

$$\begin{aligned} X_1 &= \frac{1}{2}\int_0^h B_2\left(\frac{y}{h}\right) f''(x+y)\,dy = \frac{1}{2}\int_0^h \left(\frac{y^2}{h^2} - \frac{y}{h}\right) f''(x+y)\,dy \\ &= -\int_0^h \left(\frac{y}{h^2} - \frac{1}{2h}\right) f'(x+y)\,dy \\ &= -\frac{1}{2h}[f(x+h) + f(x)] + \frac{1}{h^2}\int_0^h f(x+y)\,dy \\ &= -\frac{1}{2h}[f(x+h) + f(x)] + \frac{1}{h^2}\int_x^{x+h} f(y)\,dy \end{aligned} \tag{4.13-11}$$

Using (4.13-10) and (4.13-11), we have

$$\begin{aligned} \tfrac{1}{2}[f(x+h) + f(x)] &= \frac{1}{h}\int_x^{x+h} f(y)\,dy - hX_1 \\ &= \frac{1}{h}\int_x^{x+h} f(y)\,dy + \frac{hB_2}{2!}[f'(x+h) - f'(x)] - h^3 X_2 \\ &= \frac{1}{h}\int_x^{x+h} f(y)\,dy + \frac{hB_2}{2!}[f'(x+h) - f'(x)] \\ &\quad + \frac{h^3 B_4}{4!}[f'''(x+h) - f'''(x)] - h^5 X_3 \\ &= \cdots\cdots\cdots\cdots\cdots\cdots \\ &= \frac{1}{h}\int_x^{x+h} f(y)\,dy + \sum_{k=1}^{m} \frac{B_{2k}}{(2k)!} h^{2k-1}[f^{(2k-1)}(x+h) - f^{(2k-1)}(x)] \\ &\quad - h^{2m+1} X_{m+1} \end{aligned} \tag{4.13-12}$$

We define

$$\bar{B}_k(x) = \begin{cases} B_k(x) & 0 \le x < 1 \\ \bar{B}_k(x-1) & \text{otherwise} \end{cases} \tag{4.13-13}$$

Now consider (4.13-9) with the limits going from jh to $(j+1)h$ and with $B_{2k}(y/h)$ replaced by $\bar{B}_{2k}(y/h)$. We can repeat the derivation of (4.13-12) for $j = 1, \ldots, n-1$. Then summing the results, replacing x by x_0, and noting that the series on the right-hand side of (4.13-12) telescopes, we get

$$\begin{aligned}\sum_{j=0}^{n} f(x_0 + jh) = {} & \frac{1}{h}\int_{x_0}^{x_0+nh} f(y)\,dy \\ & + \tfrac{1}{2}[f(x_0+nh) + f(x_0)] \\ & + \sum_{k=1}^{m} \frac{B_{2k}}{(2k)!} h^{2k-1} \\ & \times [f^{(2k-1)}(x_0+nh) - f^{(2k-1)}(x_0)] + E_m \end{aligned} \tag{4.13-14}$$

where, using (4.13-9),

$$E_m = -\frac{h^{2m+1}}{(2m+2)!}\int_0^{nh} \bar{B}_{2m+2}\left(\frac{y}{h}\right) f^{(2m+2)}(x_0 + y)\,dy \tag{4.13-15}$$

By using the properties of the Bernoulli polynomials {59} (4.13-15) can be simplified to

$$E_m = \frac{nh^{2m+2}B_{2m+2}}{(2m+2)!} f^{(2m+2)}(\xi) \tag{4.13-16}$$

where $x_0 < \xi < x_0 + nh$. Equation (4.13-14) is the *Euler-Maclaurin sum formula.* It is useful even when the upper limit in the summation is infinite, although in this case the error can no longer be expressed in the form (4.13-16). A useful error estimate in the infinite case (and when n is finite) is that the error is less than the magnitude of the first neglected term in the summation on the right-hand side of (4.13-14) if $f^{(2m+2)}(x)$ and $f^{(2m+4)}(x)$ do not change sign and are of the same sign for $x_0 < x < x_0 + nh$ {60}. If just $f^{(2m+2)}(x)$ does not change sign in the interval, the error is less than twice the first neglected term {61}.

Example 4.11 Use (4.13-14) to approximate

$$\sum_{j=0}^{\infty} \frac{1}{(1+j)^2}$$

We use $f(x) = 1/x^2$ with $x_0 = h = 1$. The true sum is $\pi^2/6 \approx 1.644934067$. Using (4.13-14), our approximation for the sum is

$$\sum_{j=0}^{\infty} \frac{1}{(1+j)^2} \approx \int_1^{\infty} \frac{1}{x^2}\,dx + \frac{1}{2} + \sum_{k=1}^{m} B_{2k} = \frac{3}{2} + \sum_{k=1}^{m} B_{2k}$$

The following table indicates the error between the true result and the approximation as a function of m:

m	Error	m	Error
0	.145	6	.198
1	−.022	7	−.968
2	.012	8	6.124
3	−.012	9	−48.847
4	.021	10	480.277
5	−.055		

The approximations, after initially converging, are clearly diverging. Suppose now we try to approximate

$$\sum_{j=0}^{\infty} \frac{1}{(10+j)^2}$$

Then we have from (4.13-14), now with $x_0 = 10$,

$$\sum_{j=0}^{\infty} \frac{1}{(10+j)^2} \approx \frac{1}{10} + \frac{1}{2}\frac{1}{10^2} + \sum_{k=1}^{m} \frac{B_{2k}}{10^{2k+1}}$$

With $m = 2$ we get

$$\frac{1}{10} + \frac{1}{200} + \frac{1}{6}\frac{1}{(10)^3} - \frac{1}{30}\frac{1}{(10)^5} \approx .1051663333$$

Added to the directly calculated sum

$$\sum_{j=1}^{9} \frac{1}{j^2} \approx 1.5397677311,$$

this gives 1.6449340644, which is correct to almost nine decimals, a remarkable change from the previous case. To account for this, we note that all derivatives of $f(x)$ are of constant sign in $(0, \infty)$, so that the error is less than the first neglected term. In the first case this is B_{2m+2}, and in the second it is $B_{2m+2}/10^{2m+3}$. Now it is known [see Steffensen (1950)] that ultimately the Bernoulli numbers B_{2i}, grow as $(2i)!$, so that $B_{2m+2}/x_0^{2m+3} \to \infty$ as $m \to \infty$ for any x_0. Thus the Euler-Maclaurin sum formula diverges for $f(x) = 1/(x_0 + x)^2$ for all x_0. In particular, it diverges for both the above cases. But the series on the right-hand side of (4.13-14), although divergent when $f(x) = 1/(x_0 + x)^2$, is asymptotically convergent; i.e., it converges for a while before it starts to diverge. When $x_0 = 1$, the convergence never gets very far, but for $x_0 = 10$, when $m = 2$, $B_{2m+2}/10^{2m+3} = B_6/10^7 = 2.4 \times 10^{-9}$, so that the asymptotic convergence is very good. To get similar accuracy by direct summation of the series, we would need to sum over 10^9 terms!

This example illustrates the power of the Euler-Maclaurin formula if used judiciously. It is worth remarking that judiciousness is necessary in dealing with most asymptotic series. One exception to this rule is interpolation series (see Prob. 19, Chap. 3), which are in fact asymptotic series. In practical application of interpolation formulas we are virtually always in the region of convergence of the interpolation series.

Rearranging terms makes (4.13-14) a quadrature formula

$$\int_{x_0}^{x_0+nh} f(y)\,dy = h\sum_{j=0}^{n} f(x_0+jh) - \frac{h}{2}[f(x_0+nh)+f(x_0)] - \sum_{k=1}^{m} \frac{B_{2k}}{(2k)!} h^{2k} \times [f^{(2k-1)}(x_0+nh) - f^{(2k-1)}(x_0)] - hE_m \qquad (4.13\text{-}17)$$

The first two terms on the right-hand side of (4.13-17) are just the trapezoidal-rule approximation to the integral. Equation (4.13-17) can then be looked at as the trapezoidal rule with correction terms. If we use Eq. (4.1-7) to approximate $f^{(2k-1)}(x_0)$ in terms of forward differences, and if we use the analog of (4.1-7) with backward differences of $f(x_0+nh)$ [found by differentiating (3.3-15)] to approximate $f^{(2k-1)}(x_0+nh)$, the result is *Gregory's formula*

$$\int_{x_0}^{x_0+nh} f(y)\,dy = h(\tfrac{1}{2}f_0 + f_1 + \cdots + f_{n-1} + \tfrac{1}{2}f_n) + \frac{h}{12}(\Delta f_0 - \nabla f_n) - \frac{h}{24}(\Delta^2 f_0 + \nabla^2 f_n) + \frac{19h}{720}(\Delta^3 f_0 - \nabla^3 f_n) - \frac{3h}{160}(\Delta^4 f_0 + \nabla^4 f_n) + \cdots \qquad (4.13\text{-}18)$$

where we have written f_j for $f(x_0+jh)$. This formula is one of the few examples where it is convenient to use differences on a digital computer. For if we use (4.13-17), we must program the computer to calculate not only $f(x)$ but also m derivatives. Moreover, we may not know how large m need be a priori. But we can add terms to (4.13-18) merely by computing higher differences, and these differences will generally be much easier to compute than the derivatives.

Example 4.12 Use (4.13-18) to improve the result of Example 4.10 for 16 subintervals.

Using a difference table for $1/x$, we calculate with $a_0 = 1$ and $h = \frac{1}{8}$

$$\Delta f_0 = -\tfrac{1}{9} \qquad \nabla f_{16} = -\tfrac{1}{69}$$

and the first difference correction term in (4.13-18) is

$$\tfrac{1}{96}(\tfrac{1}{69} - \tfrac{1}{9}) = -.001006$$

which, added to 1.099768, the result for 16 subintervals, gives 1.098762. Similarly, we calculate

$$\Delta^2 f_0 = \tfrac{1}{45} \qquad \nabla^2 f_{16} = \tfrac{1}{759}$$

so that the second difference correction term is

$$-\tfrac{1}{192}(\tfrac{1}{45} + \tfrac{1}{759}) = -.000123$$

which, added to the above result, gives 1.098639. These corrections give successively better results than the trapezoidal rule. Care must be taken not to carry the correction process too far because Gregory's formula is also only asymptotically convergent in general. Moreover, of course, roundoff becomes a factor with higher differences. In {63} a direct connection between Gregory's formula and higher-order Newton-Cotes formulas is considered.

4.13-2 Summation of Rational Functions; Factorial Functions

In this section we confine ourselves to the case where $f(x)$ in (4.13-1) is a rational function. In Example 4.11 we applied the Euler-Maclaurin sum formula to such a function. Our technique there—as it will be with all slowly convergent series whose sum cannot be expressed in closed form—was to replace the sum of the slowly convergent series by the sum of a rapidly convergent series. In this section, we first consider a class of rational functions whose sum can be expressed in closed form and then use this class to convert series of rational functions to more rapidly convergent ones.

Our basic tool here will be the sequence of *factorial functions* denoted by $x^{(n)}$ and defined by

$$x^{(n)} = x(x-1)(x-2)\cdots(x-n+1) \qquad n \text{ a positive integer} \tag{4.13-19}$$

$$x^{(0)} = 1$$

so that $n^{(n)} = n!$.

Consider first the forward difference of $x^{(n)}$ with unit spacing:

$$\begin{aligned}\Delta x^{(n)} &= (x+1)^{(n)} - x^{(n)} \\ &= x(x-1)\cdots(x-n+2)[(x+1)-(x-n+1)] = nx^{(n-1)}\end{aligned} \tag{4.13-20}$$

Therefore, the factorial functions play the same role with respect to differences that the powers of x do with respect to differentiation. Thus we have

$$\sum_{x=M}^{N-1} x^{(n)} = \frac{1}{n+1}\sum_{x=M}^{N-1} \Delta x^{(n+1)} = \frac{N^{(n+1)} - M^{(n+1)}}{n+1} \tag{4.13-21}$$

since

$$\sum_{x=M}^{N-1} \Delta f(x) = f(N) - f(M) \tag{4.13-22}$$

for any $f(x)$. When n is negative, we define $x^{(n)}$ by

$$x^{(n)} = \frac{1}{(x-n)^{(-n)}} \qquad n \text{ a negative integer} \tag{4.13-23}$$

For this case, too, we can show that (4.13-20) holds and, when $n \neq -1$, that (4.13-21) also holds. Note, in particular, that $0^{(n)} \neq 0$ when n is a negative integer (why?).

We shall now use (4.13-21) to convert slowly convergent infinite series of rational functions to more rapidly convergent ones. Note that in such infinite series the degree of the denominator must exceed that of the numerator by 2 or more (why?). The basic idea of this method, which is known as *Kummer's method*, bears an analogy to the comparison test used to determine convergence or divergence of infinite series. Suppose the series we wish to sum is

$$S = \sum_{x=M}^{\infty} R(x) \tag{4.13-24}$$

where $R(x)$ is a rational function. If there is a series

$$\bar{S} = \sum_{x=M}^{\infty} \bar{R}(x) \tag{4.13-25}$$

whose sum we know and such that $\bar{R}(x)$ and $R(x)$ have the same difference in degree d between numerator and denominator, then we can use $\bar{S}$ to convert S to a more rapidly convergent series. This technique is most easily illustrated by an example.

Example 4.13 Evaluate

$$S = \sum_{x=2}^{\infty} \frac{x}{x^4 + 1}$$

Let
$$\bar{R} = \frac{1}{(x-1)(x)(x+1)} = \frac{1}{(x+1)^{(3)}} = (x-2)^{(-3)}$$

Then, from (4.13-21), $\bar{S} = \frac{1}{4}$ {68}. Now

$$S = \bar{S} + (S - \bar{S}) = \frac{1}{4} - \sum_{2}^{\infty} \frac{x^2 + 1}{(x^4 + 1)x(x^2 - 1)}$$

which converges as $1/x^5$ and therefore much more rapidly than (4.13-24). We could also have used

$$\bar{R} = \frac{1}{x(x+1)(x+2)} = (x-1)^{(-3)}$$

In this case $\bar{S} = \frac{1}{12}$ and

$$S = \frac{1}{12} - \sum_{2}^{\infty} \frac{1 - 2x^2 - 3x^3}{x(x+1)(x+2)(x^4+1)}$$

which converges as $1/x^4$.

The general result is that if $R(x)$ and $\bar{R}(x)$ both have a denominator degree which exceeds the numerator degree by d, and if

$$\lim_{x \to \infty} x^d R(x) = \lim_{x \to \infty} x^d \bar{R}(x) \tag{4.13-26}$$

then $R(x) - \bar{R}(x)$ has a difference in degree between numerator and denominator of *at least* $d + 1$ {69}. In the above example we saw how a judicious choice of $\bar{R}(x)$ made the resulting difference $d + 2$. The procedure used in the above example can be applied again and again to make the summation even more rapidly convergent, but the algebra usually becomes tedious quite rapidly.

4.13-3 The Euler Transformation

The Euler transformation is a very useful device for accelerating the convergence of an oscillating series. We shall derive here a generalization useful in summing a finite oscillating sum which reduces to the Euler transformation in the case of an infinite series. Let

$$S = v_0 - v_1 + v_2 - \cdots + (-1)^n v_n \tag{4.13-27}$$

where the v_i are usually, but not necessarily, all positive. Write

$$S(x) = v_0 - v_1 x + v_2 x^2 - \cdots + (-1)^n v_n x^n \tag{4.13-28}$$

Then

$$\begin{aligned}(1 + x)S(x) &= v_0 - (v_1 - v_0)x + (v_2 - v_1)x^2 - \cdots \\ &\quad + (-1)^n(v_n - v_{n-1})x^n + (-1)^n v_n x^{n+1} \\ &= v_0 - (\Delta v_0)x + (\Delta v_1)x^2 - \cdots + (-1)^n(\Delta v_{n-1})x^n \\ &\quad + (-1)^n v_n x^{n+1}\end{aligned} \tag{4.13-29}$$

From (4.13-29) we obtain

$$\begin{aligned}S(x) &= \frac{v_0 + (-1)^n v_n x^{n+1}}{1 + x} \\ &\quad - y[\Delta v_0 - (\Delta v_1)x + (\Delta v_2)x^2 - \cdots + (-1)^{n-1}(\Delta v_{n-1})x^{n-1}]\end{aligned} \tag{4.13-30}$$

where

$$y = \frac{x}{1 + x} \tag{4.13-31}$$

Applying this transformation again to the bracketed series in (4.13-30), we obtain

$$\begin{aligned}S(x) &= \frac{v_0 + (-1)^n v_n x^{n+1}}{1 + x} - \frac{\Delta v_0 + (-1)^{n-1}(\Delta v_{n-1})x^n}{1 + x} y \\ &\quad + y^2[\Delta^2 v_0 - (\Delta^2 v_1)x + (\Delta^2 v_2)x^2 - \cdots + (-1)^{n-2}(\Delta^2 v_{n-2})x^{n-2}]\end{aligned} \tag{4.13-32}$$

If $p \le n$, then p applications yield

$$S(x) = \frac{v_0 - y\,\Delta v_0 + y^2\,\Delta^2 v_0 - \cdots + (-1)^{p-1} y^{p-1}\,\Delta^{p-1} v_0}{1+x}$$

$$+ \frac{(-1)^n[v_n x^{n+1} + (\Delta v_{n-1})x^n y + \cdots + (\Delta^{p-1} v_{n-p+1})x^{n-p+2} y^{p-1}]}{1+x}$$

$$+ (-1)^p y^p[\Delta^p v_0 - (\Delta^p v_1)x + (\Delta^p v_2)x^2 - \cdots + (-1)^{n-p}(\Delta^p v_{n-p})x^{n-p}] \tag{4.13-33}$$

Set $x = 1$, and we obtain the identity

$$S = \tfrac{1}{2}v_0 - \tfrac{1}{4}\,\Delta v_0 + \tfrac{1}{8}\,\Delta^2 v_0 - \cdots + (-1)^{p-1} 2^{-p}\,\Delta^{p-1} v_0$$

$$+ (-1)^n(\tfrac{1}{2}v_n + \tfrac{1}{4}\,\Delta v_{n-1} + \tfrac{1}{8}\,\Delta^2 v_{n-2} + \cdots + 2^{-p}\,\Delta^{p-1} v_{n-p+1})$$

$$+ 2^{-p}(-1)^p[\Delta^p v_0 - \Delta^p v_1 + \Delta^p v_2 - \cdots + (-1)^{n-p}\,\Delta^p v_{n-p}] \qquad p \le n \tag{4.13-34}$$

Assuming now that n and p are large and that the high-order differences are small, we neglect the last bracket and obtain

$$S \approx \tfrac{1}{2}v_0 - \tfrac{1}{4}\,\Delta v_0 + \tfrac{1}{8}\,\Delta^2 v_0 - \cdots$$

$$+ (-1)^n[\tfrac{1}{2}v_n + \tfrac{1}{4}\,\Delta v_{n-1} + \tfrac{1}{8}\,\Delta^2 v_{n-2} + \cdots] \tag{4.13-35}$$

Equation (4.13-35) is useful for summing a finite oscillating sum and has been applied in evaluating the integral over a finite interval of a rapidly oscillating function.

If we now let n go to infinity, so that we have an infinite series

$$S = \sum_{j=0}^{\infty} (-1)^j v_j \tag{4.13-36}$$

which, for now, we assume to be convergent, then the terms in the brackets in (4.13-35) all tend to zero for a fixed p and (4.13-36) becomes

$$S \approx \tfrac{1}{2}v_0 - \tfrac{1}{4}\,\Delta v_0 + \tfrac{1}{8}\,\Delta^2 v_0 - \cdots + (-1)^{p-1} 2^{-p}\,\Delta^{p-1} v_0 \tag{4.13-37}$$

Letting p go to infinity, we get the *Euler transformation:*

$$\sum_{j=0}^{\infty} (-1)^j v_j = \tfrac{1}{2}v_0 - \tfrac{1}{4}\,\Delta v_0 + \tfrac{1}{8}\,\Delta^2 v_0 - \cdots \tag{4.13-38}$$

A more flexible form of this transformation involves choosing an approximate index m from which to start, so that we have

$$\sum_{j=0}^{\infty} (-1)^j v_j = \sum_{j=0}^{m-1} (-1)^j v_j$$

$$+ (-1)^m(\tfrac{1}{2}v_m - \tfrac{1}{4}\,\Delta v_m + \tfrac{1}{8}\,\Delta^2 v_m - \cdots) \tag{4.13-39}$$

Example 4.14 Let $S = \sum_{j=0}^{\infty} (-1)^j/[\log (j+2)]$. S converges very slowly. Apply (4.13-39) with $m = 2$. We have

$$v_2 = .721348 \qquad \Delta v_2 = -.100013 \qquad \Delta^2 v_2 = .036788$$

$$\Delta^3 v_2 = -.01778 \qquad \Delta^4 v_2 = .00998 \qquad \Delta^5 v_2 = -.00617$$

Therefore, to four figures

$$S = (1.442695 - .910239) + \tfrac{1}{2}(.721348) + \tfrac{1}{4}(.100013)$$
$$+ \tfrac{1}{8}(.036788) + \tfrac{1}{16}(.01778) + \tfrac{1}{32}(.00998)$$
$$+ \tfrac{1}{64}(.00617) + \cdots = .9243$$

Had we started with $m = 0$, we would have required 11 terms to achieve the same accuracy.

The Euler transformation is a sequence-to-sequence transformation which takes partial sums of one series into partial sums of another. It can be shown that if the original series converges, so does the transformed one, and in the case of oscillating series, the transformed series usually converges more rapidly. On the other hand, there exist many divergent series which are transformed into convergent series. In such cases, the sum of the transformed series can be thought to represent in some way the "average value" of the partial sums of the divergent series. The Euler transformation has been applied to the evaluation of integrals over infinite intervals of rapidly oscillating functions.

We have at best touched the surface of the existing mine of methods to sum series. References to others will be found in the Bibliographic Notes.

BIBLIOGRAPHIC NOTES

Sections 4.1 to 4.2 Numerical differentiation is discussed in most of the standard numerical analysis texts. In particular, Kopal (1961) and Hildebrand (1974) have extensive discussions. In these two books, as well as in Steffensen (1950), the error term is derived using divided differences. The proof of formula (4.1-3) appears in Isaacson and Keller (1966). Coefficients for the Lagrangian differentiation formulas based on differences have been extensively tabulated. Some tables and references to others are given by Kopal (1961). An example of analytic differentiation by the computer appears in Rall (1969). Following the original paper by Richardson and Gaunt (1927), many authors have discussed Richardson extrapolation; see the excellent survey by Joyce (1971). The application of Richardson extrapolation to numerical differentiation was suggested by Rutishauser (1963), from whom Example 4.2 is taken.

Sections 4.3 to 4.12 The most comprehensive treatment of numerical quadrature appears in Davis and Rabinowitz (1975), which contains an extensive bibliography and a collection of computer programs. Krylov (1962), which is an excellent book but slightly dated, has discussions of much of the material of this chapter. For the insight provided by other points of view and for some topics not covered here, see Kopal (1961), Hartree (1958), and Mineur (1952), as well as Krylov. Stroud (1961) gives an extensive bibliography on quadrature methods, which has been brought up to date by Davis and Rabinowitz.

Sections 4.4 to 4.8 More complete accounts of the properties of the Gaussian weights and abscissas will be found in Winston (1934). The most complete reference available on orthogonal polynomials is Szegö (1975), but most of the results we have used here are available in a more accessible form in Jackson (1941). Much of the material in Sec. 4.8-3 is from Mineur (1952).

An extensive collection of tables of abscissas and weights for Gaussian quadrature formulas is contained in Stroud and Secrest (1966).

Sections 4.9 and 4.10 The error term in the Newton-Cotes quadrature formulas has been considered by a number of authors; see Steffensen (1950), Barrett (1952), and Sard (1948). Davis (1959) gives an interesting comparison of the trapezoidal rule and Gaussian formulas in the case where we integrate a periodic function over a full period. A good discussion of Romberg's method can be found in Bauer, Rutishauser, and Stiefel (1963). Newton-Cotes formulas with weights other than $w(x) = 1$ have been considered by Kaplan (1952) and Luke (1952). An interesting treatment of the weight function sin kx is given by Filon (1928). A quadrature formula which is particularly efficient in composite rules is considered by Ralston (1959); see also Prob. 53.

Section 4.11 The adaptive Simpson scheme is treated fully in Lyness (1969). Further schemes, examples, computer programs, and references are given in Davis and Rabinowitz (1975). The various components that make up an adaptive quadrature routine are given by Rice (1975). Different strategies in applying an adaptive-quadrature scheme are compared by Malcolm and Simpson (1975).

Section 4.12 The proof that arbitrarily high accuracy can be obtained with Gaussian quadrature formulas is from Chap. 3 of Todd (1962).

Section 4.13 Steffensen (1950) discusses the Bernoulli polynomials and numbers in detail, but his notation is different from ours. Hamming (1973) considers a number of miscellaneous methods of summation; see also Cherry (1950), Rosser (1951), Shanks (1955), and Szasz (1949).

BIBLIOGRAPHY

Barrett, W. (1952): On the Remainders of Numerical Formulae with Special Reference to Differentiation, *J. Lond. Math. Soc.*, vol. 27, pp. 456–464.

Bauer, F. L., H. Rutishauser, and E. Stiefel (1963): New Aspects in Numerical Quadrature, pp. 199–218 in Experimental Arithmetic, High Speed Computing and Mathematics, vol. 15, *Proc. Symp. Appl. Math.*, Am. Math. Soc., Providence, R.I.

Bernstein, S. (1937): Sur les formules de quadrature de Cotes et de Tchebycheff, *C. R. Acad. Sc. URSS*, vol. 14, pp. 323–326.

Bromwich, T. J. (1947): *An Introduction to the Theory of Infinite Series*, 2d ed., The Macmillan Company, New York.

Cherry, T. M. (1950): Summation of Slowly Convergent Series, *Proc. Cambr. Phil. Soc.*, vol. 46, pp. 436–449.

Dahlquist, G., and Å. Björck (1974): *Numerical Methods* (trans. N. Anderson), Prentice-Hall, Inc., Englewood Cliffs, N.J.

Davis, P. J. (1959): On the Numerical Integration of Analytic Functions, pp. 45–59 in *On Numerical Approximation* (R. E. Langer, ed.), The University of Wisconsin Press, Madison, Wis.

———, and P. Rabinowitz (1975): *Methods of Numerical Integration*, Academic Press, Inc., New York.

Filon, L. N. G. (1928): On a Quadrature Formula for Trigonometric Integrals, *Proc. R. Soc. Edinb.*, vol. 49, pp. 38–47.

Hamming, R. W. (1973): *Numerical Methods for Scientists and Engineers*, 2d ed., McGraw-Hill Book Company, New York.

Hardy, G. H., J. E. Littlewood, and G. Polya (1952): *Inequalities*, 2d ed., Cambridge University Press, New York.

Hartree, D. R. (1958): *Numerical Analysis*, 2d ed., Oxford University Press, Fair Lawn, N.J.

Hildebrand, F. B. (1974): *Introduction to Numerical Analysis*, 2d ed., McGraw-Hill Book Company, New York.

Isaacson, E., and H. B. Keller (1966): *Analysis of Numerical Methods*, John Wiley & Sons, Inc., New York.

Jackson, D. (1941): *Fourier Series and Orthogonal Polynomials*, Math. Ass. Am. Carus Math. Monogr., Oberlin, Ohio.

Joyce, D. C. (1971): Survey of Extrapolation Processes in Numerical Analysis, *SIAM Rev.*, vol. 13, pp. 435–490.

Kaplan, E. L. (1952): Numerical Integration near a Singularity, *J. Math. and Phys.*, vol. 31, pp. 1–28.

Kopal, Z. (1961): *Numerical Analysis*, 2d ed., John Wiley & Sons, Inc., New York.

Krylov, V. I. (1962): *Approximate Calculation of Integrals* (trans. by A. H. Stroud), The Macmillan Company, New York.

Luke, Y. L. (1952): Mechanical Quadrature near a Singularity, *MTAC*, vol. 6, pp. 215–219.

Lyness, J. M. (1969): Notes on the Adaptive Simpson Quadrature Routine, *J. Ass. Comput. Mach.*, vol. 16, pp. 483–495.

Malcolm, M. A., and R. B. Simpson (1975): Local versus Global Strategies for Adaptive Quadrature, *ACM Trans. Math. Software*, vol. 1, pp. 129–146.

Mineur, H. (1952): *Techniques de calcul numerique*, Librarie Polytechnique, Paris.

Oliver, J. (1971): The Efficiency of Extrapolation Methods for Numerical Integration, *Numer. Math.*, vol. 17, pp. 17–32.

Rall, L. B. (1969): *Computational Solution of Nonlinear Operator Equations*, John Wiley & Sons, Inc., New York.

Ralston, A. (1959): A Family of Quadrature Formulas Which Achieve High Accuracy in Composite Rules, *J. Ass. Comput. Mach.*, vol. 6, pp. 384–394.

——— (1963): On Differentiating Error Terms, *Am. Math. Mon.*, vol. 20, pp. 187–188.

Rice, J. R. (1975): A Metalgorithm for Adaptive Quadrature, *J. Ass. Comput. Mach.*, vol. 22, pp. 61–82.

Richardson, L. F., and J. A. Gaunt (1927): The Deferred Approach to the Limit, *Trans. R. Soc. Lond.*, vol. 226A, pp. 299–361.

Rosser, J. B. (1951): Transformations to Speed the Convergence of Series, *J. Res. Natl. Bur. Std.*, vol. 46, pp. 56–64.

Rutishauser, H. (1963): Ausdehnung des Rombergschen Prinzips, *Numer. Math.*, vol. 5, pp. 48–54.

Sard, A. (1948): Integral Representation of Remainders, *Duke Math. J.*, vol. 15, pp. 333–345.

Shanks, D. (1955): Nonlinear Transformations of Divergent and Slowly Convergent Sequences, *J. Math. and Phys.*, vol. 34, pp. 1–42.

Steffensen, J. F. (1950): *Interpolation*, 2d ed., Chelsea Publishing Company, New York.

Stroud, A. H. (1961): A Bibliography on Approximate Integration, *Math. Comput.*, vol. 15, pp. 52–80.

——— and D. H. Secrest (1966): *Gaussian Quadrature Formulas*, Prentice-Hall, Inc., Englewood Cliffs, N.J.

Szasz, O. (1949): Summation of Slowly Convergent Series, *J. Math. and Phys.*, vol. 28, pp. 272–279.

Szegö, G. (1975): *Orthogonal Polynomials*, 4th ed., vol. 23, American Mathematical Society Colloquium Publications, New York.

Todd, J. (ed.): *Survey of Numerical Analysis*, McGraw-Hill Book Company, New York.

Winston, C. (1934): On Mechanical Quadrature Involving the Classical Orthogonal Polynomials, *Ann. Math.*, vol. 35, pp. 658–677.

Zygmund, A. (1952): *Trigonometric Series*, 2d ed., Chelsea Publishing Company, New York.

PROBLEMS

Section 4.1

1 In (4.1-4) let $m_j = 0$ and $a_j = a_1 + (j-1)h$. Determine the values of the coefficients $w_j^{(k)}(x)$ in the formula

$$f^{(k)}(x) \approx \sum_{j=1}^{n} w_j^{(k)}(x) f(a_j)$$

for $x = a_i$, $i = 1, \ldots, n$, $k = 1, 2$, and $n = 3, 5$.

2 (*a*) With $n = 5$ use (4.1-7) to generate approximations to the first four derivatives of $f(x)$ at $x = a_0$.

(*b*) Use the results of Prob. 13, Chap. 3 to do the same thing in terms of backward and central differences, retaining differences through order 5.

3 (*a*) Using the notation of Prob. 9, Chap. 3, show that the Taylor-series expansion of $f(x)$ can be written $Ef(x) = e^{hD}f(x)$, where D is the operator with the property $Df(x) = f'(x)$.

(*b*) From this deduce the relations $hD = \ln E = \ln(1 + \Delta) = -\ln(1 - \nabla)$.

4 Use a numerical differentiation formula with differences through order 4 to find the coefficients of the differential equation $y'' + ay' + by = x$ satisfied by the function $f(x)$ in the tabulation below; then determine $f(x)$ itself.

x	.6	.7	.8	.9	1.0	1.1
$f(x)$	1.164642	1.344218	1.517356	1.683327	1.841471	1.991207

5 (*a*) Find the first derivative of $f(x) = 1/(1 + x)$ at $x = .005$ by using a Lagrangian formula with $n = 3$. Use equally spaced intervals with the middle point at $x = .005$ and $h = 1.0$, .1, .01, respectively. Round all values of $f(x)$ to four decimal places.

(*b*) Repeat the calculations of part (*a*) with $h = .01$ using (4.1-7) and retaining terms through the second difference. Compare these results with those of part (*a*).

Section 4.2

***6** Consider Eq. (4.2-6) rewritten in the form

$$\Phi(h_i) \equiv \phi_i = \phi - \sum_{j=1}^{\infty} a_j h_i^{\gamma_j}$$

The process of Richardson extrapolation can be thought of as that of finding functions

$$\phi_m^i(h) = b_{0,m}^i - \sum_{j=1}^{m} b_{j,m}^i h^{\gamma_j}$$

which agree with the truncated series $\Phi_m(h) = \phi - \sum_{j=1}^{m} a_j h^{\gamma_j}$ at the $m + 1$ points h_{i+k}, $k = 0, \ldots, m$, and then evaluating these functions at $h = 0$ (extrapolation) to yield the approximations to ϕ, $b_{0,m}^i \equiv T_m^i$ for successive values of m. In the case $\gamma_j = j\gamma$, $\phi_m^i(h)$ is a polynomial in h^γ, and we can use polynomial interpolation to find T_m^i. Show that if the T_m^i are computed using the iterated interpolation scheme of Prob. 23, Chap. 3, we get formula (4.2-10).

*7 Consider now the case when $\gamma_j = j\gamma + \delta$.

(*a*) Using the notation of the previous problem, show that

$$\phi_m^i(h) = b_{0,m}^i - h^{\delta+\gamma} P_{m-1}^i(h^\gamma)$$

where P_{m-1}^i is a polynomial of degree $m - 1$.

(*b*) Let $E_m^i(f)$ be the value at $h = 0$ of the polynomial of degree m through the points $(h_{i+k}^\gamma, f(h_{i+k}^\gamma))$, $k = 0, \ldots, m$. Show that $E_m^i(f)$ is a linear functional.

(*c*) If we now set $\phi_m^i(h_{i+k}) = \phi_{i+k} \equiv \Phi(h_{i+k})$, $k = 0, \ldots, m$, show that $E_m^i(b_{0,m}^i h^{-\delta}) = E_m^i(h^{-\delta}\Phi)$, so that

$$T_m^i = \frac{E_m^i(h^{-\delta}\Phi)}{E_m^i(h^{-\delta})}$$

and both numerator and denominator can be computed using (4.2-10) with T replaced by E.

(*d*) Show finally that T_m^i can be computed using the following scheme:

$$T_0^i = \phi_i \qquad H_0^i = h_i^{-\delta}$$

$$T_m^i = T_{m-1}^{i+1} + \frac{T_{m-1}^{i+1} - T_{m-1}^i}{D_m^i - 1} \qquad \text{where} \quad D_m^i = \frac{h_i^\gamma H_{m-1}^{i+1}}{h_{i+m}^\gamma H_{m-1}^i}$$

$$H_m^i = H_{m-1}^{i+1} + \frac{H_{m-1}^{i+1} - H_{m-1}^i}{(h_i/h_{i+m})^\gamma - 1}$$

[Ref. Joyce (1971) p. 474.]

8 If $f(x)$ is defined and has a Taylor expansion for $x \in I = [a, b]$ but is not defined outside I, then at $x = a$ (4.2-1) does not hold, so that we cannot use (4.2-3) to approximate $f'(a)$ nor can we apply Richardson extrapolation to (4.2-1).

(*a*) Derive formulas analogous to (4.2-1) and (4.2-3) for this case using only points in $[a, b]$.

(*b*) Apply Richardson extrapolation to compute approximations to the value of the first derivative of $\cos(\sin x/\sqrt{x})$ at $x = 0$ correct to four decimal places, using the two sequences of h given in Example 4.2.

9 Instead of numerically differentiating the tabulation of Prob. 5, approximate the first and second derivatives at .6 and .7 using, respectively, the first and second differences of $f(x)$. How do these compare with the derivatives calculated in Prob. 4? Show how the error in the approximation can be estimated using the $(k + 1)$st difference.

Sections 4.3 to 4.5

10 (*a*) Let $c_0, \ldots, c_{n-1}$ be a solution of the system of linear equations

$$\alpha_{i+n} + \sum_{k=0}^{n-1} c_k \alpha_{i+k} = 0 \qquad i = 0, \ldots, n-1$$

with α_k given by (4.4-2). By writing (4.4-1) in the form

$$\alpha_{i+k} = \sum_{j=1}^{n} H_j a_j^{i+k} \qquad \begin{matrix} i = 0, \ldots, n-1 \\ k = 0, \ldots, n \end{matrix}$$

and substituting into the above linear system show that the zeros of the polynomial

$$x^n + \sum_{k=0}^{n-1} c_k x^k$$

may be used as the a_j's. Then deduce that the weights can be found by solving the linear system consisting of the first n equations of (4.4-1).

(*b*) Apply this technique to try to find the abscissa and weight of the quadrature formula

$$\int_{-1}^{1} xf(x)\,dx \approx H_1 f(a_1)$$

which is to be exact for 1 and x. Here $\alpha_k = \int_{-1}^{1} x^{k+1}\,dx$. How do you explain the result? Use the result to state an assumption which is implicit in the statement of part (*a*).

11 By making use of the function $f(x) = p_n(x)u_{n-1}(x)$, where $u_{n-1}(x)$ is an arbitrary polynomial of degree $n-1$ or less, prove that it is necessary for $p_n(x)$ in (4.4-7) to be orthogonal to all polynomials of lesser degree if the Gaussian quadrature formula is to be exact for polynomials of degree $2n-1$ or less. Does this argument have to be modified when there is a weight function?

12 (*a*) Use Gauss-Legendre quadrature with $n = 2, 3, 4, 5$ to approximate the integral

$$\int_{-4}^{4} \frac{dx}{1+x^2}$$

and compare the results with the true value (see also Sec. 4.12).

(*b*) Repeat part (*a*) for $\int_0^1 e^{-10x} \sin x\,dx$.

(*c*) Repeat part (*a*) for $\int_0^5 xe^{-3x^2}\,dx$.

(*d*) Repeat part (*a*) for $\int_{-1}^{1} (1-x^2)^{3/2} \cos x\,dx = 3\pi J_2(1) \approx 1.08294$.

(*e*) Repeat part (*a*) for $\int_{-1}^{1} \frac{\cos x}{(1-x^2)^{1/2}}\,dx = \pi J_0(1) \approx 2.40394$.

Section 4.6

13 Orthogonal polynomials. Let $\{\phi_r(x)\}$ be a sequence of polynomials such that $\phi_r(x)$ is of degree r and is orthogonal to all polynomials of lesser degree over the interval $[a, b]$ with respect to the weight function $w(x)$. In this and the following problems, we assume that $w(x)$ is of constant sign in $[a, b]$.

(*a*) By letting

$$w(x)\phi_r(x) = \frac{d^r U_r(x)}{dx^r}$$

show that $U_r(x)$ must satisfy the equation

$$[U_r^{(r-1)} q_{r-1} - U_r^{(r-2)} q'_{r-1} + U_r^{(r-3)} q''_{r-1} + \cdots + (-1)^{r-1} U_r q_{r-1}^{(r-1)}]_a^b = 0$$

where $q_{r-1}(x)$ is any polynomial of degree $r-1$ or less, superscripts denote differentiation, and the notation $]_a^b$ means that the value of the expression in brackets at $x = a$ is to be subtracted from that at $x = b$.

(*b*) Show that the boundary conditions

$$U_r(a) = U'_r(a) = U''_r(a) = \cdots = U_r^{(r-1)}(a) = 0$$

$$U_r(b) = U'_r(b) = U''_r(b) = \cdots = U_r^{(r-1)}(b) = 0$$

satisfy the requirements of part (*a*) and serve to determine $\phi_r(x)$ uniquely.

(*c*) Finally deduce that

$$\phi_r(x) = \frac{1}{w(x)} \frac{d^r U_r(x)}{dx^r}$$

where $U_r(x)$ satisfies the differential equation

$$\frac{d^{r+1}}{dx^{r+1}} \left[\frac{1}{w(x)} \frac{d^r U_r(x)}{dx^r} \right] = 0$$

subject to the boundary conditions of part (*b*). Such a representation of an orthogonal polynomial is called a *Rodrigues formula.*

14 (*a*) By writing

$$\phi_r(x) = A_r x^r + q_{r-1}(x)$$

show that
$$\gamma_r = \int_a^b w(x)\phi_r^2(x)\,dx = A_r \int_a^b x^r w(x)\phi_r(x)\,dx$$

(*b*) With $U_r(x)$ defined as in Prob. 13*a*, use the results of Prob. 13*b* to show that

$$\gamma_r = (-1)^r r!\, A_r \int_a^b U_r(x)\,dx$$

15 Prove that if $w(x)$ does not change sign in $[a, b]$, the zeros of $\phi_r(x)$ are real, distinct, and lie within $[a, b]$.

16 Recurrence relations. (*a*) Show that

$$\phi_{r+1}(x) - \alpha_r x\phi_r(x) = b_r\phi_r(x) + b_{r-1}\phi_{r-1}(x) + \cdots + b_0\phi_0(x)$$

where $\alpha_r = A_{r+1}/A_r$ and the b_i's are constants.

(*b*) Use the orthogonality property of the polynomials to show that $b_i = 0$, $i = 0, \ldots, r-2$, and thus deduce that the polynomials satisfy the recurrence relation

$$\phi_{r+1}(x) = (\alpha_r x + b_r)\phi_r(x) + b_{r-1}\phi_{r-1}(x)$$

(*c*) Show that
$$b_r = \alpha_r\left(\frac{B_{r+1}}{A_{r+1}} - \frac{B_r}{A_r}\right)$$

where B_r is the coefficient of x^{r-1} in $\phi_r(x)$.

(*d*) Show that
$$b_{r-1} = -\frac{A_{r+1}A_{r-1}}{A_r^2}\frac{\gamma_r}{\gamma_{r-1}}$$

by multiplying the recurrence relation in part (*b*) first by $w(x)\phi_{r+1}(x)$ and then by $w(x)\phi_{r-1}(x)$, integrating each equation over $[a, b]$, and solving for b_{r-1} in the resulting set of equations.

*(*e*) Use the recurrence relation to show that if $r_1 < r_2 < \cdots < r_n$ are the zeros of $\phi_n(x)$ and $s_1 < s_2 < \cdots < s_{n+1}$ are the zeros of $\phi_{n+1}(x)$, then $a < s_1 < r_1 < s_2 < r_2 < s_3 < \cdots < s_n < r_n < s_{n+1} < b$. *Hint:* Assume that $A_r > 0$ for all r.

17 The Christoffel-Darboux identity. (*a*) Divide the recurrence relation of part (*b*) of Prob. 16 by $\alpha_r\gamma_r$, multiply the result by $\phi_r(y)$, subtract from this the same equation with x and y interchanged, and use part (*d*) of Prob. 16 to obtain

$$(x-y)\frac{\phi_r(x)\phi_r(y)}{\gamma_r} = \frac{\phi_{r+1}(x)\phi_r(y) - \phi_r(x)\phi_{r+1}(y)}{\alpha_r\gamma_r} - \frac{\phi_r(x)\phi_{r-1}(y) - \phi_{r-1}(x)\phi_r(y)}{\alpha_{r-1}\gamma_{r-1}}$$

(*b*) Sum this result from $r = 0$ to $r = n$ to obtain the Christoffel-Darboux identity (4.6-3).

18 (*a*) When $[a, b] = [0, 1]$ and $w(x) = x^k$ with k a positive integer, derive $U_r(x)$ and from this an expression for $\phi_r(x)$ under the condition $\phi_r(0) = 1$.

(*b*) Use this result to calculate γ_r and A_r.

(*c*) Thus derive the form of the weights and error in the corresponding Gaussian quadrature formula as given by (4.6-8) and (4.6-9).

19 Legendre polynomials. (*a*) When $w(x) = 1$ and $[a, b] = [-1, 1]$, use the results of Prob. 13 to show that

$$U_r = C_r(x^2 - 1)^r$$

where C_r is an arbitrary constant.

(*b*) With $C_r = 1/2^r r!$, show that

$$A_r = \frac{(2r)!}{2^r(r!)^2} \qquad \text{and} \qquad \gamma_r = \frac{2}{2r+1}$$

(*c*) Show that the Legendre polynomials, denoted by $P_r(x)$, satisfy the recurrence relation

$$P_{r+1}(x) = \frac{2r+1}{r+1} x P_r(x) - \frac{r}{r+1} P_{r-1}(x)$$

(*d*) Show that $P_0(x) = 1$, $P_1(x) = x$ and use part (*c*) to generate the next six Legendre polynomials.

Section 4.7

20 Laguerre polynomials. (*a*) When $w(x) = x^\beta e^{-\alpha x}$, $\alpha > 0$, $\beta > -1$, and $(a, b) = (0, \infty)$, show that

$$U_r(x) = C_r e^{-\alpha x} x^{\beta + r}$$

(*b*) With $C_r = 1$, show that

$$A_r = (-1)^r \alpha^r \qquad \text{and} \qquad \gamma_r = \frac{r!\,\Gamma(r + \beta + 1)}{\alpha^{1+\beta}}$$

(*c*) When $\alpha = 1$, $\beta = 0$, these polynomials are called Laguerre polynomials and are denoted by $L_r(x)$. Show that they satisfy the recurrence relation

$$L_{r+1}(x) = (1 + 2r - x)L_r(x) - r^2 L_{r-1}(x)$$

(*d*) Show that $L_0(x) = 1$, $L_1(x) = 1 - x$, and use part (*c*) to generate the next six Laguerre polynomials.

(*e*) When $\alpha = 1$ and $\beta > -1$, the polynomials are called generalized Laguerre polynomials and are denoted by $L_r^\beta(x)$. Show that they satisfy the recurrence relation

$$L_{r+1}^\beta(x) = (1 + 2r + \beta - x)L_r^\beta(x) - r(r + \beta)L_{r-1}^\beta(x)$$

(*f*) Repeat part (*d*) for the generalized Laguerre polynomials with $L_0^\beta(x) = 1$, $L_1^\beta(x) = 1 + \beta - x$.

21 Hermite polynomials. (*a*) When $w(x) = e^{-\alpha^2 x^2}$ and $(a, b) = (-\infty, \infty)$, show that

$$U_r(x) = C_r e^{-\alpha^2 x^2}$$

(*b*) With $C_r = (-\alpha)^{-r}$, show that

$$A_r = (2\alpha)^r \qquad \text{and} \qquad \gamma_r = \frac{2^r r!}{\alpha} \sqrt{\pi}$$

(*c*) When $\alpha = 1$, these polynomials are called Hermite polynomials and are denoted by $H_r(x)$. Show that they satisfy the recurrence relation

$$H_{r+1}(x) = 2xH_r(x) - 2rH_{r-1}(x)$$

(*d*) Show that $H_0(x) = 1$, $H_1(x) = 2x$, and use part (*c*) to generate the next six Hermite polynomials.

22 (*a*) Use the recurrence relation of part (*b*) of Prob. 16 to show that (4.6-8) can be written

$$H_j = \frac{A_n \gamma_{n-1}}{A_{n-1} \phi_n'(a_j) \phi_{n-1}(a_j)}$$

(*b*) Apply this result to get new expressions for the weights in the Gauss-Legendre, Gauss-Laguerre, and Gauss-Hermite quadrature formulas.

(*c*) Use the following relations from Szegö (1975):

$$(1 - x^2)P_n'(x) = (n + 1)xP_n(x) - (n + 1)P_{n+1}(x)$$

$$xL_n'(x) = (x - n - 1)L_n(x) + L_{n+1}(x)$$

$$H_n'(x) = 2xH_n(x) - H_{n+1}(x)$$

to express the weights in terms of the polynomials of degree $n + 1$ only.

23 (*a*) Use Gauss-Laguerre quadrature to approximate the integral

$$\int_0^\infty e^{-10x} \sin x \, dx$$

using $n = 2, 3, 4, 5$, and compare the results with the true value.

(*b*) Approximate the integral of part (*a*) by using the result of part (*b*) of Prob. 12 and finding a bound on the integral from 1 to ∞. Compare the two results.

(c) Repeat part (*a*) for $\int_0^\infty e^{-x}/(1 + e^{-2x})\, dx$.

(*d*) Use Gauss-Hermite quadrature to approximate the integral $\int_{-\infty}^\infty |x| e^{-3x^2} dx$, with $n = 2, 3, 4, 5$, and compare the results with the true value.

(*e*) Approximate the integral of part (*d*) using the result of part (*c*) of Prob. 12 and finding a bound on the integral outside of $[-5, 5]$. Compare the two results.

(*f*) Repeat part (*d*) for $\int_{-\infty}^\infty e^{-x^2} \cos x \, dx$.

Section 4.8

24 Jacobi polynomials.

(*a*) When $w(x) = (1 - x)^\alpha (1 + x)^\beta$, $\alpha, \beta > -1$, and $[a, b] = [-1, 1]$, show that

$$U_r(x) = C_r(1 - x)^{\alpha + r}(1 + x)^{\beta + r}$$

(*b*) With $C_r = (-1)^r/(2^r r!)$, derive the relations (4.8-1) and (4.8-2). [For a recurrence relation for the Jacobi polynomials, see Jackson (1941), p. 173.]

(*c*) Thus verify (4.8-4) and (4.8-5).

(*d*) When $\alpha = \beta = 0$, verify that Gauss-Jacobi quadrature is identical to Gauss-Legendre quadrature.

***25** (*a*) When $\alpha = \beta = 1$, show that

$$J_n(x; 1, 1) = \frac{2}{n + 2} P_{n+1}'(x) + p_{n-1}(x)$$

where $P_{n+1}(x)$ is the Legendre polynomial of degree $n + 1$ and $p_{n-1}(x)$ is some polynomial of degree $n - 1$.

(*b*) Use the relationship given in Prob. 22*c* and pertinent orthogonality relationships to deduce that $p_{n-1}(x)$ must be identically zero.

(*c*) Thus derive expressions for the weights and error term in the quadrature formula

$$\int_{-1}^{1} (1 - x^2) f(x)\, dx = \sum_{j=1}^{n} H_j f(a_j) + E$$

What are the abscissas?

26 Chebyshev polynomials.

(*a*) With $\alpha = \beta = -\frac{1}{2}$ in Prob. 24 and using $C_r = (-2)^r r!/(2r)!$, use (4.8-1) and (4.8-2) to show that

$$A_r = 2^{r-1} \qquad \gamma_r = \frac{\pi}{2}$$

(*b*) Use these results to derive (4.8-10) and (4.8-11).

(*c*) By making the substitution $x = \cos\theta$ in the integral expressing the orthogonality of $T_r(x)$ to polynomials of lesser degree, show that this integral implies that

$$\int_0^{\pi} T_r(\cos\theta) \cos k\theta \, d\theta = 0 \qquad k = 1, \ldots, r - 1$$

(*d*) Thus deduce that $\qquad T_r(\cos\theta) = c_r \cos r\theta$

(*e*) Use part (*a*) to deduce that $c_r = 1$ and thus deduce (4.8-12).

(*f*) Show that the Chebyshev polynomials $T_r(x)$ satisfy the recurrence relation

$$T_{r+1}(x) = 2xT_r(x) - T_{r-1}(x)$$

27 (*a*) Verify (4.8-13) and (4.8-14). (*b*) Thus verify (4.8-15).

28 Let $F(\theta)$ be an even, periodic function of θ of period 2π. Show how the Fourier series for this function can be converted into a series of Chebyshev polynomials (cf. Sec. 7.6).

29 (*a*) Use Gauss-Chebyshev quadrature with $n = 2, 3, 4, 5$ to approximate the integral $\int_{-1}^{1} \cos x/(1 - x^2)^{1/2}\, dx$ and compare the results with the true value and the results of Prob. 12*e*.

(*b*) Repeat part (*a*) for $\int_{-1}^{1} |x|\, dx/(1 - x^4)^{1/2}$.

***30** (*a*) Derive (4.8-19) by integrating by parts the orthogonality integral for $T_{r+1}(x)$ and requiring that $A_r = 2^{r-1}$.

(*b*) Use (4.8-19) to calculate γ_n and then derive (4.8-20) to (4.8-22).

***31** (*a*) Use the change of variable $y = \sqrt{x}$ in (4.6-1) to derive (4.8-23).

(*b*) Then use (4.6-10) and the recurrence relation in Prob. 19*c* to derive (4.8-25) and (4.8-26).

(*c*) Similarly, derive (4.8-27), (4.8-29), and (4.8-30).

***32** (*a*) Use the change of variable $y = \sqrt{x}$ to derive (4.8-31).

(*b*) Then use Sec. 4.8-2 and the recurrence relation in Prob. 26*f* to derive (4.8-32) to (4.8-34).

33 (*a*) Approximate the integral $\int_0^1 \sqrt{x} \cos x\, dx$ using (4.8-18) with $n = 2$. Find a bound on the error in your approximation.

(*b*) Approximate the integral $\int_0^1 (x/1 - x)^{1/2} e^x\, dx$ using (4.8-18) with $n = 2$. Again find a bound on the error.

34 (*a*) Remove the singularity in $\int_1^a e^{-ax}/(x^2 - 1)^{1/2}\, dx$ by making a change of variable.

(*b*) Consider the integral $\int_0^a x^p f(x)\, dx$ for $p > -1$ and not an integer where $f(x)$ is regular at $x = 0$ so that the integral has a singularity at 0. By writing

$$\int_0^t x^p f(x)\,dx = \frac{1}{p+1} t^{p+1} h(t)$$

and differentiating derive a differential equation for $h(t)$. [This differential equation can be integrated numerically by one of the methods of the next chapter to find $h(a)$, which in turn gives the value of the integral.] At $t = 0$ find initial conditions for $h(t)$ and its first two derivatives. [Ref.: Hartree (1958), pp. 110–111.]

35 Consider the problem of calculating

$$g(y) = \frac{d}{dy} \int_0^y \frac{f(x)}{(y-x)^{1/2}}\,dx \qquad 0 \le y \le 1$$

(a) Why can't we differentiate under the integral sign?

(b) Suppose $f(x)$ is given at $x = 0, h, 2h, \ldots, 1$. Show how numerical quadrature and numerical differentiation can be combined to find an approximation for $g(y)$.

(c) By making the change of variable $x = y \sin^2 \theta$ and using Lagrangian interpolation to get a polynomial approximation for $f(x)$, derive another method for approximating $g(y)$. [Ref.: Hamming (1973), pp. 164–165.]

Section 4.9

36 (a) In a Gauss-Legendre composite formula if m instead of two subintervals are used, show that the only change in (4.9-12) is to replace the $1/2^{2n}$ factor by $1/m^{2n}$.

(b) Derive the equations corresponding to (4.9-13) and (4.9-14) when $n = 3$ and m subintervals are used.

37 (a) Use a Gauss-Legendre composite formula with $n = 1$ and $m = 2, 4, 8$ to approximate the integral of Prob. 12a. Compare the results with those obtained in Prob. 12.

(b) Repeat part (a) for $n = 2$ and $m = 2, 4, 8$, that is, use (4.9-13).

(c) Repeat parts (a) and (b) for the integral of part (e) of Prob. 12. Compare the results with those obtained in Probs. 12 and 29.

Section 4.10

38 Use an argument based on symmetry to prove that when $w(x)$ is symmetric on the interval $[a_0, a_n]$, the closed Newton-Cotes quadrature formula when n is even is exact for polynomials of degree $n + 1$.

39 In order to derive (4.10-7), consider the integral

$$I_j = \int_{a_j}^{a_{j+1}} p_{n+1}(x)\,dx \qquad n \text{ even}$$

(a) Show that $I_j = -I_{n-1-j}$.

(b) Use the mean-value theorem for integrals to show that

$$I_{j-1} = \frac{\xi - a_n - h}{\xi - a_0} I_j \qquad a_j < \xi < a_{j+1} \quad j < \frac{n}{2}$$

(c) Thus, deduce that $|I_{j-1}| \ge |I_j|, j < n/2$ and from this deduce that $q(x)$ is of constant sign in $[a_0, a_n]$.

(d) Finally, derive both forms of the error in (4.10-7). [Ref.: Steffensen (1950), pp. 155–157.]

***40** When n is odd, the error has the form

$$E = \frac{1}{(n+1)!} \int_{a_0}^{a_n} p_{n+1}(x) f^{(n+1)}(\xi)\, dx$$

(*a*) Write this integral as the sum of two integrals, one from a_0 to a_{n-1} and the other from a_{n-1} to a_n, and apply the mean-value theorem for integrals to express the latter integral in the form (4.10-8).

(*b*) Show that

$$\frac{1}{(n+1)!}(x - a_n) f^{(n+1)}(\xi) = \frac{1}{n!} f^{(n)}(\eta) + c$$

where c is a constant and η lies in the interval spanned by $a_0, \ldots, a_{n-1}$ and x. *Hint:* Consider the Lagrangian interpolation formula with n and $n+1$ points, divide both by $p_n(x)$, and subtract.

(*c*) Use this result to write the former of the two integrals in part (*a*) in the form

$$\frac{1}{n!} \int_{a_0}^{a_{n-1}} p_n(x) f^{(n)}(\eta)\, dx$$

where $p_n(x) = (x - a_0) \cdots (x - a_{n-1})$.

(*d*) Finally, use the result of the previous problem to derive (4.10-8). [Ref.: Steffensen (1950), pp. 162–165.]

41 (*a*) Let the quadrature formula

$$\int_{-1}^{1} f(x)\, dx \approx H_0 f(-1) + \sum_{j=1}^{n} H_j f(a_j)$$

be exact for polynomials of degree $2m - 1$ or less ($m < n$). Prove that if all the weights are positive and all the a_j's are in $[-1, 1]$, then there exists at least one a_j in the interval $[-1, \beta_1]$ where β_1 is the smallest of the zeros of $P'_m(x)$, the derivative of the Legendre polynomial of degree m. *Hint:* Use a quadrature formula where -1 and $+1$ are required to be abscissas and consider the function

$$f(x) = \frac{[P'_m(x)]^2(1 - x^2)}{x - \beta_1}$$

(*b*) Thus deduce that in the quadrature formula (4.10-1) with $[a, b] = [-1, 1]$

$$\frac{2 - n}{n} < \beta_1$$

if (4.10-1) is to be exact for polynomials of degree $2m - 1$ and all the weights are positive.

(*c*) It can be shown that

$$\beta_1 < \frac{8}{(m-1)(m+3)} - 1$$

for $m > 1$. Use this and the fact that Newton-Cotes closed formulas are exact for polynomials of degree $n + 1$ when n is even and n when n is odd to prove that the weights of such formulas are positive only when $n \leq 7$ and $n = 9$. [Ref.: Bernstein (1937).]

42 (*a*) Show that the Newton-Cotes open formulas with n even are exact for polynomials of degrees $n - 1$.

(*b*) When $w(x) = 1$, derive the error terms for the open formulas by using the techniques of Probs. 39 and 40.

***43** (*a*) Show that the Newton-Cotes quadrature formulas without the error term may be written as

$$\begin{vmatrix} \int_a^b f(x)\,dx & b-a & \int_a^b x\,dx & \cdots & \int_a^b x^n\,dx \\ f(a_0) & 1 & a_0 & \cdots & a_0^n \\ f(a_1) & 1 & a_1 & \cdots & a_1^n \\ \cdots & \cdots & \cdots & \cdots & \cdots \\ f(a_n) & 1 & a_n & \cdots & a_n^n \end{vmatrix} = 0$$

(*b*) Using an equation analogous to (4.4-1), deduce that the Cotes numbers can be calculated using the inverse of the matrix

$$A_n = \begin{bmatrix} 1 & 1 & \cdots & 1 \\ a_0 & a_1 & \cdots & a_n \\ a_0^2 & \cdots & \cdots & a_n^2 \\ \cdots & \cdots & \cdots & \cdots \\ a_0^n & \cdots & \cdots & a_n^n \end{bmatrix}$$

(*c*) Let

$$\pi_i(x) = (x - a_0)(x - a_1) \cdots (x - a_{i-1})(x - a_{i+1}) \cdots (x - a_n)$$
$$= \sum_{k=0}^{n} C_{ik} x^k \qquad i = 0, \ldots, n$$

Show that the element in the mth row and kth column of $S_n = A_n^{-1}$ is

$$\frac{C_{m-1,k-1}}{\pi_{m-1}(a_{m-1})}$$

(*d*) Derive $S_1(-\frac{1}{2}, \frac{1}{2})$ and $S_2(-1, 0, 1)$ where the values in parentheses are the abscissas.

(*e*) Show how the matrix $S_2(-1, 0, 1)$ can be used to generate any quadrature formula of the form

$$\int_a^b w(x) f(x)\,dx \approx H_0 f(-1) + H_1 f(0) + H_2 f(1)$$

Apply this to the cases (i) $w(x) = x$ and $[a, b] = [0, 1]$ and (ii) $w(x) = x^2$ and $[a, b] = [-1, 1]$. [Ref.: Hamming (1973), chap. 15.]

***44** (*a*) Use the Lagrangian interpolation formula to prove that

$$\sum_{j=-k}^{k} (-1)^j \binom{2k}{k+j} f(x + jh) = (-1)^k h^{2k} f^{(2k)}(\xi) \qquad x - kh < \xi < x + kh$$

(*b*) Use this result with $k = 2$ and the Newton-Cotes closed-type five-point formula ($n = 4$) to derive the formula

$$\int_{x-2h}^{x+2h} f(x)\,dx = \frac{2h}{9}[f(x-2h) + 8f(x-h) + 8f(x+h) + f(x+2h)]$$
$$+ \frac{4h^5}{945}[21f^{\text{iv}}(\eta_1) - 2h^2 f^{\text{vi}}(\eta_2)]$$

(*c*) Similarly, using the Newton-Cotes closed-type seven-point formula, derive *Weddle's rule*

$$\int_{x-3h}^{x+3h} f(x)\,dx = \frac{3h}{10}[f(x-3h) + 5f(x-2h) + f(x-h) + 6f(x) + f(x+h)$$

$$+ 5f(x+2h) + f(x+3h)] - \frac{h^7}{1400}[10f^{\text{vi}}(\eta_1) + 9h^2 f^{\text{viii}}(\eta_2)]$$

(*d*) Using the same seven-point formula, derive *Hardy's rule*

$$\int_{x-3h}^{x+3h} f(x)\,dx = \frac{h}{100}[28f(x-3h) + 162f(x-2h) + 220f(x) + 162f(x+2h)$$

$$+ 28f(x+3h)] + \frac{9h^7}{1400}[2f^{\text{vi}}(\eta_1) - h^2 f^{\text{viii}}(\eta_2)]$$

45 (*a*) Derive the equation analogous to (4.10-15) for the parabolic rule.

(*b*) Use the parabolic rule with $m = 4$ and 8 to approximate the integral of Example 4.3.

(*c*) Use the result of part (*a*) to get a third approximation. Compare the results with those of part (*b*) and Example 4.3.

46 By considering the significance of the derivative term in quadrature formula errors, discuss why the efficiency of Richardson extrapolation depends on the monotonic convergence of successive approximations using an increasing number of subintervals.

***47** (*a*) Show that $T_{1,k}$ in Romberg integration is the parabolic rule for 2^k subintervals.

(*b*) Similarly, show that $T_{2,k}$ is the composite rule formed using the Newton-Cotes closed formula with $n = 4$ (five points) and 2^k subintervals.

(*c*) Show that the leading term in the error of $T_{m,k}$ is of the order of $[(b-a)/2^k]^{2(m+1)}$.

(*d*) Use this and the number of abscissas that $T_{m,k}$ uses in each subinterval to show that $T_{m,k}$ corresponds to no Newton-Cotes composite rule for $m \geq 3$.

***48** (*a*) Use induction, (4.10-23), and (4.10-24) to derive (4.10-27).

(*b*) Use (4.10-28) to show that $\sum_{j=0}^{m} |c_{mj}|$ is bounded and, in fact, less than $e^{8/3}$. [The true value of the infinite product in (4.10-28) is approximately 1.969.]

(*c*) Use (4.10-27) and the boundedness of the c_{mj} to show that $c_{m,m-j} \to 0$ as $m \to \infty$ for any j.

***49** (*a*) Show that d_{jm} in (4.10-29) is given by

$$d_{jm} = c_{m0} + 2c_{m1} + 4c_{m2} + \cdots + 2^p c_{mp} \qquad j = 1, \ldots, 2^{m+k} - 1$$

where $p < m$ and 2^p is the greatest power of 2 which divides j. For $j = 0$ and 2^{m+k}, show that d_{jm} is given by one-half the above.

(*b*) Use (4.10-27) to show that

$$|c_{mj}| > 3|c_{m,j+1}|$$

(*c*) Deduce then that $d_{jm} > 0$ for all j and m and that

$$\tfrac{4}{9} < d_{jm} < \alpha$$

where $\alpha = \lim_{m\to\infty} t_m(0) \approx 1.452$.

50 Let $\phi(x)$ be a function such that

$$\phi\left(\frac{1}{4^m}\right) = T_{0,m}$$

Use (4.10-23) to show that Romberg's method is equivalent to using Neville's method of iterated interpolation (see Prob. 23, Chap. 3) to approximate $\phi(0)$.

51 (*a*) Use Romberg integration to approximate the integral of Prob. 12*a*. In particular, calculate $T_{m,0}$, $m = 1, \ldots, 6$. Compare this result with the results of Probs. 12 and 37 and Table 4.8.

(*b*) Repeat part (*a*) for the integrals of parts (*b*), (*c*), and (*d*) of Prob. 12.

52 (*a*) For $n = 1, 2$ derive the weights for the closed Newton-Cotes quadrature formula

$$\int_0^1 \sqrt{x}\, f(x)\, dx \approx \sum_{j=0}^{n} H_j f\left(a + \frac{j}{n}\right)$$

so that the formula will be exact for polynomials of degree n or less.

(*b*) Repeat part (*a*) with $\sqrt{x}$ replaced by $1/\sqrt{x}$.

(*c*) Use the results of parts (*a*) and (*b*) to find approximate values of the integrals of Prob. 33*a* and Example 4.6, respectively, and compare the corresponding results.

***53** Consider the quadrature formula

$$\int_a^b f(x)\, dx \approx H_0[f(b) - f(a)] + \sum_{j=1}^{n} H_j f(a_j)$$

(*a*) Show that

$$p_n(x) = \frac{C}{(x-a)(x-b)} \frac{d^{n-1}}{dx^{n-1}} [(x-a)^n (x-b)^n (x-\alpha)]$$

is orthogonal over $[a, b]$ to all polynomials of degree $n - 2$ or less with respect to the weight function $(x - a)(x - b)$ where C and α are arbitrary constants.

(*b*) Let the abscissas be the zeros of $p_n(x)$. Derive two simultaneous equations for H_0 and α by requiring the quadrature formula to be exact when $f(x) = p_n(x)$ and $xp_n(x)$.

(*c*) Find an expression for the H_j's by requiring the formula to be exact when

$$f(x) = l_j(x) = \frac{p_n(x)}{(x - a_j)p_n'(a_j)}$$

(*d*) By using the expression $x^k = \sum_{j=1}^{n} l_j(x) a_j^k$, $k = 0, 1, \ldots, n - 1$, show that the quadrature formula is exact for polynomials of degree $n - 1$ or less. Then use this result, the result of part (*b*), and an inductive argument to show that the quadrature formula is exact for polynomials of degree $2n$ or less. *Hint:* Write a general polynomial of degree $2n$ as $\sum_{j=0}^{n} c_j x^j p_n(x) + q_{n-1}(x)$, where $q_{n-1}(x)$ is a polynomial of degree $n - 1$.

(*e*) Prove that the zeros of $p_n(x)$ are real and distinct. (They can also be shown to lie within the interval $[a, b]$.) For $n = 1$ and $[a, b] = [0, h]$, determine H_0, H_1, α, and a_1. What advantage does this quadrature formula have in composite rules? [Ref.: Ralston (1959).]

Section 4.11

54 Verify that if (4.11-10) holds for $i = 0, \ldots, n - 1$, then $I_{as}(f)$ as given by (4.11-4) satisfies (4.11-2).

55 Apply adaptive integration to each of the integrals of Prob. 12 with $\epsilon = .001$ for part (*a*), $\epsilon = .0000001$ for part (*b*), and $\epsilon = .00001$ for the others.

Section 4.12

56 Let $\phi(x) = \exp[-(a/x^p) - (b/(1 - x)^q)]$, where $a, b > 0$ and p and q are positive integers. Define

$$g(x) = \frac{1}{k} \int_0^x \phi(t)\, dt \qquad \text{where } k = \int_0^1 \phi(t)\, dt$$

(*a*) Show that $\int_0^1 f(x)\, dx = 1/k \int_0^1 f(g(x))\phi(x)\, dx \equiv 1/k \int_0^1 s(x)\, dx$.

(*b*) Show that if $f(x)$ is infinitely differentiable in $I = [0, 1]$, then so is $s(x)$ and, in addition, $s^{(k)}(x)$ vanishes at $x = 0$ and $x = 1$ for $k = 0, 1, \ldots$.

(*c*) By using the Euler-Maclaurin formula in the form (4.13-17) show that the trapezoidal rule should give good accuracy when applied to integrate $s(x)$ over I even if $f(x)$ exhibits singular behavior at the endpoints of I.

(*d*) Finally, derive the quadrature rules

$$\int_0^1 f(x)\,dx \approx \frac{1}{nk}\sum_{j=1}^{n-1} \phi\left(\frac{j}{n}\right) f\left(g\left(\frac{j}{n}\right)\right) = \sum_{j=1}^{n-1} H_j f(a_j)$$

where the H_j and a_j, $j = 1, \ldots, n-1$, can be precomputed for various values of n, and indicate how these rules can be used to evaluate integrals of functions with endpoint singularities.

Section 4.13

57 (*a*) Use the definitions of the Bernoulli polynomials and numbers and (4.13-6) to derive (4.13-7) and (4.13-8).

(*b*) Use these identities to derive (4.13-10).

(*c*) Use (4.13-2) to generate $B_k(x)$, $k = 0, 1, 2, 3$.

***58** (*a*) Show that

$$\frac{t}{e^t - 1} + \frac{t}{2} = \frac{t}{2}\coth\frac{t}{2}$$

and from this deduce the result (4.13-6).

(*b*) Similarly, show that

$$\frac{t}{e^{t/2} + 1} - \frac{t}{2} = -\frac{t}{2}\tanh\tfrac{1}{4}t$$

and from this deduce that $B_{2k+1}(\frac{1}{2}) = 0$, $k > 0$.

(*c*) Use (4.13-2) to show that

$$B_k(1 - x) + B_k = (-1)^k[B_k(x) + B_k]$$

and thus deduce that for $k \geq 1$, $B_{2k}(x - \frac{1}{2})$ is an even function and $B_{2k+1}(x - \frac{1}{2})$ is an odd function.

(*d*) Use parts (*b*) and (*c*), (4.13-7) and (4.13-8) to deduce that, if $B_{2k+1}(x)$ vanishes anywhere in $[0, 1]$ other than at $0, \frac{1}{2}$, and 1, then $B_{2k+1}(x)$ has at least five zeros on $[0, 1]$ and $B_{2k}(x) + B_{2k}$ has at least four zeros. Thus show that $B_{2k-1}(x)$ must also have five zeros. By continuing this argument, find a contradiction and thus deduce that $B_{2k}(x)$ can vanish only at the endpoints of $[0, 1]$ and that $B_{2k}(x)$ takes on its extreme value in $[0, 1]$ at $x = \frac{1}{2}$.

59 (*a*) Use Prob. 58 to deduce that (4.13-15) can be written

$$E_m = -\frac{nh^{2m+1}f^{(2m+2)}(\eta)}{(2m+2)!}\int_0^h B_{2m+2}\left(\frac{y}{h}\right) dy$$

when $x_0 < \eta < x_0 + nh$.

(*b*) Use (4.13-8) and Prob. 58 to show that

$$\int_0^1 B_{2m+2}(t)\,dt = -B_{2m+2}$$

(*c*) Use this result to deduce (4.13-16).

***60** (*a*) Use (4.13-7) and (4.13-8) to show that

$$B''_{2k+2}(x) = (2k+1)(2k+2)[B_{2k}(x) + B_{2k}]$$

(*b*) Use (4.13-3), Prob. 59*b*, and part (*a*) to deduce that $B''_{2k+2}(0)$ has a sign different from that of $B_{2k}(x)$ on [0, 1].

(*c*) Deduce from this that $B_{2k}(x)$ and $B_{2k+2}(x)$ have opposite signs on [0, 1] and therefore that B_{2k} and B_{2k+2} have opposite signs.

(*d*) Use this result and (4.13-16) to deduce that, if $f^{(2m+2)}(x)$ and $f^{(2m+4)}(x)$ do not change sign and have the same sign on $[x_0, x_0 + nh]$, then the error in the Euler-Maclaurin sum formula is less than the magnitude of the first neglected term.

***61** (*a*) Use (4.13-2) and (4.13-5) to show that

$$\frac{t}{e^{t/2} - e^{-t/2}} = \sum_{k=0}^{\infty} [B_k(\tfrac{1}{2}) + B_k] \frac{t^k}{k!}$$

(*b*) Use (4.13-5) to show that

$$\frac{t}{e^{t/2} - e^{-t/2}} = \sum_{k=0}^{\infty} (2^{-k+1} - 1) B_k \frac{t^k}{k!}$$

(*c*) Deduce then that

$$B_k(\tfrac{1}{2}) = 2(2^{-k} - 1)B_k$$

(*d*) Use this result, Eq. (4.13-15), and Prob. 58*d* to show that if $f^{(2m+2)}(x)$ does not change sign on $[x_0, x_0 + nh]$, then the error in the Euler-Maclaurin sum formula is less than twice the first neglected term.

62 If we sum the ordinates halfway between the successive ordinates in (4.13-14), we can derive the *second Euler-Maclaurin sum formula* [see Steffenson (1950), pp. 134–135]:

$$\sum_{j=0}^{n-1} f[x_0 + (j + \tfrac{1}{2})h] = \frac{1}{h} \int_{x_0}^{x_0+nh} f(y)\,dy - \sum_{k=1}^{m} \frac{(1 - 2^{1-2k})B_{2k} h^{2k-1}}{(2k)!} \times [f^{(2k-1)}(x_0 + nh) - f^{(2k-1)}(x_0)] + E_m$$

where

$$E_m = -n \frac{(1 - 2^{-1-2m})B_{2m+2} h^{2m+2}}{(2m+2)!} f^{(2m+2)}(\xi)$$

where $x_0 < \xi < x_0 + nh$. Show that when this formula is used as a quadrature formula, it corresponds to a composite rule based on the midpoint rule with correction terms. From this formula, derive an analog of Gregory's formula.

63 (*a*) Show that if Gregory's formula (4.13-18) is written to include differences through order *n*, it is exact for polynomials of degree *n*. Thus deduce that Gregory's formula with differences through order *n* is equivalent to a Newton-Cotes closed formula with $n + 1$ points.

(*b*) Write the equivalent to Gregory's formula using central differences. This formula is called *Gauss' formula*. (Use the notation of Prob. 13*c*, Chap. 3.) [Ref.: Hamming (1973), p. 344.]

64 Euler's constant γ is defined as

$$\gamma = \lim_{n \to \infty} \left(\sum_{i=1}^{n} \frac{1}{i} - \ln n \right) \approx .57721566$$

(*a*) Use the Euler-Maclaurin sum formula with $x_0 = 1$ to try to approximate γ to eight decimal places. Explain your results.

(*b*) Repeat part (*a*) with $x_0 = 10$ and again explain your results.

65 (*a*) Use Gregory's formula with $n = 16$ and retaining differences through the second to approximate the integral of part (*a*) of Prob. 12. (Use the results of Prob. 51*a* to get the trapezoidal-rule approximation.) Compare the results with those of Probs. 12, 37, and 51 and Table 4.8.

(*b*) Repeat part (*a*) with $n = 8$. How do you explain the result?

66 (*a*) Let $f(x)$ be a function defined on $I = [a, b]$ such that $f^{(2k-1)}(a) = f^{(2k-1)}(b)$, $k = 1, \ldots, m$. Use the Euler-Maclaurin formula to determine the error when the trapezoidal rule is used to integrate $f(x)$ over I.

(*b*) Show that there are m bounds on this error and give their form.

(*c*) Show why the trapezoidal rule is a good rule to use when integrating a smooth periodic function over a full period.

(*d*) Why should we not expect Romberg integration to improve on the results of the trapezoidal rule for the functions mentioned in parts (*a*) and (*c*)?

(*e*) Evaluate the integral

$$\int_0^1 \frac{2}{2 + \sin 10\pi x}\, dx = 1.15470054$$

by the trapezoidal rule using 2, 4, 8, and 16 intervals. Apply Richardson extrapolation to these values.

(*f*) Evaluate the integral of part (*e*) using Gauss-Legendre rules with the same number of function evaluations as in part (*e*) and compare the results.

[Ref.: Davis and Rabinowitz (1975), pp. 76 and 112.]

67 (*a*) Show that x^n can be written as a linear combination of the $x^{(k)}$, $k = 0, 1, \ldots, n$. The coefficients $s_n^{(k)}$ in this linear combination are called Stirling numbers of the second kind.

(*b*) How would you use part (*a*) to evaluate

$$\sum_{x=M}^{N-1} p(x)$$

where $p(x)$ is any polynomial?

(*c*) Show that $s_n^{(n)} = 1$ for any n and $s_n^{(0)} = 0$ for $n > 0$.

(*d*) Derive the recurrence relation

$$s_{n+1}^{(k)} = s_n^{(k-1)} + ks_n^{(k)}$$

and with this tabulate the Stirling numbers of the second kind for $n = 1, 2, 3, 4, 5$.

68 (*a*) Verify that $\bar{S}$ in Example 4.13 equals $\frac{1}{4}$.

(*b*) Use the technique of Sec. 4.13-2 to convert the problem of summing

$$S = \sum_{x=1}^{\infty} \frac{x}{(x^2 + 1)^2}$$

to one of summing a more rapidly convergent series.

(*c*) Repeat part (*b*) for

$$S = \sum_{k=1}^{\infty} \frac{1}{x^4}$$

By summing a few terms of S and the more rapidly convergent series, compare the convergence of both to the true result $\pi^4/90$. [Ref.: Hamming (1973), p. 196.]

69 Prove that, if (4.13-26) is satisfied when $R(x)$ and $\bar{R}(x)$ both have denominator degrees which exceed the numerator degrees by d, then $R(x) - \bar{R}(x)$ has a difference in degree between numerator and denominator of at least $d + 1$.

70 (*a*) Let $F_{j0} = (-1)^j v_j$, so that $S = \sum_{j=0}^{\infty} (-1)^j v_j = \sum_{j=0}^{\infty} F_{j0}$. Define

$$F_{jk} = \tfrac{1}{2}[F_{j,k-1} + F_{j+1,k-1}] \qquad k = 1, 2, \ldots$$

Show that
$$\frac{1}{2}\sum_{k=0}^{m} F_{0k} = \tfrac{1}{2}v_0 - \tfrac{1}{4}\Delta v_0 + \tfrac{1}{8}\Delta^2 v_0 - \cdots + (-1)^m \Delta^m v_0/2^{m+1}$$

So that we have an alternative formulation of the Euler transformation in terms of the elements of the original series and successive averages.

(*b*) Show that

$$\frac{1}{2}\sum_{k=0}^{m} F_{0k} + F_{0,m+1} = F_{00} + \frac{1}{2}\sum_{k=0}^{m} F_{1k}$$

and in general $$\frac{1}{2}\sum_{k=0}^{m} F_{jk} + F_{j,m+1} = F_{j0} + \frac{1}{2}\sum_{k=0}^{m} F_{j+1,k}$$

(*c*) Consider the following algorithm:

$S \leftarrow \frac{1}{2}F_{00};\ j \leftarrow 0;\ k \leftarrow 1$

if $|F_{jk}| < |F_{j+1,k-1}|$ **then** $S \leftarrow S + \frac{1}{2}F_{jk};\ k \leftarrow k+1$ **else** $S \leftarrow S + F_{jk};\ j \leftarrow j+1$
repeat

Show that this chooses an appropriate index m in the modified form of the Euler transformation given by (4.13-39).

(*d*) Apply this algorithm to Example 4.14.

71 (*a*) Let G_{j0} be the $(j+1)$st partial sum of the series S, that is, $G_{j0} = \sum_{i=0}^{j} F_{i0}$, where we use the notation of Prob. 70. Define $G_{jk} = \frac{1}{2}(G_{j,k-1} + G_{j+1,k-1})$, $k = 1, 2, \ldots$. Show that

$$G_{jk} = \sum_{i=0}^{j-1} F_{i0} + \frac{1}{2}\sum_{l=0}^{k} F_{jl}$$

so that G_{0n} is equal to the Euler transformation with $n+1$ terms and G_{mn} is equal to the modified Euler transformation given by (4.13-39).

(*b*) Apply this method of averaging to the series in Example 4.14.

(*c*) Apply all three versions of the Euler transformation to compute the value of the series $S = \sum_{j=0}^{\infty} (-1)^j(2j+1)^{-1}$ to six decimal places.

[Ref.: Dahlquist and Björck (1974), pp. 71–72.]

CHAPTER

FIVE

THE NUMERICAL SOLUTION OF ORDINARY DIFFERENTIAL EQUATIONS

5.1 STATEMENT OF THE PROBLEM

Our concern in this chapter is the solution of the first-order differential equation

$$\frac{dy}{dx} = f(x, y) \qquad y(x_0) = y_0 \tag{5.1-1}$$

We assume that

1. $f(x, y)$ is defined and continuous in the strip $x_0 \le x \le b$, $-\infty < y < \infty$ with x_0 and b finite.
2. There exists a constant L such that for any x in $[x_0, b]$ and any two numbers y and y^*

$$|f(x, y) - f(x, y^*)| \le L|y - y^*|$$

These conditions are sufficient to prove that there exists on $[x_0, b]$ a unique continuous, differentiable function $y(x)$ satisfying (5.1-1).† The Lipschitz condition (2) is weaker than the assumption that $\partial f/\partial y$ is continuous and bounded (why?).

† The reader who has not recently looked at the basic existence theory for the solution of (5.1-1) would be well advised to do so now; see, for example, Henrici (1962*a*, pp. 15–26).

More generally, we may consider (5.1-1) to be a system of N first-order differential equations in which y, y_0, and f are vectors with N components. Much of what we shall develop in this chapter will be equally true for single equations and for systems. However, some of our results, particularly those on stability, will be developed for single equations only, because with systems the algebraic problems become intractable. Considered as a system, the formulation (5.1-1) is quite general in the sense that any higher-order equation or system of higher-order equations can be reduced to (5.1-1) if (and only if) the system can be rewritten with the highest-order derivative in each dependent variable appearing as the left-hand side of one equation and appearing nowhere else {1}. For example, the equation

$$\frac{d^2y}{dx^2} = f(x, y, y') \tag{5.1-2}$$

can be written

$$\frac{dz}{dx} = f(x, y, z) \qquad \frac{dy}{dx} = z \tag{5.1-3}$$

Our object in solving (5.1-1) will be to find y at a sequence of values of x, $\{x_i\}$. We distinguish here between two types of methods for effecting this solution:

1. Those in which in order to compute an approximation y_i to the true solution at the point x_i information about the solution is required solely at the previous point x_{i-1}. Such methods are called *single-step methods.* Two examples of this class of methods are Runge-Kutta methods (see Sec. 5.8) and Taylor-series methods (see Sec. 5.6-1).
2. Those in which information is needed at several previous points. These are called *multistep methods.* The case in which the points x_i are equally spaced has been extensively studied and will be treated here in depth.

There are also methods which do not clearly fall into either category, e.g., the *extrapolation methods* to be discussed in Sec. 5.9-2 and the so-called hybrid methods, which use both information from previous points and additional newly generated information. These latter will not be discussed here.

For reasons which will become clear later in the chapter, our emphasis will be on multistep methods, although Runge-Kutta methods have an important role to play, as we discuss in Sec. 5.8.

In deriving methods for the numerical solution of differential equations, the following considerations will be important:

1. How much error is incurred at each step of the computation (truncation and roundoff error) and how this error affects the results in subsequent

steps. This is the first instance in which we have had to consider the *propagation* of error incurred at one stage of a calculation into later stages. This extremely important phenomenon, which occurs in many areas of numerical analysis (see, in particular, Chap. 9), is generally discussed under the heading of stability, a stable method (see Sec. 1.7) being one in which errors incurred at one step do not tend to be magnified in later steps (we shall formalize this intuitive notion in Sec. 5.4).

2. Related to the problem of errors and error propagation is the problem of being able to estimate the error at a given stage of the computation as a function of computed results.
3. How the solution can be started. Equation (5.1-1) contains an initial condition at $x = x_0$. But multistep methods require values of y at more than one point to compute another point. Thus auxiliary means of starting the computation will often be required. Closely related to this problem is that of changing the interval between successive x_i's during the course of the computation.
4. The speed of the method. In the solution of large systems of equations ($N > 100$), the time required for the computation—even on the fastest of computers—can be considerable. Since no reasonable discussion of problems in numerical analysis can avoid such a practical consideration, speed of computation will affect our evaluation of the methods to be derived here.

5.2 NUMERICAL INTEGRATION METHODS

First we introduce some notation. Let $Y(x)$ be the true solution of (5.1-1) and $y(x)$ the calculated solution. Further, let

$$\begin{aligned} Y_i &= Y(x_i) & y_i &= y(x_i) \\ Y_i' &= \left.\frac{dY}{dx}\right|_{x=x_i} = f(x_i, Y_i) & y_i' &= f(x_i, y_i) \\ h_i &= x_{i+1} - x_i \end{aligned} \tag{5.2-1}$$

Note that since Y is the true solution, $f(x_i, Y_i)$ is equal to $dY/dx|_{x=x_i}$. However, the function $y(x)$ "exists" only at the points x_i, $i = 1, 2, 3, \ldots$. Thus, when we replace $f(x_i, y_i)$ by y_i', this is a notational convenience.

A quite general equation for computing y_i, which includes both multistep methods and Taylor-series methods, can be derived from the equation

$$Y(x_i) = \sum_{j=0}^{n} \sum_{k=0}^{m_j} A_{jk} Y^{(k)}(x_{i-j}) + T \tag{5.2-2}$$

where $A_{00} = 0$ and where we use T instead of E for the error to emphasize that this is truncation error. Actually, T is usually called the *local truncation error* because it is the error incurred in one step of the integration (as opposed to the global error over many steps; see Sec. 5.4). The equation with which we calculate is then

$$y_i = \sum_{j=0}^{n} \sum_{k=0}^{m_j} A_{jk} y_{i-j}^{(k)} \tag{5.2-3}$$

with at least one A_{j0} required to be nonzero so that each calculated value depends upon at least one previous value of y. Also at least one $A_{j1} \neq 0$, so that (5.2-3) indeed depends on the differential equation. By analogy with (5.2-1), we can calculate higher derivatives of y than the first. For example,

$$y''(x_l) = \frac{d}{dx} f(x, y) \bigg|_{(x_l, y_l)} \tag{5.2-4}$$

But because this tends to be quite tedious if $f(x, y)$ is at all complicated, and because the analysis of methods using higher derivatives can become very difficult, we shall restrict ourselves mainly to the case $m_j = 1$. In Sec. 5.9-1, we shall consider briefly methods with $m_j > 1$.

In this section we shall not only assume that $m_j = 1$ but also that the interval h_i between steps will be a constant, h, at least over a number of steps, although it can be changed whenever error considerations make this desirable; see Sec. 5.6-3. Methods conforming to these assumptions are called *numerical integration methods*. Using these assumptions in (5.2-3), we can rewrite this equation as

$$y_{n+1} = \sum_{i=0}^{p} a_i y_{n-i} + h \sum_{i=-1}^{p} b_i y'_{n-i} \tag{5.2-5}$$

where we have further changed the notation to let the last value of x at which y was calculated be x_n and to let the number of past values used to compute y_{n+1} be $p + 1$. The interval h has been introduced for later notational convenience.

The following points about (5.2-5) should be noted:

1. In any particular specialization of (5.2-5) any of the a_i's or b_i's may be zero, but we assume that either a_{n-p} or b_{n-p} is not zero.
2. If $b_{-1} = 0$, then y_{n+1} is expressed as a linear combination of (computationally) known past values of y_{n-i} and functions of the y_{n-i} and is thus easily computed. Formulas with $b_{-1} = 0$ are called explicit or *forward-integration formulas*.
3. If $b_{-1} \neq 0$, (5.2-5) is only an implicit equation for y_{n+1} since $y'_{n+1} = f(x_{n+1}, y_{n+1})$ and will generally be solvable only by an iterative procedure. It will probably come as no surprise to the reader that the greater difficulty of using formulas with $b_{-1} \neq 0$ is made up for by their more desirable properties (otherwise they would hardly be worth a

mention). Formulas with $b_{-1} \neq 0$ are called implicit or *iterative formulas*. [Analogously, (5.2-3) is said to be implicit unless all $A_{0k} = 0$, in which case it is called explicit.] Note, however, that if the differential equation (5.1-1) is linear, (5.2-5) can be solved explicitly for y_{n+1} {3}.

When $b_{-1} = 0$, (5.2-5) is an extrapolation equation in the sense that it estimates a value of y at a point x_{n+1} outside the interval spanned by x_{n-i}, $i = 0, \ldots, p$. When $b_{-1} \neq 0$, (5.2-5) still defines y_{n+1} as some function of $y_n, \ldots, y_{n-p}, y'_n, \ldots, y'_{n-p}$ and is thus also an extrapolation equation. Thus we can say that the numerical solution of ordinary differential equations is essentially a process of successive extrapolations.

5.2-1 The Method of Undetermined Coefficients

We are ready now to consider specifying the coefficients in (5.2-5). As in the previous two chapters, we shall do this so as to make (5.2-5) exact if $Y(x)$ is a polynomial of some specified degree. By "exact" here we mean that if $Y(x)$ is a polynomial of the specified degree, and if the values on the right-hand side of (5.2-5) are true values of the solution, then, except for roundoff, y_{n+1} will also be a true value. In contrast to the previous two chapters, however, we shall often not specify the coefficients so as to achieve the highest possible order of accuracy. If, for example, we have five coefficients and determine them so that (5.2-5) is exact for polynomials of degree 3 or less only, then in general we shall have one free parameter which we may use to do one or more of the following:

1. Make the coefficient in the error term small
2. Make the error-propagation properties of the formula as desirable as possible
3. Give the formula certain other desirable computational properties such as zero coefficients

Let us suppose that we wish to make (5.2-5) exact for polynomials of degree r. As in the previous chapter, we shall say that such a formula has an accuracy of order r or is of order r. Then the method of undetermined coefficients consists in considering the $r + 1$ equations derived by letting $y_i = x_i^j$, $j = 0, \ldots, r$, in (5.2-5). Before actually doing this, we note that (1) there is no loss of generality in setting $h = 1$ because putting $y_i = x_i^j$ in (5.2-5) results in the cancellation of h (why?) and (2) there is no loss of generality in letting $x_n = 0$ since the coefficients will be independent of the origin of the coordinate system (why?).

With $h = 1$ and $x_n = 0$, we have

$$
\begin{aligned}
1 &= \sum_{i=0}^{p} a_i && j = 0 \\
1 &= -\sum_{i=0}^{p} i a_i + \sum_{i=-1}^{p} b_i && j = 1 \qquad (5.2\text{-}6) \\
1 &= \sum_{i=0}^{p} (-i)^j a_i + j \sum_{i=-1}^{p} (-i)^{j-1} b_i && j = 2, \ldots, r
\end{aligned}
$$

These are $r + 1$ equations for the $2p + 3$ or fewer coefficients in (5.2-5) (some of the coefficients, for example, b_{-1}, may be postulated equal to zero). If the number of coefficients is $r + 1$, then, generally, (5.2-6) can be solved for the a_i's and b_i's. If the number of coefficients is greater than $r + 1$, we shall have free parameters in general, and if the number is less than $r + 1$, there will in general be no solution.

If the first two equation of (5.2-6) are satisfied, we say that the associated numerical integration method is *consistent*. Consistency is thus equivalent to (5.2-5) being exact for linear polynomials. All the numerical integration procedures we shall consider will be consistent.

Example 5.1 Determine the coefficients in

$$y_{n+1} = a_0 y_n + h(b_{-1} y'_{n+1} + b_0 y'_n) \qquad (5.2\text{-}7)$$

if the formula is to be exact for polynomials of degree 2, that is, $r = 2$, $p = 0$.

The equations (5.2-6) are

$$1 = a_0 \qquad 1 = b_{-1} + b_0 \qquad 1 = 2b_{-1} \qquad (5.2\text{-}8)$$

so that the equation is

$$y_{n+1} = y_n + \frac{h}{2}(y'_{n+1} + y'_n) \qquad (5.2\text{-}9)$$

If instead of polynomials of degree 2 we had required exactness only for polynomials of degree 1, then only the first two equations of (5.2-8) would have to be satisfied and (5.2-7) would then be

$$y_{n+1} = y_n + h[(1 - b_0)y'_{n+1} + b_0 y'_n]$$

with b_0 a free parameter.

Equation (5.2-9) is precisely the trapezoidal rule, as is easily seen by replacing $f(x)$ by $y'(x)$ in (4.10-11). Since it is the trapezoidal rule, we know that the truncation error incurred at each step is $-h^3 Y'''(\eta)/12$, $x_n < \eta < x_{n+1}$. This truncation error needs to be properly interpreted. It would be the difference between y_n and the true solution (exclusive of round-off) if the values on the right-hand side of (5.2-9) were *true values*. It is also worth noting at this point that the truncation error in formulas of the form

(5.2-5) will not always be simply determinable as in this case; see the next section.

Example 5.1 is a special case of the general result that any Newton-Cotes quadrature formula becomes a numerical integration formula if $f(x)$ is replaced by $y'(x)$. More generally, if we replace $f(x)$ by $y'(x)$ in the Lagrangian interpolation formula at equal intervals for the points $x_n, x_{n-1}, \ldots, x_{n-p}$ and then integrate between x_{n-j} and x_{n+1} for any j, the result is a formula of the form (5.2-5) {5}. We shall not consider this way of generating numerical integration formulas any further because the method of undetermined coefficients enables us to derive all numerical integration methods, including many of interest which are not derivable from Newton-Cotes formulas. We shall, however, use the idea of integrating interpolation polynomials to derive here two special cases of (5.2-3) and their specialization to numerical integration formulas, expressed in a different notation.

The first class of methods is the *Adams-type* methods, which have recently become popular again because of their good stability properties. These methods are based on the formula

$$Y(x_{n+1}) = Y(x_n) + \int_{x_n}^{x_{n+1}} f(x, Y(x))\, dx \qquad f(x, Y(x)) = Y'(x) \quad (5.2\text{-}10)$$

If we have approximations $y_n, y_{n-1}, \ldots, y_{n-p}$ to $Y(x)$ at the points $x_n, x_{n-1}, \ldots, x_{n-p}$, not necessarily equally spaced, then we also know $y'_{n-i} = f(x_{n-i}, y_{n-i})$, $i = 0, \ldots, p$, and we can approximate $f(x, Y(x))$ by an interpolating polynomial $Q_p(x)$. This yields the explicit Adams formula

$$y_{n+1} = y_n + \int_{x_n}^{x_{n+1}} Q_p(x)\, dx = y_n + \sum_{i=0}^{p} d_i y'_{n-i} \qquad (5.2\text{-}11)$$

where the d_i are functions of the points $x_n, \ldots, x_{n-p}$. If, on the other hand, we approximate $f(x, Y(x))$ by an interpolating polynomial $Q_{p+1}(x)$ involving also the unknown value $y'_{n+1} = f(x_{n+1}, y_{n+1})$, we get the implicit Adams formula

$$y_{n+1} = y_n + \sum_{i=-1}^{p} \hat{d}_i y'_{n-i} \qquad (5.2\text{-}12)$$

For equally spaced points, it is convenient to write $Q_p(x)$ and $Q_{p+1}(x)$ using the Newton backward-interpolation formulas (3.3-15) based on f_n and f_{n+1}, respectively, where it is convenient to use f_i instead of y'_i. This gives the *Adams-Bashforth* explicit formula

$$y_{n+1} = y_n + h(f_n + \tfrac{1}{2} \nabla f_n + \tfrac{5}{12} \nabla^2 f_n + \cdots) \qquad (5.2\text{-}13)$$

and the *Adams-Moulton* implicit formula

$$y_{n+1} = y_n + h(f_{n+1} - \tfrac{1}{2} \nabla f_{n+1} - \tfrac{1}{12} \nabla^2 f_{n+1} - \cdots) \qquad (5.2\text{-}14)$$

These formulas can also be written in the Lagrangian form (5.2-5).

The second class of methods consists of implicit formulas based on numerical differentiation. They are important for stiff problems (see Sec. 5.10). The idea here is as follows. If we approximate $Y(x)$ by a polynomial $Q_{p+1}(x)$ taking on known values $y_n, \ldots, y_{n-p}$ at the points $x_n, \ldots, x_{n-p}$ and the unknown value y_{n+1} at x_{n+1}, differentiate $Q_{p+1}(x)$, and set the value of $Q'_{p+1}(x_{n+1})$ equal to $y'_{n+1} = f(x_{n+1}, y_{n+1})$, we get an implicit equation for y_{n+1} of the form

$$y_{n+1} = \sum_{i=0}^{p} c_i y_{n-i} + d y'_{n+1} \tag{5.2-15}$$

where the coefficients depend only on the points $x_{n+1}, \ldots, x_{n-p}$. In particular, if the points are equally spaced, and if we write $Q_{p+1}(x)$ using the Newton backward formula (3.3-15) and differentiate, we get the implicit formula

$$\nabla y_{n+1} + \tfrac{1}{2} \nabla^2 y_{n+1} + \cdots + \frac{1}{m} \nabla^m y_{n+1} = h y'_{n+1} \tag{5.2-16}$$

which can also be written in form (5.2-5). In this case we can derive the coefficients a_i and b_{-1} by the method of undetermined coefficients.

5.3 TRUNCATION ERROR IN NUMERICAL INTEGRATION METHODS

As we noticed in the previous section, when a numerical integration method is equivalent to one of the numerical quadrature methods of Chap. 4, the truncation-error term can be easily written down. In general, however, numerical integration methods of the form (5.2-2) are not equivalent to numerical quadrature methods. Our object here is to determine the truncation error T in (5.2-2) for formulas of the specific type (5.2-5), although the method we present is also applicable to the general case (5.2-2). If the numerical integration method is of the form (5.2-5), the true solution satisfies the equation

$$Y_{n+1} = \sum_{i=0}^{p} a_i Y_{n-i} + h \sum_{i=-1}^{p} b_i Y'_{n-i} + T_n \tag{5.3-1}$$

where the notation T_n denotes the truncation error at the step from x_n to x_{n+1}.

In order to find an expression for T_n, it is tempting to assume that

$$T_n = ch^{r+1} Y^{(r+1)}(\eta) \tag{5.3-2}$$

where c is a constant and r is the order of accuracy of (5.3-1), in analogy with the trapezoidal-rule error and the general form of the error in numerical quadrature methods. If T_n has the form (5.3-2), c can be found by letting $Y(x)$ be x^{r+1} and substituting this into (5.3-1) (why?). In fact, most numerical integration as well as most quadrature formulas of interest do have errors of the form (5.3-2). However, as can be seen from Theorem 2.3, the truncation error cannot always be written in the form (5.3-2). Indeed, let us assume that (5.3-1) is exact when $Y(x)$ is a polynomial of degree r or less. Then we have

$$T_n = \int_{x_{n-p}}^{x_{n+1}} Y^{(n+1)}(s)G(s)\, ds \tag{5.3-3}$$

where, in this case, the Peano kernel or influence function $G(s)$ satisfies

$$G(s) = \frac{1}{r!}\left[(x_{n+1} - s)_+^r - \sum_{i=0}^{p} a_i(x_{n-i} - s)_+^r - hr \sum_{i=-1}^{p} b_i(x_{n-i} - s)_+^{r-1}\right] \tag{5.3-4}$$

where, as in (2.3-6),

$$(x - s)_+^k = \begin{cases} (x - s)^k & x \geq s \\ 0 & x < s \end{cases} \tag{5.3-5}$$

If $G(s)$ is of constant sign in $[x_{n-p}, x_{n+1}]$, we can apply the second law of the mean to get

$$T_n = \frac{Y^{(r+1)}(\eta)}{r!} \int_{x_{n-p}}^{x_{n+1}} G(s)\, ds \tag{5.3-6}$$

where $x_{n-p} < \eta < x_{n+1}$. Equation (5.3-6) is indeed of the form (5.3-2) (why?). But if $G(s)$ does change sign in $[x_{n-p}, x_{n+1}]$, the error cannot be expressed in the form (5.3-2) {7}, although we can still bound the error as

$$|T_n| \leq \frac{|Y^{(r+1)}(\eta)|}{r!} \int_{x_{n-p}}^{x_{n+1}} |G(s)|\, ds \tag{5.3-7}$$

Some examples of the use of the influence function to find errors are considered in the problems {8, 9, 10}.

Example 5.2 Consider the numerical integration method

$$y_{n+1} = (1 - a)y_n + ay_{n-1} + \frac{h}{12}[(5 - a)y'_{n+1} + (8 + 8a)y'_n + (5a - 1)y'_{n-1}] \tag{5.3-8}$$

where the parameter a is to be specified. It can be easily verified that (5.3-8) is exact for polynomials of degree 3 or less for all a (and for polynomials of degree 4 when $a = 1$). Using $r = 3$ in (5.3-4), we get for the influence function

$$3!\,G(s) =$$

$$\begin{cases} (x_{n+1} - s)^3 - \frac{h}{4}(5 - a)(x_{n+1} - s)^2 & x_n \leq s \leq x_{n+1} \\ (x_{n+1} - s)^3 - (1 - a)(x_n - s)^3 - \frac{h}{4}[(5 - a)(x_{n+1} - s)^2 + (8 + 8a)(x_n - s)^2] & x_{n-1} \leq s \leq x_n \end{cases} \tag{5.3-9}$$

from which it is verifiable that for $a \le \frac{1}{5}$, $G(s) \le 0$ in $[x_{n-1}, x_{n+1}]$ and for $a \ge 5$, $G(s) \ge 0$ in this interval but that for any other value of a, $G(s)$ changes sign in the interval. For $a \le \frac{1}{5}$ and $a \ge 5$, we can use (5.3-6) to express the error as

$$T_n = \frac{-(1-a)h^4}{24} Y^{\text{iv}}(\eta) \tag{5.3-10}$$

(The case $a = 1$ will be considered in Example 5.5.)

The method we have used here to find the error in numerical integration methods can also be fruitfully applied to finding the error in numerical quadrature methods. For consider the general quadrature formula (4.5-1). If the accuracy of this formula is r, then by Theorem 2.3,

$$E = \int_a^b f^{(r+1)}(s)G(s)\, ds \tag{5.3-11}$$

where
$$G(s) = \int_a^b w(x)(x-s)_+^r\, ds - \sum_{j=1}^{n} H_j(a_j - s)_+^r \tag{5.3-12}$$

We could have used this technique to calculate most of the error terms in the quadrature formulas of Chap. 4 as well as the errors in interpolation and numerical differentiation.

5.4 STABILITY OF NUMERICAL INTEGRATION METHODS

In this section we shall be interested (1) in considering how well the solution of the difference equation (5.2-5) approximates that of the differential equation and (2) how we can derive bounds for the accumulated error at any stage. We shall limit ourselves to the case $N = 1$, that is, a single first-order ordinary differential equation. Some of the results we shall obtain are directly applicable to systems of equations. For others, the extension to systems leads to generally intractable algebraic problems {11}.

To get an intuitive feeling for this problem, it is instructive to consider the differential equation

$$y' = -Ky \qquad y(x_0) = y_0 \tag{5.4-1}$$

whose solution is

$$Y = y_0 e^{-K(x-x_0)} \tag{5.4-2}$$

For this differential equation, (5.2-5) becomes

$$y_{n+1}(1 + hKb_{-1}) = \sum_{i=0}^{p} (a_i - hKb_i)y_{n-i} \tag{5.4-3}$$

This is a linear difference equation with constant coefficients and has the solution

$$y_n = \sum_{i=0}^{p} c_i r_i^n \tag{5.4-4}$$

where the c_i's are constants and the r_i's are the roots of

$$(1 + hKb_{-1})r^{p+1} = \sum_{i=0}^{p} (a_i - hKb_i)r^{p-i} \tag{5.4-5}$$

We have assumed in (5.4-4) that the roots in (5.4-5) are distinct. If the roots are not distinct, then terms of the form $c_i n^\alpha r_i^n$ will appear in (5.4-4) (where α is less than the multiplicity of the root). Such multiple roots will affect the details but not the substance of the development that follows.

The first thing we wish to show is that one of the roots of (5.4-5), say r_0, has the form†

$$r_0 = 1 - Kh + O(h^2) \tag{5.4-6}$$

Since we are assuming that (5.2-5) is a consistent method, we get from the first equation of (5.2-6) that when $h = 0$, (5.4-5) has a root $r_0 = 1$. For $h \neq 0$, suppose we write this root as

$$r_0 = 1 + \sum_{i=1}^{\infty} \beta_i h^i \tag{5.4-7}$$

Substituting this into (5.4-5) and using the second equation of (5.2-6), we can show {12} that $\beta_1 = -K$, which establishes (5.4-6).‡ Now since

$$1 - Kh = e^{-Kh} + O(h^2) \tag{5.4-8}$$

we can write

$$\begin{aligned} r_0^k &= [1 - Kh + O(h^2)]^k \\ &= e^{-Khk} + O(h^2) \\ &= e^{-K(x_k - x_0)} + O(h^2) \end{aligned} \tag{5.4-9}$$

The constants c_i are determined by the $p + 1$ initial conditions required to solve (5.4-3). We have noted previously that, in general, numerical integration methods require starting values that are not available from the statement of the problem, and in Sec. 5.6 [see also Sec. 5.8] we shall consider how to obtain these values. Here we are interested in calculating c_0, the coefficient of r_0^n. Let $y_0, \ldots, y_p$ be the initial values of y, with y_0 the given

† If $f(x) = O[g(x)]$ as $x \to x_0$, there exists a positive constant c such that $|f(x)| \leq c|g(x)|$ for x sufficiently close to x_0. Except where specified otherwise, we shall always have $x_0 = 0$.

‡ In fact, the $O(h^2)$ term is actually $O(h^{r+1})$, where r is the order of the numerical integration method (why?).

initial condition of the differential equation. Then (5.4-4) is a system of $p + 1$ equations for the c_i's. Using Cramer's rule, we have

$$c_0 = \frac{\begin{vmatrix} y_0 & 1 & \cdots & 1 \\ y_1 & r_1 & \cdots & r_p \\ \cdots & \cdots & \cdots & \cdots \\ y_p & r_1^p & \cdots & r_p^p \end{vmatrix}}{\begin{vmatrix} 1 & \cdots & \cdots & 1 \\ r_0 & \cdots & \cdots & r_p \\ r_0^2 & \cdots & \cdots & r_p^2 \\ \cdots & \cdots & \cdots & \cdots \\ r_0^p & \cdots & \cdots & r_p^p \end{vmatrix}} \tag{5.4-10}$$

The initial conditions $y_1, \ldots, y_p$ will generally all have errors in them, but it is reasonable to assume that as $h \to 0$, these values approach true values. This is equivalent to saying that whatever process we use to get the initial values, the error will be the order of some power of h; see Sec. 5.6. Thus the first column of the numerator approaches y_0 times the first column of the denominator as $h \to 0$, which in turn means that $c_0 \to y_0$ as $h \to 0$. This, together with our result on r_0, means that the first term of (5.4-4) approximates the true solution of the differential equation and, in fact, approaches it as $h \to 0$.

What then of the remaining terms of (5.4-4), the so-called *parasitic solution?* Note that the parasitic solution arises because the order of the difference equation $(p + 1)$ is greater than the order of the differential equation (1). The same argument used to estimate c_0 above can be used to show that as $h \to 0$, all the other c_i approach zero {12}. Thus, for small h, we expect the coefficients $c_1, \ldots, c_p$ to be small. If the solution of the difference equation is to be a useful approximation to the solution of the differential equation, each of the terms $c_i r_i^n$ in (5.4-4) must remain small with respect to $c_0 r_0^n$. This requires that

$$|r_i| \leq |r_0| \qquad i = 1, \ldots, p \tag{5.4-11}$$

The roots r_i, $i = 0, \ldots, p$, are functions of h. As $h \to 0$, we know that $r_0 \to 1$. Thus, for (5.4-11) to hold as $h \to 0$, it is necessary *that all roots of* (5.4-5) *lie on or within the unit circle.* If r_i, $i > 0$ lies on the unit circle when $h = 0$, then (5.4-11) may not hold for any positive h, but if all r_i, $i > 0$, lie within the unit circle, then for some range of positive h (5.4-11) will hold (why?). Notice that for Adams-type methods, there is only one root $r_0 = 1$ when $h = 0$. Thus we are always assured an interval in h in which all the roots r_i, $i > 0$, satisfy (5.4-11). Our first conclusion, then, is that if the solution of (5.4-3) is to be a good approximation to the solution of (5.4-1), then (5.4-11) must hold.

Now let us consider the general case when the differential equation has the form (5.1-1). Because of the derivative terms, we cannot solve (5.2-5) explicitly. But as $h \to 0$, the solution of (5.2-5) must approach that of (5.4-3). Thus, for sufficiently small h, we would expect to get results analogous to those above. We are now ready to formalize the notion of stability as it applies to a general equation of the form (5.1-1).

5.4-1 Convergence and Stability

Definition 5.1 Let the initial conditions $y_k = y_k(h)$, $k = 0, \ldots, p$, used in solving the difference equation (5.2-5) be such that

$$\lim_{h \to 0} y_k(h) = y_0 \qquad k = 0, \ldots, p \tag{5.4-12}$$

where y_0 is the given initial condition in the differential equation (5.1-1). Then the numerical integration method (5.2-5) is said to be *convergent* if for any initial-value problem (5.1-1) such that $f(x, y)$ satisfies the conditions of Sec. 5.1 the solution of (5.2-5) is such that

$$\lim_{\substack{h \to 0 \\ (n \to \infty)}} y_n = Y(x) \qquad hn = x - x_0 \tag{5.4-13}$$

for all $x \in [x_0, b]$.

The condition (5.4-12) is, of course, required because the initial conditions used in (5.2-5) will generally not be true solutions of (5.1-1). Our discussion of Eq. (5.4-1) then implies the following theorem.

Theorem 5.1 A necessary condition for a numerical integration method to be convergent is that no root of

$$r^{p+1} = \sum_{i=0}^{p} a_i r^{p-i} \tag{5.4-14}$$

that is, (5.4-5) with $h = 0$, lies outside the unit circle and that roots of magnitude 1 are simple.

PROOF Consider the equation $y' = 0$, $y(0) = 0$ whose exact solution is $Y(x) = 0$. Let the roots of (5.4-14) be $r_0, r_1, \ldots, r_p$ and suppose that they are real and simple. Then for this equation $y_k = \sum_{i=0}^{p} hr_i^k$ is a solution of (5.2-5) for $k = p + 1, p + 2, \ldots$ (why?). Moreover, for $k = 0, \ldots, p$, y_k satisfies (5.4-12) (why?). In order for y_n to satisfy (5.4-13), it is clear that the magnitude of each r_i must be less than or equal to 1. This proves the theorem in this case. In the case of complex or multiple roots, some more proof is required; we leave this to a problem {13}.

For the equation $y' = 0$, $y(0) = 0$, the sufficiency of the condition in Theorem 5.1 is not hard to prove {13}. In fact, for consistent methods the condition is sufficient in general, but the proof of this is beyond the scope of this book [see Henrici (1962*a*), pp. 244–246].

From the first equation of (5.2-6) it follows that $r = 1$ is always a root of (5.4-14). Our discussion of stability later in this section will indicate that it is in fact desirable to have the other roots of (5.4-14) as small as possible in magnitude.

Example 5.3 Determine for what values of a the method of Example 5.2 is convergent.

Equation (5.4-14) is, since $p = 1$,

$$r^2 = (1 - a)r + a$$

whose solutions are

$$r = 1 \qquad \text{and} \qquad r = -a$$

Thus, for convergence $-1 < a \leq 1$. Note that a must be greater than -1 to avoid a double root at 1.

Another necessary condition for convergence is given by the following theorem.

Theorem 5.2 A necessary condition for the convergence of a numerical integration method is that the first two equations of (5.2-6) must be satisfied; i.e., consistency is necessary for convergence.

PROOF Consider the equation $y' = 0$, $y(0) = 1$, whose exact solution is $Y(x) = 1$. For this equation, the numerical integration method (5.2-5) becomes

$$y_{n+1} = \sum_{i=0}^{p} a_i y_{n-i} \tag{5.4-15}$$

Let the starting values $y_0, \ldots, y_p$ be exact, i.e., equal to 1. Now letting $h \to 0$ and $n \to \infty$ $(nh = x)$ means that all values of y in (5.4-15) must approach 1 if the method is convergent. This proves that the first equation of (5.2-6) must be satisfied. Now consider $y' = 1$, $y(0) = 0$, whose solution is $Y(x) = x$. The difference equation now is

$$y_{n+1} = \sum_{i=0}^{p} a_i y_{n-i} + h \sum_{i=-1}^{p} b_i \tag{5.4-16}$$

Now consider the sequence defined by

$$y_n = nhA \qquad n = 0, 1, \ldots \tag{5.4-17}$$

where

$$A = \frac{\sum_{i=-1}^{p} b_i}{1 + \sum_{i=0}^{p} ia_i} \tag{5.4-18}$$

This sequence satisfies the restrictions (5.4-12) on the initial conditions and also satisfies the difference equation (5.4-16) {14}. Since the solution of (5.4-16) must approach the true solution as $h \to 0$, $n \to \infty$ $(nh = x)$, we conclude from (5.4-17) that $A = 1$. Then (5.4-18) is the second equation of (5.2-6), which completes the proof.

We have defined convergence in the limit as $h \to 0$. But in practice we are interested in what happens for finite values of h. First, we should like to know when the parasitic solutions of (5.2-5) are small in relation to the solution which approximates the solution of the differential equation, and when this is so, we should like to be able to estimate or bound the error in the computed solution. In order to discuss these matters, it is convenient to introduce the *accumulated* error after n steps, which is the difference between the true solution and the computed solution. We define

$$\epsilon_n = Y_n - y_n \tag{5.4-19}$$

Before we derive an equation for ϵ_n, we correct (5.2-5) by introducing the roundoff error R_n at each step

$$y_{n+1} = \sum_{i=0}^{p} a_i y_{n-i} + h \sum_{i=-1}^{p} b_i y'_{n-i} + R_n \tag{5.4-20}$$

This roundoff error R_n consists of a combination of roundoff errors in the y_{n-i}, roundoff errors in the computation of $y'_{n-i} = f(x_{n-i}, y_{n-i})$, and roundoff errors in the machine computation of (5.2-5). In most solutions of ordinary differential equations on digital computers, truncation error is far larger than roundoff error. However, in certain applications, particularly in the case of real-time† computations, roundoff error may be significant. Moreover, roundoff error limits the values of h which can be used. For example, suppose we take a simple-minded scheme such as *Euler's method*, which is derived by dropping all terms involving differences of f_n in (5.2-13):

$$y_{n+1} = y_n + hf(x_n, y_n) \qquad y_0 = y(x_0) \tag{5.4-21}$$

It can be shown that this method yields a solution which converges to the true solution $Y(x)$ as $h \to 0$. However, if we compute with a sequence of values of h tending to zero, we shall find that because of roundoff error there is a threshold below which we cannot reduce h and still retain a meaningful computation.

Subtracting (5.4-20) from (5.3-1), we get

$$\epsilon_{n+1} = \sum_{i=0}^{p} a_i \epsilon_{n-i} + h \sum_{i=-1}^{p} b_i \epsilon'_{n-i} + E_n \tag{5.4-22}$$

† In real-time applications the computer is intimately tied to an operative physical system, e.g., missile tracking or on-line control of a chemical plant. In such cases because of inherent physical inaccuracies the number of digits used in the computation may be small, and thus the roundoff may be large.

where $E_n = T_n - R_n$ is the error *introduced* at the step from x_n to x_{n+1} (in contrast to ϵ_n, which is the total error that has *accumulated* after the n steps). Using the mean-value theorem, we write

$$\begin{aligned}\epsilon'_{n-i} &= Y'_{n-i} - y'_{n-i} = f(x_{n-i}, Y_{n-i}) - f(x_{n-i}, y_{n-i}) \\ &= (Y_{n-i} - y_{n-i})f_y(x_{n-i}, \eta_{n-i}) = \epsilon_{n-i} f_y(x_{n-i}, \eta_{n-i}) \qquad (5.4\text{-}23)\end{aligned}$$

where the subscript y denotes partial differential and η_{n-i} lies between y_{n-i} and Y_{n-i}. Substituting (5.4-23) into (5.4-22), we get

$$\epsilon_{n+1}[1 - hb_{-1} f_y(x_{n+1}, \eta_{n+1})] = \sum_{i=0}^{p} [a_i + hb_i f_y(x_{n-i}, \eta_{n-i})]\epsilon_{n-i} + E_n \qquad (5.4\text{-}24)$$

When the differential equation is (5.4-1), and we use the simplifying assumption that the per step error E_n is a constant E, this difference equation becomes

$$\epsilon_{n+1}(1 + hb_{-1}K) = \sum_{i=0}^{p} (a_i - hb_i K)\epsilon_{n-i} + E \qquad (5.4\text{-}25)$$

which has the same form as (5.4-3) except for the inhomogeneous term E. If no roots are multiple, the solution of (5.4-25) is

$$\epsilon_n = \sum_{i=0}^{p} d_i r_i^n + \frac{E}{hK \sum_{i=-1}^{p} b_i} \qquad (5.4\text{-}26)$$

where we have used the first equation of (5.2-6) in getting the particular solution. The r_i's are the same as those in (5.4-4), but the d_i's depend on the initial conditions on the error for $n = 0, 1, \ldots, p$. For example, d_0 is given by an equation analogous to (5.4-10) with the first column in the numerator replaced by the initial errors. We assume that the initial errors are such that $|d_0| \ll |c_0|$, for otherwise the computed solution will be of no use, quite aside from the errors caused by the parasitic solution.

We note that if $|r_0| > 1$ for small h, which corresponds to $-K > 0$ [cf. (5.4-6)], the solution is an increasing exponential. Since we cannot expect to keep the error bounded when the solution is unbounded, it is not surprising that the $d_0 r_0^n$ term in (5.4-26) is also unbounded in this case. What we can hope to do is to keep the error *small relative to the true solution*, in which case we shall say the method of solution is stable. Since $|d_0| \ll |c_0|$, this will be true if the terms in (5.4-26) for $i = 1, \ldots, p$, which correspond to the parasitic solution, remain small relative to the r_0 term. This is equivalent to saying that an error introduced in the initial conditions or at a later stage of the computation will not propagate with a magnitude which increases relative to the magnitude of the true solution. This brings us back to the condition (5.4-11). But before we give a formal definition of stability, let us consider again the general case of Eq. (5.4-24).

This equation is not tractable as it stands, so that some simplifying assumptions are necessary. The most natural way of modifying (5.4-24) is to replace each f_y by some constant value $-K$ and E_n by E and thus get an equation of the form (5.4-25). Let us consider two ways of choosing the constants K and E:

1. Suppose K and E are such that

$$|f_y(x, y)| < -K \qquad |E_n| < E \tag{5.4-27}$$

for all n and for all points which occur in the solution of (5.4-24). Consider a new equation formed from (5.4-24) by replacing f_y by $-K$, E_n by E, b_i by $|b_i|$, and a_i by $|a_i|$. Suppose also that the initial conditions used in the solution of this new equation are all greater than or equal to the magnitudes of the corresponding initial conditions in (5.4-24). We leave to a problem {15} the proof of the result that if $|hb_{-1}K| < 1$, the solution of the new equation is greater than or equal to the magnitude of the solution of (5.4-24) *for all n*. This approach leads to a bound on ϵ_n which tends to be very conservative.

In order to elucidate how the error propagates from one step to the next, which is the essence of stability and which is a *local* behavior, we use a different approach.

2. Consider (5.4-24) for any value of n. For this n let $-K$ be a *characteristic* or average value of f_y for points in the neighborhood of those in (5.4-24). Similarly, let E be a characteristic value of E_n. Then, since in practice both f_y and E_n change slowly with n, we expect that locally the solution of (5.4-25) will behave like that of (5.4-24).

In what follows we shall use the latter approach. The stability of a numerical integration method will be defined in terms of the solution of the characteristic equation (5.4-5) of the difference equation (5.4-25). In (5.4-5), therefore, $-K$ is to be taken as a characteristic value of $f_y(x, y)$. Thus, whether or not a method is stable will depend upon the particular equation (5.1-1) to which it is being applied.

Definition 5.2 Let (5.2-5) be a consistent numerical integration method. Let r_i, $i = 0, \ldots, p$, be the roots of (5.4-5), with r_0 the root corresponding to the term in (5.4-4) which approximates the solution of the differential equation. Then this method is said to be (*relatively*) *stable* on an interval $[\alpha, \beta]$, which must include zero if, for all hK in this interval,

$$\left|\frac{r_i}{r_0}\right| \le 1 \qquad i = 1, \ldots, p \tag{5.4-28}$$

and if, when $|r_i| = |r_0|$, r_i is a simple root. It is said to be *absolutely stable* on an interval $[\gamma, \delta]$ if for all hk in this interval

$$|r_i| < 1 \qquad i = 0, \ldots, p \tag{5.4-29}$$

Remarks

1 With $[\alpha, \beta]$ required to include zero, we are assured that for any K we can make the solution stable by choosing h sufficiently small. Since (5.4-28) must hold for $h = 0$, a necessary condition for stability is that when $h = 0$, no r_i, $i = 1, \ldots, p$, lie outside the unit circle and that those roots on the unit circle must be simple. Thus, from Theorem 5.1, we conclude that *convergence is necessary for stability.*

2. We allow roots of magnitude equal to r_0 because if the errors in the initial conditions are small, we shall have $|d_i| \ll |d_0|$, $i = 1, \ldots, p$. Thus if $|r_i| = |r_0|$, although the parasitic solution will not decrease in magnitude relative to the r_0 term, it will remain small relative to the r_0 term. The requirement of no multiple roots of magnitude $|r_0|$ is necessary because of the factor $n^\alpha r_i^n$ introduced into (5.4-26) by a multiple root. Definition 5.2 in fact requires that r_0 be real. This is reasonable because we cannot expect the term in (5.4-4) in r_0 to be a good approximation to the solution of a real differential equation when r_0 is complex.

3. From (5.4-6) we see that it makes sense to consider absolute stability only for the case $hK > 0$. We then see that the condition of absolute stability is equivalent to requiring that when the solution of (5.1-1) is decreasing in magnitude, all the parasitic solutions also decrease in magnitude (why?). The determination of those values of hK for which a method is absolutely stable is easier than the determination of the interval of relative stability, {16, 17}.

4. In order that a numerical integration method may be stable for as large a class of differential equations as possible, it is desirable that $[\alpha, \beta]$ be as large as possible. It follows then that it is desirable to have all the roots of the convergence equation (5.4-14), except the one at $r = 1$, as small as possible. Since the roots of (5.4-3) are continuous functions of the coefficients, the range of Kh for which (5.4-28) is satisfied will tend to be larger the smaller the roots r_i, $i = 1, \ldots, p$. In particular, therefore, we would like to minimize the maximum magnitude of the r_i, $i = 1, \ldots, p$. The Adams-type methods are thus optimal from this point of view since all $r_i = 0$, $i = 1, \ldots, p$.

Example 5.4 Determine for what values of hK the consistent method of Example 5.1, that is, (5.2-9), is stable.

Since $p = 0$, Eq. (5.4-5) becomes

$$(1 + \tfrac{1}{2}hK)r = 1 - \tfrac{1}{2}hK$$

Thus there is only one root

$$r = \frac{1 - \frac{1}{2}hK}{1 + \frac{1}{2}hK}$$

When $h = 0$, $r = 1$, so that by Theorem 5.1 the method is convergent. Since there is just one root, Definition 5.2 is trivially satisfied and so this method is stable on the interval $(-\infty, \infty)$.

Example 5.5 For $a = 1$, determine the values of hK for which the consistent method of Example 5.2 is stable.

The equation for the roots is

$$(1 + \tfrac{1}{3}hK)r^2 = -(\tfrac{4}{3}hK)r + (1 - \tfrac{1}{3}hK)$$

and its solution is

$$r = \frac{1}{2(1 + \frac{1}{3}hK)}\{-\tfrac{4}{3}hK \pm [4 + \tfrac{4}{3}(hK)^2]^{1/2}\}$$

The plus sign corresponds to r_0 since this root approaches 1 as $h \to 0$. As $h \to 0$, $r_1 \to -1$, so that the method is convergent. The magnitude of the ratio of the roots is

$$\left|\frac{r_1}{r_0}\right| = \left|\frac{-\frac{4}{3}hK - [4 + \frac{4}{3}(hK)^2]^{1/2}}{-\frac{4}{3}hK + [4 + \frac{4}{3}(hK)^2]^{1/2}}\right|$$

For $hK < 0$ this magnitude is always less than 1, but for $hK > 0$ it is always greater than 1. When $hK = 0$, $r_0 = 1$ and $r_1 = -1$. Therefore, this method is stable only on intervals of the form $[\alpha, 0]$, where α is any negative number. Whenever f_y is negative, that is, $hK > 0$, r_1 has magnitude greater than r_0 and we would expect this method to exhibit bad error behavior; that this is so we shall indicate in Sec. 5.7-2. Since generally any differential equation or system of differential equations is such that f_y takes on both positive and negative values, this method should be avoided.

Because convergence is necessary for stability, it is an obvious first step in testing the stability of a numerical integration method to see whether the roots of (5.4-5) lie on or within the unit circle when $h = 0$. Two ways of doing this which do not require calculating the roots of (5.4-5) are considered in {16, 17}. If this necessary condition is satisfied, we may proceed to determine the range of values of hK for which the method is stable; see Sec. 5.5-4.

5.4-2 Propagated-Error Bounds and Estimates

If a stable method is used to compute the solution of (5.4-1), the main contribution to the accumulated or *propagated* error in the summation in (5.4-26) is given by the r_0 term. To estimate d_0 we use Cramer's rule and get, analogous to (5.4-10),

$$d_0 = \frac{\begin{vmatrix} e+\gamma & 1 & \cdots & 1 \\ e+\gamma & r_1 & \cdots & r_p \\ \cdots & \cdots & \cdots & \cdots \\ e+\gamma & r_1^p & \cdots & r_p^p \end{vmatrix}}{\begin{vmatrix} 1 & 1 & \cdots & 1 \\ r_0 & r_1 & \cdots & r_p \\ \cdots & \cdots & \cdots & \cdots \\ r_0^p & r_1^p & \cdots & r_p^p \end{vmatrix}} \tag{5.4-30}$$

where $\gamma = (-E)/hK \sum_{i=-1}^{p} b_i$ and we have assumed for simplicity that the errors in the initial conditions $y_0, \ldots, y_p$ are all equal to e.† The root r_0, we have seen, is close to 1; so, making the further simplifying assumption that it is 1, we get

$$d_0 \approx e + \gamma \tag{5.4-31}$$

An estimate of the propagated error is then given by the first term in the summation (5.4-26) plus the particular solution

$$\epsilon_n \approx d_0 r_0^n + \frac{E}{hK \sum_{i=1}^{p} b_i} \approx e - \frac{E}{hK \sum_{i=-1}^{p} b_i} e^{-Khn} + \frac{E}{hK \sum_{i=-1}^{p} b_i} \tag{5.4-32}$$

where we have replaced r_0 by its approximate value e^{-Kh}. [Note the different approximations for r_0 which led to (5.4-31) and (5.4-32); why are they both reasonable?] To get an estimate of the propagated error in the general case of Eq. (5.1-1), we replace f_y by $-K$, as in the previous section. If E and K are such that the conditions (5.4-27) are satisfied, then (5.4-32) will usually be a bound on the error and a quite conservative one [cf. {15}].

To use (5.4-32) we need estimates of e, E, and K, and the latter, at least, is often very difficult to get. As we shall see in the next section, there is an effective way of estimating the error of each step, and for a stable method, this may be used to control the overall error in the computation. However, (5.4-32) does indicate what quantities affect the error and how they affect it.

The definition and discussion of convergence in this section did not depend on the fact that only one equation was being considered. Definition 5.1 and Theorems 5.1 and 5.2 are all valid if the relevant quantities are vectors. The discussion of stability, however, does require some modifications in the case of systems of equations. By careful application of the mean-value theorem in (5.4-23), we arrive at a system of equations (5.4-25). For nonzero values of h, this system is cross-coupled; i.e., errors corresponding to different dependent variables appear in the same equation. The resultant system of polynomial equations corresponding to (5.4-5) is generally somewhat intractable {11}.

5.5 PREDICTOR-CORRECTOR METHODS

Consider the two numerical integration methods

$$y_{n+1} = y_n + \frac{h}{2}(y'_{n+1} + y'_n) \tag{5.5-1}$$

† The error in y_0 is caused by the necessity of rounding the given initial condition when inserting it in the computer. The other y_i's also are in error because of the truncation error in the method used to calculate them; see Sec. 5.6.

$$y_{n+1} = y_{n-2} + \frac{3h}{2}(y'_n + y'_{n-1}) \tag{5.5-2}$$

Both these equations are of order 2 [see Example 5.1 for (5.5-1)], and their truncation errors are immediately determinable since they are equivalent to a Newton-Cotes closed and open formula, respectively. These truncation errors are, respectively, $-(h^3/12)Y'''(\eta_1)$ and $(3h^3/4)Y'''(\eta_2)$. Equation (5.5-1), which is an iterative formula, is substantially more accurate (by a factor of 9 in general) than (5.5-2), which is a forward-integration formula. This is an illustration of the general rule that for formulas of corresponding order, iterative formulas are substantially more accurate than forward formulas. Thus, despite their added difficulty in use, it is worthwhile to use them. In this section, we consider methods by which this can most efficiently be done. Again for convenience we let the number of equations be 1, but the extension to systems is straightforward.

5.5-1 Convergence of the Iterations

In order to solve (5.2-5) for y_{n+1} when $b_{-1} \neq 0$, it will be necessary in general to use an iterative procedure. That is, we guess or somehow estimate an initial value of y_{n+1}, call it $y_{n+1}^{(0)}$, calculate $f(x_{n+1}, y_{n+1}^{(0)})$, insert this on the right-hand side of (5.2-5) to get $y_{n+1}^{(1)}$, and continue this process until convergence to some desired degree of accuracy is obtained. But first we must be sure that the process will converge. To derive the condition for convergence, we rewrite (5.2-5) as

$$y_{n+1}^{(j+1)} = \sum_{i=0}^{p} (a_i y_{n-i} + hb_i y'_{n-i}) + hb_{-1}(y_{n+1}^{(j)})' \tag{5.5-3}$$

where $y_{n+1}^{(j)}$ is the jth approximation to y_{n+1}. The correct value,† y_{n+1}, satisfies (5.2-5), which we rewrite for convenience as

$$y_{n+1} = \sum_{i=0}^{p} (a_i y_{n-i} + hb_i y'_{n-i}) + hb_{-1} y'_{n+1} \tag{5.5-4}$$

Subtracting (5.5-3) from (5.5-4), we get

$$y_{n+1} - y_{n+1}^{(j+1)} = hb_{-1}[y'_{n+1} - (y_{n+1}^{(j)})'] \tag{5.5-5}$$

which, using the mean-value theorem, becomes

$$y_{n+1} - y_{n+1}^{(j+1)} = hb_{-1} f_y(x_{n+1}, \eta^{(j)})(y_{n+1} - y_{n+1}^{(j)}) \tag{5.5-6}$$

† That is, the true solution of (5.2-5), not the true solution of the differential equation.

where $\eta^{(j)}$ lies between y_{n+1} and $y_{n+1}^{(j)}$. If in a neighborhood of (x_{n+1}, y_{n+1}) which includes all points $(x_{n+1}, y_{n+1}^{(j)})$

$$|f_y(x, y)| < K \tag{5.5-7}$$

then†
$$|y_{n+1} - y_{n+1}^{(j+1)}| < hb_{-1}K\,|y_{n+1} - y_{n+1}^{(j)}| \tag{5.5-8}$$

and, by induction, $|y_{n+1} - y_{n+1}^{(j+1)}| < (hb_{-1}K)^{j+1}\,|y_{n+1} - y_{n+1}^{(0)}|$ (5.5-9)

Thus, if

$$hb_{-1}K < 1 \tag{5.5-10}$$

then, as $j \to \infty$, $y_{n+1}^{(j+1)} \to y_{n+1}$ and the iteration converges. Moreover, the difference $|y_{n+1} - y_{n+1}^{(j)}|$ is monotonically decreasing for all j. In most cases in which we iterate to convergence, we shall assume that h has been chosen so that (5.5-10) is satisfied. The magnitude of $hb_{-1}K$ then determines the rate at which the iteration converges, so that for rapid convergence we should have the $hb_{-1}K \ll 1$ {20}.

For stiff equations (Sec. 5.10), (5.5-10) will not usually hold, so that if we must iterate to convergence, we must use some method other than simple iteration to solve the nonlinear equations (see Sec. 8.8).

5.5-2 Predictors and Correctors

Suppose we are going to use an iterative method to solve the differential equation (5.1-1). The only term in (5.5-3) that changes from one iteration to the next is the last term on the right-hand side. Thus the major calculation in each iteration is the value of $(y_{n+1}^{(j)})' = f(x_{n+1}, y_{n+1}^{(j)})$. In practice, the evaluation of $f(x, y)$ will generally be much more time-consuming than evaluation of the whole right-hand side of (5.5-3). Thus, in comparing the computational efficiency of various methods, we shall be interested particularly in how many evaluations of $f(x, y)$ are necessary at each step.

In many cases, we shall perform the iteration of the previous section until two successive iterates differ by less than some tolerance. We shall then accept the final iterate as y_{n+1}. The number of iterations required [each of which requires one evaluation of $f(x, y)$] will depend upon

1. The accuracy of the initial guess or estimation. Clearly, the nearer $y_{n+1}^{(0)}$ is to y_{n+1}, the faster the iteration will converge.
2. The accuracy desired in the final value of y_{n+1}. For example, if the error E_n at each step is of the order 10^{-5}, there is no reason to require y_{n+1} to be correct to 10 decimals. This question is considered in more detail in Sec. 5.7-2.
3. The value of $hb_{-1}K$.

† In what follows, we assume that b_{-1} is positive, as in all practical cases it is.

Our first object here is to consider ways of *predicting* $y_{n+1}^{(0)}$ as accurately as possible consistent with other properties that are desirable in such *predictors*. Using the predicted value, we shall then use an iterative formula to *correct* the prediction, hence the name *predictor-corrector methods*.

The best way to predict $y_{n+1}^{(0)}$ is to use a forward-integration formula since such a formula expresses y_{n+1} in terms of known past values of y and y'. Thus, for example, (5.5-2) could be used as a predictor for (5.5-1). The predictor-corrector system would then be

$$\text{Predictor:} \qquad y_{n+1}^{(0)} = y_{n-2} + \frac{3h}{2}(y_n' + y_{n-1}')$$

$$(y_{n+1}^{(0)})' = f(x_{n+1}, y_{n+1}^{(0)}) \tag{5.5-11}$$

$$\text{Corrector:} \qquad y_{n+1}^{(j+1)} = y_n + \frac{h}{2}[(y_{n+1}^{(j)})' + y_n'] \qquad j = 0, 1, \ldots$$

where the corrector is iterated until the desired degree of convergence is achieved. We noted previously that the truncation error of both predictor and corrector involves a third derivative. As we shall see in Sec. 5.5-3, it is important that both predictor and corrector have error terms with derivatives of the same order.

Equation (5.5-11) is a second-order predictor-corrector system. A fourth-order system using corresponding open and closed Newton-Cotes integration formulas is

$$\text{Predictor:} \qquad y_{n+1}^{(0)} = y_{n-3} + \frac{4h}{3}(2y_n' - y_{n-1}' + 2y_{n-2}')$$

$$(y_{n+1}^{(0)})' = f(x_{n+1}, y_{n+1}^{(0)}) \tag{5.5-12}$$

$$\text{Corrector:} \qquad y_{n+1}^{(j+1)} = y_{n-1} + \frac{h}{3}[(y_{n+1}^{(j)})' + 4y_n' + y_{n-1}'] \qquad j = 0, 1, \ldots$$

where the truncation-error terms of predictor and corrector are, respectively, $\frac{14}{45}h^5 Y^{\mathrm{v}}(\eta_1)$ and $-\frac{1}{90}h^5 Y^{\mathrm{v}}(\eta_2)$. The corrector is, of course, just Simpson's rule. This predictor-corrector method is known as *Milne's method*. In Example 5.5 we showed that the corrector in (5.5-12) is not stable for any positive value of hK, that is, $\partial f/\partial y$ negative. For this reason, the use of (5.5-12) is not advisable unless the number of steps of the integration to be carried out is too small to allow the parasitic solution to achieve a substantial magnitude or unless it is known that $\partial f/\partial y$ is positive. In Sec. 5.5-4, we shall consider a modification of (5.5-12) which is stable for sufficiently small values of hK.

In Fig. 5.1, we have indicated the sequence of steps required to use either (5.5-11) or (5.5-12) in proceeding from x_n to x_{n+1}. The use of convergence tolerances and other computational matters alluded to in Fig. 5.1 will be discussed in some detail in Sec. 5.7-2.

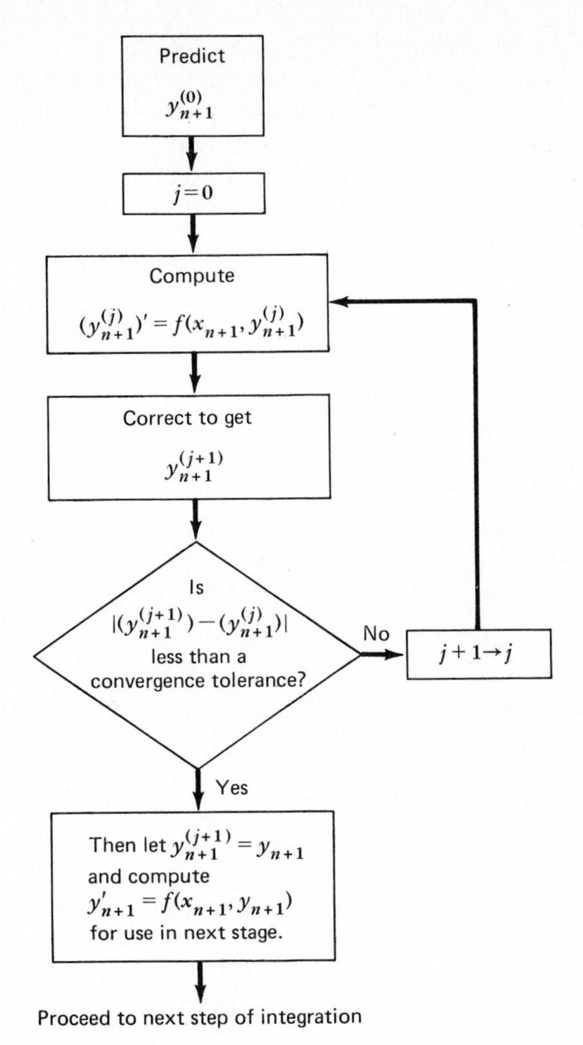

Figure 5.1 The use of predictor-corrector methods.

We are, of course, not restricted to using a Newton-Cotes open formula as predictor or a closed formula as corrector. Another choice for predictors is the Adams-Bashforth formulas (5.2-13). For example, we have

$$p = 0: \qquad y_{n+1} = y_n + hy'_n \tag{5.5-13}$$

$$p = 1: \qquad y_{n+1} = y_n + \frac{h}{2}(3y'_n - y'_{n-1}) \tag{5.5-14}$$

$$p = 2: \qquad y_{n+1} = y_n + \frac{h}{12}(23y'_n - 16y'_{n-1} + 5y'_{n-2}) \tag{5.5-15}$$

The case $p = 0$ is called *Euler's method* or the *point-slope method*. As we have seen, Adams-Bashforth predictors can all be generated by integrating Newton's backward-interpolation formula, and thus the truncation-error terms in (5.5-13) to (5.5-15) are easily found {21}. A similar situation holds with correctors given by the Adams-Moulton formulas (5.2-14).

If the point-slope method is used to integrate (5.4-1), the error for the case $K < 0$ is indicated graphically in Fig. 5.2. We see that the calculated solution always lags behind the true solution. Clearly, this will be the case for any exponentially increasing solution. A formula that attempts to avoid this difficulty is the *midpoint method*

$$y_{n+1} = y_{n-1} + 2hy'_n \tag{5.5-16}$$

where the derivative used is at the midpoint between the two abscissas. The midpoint method is a special case of a method known as *Nystrom's method* {23}.

Another class of predictors can be derived from the Hermite or modified Hermite interpolation formulas by letting $x = x_{n+1}$ and replacing $f(x)$ by $y(x)$. A fifth-order example of such a formula derived from (3.7-17) with $n = 3$ is

$$y_{n+1} = -18y_n + 9y_{n-1} + 10y_{n-2} + h(9y'_n + 18y'_{n-1} + 3y'_{n-2}) \tag{5.5-17}$$

Note that the fluctuation in coefficients means that this formula has bad roundoff properties. Other formulas of this type are considered in {24}; see also P_6 in Table 5.1.

On a digital computer it is common to use fourth-order predictor-corrector methods since methods of this order provide sufficient accuracy for most problems and are reasonably straightfoward to derive and use. However, there is a tendency now to use higher-order methods at a point where the function is assumed to be smooth on the basis of the results of the calculation up to that point. Conversely, low-order methods are used when the function starts to show singular behavior. Such a *variable-order method* is thus seen to be adaptive in the sense of Sec. 4.11. When combined with the choice of step length based on the best estimate consistent with error con-

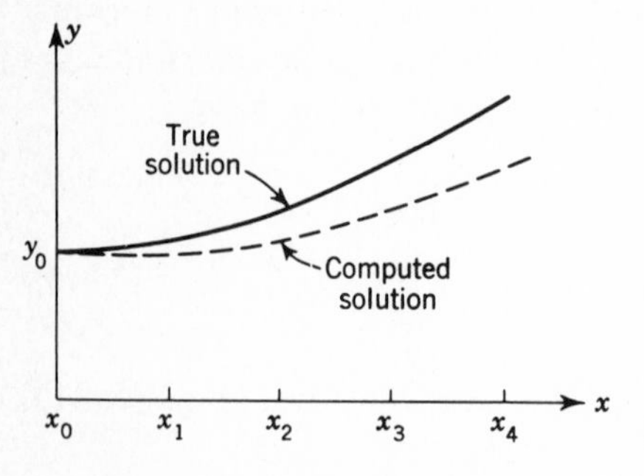

Figure 5.2 The point-slope method.

siderations, this leads to the very efficient and accurate *variable-order–variable-step* integration methods (see Sec. 5.7-1). Since the implementation of these methods is quite complicated, it is worth using them only for large systems of equations or for integration over a large interval. For small systems, the fourth-order methods studied here are adequate.

In Table 5.1 we have listed a number of fourth-order predictors. The error coefficient is the coefficient of $h^5 Y^{\text{v}}(\eta)$ in the error term. The choice of a predictor is not nearly so critical as the choice of a corrector. By far the most important factor is low truncation error in order to assure as good a prediction as possible. Other factors of some importance are (1) ease of computation; e.g., zero coefficients make the evaluation of the predictor easier, and (2) roundoff properties; note that the extremely bad roundoff properties of P_6 lower its value despite its extremely good truncation error.

The choice of a corrector depends importantly on stability. For this reason we defer discussion of the choice of a corrector until Sec. 5.5-4, in which corrector stability is considered.

5.5-3 Error Estimation

At each step in the use of predictor-corrector methods, we get two estimates of the solution at x_{n+1}, the predicted value and the corrected value. This enables us to obtain an *estimate*, called a *Milne-type estimate*, of the error incurred at *each step*. This estimation, which gives us an idea of the order of magnitude of the error, enables us to judge whether to accept the corrected value or to reject it. If we reject it, we must repeat the step with a smaller value of h. If we accept it, the estimate enables us to determine whether to (1) decrease the interval size h if the error is too large or (2) increase the interval size and thereby speed up the computation if the error is smaller than needed for the accuracy we desire or (3) leave the interval

Table 5.1 Fourth-order predictors

	P_1 (Adams)	P_2 (Milne)	P_3	P_4	P_5	P_6
a_0	1	0	0	0	$\frac{1}{3}$	-9
a_1	0	0	0	1	$\frac{1}{3}$	9
a_2	0	0	1	0	$\frac{1}{3}$	1
a_3	0	1	0	0	0	0
b_0	$\frac{55}{24}$	$\frac{8}{3}$	$\frac{21}{8}$	$\frac{8}{3}$	$\frac{91}{36}$	6
b_1	$-\frac{59}{24}$	$-\frac{4}{3}$	$-\frac{9}{8}$	$-\frac{5}{3}$	$-\frac{63}{36}$	6
b_2	$\frac{37}{24}$	$\frac{8}{3}$	$\frac{15}{8}$	$\frac{4}{3}$	$\frac{57}{36}$	0
b_3	$-\frac{9}{24}$	0	$-\frac{3}{8}$	$-\frac{1}{3}$	$-\frac{13}{36}$	0
Error coefficient	$\frac{251}{720}$	$\frac{224}{720}$	$\frac{243}{720}$	$\frac{232}{720}$	$\frac{242}{720}$	$\frac{72}{720}$

unchanged. Implicit in the above is an assumption to be used throughout this section, namely that truncation error is the dominant source of error.

We shall use the method (5.5-12) to illustrate error estimation. When (5.3-1) and the definition of ϵ_n in (5.4-19) are used, the predictor in (5.5-12) can be written

$$y_{n+1}^{(0)} = Y_{n-3} + \frac{4h}{3}(2Y'_n - Y'_{n-1} + 2Y'_{n-2}) - \epsilon_{n-3}$$
$$- \frac{4h}{3}(2\epsilon'_n - \epsilon'_{n-1} + 2\epsilon'_{n-2}) = Y_{n+1} - \epsilon_{n-3}$$
$$- \frac{4h}{3}(2\epsilon'_n - \epsilon'_{n-1} + 2\epsilon'_{n-2}) - \tfrac{14}{45}h^5 Y^{\text{v}}(\eta_1) \qquad (5.5\text{-}18)$$

Similarly, removing the superscripts, we can write the corrector as

$$y_{n+1} = Y_{n+1} - \epsilon_{n-1} - \frac{h}{3}(\epsilon'_{n+1} + 4\epsilon'_n + \epsilon'_{n-1}) + \tfrac{1}{90}h^5 Y^{\text{v}}(\eta_2) \quad (5.5\text{-}19)$$

where y_{n+1} is the value that would be obtained if the corrector were iterated to convergence. Subtracting (5.5-18) from (5.5-19), we get

$$y_{n+1} - y_{n+1}^{(0)} = \tfrac{1}{90}h^5 Y^{\text{v}}(\eta_2) + \tfrac{14}{45}h^5 Y^{\text{v}}(\eta_1)$$
$$- \frac{h}{3}(\epsilon'_{n+1} - 4\epsilon'_n + 5\epsilon'_{n-1} - 8\epsilon'_{n-2}) + \epsilon_{n-3} - \epsilon_{n-1} \qquad (5.5\text{-}20)$$

If we assume (1) that ϵ_i changes slowly from step to step and (2) that $(h/3)\epsilon'_i$ is small compared with the truncation error, we may drop the ϵ_i and ϵ'_i terms from (5.5-20). Then, if we further assume that $Y^{\text{v}}(x)$ does not change greatly between η_1 and η_2, we can write

$$y_{n+1} - y_{n+1}^{(0)} \approx \tfrac{29}{90}h^5 Y^{\text{v}}(\eta) \qquad (5.5\text{-}21)$$

Thus an estimate of $h^5 Y^{\text{v}}(\eta)$ is given by

$$h^5 Y^{\text{v}}(\eta) \approx \tfrac{90}{29}(y_{n+1} - y_{n+1}^{(0)}) \qquad (5.5\text{-}22)$$

The truncation error incurred at each step T_n can then be estimated as

$$T_n = -\tfrac{1}{90}h^5 Y^{\text{v}}(\eta) \approx -\tfrac{1}{29}(y_{n+1} - y_{n+1}^{(0)}) \qquad (5.5\text{-}23)$$

Therefore, we can use the difference in the predicted and corrected values to estimate how much error is being made at each step.

The truncation error in the predictor can be estimated as

$$T_n^{(0)} = \tfrac{14}{45}h^5 Y^{\text{v}}(\eta) \approx \tfrac{28}{29}(y_{n+1} - y_{n+1}^{(0)}) \qquad (5.5\text{-}24)$$

This not only enables us to estimate how good our predictions are but also lets us improve the prediction. For, assuming that the difference between the predicted and corrected values at each step changes slowly, we can estimate $T_n^{(0)}$ as

$$T_n^{(0)} \approx \tfrac{28}{29}(y_n - y_n^{(0)}) \tag{5.5-25}$$

Therefore,
$$\bar{y}_{n+1}^{(0)} = y_{n+1}^{(0)} + \tfrac{28}{29}(y_n - y_n^{(0)}) \tag{5.5-26}$$

will, in general, be an improved value of the prediction. The complete predictor-corrector method (5.5-12) then is

$$\text{Predictor:} \quad y_{n+1}^{(0)} = y_{n-3} + \frac{4h}{3}(2y'_n - y'_{n-1} + 2y'_{n-2})$$

$$\text{Modifier:} \quad \bar{y}_{n+1}^{(0)} = y_{n+1}^{(0)} + \tfrac{28}{29}(y_n - y_n^{(0)}) \tag{5.5-27}$$

$$(\bar{y}_{n+1}^{(0)})' = f(x_{n+1}, \bar{y}_{n+1}^{(0)})$$

$$\text{Corrector:} \quad y_{n+1}^{(j+1)} = y_{n-1} + \frac{h}{3}[(y_{n+1}^{(j)})' + 4y'_n + y'_{n-1}] \qquad j = 0, 1, \ldots$$

with $(\bar{y}_{n+1}^{(0)})'$ being used in the corrector initially.†

This procedure that we have illustrated for (5.5-12) can clearly be used for any predictor-corrector method as long as the *order of predictor and corrector are the same*. The corresponding case for (5.5-11) is considered in {25}.

A reasonable question to ask at this point is why we do not use the estimate of the corrector truncation error (5.5-23) to improve the corrected value. In fact this can be done, but, as shown in {26}, doing this is equivalent to using a system of one higher order (in this case, 5). Moreover, correcting the corrector affects the stability properties of the corrector. Therefore, rather than correcting the corrector, it is probably a better idea to use a higher-order system in the first place.

5.5-4 Stability

Two basic factors determine the value of a given corrector formula in comparison with others of the same order: (1) the coefficient in the error term and (2) its stability properties. That these two properties tend to work against each other will probably not surprise the reader. Other factors of subsidiary importance are (3) the roundoff properties and (4) the ease with which it can be computed (zero coefficients, "simple" coefficients, etc.). Our aim here is to develop a fourth-order corrector with as desirable properties as possible. To do this, we consider a corrector of the form

$$y_{n+1} = a_0 y_n + a_1 y_{n-1} + a_2 y_{n-2} + h(b_{-1} y'_{n+1} + b_0 y'_n + b_1 y'_{n-1}) \tag{5.5-28}$$

† At the first predictor-corrector step, i.e., after starting values have been computed (see Sec. 5.6), there will be no previous value of $y_n^{(0)}$ to use in the modifier, which therefore should be omitted at this step.

which uses data at only the last three points and contains six coefficients, one more than is necessary to achieve an order of 4. We shall use this extra degree of freedom to give the corrector desirable stability properties. We could also include a term $b_2 y'_{n-2}$ in (5.5-28) and thus achieve another degree of freedom, still using data at only three past points. This other degree of freedom could, for example, be used to achieve good roundoff properties. This is considered in {27}.

The requirement that (5.5-28) be exact for $y(x) = x^j, j = 0, \ldots, 4$, leads to the equations {5}

$$\begin{aligned} &a_0 = \tfrac{1}{8}(9 - 9a_1) \qquad a_1 = a_1 \qquad a_2 = -\tfrac{1}{8}(1 - a_1) \\ &b_{-1} = \tfrac{1}{24}(9 - a_1) \qquad b_0 = \tfrac{1}{12}(9 + 7a_1) \qquad b_1 = \tfrac{1}{24}(-9 + 17a_1) \end{aligned} \tag{5.5-29}$$

where a_1 is the free parameter. The influence function in the truncation error is given by

$$\begin{aligned} G(s) = (x_{n+1} - s)_+^4 &- a_0(x_n - s)_+^4 - a_1(x_{n-1} - s)_+^4 \\ &- a_2(x_{n-2} - s)_+^4 - 4h[b_{-1}(x_{n+1} - s)_+^3 \\ &+ b_0(x_n - s)_+^3 + b_1(x_n - 1)_+^3] \end{aligned} \tag{5.5-30}$$

It is naturally of interest to determine when $G(s)$ is of constant sign in $[x_{n-2}, x_{n+1}]$ as a function of a_1. We leave determination of this to a problem {10}, but note here that for a_1 in $[-.6, 1.0]$ $G(s)$ is indeed of constant sign.

The stability equation for (5.5-28), that is, the equation corresponding to (5.4-5), is

$$(1 + hKb_{-1})r^3 = (a_0 - hKb_0)r^2 + (a_1 - hKb_1)r + a_2 \tag{5.5-31}$$

To determine the value of hK for which (5.4-28) holds as a function of a_1 is quite difficult for this cubic. We start by considering the case $h = 0$. As a function of a_1, the three roots are shown in Fig. 5.3. We conclude that only in the interval $-.6 \le a_1 \le 1.0$ can there be stability. Because the roots of (5.5-31) are continuous functions of hK, for any value of a_1 interior to $[-.6, 1.0]$ there will be some range of hK for which the method is stable. In order to get some insight into the best value of a_1 to choose, consider the data in Table 5.2. The error coefficient is the coefficient of $h^5 Y^{\text{v}}(\eta)$ in the error term. We see that the error coefficient steadily decreases from 1.0 to $-.6$ [note that $a_1 = 1$ is Milne's corrector in (5.5-12)]. Also the roundoff properties, judged by the sum of the squares of a_0, a_1, and a_2, become steadily worse in this direction from $\frac{9}{17}$ on down. A reasonable choice would seem to be $a_1 = 0$ since it is near the center of the interval [and thus is likely to be stable for a greater range of hK than values near the ends of the interval (why?)], has one zero coefficient, a reasonable error-term coefficient, and reasonable roundoff properties. In Fig. 5.4 we have plotted the roots of

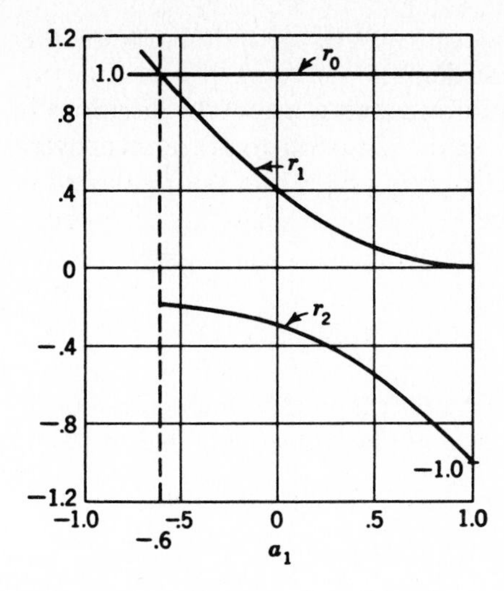

Figure 5.3 Roots of (5.5-31) with $h = 0$.

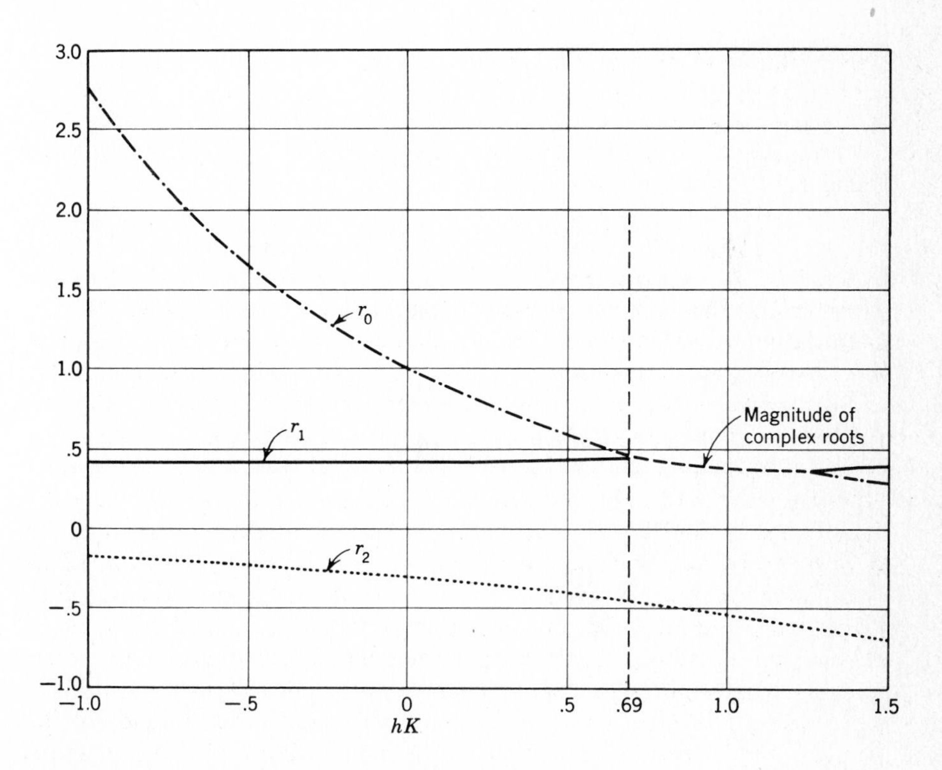

Figure 5.4 Roots of (5.5-31) for $a_1 = 0$.

(5.5-31) for $a_1 = 0$. For $hK \leq +.69$ and all negative hK of importance, the stability condition (5.4-28) is always satisfied.

When $a_1 = 0$, the condition (5.5-10) for convergence of the iterations is that $|hK|$ be less than $\frac{8}{3}$ since $b_{-1} = \frac{3}{8}$. In fact, in order to get rapid convergence of the corrector, we would want $|hK| \ll \frac{8}{3}$. Thus the requirement $hK \leq +.69$ is not restrictive in practice.

Our conclusion then is that the corrector (5.5-28) with $a_1 = 0$

$$y_{n+1} = \tfrac{1}{8}(9y_n - y_{n-2}) + \frac{3h}{8}(y'_{n+1} + 2y'_n - y'_{n-1}) \tag{5.5-32}$$

is a desirable fourth-order corrector to use in place of Milne's corrector in (5.5-12). Using this corrector and Milne's predictor, we obtain *Hamming's method* {29}

$$\begin{aligned} &\text{Predictor:} && y_{n+1}^{(0)} = y_{n-3} + \frac{4h}{3}(2y'_n - y'_{n-1} + 2y'_{n-2}) \\ &\text{Modifier:} && \bar{y}_{n+1}^{(0)} = y_{n+1}^{(0)} + \tfrac{112}{121}(y_n - y_n^{(0)}) \\ & && (\bar{y}_{n+1}^{(0)})' = f(x_{n+1}, \bar{y}_{n+1}^{(0)}) \\ &\text{Corrector:} && y_{n+1}^{(j+1)} = \tfrac{1}{8}(9y_n - y_{n-2}) \\ & && \qquad + \frac{3h}{8}[(y_{n+1}^{(j)})' + 2y'_n - y'_{n-1}] \qquad j = 0, 1, \ldots \end{aligned} \tag{5.5-33}$$

Using Table 5.2 we get, analogous to (5.5-23),

$$T_n = -\tfrac{1}{40}h^5 Y^{\text{v}}(\eta) \approx \tfrac{9}{121}(y_{n+1} - y_{n+1}^{(0)}) \tag{5.5-34}$$

In Sec. 5.7-2 we shall give some numerical examples comparing this method with (5.5-27). An analysis similar to that in this section is possible for methods of any order {29}.

Table 5.2 Correctors for sample values of a_1

a_1	1 (Milne)	$\frac{9}{17}$	$\frac{1}{9}$	0 (Hamming)	$-\frac{1}{7}$	$-\frac{9}{31}$	$-\frac{3}{5}$
a_0	0	$\frac{9}{17}$	1	$\frac{9}{8}$	$\frac{9}{7}$	$\frac{45}{31}$	$\frac{9}{5}$
a_1	1	$\frac{9}{17}$	$\frac{1}{9}$	0	$-\frac{1}{7}$	$-\frac{9}{31}$	$-\frac{3}{5}$
a_2	0	$-\frac{1}{17}$	$-\frac{1}{9}$	$-\frac{1}{8}$	$-\frac{1}{7}$	$-\frac{5}{31}$	$-\frac{1}{5}$
b_{-1}	$\frac{1}{3}$	$\frac{6}{17}$	$\frac{10}{27}$	$\frac{3}{8}$	$\frac{8}{21}$	$\frac{12}{31}$	$\frac{2}{5}$
b_0	$\frac{4}{3}$	$\frac{18}{17}$	$\frac{22}{27}$	$\frac{3}{4}$	$\frac{2}{3}$	$\frac{18}{31}$	$\frac{2}{5}$
b_1	$\frac{1}{3}$	0	$-\frac{8}{27}$	$-\frac{3}{8}$	$-\frac{10}{21}$	$\frac{18}{31}$	$-\frac{4}{5}$
Error coefficient	$-\frac{1}{90}$	$-\frac{3}{170}$	$-\frac{19}{810}$	$-\frac{1}{40}$	$-\frac{17}{630}$	$-\frac{9}{310}$	$-\frac{1}{30}$

Another useful fourth-order corrector, though not of the form (5.5-28), is the Adams-Moulton corrector

$$y_{n+1} = y_n + \frac{h}{720}(251y'_{n+1} + 646y'_n - 264y'_{n-1} + 106y'_{n-2} - 19y'_{n-3})$$

with error coefficient $-\frac{3}{160}$. The stability characteristics of this formula are much better that those of (5.5-32) in that the interval of absolute stability is (0, 3.0) compared with (0, .69) for formula (5.5-32). In general, the Adams-Moulton formulas (5.2-14) have good stability characteristics. One reason is that for $h = 0$, all extraneous roots vanish. In addition, the value of y_{n+1} is directly coupled to the previous value y_n by an integral, and integration is a smoothing operation.

5.6 STARTING THE SOLUTION AND CHANGING THE INTERVAL

We come now to two problems we have thus far put off:

1. How do we obtain the starting values (initial conditions) for (5.2-5) that are required besides the initial condition of the differential equation?
2. How do we change the interval h during the computation if our error estimate indicates that this is desirable?

There are, of course, some *self-starting methods*—Eq. (5.5-13) is an example–which require no starting values other than that provided by the differential equation. While these methods are of low order and therefore not sufficiently accurate for most problems, they are very useful in variable-order–variable-step methods (Sec. 5.7-1). In this section, however, we shall be concerned with starting values and interval changes for higher-order methods which are to be used over the whole range of integration.

5.6-1 Analytic Methods

Taylor series The Taylor-series expansion of $y(x)$ about x_0 can be written

$$y(x_0 + hs) = y_0 + hsy'_0 + \frac{h^2s^2}{2!}y''_0 + \cdots \tag{5.6-1}$$

Using the given initial condition y_0, we can, using (5.1-1), calculate y'_0. Then by differentiating the differential equation, higher derivatives of y at x_0 can be calculated. Thus (5.6-1) can be used to approximate $y(x_0 + hs)$ for any s for which the series converges.

Example 5.6 Use (5.6-1) to calculate initial values at x_1, x_2, and x_3 for the differential equation $y' = y^2$, $y(0) = 1$ with $h = .01$.

We calculate as follows:

$$y'(0) = 1 \qquad y'' = 2yy' \qquad y''(0) = 2$$

$$y''' = 2(y')^2 + 2yy'' \qquad y'''(0) = 2 + 4 = 6 \qquad \text{etc.}$$

Then $$y(hs) = 1 + .01s + 10^{-4}s^2 + 10^{-6}s^3 + \cdots$$

which can be used with $s = 1, 2, 3$, to approximate y_1, y_2, y_3 {35}.

If the Taylor series converges for the requisite values of x, this process can be used to get initial values of any desired accuracy. Customarily, one would desire the accuracy of the initial values to be at least as great as that of the numerical integration procedure to be used. An alternative approach to the above is considered in {34}.

The method of successive approximations (Picard's method) Equation (5.1-1) can be written

$$y(x) = y_0 + \int_{x_0}^{x} f(x, y)\,dx \tag{5.6-2}$$

Assuming an initial approximation $y(x) = y_0(x)$ and inserting it on the right-hand side of (5.6-2), we generate an approximation $y_1(x)$. This process can be iterated, and if $f(x, y)$ satisfies the conditions of page 164 [see Ince (1926)], the process will converge in a neighborhood of x_0. Thus we can obtain approximations to $Y(x)$ at the desired values of x. This technique is of great importance in the theory of the existence of solutions of (5.1-1), but the difficulty of evaluating the integral in (5.6-2) makes it impractical for numerical computations.

While both these methods involve *analytic* operations, the Taylor-series method involves only analytic differentiation and can be mechanized quite readily on a digital computer. In fact, the Taylor-series method has been proposed as a general-purpose numerical integration method, and programs exist for solving systems of differential equations by Taylor series using analytic continuation methods. However, if analytic differentiation capabilities are available, the Hermite methods discussed in Sec. 5.9-1 must also be considered. Picard's method, on the other hand, involves indefinite integration. While programs have been written to mechanize this process, they do not always work, even when the integral can be expressed in terms of elementary functions. Moreover, there exist many functions expressed in terms of elementary functions for which the indefinite integral cannot be so expressed. A simple example of this is the function e^{x^2}. Hence the iteration process may break down very quickly.

5.6-2 A Numerical Method

Suppose, as in Example 5.6, that three values of y besides y_0 are needed. Let us approximate $y'(x)$ by a Lagrangian interpolation formula using the four equally spaced abscissas x_0, x_1, x_2, and x_3. Then we insert the result of this on the right-hand side of (5.6-2) and integrate from x_0 to, respectively, x_1, x_2, and x_3. So doing, we get {36}.

$$
\begin{aligned}
y_1 &= y_0 + \frac{h}{24}(9y'_0 + 19y'_1 - 5y'_2 + y'_3) \\
y_2 &= y_0 + \frac{h}{3}(y'_0 + 4y'_1 + y'_2) \\
y_3 &= y_0 + \frac{h}{8}(3y'_0 + 9y'_1 + 9y'_2 + 3y'_3)
\end{aligned}
\tag{5.6-3}
$$

where the error term in all three equations can be shown to be $O(h^5)$ {36}. To use (5.6-3), we make an initial estimate of y_1, y_2, and y_3, use (5.1-1) to calculate y'_1, y'_2, and y'_3, and then use (5.6-3) to get new values of y_1, y_2, and y_3. This process can be iterated and if it converges it will give the four required starting values. Note, for example, that the method (5.5-33) requires precisely four starting values.

Two disadvantages of this method are its possible lack of convergence and, in any case, the tedious computation involved. It can, of course, be mechanized but has no advantage in this respect over the Runge-Kutta methods to be discussed in Sec. 5.8. As we shall see, these methods are not only the most desirable methods to use in generating starting values, but they have their proponents as a general method for numerical solution of differential equations competitive with predictor-corrector methods.

5.6-3 Changing the Interval

Usually the solution of (5.1-1) is desired for some final value of x, say x_F, and at this value of x it is desired that the value of y should be in error by no more than some predetermined tolerance. Normally the initial value of h will be chosen so that if the a priori estimate of the per step error is correct, the solution will have the desired accuracy. However, as the computation proceeds, the error-estimation procedure of Sec. 5.5-3 may indicate that (1) the per step error is larger at each step than is allowable if the final value of y is to have the desired accuracy or (2) the error is significantly smaller than is necessary. Assuming, as we shall here, that truncation error is the dominant factor, the indicated action in the first case above is to decrease the step size h since truncation error depends on a power of h. Conversely, in the second case, we increase the step size so that the computation will proceed more rapidly. Since the predictor-corrector methods we have been considering require a constant step size, whenever the interval is changed, more starting values must be generated. For example, suppose the second-order method (5.5-11) is being used and after y_n has been computed, it is desired to halve the value of h. Then in calculating $y(x_n + \frac{1}{2}h)$ the predictor requires values of y at x_n, $x_n - \frac{1}{2}h$, and x_{n-1}. Thus, in order to proceed with the computation, we need the value of y at $x_n - \frac{1}{2}h$. An effective method for doing this would be to use a single-step method such as a second-order Runge-Kutta method (see Sec. 5.8-2) [because (5.5-11) is a second-order method] starting from x_{n-1} with a step size $h/2$. An alternative procedure would be to use such a method starting from x_n to generate values of y at $x_n + \frac{1}{2}h$ and $x_n + h$ and then switch back to the use of the predictor-corrector method. The use of a single-step method such as a Runge-Kutta method to change the interval h is particularly simple on a digital computer when the method has been used to calculate starting values and is therefore part of the computer program. These methods are, in fact, easily used when the interval h is to be increased or decreased by any factor whatsoever.

Another approach to changing the interval is to use one of the interpolation formulas of Chap. 3 and previously computed values of $y(x)$ and $y'(x)$ to interpolate or extrapolate to get those values of y [such as $y(x_n - \frac{1}{2}h)$] which are required to continue the computation at the new interval. Since $y'(x) = f(x, y(x))$ is also available at the grid points, the Hermite interpolation formulas (Sec. 3.7) are most appropriate. A particular example of this is considered in {37}. With such interpolation or extrapolation procedures, it is convenient to halve the interval when decreasing it and to double it when increasing it. But when the interval is to be doubled, it is possible to avoid the use of Runge-Kutta methods or interpolation methods entirely {37}. In Sec. 5.7-2, we shall give an example of changing the interval in practice.

5.7 USING PREDICTOR–CORRECTOR METHODS

One computational problem that we have deferred thus far is the determination of when to stop iterating the corrector. Let us assume that because of the final accuracy that we desire in our solution we have determined a bound on the per step error or the per step relative error which if not exceeded during the computation will enable us to achieve the desired accuracy. It is reasonable then to require that any error made by not iterating the corrector to convergence be small compared with the allowable error, or to put it another way, this error should be small compared with truncation and roundoff errors.

Let y be the value that would be obtained if the corrector were iterated to convergence, and let $y^{(i)}$ be the value of the ith iterate (with $y^{(0)}$ the predicted value) where, for convenience, we have dropped the subscript n. As in Sec. 5.5-1 we assume that b_{-1} is positive. Then, since only the b_{-1} term in the numerical integration method (5.2-5) changes from one iteration to the next,

$$y^{(i+1)} - y^{(i)} = hb_{-1}[(y^{(i)})' - (y^{(i-1)})'] \tag{5.7-1}$$

Let $|(y^{(i)})' - (y^{(i-1)})'| = \delta_i$. Then

$$|y^{(i+1)} - y^{(i)}| = hb_{-1}\,\delta_i \tag{5.7-2}$$

Using the mean-value theorem,

$$\begin{aligned}\delta_{i+1} = |(y^{(i+1)})' - (y^{(i)})'| &= |f(x, y^{(i+1)}) - f(x, y^{(i)})| \\ &\leq |y^{(i+1)} - y^{(i)}|K = hb_{-1}\,\delta_i K \end{aligned} \tag{5.7-3}$$

if $|\partial f/\partial y| \leq K$ in the region of interest. Then from (5.7-2)

$$|y^{(i+2)} - y^{(i+1)}| \leq (hb_{-1})^2\,\delta_i K \tag{5.7-4}$$

and in this way the differences of successive iterates can be bounded. Now

$$|y - y^{(i+1)}| \leq |y^{(i+2)} - y^{(i+1)}| + |y^{(i+3)} - y^{(i+2)})| + \cdots$$

and using the above results, we get

$$\begin{aligned}|y - y^{(i+1)}| &\leq h^2b_{-1}^2\,\delta_i K(1 + hb_{-1}K + h^2b_{-1}^2K^2 + \cdots) \\ &= \frac{h^2b_{-1}^2\,\delta_i K}{1 - hb_{-1}K}\end{aligned} \tag{5.7-5}$$

if $|hb_{-1}K| < 1$ which, in fact, it must be for convergence of the corrector iterations. Indeed, in order to get rapid convergence of the corrector, we have noted that we should have $|hb_{-1}K| \ll 1$. Thus we have the result that $h^2b_{-1}^2\,\delta_i K$ should be a good approximation to the right-hand side of (5.7-5). In fact, since we have considered only the maximum values of quantities in this derivation, we expect that $h^2b_{-1}^2\,\delta_i K$ will be a quite conservative bound on the error incurred by stopping the iteration after the computation of $(y^{(i)})'$ and $y^{(i+1)}$. This suggests the following procedure. After each corrector iteration, compute δ_i and compare it with a *convergence factor* chosen so that if δ_i is less than the convergence factor, termininating the iteration with $(y^{(i)})'$ and $y^{(i+1)}$ will result in a value of $h^2b_{-1}^2\,\delta_i K$ which is small compared with the allowable per step error. We would compare δ_i with the product of the convergence factor and $y^{(i)}$ if we were interested in controlling the relative error. In either case the test requires some estimate of K, the bound on $|\partial f/\partial y|$. Since

$$\left|\frac{\partial f}{\partial y}\right| \approx \left|\frac{(y^{(i)})' - (y^{(i-1)})'}{y^{(i)} - y^{(i-1)}}\right| = \frac{\delta_i}{|y^{(i)} - y^{(i-1)}|} \tag{5.7-6}$$

and since, as we have noted, $h^2 b_{-1}^2 \delta_i K$ is a conservative bound on the error incurred by terminating the iteration, equation (5.7-6) can be used as the estimate of K. Note that this estimate is appropriate only for a single differential equation. For a system of equations the situation is much more complicated.

Generally in the numerical solution of differential equations, the test described above will be satisfied after the *first* application of the corrector, that is, $i = 1$, because the modified predicted value itself will usually be quite accurate. Thus generally only *two* evaluations of $f(x, y)$ are required at each step, one after computing the modified predicted value and one after computing the first corrected value. This is why fourth-order predictor-corrector methods are generally substantially faster than fourth-order Runge-Kutta methods, which require four evaluations of $f(x, y)$ per step (see Sec. 5.8).

There is another way of using predictor-corrector pairs, namely to determine beforehand a positive integer m, usually taken to be 1, and to iterate the corrector exactly m times. This leads to an integration method which has different properties from that method in which the corrector is iterated to convergence. In the latter case, the properties of the method are independent of which predictor is used, while in the former case, the choice of predictor has a considerable influence on the properties of the method, especially its stability properties. Thus, if we correct only once, the stability equation takes the form

$$(1 + hKb_{-1})r^{p+1} = \sum_{i=0}^{p} (a_i - hKb_i)r^{p-i} - hKb_{-1} \sum_{i=0}^{p} (a_i^* - hKb_i^*)r^{p-i} \tag{5.7-7}$$

where the predictor has the form

$$y_{n+1} = \sum_{i=0}^{p} a_i^* y_{n-i} + h \sum_{i=0}^{p} b_i^* y'_{n-i} \tag{5.7-8}$$

and the corrector is given by (5.2-5) with $b_{-1} > 0$. In the usual fourth-order case, the Adams-Bashforth-Moulton pair has an absolute stability interval of (0, 1.3) while the Adams-Moulton formula corrected to convergence has an interval of (0, 3.0).

A curious situation can arise in using this mode of operation, as is illustrated by the Milne predictor-corrector pair (5.5-12). If the corrector is applied only once, the absolute stability interval becomes (.3, .8). This means that by reducing h, one can enter a region of instability! In this mode, we can also modify the predictor but at the cost of changing the stability equation so that it is almost impossible to analyze it. Similarly, we can use a Milne-type estimate to monitor the calculation so as to determine when to change the step size. The estimate is, of course, the difference between the predictor and the mth corrector.

5.7-1 Variable-Order–Variable-Step Methods

In these methods, the computation is started with a low-order rule and a small step size so that they are self-starting. As the computation proceeds, both the order and the step size are modified as the circumstances demand so that at each point the optimal order and step size are chosen, subject to certain constraints. The goal is to reach the endpoint b of the interval of interest $[a, b]$ with a solution satisfying a given accuracy criterion using the least amount of computation. Of course, stability considerations must not be neglected in the course of the computation, but the stability theory in these methods is not yet fully understood and decisions are based on experimental evidence. One decision which is usually made is to use Adams-type methods since they have the best stability properties among the constant-order–constant-step methods.

There is a variety of variable-order–variable-step methods, each employing some of the above techniques. Thus, in some methods the corrector is iterated to convergence, while in others it is applied only once. In some methods, Milne-type estimates are used for the local truncation error while in others what corresponds to the first neglected difference is used. In some methods constraints are imposed on the changing of the order and in others on the changing of the step size. This is necessary since experimental evidence indicates that changing both the order and the step size too frequently introduces instabilities into the computation in a way which is not too well understood theoretically. We present here one of several strategies for step size and order adjustment. Other approaches can be found in the references cited in the Bibliography.

Let us assume now that we are integrating from the point x_n, using the step size h_n, and a method of order q. We thus compute a value y_{n+1} at the point $x_{n+1} = x_n + h_n$ and an estimate T_q of the local error (see, for example, Sec. 5.5-3), where

$$T_q \approx \gamma_q h_n^{q+1} \tag{5.7-9}$$

We now check whether this integration step has been successful by testing if

$$|T_q| \le \bar{\epsilon} h_n \tag{5.7-10}$$

where $\bar{\epsilon} = \epsilon/H$ and $H = b - a$ is the length of the interval over which errors of size ϵ are to be allowed (cf. Sec. 5.8-7). If (5.7-10) does not hold, we repeat the integration with the new step size $\bar{h}_n = \eta_q h_n$, where

$$\eta_q = \left(\frac{\lambda \bar{\epsilon} h_n}{|T_q|}\right)^{1/q} \tag{5.7-11}$$

and λ is a safety factor, $0 < \lambda \le 1$, typically taken as .8. With this choice of $\bar{h}_n$ we have that

$$|\bar{T}_q| \simeq |\gamma_q \bar{h}_n^{q+1}| = |\gamma_q \bar{h}_n \eta_q^q h_n^q| \simeq \lambda \bar{\epsilon} \bar{h}_n \tag{5.7-12}$$

so that the integration with step size $\bar{h}_n$ will be satisfactory.

If (5.7-10) does hold, we proceed to the next stage in the integration. At this new stage, we must decide on a new step size h_{n+1} and a new order. Since we are changing the step size freely, we must restrict changes in order to avoid instability. Hence, we permit a change in order from q to $q-1$ or $q+1$ only if $q+1$ steps have been taken with order q. If this is not the case, we retain the previous order q and set $h_{n+1} = \bar{h}_n$, based on the assumption that γ_q does not vary too much as we proceed from x_n to x_{n+1}.

If it is permissible to change the order by 1, we test whether such a change will be beneficial in that it will allow a choice of a larger step size. To determine this, we compute estimates of T_{q-1} and T_{q+1}, that is, the local truncation errors had we integrated from x_n with methods of orders $q-1$ or $q+1$, respectively. The information needed to compute these estimates is either already available or can easily be generated. How this information is computed depends on the particular details of the method used. Since we are only interested in the basic ideas of these methods, we shall not discuss this aspect of the computation. Having computed T_{q-1} and T_{q+1}, we compute η_{q-1} and η_{q+1} by formulas analogous to (5.7-11). We now have three potential step sizes at our disposal, $\eta_{q-1} h_n$, $\eta_q h_n$, and $\eta_{q+1} h_n$. As before, the estimated local error, taking h_{n+1} to be any of these step sizes and using a method of corresponding order, is approximately equal to $\lambda \bar{\epsilon} h_{n+1}$. Hence, we choose that order which gives a maximum step size, thus defining h_{n+1}. In this way, we step along the integration interval $[a, b]$ until we reach b, using an optimal number of steps consistent with stability considerations, which preclude changing the order too frequently.

5.7-2 Some Illustrative Examples

Our main object in this section is to compare the use of two predictor-corrector methods, that of Milne (5.5-27) and that of Hamming (5.5-33), in the solution of some differential equations in order to illustrate a number of the points made in this chapter.

First example The first equation we consider is

$$\frac{dy}{dx} = -y \qquad y(0) = 1 \tag{5.7-13}$$

whose solution is $Y = e^{-x}$. For a value of $h = .1$, the results of this computation for various values of x are shown in Table 5.3. All the computation was carried out on a digital computer using floating-point decimal arithmetic with eight significant figures. For both predictor-corrector methods, the

Table 5.3 Results of numerical solution of $y' = -y$, $y(0) = 1$†

		Hamming's method		Milne's method		Hermite method‡	
x	e^{-x}	y	Error	y	Error	y	Error
.1	.90483742	.90483753	−1.1E − 7	.90483753	−1.1E − 7	.90483753	−1.1E − 7
.2	.81873075	.81873093	−1.8E − 7	.81873093	−1.8E − 7	.81873093	−1.8E − 7
.3	.74081822	.74081846	−2.4E − 7	.74081846	−2.4E − 7	.74081846	−2.4E − 7
.4	.67032005	.67032033	−2.8E − 7	.67032033	−2.8E − 7	.67032033	−2.8E − 7
.5	.60653066	.60653078	−1.2E − 7	.60653078	−1.2E − 7	.60653092	−2.6E − 7
.6	.54881164	.54881160	4.0E − 8	.54881183	−1.9E − 7	.54881188	−2.4E − 7
.7	.49658530	.49658511	1.9E − 7	.49658534	−4.0E − 8	.49658553	−2.3E − 7
.8	.44932896	.44932865	3.1E − 7	.44932910	−1.4E − 7	.44932918	−2.2E − 7
.9	.40656966	.40656924	4.2E − 7	.40656964	2.0E − 8	.40656986	−2.0E − 7
1.0	.36787944	.36787894	5.0E − 7	.36787953	−9.0E − 8	.36787963	−1.9E − 7
2.0	.13533528	.13533469	5.9E − 7	.13533534	−6.0E − 8	.13533536	−8.0E − 8
3.0	4.9787068E − 2	4.9786674E − 2	3.9E − 7	4.9787174E − 2	−1.1E − 7	4.9787110E − 2	−4.2E − 8
4.0	1.8315639E − 2	1.8315434E − 2	2.1E − 7	1.8315820E − 2	−1.8E − 7	1.8315655E − 2	−1.6E − 8
5.0	6.7379470E − 3	6.7378495E − 3	9.8E − 8	6.7382209E − 3	−2.7E − 7	6.7379539E − 3	−6.9E − 9
6.0	2.4787522E − 3	2.4787080E − 3	4.4E − 8	2.4791447E − 3	−3.9E − 7	2.4787550E − 3	−2.8E − 9
7.0	9.1188197E − 4	9.1186306E − 4	1.9E − 8	9.1243460E − 4	−5.5E − 7	9.1188321E − 4	−1.2E − 9
8.0	3.3546263E − 4	3.3545448E − 4	8.2E − 9	3.3623582E − 4	−7.7E − 7	3.3546314E − 4	−5.1E − 10
9.0	1.2340980E − 4	1.2340648E − 4	3.3E − 9	1.2448957E − 4	−1.1E − 6	1.2341002E − 4	−2.2E − 10
10.0	4.5399930E − 5	4.5398555E − 5	1.4E − 9	4.6906894E − 5	−1.5E − 6	4.5400018E − 5	−8.8E − 11
11.0	1.6701701E − 5	1.6701140E − 5	5.6E − 10	1.8804490E − 5	−2.1E − 6	1.6701734E − 5	−3.3E − 11
12.0	6.1442124E − 6	6.1439854E − 6	2.3E − 10	9.0782370E − 6	−2.9E − 6	6.1442251E − 6	−1.3E − 11
13.0	2.2603294E − 6	2.2602384E − 6	9.1E − 11	6.3541098E − 6	−4.1E − 6	2.2603344E − 6	−5.0E − 12
13.5	1.3709591E − 6			−3.4646962E − 6	4.8E − 6		
14.0	8.3152872E − 7	8.3149279E − 7	3.6E − 11	6.5434628E − 6	−5.7E − 6	8.3153061E − 7	−1.9E − 12
14.5	5.0434766E − 7			−6.2426818E − 6	6.7E − 6		
15.0	3.0590232E − 7	3.0588805E − 7	1.4E − 12	8.2755880E − 6	−8.0E − 6	3.0590305E − 7	−7.3E − 13

† The notation E − 2, etc., is shorthand for $\times 10^{-2}$, etc. ‡ The Hermite method is discussed in Sec. 5.9-1.

fourth-order Runge-Kutta method (5.8-46) was used to calculate the values through $x = .4$.† The truncation errors for the two methods are

$$T_n = \begin{cases} -\frac{1}{40}h^5 Y^{\text{v}}(\eta) = 2.5 \times 10^{-7}e^{-\eta} & \text{Hamming} \\ -\frac{1}{90}h^5 Y^{\text{v}}(\eta) \approx 1.1 \times 10^{-7}e^{-\eta} & \text{Milne} \end{cases} \tag{5.7-14}$$

Since the solution of (5.7-13) is rapidly decreasing, we used a relative-error criterion in determining when to terminate the corrector iteration. The convergence factor used was 1.0×10^{-5}. For if δ_i is less than this convergence factor, then with $K = 1$

$$h^2 b_{-1}^2\, \delta_i K < \begin{cases} \frac{9}{64} \times 10^{-7} & \text{Hamming} \\ \frac{1}{9} \times 10^{-7} & \text{Milne} \end{cases} \tag{5.7-15}$$

Since we are using relative error, these must be compared with the coefficients of $e^{-\eta}$ in (5.7-14). For Hamming's and Milne's method, the bounds in (5.7-15) are, respectively, about $\frac{1}{18}$ and $\frac{1}{10}$ times the coefficients in (5.7-14). Therefore, we conclude that if δ_i is less than 1.0×10^{-5}, the error incurred by not iterating the corrector to convergence will not be serious.

During the early stages of the computation until $x = 4.0$, the smaller truncation error of Milne's method results in smaller errors using this method. From $x = 4.0$ on, however, the superiority of Hamming's method becomes manifest. At $x = 15.0$ there is a relative error of 10^{-5} in Hamming's method. Correspondingly, the relative error in Milne's method at $x = 15.0$ is about -25. The reason why the early good behavior of Milne's method is not continued is, of course, the instability of Milne's method for all positive K that we discussed in Sec. 5.5-4. (Here $\partial f/\partial y = -1$, so that $K = 1$.) In the early stages of the computation, the coefficient of r_1 in (5.4-26) is small because of the accuracy of the initial conditions generated using (5.8-46). During this part of the computation, the per step truncation error determines the error, and thus Milne's method gives more accurate results than Hamming's. But sooner or later the term in r_1^n must predominate in the error in Milne's method, thereby producing the instability which is evident later in the computation. In fact, late in the computation the sign of y alternates from one step to the next, e.g., note the entries for $x = 13.5$ and 14.5. The stability of Hamming's method is best illustrated by noticing the quite slow growth in the relative error as the computation proceeds. This example is a good illustration of the necessity of using a stable method for any solution of a differential equation that is going to proceed over more than just a few steps in h.

For each of the 146 steps of the computation using (5.5-33), the test of the convergence factor was satisfied after one application of the corrector, so that just two evaluations of $f(x, y)$ were needed at each step. The instability

† Actually, only values through $x = .3$ are required to start Milne's or Hamming's methods.

of Milne's method, however, necessitated an average of three corrector iterations per step, one near the beginning but five toward the end. Finally, we note that in this example the truncation error is substantially greater than roundoff; for a contrast, see the third example below.

Second example Consider the equation

$$\frac{dy}{dx} = y \qquad y(0) = 1 \tag{5.7-16}$$

whose solution is $Y = e^x$. In Table 5.4, we have the results of this computation tabulated. Again an interval of $h = .1$ was used, the first four steps were calculated using (5.8-46), and floating-point arithmetic was used throughout with a convergence factor of 1.0×10^{-5} and a relative-error criterion. This time the superior per step error of Milne's method causes the solution with

Table 5.4 Results of numerical solution of $y' = y$, $y(0) = 1$†

		Hamming's method		Milne's method		Hermite method‡	
x	e^x	y	Error	y	Error	y	Error
.1	1.1051709	1.1051708	1.0E−7	1.1051708	1.0E−7	1.1051708	1.0E−7
.2	1.2214028	1.2214024	4.0E−7	1.2214024	4.0E−7	1.2214024	4.0E−7
.3	1.3498588	1.3498582	6.0E−7	1.3498582	6.0E − 7	1.3498582	6.0E−7
.4	1.4918247	1.4918238	9.0E−7	1.4918238	9.0E − 7	1.4918238	9.0E−7
.5	1.6487213	1.6487205	8.0E−7	1.6487206	7.0E−7	1.6487203	1.0E−6
.6	1.8221188	1.8221184	4.0E−7	1.8221179	9.0E−7	1.8221178	1.0E−6
.7	2.0137527	2.0137529	−2.0E−7	2.0137520	7.0E−7	2.0137516	1.1E−6
.8	2.2255409	2.2255418	−9.0E−7	2.2255400	9.0E−7	2.2255397	1.2E−6
.9	2.4596031	2.4596048	−1.7E−6	2.4596024	7.0E−7	2.4596017	1.4E−6
1.0	2.7182818	2.7182845	−2.7E−6	2.7182810	8.0E−7	2.7182803	1.5E−6
2.0	7.3890561	7.3890860	−3.0E−5	7.3890573	−1.2E−6	7.3890511	5.0E−6
3.0	20.085537	20.085675	−1.4E−4	20.085548	−1.1E−5	20.085520	1.7E−5
4.0	54.598150	54.598685	−5.4E−4	54.598204	−5.4E−5	54.598097	5.3E−5
5.0	148.41316	148.41509	−1.9E−3	148.41337	−2.1E−4	148.41301	1.5E−4
6.0	403.42879	403.43529	−6.5E−3	403.42952	−7.3E−4	403.42837	4.2E−4
7.0	1096.6332	1096.6542	−2.1E−2	1096.6357	−2.5E−3	1096.6319	1.3E−3
8.0	2980.9580	2981.0243	−6.6E−2	2980.9661	−8.1E−3	2980.9541	3.9E−3
9.0	8103.0839	8103.2885	− .20	8103.1094	−2.6E−2	8103.0722	1.2E−2
10.0	22026.466	22027.089	− .62	22026.544	−7.8E−2	22026.432	3.4E−2
11.0	59874.142	59876.025	−1.9	59874.381	− .24	59874.046	.096
12.0	162754.79	162760.39	−5.6	162755.51	− .72	162754.51	.28
13.0	442413.39	442429.93	−16.5	442415.51	−2.1	442412.57	.82
14.0	1202604.3	1202652.9	−48.6	1202610.6	−6.3	1202602.0	2.3
15.0	3269017.4	3269159.6	−142.2	3269035.8	−18.4	3269011.1	6.3

† The notation E−7, etc., is shorthand for $\times 10^{-7}$, etc.

‡ The Hermite method is discussed in Sec. 5.9-1.

that method to be more accurate throughout the computation. For with $\partial f/\partial y = 1$ the magnitude of r_1 is less than that of r_0 (see Example 5.5). The smaller error in Hamming's method for $x = .5$ to $x = .9$ is a result of the errors in the values calculated using (5.8-46). These errors are positive, as we see from Table 5.4, but the truncation errors of both Milne's and Hamming's method are negative (why?), so that the *larger* negative error in Hamming's method overcomes the positive error in the initial values more rapidly than the error in Milne's method. For both methods only one application of the corrector was needed at each step.

The growth of the error as x increases makes this example a good one with which to illustrate change of interval. The quantity $y_n - y_n^{(0)}$ grows from 3.9×10^{-6} at $x = .5$ to 7.9×10^{-2} at $x = 10.0$ using Hamming's method and from 4.0×10^{-6} to 5.9×10^{-2} using Milne's method. For illustrative purposes in Table 5.5, we give the results of changing the interval to $h = .05$ at $x = 10.0$. The values of y at $x = 10.05, 10.1, 10.15, 10.2$ were calculated using (5.8-46). The percentage improvement in the error is substantially greater for Milne's than for Hamming's method. The reason for this is that the propagated-error terms in the solution are not negligible at $x = 10.0$ in Hamming's method, and the reduction in h (and therefore in the truncation error) does not prevent the continuation of this error propagation. In Milne's method the error-propagation terms are very small when $x = 10.0$, and thus the reduction in truncation error has more effect on the overall error.

This second example illustrates the general rule, that between two *stable* methods, the one with the smaller truncation error should usually be chosen. It is well to emphasize at this point that a priori, in most numerical solutions of differential equations, we do not know much about the behavior of $\partial f/\partial y$ and, moreover, that in most cases $\partial f/\partial y$ will take on both positive and negative values during the computation (with systems the behavior will be even more complex). Thus we must generally choose a method such as Hamming's, which is stable for both positive and negative $\partial f/\partial y$, rather than Milne's.

Table 5.5 Change of interval in the solution of $y' = y$, $y(0) = 1$

		Hamming's method		Milne's method		Hermite method	
x	e^x	y	Error	y	Error	y	Error
11.0	59874.142	59875.835	−1.7	59874.339	−.20	59874.043	.099
12.0	162754.79	162759.51	−4.7	162755.29	−.50	162754.53	.26
13.0	442413.39	442426.39	−13.0	442414.63	−1.2	442412.69	.70
14.0	1202604.3	1202640.2	−35.9	1202607.5	−3.2	1202602.3	2.0
15.0	3269017.4	3269117.1	−99.7	3269025.2	−7.8	3269012.1	5.3

Table 5.6 Results of numerical solution of $y' = 1(1 + \tan^2 y)$, $y(0) = 0$†

		Hamming's method		Milne's method		Hermite method‡	
x	$\tan^{-1} x$	y	Error	y	Error	y	Error
.1	9.668652E−2	9.9668686E − 2	−3.4E−8	9.9668686E−2	−3.4E−8	9.9668686E−2	−3.4E−8
.2	.19739556	.19739560	−4.0E−8	.19739560	−4.0E−8	.19739560	−4.0E−8
.3	.29145679	.29145683	−4.0E−8	.29145683	−4.0E−8	.29145683	−4.0E−8
.4	.38050638	.38050639	−1.0E−8	.38050639	−1.0E−8	.38050639	−1.0E−8
.5	.46364761	.46364675	8.6E−7	.46364707	5.4E−7	.46364772	−1.1E−7
.6	.54041950	.54041660	2.9E−6	.54041858	9.2E−7	.54041974	−2.4E−7
.7	.61072596	.61072066	5.3E−6	.61072458	1.4E−6	.61072632	−3.6E−7
.8	.67474094	.67473336	7.6E−6	.67473938	1.6E−6	.67474137	−4.3E−7
.9	.73281510	.73280581	9.3E−6	.73281330	1.8E−6	.73281556	−4.6E−7
1.0	.78539816	.78538785	1.0E−5	.78539642	1.7E−6	.78539862	−4.6E−7
2.0	1.1071487	1.1071434	5.3E−6	1.1071478	9.0E−7	1.1071490	−3.0E−7
3.0	1.2490458	1.2490442	1.6E−6	1.2490451	7.0E−7	1.2490458	0.0
4.0	1.3258177	1.3258166	1.1E−6	1.3258169	8.0E−7	1.3258177	0.0
5.0	1.3734008	1.3733999	9.0E−7	1.3734000	8.0E−7	1.3734009	−1.0E−7
6.0	1.4056476	1.4056473	3.0E−7	1.4056468	8.0E−7	1.4056478	−2.0E−7
7.0	1.4288993	1.4288989	4.0E−7	1.4288984	9.0E−7	1.4288993	0.0
8.0	1.4464413	1.4464410	3.0E−7	1.4464403	1.0E−6	1.4464413	0.0
9.0	1.4601391	1.4601390	1.0E−7	1.4601379	1.2E−6	1.4601393	−2.0E−7
10.0	1.4711277	1.4711278	−1.0E−7	1.4711264	1.3E−6	1.4711280	−3.0E−7
11.0	1.4801364	1.4801369	−5.0E−7	1.4801350	1.4E−6	1.4801367	−3.0E−7
12.0	1.4876551	1.4876561	−1.0E−6	1.4876535	1.6E−6	1.4876551	0.0
13.0	1.4940244	1.4940256	−1.2E−6	1.4940227	1.7E−6	1.4940242	2.0E−7
14.0	1.4994889	1.4994904	−1.5E−6	1.4994870	1.9E−6	1.4994882	7.0E−7
15.0	1.5042282	1.5042300	−1.8E−6	1.5042262	2.0E−6	1.5042272	1.0E−6
16.0	1.5083775	1.5083790	−1.5E−6	1.5083754	2.1E−6	1.5083761	1.4E−6
17.0	1.5120405	1.5120423	−1.8E−6	1.5120382	2.3E−6	1.5120388	1.7E−6
18.0	1.5152978	1.5152994	−1.6E−6	1.5152954	2.4E−6	1.5152958	2.0E−6
19.0	1.5182133	1.5182148	−1.5E−6	1.5182107	2.6E−6	1.5182110	2.3E−6
20.0	1.5208379	1.5208395	−1.6E−6	1.5208353	2.6E−6	1.5208354	2.5E−6

† The notation E−2, etc., is shorthand for $\times 10^{-2}$, etc.

‡ The Hermite method is discussed in Sec. 5.9-1.

We did not explicitly use the error-estimation ability of Sec. 5.5-3 in either of these examples. In fact, the change of interval discussed above could have been done automatically using error estimation {39}.

Third example Consider the equation

$$y' = \frac{1}{1 + \tan^2 y} \qquad y(0) = 0 \tag{5.7-17}$$

whose solution is $Y = \tan^{-1} x$. Again with $h = .1$, using (5.8-46) for the first four values and using floating-point arithmetic, some values for the computation are shown in Table 5.6. Since the magnitude of the solution does not change greatly, we used an absolute-error criterion with a convergence factor of 5.0×10^{-5}. In both cases, only one application of the corrector per step was needed. As in the second example, $\partial f/\partial y$ is always positive, so that again Milne's method does not lead to instability. Thus, again the smaller truncation error in Milne's method causes that method to give higher accuracy until late in the computation, when the errors in the two computations become almost equal in magnitude although opposite in sign. This is because, early in the computation, the roundoff error becomes a significant part of the total error and late in the computation the roundoff dominates. This occurs because $\tan^2 y$ gets very large as y approaches $\pi/2$, thus making the truncation error almost zero, so that the roundoff error, even though occurring in the eighth figure, is greater than the truncation error. The roundoff properties of Milne's method are slightly better than those of Hamming's {40}, but because of the statistical nature of the roundoff, this is not enough to show up significantly in the results tabulated in Table 5.6. Proper use of the error-estimation ability of predictor-corrector methods would have led us to increase the value of h when the roundoff error became dominant. In fact, h should always be increased in the numerical solution of differential equations when roundoff becomes dominant if values of the solution at the smaller spacing are not required (why?).

5.8 RUNGE-KUTTA METHODS

These methods can be used to generate not only starting values but, in fact, the whole solution. They are self-starting and easy to program for a digital computer, but these advantages do not overcome their disadvantages in error-estimation ability and speed relative to predictor-corrector methods. However, their value in starting the solution and in changing the interval is great. Furthermore, because of their simplicity, they are competitive in terms of speed with predictor-corrector methods whenever the derivative function $f(x, y)$ is simple. In this case, the time spent on "housekeeping" operations may be greater than that spent on pure computation, so that a simple algorithm performs better than a more complicated one even if more function evaluations are required. In addition, as we shall see, we can achieve an error estimate at the cost of several additional function evaluations. Hence, the second advantage of predictor-corrector methods also vanishes when function-evaluation time is not critical. This situation is particularly common with large systems of differential equations.

The basis of all Runge-Kutta methods is to express the difference between the values of y at x_{n+1} and x_n as

$$y_{n+1} - y_n = \sum_{i=1}^{p} w_i k_i \tag{5.8-1}$$

where the w_i's are constants and

$$k_i = h_n f\left(x_n + \alpha_i h_n,\, y_n + \sum_{j=1}^{i-1} \beta_{ij} k_j\right) \tag{5.8-2}$$

with $h_n = x_{n+1} - x_n$ and $\alpha_1 = 0$. We use h_n instead of h since it is possible to vary the interval at each stage of the use of a Runge-Kutta method. Clearly, given the w_i's, α_i's, and β_{ij}'s, (5.8-2) is a single-step method for the solution of (5.1-1). In this section we shall again consider a single equation, but the extension to N equations is straightforward {50}.

Our object is to determine the w_i's, α_i's, and β_{ij}'s so that (5.8-1) has the properties we desire. In particular, our object is to make the coefficients of h_n^r in the Taylor-series expansion of both sides of (5.8-1) about (x_n, y_n) identical for $r = 1, 2, \ldots, m$. As we shall see, we can do no better than $m = p$. The resulting formula will be called a *Runge-Kutta method of order m*. For convenience throughout this section we shall not distinguish notationally between the true and calculated solutions. In particular, we shall differentiate y as if it were the true solution.

The expansion of the left-hand side of (5.8-1) is

$$y_{n+1} - y_n = \sum_{t=1}^{\infty} h_n^t y_n^{(t)}/t! \tag{5.8-3}$$

From (5.1-1)

$$\begin{aligned} y_n^{(t)} &= \frac{d^{t-1}}{dx^{t-1}} f(x_n, y_n) = \left(\frac{\partial}{\partial x} + \frac{dy}{dx}\frac{\partial}{\partial y}\right)^{t-1} f(x_n, y_n) \\ &= \left(\frac{\partial}{\partial x} + f\,\frac{\partial}{\partial y}\right)^{t-1} f(x_n, y_n) \end{aligned} \tag{5.8-4}$$

where $f = f(x, y)$. Using this, (5.8-3) becomes

$$y_{n+1} - y_n = \sum_{t=0}^{\infty} \frac{h_n^{t+1}}{(t+1)!}\left(\frac{\partial}{\partial x} + f\,\frac{\partial}{\partial y}\right)^t f(x_n, y_n) \tag{5.8-5}$$

We define

$$D = \frac{\partial}{\partial x} + f_n \frac{\partial}{\partial y} \qquad f_n = f(x_n, y_n) \tag{5.8-6}$$

and so can write, for example,

$$\left[\frac{\partial}{\partial x} + f\frac{\partial}{\partial y}\right]^2 f(x_n, y_n) = D^2 f + f_y Df \Big|_n \tag{5.8-7}$$

where the notation $|_n$ means that all quantities should be evaluated at (x_n, y_n) (and h should be replaced by h_n).

When (5.8-6) is used, the first few terms of (5.8-5) are

$$\begin{aligned} y_{n+1} - y_n = \Big[& hf + \frac{h^2}{2!}Df + \frac{h^3}{3!}(D^2 f + f_y Df) \\ & + \frac{h^4}{4!}(D^3 f + f_y D^2 f + f_y^2 Df + 3Df\,Df_y) + \frac{h^5}{5!}(D^4 f \\ & + 6Df\,D^2 f_y + 4D^2 f\,Df_y + D^2 f f_y^2 + Dff_y^3 + 3(Df)^2 f_{yy} \\ & + D^3 ff_y + 7f_y\,Df\;Df_y)\Big]\Big|_n + O(h_n^6) \end{aligned} \tag{5.8-8}$$

To get the expansion of the right-hand side of (5.8-1), we first use the Taylor-series expansion for two variables to write

$$f\left[x_n + \alpha_i h_n, y_n + \left(\sum_{j=1}^{i-1} \beta_{ij}\right) h_n f_n\right] = \sum_{t=0}^{\infty} h_n^t D_i^t f(x_n, y_n)/t! \tag{5.8-9}$$

where

$$D_i = \alpha_i \frac{\partial}{\partial x} + \left(\sum_{j=1}^{i-1} \beta_{ij}\right) f_n \frac{\partial}{\partial y} \tag{5.8-10}$$

Using (5.8-9) and (5.8-10), we can get expansions for each k_i on the right-hand side of (5.8-1).

Since $\alpha_1 = 0$,

$$k_1 = h_n f_n \tag{5.8-11}$$

For k_2 we have

$$\begin{aligned} k_2 &= h_n f(x_n + \alpha_2 h_n, y_n + \beta_{21} k_1) = h_n f(x_n + \alpha_2 h_n, y_n + \beta_{21} h_n f_n) \\ &= h_n \sum_{t=0}^{\infty} h_n^t D_2^t f(x_n, y_n)/t! \end{aligned} \tag{5.8-12}$$

where D_2 is given by (5.8-10). For k_3 we proceed as follows, using (5.8-11) and (5.8-12):

$$\begin{aligned} k_3 &= h_n f(x_n + \alpha_3 h_n, y_n + \beta_{31} k_1 + \beta_{32} k_2) \\ &= h_n f[x_n + \alpha_3 h_n, y_n + (\beta_{31} + \beta_{32}) h_n f_n + \beta_{32}(k_2 - h_n f_n)] \\ &= h_n \sum_{t=0}^{\infty} \left[h_n D_3 + \beta_{32}(k_2 - h_n f_n)\frac{\partial}{\partial y}\right]^t f(x_n, y_n)/t! \end{aligned} \tag{5.8-13}$$

By using (5.8-12) in (5.8-13) we can consider (5.8-13) to be an expansion in powers of h_n. The procedure for k_3 suggests the general procedure for k_i. We write

$$\begin{aligned} k_i &= h_n f\left(x_n + \alpha_i h_n, y_n + \sum_{j=1}^{i-1} \beta_{ij} k_j\right) \\ &= h_n f\left\{x_n + \alpha_i h_n, y_n + \sum_{j=1}^{i-1} [\beta_{ij} h_n f_n + \beta_{ij}(k_j - h_n f_n)]\right\} \\ &= h_n \sum_{t=0}^{\infty} \left[h_n D_i + \sum_{j=2}^{i-1} \beta_{ij}(k_j - h_n f_n) \frac{\partial}{\partial y}\right]^t f(x_n, y_n)/t! \end{aligned} \tag{5.8-14}$$

and then use the results for $k_j, j < i$ to write k_i as an expansion in powers of h_n. Leaving the algebra to a problem {42}, we give the results for $i = 1, 2, 3, 4$, retaining terms through h_n^5:

$$k_1 = hf \Big|_n \tag{5.8-15}$$

$$k_2 = hf + h^2 D_2 f + \frac{h^3}{2!} D_2^2 f + \frac{h^4}{3!} D_2^3 f + \frac{h^5}{4!} D_2^4 f \Big|_n + O(h_n^6) \tag{5.8-16}$$

$$\begin{aligned} k_3 = hf &+ h^2 D_3 f + h^3(\tfrac{1}{2} D_3^2 f + \beta_{32} f_y D_2 f) + h^4 \Big[\tfrac{1}{6} D_3^3 f \\ &+ \frac{\beta_{32}}{2} f_y D_2^2 f + \beta_{32} D_2 f D_3 f_y\Big] + h^5 \Big[\tfrac{1}{24} D_3^4 f + \frac{\beta_{32}}{6} f_y D_2^3 f \\ &+ \frac{\beta_{32}}{2} D_2^2 f D_3 f_y + \frac{\beta_{32}^2}{2} f_{yy}(D_2 f)^2 + \frac{\beta_{32}}{2} D_2 f D_3^2 f_y\Big]\Big|_n \\ &+ O(h_n^6) \end{aligned} \tag{5.8-17}$$

$$\begin{aligned} k_4 = hf &+ h^2 D_4 f + h^3(\tfrac{1}{2} D_4^2 f + \beta_{42} f_y D_2 f + \beta_{43} f_y D_3 f) \\ &+ h^4[\tfrac{1}{6} D_4^3 f + \tfrac{1}{2}\beta_{42} f_y D_2^2 f + \beta_{32}\beta_{43}(f_y)^2 D_2 f + \tfrac{1}{2}\beta_{43} f_y D_3^2 f \\ &+ \beta_{42} D_2 f D_4 f_y + \beta_{43} D_3 f D_4 f_y] + h^5[\tfrac{1}{24} D_4^4 f + \tfrac{1}{6}\beta_{42} f_y D_2^3 f \\ &+ \tfrac{1}{2}\beta_{32}\beta_{43}(f_y)^2 D_2^2 f + \beta_{32}\beta_{43} f_y D_2 f D_3 f_y + \tfrac{1}{6}\beta_{43} f_y D_3^3 f \\ &+ \tfrac{1}{2}\beta_{42} D_4 f_y D_2^2 f + \tfrac{1}{2}\beta_{43} D_4 f_y D_3^2 f + \tfrac{1}{2}\beta_{42}^2 f_{yy} D_2^2 f \\ &+ \beta_{42}\beta_{43} f_{yy} D_2 f D_3 f + \tfrac{1}{2}\beta_{43}^2 f_{yy} D_3^2 f + \tfrac{1}{2}\beta_{42} D_2 f D_4^2 f_y \\ &+ \tfrac{1}{2}\beta_{43} D_3 f D_4^2 f_y + \beta_{43}\beta_{32} f_y D_2 f D_4 f_y]\Big|_n + O(h_n^6) \end{aligned} \tag{5.8-18}$$

Equations (5.8-15) to (5.8-18) will enable us to develop all Runge-Kutta methods through order 4, and the terms in h_n^5 will facilitate discussion of the error term.

Substituting (5.8-15) to (5.8-18) and (5.8-8) into (5.8-1) and matching powers of h_n through h_n^4, we get

$$h_n: \quad w_1 + w_2 + w_3 + w_4 = 1 \tag{5.8-19}$$

$$h_n^2: \quad w_2 D_2 f + w_3 D_3 f + w_4 D_4 f = \frac{1}{2!} Df \tag{5.8-20}$$

$$\begin{aligned} h_n^3: \quad & \tfrac{1}{2}[w_2 D_2^2 f + w_3 D_3^2 f + w_4 D_4^2 f] + f_y[w_3 \beta_{32} D_2 f \\ & + w_4(\beta_{42} D_2 f + \beta_{43} D_3 f)] = \frac{1}{3!}(D^2 f + f_y Df) \end{aligned} \tag{5.8-21}$$

$$\begin{aligned} h_n^4: \quad & \tfrac{1}{6}[w_2 D_2^3 f + w_3 D_3^3 f + w_4 D_4^3 f] + \tfrac{1}{2} f_y[w_3 \beta_{32} D_2^2 f \\ & + w_4(\beta_{42} D_2^2 f + \beta_{43} D_3^2 f)] + [w_3 \beta_{32} D_2 f D_3 f_y \\ & + w_4(\beta_{42} D_2 f D_4 f_y + \beta_{43} D_3 f D_4 f_y)] \\ & + [w_4 \beta_{32} \beta_{43} (f_y)^2 D_2 f] = \frac{1}{4!}[D^3 f + f_y D^2 f \\ & + 3Df\, Df_y + f_y^2 Df] \end{aligned} \tag{5.8-22}$$

These are in reality not four but eight equations, since if the values of the w_i's, α_i's, and β_{ij}'s are to be independent of $f(x, y)$, as they must be to be useful, the expressions in square brackets on the left-hand sides of (5.8-21) and (5.8-22), which are homogeneous in the operators, must equal the corresponding terms on the right-hand sides {42}. Moreover, if the resulting eight equations are to be independent of $f(x, y)$, the ratios

$$\frac{D_j f}{Df} \quad j = 2, 3, 4 \qquad \text{and} \qquad \frac{D_i f_y}{Df_y} \quad j = 3, 4 \tag{5.8-23}$$

must be constant. This will be true if

$$\alpha_i = \sum_{j=1}^{i-1} \beta_{ij} \qquad i = 2, 3, 4 \tag{5.8-24}$$

for then

$$D_i = \alpha_i D \tag{5.8-25}$$

Finally, then, the eight equations are

$$\begin{aligned} w_1 + w_2 + w_3 + w_4 &= 1 \\ w_2 \alpha_2 + w_3 \alpha_3 + w_4 \alpha_4 &= \tfrac{1}{2} \\ w_2 \alpha_2^2 + w_3 \alpha_3^2 + w_4 \alpha_4^2 &= \tfrac{1}{3} \\ w_3 \alpha_2 \beta_{32} + w_4(\alpha_2 \beta_{42} + \alpha_3 \beta_{43}) &= \tfrac{1}{6} \\ w_2 \alpha_2^3 + w_3 \alpha_3^3 + w_4 \alpha_4^3 &= \tfrac{1}{4} \\ w_3 \alpha_2^2 \beta_{32} + w_4(\alpha_2^2 \beta_{42} + \alpha_3^2 \beta_{43}) &= \tfrac{1}{12} \\ w_3 \alpha_2 \alpha_3 \beta_{32} + w_4(\alpha_2 \beta_{42} + \alpha_3 \beta_{43})\alpha_4 &= \tfrac{1}{8} \\ w_4 \alpha_2 \beta_{32} \beta_{43} &= \tfrac{1}{24} \end{aligned} \tag{5.8-26}$$

where the first equation corresponds to (5.8-19), the second to (5.8-20), the next two to (5.8-21), and the last four to (5.8-22). The system (5.8-24) and (5.8-26) has 11 equations and 13 unknowns, which will generally be sufficient to determine the parameters with two degrees of freedom. We note that because of the last equation, it is *necessary* to include the k_4 term in order to achieve accuracy through h_n^4 (why?), thus verifying that p must be at least m. For the cases $m = 2$ and 3, we can also show that $p = m$. The general result that $p \geq m$ is not hard to prove {43}. Since a treatment of the cases $m > 4$ is quite involved, we shall be considering in detail only the cases $m = 2, 3, 4$. First, however, it will be convenient to consider the errors in Runge-Kutta methods.

5.8-1 Errors in Runge-Kutta Methods

Here alone in our discussion of Runge-Kutta methods is it important to consider a single differential equation. For systems of equations, the algebra becomes intractable.

Truncation error Eqution (5.8-1) is to be exact for powers of h_n through h_n^m. Therefore, the truncation error T_m can be written

$$T_m = \gamma_m h_n^{m+1} + O(h_n^{m+2}) \tag{5.8-27}$$

where, of course, both γ_m and T_m really depend on $f(x, y)$. To estimate T_m, we are forced to consider only γ_m because consideration of the higher-order terms is algebraically intractable. The bounds on γ_m that we shall obtain will be very conservative; i.e., the true magnitude of γ_m will generally be much less than the bound. Thus, if the $O(h_n^{m+2})$ term is small compared with $\gamma_m h_n^{m+1}$, as we expect it will be if h_n is small, then the bound on $\gamma_m h_n^{m+1}$ will usually be a bound on the error as a whole.

Using (5.8-8), (5.8-15) to (5.8-18), and (5.8-25), we calculate

$$\gamma_2 = \left(\frac{1}{6} - \frac{\alpha_2^2 w_2}{2}\right) D^2 f + \tfrac{1}{6} f_y Df \tag{5.8-28}$$

$$\begin{aligned}\gamma_3 = & \left[\frac{1}{4!} - \frac{1}{3!}(\alpha_2^3 w_2 + \alpha_3^3 w_3)\right] D^3 f \\ & + \left(\frac{1}{4!} - \frac{1}{2!}\alpha_2^2 \beta_{32} w_3\right) f_y D^2 f + \left(\frac{3}{4!} - \alpha_2 \alpha_3 \beta_{32} w_3\right) Df Df_y \\ & + \frac{1}{4!} f_y^2 Df \end{aligned} \tag{5.8-29}$$

$$\gamma_4 = \left(\frac{1}{120} - \frac{w_2\alpha_2^4 + w_3\alpha_3^4 + w_4\alpha_4^4}{24}\right)D^4 f$$

$$+ \left[\frac{1}{20} - \frac{w_3\alpha_2\alpha_3^2\beta_{32} + w_4\alpha_4^2(\alpha_2\beta_{42} + \alpha_3\beta_{43})}{2}\right]D^2 f_y Df$$

$$+ \left[\frac{1}{30} - \frac{w_3\beta_{32}\alpha_2^2\alpha_3 + w_4\alpha_4(\beta_{42}\alpha_2^2 + \beta_{43}\alpha_3^2)}{2}\right]Df_y D^2 f$$

$$+ \left(\frac{1}{120} - \frac{w_4\beta_{43}\beta_{32}\alpha_2^2}{2}\right)f_y^2 D^2 f$$

$$+ \left[\frac{1}{40} - \frac{w_3\beta_{32}^2\alpha_2^2 + w_4(\beta_{43}\alpha_3 + \beta_{42}\alpha_2)^2}{2}\right]f_{yy}D^2 f$$

$$+ \left[\frac{1}{120} - \frac{w_3\beta_{32}\alpha_2^3 + w_4(\beta_{43}\alpha_3^3 + \beta_{42}\alpha_2^3)}{6}\right]f_y D^3 f$$

$$+ [\tfrac{7}{120} - w_4\beta_{43}\beta_{32}\alpha_2(\alpha_3 + \alpha_4)]f_y Df_y Df + \tfrac{1}{120}f_y^3 Df \qquad (5.8\text{-}30)$$

In order to bound γ_m, we assume the following bounds for $f(x, y)$ and its derivatives in a region R about (x_n, y_n) containing all points in (5.8-2):

$$|f(x, y)| < M \qquad \left|\frac{\partial^{i+j} f}{\partial x^i\,\partial y^j}\right| < \frac{L^{i+j}}{M^{j-1}} \qquad i + j \le m \qquad (5.8\text{-}31)$$

the latter being chosen because it leads to the convenient forms below. Using these bounds and (5.8-6), we can, for example, bound $D^2 f$ as

$$|D^2 f| = \left|\frac{\partial^2 f}{\partial x^2} + 2f_n\frac{\partial^2 f}{\partial x\,\partial y} + f_n^2\frac{\partial^2 f}{\partial y^2}\right| < 4ML^2$$

Bounding the other derivative terms in (5.8-28) to (5.8-30) similarly and using $\alpha_4 = 1$, we get {44}

$$|\gamma_2| < \left(4\left|\frac{1}{6} - \frac{\alpha_2^2 w_2}{2}\right| + \frac{1}{3}\right)ML^2 \qquad (5.8\text{-}32)$$

$$|\gamma_3| < [8\,|\tfrac{1}{24} - \tfrac{1}{6}(\alpha_2^3 w_2 + \alpha_3^3 w_3)| + 4\,|\tfrac{1}{24} - \tfrac{1}{2}\alpha_2^2\beta_{32}w_3|$$
$$+ 4\,|\tfrac{1}{8} - \alpha_2\alpha_3\beta_{32}w_3| + \tfrac{1}{12}]ML^3 \qquad (5.8\text{-}33)$$

$$|\gamma_4| < (16\,|b_1| + 4\,|b_2| + |b_2 + 3b_3| + |2b_2 + 3b_3|$$
$$+ |b_2 + b_3| + |b_3| + 8\,|b_4| + |b_5| + |2b_5 + b_7|$$
$$+ |b_5 + b_6 + b_7| + |b_6| + |2b_6 + b_7|$$
$$+ |b_7| + 2\,|b_8|)ML^4 \qquad (5.8\text{-}34)$$

where

$$\begin{aligned}
b_1 &= \tfrac{1}{120} - \tfrac{1}{24}(\alpha_2^4 w_2 + \alpha_3^4 w_3 + w_4) \\
b_2 &= \tfrac{1}{20} - \tfrac{1}{2}[\alpha_2 \alpha_3^2 \beta_{32} w_3 + (\alpha_2 \beta_{42} + \alpha_3 \beta_{43}) w_4] \\
b_3 &= \tfrac{1}{120} - \tfrac{1}{6}[\alpha_2^3 \beta_{32} w_3 + (\alpha_2^3 \beta_{42} + \alpha_3^3 \beta_{43}) w_4] \\
b_4 &= \tfrac{1}{30} - \tfrac{1}{2}[\alpha_2^2 \alpha_3 \beta_{32} w_3 + (\alpha_2^2 \beta_{42} + \alpha_3^2 \beta_{43}) w_4] \\
b_5 &= \tfrac{1}{120} - \tfrac{1}{2}\alpha_2^2 \beta_{32} \beta_{43} w_4 \\
b_6 &= \tfrac{1}{40} - \tfrac{1}{2}[\alpha_2^2 \beta_{32}^2 w_3 + (\alpha_2 \beta_{42} + \alpha_3 \beta_{43})^2 w_4] \\
b_7 &= \tfrac{7}{120} - \alpha_2(1 + \alpha_3)\beta_{32} \beta_{43} w_4 \\
b_8 &= \tfrac{1}{120}
\end{aligned} \tag{5.8-35}$$

We have noted that, for $m = 4$, the system (5.8-26) is underdetermined. As we shall see, the corresponding systems for $m = 2$ and 3 are also underdetermined. In all three cases the extra parameters can be used to minimize the above bound on γ_m, as we shall indicate in Secs. 5.8-2 to 5.8-4.

Propagated-error bounds If Runge-Kutta methods are used for the complete solution of (5.1-1), bounds on the propagated error will also be important. We shall content ourselves here with stating without proof [see Galler and Rozenberg (1960)] the following theorem on propagated error bounds, which covers many of the specific Runge-Kutta methods we shall consider [but cf. (5.8-46)].

Theorem 5.3 If $w_i > 0$, $\alpha_i > 0$, $\beta_{ij} > 0$ $(i = 1, \ldots, 4, j = 1, \ldots, 3\ ij \neq 42)$, $\beta_{42} < 0$, if $\partial f/\partial y$ is continuous, negative, and bounded from above and below in a region D in the xy plane:

$$-M_2 < \frac{\partial f}{\partial y} < -M_1 < 0$$

if the maximum error (truncation and roundoff) committed in any step is less than E in magnitude, and if the solution remains in a region D^*, approaching no closer to the boundary of D than $Qh + |\epsilon_i|$, where $Q = \max_{(x,\, y) \in D} f(x, y)$, then the total error at the ith step ϵ_i satisfies the inequality

$$|\epsilon_i| < \frac{2E}{hM_1}$$

where h is constant for all steps and must be such that

$$h < \min\left(\frac{4M_1^3}{M_2^4}, \frac{M_1}{M_2^2 - 2w_4 \beta_{42} M_2^2 - 2w_4 \beta_{42} \alpha_2 M_1 M_2}\right)$$

This indicates the difficulty of obtaining definitive results about Runge-Kutta methods. The bound can be expected to be very conservative.

Roundoff error We could choose the free parameters to minimize the roundoff in (5.8-1), i.e., to make the w_i as nearly equal as possible, but, as with numerical integration methods, roundoff is generally not significant compared with truncation, and so we shall ignore it when choosing the free parameters.

5.8-2 Second-Order Methods

For the case $m = 2$ the system (5.8-26) retains only the equations pertaining to h_n^2. These, together with (5.8-24) for $i = 2$, are

$$w_1 + w_2 = 1 \qquad \alpha_2 w_2 = \tfrac{1}{2} \qquad \beta_{21} = \alpha_2 \tag{5.8-36}$$

Three second-order methods of interest correspond to $\alpha_2 = \frac{1}{2}, \frac{2}{3}, 1$ for which (5.8-1) becomes, respectively,

$$y_{n+1} - y_n = h_n f(x_n + \tfrac{1}{2}h_n, y_n + \tfrac{1}{2}h_n f_n) \tag{5.8-37}$$

$$y_{n+1} - y_n = \tfrac{1}{4}h_n[f(x_n, y_n) + 3f(x_n + \tfrac{2}{3}h_n, y_n + \tfrac{2}{3}h_n f_n)] \tag{5.8-38}$$

$$y_{n+1} - y_n = \tfrac{1}{2}h_n[f(x_n, y_n) + f(x_n + h_n, y_n + h_n f_n)] \tag{5.8-39}$$

When $f(x, y)$ is a function of x only, the method (5.8-37) corresponds to the midpoint rule and (5.8-39) to the trapezoidal rule. Equation (5.8-38) is that for which the bound on γ_2 in (5.8-32) is minimized {46}. The bounds for (5.8-37) to (5.8-39) are, respectively, $\frac{1}{2}ML^2$, $\frac{1}{3}ML^2$, and $\frac{2}{3}ML^2$.

5.8-3 Third-Order Methods

The equations are

$$w_1 + w_2 + w_3 = 1 \qquad \alpha_2 w_2 + \alpha_3 w_3 = \tfrac{1}{2} \qquad \alpha_2^2 w_2 + \alpha_3^2 w_3 = \tfrac{1}{3}$$

$$\alpha_2 \beta_{32} w_3 = \tfrac{1}{6} \qquad \alpha_2 = \beta_{21} \qquad \alpha_3 = \beta_{31} + \beta_{32} \tag{5.8-40}$$

a two-parameter family, which can be written

$$\begin{aligned}
w_1 &= 1 + \frac{2 - 3(\alpha_2 + \alpha_3)}{6\alpha_2\alpha_3} \\
w_2 &= \frac{3\alpha_3 - 2}{6\alpha_2(\alpha_3 - \alpha_2)} \\
w_3 &= \frac{2 - 3\alpha_2}{6\alpha_3(\alpha_3 - \alpha_2)} && \alpha_2 \neq \alpha_3 \\
& && \alpha_2, \alpha_3 \neq 0 \\
\beta_{21} &= \alpha_2 && \alpha_2 \neq \tfrac{2}{3} \\
\beta_{31} &= \frac{3\alpha_2\alpha_3(1 - \alpha_2) - \alpha_3^2}{\alpha_2(2 - 3\alpha_2)} \\
\beta_{32} &= \frac{\alpha_3(\alpha_3 - \alpha_2)}{\alpha_2(2 - 3\alpha_2)}
\end{aligned} \tag{5.8-41}$$

The cases $\alpha_2 = \alpha_3$ and α_2 or $\alpha_3 = 0$ we leave to a problem {45}. Two third-order methods of interest are

$$\begin{aligned} y_{n+1} - y_n &= \tfrac{2}{9}k_1 + \tfrac{1}{3}k_2 + \tfrac{4}{9}k_3 \\ k_1 &= h_n f(x_n, y_n) \\ k_2 &= h_n f(x_n + \tfrac{1}{2}h_n, y_n + \tfrac{1}{2}k_1) \\ k_3 &= h_n f(x_n + \tfrac{3}{4}h_n, y_n + \tfrac{3}{4}k_2) \end{aligned} \tag{5.8-42}$$

$$\begin{aligned} y_{n+1} - y_n &= \tfrac{1}{6}(k_1 + 4k_2 + k_3) \\ k_1 &= h_n f(x_n, y_n) \\ k_2 &= h_n f(x_n + \tfrac{1}{2}h_n, y_n + \tfrac{1}{2}k_1) \\ k_3 &= h_n f(x_n + h_n, y_n - k_1 + 2k_2) \end{aligned} \tag{5.8-43}$$

When $f(x, y)$ is a function of x only (5.8-43) is Simpson's rule. The method (5.8-42) is that for which the bound on γ_3 is minimized {46}. For the methods (5.8-42) and (5.8-43), the bounds are, respectively, $\frac{1}{9}ML^3$ and $\frac{13}{36}ML^3$.

5.8-4 Fourth-Order Methods

The two-parameter system (5.8-26) can be solved to give

$$w_1 = \frac{1}{2} + \frac{1 - 2(\alpha_2 + \alpha_3)}{12\alpha_2\alpha_3} \qquad w_2 = \frac{2\alpha_3 - 1}{12\alpha_2(\alpha_3 - \alpha_2)(1 - \alpha_2)}$$

$$w_3 = \frac{1 - 2\alpha_2}{12\alpha_3(\alpha_3 - \alpha_2)(1 - \alpha_3)} \qquad w_4 = \frac{1}{2} + \frac{2(\alpha_2 + \alpha_3) - 3}{12(1 - \alpha_2)(1 - \alpha_3)}$$

$$\beta_{32} = \frac{\alpha_3(\alpha_3 - \alpha_2)}{2\alpha_2(1 - 2\alpha_2)} \qquad \alpha_4 = 1 \tag{5.8-44}$$

$$\beta_{42} = \frac{(1 - \alpha_2)[\alpha_2 + \alpha_3 - 1 - (2\alpha_3 - 1)^2]}{2\alpha_2(\alpha_3 - \alpha_2)[6\alpha_2\alpha_3 - 4(\alpha_2 + \alpha_3) + 3]}$$

$$\beta_{43} = \frac{(1 - 2\alpha_2)(1 - \alpha_2)(1 - \alpha_3)}{\alpha_3(\alpha_3 - \alpha_2)[6\alpha_2\alpha_3 - 4(\alpha_2 + \alpha_3) + 3]}$$

except when $\alpha_2, \alpha_3 = 0$, $\alpha_2, \alpha_3 = 1$, $\alpha_2 = \alpha_3$ or the denominators of β_{32}, β_{42}, or β_{43} vanish. These special cases are considered in {47} and {48}. The most commonly used fourth-order Runge-Kutta method is one in which $\alpha_2 = \alpha_3 = \frac{1}{2}$ and its equations are

$$\begin{aligned} y_{n+1} - y_n &= \tfrac{1}{6}(k_1 + 2k_2 + 2k_3 + k_4) \\ k_1 &= h_n f(x_n, y_n) \\ k_2 &= h_n f(x_n + \tfrac{1}{2}h_n, y_n + \tfrac{1}{2}k_1) \\ k_3 &= h_n f(x_n + \tfrac{1}{2}h_n, y_n + \tfrac{1}{2}k_2) \\ k_4 &= h_n f(x_n + h_n, y_n + k_3) \end{aligned} \tag{5.8-45}$$

The method for which the error bound on γ_4 is minimized corresponds to $\alpha_2 = .4$, $\alpha_3 = \frac{7}{8} - \frac{3}{16}\sqrt{5}$ and has the equations

$$\begin{aligned}
y_{n+1} - y_n &= .17476028k_1 - .55148066k_2 + 1.20553560k_3 + .17118478k_4 \\
k_1 &= h_n f(x_n, y_n) \\
k_2 &= h_n f(x_n + .4h_n, y_n + .4k_1) \\
k_3 &= h_n f(x_n + .45573725h_n, y_n + .29697761k_1 + .15875964k_2) \\
k_4 &= h_n f(x_n + h_n, y_n + .21810040k_1 - 3.05096516k_2 + 3.83286476k_3)
\end{aligned} \tag{5.8-46}$$

The error bounds on the methods (5.8-45) and (5.8-46) are, respectively, $\frac{73}{720}ML^4$ and $5.4627 \times 10^{-2}ML^4$. A third method which is useful for error estimation (see Sec. 5.8-6) has the equations

$$\begin{aligned}
y_{n+1} - y_n &= \tfrac{1}{6}(k_1 + 4k_3 + k_4) \\
k_1 &= h_n f(x_n, y_n) \\
k_2 &= h_n f(x_n + \tfrac{1}{2}h_n, y_n + \tfrac{1}{2}k_1) \\
k_3 &= h_n f(x_n + \tfrac{1}{2}h_n, y_n + \tfrac{1}{4}k_1 + \tfrac{1}{4}k_2) \\
k_4 &= h_n f(x_n + h_n, y_n - k_2 + 2k_3)
\end{aligned} \tag{5.8-47}$$

5.8-5 Higher-Order Methods

For $m > 4$, it turns out that $p > m$. Thus, to achieve fifth-order accuracy, six stages are necessary and for sixth-order, seven is the minimum number required. For $m \geq 7$, it has been shown that $p_m \geq m + 2$. This does not detract from the usefulness of higher-order methods, since under proper conditions of smoothness the use of such methods allows one to take a much larger step length h and thus cover the interval of integration using fewer function evaluations. Since the algebra involved in computing the formulas for such methods is formidable, it is only recently that these methods have been generated using computerized algebraic-manipulation systems. A wide variety of such methods exist since there are many free parameters in the governing equations. One of the considerations in the choice of these parameters is that of absolute stability, which we discuss below; another is connected with local error estimation. As an example of such methods, we give the following equations for a particular fifth-order method which has certain valuable features considered below.

$$y_{n+1} - y_n = \tfrac{1}{24}k_1 + \tfrac{5}{48}k_4 + \tfrac{27}{56}k_5 + \tfrac{125}{336}k_6$$

$$\begin{aligned}
k_1 &= h_n f(x_n, y_n) \\
k_2 &= h_n f(x_n + \tfrac{1}{2}h_n, y_n + \tfrac{1}{2}k_1) \\
k_3 &= h_n f(x_n + \tfrac{1}{2}h_n, y_n + \tfrac{1}{4}k_1 + \tfrac{1}{4}k_2) \\
k_4 &= h_n f(x_n + h_n, y_n - k_2 + 2k_3) \\
k_5 &= h_n f(x_n + \tfrac{2}{3}h_n, y_n + \tfrac{7}{27}k_1 + \tfrac{10}{27}k_2 + \tfrac{1}{27}k_4) \\
k_6 &= h_n f(x_n + \tfrac{1}{5}h_n, y_n + \tfrac{28}{625}k_1 - \tfrac{1}{5}k_2 \\
&\quad + \tfrac{546}{625}k_3 + \tfrac{54}{625}k_4 - \tfrac{378}{625}k_5)
\end{aligned} \tag{5.8-48}$$

5.8-6 Practical Error Estimation

The error bounds derived in Sec. 5.8-1 are not very practical, since it is very difficult to find values of L and M for which (5.8-31) holds. Consequently, we seek other ways to estimate the local truncation error T_m or more specifically γ_m, where, we recall from (5.8-27),

$$T_m = \gamma_m h_n^{m+1} + O(h_n^{m+2})$$

One way, similar to that used in adaptive-integration schemes (cf. Sec. 4.11), is to integrate over two successive intervals with the same step size h_n, then integrate over the double interval with step size $2h_n$ and compare the results. In the first integration, we have

$$y_{n+1} \approx Y(x_{n+1}) + \gamma_m h_n^{m+1}$$

$$y_{n+2} \approx Y(x_{n+2}) + 2\gamma_m h_n^{m+1}$$

where we have assumed that γ_m does not vary much over the interval $[x_n, x_{n+2}]$ and that we have started with the exact value $Y(x_n)$ at x_n. There is a further assumption that using y_{n+1} in place of $Y(x_{n+1})$ in going from x_{n+1} to x_{n+2} does not affect the form of the local truncation error. If we now integrate directly from x_n to x_{n+2} with step size $2h$, to yield the value $\bar{y}_{n+2}$, we have that

$$\bar{y}_{n+2} \approx Y(x_{n+2}) + \gamma_m(2h_n)^{m+1}$$

Hence $y_{n+2} - \bar{y}_{n+2} \approx \gamma_m(2 - 2^{m+1})h_n^{m+1}$, so that

$$T_m \approx \gamma_m h_n^{m+1} \approx \frac{y_{n+2} - \bar{y}_{n+2}}{2 - 2^{m+1}} \tag{5.8-49}$$

Thus, if the method requires p_m function evaluations, at the cost of $p_m - 1$ additional function evaluations (because k_1 is the same for both step sizes) every two steps, we can monitor the calculation to see whether the local truncation error remains acceptable. Furthermore, as we shall see, this estimate of T_m can be used to determine a good value of the step size h_{n+2}.

A second method for estimating T_m is to compare the value of y_{n+1} with a second value $\bar{y}_{n+1}$ computed using a method of order $m+1$. For then we have, assuming again that $y_n = Y(x_n)$,

$$\bar{y}_{n+1} \approx Y(x_{n+1}) + \gamma_{m+1} h_n^{m+2} + O(h_n^{m+3})$$

so that $$y_{n+1} - \bar{y}_{n+1} \approx \gamma_m h_n^{m+1} + O(h_n^{m+2}) \approx T_m \tag{5.8-50}$$

In general, this is a very expensive way to estimate T_m since it requires $p_m + p_{m+1} - 1$ function evaluations per step. However, by a clever choice of the parameters β_{ij} it is possible, for some values of m, to determine a method of order $m+1$ in which a method of order m is embedded; that is, k_i, $i = 1, \ldots, p_m$, are the same for both methods.

Then $$\bar{y}_{n+1} = y_n + \sum_{i=1}^{p_{m+1}} w_i k_i \qquad \text{and} \qquad y_{n+1} = y_n + \sum_{i=1}^{p_m} \bar{w}_i k_i$$

Thus, at the cost of $p_{m+1} - p_m$ additional function evaluations, we get an estimate of the local truncation error. The fifth-order method given in (5.8-48) is an example of this, since the fourth-order method (5.8-47) is embedded in it. Thus, (5.8-47) is a fourth-order method with the local truncation-error estimate

$$T_4 \approx \tfrac{1}{8}k_1 + \tfrac{2}{3}k_2 + \tfrac{1}{16}k_4 - \tfrac{27}{56}k_5 - \tfrac{125}{336}k_6 \tag{5.8-51}$$

which results by formally subtracting the first equation in (5.8-48) from the first equation in (5.8-47).

5.8-7 Step-Size Strategy

Once we have an estimate of the local truncation error in one or two integration steps, we can use this information to decide whether to accept or reject the computed values and to decide on a new value of h to be used either to repeat the computation or to continue. If the step size is changed, the choice of the new h should be such that the estimated local truncation error with the new value will be less than the allowable error over the new step, $[\tilde{x}, \tilde{x} + h]$, where $\tilde{x} = x_n$ in case the computation is repeated and $\tilde{x} = x_{n+1}$ or $\tilde{x} = x_{n+2}$ otherwise. The details are as follows.

Let us assume that we are integrating over the interval $[a, b]$, where $H = b - a$, and that we want the total error at $x = b$ to be less than ϵ. We now make the assumption that the total error equals the sum of the local truncation errors over all subintervals. This assumption may not hold when $|f_y|$ is large, for then small changes in y cause large changes in $f(x, y)$ and there is a considerable amount of propagated error. In such cases, the best we can do is to work with a tolerance ϵ which is smaller than the accuracy we desire in the hope that the deleterious effect of error propagation will be controlled in this way. Returning now to our original assumption, we see that if we integrate over the subinterval $[x_n, x_n + h_n]$, we require that $|T_m^{(n)}| < \epsilon h_n/H = \bar{\epsilon}h_n$, where $\bar{\epsilon} = \epsilon/H$, and where $T_m^{(n)}$ is the local truncation error over $[x_n, x_n + h_n]$. If our estimate of $|T_m^{(n)}|$ is $< \bar{\epsilon}h_n$ after one step of length h_n, or if our estimate of $|T_m^{(n)}| + |T_m^{(n+1)}|$ is $< 2\bar{\epsilon}h_n$ after two steps of total length $2h_n$, we accept the results and continue from x_{n+1} or x_{n+2}, respectively. Otherwise, we must recompute from x_n with a smaller value of h_n. A new value of h_n may also be desirable even when we accept the results of the computation, as we shall shortly see.

In either case, we have

$$T_m^{(n)} \approx \gamma_m h_n^{m+1}$$

If $|T_m^{(n)}| > \bar{\epsilon}h_n$, we must choose $\bar{h}_n$, the new value of h, so that for the new local truncation error, $|\bar{T}_m^{(n)}| < \bar{\epsilon}\bar{h}_n$. Since $\bar{T}_m^{(n)} \approx \gamma_m \bar{h}_m^{m+1}$ and $\gamma_m \approx T_m^{(n)}/h_n^{m+1}$, we require that

$$\frac{|T_m^{(n)}|}{h_n^{m+1}} \bar{h}_n^{m+1} < \bar{\epsilon}\bar{h}_n$$

from which we deduce that any choice $\bar{h}_n$ such that

$$\bar{h}_n < \tilde{h} = \left(\frac{\bar{\epsilon}h_n^{m+1}}{|T_m^{(n)}|}\right)^{1/m}$$

will be satisfactory. However, to be on the safe side, we choose $\bar{h}_n = .8\tilde{h}$.

Assuming that γ_m does not change much as we proceed from one interval to the next, we see that even when we accept the computation at $x_n + h_n$ or $x_n + 2h_n$, we can use the same reasoning as above to choose the next step length h_{n+1} or h_{n+2} as equal to $.8\tilde{h}$. However, we must make the proviso that this new value be neither too large nor too small, both in an absolute sense and relative to the previous value of h. When the proposed value of h is not within some preset bounds, we take the new value of h as the extreme point of the permissible interval of values of h.

5.8-8 Stability

Since the Runge-Kutta methods are single-step methods, it is meaningless to apply the definition of relative stability to them since there is only one root to the (generally) nonlinear difference equation (5.8-1). However, we can define absolute stability for these methods in terms of equation (5.4-1) with $K > 0$. We know that the solution of this equation goes to zero as $x \to \infty$. Hence, it is only natural to require that the numerical solution should also have this property. Assume a constant step size h. We can then define a

Runge-Kutta method to be absolutely stable for a certain value $\bar{h} = hK$, $K > 0$, if the numerical solution using this particular method applied to (5.4-1) goes to zero as $n \to \infty$.

Example 5.7 Apply the Runge-Kutta method of order 2, (5.8-39), to (5.4-1). We obtain

$$y_{n+1} = y_n + \frac{h}{2}[-Ky_n - K(y_n - hKy_n)] = \left(1 - hK + \frac{h^2}{2}K^2\right)y_n$$

so that, by induction, $y_{n+1} = [1 - hK + (h^2/2)K^2]^{n+1}y_0$. In order that $y_n \to 0$ as $n \to \infty$ for $y_0 \neq 0$, it is necessary and sufficient that $[1 - \bar{h} + (\bar{h}^2/2)] < 1$, from which it follows that, for $0 < \bar{h} < 2$, the method (5.8-39) is absolutely stable. In fact, for any second-order method using two function evaluations, the interval of absolute stability is (0, 2).

In general, for a p-stage method of order m, we have that {52} $y_{n+1} = ry_n$, where

$$r = 1 - \bar{h} + \tfrac{1}{2}\bar{h}^2 - \cdots + \frac{1}{m!}(-\bar{h})^m + \sum_{i=m+1}^{p} \gamma_i(-\bar{h})^i \qquad (5.8\text{-}52)$$

and the coefficients γ_i depend on the constants of the particular method. If $p = m$, the value of the sum is zero, so that for all methods of order m for which $p = m$, the interval of stability is the same. Thus for $m = 3$, the interval is (0, 2.51), and for $m = 4$, it is (0, 2.78). For $m > 4$, $p > m$, and one of the criteria in choosing the constants of such a method is to maximize the interval of absolute stability.

For the general equation $y' = f(x, y)$, we approximate K by $-f_y(x, y)$ at some point in the interval $(x, x + h)$. For a single equation, we can approximate f_y quite easily provided two different values of k use the same value of α. Thus if

$$k_r = h_n f\left(x_n + \alpha h_n, y_n + \sum_{j=1}^{r-1} \beta_{rj}k_j\right)$$

and

$$k_s = h_n f\left(x_n + \alpha h_n, y_n + \sum_{j=1}^{s-1} \beta_{sj}k_j\right)$$

then

$$K \approx f_y(x + \alpha h_n, \bar{y}_n) \approx \frac{k_r - k_s}{\sum_{j=1}^{r-1} \beta_{rj}k_j - \sum_{j=1}^{s-1} \beta_{sj}k_j}$$

where $\bar{y}_n$ is some intermediate value. We see from this that another criterion in choosing the free parameters of a method is that $\alpha_r = \alpha_s$ for some pair of indices $0 < r < s \leq p$. This occurs, for example, in the classical fourth-order method (5.8-45) as well as in the fifth-order method (5.8-48).

We stress that this estimate can only be obtained for a single equation. However, in general, the problem of stability does not arise in the integration of a single equation since accuracy considerations usually demand a choice of step size h which keeps the method stable. The problems arise in systems of equations. Here, the test system has the form

$$\mathbf{y}' = A\mathbf{y} \qquad \mathbf{y}(x_0) = \mathbf{y}_0$$

where $\mathbf{y}$ is an n vector and A an $n \times n$ matrix. The situation corresponding to $K > 0$ for the single equation (5.4-1) is that all eigenvalues λ_i of A have negative real part. A method is then absolutely stable for all values of h such that $\bar{h}_i = |h\lambda_i| < 1$ when Re $\lambda_i < 0$ for all i. We shall return to this in Sec. 5.10 on stiff equations.

5.8-9 Comparison of Runge-Kutta and Predictor-Corrector Methods

In order to compare Runge-Kutta with corresponding-order predictor-corrector methods, we note the following points:

1. Runge-Kutta methods are self-starting, the interval between steps may be changed at will, and, in general, they are particularly straightforward to apply on a digital computer.
2. They are comparable in accuracy (often more accurate) than corresponding-order predictor-corrector methods {49}. However, if we do not monitor the per step error by using additional function evaluations, we shall generally be required to choose the step size h conservatively, i.e., smaller than is actually necessary to achieve the desired accuracy.
3. Further, they require a number of evaluations of $f(x, y)$ at each step at least equal to the order of the method. As we have seen, predictor-corrector methods generally require only two evaluations per step. Since evaluation of $f(x, y)$ is usually the most time-consuming part of solving (5.1-1), this means that predictor-corrector methods are generally faster than Runge-Kutta methods; e.g., fourth-order predictor-corrector methods are nearly twice as fast as fourth-order Runge-Kutta methods.
4. Finally, monitoring the local truncation error does not involve any additional function evaluations using predictor-corrector methods, whereas it is quite expensive for Runge-Kutta methods.

On a digital computer reasons 2 to 4 are much more compelling than 1, and thus predictor-corrector methods are the indicated methods to use, except when $f(x, y)$ is simple to compute, in which case, point 1 becomes important.

The self-starting characteristic of Runge-Kutta methods makes them an ideal adjunct to the usual predictor-corrector methods for starting the solution. Since they will be used for only a few steps of the computation, truncation error and not stability is the key consideration. Therefore, for this purpose, the minimum-error-bound Runge-Kutta methods should be used.

The methods we have derived may be used for systems of equations {50}, although the error bounds were derived for single equations. It is reasonable to assume, however, that methods which are best in this sense for single equations will be at least nearly best for systems.

5.9 OTHER NUMERICAL INTEGRATION METHODS

5.9-1 Methods Based on Higher Derivatives

In this section we shall use derivatives higher than the first in formulas for the solution of first-order differential equations. One example of such a formula is readily derivable from the Euler-Maclaurin sum formula in the form (4.13-17). Replacing $f(x)$ with $y'(x)$, we get, after making some simple changes in notation,

$$y_{n+1} = y_1 + h \sum_{i=-1}^{n-1} y'_{n-i} - \frac{h}{2}(y'_{n+1} + y'_1) - \sum_{k=1}^{m} \frac{B_{2k}}{(2k)!} h^{2k-1}(y_{n+1}^{(2k)} - y_1^{(2k)}) \tag{5.9-1}$$

with the truncation error given by (4.13-16)

$$E_m = \frac{nh^{2m+2}B_{2m+2}}{(2m+2)!} Y^{(2m+3)}(\xi) \tag{5.9-2}$$

Our main interest in this section, however, is in a class of predictor and corrector formulas which use the second as well as the first derivative. The corrector formulas are generated by taking the Hermite interpolation formula (3.7-17) (or the modified Hermite formula) based on the points x_{n+1}, $x_n, \ldots, x_{n-p}$ and integrating between x_n and x_{n+1} after replacing $f(x)$ by $y'(x)$. The result of this is

$$y_{n+1} = y_n + \sum_{i=-1}^{p} \left[\int_{x_n}^{x_{n+1}} h_{n-i}(x)\, dx\right] y'_{n-i} + \sum_{i=-1}^{p} \left[\int_{x_n}^{x_{n+1}} \bar{h}_{n-i}(x)\, dx\right] y''_{n-i} \tag{5.9-3}$$

The truncation error is given, using (4.4-12), as

$$T_n = \frac{Y^{(2p+5)}(\eta)}{(2p+4)!} \int_{x_n}^{x_{n+1}} [(x - x_{n+1}) \cdots (x - x_{n-p})]^2\, dx \tag{5.9-4}$$

With $p = 0$ in (5.9-3), we get the fourth-order corrector

$$y_{n+1} = y_n + \frac{h}{2}(y'_{n+1} + y'_n) + \frac{h^2}{12}(-y''_{n+1} + y''_n) \tag{5.9-5}$$

which has an error term

$$T_n = \frac{h^5}{720} Y^{\text{v}}(\eta) \tag{5.9-6}$$

The stability equation for (5.9-3) has only one root, $r_0 = 1$ at $h = 0$. Therefore, for some range of values of Kh, this class of methods is stable. In particular, (5.9-5) is stable for all values of Kh {54}. Moreover, note that the truncation-error term (5.9-6) is substantially smaller than that given by any of the correctors in Table 5.2. Thus, if the second derivative of y can be calculated easily, (5.9-5) may indeed be a good choice for a corrector. Indeed, when (5.1-1) is a single equation, the second derivative is given by

$$y'' = \frac{d}{dx} f(x, y) = \frac{\partial f}{\partial y} y' + \frac{\partial f}{\partial x} \tag{5.9-7}$$

It is often true that, having computed $f(x, y)$, the calculation of the partial derivatives of $f(x, y)$ can be done very quickly on a digital computer, in which case the second derivative as given by (5.9-7) is indeed easily calculable. Of course, for systems of equations, the calculation of the second derivative becomes substantially more complicated.

To get a class of predictors, we use a Hermite formula based on the points $x_n, \ldots, x_{n-p}$ and proceed as above {53}. A convenient predictor to use with (5.9-5) is

$$y_{n+1} = y_n + \frac{h}{2}(-y'_n + 3y'_{n-1}) + \frac{h^2}{12}(17y''_n + 7y''_{n-1}) \tag{5.9-8}$$

which has a truncation error

$$T_n = \frac{31h^5}{720} Y^{\text{v}}(\eta) \tag{5.9-9}$$

The set of equations analogous to (5.5-27) and (5.5-33) is

Predictor:

$$y^{(0)}_{n+1} = y_n + \frac{h}{2}(-y'_n + 3y'_{n-1}) + \frac{h^2}{12}(17y''_n + 7y''_{n-1})$$

Modifier:

$$\bar{y}^{(0)}_{n+1} = y^{(0)}_{n+1} + \tfrac{31}{30}(y_n - y^{(0)}_n) \tag{5.9-10}$$

$$(\bar{y}^{(0)}_{n+1})' = f(x_{n+1}, \bar{y}^{(0)}_{n+1})$$

Corrector:

$$y^{(j+1)}_{n+1} = y_n + \frac{h}{2}[(y^{(j)}_{n+1})' + y'_n] + \frac{h^2}{12}[-(y^{(j)}_{n+1})'' + y''_n] \qquad j = 0, 1, \ldots$$

Example 5.8 Use (5.9-10) to perform the same calculations as in Sec. 5.7-2 on the same three examples.

The results for these calculations corresponding to those of Sec. 5.7-2 are given in the last two columns of Tables 5.3 to 5.6. Since the Hermite method is stable and has a smaller truncation error than either Hamming's or Milne's methods, we would expect it to give better results than Hamming's method for the differential equation (5.7-13) and better results than both Milne's and Hamming's methods for Eq. (5.7-16), and in fact it does. The importance of roundoff is again in evidence in the solution of the differential equation (5.7-17), where the Hermite method is initially best, but by the end of the computation, the errors in all three methods are similar in magnitude. In all the computations of this section, the same convergence factors were used as in the examples of Sec. 5.7-2, and in all cases one application of the corrector at each step was sufficient. Thus, when the second derivative is easy to calculate (it is very easy for the first two examples considered here and not quite so easy for the last), the Hermite method is to be recommended over Hamming's method.

5.9-2 Extrapolation Methods

Consider the Euler, or point-slope, method for solving a differential equation. It does not require any additional starting values, and the solution is given by the recursive formula

$$y_{n+1} = y_n + hf(x_n, y_n) \qquad y_0 = y(x_0) \qquad n = 1, \ldots, N \tag{5.9-11}$$

The sequence $\{y_n\}$ depends on the step size h, so that we can denote the computed solution by $y(x; h)$; the domain of $y(x; h)$ is the set $\{x_n\}$, where $x_n = x_0 + nh$, $n = 0, 1, \ldots, N$. For this function $y(x; h)$ it can be shown that

$$y(x; h) = Y(x) + c_1(x)h + c_2(x)h^2 + \cdots + c_p(x)h^p + O(h^{p+1}) \tag{5.9-12}$$

To apply Richardson extrapolation (Sec. 4.2) to this problem, we choose a sequence of pairs (h_i, N_i) such that $x = x_0 + h_i N_i$ is a constant, calculate $y(x; h_i)$ for each pair, and apply formula (4.2-10) with $\gamma = 1$. We shall not pursue this case further since there exists a more efficient integration method yielding an expansion like the above but containing only even powers of h. This gives better results using extrapolation. The method is based on the midpoint formula (5.5-16). Used by itself, the midpoint formula requires two starting values and also exhibits an oscillating error due to the fact that y_{n+1} is coupled to y_{n-1} and not to the immediately preceding value y_n, which only enters via the derivative. On the other hand, it is more accurate than the Euler method in that the local truncation error is $O(h^3)$ and the global error is $O(h^2)$, while in the Euler method, the errors are $O(h^2)$ and $O(h)$, respectively [cf. (5.4-32)]. In order to overcome the oscillating error, we introduce a damping process to get the *modified midpoint method* of Gragg. Set

$$\begin{aligned} y_1 &= y_0 + hy'_0 \\ y_{n+1} &= y_{n-1} + 2hy'_n \qquad n = 1, 2, \ldots, N;\ N \text{ even} \\ \hat{y}_N &= \tfrac{1}{4}(y_{N-1} + 2y_N + y_{N+1}) \end{aligned} \tag{5.9-13}$$

If we set $x = x_0 + Nh$ and $y(x; h) = \hat{y}_N$, it can be shown that

$$y(x; h) = Y(x) + c_1(x)h^2 + c_2(x)h^4 + \cdots \tag{5.9-14}$$

Hence, the convergence of Richardson extrapolation should be quicker if we calculate $y(x; h_i)$ for a sequence of pairs (h_i, N_i), $x = x_0 + h_i N_i$, and now apply (4.2-10) with $\gamma = 2$.

Example 5.9 Apply (5.9-13) and Richardson extrapolation to the differential equation $y' = y^2$, $y(0) = .25$.

Starting with $h_1 = \frac{1}{2}$ and setting $h_i = \frac{1}{2}h_{i-1}$, we calculate with $N_i = 2^i$:

h	N	$y(1; h) = T_0$	T_1	T_2
.5	2	.33225 2741		
			.33331 4614	
.25	4	.33304 9146		.33333 3213
			.33333 2051	
.125	8	.33326 1325		

The exact solution of the differential equation is $y(x) = 1/(4 - x)$, so that $Y(1) = \frac{1}{3}$. The extrapolated solution is correct to six figures, and the error is $\frac{1}{600}$ times the error in $y(1; .125)$.

Two general strategies suggest themselves in using extrapolation methods. Suppose we wish the solution at a set of points $x_0, x_1, x_2, \ldots$ such that $x_{n+1} - x_n = H$ and we use the modified midpoint method. Then to proceed from x_n to x_{n+1}, we take a succession of values of h, which might be the usual Romberg sequence $H/2$, $H/4$, $H/8$, ... (as in the example above) or a sequence such as $H/2$, $H/4$, $H/6$, $H/8$, $H/12$, ..., which, as we noted at the end of Section 4.10, is more efficient for quadrature. In any case, $N = H/h$ must be even. We then apply (4.2-10), as in the example above, until we have a satisfactory extrapolated value of x_{n+1}. Then we proceed to x_{n+2} with the accepted value at x_{n+1} as initial value. This so-called *active* extrapolation is a single-step method as defined in Sec. 5.1.

Another possibility is *passive* extrapolation. This can best be illustrated using the trapezoidal rule

$$y_{n+1} = y_n + \frac{h}{2}(y_n' + y_{n+1}') \tag{5.9-15}$$

If we assume that y_{n+1} satisfies this equation, i.e., that we have iterated to convergence, and if we integrate our differential equation over the entire integration interval to yield the function $y(x; h)$ defined on the set $\{x_n\}$ as previously, then the error at every grid point x_n, $n = 1, 2, \ldots, N$, satisfies (5.9-14). Thus, after integrating over the entire interval with various values of

h_i, $h_i = H/2^i$, $i = 0, 1, 2, \ldots$, or some other sequence h_i with $h_0 = H$, where H is the spacing we are interested in, we can extrapolate at each of the basic grid points $x_j = x_0 + jH$. While this method gives very accurate results for difficult problems, it is very costly in time because of the requirement that we iterate to convergence even when the value of h_i is not small enough to ensure rapid convergence of the iterations (5.5-3) or alternatively use a different method to solve the nonlinear equation (5.9-15).

5.10 STIFF EQUATIONS

Although the concept of stiffness is usually associated with systems of first-order differential equations, we shall first illustrate the problem with a single equation. We shall then discuss the general situation and one of the many proposed solutions to the problem.

Consider the differential equation

$$y' = 100(\sin x - y) \qquad y(0) = 0 \tag{5.10-1}$$

The exact solution is

$$y(x) = \frac{\sin x - .01 \cos x + .01e^{-100x}}{1.0001} \tag{5.10-2}$$

Since the exponential term is less than 10^{-6} for $x = .1$, one would expect that a step size of $h = .1$ would be possible for $x > .1$. However, if we compute with $h = .03$ using the Runge-Kutta method (5.8-45), we find that $y(3) = 6.7 \times 10^{11}$, whereas with $h = .025$ we get $y(3) = .150943$, which is a good result. The problem here is that while the component $.01e^{-100x}$ does not contribute anything to the solution after a short interval, i.e., it behaves like a transient, nevertheless, it influences the stability interval throughout the computation. Since the stability interval is measured in units of $\bar{h} = h|\partial f/\partial y|$ and $\partial f/\partial y = -100$, we see that we need a very small value of h to keep $\bar{h}$ reasonable. For the method given by Eqs. (5.8-45) $\bar{h}$ must be in the interval (0, 2.78) for absolute stability. This translates into an interval (0, .0278) for h, which explains why the value $h = .03$ causes instability while $h = .025$ still gives a stable computation.

The situation for systems is similar. For the system

$$\begin{aligned} y^{[1]'} &= f_1(x, y^{[1]}, \ldots, y^{[m]}) \\ &\cdots\cdots\cdots\cdots\cdots\cdots \\ y^{[m]'} &= f_m(x, y^{[1]}, \ldots, y^{[m]}) \end{aligned} \tag{5.10-3}$$

or equivalently

$$\mathbf{y}' = \mathbf{f}(x, \mathbf{y}) \tag{5.10-4}$$

where

$$\mathbf{y} = [y^{[1]}, \ldots, y^{[m]}]^T \tag{5.10-5}$$

$\bar{h} = h\rho(J)$, where J is the Jacobian matrix

$$J = \left[\frac{\partial f_i}{\partial y^{[j]}}\right] \qquad i, j = 1, \ldots, m \tag{5.10-6}$$

and $\rho(J)$ is the spectral radius of J, which equals the maximum of the magnitudes of the eigenvalues of J (see Sec. 1.3). Hence, if $\rho(J)$ is large, h will have to be very small to ensure absolute stability.

In order to determine when absolute stability is essential, we consider a special case of (5.10-4), namely the system of inhomogeneous linear equations with constant coefficients

$$\mathbf{y}' = A\mathbf{y} + \boldsymbol{\phi}(x) \tag{5.10-7}$$

where $\boldsymbol{\phi}(x)$ is a vector-valued function of the independent variable x and we assume that the $m \times m$ matrix A has distinct eigenvalues. In this case, the system (5.10-7) has a general solution of the form

$$\mathbf{y}(x) = \sum_{k=1}^{m} c_k e^{\lambda_k x}\mathbf{z}_k + \boldsymbol{\psi}(x) \tag{5.10-8}$$

where the λ_k, $k = 1, \ldots, m$, are the (distinct) eigenvalues of A and the $\mathbf{z}_k$ are the corresponding eigenvectors. The c_k depend on the initial values of the system (5.10-7). Now, if Re $\lambda_k > 0$ for some k, then generally $\mathbf{y}(x) \to \infty$ as $x \to \infty$, so that absolute stability is not relevant. However, if Re $\lambda_k < 0$ for all k, then the term

$$\sum_{k=1}^{m} c_k e^{\lambda_k x}\mathbf{z}_k \to 0 \qquad \text{as } k \to \infty \tag{5.10-9}$$

so that absolute stability becomes important. (The case Re $\lambda_k = 0$ corresponds to an oscillating term. If $|\lambda_k| \gg 0$, the oscillation is very rapid; this case is the subject of much current research activity. We shall not treat the oscillating case here.) When (5.10-9) holds, the first term in the solution (5.10-8) is called the *transient solution* and the second term $\boldsymbol{\psi}(x)$ is called the *steady-state solution.*

Let λ_μ and λ_ν be two eigenvalues of A such that

$$|\text{Re } \lambda_\mu| \leq |\text{Re } \lambda_k| \leq |\text{Re } \lambda_\nu| \qquad k = 1, \ldots, m$$

If we are interested in finding the steady-state solution $\boldsymbol{\psi}(x)$, we must integrate (5.10-7) until the slowest-decaying exponential in the solution, $e^{\lambda_\mu x}$, is negligible. (Recall that Re $\lambda_k < 0$ for all k.) Thus, the smaller $|\text{Re } \lambda_\mu|$ is, the larger the integration interval will be. On the other hand, because of stability considerations, the size of the integration step is determined by $\rho(A)$, which is greater than or equal to $|\text{Re } \lambda_\nu|$. Hence, if $|\text{Re } \lambda_\nu| \gg |\text{Re } \lambda_\mu|$, we have to take an excessively large number of steps to cover the integration interval of interest in order to find the steady-state solution. This situation is called *stiffness.* We summarize the above discussion with the following definition.

Definition 5.3 The linear system $\mathbf{y}' = A\mathbf{y} + \boldsymbol{\phi}(x)$ is said to be *stiff* if

1. $\text{Re } \lambda_k < 0$, $k = 1, \ldots, m$.
2. $\max_{1 \le k \le m} |\text{Re } \lambda_k| \gg \min_{1 \le k \le m} |\text{Re } \lambda_k|$

where the λ_k, $k = 1, \ldots, m$, are the eigenvalues of A. The ratio

$$\frac{\max_{1 \le k \le m} |\text{Re } \lambda_k|}{\min_{1 \le k \le m} |\text{Re } \lambda_k|}$$

is called the *stiffness ratio*.

For the nonlinear system $\mathbf{y}' = \mathbf{f}(x, \mathbf{y})$ stiffness is determined by the eigenvalue structure of the Jacobian J, so that it depends on the solution $\mathbf{y}$ and ultimately on the independent variable x. We say that the system (5.10-4) is stiff in an interval I if for $x \in I$ the eigenvalues of the Jacobian J satisfy conditions 1 and 2 in Definition 5.3.

If the system (5.10-4) is a model of a physical system, stiffness means that there are processes in the physical system described by the differential equations with significantly different time scales (or time constants). The usual numerical methods are inadequate since they require that we take account of the fastest process even after the corresponding component has died out in the exact solution. The only satisfactory methods are those which are absolutely stable in a good portion of $\mathcal{N}$, that part of the complex plane with negative real part.

The trapezoidal rule

$$\mathbf{y}_{n+1} = \mathbf{y}_n + \frac{h}{2}[\mathbf{f}(x_n, \mathbf{y}_n) + \mathbf{f}(x_{n+1}, \mathbf{y}_{n+1})] \tag{5.10-10}$$

is absolutely stable in all of $\mathcal{N}$. However, in applying it, we must make sure that this equation is satisfied exactly to ensure stability; i.e., we must iterate to convergence. However, if we use a relatively large h, and this is the whole point of using a rule absolutely stable in $\mathcal{N}$, we cannot expect convergence if we use the iteration (5.5-3) since for systems of equations this requires that $\frac{1}{2}h\rho(J) < 1$. Thus, viewing (5.10-10) as a nonlinear system of equations for $y_{n+1}^{[1]}, \ldots, y_{n+1}^{[m]}$, we must use the Newton-Raphson method or one of its variants to solve (5.10-10). As this implies, the problems involved in solving stiff systems are quite formidable.

A family of methods proposed by Gear has proved successful for stiff equations in a great number of cases. These methods are based on the implicit formulas (5.2-15) and (5.2-16) derived from numerical differentiation. Their properties are such that they are stable for all values of $\bar{h}$ such that $\text{Re } \bar{h} < -a$ for some relatively small positive value of a and accurate for values of $\bar{h}$ in the rectangle $-a \leq x \leq a$, $-b \leq y \leq b$, b some small positive value. Thus for values of $\bar{h}$ where stability is the prime consideration, that is, Re $\bar{h} < -a$, these methods are stable, and where accuracy is more important, that is, Re $\bar{h} > -a$, they are accurate. Of course, whenever we try to achieve higher accuracy, stability deteriorates, but we should expect this. Gear's method can be applied in the fixed-step mode (5.2-16) and in the variable-step–variable-order mode described in Sec. 5.7-1. The order must be limited to 5 because of stability considerations. The actual implementation of these methods is quite complicated, but the results attained justify the effort invested. Details can be found in the references given in the Bibliographic Notes, where references to other methods for solving stiff systems can also be found.

Example 5.10 Solve the following nonlinear stiff system, which arises in a problem of reaction kinetics:

$$u' = .01 - (.01 + u + v)[1 + (u + 1000)(u + 1)] \equiv s(u, v)$$

$$v' = .01 - (.01 + u + v)(1 + v^2) \equiv t(u, v) \qquad (5.10\text{-}11)$$

$$u(0) = v(0) = 0 \qquad 0 \leq x \leq 100$$

The eigenvalues of the Jacobian of this system depend on x. At $x = 0$ they are -1012 and $-.01$, while at $x = 100$ they are -21.7 and $-.089$, so that the system is initially very stiff (stiffness ratio $\approx 10^5$) but much less so for large x (stiffness ratio ≈ 245).

No theoretical solution is known for his system, so that its exact solution was computed using the classical Runge-Kutta fourth-order method (5.8-45). Since $\rho(J) = 1012$ for $x = 0$, and since the absolute stability interval for this method is (0, 2.78), the step size h must be less than .00278 and indeed, a computation with $h = .005$ blew up at the second stage in the integration while $h = .004$ and $h = 1/300$ also gave incorrect results. For smaller values of h, we got the following values at $x = 100$:

h	$u(100)$	$v(100)$
.0025	−.99164 24297	.98333 67696
.0020	−.99164 21286	.98333 64258
.0010	−.99164 20711	.98333 63603
.0005	−.99164 20699	.98333 63589

so that to eight figures, we have

$$u(100) = -.99164207 \qquad v(100) = .98333636$$

Note that the use of (5.8-45) required $400/h$ function evaluations, so that for $h = .001$ the right-hand side of (5.10-11) was evaluated 400,000 times.

The system (5.10-11) was also solved using the implicit methods of Gear of orders 3 and 4. The equations for these methods, derived from (5.2-16), are, respectively,

$$\mathbf{y}_{n+1} = \tfrac{1}{11}[18\mathbf{y}_n - 9\mathbf{y}_{n-1} + 2\mathbf{y}_{n-2} + 6h\mathbf{f}(x_{n+1}, \mathbf{y}_{n+1})]$$

and

$$\mathbf{y}_{n+1} = \tfrac{1}{25}[48\mathbf{y}_n - 36\mathbf{y}_{n-1} + 16\mathbf{y}_{n-2} - 3\mathbf{y}_{n-3} + 12h\mathbf{f}(x_{n+1}, \mathbf{y}_{n+1})]$$

These equations were solved using a step length $h = .1$ and starting values obtained from the Runge-Kutta solution. The implicit equations were solved both by Newton's method and by the modified Newton method (sec Sec. 8.8). The details of the computation for the third-order case are as follows (those for the fourth order case are similar).

The implicit equations we had to solve had the form

$$c(u_{n+1}, v_{n+1}) \equiv u_{n+1} - \tfrac{18}{11}u_n + \tfrac{9}{11}u_{n-1} - \tfrac{2}{11}u_{n-2} - \hat{h}s(u_{n+1}, v_{n+1}) = 0$$

$$d(u_{n+1}, v_{n+1}) \equiv v_{n+1} - \tfrac{18}{11}v_n + \tfrac{9}{11}v_{n-1} - \tfrac{2}{11}v_{n-2} - \hat{h}t(u_{n+1}, v_{n+1}) = 0 \quad (5.10\text{-}12)$$

where $\hat{h} = \frac{6}{11}h$, and where u_n, u_{n-1}, u_{n-2}, v_n, v_{n-1}, and v_{n-2} remain constant during the iterations for the solution of (5.10-12).

The Jacobian of this system is

$$J(u, v) = \begin{bmatrix} 1 - \hat{h}\dfrac{\partial s}{\partial u} & -\hat{h}\dfrac{\partial s}{\partial v} \\ -\hat{h}\dfrac{\partial t}{\partial u} & 1 - \hat{h}\dfrac{\partial t}{\partial v} \end{bmatrix}$$

$$= \begin{bmatrix} 1 - \hat{h}(g(u, v)(2u + 1000) + r(u)) & -\hat{h}r(u) \\ -\hat{h}(1 + v^2) & 1 - \hat{h}(g(u, v)(2v) + (1 + v^2)) \end{bmatrix}$$

where $g(u, v) = .01 + u + v$, $r(u) = 1 + (u + 1000)(u + 1)$, and

$$J^{-1}(u, v) = \frac{1}{D}\begin{bmatrix} 1 - \hat{h}\dfrac{\partial t}{\partial v} & \hat{h}\dfrac{\partial s}{\partial v} \\ \hat{h}\dfrac{\partial t}{\partial u} & 1 - \hat{h}\dfrac{\partial s}{\partial u} \end{bmatrix}$$

where

$$D = \left(1 - \hat{h}\frac{\partial s}{\partial u}\right)\left(1 - h\frac{\partial t}{\partial v}\right) - \hat{h}^2\frac{\partial s}{\partial v}\frac{\partial t}{\partial u}$$

As initial approximations to u_{n+1} and v_{n+1}, we took $u_{n+1}^{(0)} = 2u_n - u_{n-1}$, $v_{n+1}^{(0)} = 2v_n - v_{n-1}$. The iteration using Newton's method was

$$\begin{bmatrix} u_{n+1}^{(i+1)} \\ v_{n+1}^{(i+1)} \end{bmatrix} = \begin{bmatrix} u_{n+1}^{(i)} \\ v_{n+1}^{(i)} \end{bmatrix} - J^{-1}(u_{n+1}^{(i)}, v_{n+1}^{(i)})\begin{bmatrix} c(u_{n+1}^{(i)}, v_{n+1}^{(i)}) \\ d(u_{n+1}^{(i)}, v_{n+1}^{(i)}) \end{bmatrix}$$

while in the modified Newton method, the same formula was used except that $J^{-1}(u_{n+1}^{(i)}, v_{n+1}^{(i)})$ was replaced by $J^{-1}(u_{n+1}^{(0)}, v_{n+1}^{(0)})$.

The results obtained with two iterations using Newton's method and the modified Newton method were identical to more than eight figures and are as follows:

Order	$u(100)$	$v(100)$
3	−.99164187	.98333613
4	−.99164208	.98333637

We thus see that in this stiff system comparable accuracy was attained with much less work.

BIBLIOGRAPHIC NOTES

An excellent, general reference on the numerical solution of ordinary differential equations is Henrici (1962*a*). A more theoretical treatment of the material in Henrici and of the more recent developments in the field is given by Stetter (1973). Other recent treatments are those of Gear (1971), Lambert (1973), and Lapidus and Seinfeld (1971), while the more classical references are Collatz (1960) and Milne (1953). Unconventional approaches to the subject are given by Miller (1966) and Daniel and Moore (1970). Fox (1962) surveys the field of numerical solution of ordinary differential equations as well as integral and partial differential equations. This is only one of many conference proceedings dealing principally with the numerical solution of ordinary differential equations.

Section 5.1 No student should approach the numerical solution of ordinary differential equations without a firm grounding in basic existence theory. Henrici (1962a) contains an introduction to this subject. More extensive references are Coddington and Levinson (1955) and Ince (1926). Discussions of hybrid methods will be found in Lambert (1973) and Stetter (1973).

Section 5.2 The method of undetermined coefficients, which is now widely used in the development of numerical integration formulas, was introduced by Dahlquist (1956). A more accessible reference is Hamming (1959). For other examples of its use, see Hull and Newbery (1959), Ralston (1961), and Crane and Lambert (1962).

The Adams-Bashforth and Adams-Moulton formulas are classical and are discussed in all the references mentioned above. The implicit formulas based on numerical differentiation originate with Gear; see Gear (1971).

Section 5.3 The method of this section, particularly the use of the influence function, is due to Milne (1949). Other examples of its use can be found in Hamming (1973) and Hildebrand (1974). Hildebrand also considers some more sophisticated methods of deriving error terms. Barrett (1952) considers some matters related to the material of this section.

Section 5.4 It is possible to find almost as many definitions of stability as there are references on numerical integration methods. Relative stability is considered by Hamming (1959) and Hull and Newbery (1962) and is discussed in detail by Ralston (1965). Many authors define stability only for the case $h = 0$. Henrici (1962*a*, *b*) has excellent discussions of this case; see also Hildebrand (1974) and Lambert (1973). Much of the material of Sec. 5.4-2 is from Hull and Newbery (1961).

Section 5.5 All the books on the numerical solution of ordinary differential equations mentioned above, as well as the book by Hamming (1973) and, in particular, the paper by Hamming (1959), are good references on predictor-corrector methods. The last is the source of most of Sec. 5.5-4. Other papers on the use of predictor-corrector methods are those of Hull and Newbery (1959, 1961), Ralston (1961), and Crane and Lambert (1962). A technique to remove the instability of Milne's method is considered by Milne and Reynolds (1959, 1960).

Section 5.6 For the use of Picard's method in existence theory, see Ince (1926). A recent discussion of the use of Taylor-series methods is given by Barton et al. (1971). There is a good discussion of non-Runge-Kutta methods for starting the solution in Hildebrand (1974).

Section 5.7 Other examples of the use of predictor-corrector methods are given by Milne (1953), Henrici (1962*a*), Hildebrand (1974), and Hamming (1973); see also Ralston (1961). Nordsieck (1962) discusses the numerical solution of ordinary differential equations in an overall computational sense. His approach has been extended and implemented in a very successful computer program, applicable also to stiff systems, by Gear (1971).

The variable-order–variable-step method described here is based on that of Byrne and Hindmarsh (1975). Another approach is the subject of a book by Shampine and Gordon (1975), which gives a fully documented computer program implementing their method.

Most of the generally accepted methods for the numerical solution of ordinary differential equations are compared in Hull et al. (1972).

Section 5.8 Much of the discussion of Runge-Kutta methods is taken from Ralston (1962*b*). Kopal (1961), among many others, also has an extensive discussion of Runge-Kutta methods. The pair of Runge-Kutta formulas (5.8-47) and (5.8-48) is given by England (1969). Other such pairs for various orders m are given by Fehlberg (1969, 1970). A slightly different approach in the same spirit appears in Zonneveld (1970).

The advantages and disadvantages of high-order Runge-Kutta methods are discussed by Curtis (1975). The step-size strategy in Sec. 5.8-7 is essentially that of Shampine and Allen (1973). Lambert (1973) is the source of Sec. 5.8-8 on stability of Runge-Kutta methods, including Example 5.7.

An interesting generalization of the Runge-Kutta idea to integrate forward N steps at a time instead of a single step has been proposed by Rosser (1967).

Section 5.9 Henrici (1962*a*), Hamming (1973), and Hildebrand (1974), among others, have discussions of methods based on higher derivatives.

Extrapolation methods for ordinary differential equations were first suggested by Gragg, who developed the formulas (5.9-15); see Gragg (1965). A computer program implementing one of these methods is given by Fox (1971). Example 5.9 is from Dahlquist and Björck (1974), from whom the terms active and passive extrapolation are taken.

Section 5.10 The definition of stiffness is that given in Lambert (1973), which is also the source of Example 5.10. A full description of Gear's method appears in Gear (1971). A survey of numerical methods for stiff systems appears in Lapidus and Seinfeld (1971). A variable-order–variable-step method applicable to stiff systems is described in Byrne and Hindmarsh (1975). A comparison of various methods for solving stiff systems is given by Enright et al. (1975).

BIBLIOGRAPHY

Barrett, W. (1952): On the Remainders of Numerical Formulae with Special References to Differentiation, *J. Lond. Math. Soc.*, vol. 27, pp. 456–464.

Barton, D., I. M. Willers, and R. V. M. Zahar (1971): Taylor Series Methods for Ordinary Differential Equations: An Evaluation, pp. 369–390 in *Mathematical Software* (J. R. Rice, ed.), Academic Press, Inc., New York.

Byrne, G. D., and A. C. Hindmarsh (1975): A Polyalgorithm for the Numerical Solution of Ordinary Differential Equations, *ACM Trans. Math. Software*, vol. 1, pp. 71–96.

Ceschino, F., and J. Kuntzmann (1966): *Numerical Solution of Initial Value Problems*, Prentice-Hall, Inc., Englewood Cliffs, N.J.

Coddington, E. A., and N. Levinson (1955): *Theory of Ordinary Differential Equations*, McGraw-Hill Book Company, New York.

Collatz, L. (1960): *The Numerical Treatment of Differential Equations* (transl. P. G. Williams), 3d ed., Springer-Verlag OHG, Berlin.

Crane, R. L., and R. J. Lambert (1962): Stability of a Generalized Corrector Formula, *J. Ass. Comput. Mach.*, vol. 9, pp. 104–117.

Curtis, A. R. (1975): High-order Explicit Runge-Kutta Formulae, Their Uses, and Limitations, *J. Inst. Math. Appl.*, vol. 16, pp. 35–55.

Dahlquist, G. (1956): Convergence and Stability in the Numerical Solution of Ordinary Differential Equations, *Math. Scand.*, vol. 4, pp. 33–53.

——— and Å. Björck (1974): *Numerical Methods* (trans. N. Anderson), Prentice-Hall, Inc., Englewood Cliffs, N.J.

Daniel, J. W., and R. E. Moore, (1970): *Computation and Theory in Ordinary Differential Equations*, W. H. Freeman and Company, San Francisco.

Emanuel, G. (1963): The Wilf Stability Criterion for Numerical Integration, *J. Ass. Comput. Mach.*, vol. 10, pp. 557–561.

England, R. (1969): Error Estimates for Runge-Kutta Type Solutions to Systems of Ordinary Differential Equations, *Comput. J.*, vol. 12, pp. 166–170.

Enright, W. H., T. E. Hull, and B. Lindberg (1975): Comparing Numerical Methods for Stiff Systems of O.D.E's, *BIT*, vol. 15, pp. 10–48.

Fehlberg, E. (1969): Klassische Runge-Kutta Formeln fünfter und siebenter Ordnung mit Schrittweiter-Kontrolle, *Computing*, vol. 4, pp. 93–106.

——— (1970): Klassische Runge-Kutta Formeln vierter und niedriger Ordnung mit Schrittweiter-Kontrolle und ihre Anwendung auf Wärmeleitungsprobleme, *Computing*, vol. 6, pp. 61–71.

Fox, L. (1962): *Numerical Solution of Ordinary and Partial Differential Equations*, Addison-Wesley Publishing Company, Inc., Reading, Mass.

Fox, P. A. (1971): DESUB: Integration of a First-Order System of Ordinary Differential Equations, pp. 477–507 in *Mathematical Software* (J. R. Rice, ed.), Academic Press, Inc., New York.

Galler, B. A., and D. P. Rozenberg (1960): A Generalization of a Theorem of Carr on Error Bounds for Runge-Kutta Procedures, *J. Ass. Comput. Mach.*, vol. 7, pp. 57–60.

Gear, C. W. (1971): *Numerical Initial Value Problems in Ordinary Differential Equations*, Prentice-Hall, Inc., Englewood Cliffs, N.J.

Gragg, W. B. (1965): On Extrapolation Algorithms for Ordinary Initial Value Problems, *SIAM J. Numer. Anal.*, vol. 2, pp. 384–403.

Hamming, R. W. (1959): Stable Predictor-Corrector Methods for Ordinary Differential Equations, *J. Ass. Comput. Mach.*, vol. 6, pp. 37–47.

——— (1973): *Numerical Methods for Scientists and Engineers*, 2d ed., McGraw-Hill Book Company, New York.

Henrici, P. (1962*a*): *Discrete Variable Methods in Ordinary Differential Equations*, John Wiley & Sons, Inc., New York.

——— (1962*b*): *Error Propagation for Difference Methods*, John Wiley & Sons, Inc., New York.

Hildebrand, F. B. (1974): *Introduction to Numerical Analysis*, 2d ed., McGraw-Hill Book Company, New York.

Hull, T. E., and A. L. Creemer (1963): The Efficiency of Predictor-Corrector Procedures, *J. Ass. Comput. Mach.*, vol. 10, pp. 291–301.

———, W. H. Enright, B. M. Fellen, and A. E. Sedgwick (1972): Comparing Numerical Methods for Ordinary Differential Equations, *SIAM J. Numer. Anal.*, vol. 9, pp. 603–637.

——— and A. C. R. Newbery (1959): Error Bounds for a Family of Three-Point Integration Procedures, *J. Soc. Ind. Appl. Math.*, vol. 7, pp. 402–412.

——— and ——— (1961) Integration Procedures Which Minimize Propagated Errors, *J. Soc. Ind. Appl. Math.*, vol. 9, pp. 31–47.

——— and ——— (1962): Corrector Formulas for Multi-step Integration Formulas, *J. Soc. Ind. Appl. Math.*, vol. 10, pp. 351–369.

Ince, E. L. (1926): *Ordinary Differential Equations*, Dover Publications, Inc., New York.

Kopal, Z. (1961): *Numerical Analysis*, 2d ed., John Wiley & Sons, Inc., New York.

Lambert, J. D. (1973): *Computational Methods in Ordinary Differential Equations*, John Wiley & Sons, Inc., New York.

Lapidus, L., and J. H. Seinfeld (1971): *Numerical Solution of Ordinary Differential Equations*, Academic Press, Inc., New York.

Miller, J. C. P. (1966): The Numerical Solution of Ordinary Differential Equations, pp. 63–98 in *Numerical Analysis, An Introduction* (J. Wash, ed.), Academic Press, Inc., London.

Milne, W. E. (1949): The Remainder in Linear Methods of Approximation, *J. Res. Bur. Std.*, vol. 43, pp. 501–511.
——— (1953): *Numerical Solution of Ordinary Differential Equations*, John Wiley & Sons, Inc., New York.
——— and R. R. Reynolds (1959): Stability of a Numerical Solution of Differential Equations, I, *J. Ass. Comput. Mach.*, vol. 6, pp. 196–203.
——— and ——— (1960): Stability of a Numerical Solution of Differential Equations, II, *J. Ass. Comput. Mach.*, vol. 7, pp. 46–56.
Nordsieck, A. (1962): On Numerical Integration of Ordinary Differential Equations, *Math. Comput.*, vol. 16, pp. 22–49.
Ralston, A. (1961): Some Theoretical and Computational Matters Relating to Predictor-Corrector Methods of Numerical Integration, *Comput. J.*, vol. 4, pp. 64–67.
——— (1962*a*): A Symmetric Matrix Formulation of the Hurwitz-Routh Stability Criterion, *IRE Trans. Autom. Control*, vol. AC-7, pp. 50–51.
——— (1962*b*): Runge-Kutta Methods with Minimum Error Bounds, *Math. Comput.*, vol. 16, pp. 431–437.
——— (1965): Relative Stability in the Numerical Solution of Ordinary Differential Equations, *SIAM Rev.*, vol. 7, pp. 114–125.
Rosser, J. B. (1967): A Runge-Kutta for All Seasons, *SIAM Rev.*, vol. 9, pp. 417–452.
Shampine, L. F., and R. C. Allen, Jr. (1973): *Numerical Computing: An Introduction*, W. B. Saunders Company, Philadelphia.
——— and M. K. Gordon (1975): *Computer Solution of Ordinary Differential Equations: The Initial Value Problem*, W. H. Freeman and Company, San Francisco.
Stetter, H. J. (1973): *Analysis of Discretization Methods for Ordinary Differential Equations*, Springer-Verlag, Berlin.
Wilf, H. S. (1957): An Open Formula for the Numerical Integration of First Order Differential Equations, I, *MTAC*, vol. 11, pp. 201–203.
——— (1958): An Open Formula for the Numerical Integration of First Order Differential Equations, II, *MTAC*, vol. 12, pp. 55–58.
——— (1959): A Stability Criterion for Numerical Integration. *J. Ass. Comput. Mach.*, vol. 6, pp. 363–365.
——— (1960): Maximally Stable Numerical Integration, *J. Soc. Ind. Appl. Math.*, vol. 8, pp. 537–540.
Zonneveld, J. A. (1970): *Automatic Numerical Integration*, 2d ed., Mathematisch Centrum, Amsterdam.

PROBLEMS

Section 5.1

1 (*a*) Prove that a system of ordinary differential equations can be written in the form (5.1-1) if and only if the system can be rewritten with the highest-order derivative in each variable appearing as the left-hand side of one equation and nowhere else.

(*b*) Write the following system in the form (5.1-1):

$$\frac{d^3y}{dx^3} + \frac{dy}{dx}\frac{dz}{dx} + xy^2z\frac{d^2y}{dx^2} = \sin y$$

$$\frac{d^2z}{dx^2}\frac{dy}{dx} + y^3\frac{d^2y}{dx^2} + \left(\frac{dz}{dx}\right)^{3/2}\frac{dy}{dx} = e^{yz}$$

Section 5.2

2 (*a*) Derive a numerical integration method of the form (5.2-3) for $n = 1$, $m_0 = 0$, and any m_1 by using a truncated Taylor series.

(*b*) Explain how you would use the method of part (*a*) to find the solution of $y' = xy^2$, $y(0) = 1$. With $m = 4$ and $h = .2$, find an approximate solution at $x = .2$ and $.4$ and compare with the true solution.

3 When the system (5.1-1) is linear, show that there is no computational difference between using a forward-integration formula and an iterative formula.

4 (*a*) Derive a fifth-order iterative formula of the form

$$y_{n+1} = ay_n + by_{n-1} + cy_{n-2} + dy_{n-3} + h(ey'_{n+1} + fy'_n + gy'_{n-1})$$

Express each coefficient in terms of b. Find a value $0 < b \le 1$ that leads to a simple set of coefficients.

(*b*) Repeat part (*a*) with the dy_{n-3} term replaced by a term dy'_{n-2}. [Ref.: Ralston (1961).]

5 (*a*) For $j = 0, 1, 2, 3$ use the method of undetermined coefficients to derive fourth-order forward-integration methods of the form

$$y_{n+1} = a_j y_{n-j} + h(by'_n + cy'_{n-1} + dy'_{n-2} + ey'_{n-3})$$

Show how each of these methods could have been derived by integrating the Lagrangian interpolation formula for $y'(x)$.

(*b*) Verify that Eqs. (5.5-29) are such that (5.5-28) will be fourth order.

6 (*a*) Derive the Lagrangian form (5.2-5) of the Adams-Bashforth formula (5.2-13) for $p = 0, 1, 2, 3$.

(*b*) Repeat part (*a*) for the Adams-Moulton formula (5.2-14).

(*c*) Repeat part (*a*) for the implicit formula (5.2-16).

Section 5.3

7 Prove that if the influence function $G(s)$ does change sign over the interval of integration $[a, b]$, then there exists some continuous function $y(s)$ for which

$$\int_a^b G(s)y(s)\, ds \neq y(\eta) \int_a^b G(s)\, ds$$

for any η in $[a, b]$.

***8** (*a*) Use (5.3-12) to derive the error terms for the Newton-Cotes closed formulas with $n = 1, 2$ on an interval $[0, nh]$ and with $w(x) = 1$.

(*b*) Consider the quadrature method

$$\int_{-1}^{1} f(x)\, dx = f(\alpha) + f(-\alpha) + E \qquad 0 \le \alpha \le 1$$

which is exact for 1 and x. For what values of α is the influence function of constant sign in $[-1, 1]$? Find E for those values of α. Explain the behavior at $\alpha = 1/\sqrt{3}$. [Ref.: Hildebrand (1974), pp. 212–213.]

***9** (*a*) By calculating $G(s)$ at x_{n-3}, x_{n-2}, x_{n-1}, x_n, and x_{n+1}, surmise for which values of b in Prob. 4*a* the error can be expressed in the form (5.3-2). Why is this technique not sufficient to *prove* that the error can be expressed in this form?

(*b*) For the values of b found in part (*a*), calculate the explicit form of the error term.

***10** (*a*) For each value of j in Prob. 5*a*, use the influence function to determine whether the error can be expressed in the form (5.3-2), and if so, find the error.

(*b*) Do both parts of Prob. 9 for the iterative formula (5.5-28), whose coefficients are given by (5.5-29).

Section 5.4

11 (*a*) Suppose that in place of (5.4-1) we have the system

$$(y^{(1)})' = -K_{11}y^{(1)} - K_{12}y^{(2)} \qquad y^{(1)}(x_0) = y_{01}$$

$$(y^{(2)})' = -K_{21}y^{(1)} - K_{22}y^{(2)} \qquad y^{(2)}(x_0) = y_{02}$$

Carry through an analysis similar to that in Eqs. (5.4-3) to (5.4-5) to show that the stability equation is now two coupled polynomial equations. Thus deduce that the analysis of Sec. 5.4 leads to generally intractable algebraic problems when more than one equation is being considered.

(*b*) However, show that the *convergence* of a numerical integration method is independent of whether a single equation or a system is being solved.

12 (*a*) By substituting (5.4-7) into (5.4-5) and using (5.2-6), show that $\beta_1 = -K$.

(*b*) Use Cramer's rule to express c_i, $i \neq 0$, in (5.4-4) in a form similar to (5.4-10), and use this result to show that if $y_j \to y_0$, $j = 1, \ldots, p$, as $h \to 0$, then $c_i \to 0$ as $h \to 0$ for $i \neq 0$.

13 (*a*) Prove Theorem 5.1 in the case where (5.4-14) has simple complex roots. *Hint:* If $r = Re^{i\theta}$ consider $w_k = hR^k \cos k\theta$ and $w_k^2 - w_{k+1}w_{k-1}$.

(*b*) Use a similar technique to prove Theorem 5.1 in the case of multiple roots.

(*c*) Show that for the equation $y' = 0$, $y(0) = 0$ the condition of Theorem 5.1 is also sufficient for convergence. [Ref.; Henrici (1962*a*), pp. 218–219.]

14 (*a*) Show that the sequence (5.4-17) with A given by (5.4-18) satisfies (5.4-16).

(*b*) Give an example of a numerical integration method which is convergent and satisfies only the first two equations of (5.2-6).

***15** Suppose that $|f_y(x, y)| \leq -K$ and $|E_n| < E$ for all n and all $f(x, y)$ involved in calculating the solution of (5.1-1). Suppose further than $|Khb_{-1}| < 1$.

(*a*) Show that the solution of the difference equation

$$e_{n+1}(1 + h|b_{-1}|K) = \sum_{i=0}^{p} (|a_i| - h|b_i|K)e_{n-i} + E$$

is such that

$$|\epsilon_i| < e_i \qquad i = p+1, p+2, \ldots$$

where ϵ_i is the solution of (5.4-24), if the initial conditions for the difference equations are such that

$$|\epsilon_i| < e_i \qquad i = 0, \ldots, p$$

Thus deduce that the solution of the above difference equation dominates that of (5.4-24).

(*b*) Show that if all the a_i's are positive, the difference equation of part (*a*) has a particular solution $E/Kh\sigma$, where $\sigma = \sum_{i=-1}^{p} |b_i|$.

(*c*) Show that if again all the a_i's are positive, one solution of the homogeneous difference equation formed by setting $E = 0$ in part (*a*) is given by $e_n = r_0^n$, where

$$r_0 = 1 - \frac{Kh\sigma}{\sigma'} + O(h^2) \qquad \sigma' = \sum_{i=-1}^{p} b_i$$

(*d*) Suppose $|\epsilon_i| < \lambda$, $i = 0, \ldots, p$. Show that

$$e_n = \lambda r_0^n - \frac{E}{Kh\sigma}(r_0^n - 1)$$

is a solution of the difference equation of part (*a*) which is greater in magnitude than ϵ_n for all n, and thus deduce a bound for the propagated error. [Ref.: Hildebrand (1974), pp. 266–268, and Hull and Newbery (1961).]

***16** (*a*) Show that the transformation

$$w = \frac{z-1}{z+1}$$

maps the unit circle in the z plane into the left half of the w plane.

(*b*) Show that the result of applying this transformation to the polynomial equation

$$c_0 z^n + c_1 z^{n-1} + \cdots + c_{n-1} z + c_n = 0$$

is a polynomial equation

$$d_0 w^n + d_1 w^{n-1} + \cdots + d_{n-1} w + d_n = 0$$

where

$$d_j = \sum_{i=0}^{n} c_i r_{n-i,\,n-j}^{(n)}$$

with the $r_{ij}^{(n)}$ defined by

$$(1+x)^i(1-x)^{n-i} = \sum_{j=0}^{n} r_{ij}^{(n)} x^j$$

(*c*) Deduce then that (5.4-5) has all its roots on or inside the unit circle if and only if the polynomial in w in part (*b*) has all its roots in the left half plane or on the imaginary axis where

$$d_j = \sum_{i=0}^{p+1} (-a_{i-1} + hKb_{i-1}) r_{p+1-i,\,p+1+j}^{(p+1)} \qquad a_{-1} = 1$$

(*d*) The Hurwitz-Routh criterion states that the polynomial in w in part (*b*) has all its zeros in the left half plane or on the imaginary axis if and only if when $d_0 > 0$ (< 0), all the principal minors of the matrix $D = [D_{ij}]$ are nonnegative (nonpositive), where

$$D_{ij} = d_{2i+1-j} \qquad i, j = 0, 1, \ldots, n-1; \qquad d_k = 0 \text{ if } k < 0 \text{ or } k > n$$

Use this criterion and the results of parts (*b*) and (*c*) to find those values of b for which the method of Prob. 5*b* is convergent.

(*e*) For $b = 0, \frac{1}{2}$, find those values of Kh for which the roots of the stability equation (5.4-5) lie within the unit circle. Does this help determine when the method is stable? Why? [Ref.: Ralston (1962*a*).]

***17** (*a*) A theorem of Schur states that the roots of the polynomial

$$c_0 z^n + c_1 z^{n-1} + \cdots + c_n = 0$$

will be on or within the unit circle if and only if the quadratic form

$$\sum_{j=0}^{n-1} [(c_0 x_j + c_1 x_{j+1} + \cdots + c_{n-j-1} x_{n-1})^2 - (c_n x_j + c_{n-1} x_{j+1} + \cdots + c_{j+1} x_{n-1})^2]$$

is nonnegative definite. From this derive the result that a sufficient condition for a numerical integration method (5.2-5) to be convergent is that the matrix with elements

$$A_{rs} = \sum_{l=0}^{\min(r,\,s)} (c_{r-l} c_{s-l} - c_{p+l+1-r} c_{p+l+1-s}) \qquad r, s = 0, 1, \ldots, p$$

where

$$c_j = -a_{j-1} \qquad j = 0, 1, \ldots, p+1 \qquad a_{-1} = -1$$

be nonnegative definite. Why is this condition not necessary?

(*b*) Use this criterion to repeat the calculation of part (*d*) of the previous problem.

(*c*) If we wish to know for what values of Kh the roots of the polynomial equation related to (5.4-25) lie within the unit circle, what do we have to replace c_j by in part (*a*)?

(*d*) Repeat Prob. 16*e* using this criterion. Why is this method more difficult to use than that of the previous problem? [But, see Emanuel (1963).] [Ref.: Wilf (1959).]

***18** (*a*) For $b = \frac{2}{3}$ determine for what values of Kh the stability polynomial for the method of Prob. 4*a* has all its roots on or inside the unit circle.

(*b*) Show that, independent of the value of b, the iterative formula of Prob. 4*b* is never stable. [Ref.: Ralston (1961).]

19 (*a*) Show that when $K < 0$, at least one root of the stability equation (5.4-5) always lies outside the unit circle for sufficiently small values of h.

(*b*) Discuss the merits and limitations of a definition of stability which requires only that the roots of (5.4-5) lie within the unit circle when $K > 0$.

Section 5.5

20 (*a*) With $b = \frac{2}{3}$ in the iterative formula of Prob. 4*a*, determine how large h may be for (5.5-10) to be satisfied for the differential equation $y' = -y$, $y(0) = 1$.

(*b*) For this value of h and $b = \frac{2}{3}$, use the result of Prob. 9*b* to determine the truncation error of the method for the same differential equation.

(*c*) Repeat parts (*a*) and (*b*) for the iterative formula of Prob. 5*b* with $b = 0, 1$ using the result of Prob. 10*b*. Show that when $b = 1$, this is Milne's corrector and when $b = 0$, Hamming's corrector.

(*d*) Give an intuitive argument to justify the assertion that unless $|hK| \ll 1$, the truncation error will not be very small.

21 (*a*) Using the notation of (5.2-5), show that by integrating Newton's backward formula we can obtain a class of predictors of order $p + 1$ with $a_0 = 1$, $b_{-1} = 0$, and $a_i = 0$, $i \neq 0$.

(*b*) Using the error term in Newton's formula, derive the truncation errors for (5.5-13) to (5.5-15).

22 (*a*) Show how Newton's backward formula can be used to derive the class of correctors known as *Adams-Moulton correctors*

$$y_{n+1} = y_n + \sum_{i=-1}^{p} b_i y'_{n-i}$$

by suitably varying the limits of integration from those of the previous problem.

(*b*) Find an expression for the error term for this class of formulas.

(*c*) Display the error terms for $p = 0, 1, 2, 3$.

23 (*a*) Show how Newton's backward formula can be used to generate *Nystrom's predictors*

$$y_{n+1} = y_{n-1} + \sum_{i=0}^{p} b_i y'_{n-i}$$

(*b*) Find an expression for the error term. Why does the problem of finding the error term differ from that in Probs. 21 and 22?

(*c*) Display the formulas of this class for $p = 0, 1, 2$. For $p = 0, 1$ display the error term by making use of open Newton-Cotes quadrature formulas.

24 (*a*) Show how the modified Hermite interpolation formula can be used to generate formulas of the form

$$y_{n+1} = \sum_{i=0}^{p} a_i y_{n-i} + \sum_{i=s}^{q} b_i y'_{n-i} \qquad q \leq p$$

which have an order of accuracy $p + q + 1 - s$.

(*b*) In particular, derive the predictor P_6 of Table 5.1.

(*c*) With $p = 2$, $q = 2$, $s = 1$, derive the predictor and compare its error-term coefficient with that of P_6.

25 Use the error terms of the predictor and corrector of (5.5-11) to derive an estimate of the truncation error in both predictor and corrector. Use the estimate of the predictor error to get a method analogous to (5.5-27).

26 (*a*) Show that if the estimate of the corrector error in (5.5-27) were used to modify the corrector, the resulting method would be of fifth order.

(*b*) Generalize this result to prove that if the estimate of the corrector error is used to modify the corrector, the resulting predictor-corrector method always has an order of 1 greater than that of the predictor and corrector.

***27** (*a*) If a term $b_2 y'_{n-2}$ is added to the corrector (5.5-28), derive the equations analogous to (5.5-29) if the corrector is still to be of order 4. Let a_1 and a_2 be the free parameters.

(*b*) If this corrector is used to compute the solution of $y' = 0$, $y(0) = 0$ but the initial conditions of the difference equation are $y_{-2} = y_{-1} = 0$, $y_0 = \epsilon$ (a roundoff error), show that the solution of the difference equation is

$$y_n = C_1 r_1^n + C_2 r_2^n + C_3 \qquad \text{where } C_3 = \frac{\epsilon}{1 + a_1 + 2a_2}$$

(*c*) Thus deduce that if the corrector is stable, the growth of this single roundoff error (in fact, of course, a new roundoff error will be introduced at each step) is determined by $1 + a_1 + 2a_2$. If $1 + a_1 + 2a_2$ is positive (it usually is; see Table 5.2), show that there will be no magnification of the roundoff error if $a_2 \geq (-a_1/2)$. Which methods in Table 5.2 satisfy this condition?

(*d*) The above indicates that the accumulated roundoff error is correlated (not completely random) because of the relation between successive values of y_n through the corrector. What factor should be kept small to control the random component of the error? Compare this quantity for the correctors of Table 5.2.

(*e*) Assuming that the influence function $G(s)$ for the corrector of part (*a*) is of constant sign over the necessary interval, find the form of the error term.

(*f*) Use the result of Prob. 10*b* to deduce that, with $b_2 = 0$ and $-.6 \leq a_1 \leq 1$, $G(s)$ is of constant sign, and thus calculate the last row in Table 5.2. [Ref.: Hamming (1973), pp. 369-371, 395-401].

***28** Consider the corrector (5.3-8).

(*a*) Use Prob. 16 to show that if the corrector is convergent, then $-1 \leq a \leq 1$.

(*b*) For these values of a show that if the roots of (5.4-5) are to lie within the unit circle when $K > 0$, then $Kh \leq (6 - 6a)/(1 + a)$.

(*c*) Find the value of a which maximizes the range of Kh in part (*b*) and is such that part (*a*) and (5.5-10) are satisfied. For this value of a, determine the other coefficients. This method has been called the *maximally stable formula* of this type of third order. (Note, however, that here stability is defined as in Prob. 19*b*.) [Ref.: Wilf (1960).]

29 (*a*) Use the errors of the predictor and corrector of (5.5-33) to derive the modifier equation in (5.5-33).

(*b*) Find the expression for the truncation error in (5.5-17).

(*c*) Display the system of equations analogous to (5.5-27) and (5.5-33) for the corrector of Prob. 4*a* with $b = \frac{2}{3}$, using as a predictor (5.5-17). Use the results of Prob. 9*b*. [Ref.: Hamming (1959) and Ralston (1961).]

30 (*a*) Derive (5.5-1) and (5.5-2) using the method of undetermined coefficients.

(*b*) Derive the same formulas by integrating the proper Lagrangian interpolation formula.

(*c*) Can integration of the Lagrangian interpolation formula be used to derive all formulas of the form (5.2-5) with only one value of y_i on the right-hand side? Why?

***31** Obrechkoff's method. Suppose we wish to find an equation for the numerical integration of $y' = f(x, y)$ which has an error of the form

$$E = \frac{1}{(2r)!} \int_0^h x^r (x-h)^r Y^{(2r+1)}(x)\, dx$$

where h is the step size.

(*a*) Integrate by parts r times to obtain

$$E = \frac{1}{(2r)!} \int_0^h Y^{(r+1)}(x) \frac{d^r}{dx^r} [x^r (h-x)^r]\, dx$$

(*b*) Integrate by parts r additional times and translate the origin to x_n to obtain the desired numerical integration formula

$$y_{n+1} = y_n + \frac{r!}{(2r)!} \sum_{j=1}^{r} (-1)^{j-1} \frac{(2r-j)!\, h^j}{(r-j)!\, j!} [y_{n+1}^{(j)} + (-1)^{j-1} y_n^{(j)}] + E$$

where $x_{n+1} = x_n + h$.

(*c*) Show that the error term can be expressed in the form

$$E = (-1)^r \frac{h^{2r+1}}{2r+1} \left[\frac{r!}{(2r)!}\right]^2 Y^{(2r+1)}(\eta)$$

with $x_n < \eta < x_{n+1}$. [Ref.: Hildebrand (1974), pp. 284–285.]

32 In order to avoid a calculation of $f(x, y)$, it is possible to predict y' directly.

(*a*) Display an extrapolation formula which predicts y'_{n+1} using $y_n, y_{n-1}, \ldots, y_{n-r}$.

(*b*) What is the error in this formula?

(*c*) Discuss the relative merits of this type of prediction and the prediction of y as discussed in Sec. 5.5. Can an estimate of the error at each step be found when y' is being predicted? [Ref.: Ralston (1961).]

33 (*a*) Use the results of Prob. 1, Chap. 4, for $k = 1$ and $n = 3$ to derive the corrector of (5.5-11) by solving the second equation for $f(a_3)$, inserting the result in the first equation and making the appropriate changes in notation.

(*b*) Derive formulas for y'_i, $i = 0, 1, 2, 3$, in terms of y_i, $i = 0, 1, 2, 3$. For each equation also derive the error term.

(*c*) Eliminate y_2 and y_3 from the first three of the equations derived in part (*b*) to get

$$y_1 - y_0 = \frac{h}{12}(5y'_0 + 8y'_1 - y'_2)$$

(*d*) Eliminate y_3 from the first two equations in part (*b*) to get

$$y_2 = 5y_0 - 4y_1 + 2h(y'_0 + 2y'_1)$$

(*e*) Show how the equations of parts (*c*) and (*d*) together form a self-starting numerical integration method by using part (*d*) to "guess" the value of y'_2 required in part (*c*). Show that the error term of the overall method is $O(h^4)$. [Ref.: Wilf (1957).]

Section 5.6

34 (*a*) Show that starting values for a numerical integration method can be found by letting

$$y(x) = \sum_{k=0}^{\infty} A_k (x - x_0)^k$$

substituting this in (5.1-1), letting $x - x_0 = hs$, and getting a recurrence relation for the A_k's, A_0 being determined from the initial condition. For what types of functions $f(x, y)$ is this method most applicable?

(*b*) Show that this method gives results identical to the Taylor-series method of Sec. 5.6-1.

35 (*a*) Complete the calculation of Example 5.6 to get values of $y(.01s)$, $s = 1, 2, 3$ using terms through the third derivative. Compare these results with the true values.

(*b*) Repeat the calculation of part (*a*) for $h = .1$.

36 (*a*) Derive Eqs. (5.6-3).

(*b*) Show that the truncation error in all the equations is $O(h^5)$. For which equation can a truncation error of the form (5.3-2) be calculated easily? Why?

(*c*) Use these equations to find initial values for the differential equation $y' = y$, $y(0) = 1$ with $h = .1$. Use 1.1, 1.2, and 1.3 as initial guesses of y_1, y_2, and y_3, respectively. Carry eight decimal places and compare the answers with the true values.

37 (*a*) If a Hermite interpolation formula is used to halve the interval in a numerical integration using the predictor-corrector method (5.5-33), how many points should be used in the interpolation? If $y(x_n)$ has been computed using the interval h and $y(x_n + \frac{1}{2}h)$ is to be calculated next using (5.5-33), what interpolated values have to be computed?

(*b*) If the interval is to be doubled, show that no restarting is necessary if enough past values have been saved. For the method (5.5-33), how many past values, that is, y_{n-1}, y_{n-2}, etc., must be available for this method?

38 For the differential equations

$$y' + y = 0 \qquad y(0) = 1 \tag{1}$$

$$y' + 2xy = 2x^3 \qquad y(0) = 0 \tag{2}$$

$$y' + y + xy^2 = 0 \qquad y(0) = 1 \tag{3}$$

$$y' - 2y \tan x = 2 \tan x \qquad y(0) = 1 \tag{4}$$

(*a*) Use five terms of a Taylor series to obtain values of y at $x = 0.1, 0.2, 0.3$.

(*b*) Solve (1) using Picard's method.

(*c*) Compute starting values for equation (1) at $x = 0.1, 0.2, 0.3$ using the method of Sec. 5.6-2. Use initial values $y_1 = .9$, $y_2 = .8$, $y_3 = .7$, and do three iterations.

(*d*) Repeat part (*a*) using the Runge-Kutta method of (i) second order (5.8-38), (ii) third order (5.8-42), (iii) fourth order (5.8-46).

Section 5.7

39 (*a*) Use the data of Table 5.3 for $x = .1, .2, .3, .4$ to calculate y at $x = .5$ using Hamming's and Milne's methods (skipping the modifier step in both cases since this is the first predictor-corrector step.). Estimate the error using the technique of Sec. 5.5-3 and compare with the actual error.

(*b*) As in part (*a*), calculate y at $x = .6$, this time including the modifier step.

40 (*a*) Compare the roundoff properties of Milne's and Hamming's methods (cf. Prob. 27). Would you expect the difference to be significant in the calculations of Table 5.6?

(*b*) Show that roundoff error dominates in the calculation of Table 5.6 for large x.

(*c*) Why should the interval be increased if possible when roundoff is the dominant error?

41 For the four differential equations of Prob. 38, using the results of (iii) of Prob. 38*d*, continue the solution from $x = 0.3$ to $x = 1.0$ with $h = 0.1$, using (*a*) the trapezoidal rule method (5.5-11); (*b*) Euler's method (5.5-13); (*c*) Adams's method with $p = 2$(5.5-15); (*d*) Milne's method (5.5-27); (*e*) Hamming's method (5.5-33). In (*a*), (*d*), and (*e*) estimate the truncation error at each step. In all five cases compare the calculated and true values. How do you account for the uniformly poor results for equation (4)?

Section 5.8

***42** (*a*) Derive Eqs. (5.8-15) to (5.8-18).

(*b*) If the w_i's, α_i's, and β_{ij}'s are to be independent of $f(x, y)$, why must the expressions in brackets on the left-hand sides of (5.8-21) and (5.8-22) equal the corresponding operators on the right-hand sides?

43 (*a*) Verify that for Runge-Kutta methods of order 2 and 3, m must be at least 2 and 3, respectively.

(*b*) Prove that $m \geq M$ for all orders by showing that the coefficient of h_n^M on the right-hand side of (5.8-1) always contains a term $cw_M(f_y)^{M-2}D_2 f$, where c is a constant, and no other terms of this form in w_j, $j < M$.

44 (*a*) Starting from (5.8-16) to (5.8-18), derive the error bounds (5.8-32) to (5.8-34).

(*b*) Show that $\alpha_2 = \frac{2}{3}$ minimizes γ_2.

(*c*) Show that $\alpha_2 = \frac{1}{2}$, $\alpha_3 = \frac{3}{4}$ minimizes γ_3. [Ref.: Ralston (1962*b*).]

45 (*a*) Verify (5.8-41).

(*b*) Show that when $\alpha_2 = 0$, there are no Runge-Kutta third-order methods.

(*c*) Derive the one-parameter family of third-order Runge-Kutta methods when $\alpha_2 = \alpha_3$ and when $\alpha_3 = 0$. [Ref.: Ralston (1962*b*).]

46 (*a*) Use the results of Prob. 44 to verify that (5.8-38) is that second-order Runge-Kutta method for which the bound on γ_2 is minimized.

(*b*) Verify that (5.8-42) is that third-order Runge-Kutta method for which the bound on γ_3 is minimized. [Ref.: Ralson (1962b).]

47 For fourth-order Runge-Kutta methods:

(*a*) Verify (5.8-44).

(*b*) Show that no solutions are possible when $\alpha_2 = 0$ or $\alpha_3 = 1$. Why is it reasonable to require $\alpha_2 \neq 0$ in any Runge-Kutta method?

(*c*) Find the one-parameter family of solutions when $\alpha_2 = \alpha_3, \alpha_2 = 1, \alpha_3 = 0$. [Ref.: Kopal (1961), pp. 206–209.].

(*d*) Verify that (5.8-47) is of fourth order.

48 (*a*) For what values of α_2 and α_3 other than the special cases of the previous problem can β_{32}, β_{42}, or β_{43} be infinite?

(*b*) What happens to the corresponding weights in these cases?

(*c*) Even if the corresponding weight is zero and this term is dropped in (5.8-1), will the method be of fourth order?

49 (*a*) Compare the errors in the second-order Runge-Kutta method (5.8-38) and the second-order corrector of (5.5-11) when $f(x, y) = x^2$ and when $f(x, y) = y$. Use (5.8-28) for the Runge-Kutta error.

(*b*) Do the same for the fourth-order Runge-Kutta method (5.8-46) and the fourth-order corrector in (5.5-33) for $f(x, y) = x^4$ and y.

(*c*) Why do the results of part (*b*) indicate that it is very difficult to assess the truncation error in a particular Runge-Kutta method using simple examples of this type? How does the highest-order term in (5.8-27) affect the comparative error estimates? Which of the two functions $f(x, y) = y$ and $f(x, y) = x^m$ is a more realistic test of the value of a method?

***50** For a system of two simultaneous differential equations

$$\frac{dy}{dx} = f(x, y, z) \qquad \frac{dz}{dx} = g(x, y, z)$$

the Runge-Kutta equations corresponding to (5.8-1) are

$$y_{n+1} - y_n = \sum_{i=1}^{m} w_i k_i \qquad z_{n+1} - z_n = \sum_{i=1}^{m} v_i m_i$$

where
$$k_i = h_n f\left(x_n + \alpha_i h_n,\ y_n + \sum_{j=1}^{i-1} \beta_{ij} k_j,\ z_n + \sum_{j=1}^{i-1} \gamma_{ij} m_j\right)$$
$$m_i = h_n g\left(x_n + \alpha_i h_n,\ y_n + \sum_{j=1}^{i-1} \beta_{ij} k_j,\ z_n + \sum_{j=1}^{i-1} \gamma_{ij} m_j\right)$$

(*a*) Use the analog of (5.8-14) to show that the coefficient of h_n^t in k_i and m_i contains no term which includes a product of a β_{ij} and γ_{ij} for $t \leq 4$.

(*b*) Deduce then that a Runge-Kutta method of order $m \leq 4$ for the above system is given by using w_i, α_i, β_{ij} as for a single equation and by having $v_i = w_i$, $\beta_{ij} = \gamma_{ij}$.

(*c*) Can this result be generalized for a system of N equations? Why?

51 For the equation $d^2y/dx^2 = f(x, y, y')$ use the equations in the previous problem to write the equations for the fourth-order Runge-Kutta method corresponding to (5.8-45). How are these equations simplified when f is independent of y'?

52 (*a*) Verify (5.8-52).

(*b*) Determine the values of γ_i for the six-stage method of order 5 given by (5.8-48).

(*c*) Verify that the interval of stability for a three-stage method of order 3 is (0, 2.51).

(*d*) Repeat part (*c*) for $m = 4$ and (0, 2.78).

Section 5.9

53 (*a*) Verify (5.9-5) and (5.9-6) using (5.9-3) and (5.9-4).

(*b*) Use the Hermite interpolation formula to derive (5.9-8) and (5.9-9).

(*c*) Verify the modifier equation in (5.9-10) and find an estimate of T_n.

54 (*a*) Derive the equation analogous to (5.4-5) for Eq. (5.9-3).

(*b*) Deduce that any method of the form (5.9-3) is stable for some range of values of Kh.

(*c*) Deduce in particular that (5.9-5) is stable for all values of Kh.

55 For the differential equation $y'' = y$, $y(0) = 1$, $y'(0) = -1$ use Hamming's method with $h = 0.1$ applied to two first-order equations to take the solution to $x = 1.0$. Use (5.8-46) to get starting values.

56 Use (5.9-10) to calculate the solutions of the four differential equations in Prob. 38. Use $h = .1$ and the results of (iii) of Prob. 38*d* to carry the solution from $x = .3$ to $x = 1.0$. Compare the results with those of Prob. 41. Instead of starting the computation at $x = .4$, where could you have started it?

57 Solve the differential equation of Example 5.9 using the sequence $H/2$, $H/4$, $H/6$, $H/8$, $H/12, \ldots$.

58 (*a*) Apply (5.9-13) and Richardson extrapolation to calculate the solutions of the four differential equations in Prob. 38 at $x = 1$ using the sequence $H/2$, $H/4$, $H/8, \ldots$.

(*b*) Repeat part (*a*) using the sequence of Prob. 57.

59 Using passive extrapolation and the trapezoidal rule (5.9-15), solve the following three differential equations at the values $x = 1(1)10$, and compare with the values in Tables 5.3, 5.4 and 5.6.

$$y' = -y \qquad y(0) = 1 \tag{1}$$

$$y' = y \qquad y(0) = 1 \tag{2}$$

$$y' = \frac{1}{1 + \tan^2 y} \qquad y(0) = 0 \tag{3}$$

Use $h_i = H/2^i$.

Section 5.10

60 (*a*) Compute the solution of (5.10-1) at $x = 3$ using (5.9-13) and Richardson extrapolation with the sequence $H/2$, $H/4$, $H/8$, ...

(*b*) Repeat part (*a*) using the sequence of Prob. 57.

(*c*) Repeat part (*a*) using Hamming's method.

(*d*) Repeat part (*a*) using the Hermite method.

61 Prove that the trapezoidal rule (5.10-10) is absolutely stable in the entire left-hand side of the complex plane.

CHAPTER

SIX

FUNCTIONAL APPROXIMATION: LEAST-SQUARES TECHNIQUES

6.1 INTRODUCTION

Polynomial interpolation is a method of approximating the value of a function at a point by means of a polynomial passing through known functional values. A major virtue of this method of approximation is its ease of implementation. Another virtue is that it leads to an expression for the truncation error in the approximation which can often be estimated or bounded. Implicit in our discussion of polynomial interpolation in Chap. 3 was the assumption that truncation and not roundoff error was the major source of error.

In this chapter we consider the problem of approximating a function whose values at a sequence of points are generally known only empirically and thus are subject to inherent errors which may be large. Thus roundoff will be a serious source of error and often the controlling source. Moreover, it is often the case that an approximation to such a function is desired which can be manipulated analytically—in particular differentiated—with a reasonable degree of accuracy. This, we saw in Chap. 4, is generally difficult with exact polynomial approximations. In fact, the $1/h^k$ factor in the numerical differentiation formula (4.1-15) means that if the inherent error in the functional value is large, differentiation will cause a serious "noise" problem. This is in contrast to the case where the functional value can be calculated to the full word length (say 10 decimals) of a digital computer, so that the roundoff error will be small. The subject of this chapter, least-squares approximations, is concerned with a technique by which noisy functional values can be used to generate a smooth approximation to the function. This smooth approximation can then, for example, be used to approximate the derivative of the function more accurately than exact approximations.

The reader may be familiar with the principle of least squares as applied to continuous functions over an interval $[a, b]$. In this chapter we shall be concerned entirely with the principle of least squares as applied to functions known only at a discrete set of points. However, the derivation of the principle of least squares in the next section for a discrete set of points is precisely analogous to the derivation in the continuous case.

Our reason for emphasizing the discrete rather than the continuous case is that this is the case of interest in numerical applications. Approximating continuous functions by least-squares techniques is, of course, of great theoretical interest. In the case where the approximating functions are polynomials, such approximations of continuous functions lead naturally to the development of the orthogonal polynomials discussed in Chap. 4. As we shall see in Sec. 6.4, orthogonal polynomials also play an important role in discrete least-squares approximations.

6.2 THE PRINCIPLE OF LEAST SQUARES

We are now ready to make precise our heuristic definition of least-squares approximations in Sec. 2.1-2. Let $f(x)$ be a function and $\{x_i\}$, $i = 1, \ldots, n$, be a sequence of *data points* at which we have observed values of $f(x)$ which generally will be in error. We denote $f(x_i)$, the true value at x_i, by f_i, and we denote the observed value at x_i by $\bar{f}_i$. We define $E_i = f_i - \bar{f}_i$. Throughout this chapter we shall assume that the errors at different data points are uncorrelated, i.e., independent.

Let $\{\phi_j(x)\}$, $j = 0, 1, \ldots$, be a (generally finite) sequence of functions defined for every x_i. Then our object is to approximate $\bar{f}_i$ by a linear combination of the $\{\phi_j(x)\}$

$$\bar{f}_i \approx \sum_{j=0}^{m} a_j^{(m)} \phi_j(x_i) \qquad i = 1, \ldots, n \tag{6.2-1}$$

with the $a_j^{(m)}$ to be determined so that

$$\begin{aligned} H(a_0^{(m)}, \ldots, a_m^{(m)}) &\equiv \sum_{i=1}^{n} w(x_i) \left[\bar{f}_i - \sum_{j=0}^{m} a_j^{(m)} \phi_j(x_i) \right]^2 \\ &= \sum_{i=1}^{n} w(x_i) R_i^2 \end{aligned} \tag{6.2-2}$$

is minimized. The function $w(x)$ is called the *weight function* and is assumed to be such that $w(x_i) \geq 0$, $i = 1, \ldots, n$. The quantity R_i is called the *residual* at x_i. The superscript m on $a_j^{(m)}$ denotes the fact that the coefficient of $\phi_j(x)$ will generally depend on m, although in Sec. 6.4 we shall see that this is not always the case. Having determined the $a_j^{(m)}$ so as to satisfy (6.2-2), we have then an approximation

$$y_m(x) = \sum_{j=0}^{m} a_j^{(m)} \phi_j(x) \tag{6.2-3}$$

which is called a least-squares approximation of $f(x)$ over $\{x_i\}$. We can use this approximation not only at the points $\{x_i\}$ but also at other values of x. In this sense, our development is analogous to that of Chap. 3, in which functional values at a discrete set of points were used to derive an approximation to be used over an interval.

If $\phi_j(x) = x^j$ and there are no more data points than parameters in the approximating polynomial, that is, $n \leq m + 1$, then by making the summation in (6.2-3) the Lagrangian interpolation polynomial corresponding to the points $\{x_i\}$, we would have $y_i = \bar{f}_i$, $i = 1, \ldots, n$, where $y_i = y(x_i)$. Since (6.2-2) would then be zero, this would be the desired minimum. In this chapter, however, we shall be concerned with the case $n > m + 1$; that is, we use a number of approximating functions less than the number of data points. Thus, for example, we might approximate a function known at five data points by a polynomial of degree 1, that is, $m = 1$. We could derive a polynomial of degree 4 passing through these five points, but such a fourth-degree polynomial will not enable us to smooth the empirical data. However, as we shall see, such smoothing is possible in general when $m + 1 < n$.

Graphically, this is illustrated by Fig. 6.1. Suppose we have empirical data at five points on a function which is in fact linear as shown. The errors in the empirical data as such are much too great to allow approximating the true function by an *exact* linear approximation using any two points (although, if we knew which two points to choose—which we never shall—

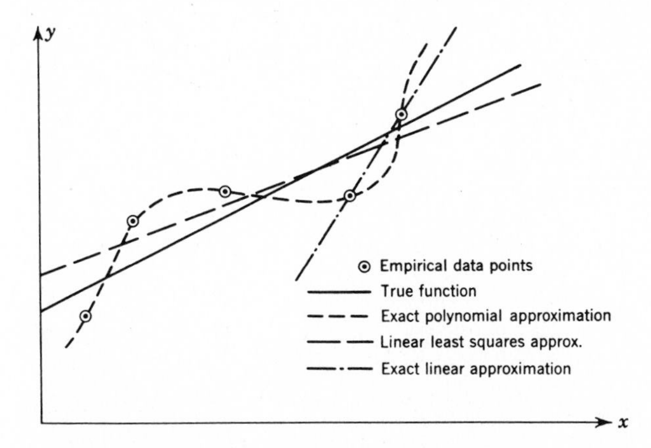

Figure 6.1 Least-squares and exact approximations.

we could get quite a good approximation this way). If we pass a fourth-degree polynomial through the five points, we get an approximation whose deviation from the true function is not too bad but whose derivatives are much different from those of the true function. On the other hand, the linear least-squares approximation not only lies close to the true function but also has a similar slope.

Although the considerations above were for the case where the set $\{\phi_j(x)\}$ is the powers of x, the points we have made are equally true for other sets of functions, in particular for the functions $\{\sin jx, \cos jx\}$, which we shall also consider in this chapter.

The use of Eq. (6.2-2) as the one to be minimized instead of

$$\sum_{i=1}^{n} w(x_i)\,|R_i| \tag{6.2-4}$$

or

$$\max_{i=1,\ldots,n} w(x_i)\,|R_i| \tag{6.2-5}$$

is motivated by an analytical consideration. If the magnitude of the error is the quantity in which we are directly interested, minimizing (6.2-4) would be more desirable than minimizing H. Minimizing (6.2-5) (the so-called *minimax* approximation) has the advantage of giving us a sure bound on the error at any data point. The desirability in some circumstances of having such a sure bound on the error, not just at a discrete set of points but over a whole interval, is the motivation behind the whole of Chap. 7. But for general application with empirical data, the thing that rules out (6.2-4) and (6.2-5) is the fact that the absolute value is not a differentiable function of x. This makes the determination of the constants $a_j^{(m)}$ substantially more difficult in general using either (6.2-4) or (6.2-5) than using (6.2-2), although linear programming methods can be used in both these cases to determine the constants (see Sec. 9.10).

To calculate the $a_j^{(m)}$'s, we take the partial derivative of H in (6.2-2) with respect to $a_k^{(m)}$ and set it equal to 0, thereby obtaining

$$\frac{\partial H}{\partial a_k^{(m)}} = -2\sum_{i=1}^{n} w_i \left[\bar{f}_i - \sum_{j=0}^{m} a_j^{(m)}\phi_j(x_i)\right]\phi_k(x_i) = 0$$

$$k = 0, \ldots, m; \; w_i = w(x_i) \tag{6.2-6}$$

Equation (6.2-6) is a system of $m + 1$ linear equations for the $m + 1$ unknown $a_j^{(m)}$'s. This system is called the *normal equations*. If the determinant of the coefficients does not vanish, we can solve for the $a_j^{(m)}$'s. By considering $H(a_0^{(m)} + \Delta a_0, \ldots, a_m^{(m)} + \Delta a_m)$ it is not hard to show that this solution is indeed a minimum.

Our basic assumption in this chapter is that for some unknown value of

m, say M, the true function $f(x)$ can be expressed as a finite linear combination of the set of functions $\{\phi_j(x)\}$; that is, we assume

$$f(x) = \sum_{j=0}^{M} a_j^{(M)} \phi_j(x) \tag{6.2-7}$$

Now clearly this assumption will not always be satisfied in practice, but if the assumption is a good approximation to reality, the results we shall derive based on this assumption will be useful.

6.3 POLYNOMIAL LEAST-SQUARES APPROXIMATIONS

In this and the next section, we consider the case in which $\phi_j(x)$ is a polynomial of degree j. In particular, in this section we shall consider the case $\phi_j(x) = x^j$ and $w(x) = 1$. For this case Eq. (6.2-6) becomes, after canceling the -2,

$$\sum_{i=1}^{n} \left(\bar{f}_i - \sum_{j=0}^{m} a_j^{(m)} x_i^j \right) x_i^k = 0 \qquad k = 0, \ldots, m \tag{6.3-1}$$

Interchanging summations, we can rewrite (6.3-1) as

$$\sum_{j=0}^{m} a_j^{(m)} \left(\sum_{i=1}^{n} x_i^{j+k} \right) = \sum_{i=1}^{n} \bar{f}_i x_i^k \qquad k = 0, \ldots, m \tag{6.3-2}$$

Using the notation

$$g_{jk} = \sum_{i=1}^{n} x_i^{j+k} \qquad \rho_k = \sum_{i=1}^{n} \bar{f}_i x_i^k \tag{6.3-3}$$

the normal equations can be written

$$\sum_{j=0}^{m} g_{jk} a_j^{(m)} = \rho_k \qquad k = 0, \ldots, m \tag{6.3-4}$$

Using matrix calculus, it can be proved that the least-squares problem and, thus the system (6.3-4), has a unique solution. We leave the proof to a problem {2}.

6.3-1 Solution of the Normal Equations

We seem at this point to have solved the least-squares problem for the case $\phi_j(x) = x^j$, $w(x) = 1$. All we need do is perform the perhaps tedious calculations required to solve the normal equations (6.3-4). And indeed for small values of m, say up to 5 or 6, experience indicates that the solution of (6.3-4) produces quite good least-squares approximations. But for greater values of

m, the solutions found by solving (6.3-4) generally lead to progressively poorer least-squares approximations. Moreover, this is quite independent of which of the many methods available for the solution of (6.3-4) (see Chap. 9) is used. An explanation of this can be found using the following argument.

For convenience let us assume that the points x_i are all in the interval (0, 1). Further, let us assume that they are distributed fairly uniformly in this interval. Then g_{jk} as defined in (6.3-3) has the form of n times a Riemann sum. For large n, then, the approximation

$$g_{jk} = \sum_{i=1}^{n} x_i^{j+k} \approx n \int_0^1 x^{j+k}\,dx = \frac{n}{j+k+1} \qquad j, k = 0, \ldots, m \qquad (6.3\text{-}5)$$

should be a good one. Let $G = [g_{jk}]$ be the matrix of coefficients in (6.3-4). Using (6.3-5), we approximate G by n times the matrix H, where

$$H = \begin{bmatrix} 1 & \frac{1}{2} & \frac{1}{3} & \cdots & \frac{1}{m+1} \\ \frac{1}{2} & \frac{1}{3} & \frac{1}{4} & \cdots & \frac{1}{m+2} \\ \frac{1}{3} & \cdots & & \cdots & \frac{1}{m+3} \\ \cdots & \cdots & \cdots & \cdots & \cdots \\ \frac{1}{m+1} & \cdots & & \cdots & \frac{1}{2m+1} \end{bmatrix} \qquad (6.3\text{-}6)$$

This matrix is the principal minor of order $m + 1$ of the infinite *Hilbert matrix*. This matrix is a classical example of an *ill-conditioned matrix* (cf. Sec. 1-7). A matrix is ill-conditioned if when it has been normalized so that its largest element has order of magnitude 1, as, for example, in (6.3-6), its inverse has very large elements. Thus, for example, when $m = 9$, the inverse of (6.3-6) has elements of magnitude 3×10^{12}. The result of this is that any roundoff error incurred in entering the coefficients of H_m into the computer (and such errors are inevitable because the elements of H generally have infinite binary or decimal expansions) will result in a matrix whose true inverse has greatly magnified errors. For quite small values of m, therefore, it becomes impossible to compute an accurate solution to a set of linear equations whose coefficient matrix is H_m. (It might be expected that roundoff errors introduced in the calculation of the inverse would make this problem much worse, but, interestingly enough, this is not so. The inverse of the matrix whose elements are rounded values of those in (6.3-6) can be calculated very accurately [see Wilkinson (1961)] although, of course, it still differs considerably from the true inverse of H_m.)

For m even as large as 9, the situation as we have presented it is so bad that even though G is only an approximation to a Hilbert matrix, we still expect to have a great deal of difficulty in solving the normal equations. Some actual examples to illustrate how hard it is to solve the system (6.3-4) with any degree of accuracy are considered in the problems {5}.

The previous argument is a cogent one against using $\phi_j(x) = x^j$ for all but very small values of m. But to make the case even stronger, we now consider this class of functions from another standpoint.

6.3-2 Choosing the Degree of the Polynomial

Given a value of n, how is one to choose m, the degree of the polynomial approximation? This problem is analogous to choosing the order of an interpolation or quadrature formula as discussed in previous chapters. But whereas in those cases we were interested in making an error term small and at the same time being able to estimate it, here the considerations are different.

Our basic hypothesis is that the true function $f(x)$ is a polynomial of degree $M < n$ or at least can be accurately represented by such a polynomial. A priori we do not know what M is; our problem is to find it. If we choose a value of $m < M$, then clearly it is impossible to get a good representation of the true function. On the other hand, choosing a value of $m > M$ also defeats our purpose. We have pointed out that by choosing $m = n - 1$ we can make

$$\delta_m^2 = \sum_{i=1}^{n} w_i R_i^2 = \sum_{i=1}^{n} w_i \left(\bar{f}_i - \sum_{j=0}^{m} a_j^{(m)} x_i^j \right)^2 \qquad w_i = w(x_i) \tag{6.3-7}$$

equal to 0. But in so doing, we shall have lost all smoothing properties of least-squares approximations. In fact, any value of $m > M$ sacrifices some smoothing.

When we are using powers of x, (6.2-7) becomes

$$f(x) = \sum_{j=0}^{M} a_j^{(M)} x^j \tag{6.3-8}$$

Therefore, if we knew M and calculated the least-squares approximation

$$y_{M+1}(x) = \sum_{j=0}^{M+1} a_j^{(M+1)} x^j \tag{6.3-9}$$

using the observed data $\{\bar{f}_i\}$, then *statistically* $a_{M+1}^{(M+1)}$ should be 0. That is, if there were no errors in the data, it would be 0, but because of these errors, it will not be 0 even if the assumption that $f(x)$ has the form (6.3-8) is correct. We should like then to test the statistical hypothesis that $a_{M+1}^{(M+1)} = 0$. In order to be able to do this, we make the one further assumption that the errors E_i are normally distributed with zero mean and variance σ^2/w_i. This assumption is reasonable because more accurate measurements, i.e., those with small variance, will usually be more heavily weighted.

This statistical hypothesis that we wish to test is often called the *null hypothesis*. It can be tested using maximum-likelihood statistical methods, a discussion of which is beyond the scope of this book [see Wilks (1962)]. Here we only state the result that if the null hypothesis is correct, then the expected value of

$$\sigma_m^2 = \frac{\delta_m^2}{n - m - 1} \tag{6.3-10}$$

will be independent of m for $m = M, M + 1, \ldots, n - 1$. Thus in practice, since we do not know M, we would wish to solve the normal equations (6.3-4) for $m = 1, 2, \ldots$, compute σ_m^2, and continue as long as σ_m^2 decreases significantly with increasing m. As soon as a value of m is reached after which no significant decrease occurs in σ_m^2, this m is that of the null hypothesis and we have the desired least-squares approximation. In order to guard against the possibility that the underlying function is odd or even and that successive values of σ_m will therefore be nearly equal before $m = M$, in practice we should stop the computation only after several σ_m are almost the same.

This means that we must compute the solution of the normal equations for a sequence of values of m.† Although some of the computations in the solution of these equations for $m = r$ can be used to compute the solution for $m = r + 1$, there is nevertheless significant additional calculation at each stage [see p. 423]. This and the problems adduced in Sec. 6.3-1 constitute a strong case against using $\phi_j(x) = x^j$. In the next section we shall indicate how both the analytic problems of Sec. 6.3-1 and the computational problems discussed in this section can be avoided by the use of orthogonal polynomials.

6.4 ORTHOGONAL-POLYNOMIAL APPROXIMATIONS

If $p_j(x)$ is a polynomial of degree j, the least-squares approximation of degree m can be written

$$y_m(x) = \sum_{j=0}^{m} b_j^{(m)} p_j(x) \tag{6.4-1}$$

In order to minimize

$$H(b_0^{(m)}, \ldots, b_m^{(m)}) \equiv \sum_{i=1}^{n} w_i[\bar{f}_i - y_m(x_i)]^2 \tag{6.4-2}$$

† In practice, using our knowledge of the problem, we would often start with a value of $m > 1$.

we proceed as in the previous section. We get, corresponding to (6.3-4),

$$\sum_{j=0}^{m} d_{jk} b_j^{(m)} = \omega_k \qquad k = 0, \ldots, m \tag{6.4-3}$$

where
$$d_{jk} = \sum_{i=1}^{n} w_i p_j(x_i) p_k(x_i) \qquad \omega_k = \sum_{i=1}^{n} w_i f_i p_k(x_i) \tag{6.4-4}$$

For arbitrary choice of the $\{p_j(x)\}$, the computational problems involved in solving these normal equations can be just as serious as before. If, however, the $\{p_j(x)\}$ are chosen so that the nondiagonal terms of the matrix $D = [d_{jk}]$ are small compared with the diagonal elements, then the matrix D, unlike G, will not be ill-conditioned. In particular, if the $\{p_j(x)\}$ are *orthogonal* over the sets of points $\{x_i\}$, the off-diagonal terms will all be 0. By definition, a set of polynomials $\{p_j(x)\}$ is *orthogonal* over a set of points $\{x_i\}$ with respect to a weight function $w(x)$ if

$$\sum_{i=1}^{n} w_i p_j^{(n)}(x_i) p_k^{(n)}(x_i) = 0 \qquad \text{if } j \neq k$$
$$w_i = w(x_i) \tag{6.4-5}$$

where the superscript denotes the fact that the polynomial will depend on the number of points n. We assume in what follows that $w_i > 0$ for all i. If the $\{p_j^{(n)}(x)\}$ are orthogonal, then, as defined in (6.4-4), $d_{jk} = 0$, $j \neq k$. The system (6.4-3) then becomes

$$d_{kk} b_k^{(m)} = \omega_k \qquad k = 0, \ldots, m \tag{6.4-6}$$

which has the immediate solution

$$b_k^{(m)} = \frac{\omega_k}{d_{kk}} \qquad k = 0, \ldots, m \tag{6.4-7}$$

thereby eliminating the problems of solving an ill-conditioned system of normal equations. Moreover, the solution with m replaced by $m + 1$ is given by

$$b_k^{(m+1)} = \frac{\omega_k}{d_{kk}} \qquad k = 0, \ldots, m + 1 \tag{6.4-8}$$

with ω_k and d_{kk} again given by (6.4-4). Thus

$$b_k^{(m)} = b_k^{(m+1)} \qquad k = 0, \ldots, m \tag{6.4-9}$$

Therefore, to compute the solution for $m + 1$, we need only compute ω_{m+1} and $d_{m+1,m+1}$. Since (6.4-9) indicates that b_k is in fact independent of m, we shall from now on drop the superscript.

The development above indicates that the use of orthogonal polynomials enables us to avoid the difficulties of both Secs. 6.3-1 and 6.3-2. We now proceed to consider the generation of polynomials orthogonal over discrete sets of points which need not be equally spaced. A convenient method for this is the *Gram-Schmidt orthogonalization* process. We begin with a set of m polynomials $q_j(x)$, $j = 0, 1, \ldots, m-1$, where $q_j(x)$ is a polynomial of degree j, which are linearly independent over the set $\{x_i\}$. That is, there exist no constants $\{c_j\}$ other than $c_j = 0, j = 1, \ldots, m-1$, such that

$$\sum_{j=0}^{m-1} c_j q_j(x_i) = 0 \qquad i = 1, \ldots, n \tag{6.4-10}$$

A convenient choice for $\{q_i(x)\}$ is often $1, x, x^2, \ldots, x^{m-1}$. In any case let

$$p_0(x) = q_0(x)$$

$$p_j(x) = q_j(x) - \sum_{r=0}^{j-1} d_{rj} p_r(x) \qquad j = 1, \ldots, m-1 \tag{6.4-11}$$

so that $p_j(x)$ is a polynomial of degree j. Suppose we have determined $p_0(x)$, $p_1(x), \ldots, p_k(x)$ so that (6.4-5) is satisfied. Then, to determine $p_{k+1}(x)$ orthogonal to all $p_j(x)$, $j \le k$, we use (6.4-11) with $j = k+1$ to write

$$\sum_{i=1}^{n} w_i p_{k+1}(x_i) p_j(x_i) = \sum_{i=1}^{n} w_i q_{k+1}(x_i) p_j(x_i) - \sum_{r=0}^{k} d_{r,k+1} \sum_{i=1}^{n} w_i p_r(x_i) p_j(x_i)$$
$$j = 0, 1, \ldots, k \tag{6.4-12}$$

We wish the left-hand side to be 0. By orthogonality all the terms in the double summation on the right-hand side are 0 except the term in $r = j$. Therefore

$$d_{j,k+1} = \frac{\sum_{i=1}^{n} w_i q_{k+1}(x_i) p_j(x_i)}{\sum_{i=1}^{n} w_i p_j^2(x_i)} \qquad j = 0, 1, \ldots, k \tag{6.4-13}$$

In this way $p_{k+1}(x)$ is determined orthogonal to all $p_j(x)$ of lower degree, and continuing this process leads to a set of m orthogonal polynomials. The Gram-Schmidt process is also commonly used to generate sets of functions orthogonal over an interval or sets of orthogonal vectors {7}.

A more convenient and efficient method for the derivation of orthogonal polynomials over discrete sets of points is the use of recurrence relations. Suppose that $\{p_j(x)\}$ is any sequence of polynomials satisfying the orthogonality relationship (6.4-5) with respect to some positive weight function $w(x)$ and some sequence of data points $\{x_i\}$. We shall show by induction that there exists a relation of the form

$$p_{j+1}(x) = (x - \alpha_{j+1}) p_j(x) - \beta_j p_{j-1}(x) \qquad j = 0, 1, \ldots$$
$$p_0(x) = 1, \; p_{-1}(x) = 0 \tag{6.4-14}$$

where α_{j+1} and β_j are constants to be determined. For $j = 0$ (6.4-14) becomes

$$p_1(x) = (x - \alpha_1) \tag{6.4-15}$$

The relation (6.4-5) requires that

$$\sum_{i=1}^{n} w_i p_0(x_i) p_1(x_i) = \sum_{i=1}^{n} w_i(x_i - \alpha_1) = 0 \tag{6.4-16}$$

from which it follows that

$$\alpha_1 = \frac{\sum_{i=1}^{n} w_i x_i}{\sum_{i=1}^{n} w_i} \tag{6.4-17}$$

Let us suppose that for $j = 0, 1, \ldots, k$ the polynomials $p_j(x)$ satisfy a relationship of the form (6.4-14) and the orthogonality relationship (6.4-5). Then we wish to show that we can choose α_{k+1} and β_k so that with $p_{k+1}(x)$ defined by (6.4-14)

$$\sum_{i=1}^{n} w_i p_j(x_i) p_{k+1}(x_i) = 0 \qquad j = 0, 1, \ldots, k \tag{6.4-18}$$

Substituting (6.14-14) with $j = k$ into (6.4-18), we have

$$\sum_{i=1}^{n} w_i x_i p_j(x_i) p_k(x_i) - \alpha_{k+1} \sum_{i=1}^{n} w_i p_j(x_i) p_k(x_i)$$
$$- \beta_k \sum_{i=1}^{n} w_i p_j(x_i) p_{k-1}(x_i) = 0 \qquad j = 0, 1, \ldots, k \tag{6.4-19}$$

For $j = 0, 1, \ldots, k - 2$ the last two terms on the left-hand side of (6.4-19) are identically 0 by the induction hypothesis. Moreover, for these values of j, $x_i p_j(x_i)$ in the first term is a polynomial of degree no greater than $k - 1$ and thus can be expressed as a linear combination of the $p_j(x)$, $j = 0, \ldots, k - 1$. Therefore, again by the induction hypothesis, the first term is also 0. For $j = 0, 1, \ldots, k - 2$, then, (6.4-5) is satisfied for *any* choice of α_{k+1} and β_k.

For $j = k - 1$ the second term is still 0, and so we get the requirement that

$$\beta_k = \frac{\sum_{i=1}^{n} w_i x_i p_{k-1}(x_i) p_k(x_i)}{\sum_{i=1}^{n} w_i [p_{k-1}(x_i)]^2} = \frac{\sum_{i=1}^{n} w_i [p_k(x_i)]^2}{\sum_{i=1}^{n} w_i [p_{k-1}(x_i)]^2} \tag{6.4-20}$$

the second form following from use of (6.4-14). For $j = k$ the third term vanishes, and we get

$$\alpha_{k+1} = \frac{\sum_{i=1}^{n} w_i x_i [p_k(x_i)]^2}{\sum_{i=1}^{n} w_i [p_k(x_i)]^2} \tag{6.4-21}$$

Thus, we have the result that with β_k and α_{k+1} given by (6.4-20) and (6.4-21), the polynomial of degree $k + 1$, $p_{k+1}(x)$, defined by (6.4-14), satisfies the orthogonality relation (6.4-5). This proves our assertion that a recurrence relation of the form (6.4-14) exists. We have assumed in the above that the denominators in (6.4-20) and (6.4-21) do not vanish. A denominator can vanish only if

$$p_j(x_i) = 0 \qquad i = 1, \ldots, n \tag{6.4-22}$$

for some j. If (6.4-22) held for no j, we could generate an unending sequence of polynomials $p_j(x)$. Given n data points, we expect to be able to generate at most n independent polynomials $p_0(x), \ldots, p_{n-1}(x)$. Therefore, it is no surprise that we can show {8} that if (6.4-14) is used to generate $p_n(x)$, then $p_n(x_i) = 0$, $i = 1, \ldots, n$.

Using (6.4-14), (6.4-20), and (6.4-21), we can generate the least-squares approximation

$$y_m(x) = \sum_{j=0}^{m} b_j p_j(x) \tag{6.4-23}$$

with

$$b_j = \frac{\omega_j}{\gamma_j} \tag{6.4-24}$$

where

$$\omega_j = \sum_{i=1}^{n} w_i \bar{f}_i p_j(x_i) \tag{6.4-25}$$

and

$$\gamma_j = \sum_{i=1}^{n} w_i [p_j(x_i)]^2 \tag{6.4-26}$$

This technique of generating orthogonal polynomials is a very powerful one, and since it is easily mechanized, it is very useful on computers. In addition, it provides a convenient method of evaluating the approximation $y_m(x)$. It is not very hard to show {9} that the recurrence

$$q_k(x) = b_k + (x - \alpha_{k+1})q_{k+1}(x) - \beta_{k+1} q_{k+2}(x) \qquad k = m, m-1, \ldots, 0$$
$$q_{m+1}(x) = q_{m+2}(x) = 0 \tag{6.4-27}$$

is such that $q_0(x) = y_m(x)$. Similar recurrences can be used to evaluate the derivatives of $y_m(x)$ {9}.

When the data points are equally spaced and for the particular case

$w(x) = 1$, the orthogonal polynomials are called *Gram polynomials*. In this case it is convenient to use an odd number of points $2L + 1$

$$x_s = x_0 + sh \qquad s = -L, \ldots, -1, 0, 1, \ldots, L \tag{6.4-28}$$

It can be shown {16} that $\alpha_{j+1} = 0$ for all j, so that the resulting recurrence relation is {16}

$$\frac{1}{\epsilon_{j+1}} p_{j+1}(s, 2L) = \frac{s}{\epsilon_j} p_j(s, 2L) - \frac{\beta_j}{\epsilon_{j-1}} p_{j-1}(s, 2L) \qquad j = 0, 1, \ldots$$

$$p_0(s, 2L) = 1 \qquad p_{-1}(s, 2L) = 0 \tag{6.4-29}$$

where $$\beta_j = \frac{j^2[(2L+1)^2 - j^2]}{4(4j^2 - 1)} \qquad \epsilon_j = \frac{(2j)!}{(j!)^2} \frac{1}{(2L)^{(j)}} \tag{6.4-30}$$

Before presenting an example of a least-squares orthogonal polynomial approximation in the next section, we develop and present here an algorithm for generating approximations of the form (6.4-23). Our technique is to use (6.4-14), (6.4-20), and (6.4-21) to generate the orthogonal polynomials or, more precisely, the values of these polynomials at the points x_i.

Input

$\{x_i, w_i, f_i\} \qquad i = 1, \ldots, n$

m

Algorithm

for $i = 1, \ldots, n$ **do** $p_0(x_i) \leftarrow 1;\ p_{-1}(x_i) \leftarrow 0;\ y_m(x_i) \leftarrow 0$ **endfor**

$\gamma_0 \leftarrow \sum_{i=1}^{n} w_i;\ \beta_0 \leftarrow 0$

for $j = 0, \ldots, m$ **do**

$\omega_j \leftarrow \sum_{i=1}^{n} w_i f_i p_j(x_i)$

$b_j \leftarrow \omega_j/\gamma_j$

for $i = 1, \ldots, n$ **do**

$y_m(x_i) \leftarrow y_m(x_i) + b_j p_j(x_i)$

endfor

if $j = m$ **then stop**

$\alpha_{j+1} \leftarrow \sum_{i=1}^{n} w_i x_i [p_j(x_i)]^2/\gamma_j$

for $i = 1, \ldots, n$ **do** $p_{j+1}(x_i) \leftarrow (x_i - \alpha_{j+1})p_j(x_i) - \beta_j p_{j-1}(x_i)$ **endfor**

$\gamma_{j+1} \leftarrow \sum_{i=1}^{n} w_i [p_j(x_i)]^2$

$\beta_{j+1} \leftarrow \gamma_{j+1}/\gamma_j$

endfor

Output

$b_j, j = 0, \ldots, m$

$\alpha_j, j = 1, \ldots, m$

$\beta_j, j = 1, \ldots, m - 1$

$y_m(x_i), i = 1, \ldots, n$ (smoothed values at the data points)

Note that the α_j and β_j are listed as outputs in order to make it possible to use the recurrence relation (6.4-27) to compute $y_m(x)$ at points other than the given x_i.

6.5 AN EXAMPLE OF THE GENERATION OF LEAST-SQUARES APPROXIMATIONS

Suppose we are given the empirical data (cf. Example 4.1)

x_i	.1	.2	.3	.4	.5	.6	.7	.8	.9
f_i	5.1234	5.3057	5.5687	5.9378	6.4370	7.0978	7.9493	9.0253	10.3627

and we wish to find the best least-squares polynomial approximation to $f(x)$ with weight function $w(x) = 1$. Since the data points are equally spaced, we could use the Gram polynomials. Instead, however, we shall use the algorithm at the end of the previous section. Using this algorithm with $m = 5$ we calculate:

j	γ_j	ω_j	b_j	α_j	β_j
0	9	62.80767	6.97863		
1	.6	3.80372	6.33953	$\frac{1}{2}$	$\frac{1}{15}$
2	$\frac{11.088}{360{,}000}$	2.05302	8.16435	$\frac{1}{2}$	$\frac{77}{1500}$
3	$\frac{32{,}076}{225 \times 10^5}$	.00711	4.98754	$\frac{1}{2}$	$\frac{81}{1750}$
4	$\frac{45{,}045}{765{,}625 \times 10^3}$	.00006	.99883	$\frac{1}{2}$	$\frac{13}{315}$
5	$\frac{13}{6{,}250{,}000}$	.0000003	.12821	$\frac{1}{2}$	

and $y_5(.1) = 5.1234 \qquad y_5(.2) = 5.3056 \qquad y_5(.3) = 5.5689$

$y_5(.4) = 5.9374 \qquad y_5(.5) = 6.4374 \qquad y_5(.6) = 7.0976$

$y_5(.7) = 7.9491 \qquad y_5(.8) = 9.0255 \qquad y_5(.9) = 10.3626$

Using the results computed with the algorithm we then, as in (6.3-7), calculate

$$\delta_m^2 = \sum_{i=1}^{9} \left[\bar{f}_i - \sum_{j=0}^{m} b_j p_j(x) \right]^2 \tag{6.5-1}$$

We obtain

$$\begin{aligned} &\delta_0^2 = 26.202 \qquad &\delta_1^2 = 2.089 \qquad &\delta_2^2 = .0355 \\ &\delta_3^2 = .000059 \qquad &\delta_4^2 = .00000049 \qquad &\delta_5^2 = .00000045 \end{aligned} \tag{6.5-2}$$

Then, using (6.3-10), we calculate

$$\begin{aligned} &\sigma_0^2 = 3.275 \qquad &\sigma_1^2 = .298 \qquad &\sigma_2^2 = .0059 \\ &\sigma_3^2 = .000012 \qquad &\sigma_4^2 = .00000012 \qquad &\sigma_5^2 = .00000015 \end{aligned} \tag{6.5-3}$$

from which we conclude that $m = 4$ gives us the best least-squares approximation. This approximation is

$$y_4(x) = \sum_{j=0}^{4} b_j p_j(x) \tag{6.5-4}$$

with the b_j's given by (6.4-24) and $p_j(x)$ given by (6.4-14). Although there is normally no computational need to do so, we can convert (6.5-4) into an approximation using powers of x, in which case we get

$$y_4(x) = .9988x^4 + 2.9898x^3 + 2.0172x^2 + .9920x + 5.0010 \tag{6.5-5}$$

In fact the values given in the table at the start of the section are perturbations of the values of

$$f(x) = x^4 + 3x^3 + 2x^2 + x + 5 \tag{6.5-6}$$

The true values of $f(x)$ at the points given in the table are

x_i	.1	.2	.3	.4	.5	.6	.7	.8	.9
f_i	5.1231	5.3056	5.5691	5.9376	6.4375	7.0976	7.9491	9.0256	10.3631

In order to show the effect of the ill condition of the matrix G, let us now repeat the above computation using powers of x instead of orthogonal polynomials. Using (6.3-3) and (6.3-4), we wish to calculate $a_j^{(4)}$, $j = 0, \ldots, 4$.

We get for G, the matrix of the g_{jk},

$$G = \begin{bmatrix} 9.0 & 4.5 & 2.85 & 2.025 & 1.5333 \\ 4.5 & 2.85 & 2.025 & 1.5333 & 1.20825 \\ 2.85 & 2.025 & 1.5333 & 1.20825 & .978405 \\ 2.025 & 1.5333 & 1.20825 & .978405 & .8080425 \\ 1.5333 & 1.20825 & .978405 & .8080425 & .67731333 \end{bmatrix}$$

$$= 9 \begin{bmatrix} 1.00000 & .50000 & .31667 & .22500 & .17037 \\ .50000 & .31667 & .22500 & .17037 & .13425 \\ .31667 & .22500 & .17037 & .13425 & .10871 \\ .22500 & .17037 & .13425 & .10871 & .08978 \\ .17037 & .13425 & .10871 & .08978 & .07526 \end{bmatrix} \tag{6.5-7}$$

From (6.5-7) we see that $\frac{1}{9}G$ is quite close to (6.3-6), as we would expect since the points x_i are all in the interval (0, 1) and are equally spaced. For the determinant of G we get

$$\det G = .000000014 \tag{6.5-8}$$

Therefore, we expect that the errors incurred in the calculation of G will cause substantial errors in the solution of the normal equations. We emphasize that the orthogonal-polynomial and powers-of-x formulations are two ways of stating precisely the same problem. Therefore, any difference between (6.5-5) and the solution of (6.3-4) will be entirely due to different computational techniques.

When (6.3-3) is used, the right-hand side of (6.3-4) is

$$\rho_0 = 62.8077 \qquad \rho_1 = 35.20757 \qquad \rho_2 = 23.944287$$

$$\rho_3 = 17.8176647 \qquad \rho_4 = 13.93266027 \tag{6.5-9}$$

If we solve the normal equations using Gaussian elimination (see Sec. 9.3-1) and carry six decimal places throughout the computation, we get, rounding the results to four decimal places,

$$a_4^{(4)} = .9672 \qquad a_3^{(4)} = 3.0522 \qquad a_2^{(4)} = 1.9763$$

$$a_1^{(4)} = 1.0020 \qquad a_0^{(4)} = 5.0003 \tag{6.5-10}$$

which in the case of $a_4^{(4)}$ and $a_3^{(4)}$ have errors far larger than those in (6.5-5).

A possible source of loss of significance when using orthogonal polynomials is in the computation of the b_j's. If, for example, the magnitude of $p_j(x_i)$ is small for all x_i, the calculation of the quotient ω_j/γ_j may result in a substantial loss of significance, particularly in fixed-point calculations {19}. This is not a problem when using the Gram polynomials with the normalized variable s, but it can be a serious problem when using the recurrence-relation technique. To avoid loss of significance in this latter case, it is desirable to scale and shift the data points from their original interval to a more convenient one; i.e., in effect to normalize the independent variable. Such a convenient interval [Forsythe (1957)] when $w(x) = 1$ is $[-2, 2]$.

It is important that the reader clearly distinguish the two types of errors considered here. On the one hand, the ill condition of G causes the difference between the calculated coefficients given by (6.5-5) and (6.5-10). On the other hand, the difference between the coefficients in (6.5-5) and the true coefficients (6.5-6) is due to the inherent empirical errors in the data.

6.6 THE FOURIER APPROXIMATION

Particularly when the data are from a real-time application, there may be physical knowledge of the function $f(x)$ which indicates that it is periodic. In this case it is advantageous to use the trigonometric (or Fourier) functions instead of polynomials as the least-squares approximating functions. In Sec. 6.6-2 we consider least-squares approximations based on the Fourier functions, but because the summations which arise in such approximations play a role in many applications besides least-squares approximations, we consider first, in Sec. 6.6-1, the evaluation of such sums by the algorithm known as the fast Fourier transform.

6.6-1 The Fast Fourier Transform

Let g_k, $k = 0, 1, \ldots, N-1$ be a set of complex numbers and let

$$\begin{aligned} G_j &= \sum_{k=0}^{N-1} g_k e^{2\pi i jk/N} \qquad j = 0, 1, \ldots, N-1 \\ &= \sum_{k=0}^{N-1} g_k w^{jk} \qquad \text{where } w = e^{2\pi i/N} \end{aligned} \tag{6.6-1}$$

Equation (6.6-1) is often called the *discrete Fourier transform* (DFT) of the sequence $\{g_k\}$ by analogy with the (continuous) Fourier transform

$$G(x) = \int_{-\infty}^{\infty} g(t) e^{2\pi i x t}\, dt \tag{6.6-2}$$

Indeed, there is a direct relationship between the discrete and continuous transforms, whose derivation we leave to a problem {21}. In the same way that the continuous Fourier transform can be inverted, so can the discrete transform, to yield

$$g_k = \frac{1}{N} \sum_{j=0}^{N-1} G_j w^{-jk} \qquad k = 0, 1, \ldots, N-1 \tag{6.6-3}$$

Using the orthogonality relationship {22}

$$\sum_{k=0}^{N-1} w^{jk} w^{-rk} = \begin{cases} N & \text{if } j \equiv r \pmod{N} \\ 0 & \text{otherwise} \end{cases} \tag{6.6-4}$$

it is not hard to show that if G_j given by (6.6-1) is substituted into (6.6-3), the right-hand side gives back g_k {22}. The G_j and g_k thus form a *transform pair*, and it is convenient to use the notation

$$G_j \leftrightarrow g_k$$

to denote this. Moreover, since both (6.6-1) and (6.6-3) are periodic with period N, we may consider G_j and g_k to be defined for all j and k with $G_{j+rN} = G_j$ and $g_{k+rN} = g_k$.

Among the properties of the DFT we state the following, leaving the proofs to a problem {22}.

Property 1 *Linearity* If $G_j \leftrightarrow g_k$ and $H_j \leftrightarrow h_k$ and α and β are any complex constants, then $\alpha G_j + \beta H_j \leftrightarrow \alpha g_k + \beta h_k$.

Property 2 *Shifting* If $G_j \leftrightarrow g_k$, then

$$w^{jr} G_j \leftrightarrow g_{k-r} \qquad \text{and} \qquad G_{j-s} \leftrightarrow w^{-ks} g_s$$

Property 3 *Convolution* If $G_j \leftrightarrow g_k$ and $H_j \leftrightarrow h_k$, then

$$\frac{1}{N} \sum_{r=0}^{N-1} G_r H_{j-r} \leftrightarrow g_k h_k \qquad \text{and} \qquad G_j H_j \leftrightarrow \sum_{r=0}^{N-1} g_k h_{r-k}$$

In addition to these properties, there are many other results about the DFT, some of which are considered in the problems {21 to 24}. Our main interest here, however, is the calculation of the DFT, which we now consider.

We begin by noting that, given the g_k's, calculation of the G_j using (6.6-1) as given would require N complex multiplications and additions for each j or N^2 for all the G_j's. Now suppose N can be factored into

$$N = r_1 r_2 \cdots r_t \tag{6.6-5}$$

Corresponding to the indices j and k we define t-tuples $(j_1, \ldots, j_t)$ and $(k_1, \ldots, k_t)$ such that

$$\begin{aligned} j &= j_1 + r_1 j_2 + r_1 r_2 j_3 + \cdots + r_1 r_2 \cdots r_{t-1} j_t \\ j_s &= 0, 1, \ldots, r_s - 1 \\ s &= 1, \ldots, t \\ k &= k_t + r_t k_{t-1} + r_t r_{t-1} k_{t-2} + \cdots + r_t \cdots r_2 k_1 \\ k_s &= 0, 1, \ldots, r_s - 1 \\ s &= 1, \ldots, t \end{aligned} \tag{6.6-6}$$

The integers $j_1, \ldots, j_t$ and $k_1, \ldots, k_t$ are called the *digits* of j and k, respectively. For example, if $N = 30$, then

$$N = 2 \cdot 3 \cdot 5 \qquad j = j_1 + 2j_2 + 6j_3 \qquad k = k_3 + 5k_2 + 15k_1 \tag{6.6-7}$$

If $j = 23$, then $j_1 = 1, j_2 = 2, j_3 = 3$, and if $k = 8$, then $k_1 = 0, k_2 = 1, k_3 = 3$.

Now we shall develop the *fast Fourier transform* (FFT) algorithm for the case $t = 3$ in order to keep the algebra relatively simple. The generalization for an arbitrary value of t will be fairly obvious {25}. Using (6.6-6), we write

$$w^{jk} = w^{j(k_3 + r_3 k_2 + r_3 r_2 k_1)} \tag{6.6-8}$$

and substitute into (6.6-1) to obtain

$$G_j = \sum_{k_3=0}^{r_3-1} \sum_{k_2=0}^{r_2-1} \sum_{k_1=0}^{r_1-1} g_k w^{j(r_3 r_2)k_1} w^{jr_3 k_2} w^{jk_3} \tag{6.6-9}$$

Now using (6.6-6), we have

$$w^{jr_3 r_2} = w^{(j_1 + r_1 j_2 + r_1 r_2 j_3)(r_3 r_2)} = w^{j_1 r_3 r_2} \tag{6.6-10}$$

since all other exponents have terms in $r_1 r_2 r_3 = N$ and $w^{\alpha N} = 1$, where α is any integer. Similarly

$$w^{jr_3} = w^{(j_1 + r_1 j_2) r_3} \tag{6.6-11}$$

Substituting (6.6-10) and (6.6-11) into (6.6-9), we obtain

$$G_{j_1, j_2, j_3} = \sum_{k_3=0}^{r_3-1} \left\{ \sum_{k_2=0}^{r_2-1} \left[\sum_{k_1=0}^{r_1-1} g_k w^{j_1 r_3 r_2 k_1} \right] w^{(j_1 + r_1 j_2) r_3 k_2} \right\} w^{(j_1 + r_1 j_2 + r_1 r_2 j_3) k_3} \tag{6.6-12}$$

Noting that the term in square brackets depends only on k_1 and the term in braces only on k_2, we describe the FFT algorithm as follows:

Input

g_k — N values stored in increasing order of the index k, that is, from $(k_1, k_2, k_3) = (0, 0, 0)$ to $(r_1 - 1, r_2 - 1, r_3 - 1)$

Algorithm

$$\begin{aligned} &f_0(k_1, k_2, k_3) \leftarrow g_k \\ &f_1(j_1, k_2, k_3) \leftarrow \sum_{k_1=0}^{r_1-1} f_0(k_1, k_2, k_3) w^{j_1 r_3 r_2 k_1} \\ &f_2(j_1, j_2, k_3) \leftarrow \sum_{k_2=0}^{r_2-1} f_1(j_1, k_2, k_3) w^{(j_1 + r_1 j_2) r_3 k_2} \\ &f_3(j_1, j_2, j_3) \leftarrow \sum_{k_3=0}^{r_3-1} f_2(j_1, j_2, k_3) w^{(j_1 + r_1 j_2 + r_1 r_2 j_3) k_3} \end{aligned} \tag{6.6-13}$$

Output

$G_j = f_3(j_1, j_2, j_3)$

From (6.6-12) it follows that this algorithm does indeed compute the G_j. Moreover, the number of complex multiplications and additions required in (6.6-13) for each value of the argument triple on the left-hand side is

$$r_1 + r_2 + r_3$$

Since there are N possible argument triples, the total number of operations is

$$N(r_1 + r_2 + r_3) \tag{6.6-14}$$

which is never greater than N^2 and when r_1, r_2 and r_3 are substantially greater than 1, is much less than N^2. Moreover, if $t > 3$, the equation corresponding to (6.6-14) is {25}

$$N(r_1 + r_2 + \cdots + r_t) \tag{6.6-15}$$

where the inequality with respect to N^2 is likely to be even more pronounced. Thus, the FFT algorithm is, at least at first glance, more efficient than direct evaluation of the DFT, and for large N the reduction in computation may be quite dramatic. There is, of course, a substantial amount of bookkeeping implied by (6.6-13), which would seem to lessen the overall computational advantage of the FFT algorithm; we shall return to this point below.

One aspect of (6.6-13) which tends to be confusing on initial exposure to the FFT is that the order of G_j's computed is different from the natural order. Consider the following case:

$$N = 12 = 2 \cdot 2 \cdot 3 \qquad j = j_1 + 2j_2 + 4j_3 \qquad k = k_3 + 3k_2 + 6k_1 \tag{6.6-16}$$

In organizing the computation of (6.6-13) we would expect to choose the natural order of k from 0 to 11, as shown in Table 6.1. Equation (6.6-13) makes it clear that if we begin with $g_0, \ldots, g_{11}$ in 12 successive memory locations, then at successive steps of the algorithm we may

Overwrite $f_1(j_1, k_2, k_3)$ on $g_k = f_0(k_1, k_2, k_3)$
Overwrite $f_2(j_1, j_2, k_3)$ on $f_1(j_1, k_2, k_3)$
Overwrite $f_3(j_1, j_2, j_3) = G_j$ on $f_2(j_1, j_2, k_3)$

But Table 6.1 indicates that the order of the G_j in these 12 locations is not the natural order and that to obtain the natural order therefore requires some unscrambling. But we do note here that if we reverse the order of the digits j_1, j_2, and j_3 and use the natural order of the reversed digits, we obtain the natural order of j as shown in Table 6.1. It can be proved that this *digit reversal* always results in the natural order of j.

Table 6.1 Correspondence between digits of k and j for case $N = 12$

k	j_1 k_1	j_2 k_2	j_3 k_3	j	j_3	j_2	j_1	j
0	0	0	0	0	0	0	0	0
1	0	0	1	4	0	0	1	1
2	0	0	2	8	0	1	0	2
3	0	1	0	2	0	1	1	3
4	0	1	1	6	1	0	0	4
5	0	1	2	10	1	0	1	5
6	1	0	0	1	1	1	0	6
7	1	0	1	5	1	1	1	7
8	1	0	2	9	2	0	0	8
9	1	1	0	3	2	0	1	9
10	1	1	1	7	2	1	0	10
11	1	1	2	11	2	1	1	11

Implementation of the FFT algorithm is most convenient and efficient when N is chosen so that

$$N = r^t$$

Then from (6.6-15) it follows that the number of operations required is

$$N(rt) = Nr \log_r N = \frac{r}{\log_2 r} N \log_2 N \tag{6.6-17}$$

For fixed N the coefficient of $N \log_2 N$ in (6.6-17) is minimized when $r = 3$, which implies that N should be chosen as a power of 3. But for various reasons, most notably that the calculation and use of the powers of w is simplified, it is most common to choose $r = 2$. Because of the importance of this case, we now consider it in some detail.

When $N = 2^t$, the digits of j and k are all either 0 or 1. Indeed, $j_t, \ldots, j_1$ and $k_1, \ldots, k_t$ are, respectively, just the bits of the binary representations of j and k. The algorithm (6.6-13), now expressed for a general t, becomes

$$f_0(k_1, k_2, \ldots, k_t) = g_k$$

$$\cdots\cdots\cdots\cdots\cdots\cdots$$

$$f_l(j_1, j_2, \ldots, j_l, k_{l+1}, \ldots, k_t) \tag{6.6-18}$$

$$= \sum_{k_l=0}^{1} f_{l-1}(j_1, \ldots, j_{l-1}, k_l, \ldots, k_t) w^{(\Sigma_{s=1}^{l} j_s 2^{s-1}) 2^{t-l} k_l}$$

$$\cdots\cdots\cdots\cdots\cdots\cdots\cdots\cdots\cdots\cdots\cdots\cdots$$

$$G_j = f_t(j_1, j_2, \ldots, j_t) = \sum_{k_t=0}^{1} f_{t-1}(j_1, \ldots, j_{t-1}, k_t) w^{jk_t}$$

The statement of this algorithm, however, can be much simplified, as follows. Define

$$J_1 = 0$$

$$J_l = j_1 + 2j_2 + \cdots + 2^{l-2}j_{l-1} \qquad l = 2, \ldots, t \tag{6.6-19}$$

and

$$K_l = k_{l+1} + 2k_{l+2} + \cdots + 2^{t-l-1}k_t \qquad l = 0, \ldots, t-1$$

$$K_t = 0 \tag{6.6-20}$$

Then we can write the index j in (6.6-18) as

$$j = J_l + 2^{l-1}j_l + 2^l K_l \qquad l = 1, \ldots, t \tag{6.6-21}$$

It is not hard to show {26} that with J_l and K_l defined as in (6.6-19) and (6.6-20), the index j in (6.6-21) goes through its full range of values from 0 to $2^t - 1$ for each l. The middle equation of (6.6-18) can then be written

$$f_l(J_l + 2^{l-1}j_l + 2^l K_l) = \sum_{k_l=0}^{1} f_{l-1}(J_l + 2^{l-1}k_l + 2^l K_l)w^{J_{l+1}2^{t-l}k_l} \tag{6.6-22}$$

where the ranges of the indices are

$$J_l = 0, 1, \ldots, 2^{l-1} - 1 \qquad K_l = 0, 1, \ldots, 2^{t-1} - 1 \tag{6.6-23}$$

When $l = t$ in (6.6-22), the index of f_t is, using (6.6-19) and (6.6-20),

$$\begin{aligned} J_t + 2^{t-1}j_t + 2^t K_t &= J_t + 2^{t-1}j_t \\ &= j_1 + 2j_2 + \cdots + 2^{t-2}j_{t-1} + 2^{t-1}j_t \end{aligned}$$

which, using (6.6-6), is precisely j. Thus, $f_t(0), f_t(1), \ldots, f_t(2^{t-1})$ correspond to $j = 0, 1, \ldots, 2^{t-1}$, and therefore the G_j's are calculated in their natural order. But this happens because

$$K_0 = k_1 + 2k_2 + \cdots + 2^{t-1}k_t$$

has its bits reversed from those of k, as can be seen from (6.6-6). Thus, in order for the G_j's to be computed in their natural order, the g_k's must be ordered in bit-reversal sequence; i.e., the initial order of the g_k must correspond to the order resulting from taking the bits of the binary expansion of K, namely $k_t, k_{t-1}, \ldots, k_1$, and computing K_0 as above. Example 6.1 will illustrate this explicitly.

Now we split (6.6-22) into two equations, one for $j_l = 0$ and one for $j_l = 1$, and write out both terms in the right-hand side sum to obtain

$$\begin{aligned} f_l(J_l + 2^l K_l) &= f_{l-1}(J_l + 2^l K_l) \\ &\quad + f_{l-1}(J_l + 2^l K_l + 2^{l-1})w^{J_l 2^{t-l}} \\ f_l(J_l + 2^l K_l + 2^{l-1}) &= f_{l-1}(J_l + 2^l K_l) \\ &\quad + f_{l-1}(J_l + 2^l K_l + 2^{l-1})w^{(J_l + 2^{l-1})2^{t-l}} \end{aligned} \tag{6.6-24}$$

Finally, we note that $w^{(2^{l-1})(2^{t-l})} = w^{2^{t-1}} = -1$, and we define

$$p = J_l + 2^l K_l \qquad q = p + 2^{l-1} \tag{6.6-25}$$

so that p and q both take on half the values from 0 to $2^t - 1$. Then we obtain the complete algorithm for this case as

$$\begin{aligned} f_0(k) &= g_k \\ &\cdots\cdots\cdots \\ f_l(p) &= f_{l-1}(p) + f_{l-1}(q)w^{J_l 2^{t-l}} \\ f_l(q) &= f_{l-1}(p) - f_{l-1}(q)w^{J_l 2^{t-l}} \\ &\cdots\cdots\cdots\cdots\cdots\cdots \\ G_p &= f_t(p) \qquad G_q = f_t(q) \end{aligned} \tag{6.6-26}$$

Thus, at each stage of the algorithm we proceed as follows:

1. Let J_l and K_l run through all possible pairs of values in (6.6-23).
2. For each pair calculate p and q as in (6.6-25).
3. Then calculate $f_l(p)$ and $f_l(q)$ from (6.6-26).

Normally in the computer implementation of the algorithm the quantities $w^{J_l 2^{t-l}}$ which appear in (6.6-26) are precalculated and stored in a table (or a set of coefficients from which they are easily calculated is stored in a table {26}). Note that after the complete calculation implied by (6.6-26), we must perform a digit reversal, as described above, to get the G_j in their natural order. In the binary case we are considering, this *bit* reversal consists only of taking the bits of the binary expansion of j and computing j' as the binary expansion of the reversal of these bits. The correct position of G_j as calculated is then $G_{j'}$. Alternatively, as in Example 6.1 below, we may reorder the g_k so that the G_j are calculated in their correct order.

These remarks and the algorithm itself imply that the bookkeeping in (6.6-26) is quite simple and straightforward. The savings achievable, therefore, by reducing the N^2 operations for the DFT to the $N \log_2 N$ are considerable and for large N can be dramatic. The FFT algorithm in the form (6.6-26) or equivalent forms {27} has become extremely popular and useful.

To conclude this section we note a couple of other features of the FFT:

1. If we are given the G_j instead of the g_k, the algorithms (6.6-13) or (6.6-26) are easily applied, the only difference being the change of sign of the exponent of w from plus to minus.
2. An important case is the one where the g_k are all real and we are interested in the cosine transform

$$G_j = \sum_{k=0}^{N-1} g_k \cos \frac{2\pi jk}{N} \tag{6.6-27}$$

The statement of the FFT algorithm for this case is easily derived from (6.6-13) or (6.6-26) {27}. Similarly we may also consider the sine transform

$$G_j = \sum_{k=1}^{N-1} g_k \sin \frac{2\pi jk}{N} \tag{6.6-28}$$

The example which follows has too small a value of N for the FFT algorithm to show any substantial efficiency over direct calculation of the DFT, but it does illustrate how the algorithm works.

Example 6.1 Given the data

k	0	1	2	3	4	5	6	7
g_k	1	$1+i$	0	$1-i$	0	$1+i$	0	$1-i$

compute G_j, $j = 0, 1, \ldots, 7$.

Since $N = 8 = 2^3$, we use (6.6-26). First we perform a bit reversal and reorder the g_k to get f_0. We do this by taking the bits of k and reversing them as in the table below:

k	k_3	k_2	k_1	$4k_1 + 2k_2 + k_3 = k'$	k	k_3	k_2	k_1	$4k_1 + 2k_2 + k_3 = k'$
0	0	0	0	0	4	1	0	0	1
1	0	0	1	4	5	1	0	1	5
2	0	1	0	2	6	1	1	0	3
3	0	1	1	6	7	1	1	1	7

Thus the order of f_0 is with k' in its natural order.

We begin by calculating the necessary powers of w. We have

$$w = e^{2\pi i/8} = \cos\frac{\pi}{4} + i \sin\frac{\pi}{4} = \frac{\sqrt{2}}{2}(1+i)$$

$$w^2 = i \qquad w^3 = \frac{\sqrt{2}}{2}(-1+i)$$

Using these and (6.6-25) and (6.6-26), we then calculate

	0	1	2	3	4	5	6	7
f_0	1	0	0	0	$1+i$	$1+i$	$1-i$	$1-i$
f_1	1	1	0	0	$2+2i$	0	$2-2i$	0
f_2	1	1	1	1	1	4	$4i$	0
$f_3 = G_j$	5	1	-3	1	-3	1	5	1

The reader who wishes to check the values of G_j using (6.6-1) will find that even in this simple case the FFT algorithm has much to recommend it over brute force.

6.6-2 Least-Squares Approximations and Trigonometric Interpolation

Throughout this section we assume that the set of points $\{x_i\}$ is equally spaced. As in Sec. 6.4, we assume that the number of points n is odd and equal to $2L + 1$. The corresponding results when n is even are considered in a problem {31}.

For convenience we set $x_i = 2\pi i/(2L + 1)$, $i = 0, \ldots, 2L$. The set of functions $\{\phi_j(x)\}$ that we shall use are the $2L + 1$ functions, 1, $\cos x, \ldots, \cos Lx$, $\sin x, \ldots, \sin Lx$. We limit ourselves to a number of functions equal to the number of data points because, as before, we expect to have no more than $2L + 1$ independent functions on $2L + 1$ points. In fact, it is not hard to show {29} that $\sin kx$ or $\cos kx$, where k is an integer greater than L, can be simply expressed in terms of one of the above functions on the set $\{x_i\}$.

Just as the Fourier functions satisfy an orthogonality relationship over the interval $[0, \pi]$, so do the above functions satisfy an orthogonality relationship over the discrete set of points $\{x_i\}$. In fact, we can show that {30}

$$\sum_{i=0}^{2L} \sin jx_i \sin kx_i = \begin{cases} 0 & j \neq k \\ \dfrac{2L+1}{2} & j = k \neq 0 \\ 0 & j = k = 0 \end{cases} \tag{6.6-29}$$

$$\sum_{i=0}^{2L} \cos jx_i \cos kx_i = \begin{cases} 0 & j \neq k \\ \dfrac{2L+1}{2} & j = k \neq 0 \\ 2L+1 & j = k = 0 \end{cases} \tag{6.6-30}$$

$$\sum_{i=0}^{2L} \cos jx_i \sin kx_i = 0 \qquad \text{all } j, k \tag{6.6-31}$$

where j and k are restricted to run between 0 and L.

With $2L + 1$ functions we would expect to be able to fit exactly the $2L + 1$ points $\{x_i\}$, but in the least-squares context, this again is just what we do not wish to do. As before, we want to use enough functions to provide a good approximation to the true function $f(x)$ but not so many that we lose smoothing. Suppose then, as before, that $\bar{f}_i$ is the observed value of $f(x)$ at x_i. We approximate $f(x)$ by

$$y_m(x) = \tfrac{1}{2}a_0 + \sum_{j=1}^{m} (a_j \cos jx + b_j \sin jx) \qquad m < L \tag{6.6-32}$$

in direct analogy with standard Fourier-series notation. Again we determine the a_j's and b_j's so that the sum of the squares of the differences $\bar{f}_i - y_m(x_i)$

will be minimized. When we use the orthogonality relations (6.6-29) to (6.6-31), the normal equations (6.2-6) yield

$$a_j = \frac{2}{2L+1}\sum_{i=0}^{2L} \bar{f}_i \cos jx_i$$

$$= \frac{2}{2L+1}\sum_{i=0}^{2L} \bar{f}_i \cos \frac{2\pi ij}{2L+1} \qquad j = 0, \ldots, m \tag{6.6-33}$$

$$b_j = \frac{2}{2L+1}\sum_{i=1}^{2L} \bar{f}_i \sin jx_i$$

$$= \frac{2}{2L+1}\sum_{i=1}^{2L} \bar{f}_i \sin \frac{2\pi ij}{2L+1} \qquad j = 1, \ldots, m \tag{6.6-34}$$

In order to compute the coefficients in (6.6-33) and (6.6-34) we can, of course, evaluate the summations directly. But referring to (6.6-27) and (6.6-28), we see that these summations have just the form of those used in the FFT with g's replaced by $\bar{f}$'s and N replaced by $2L + 1$. Although $2L$ is not usually large enough for use of the FFT to result in a great saving of time, it can conveniently be used to evaluate (6.6-33) and (6.6-34). Of course when the number of points is odd, we cannot use the power-of-2 FFT algorithm (6.6-26) but must use the more general case (6.6-13). Another algorithm which is much more efficient than direct evaluation for the summations (6.6-33) and (6.6-34) is considered in a problem {32}.

The results of Sec. 6.3 can be applied to approximations of the form (6.6-32) just as they were to polynomial approximations. In particular, we can calculate

$$\delta_m^2 = \sum_{i=0}^{2L}\left\{\bar{f}_i - \left[\tfrac{1}{2}a_0 + \sum_{j=1}^{m}(a_j \cos jx_i + b_j \sin jx_i)\right]\right\}^2$$

$$= \sum_{i=0}^{2L} \bar{f}_i^2 - \frac{2L+1}{2}\left[\frac{a_0^2}{2} + \sum_{j=1}^{m}(a_j^2 + b_j^2)\right] \tag{6.6-35}$$

by making use of the orthogonality relations (6.6-29) to (6.6-31) {30}.

Example 6.2 Use the data in the following table:

x_i	0	$2\pi/9$	$4\pi/9$	$2\pi/3$	$8\pi/9$	$10\pi/9$	$4\pi/3$	$14\pi/9$	$16\pi/9$
$\bar{f}_i$	3.0004	5.7203	3.1993	−1.0981	−.8679	2.9890	4.0985	1.1477	−.1882

to calculate Fourier least-squares approximations with $m = 1, 2, 3$.

By direct evaluation of (6.6-33) and (6.6-34) or using the algorithm of {32} we get

$$a_0 = 4.00022, \qquad a_1 = .99998 \qquad a_2 = .00011 \qquad a_3 = .00023$$

$$b_1 = .00029 \qquad b_2 = 2.99997 \qquad b_3 = .00002$$

Using the second equality in (6.6-35), we calculate

$$\delta_0^2 = 44.99932 \qquad \delta_1^2 = 40.49950 \qquad \delta_2^2 = .00031 \qquad \delta_3^2 = .00031$$

To find the value of m which results in the best least-squares approximation, we would generally use (6.3-10) with $2(L - m)$ in the denominator (why?). In this example, however, it is clear—as indeed the coefficients a_i and b_i indicated—that $m = 2$ gives the desired least-squares approximation. In fact, the data in the table are slightly perturbed values of $f(x) = 2 + \cos x + 3 \sin 2x$.

In Sec. 6.2 we noted that a least-squares approximation in which the number of functions used equals the number of data points results in an exact approximation. In particular, in the polynomial case, we get the Lagrangian interpolation formula. When our approximating functions are the trigonometric functions, we get a formula for trigonometric interpolation. In common with our approach to interpolation in Chap. 3, we shall assume that the given function values are exact (except perhaps for some roundoff error).

When $m = L$, Eq. (6.6-32) becomes the discrete Fourier series for the function defined by the set of values $\{f_i\}$

$$y_L(x) = \tfrac{1}{2}a_0 + \sum_{j=1}^{L} (a_j \cos jx + b_j \sin jx) \tag{6.6-36}$$

with the a_j's and b_j's again given by (6.6-33) and (6.6-34). Now, however, δ_L^2 as given by (6.6-35) is 0. To prove this we substitute (6.6-33) and (6.6-34) into (6.6-36) to get

$$\begin{aligned} y_L(x) &= \frac{2}{2L+1} \sum_{i=0}^{2L} f_i \left[\frac{1}{2} + \sum_{j=1}^{L} (\cos jx_i \cos jx + \sin jx_i \sin jx)\right] \\ &= \frac{2}{2L+1} \sum_{i=0}^{2L} f_i \left[\frac{1}{2} + \sum_{j=1}^{L} \cos j(x_i - x)\right] \end{aligned} \tag{6.6-37}$$

We wish to show that $y_L(x_k) = f_k$. First we note that

$$\begin{aligned} \cos j(x_i - x_k) &= \cos \frac{2\pi}{2L+1} j(i-k) \\ &= \cos \left[\frac{2\pi}{2L+1} (i-k)(2L+1-j)\right] \\ &= \cos [(2L+1-j)(x_i - x_k)] \end{aligned} \tag{6.6-38}$$

Using this, we rewrite (6.6-37) at $x = x_k$ as

$$y_L(x_k) =$$

$$\frac{2}{2L+1} \sum_{i=0}^{2L} f_i \left[\frac{1}{2} + \frac{1}{2} \sum_{j=1}^{L} \cos j(x_i - x_k) + \frac{1}{2} \sum_{j=L+1}^{2L} \cos j(x_i - x_k) \right]$$

$$= \frac{1}{2L+1} \sum_{i=0}^{2L} f_i \sum_{j=0}^{2L} \cos j(x_i - x_k) \tag{6.6-39}$$

But using (6.6-29) and (6.6-30), we can show that {33}

$$\sum_{j=0}^{2L} \cos j(x_i - x_k) = \begin{cases} 2L+1 & i = k \\ 0 & i \neq k \end{cases} \tag{6.6-40}$$

from which it follows that in (6.6-39)

$$y_L(x_k) = f_k \tag{6.6-41}$$

as we wished to prove.

Equation (6.6-36) or, equivalently, (6.6-37) is an equation for trigonometric interpolation which agrees with the observed data at $x = x_i$, $i = 0, \ldots, 2L$, and can be used to interpolate at values of $x \neq x_i$. In the form (6.6-37) the formula is the trigonometric analog to the Lagrangian interpolation formula for equally spaced data in the polynomial case. The coefficient of f_i

$$\frac{2}{2L+1} \left[\frac{1}{2} + \sum_{j=1}^{L} \cos j(x_i - x) \right] \tag{6.6-42}$$

is therefore the trigonometric analog of the Lagrangian interpolation polynomial $l_i(x)$. By suitably manipulating the term in brackets in (6.6-42), we can show that {33}

$$\frac{1}{2} + \sum_{j=1}^{L} \cos j(x_i - x) = \frac{\sin (L + \frac{1}{2})(x_i - x)}{2 \sin \frac{1}{2}(x_i - x)} \tag{6.6-43}$$

and thus (6.6-37) can be written

$$y_L(x) = \frac{1}{2L+1} \sum_{i=0}^{2L} \frac{\sin (L + \frac{1}{2})(x_i - x)}{\sin \frac{1}{2}(x_i - x)} f_i \tag{6.6-44}$$

In contrast to the polynomial case, we cannot derive any very useful closed form for the error in the approximation (6.6-44); see, however, {34} for one expression for the error.

BIBLIOGRAPHIC NOTES

There is a wide literature on methods for fitting curves to numerical data. For a general reference covering the subject matter of this chapter as well as the statistical background and many related subjects we recommend Guest (1961).

Sections 6.1 and 6.2 Most numerical analysis texts contain some material on least-squares approximation; in particular, see Isaacson and Keller (1966), Shampine and Allen (1973), and Hamming (1973). The first of these also includes a good deal of material on least-squares approximations over continuous intervals. Davis (1963) and Rice (1964) consider least-squares approximation in the context of linear spaces; the latter also considers discrete least-squares approximations and the problem of minimizing (6.2-4).

Section 6.3 Much of the material from this section is from Forsythe (1957). For more information on the ill condition of the Hilbert matrix, see Todd (1954), and for related computational problems, see Wilkinson (1961). Guest (1961) discusses methods for the solution of the normal equations; see also Chap. 9.

Section 6.4 Guest (1961) has an excellent chapter on the generation and use of orthogonal polynomials in least-squares approximations. Another good source is the paper by Birge and Weinberg (1947); see also Forsythe (1957) and Aitken (1932). A good basic treatment will be found in Shampine and Allen (1973).

Section 6.6 Three good references on the fast Fourier transform are Davis and Rabinowitz (1975), Cooley, Lewis, and Welch (1977), and Brigham (1974). For a much fuller discussion of the numerical analysis of periodic functions, see Hamming (1973). A good discussion of trigonometric interpolation will be found in Lanczos (1956); see also Lanczos (1938). The standard text on the subject of trigonometric series in general is Zygmund (1952).

BIBLIOGRAPHY

Aitken, A. C. (1932): On the Graduation of Data by the Orthogonal Polynomials of Least Squares, *Proc. Roy. Soc. Edinb.*, vol. 53, pp. 54–78.

Birge, R. T., and J. W. Weinberg (1947): Least Squares Fitting of Data by Means of Polynomials, *Rev. Mod. Phys.*, vol. 19, pp. 298–360.

Brigham, E. O. (1974): *The Fast Fourier Transform*, Prentice-Hall, Inc., Englewood Cliffs, N.J.

Cooley, J. W., P. A. W. Lewis, and P. D. Welch (1977): The Fast Fourier Transform and Its Applications to Time Series Analysis, pp. 377–423 in *Statistical Methods for Digital Computers* (K. Enslein, A. Ralston, and H. S. Wilf, eds.), John Wiley & Sons, Inc., New York.

Davis, P. J. (1963): *Interpolation and Approximation*, Blaisdell Publishing Company, New York.

——— and P. Rabinowitz (1975): *Methods of Numerical Integration*, Academic Press, Inc., New York.

Forsythe, G. E. (1957): Generation and Use of Orthogonal Polynomials for Data-fitting on a Digital Computer, *J. Soc. Ind. Appl. Math.*, vol. 5, pp. 74–88.

Goertzel, G. (1958): An Algorithm for the Evaluation of Finite Trigonometric Series, *Am. Math. Mon.*, vol. 65, pp. 34–35.

Guest, P. G. (1961): *Numerical Methods of Curve Fitting*, Cambridge University Press, New York.

Hamming, R. W. (1973): *Numerical Methods for Scientists and Engineers*, 2d ed., McGraw-Hill Book Company, New York.

Hildebrand, F. B. (1974): *Introduction to Numerical Analysis*, 2d ed., McGraw-Hill Book Company, New York.

Isaacson, E., and H. B. Keller (1966): *Analysis of Numerical Methods*, John Wiley & Sons, Inc., New York.

Lanczos, C. (1938): Trigonometric Interpolation of Empirical and Analytic Functions, *J. Math. and Phys.*, vol. 17, pp. 123–199.

——— (1956): *Applied Analysis*, Prentice-Hall, Inc., Englewood Cliffs, N.J.

Rice, J. R. (1964): *The Approximation of Functions*, vol. 1, Addison-Wesley Publishing Company, Inc., Reading, Mass.
Savage, I. R., and E. Lukacs (1954): Tables of Inverses of Finite Segments of the Hilbert Matrix in *Contributions to the Solution of Systems of Linear Equations and the Determination of Eigenvalues* (O. Taussky, ed.), vol. 39, National Bureau of Standards Applied Mathematics Series.
Shampine, L. F., and R. C. Allen, Jr. (1973): *Numerical Computing: An Introduction*, W. B. Saunders Company, Philadelphia.
Shannon, C. E. (1949): Communication in the Presence of Noise, *Proc. IRE*, vol. 37, pp. 10–21.
Todd, J. (1954): The Condition of Finite Segments of the Hilbert Matrix in *Contributions to the Solution of Systems of Linear Equations and the Determination of Eigenvalues* (O. Taussky, ed.), vol. 39, National Bureau of Standards Applied Mathematics Series.
Wilkinson, J. H. (1961): Error Analysis of Direct Methods of Matrix Inversion, *J. Ass. Comput. Mach.*, vol. 8, pp. 281–330.
Wilks, S. S. (1962): *Mathematical Statistics*, rev. ed., John Wiley & Sons, Inc., New York.
Zygmund, A. (1952): *Trigonometric Series*, 2d ed., Chelsea Publishing Company, New York.

PROBLEMS

Section 6.2

1 Given the data

x_i	-1.00	$-.75$	$-.50$	$-.25$	0	.25	.50	.75	1.00
f_i	$-.2209$	.3295	.8826	1.4392	2.0003	2.5645	3.1334	3.7061	4.2836

calculate the coefficients in the normal equations for $m = 3, 4, 5$ with (a) $\phi_j(x) = x^j$; (b) $\phi_j(x) = P_j(x)$, the Legendre polynomial of degree j. Use $w(x) = 1$.

Section 6.3

***2** Let $\mathbf{v}$ be a column vector such that $\mathbf{v}^T = (v_1, \ldots, v_n)$, where T denotes the transpose. Define the norm of $\mathbf{v}$ to be $\|\mathbf{v}\| = (v_1^2 + v_2^2 + \cdots + v_n^2)^{1/2}$.

(*a*) Show that the least-squares problem for polynomials with $w(x) = 1$ can be written in the form

$$\|\mathbf{f} - Q\mathbf{a}^{(m)}\|^2 = \text{minimum}$$

where $\mathbf{f}^T = (f_1, \ldots, f_n)$, $(\mathbf{a}^{(m)})^T = (a_0^{(m)}, \ldots, a_m^{(m)})$, and Q is an $n \times (m+1)$ matrix with columns $\mathbf{q}_j = (x_1^j, x_2^j, \ldots, x_n^j)^T$, $j = 0, \ldots, m$.

(*b*) If $n > m$ and $\mathbf{c}$ is a column vector with $m + 1$ components, show that

$$\mathbf{c}^T Q^T Q \mathbf{c} \geq 0$$

with the equality holding only if $\mathbf{c} = \mathbf{0}$.

(*c*) Show that the minimum problem of part (*a*) is equivalent to minimizing

$$(\mathbf{a}^{(m)} - G^{-1}\mathbf{g})^T G(\mathbf{a}^{(m)} - G^{-1}\mathbf{g}) + \mathbf{f}^T\mathbf{f} - \mathbf{g}^T(G^{-1})^T\mathbf{g}$$

where $G = Q^T Q$ and $\mathbf{g} = Q^T\mathbf{f}$.

(*d*) Use this result and part (*b*) to show that

$$\mathbf{a}^{(m)} = G^{-1}Q^T\mathbf{f}$$

is the unique solution of the minimum problem. [Ref.: Forsythe (1957).]

3 (*a*) If the set of points $\{x_i\}$ is symmetrically placed with respect to 0, show that the system of normal equations (6.3-4) can be "decoupled" into two sets of equations with $(m + 1)/2$ equations in each set if m is odd and $m/2$ and $(m/2) + 1$ equations in the two sets if m is even.

(*b*) Display these two sets of equations for the normal equations derived in Prob. 1*a*.

(*c*) Can the systems derived in Prob. 1*b* be decoupled? What is the general rule?

***4** A theorem of Cauchy states that if $a_1, \ldots, a_n, b_1, \ldots, b_n$ are $2n$ numbers, the determinant with elements $(a_i + b_k)^{-1}$, $i = 1, \ldots, n$, $k = 1, \ldots, n$, has the value

$$\left| \frac{1}{a_i + b_k} \right| = \frac{\prod\limits_{i>k=1}^{n} (a_i - a_k)(b_i - b_k)}{\prod\limits_{i=1}^{n} \prod\limits_{k=1}^{n} (a_i + b_k)}$$

(*a*) Use this theorem to show that Δ_n, the determinant of H_n, is given by

$$\Delta_n = \frac{\left(\prod\limits_{k=1}^{n-1} k!\right)^2}{n^n \prod\limits_{k=1}^{n-1} (n^2 - k^2)^{n-k}}$$

(*b*) If Δ_n^{ij} is the minor of the element in the ith row and jth column of H_n, show that

$$\Delta_n^{ij} = \frac{(n+i-1)!\,(n+j-1)!}{[(i-1)!\,(j-1)!]^2 n!\,(n-i)!\,(n-j)!} \frac{\left(\prod\limits_{k=1}^{n-1} k!\right)^3}{\prod\limits_{k=1}^{n-1} (n+k)!} \frac{1}{i+j-1}$$

(*c*) Use induction to show that

$$\frac{n^n}{n!} \prod_{k=1}^{n-1} \frac{(n^2 - k^2)^{n-k} k!}{(n+k)!} = 1$$

(*d*) Use the results of parts (*a*) to (*c*) to show that h_n^{ij}, the element in the ith row and jth column of H_n^{-1}, is given by

$$h_n^{ij} = \frac{(-1)^{i+j}}{i+j-1} \frac{(n+i-1)!\,(n+j-1)!}{[(i-1)!\,(j-1)!]^2 (n-i)!\,(n-j)!}$$

(*e*) Use this result to show that

$$h_{n+1}^{ij} = \frac{(n+i)(n+j)}{(n+1-i)(n+1-j)} h_n^{ij} \qquad i, j = 1, \ldots, n$$

$$h_{n+1}^{n+1,j} = h_{n+1}^{j,n+1} = \frac{(-1)^{n+j+1}}{(n+j)} \frac{(2n+1)!\,(n+j)!}{[n!\,(j-1)!]^2 (n+1-j)!}$$

(*f*) Use the results of part (*e*) to calculate H_n^{-1}, $n = 2, 3, 4, 5$. [Ref.: Savage and Lukacs (1954).]

5 Using equations derived in Prob. 1, compute the coefficients of the least-squares approximations for $m = 3, 4, 5$ for the case $\phi_j(x) = x^j$. Use any technique to solve the normal equations. Also calculate the determinant of the coefficients in the normal equations (see Prob. 10).

6 (*a*) Repeat the calculations of the previous problem for the case $\phi_j(x) = P_j(x)$.

(*b*) Convert the Legendre polynomial approximations to the form of approximations in powers of x. Compare the coefficients with those found in the previous problem. How do you account for the differences? Which coefficients do you expect to be more accurate? Why? (See Prob. 10.)

Section 6.4

7 (*a*) Given m vectors $\mathbf{v}_1, \mathbf{v}_2, \ldots, \mathbf{v}_m$, show how the Gram-Schmidt procedure can be used to generate a related set of orthogonal vectors.

(*b*) Starting with the powers of x $(1, x, x^2, \ldots)$, show how the Gram-Schmidt procedure can be used to generate the Legendre polynomials (see Prob. 19, Chap. 4).

8 (*a*) Prove that any polynomial of degree j satisfying the orthogonality relationship (6.4-5) with $w_i > 0$ for all i has j distinct zeros interior to the interval spanned by the points $\{x_i\}$.

(*b*) Show that if $p_j^{(n)}(x)$, $j = 0, 1, \ldots, n$, are a set of polynomials of degree j satisfying (6.4-5), then $p_n^{(n)}(x_i) = 0$, $i = 1, \ldots, n$.

9 (*a*) By substituting for b_k in (6.4-23) the value found by solving for b_k in (6.4-27) show that $q_0(x)$ as defined in (6.4-27) is equal to $y_m(x)$.

(*b*) Derive a similar recurrence for $y'_m(x)$.

10 (*a*) Use the data of Prob. 1 and orthogonal polynomials generated using (6.4-14) to generate a least-squares approximation for $m = 1, 2, 3, 4, 5$. Express the approximations as sums of powers of x. Compare these results with those of Probs. 5 and 6 and discuss the differences.

(*b*) From these results which, if any, of these values of m would you choose to be M in (6.2-7)?

11 (*a*) Show that

$$\Delta(u_i v_i) = u_i \, \Delta v_i + v_{i+1} \, \Delta u_i$$

(*b*) Use this to derive the formula for summation by parts

$$\sum_{i=R}^{S} u_i \, \Delta v_i = u_i v_i \Big|_R^{S+1} - \sum_{i=R}^{S} v_{i+1} \, \Delta u_i .$$

(*c*) Derive also the alternative formula

$$\sum_{i=R}^{S} u_i \, \Delta v_i = u_{i-1} v_i \Big|_R^{S+1} - \sum_{i=R}^{S} v_i \, \Delta u_{i-1}$$

12 (*a*) If the $N + 1$ points x_i, $i = 0, 1, \ldots, N$, are equally spaced at an interval h, show that the orthogonality condition (6.4-5) can be recast as

$$\sum_{i=0}^{N} q_{j-1}(i) \, \Delta^j U_j(i, N) = 0$$

where $\Delta^j U_j(i, N) = w_i p_j(x_0 + ih)$ and $q_{j-1}(x)$ is a polynomial of degree $j - 1$ or less.

(*b*) Use the results of Prob. 11 to derive the conditions

$$U_j(k, N) = 0 \qquad k = 0, 1, \ldots, j - 1$$

$$U_j(N + k, N) = 0 \qquad k = 1, \ldots, j$$

(*c*) For $w(x) = 1$ show that

$$\Delta^{2j+1} U_j(i, N) = 0$$

and then conclude that

$$U_j(i, N) = A_{jN}\, i^{(j)}(i - N - 1)^{(j)}$$

where A_{jN} is an arbitrary constant.

13 (*a*) Use the series expansions for $(1 + x)^p$ and $(1 + x)^m$ to derive the identity

$$\binom{m+p}{n} = \sum_{k=0}^{n} \binom{p}{k}\binom{m}{n-k} = \sum_{k=0}^{n} \binom{p}{n-k}\binom{m}{k}$$

(*b*) If we define

$$\binom{-p}{q} = \frac{(-p)^{(q)}}{q!}$$

where p and q are positive integers, show that

$$\binom{-p}{q} = (-1)^q \binom{p+q-1}{q}$$

(*c*) Use the results of parts (*a*) and (*b*) to derive the identity

$$\binom{m-p}{n} = \sum_{k=0}^{n} (-1)^k \binom{p+k-1}{k}\binom{m}{n-k}$$

$$= \sum_{k=0}^{n} (-1)^{n+k} \binom{p+n-k-1}{n-k}\binom{m}{k}$$

***14** (*a*) Use the result of Prob. 13 to show that

$$(i - N - 1)^{(j)} = (-1)^j j! \sum_{k=0}^{j} (-1)^k \binom{N-k}{j-k}\binom{i-j}{k}$$

(*b*) Derive the relation

$$(i)^{(n)}(i - n)^{(k)} = (i)^{(n+k)}$$

(*c*) Use the results of parts (*a*) and (*b*) and Prob. 12*c* to show that

$$U_j(i, N) = (-1)^j j!\, A_{jN} \sum_{k=0}^{j} \frac{(-1)^k}{k!} \binom{N-k}{j-k} i^{(j+k)}$$

(*d*) Then with $w(x) = 1$ use the definition of U_j in Prob. 12*a* to derive

$$p_j(i, N) = (-1)^j j!\, A_{jN} \sum_{k=0}^{j} (-1)^k \frac{(j+k)^{(j)}}{k!} \binom{N-k}{j-k} i^{(k)}$$

(*e*) Use this to derive

$$p_j(i, N) = B_{jN} \sum_{k=0}^{j} (-1)^k \frac{(j+k)^{(2k)}}{(k!)^2} \frac{i^{(k)}}{N^{(k)}}$$

[Ref.: Hildebrand (1974).]

15 (*a*) Use the results of Probs. 13 and 14 to show that $p_j(N, N) = 1$ if $B_{jN} = (-1)^j$.

(*b*) If $N = 2L$ and s is defined so that $i = L + s$, show that the formula derived in Prob. 14*e* becomes

$$p_j(s, 2L) = \sum_{k=0}^{j} (-1)^{k+j} \frac{(j+k)^{(2k)}}{(k!)^2} \frac{(L+s)^{(k)}}{(2L)^{(k)}}$$

(*c*) Use induction to prove that if s in $p_j(s, 2L)$ is replaced by sL, then

$$\lim_{L\to\infty} p_j(sL, 2L) = P_j(s)$$

16 (*a*) Display the equations corresponding to (6.4-23) to (6.4-26) for the Gram polynomials.

(*b*) If we define $\gamma_j(N) = \sum_{j=0}^{n} w_i[p_j(i, N)]^2$, show that

$$\gamma_j(N) = (-1)^j j!\, c_j \sum_{i=0}^{N} U_j(i+j, N)$$

where c_j is the coefficient of i^j in $p_j(i, N)$ and U_j is as defined in Prob. 12.

(*c*) Use this result and summation by parts to show that when $w(x) = 1$,

$$\sum_{i=0}^{N} [p_j(i, N)]^2 = \frac{(-1)^j(2j)!}{[j!\,N^{(j)}]^2} \sum_{i=0}^{N} (j+i)^{(j)}(j+i-N-1)^{(j)}$$

$$= \frac{1}{[N^{(j)}]^2} \sum_{i=0}^{N} (j+i)^{(2j)}$$

(*d*) Use this result to derive

$$\gamma_j = \frac{(2L+1+j)!\,(2L+j)!}{(2j+1)[(2L)!]^2}$$

(*e*) Use an argument based on symmetry to show that $\alpha_{j+1} = 0$ in the recurrence relation for the Gram polynomials corresponding to (6.4-14).

(*f*) Finally use this result to derive (6.4-29) and (6.4-30).

Section 6.5

17 Solve the system (6.3-4) with $m = 4$ for the data of Sec. 6.5 using any method and carrying (*a*) six decimal places; (*b*) eight decimal places. In both cases substitute the results back into the system of equations to see how well they satisfy the system.

18 For the data of Sec. 6.5:

(*a*) Use a Lagrangian five-point formula (see Sec. 4.1) to differentiate the data numerically at $x = .3, .4, .5, .6, .7$. In each case center the Lagrangian formula at the point at which the derivative is to be calculated.

(*b*) Calculate the derivatives at these points using the least-squares approximation (6.5-5).

(*c*) Calculate the derivatives using the true function (6.5-6). Compare these results with those of parts (*a*) and (*b*) and discuss the reasons for the errors.

19 Given the data

x_i	.4	.5	.6	.7	.8	.9	1.0
f_i	−.9435	−.9996	−.9362	−.7284	−.3517	.2164	.9998

(*a*) Find the best least-squares orthogonal polynomial approximation to these data.

(*b*) Calculate the residuals at the data points using the approximations of part (*a*).

20 (*a*) Use (6.4-28) to (6.4-30) to derive the first six Gram polynomials for the case $L = 4$.

(*b*) Use these and the data in Sec. 6.5 to derive a least-squares approximation to these data corresponding to that in Sec. 6.5.

(*c*) Compare your results with those in Sec. 6.5 and discuss any significant differences.

Section 6.6

21 (*a*) In (6.6-2) let $g(t)\,dt = ds(t)$ and so derive the Fourier-Stieltjes form of the Fourier transform.

(*b*) Let $s(t)$ be a function which is 0 except at the points $0, 1/N, \ldots, (N-1)/N$ such that $s(k/N) = g_k$. With $x = j$ show that the Fourier-Stieltjes transform of part (*a*) reduces to (6.6-1).

22 (*a*) Using the definition of w in Sec. 6.6-1, verify the orthogonality relationship (6.6-4).

(*b*) Use this orthogonality to verify Eq. (6.6-3) by substituting (6.6-1) into it.

(*c*) Prove the correctness of the linearity, shifting, and convolution formulas for the DFT.

23 (*a*) If $\delta(j) = 1$ when $j \equiv 0 \pmod N$ and 0 otherwise, show that

$$N\,\delta(j) \leftrightarrow 1 \qquad 1 \leftrightarrow \delta(k)$$

(*b*) Deduce from this that if each G_j is replaced by $G_j - c$ for some constant c, then all the g_k's are unchanged except g_0, which becomes $g_0 - c$.

(*c*) If the G_j's are observations of a random variable, express the mean of these observations in terms of the g_k's.

24 (*a*) If $G_j \leftrightarrow g_k$, show that $\tilde{G}_{-j} \leftrightarrow \tilde{g}_k$ and $\tilde{G}_j \leftrightarrow \tilde{g}_{-k}$ where the tilde ($\sim$) represents the complex conjugate.

(*b*) If $G_j \leftrightarrow g_k$ and $H_j \leftrightarrow h_k$, show that

$$\frac{1}{N}\sum_{r=0}^{N-1} G_r H_{r-j} = \frac{1}{N}\sum_{r=0}^{N-1} G_{j+r} H_r \leftrightarrow g_k h_{-k}$$

(*c*) Use these two results to prove Parseval's theorem, namely if $G_j \leftrightarrow g_k$, then

$$\frac{1}{N}\sum_{j=0}^{N-1} |G_j|^2 = \sum_{k=0}^{N-1} |g_k|^2$$

25 (*a*) Display the equations corresponding to (6.6-8) to (6.6-13) for an arbitrary value of t.

(*b*) Show that (6.6-15) is the equation corresponding to (6.6-14) for arbitrary t.

26 (*a*) Show that the number of operations in the $N = r^t$ case is minimized for fixed N if $r = 3$.

(*b*) Show that j in (6.6-21) takes on all values from 0 to $2^l - 1$ for each l.

(*c*) Suppose a table of $\sin(2\pi s/N)$, $s = 0, 1, \ldots, N/4$, is stored in a computer. Show how w^k, $k = 1, \ldots, N$, can be computed using only sums and differences of the stored quantities.

27 (*a*) Writing jk in (6.6-8) as $k(j_1 + r_1 j_2 + r_1 r_2 j_3)$, develop the analog of (6.6-13) which is the so-called *Sande-Tukey* form of the FFT.

(*b*) Similarly develop the Sande-Tukey form of the binary algorithm by displaying the equations corresponding to (6.6-19) to (6.6-26).

(*c*) For both the $t = 3$ case and the binary case display the FFT algorithm for the sine and cosine transform cases, (6.6-27) and (6.6-28).

28 (*a*) Let $G_j = 1$, $j = 0, 1, 2, 3$. Calculate directly

$$H_k = \sum_{j=0}^{3-k} G_j G_{j+k} \qquad k = 0, 1, 2, 3$$

(*b*) Define $G_j = 0$, $j = 4, 5, 6, 7$, and compute g_k using the FFT algorithm.

(*c*) Compute $|g_k|^2$ and then use the FFT algorithm to compute

$$H_j = 8\sum_{k=0}^{7} |g_k|^2 w^{jk}$$

(*d*) Compare these results with those of part (*a*) and use the results of Prob. 24*b* to explain the comparison.
[Ref.: Cooley, Lewis, and Welch (1977).]

29 (*a*) With $x_i = 2\pi i/(2L+1)$ show that

$$\cos kx_i = \cos (2L+1-k)x_i \qquad \text{and} \qquad \sin kx_i = -\sin (2L+1-k)x_i$$

(*b*) Thus deduce that for any $k > L$, $\cos kx_i$ and $\sin kx_i$, $i = 0, \ldots, 2L$, can be expressed in terms of, respectively, $\cos jx_i$ and $\sin jx_i$ for $j \le L$.

30 (*a*) Derive the result

$$\sum_{k=0}^{2L} e^{ik\alpha} = \begin{cases} e^{i\alpha L} \dfrac{\sin (L+\frac{1}{2})\alpha}{\sin \alpha/2} & \alpha \ne 2\pi v \\ 2L+1 & \alpha = 2\pi v \end{cases}$$

where v is any integer.

(*b*) Deduce from this that

$$\sum_{k=0}^{2L} \cos k\alpha = \begin{cases} \dfrac{\cos L\alpha \sin (L+\frac{1}{2})\alpha}{\sin \alpha/2} & \alpha \ne 2\pi v \\ 2L+1 & \alpha = 2\pi v \end{cases}$$

$$\sum_{k=0}^{2L} \sin k\alpha = \frac{\sin L\alpha \sin (L+\frac{1}{2})\alpha}{\sin \alpha/2}$$

(*c*) Then letting $\alpha = 2\pi j/(2L+1)$, deduce that

$$\sum_{k=0}^{2L} \cos jx_k = \begin{cases} 0 & j \ne v(2L+1) \\ 2L+1 & j = v(2L+1) \end{cases} \qquad \text{and} \qquad \sum_{k=0}^{2L} \sin jx_k = 0$$

where $x_k = 2\pi k/(2L+1)$.

(*d*) Use these results and the identities for the products of two sines, two cosines, and one sine and one cosine to derive (6.6-29) to (6.6-31) and (6.6-35).

31 (*a*) Derive the relations corresponding to (6.6-29) to (6.6-31) when the number of points n is even.

(*b*) Derive the relations corresponding to (6.6-33), (6.6-34), and (6.6-35) when n is even.

(*c*) When n is even consider replacing the L in (6.6-36) by $L-1$ and adding a term $\frac{1}{2}a_L \cos Lx$. Find an expression for a_L analogous to (6.6-33). Why isn't it reasonable to add a term in $\sin Lx$?

***32** (*a*) With the definition

$$V_k(x) = \sum_{i=k}^{2L} \bar{f}_i \sin (i-k+1)x \qquad k = 1, .., 2L$$

$$V_{2L+1}(x) = V_{2L+2}(x) = 0$$

derive the recurrence relation

$$\bar{f}_k \sin x + 2 \cos x \, V_{k+1}(x) - V_{k+2}(x) = V_k(x)$$

(*b*) Defining $U_{kj} \sin x_j = V_k(x_j)$, find the analogous recurrence for U_{kj}.

(*c*) Then show that a_j and b_j in (6.6-33) and (6.6-34) are given by

$$a_j = \left(\frac{2}{2L+1}\right)\left(\bar{f}_0 + U_{1j}\cos\frac{2\pi j}{2L+1} - U_{2j}\right)$$

$$b_j = \left(\frac{2}{2L+1}\right)U_{1j}\sin\frac{2\pi j}{2L+1}$$

[Ref.: Goertzel (1958).]

***33** (*a*) Verify (6.6-40) and from this deduce (6.6-41). (*b*) Derive (6.6-43) and thus deduce (6.6-44). (*c*) Derive the analog of (6.6-44) when the number of points is even. [Ref.: Hamming (1973), pp. 516–518.]

***34** (*a*) Show that the error $\epsilon_L = f(x) - y_L(x)$ in the approximation (6.6-44) can be expressed in the form

$$\epsilon_L = \frac{1}{2L+1}\sum_{i=0}^{2L}\frac{\sin(L+\frac{1}{2})(x_i - x)}{\sin\frac{1}{2}(x_i - x)}[f(x) - f_i]$$

(*b*) Deduce from this that, particularly if $f(x)$ does not change rapidly, the major contribution to the error comes from the term or terms with x_i nearest to x. [Ref.: Hamming (1973), pp. 516–518.]

35 (*a*) Suppose $f(x)$ is a periodic function of period 2π. Show that the coefficients of the discrete Fourier expansion, (6.6-33) and (6.6-34), are what would be obtained if the trapezoidal rule were used to approximate the coefficients of the continuous Fourier series for $f(x)$ on $[0, 2\pi]$.

(*b*) Show that for any periodic function the trapezoidal-rule correction terms in the Euler-Maclaurin sum formula drop out if the interval is a multiple of the period. Does this indicate that the trapezoidal rule is exact for periodic functions? Why? [Ref.: Hildebrand (1974), p. 454.]

36 Given the data

x_i	0	$2\pi/7$	$4\pi/7$	$6\pi/7$	$8\pi/7$	$10\pi/7$	$12\pi/7$
f_i	1.0004	−.1190	1.5987	.2115	−.6567	−.3514	−1.6824

(*a*) Find the best least-squares Fourier approximation to these data. (*b*) Calculate the residuals at the data points. (*c*) With $m = 3$ calculate δ_m^2 using both forms of (6.6-35), and discuss the difference.

37 Let $f(x)$ be a function whose discrete Fourier series $y_L(x)$ corresponding to a set of values $\{\bar{f}_i\}$, $i = 0, \ldots, 2L$, is given by (6.6-36). Let the continuous Fourier-series expansion of the observed function $\bar{f}(x)$ on $[0, 2\pi]$ be given by

$$\bar{f}(x) = \tfrac{1}{2}A_0 + \sum_{j=1}^{\infty}(A_j\cos jx + B_j\sin jx)$$

(*a*) By summing both sides of the above equation for $x = x_i$, $i = 0, \ldots, 2L$, and using (6.6-33), prove that

$$a_0 = A_0 + 2\sum_{j=1}^{\infty}A_{(2L+1)j}$$

(*b*) Now by multiplying the continuous Fourier series by, respectively, cos kx and sin kx and summing as above prove that

$$a_k = A_k + \sum_{j=1}^{\infty} (A_{(2L+1)j-k} + A_{(2L+1)j+k}) \qquad k > 0$$

$$b_k = B_k + \sum_{j=1}^{\infty} (B_{(2L+1)j+k} - B_{(2L+1)j-k}) \qquad k > 0$$

The importance of this result is that when the function $f(x)$ is *sampled* at the equally spaced points $\{x_i\}$, the calculated discrete Fourier series includes in its coefficients the effects of higher frequencies than the *sampling rate*. This "folding" back of the higher frequencies on the lower ones is often called *aliasing*. [Ref.: Hamming (1973), pp. 505–508.]

***38** A function $f(x)$ is called band-limited if its Fourier transform

$$F(\lambda) = \int_{-\infty}^{\infty} f(t)e^{-2\pi i\lambda t}\, dt$$

vanishes outside the open interval $(-\Omega, \Omega)$.

(*a*) Show that, if $f(x)$ is band-limited, then

$$F(\lambda) = F_1(\lambda)P(\lambda)$$

where $F_1(\lambda)$ is periodic with period Ω and

$$P(\lambda) = \begin{cases} 1 & |\lambda| < \Omega \\ 0 & |\lambda| \geq \Omega \end{cases}$$

(*b*) Show that

$$F_1(\lambda) = \sum_{k=-\infty}^{\infty} c_k e^{(i\pi/\Omega)k\lambda} \qquad \text{where } c_k = \frac{1}{2\Omega} f\left(\frac{-k}{2\Omega}\right)$$

and thus deduce that

$$F(\lambda) = \sum_{k=-\infty}^{\infty} f\left(\frac{k}{2\Omega}\right) \frac{P(\lambda)}{2\Omega} e^{-(i\pi/\Omega)k\lambda}$$

(*c*) Show that $P(\lambda)$ is the Fourier transform of

$$p(x) = \frac{\sin 2\pi\Omega x}{\pi x}$$

(*d*) Use part (*c*) to take the inverse transform of $F(\lambda)$ in part (*b*) and thus show that

$$f(x) = \sum_{k=-\infty}^{\infty} f\left(\frac{k}{2\Omega}\right) \frac{\sin \pi(2\Omega x - k)}{\pi(2\Omega x - k)}$$

This result is known as the *sampling theorem* and is due to Shannon (1949). It implies that if $f(x)$ is sampled at equal intervals with a frequency greater than the bandwidth Ω of the function, the entire function can be reconstructed from this infinite set of samples. When we approximate the infinite sum above by a finite sum, we have the analog for nonperiodic functions to the approximation (6.6-44) for periodic functions. [See Hamming (1973), pp. 557–559, for a good discussion of the sampling theorem and its importance.]

CHAPTER

SEVEN

FUNCTIONAL APPROXIMATION: MINIMUM MAXIMUM ERROR TECHNIQUES

7.1 GENERAL REMARKS

One way of evaluating a mathematical function—trigonometric, logarithmic, exponential, Bessel, etc.—on a digital computer would be to store a table of the function in the memory of the computer and to use an interpolation formula to evaluate the function at nontabulated points. Not only is this technique extremely wasteful of the memory of the computer, but also it generally has no advantages in speed or accuracy over the techniques to be discussed in this chapter. These techniques all involve approximating a function $f(x)$ by a rational function.† We noted in Chap. 2 that a rational function is the most general function of a variable x that can be evaluated directly on a digital computer. But why use rational functions rather than the more familiar polynomial approximations (which are, of course, just a special case of rational functions)? To answer this we need to consider our aims in approximating functions on a computer.

† We can, however, use a different rational function on different intervals.

The general situation is this: a computation is to be performed in which a certain mathematical function is to be evaluated many (perhaps millions of) times. It is known a priori that the arguments of this function will be in some interval (perhaps infinite), but it is not known a priori what the arguments will be. Thus the function must be approximated over the entire interval. The property of this approximation of most importance to us is the error between it and the true function, an error which will vary over the interval.

Generally, in the overall computation, it is desirable to be able to bound the error in the result. To bound this error without a priori knowledge of what the numbers involved will be, we must consider the worst possible case. Thus, in an approximation to a function to be used in such a computation, the property of the error that is of most importance is the maximum relative or absolute error (in magnitude) on the interval. Therefore, the major aim of a computer approximation to a function is to make the maximum error as small as possible, or, to use terminology introduced previously, we are interested in minimizing the L_∞ or uniform norm which is sometimes also called the Chebyshev norm. In the last section of this chapter we shall develop techniques for generating a rational approximation to a function which, among all rational approximations with the same degree polynomial in numerator and denominator, has the *minimum maximum error*. We shall call such an approximation the *Chebyshev* or, more often, the *minimax approximation*. Before this, however, we shall develop techniques for generating good, if not minimax, approximations.

If an approximation is to be evaluated millions of times, another aim of a computer approximation must certainly be to achieve maximum speed. We shall estimate the speed with which an approximation can be evaluated by considering the number of multiplications and divisions required. We shall assume for convenience that multiplication and division are equally time-consuming although on some computers division is more time-consuming than multiplication. We should also note that floating-point addition and subtraction are sometimes as (or almost as) time-consuming as multiplication and division. Nevertheless, our conclusions based only on a consideration of multiplications and divisions will be generally valid.

Our reason for preferring rational approximations to polynomial approximations is quite simple. For a given amount of computation, rational approximations lead to smaller maximum errors than polynomial approximations for the functions most commonly approximated on a digital computer. This assertion will be illustrated empirically by examples later in this chapter. It implies the need to compare the computation required to evaluate one rational function with that required to evaluate another rational function or a polynomial. Since the means by which rational functions are evaluated is of both interest and significance, we shall, in the next section, consider the computational aspects of the evaluation of rational approximations.

7.2 RATIONAL FUNCTIONS, POLYNOMIALS, AND CONTINUED FRACTIONS

Let

$$R_{mk}(x) = \frac{P_m(x)}{Q_k(x)} \tag{7.2-1}$$

be a rational approximation to a function $f(x)$, where $P_m(x)$ and $Q_k(x)$ are polynomials of degree at most m and k, respectively. We shall call $N = m + k$ the *index* of $R_{mk}(x)$. The number of coefficients at our disposal in $R_{mk}(x)$ is $N + 1$ since one of the $N + 2$ coefficients in the numerator and denominator may be chosen arbitrarily. In general it is true that the greater the index, the higher the accuracy of the approximation. Moreover, for a particular function over intervals of interest to us, all approximations with the same index, in contrast to approximation with other indices, require similar amounts of computation and achieve similar accuracy. In this section we shall consider the computational aspect and not the accuracy of rational approximations.

Let us consider first the evaluation of the polynomial

$$p_n(x) = x^n + a_{n-1}x^{n-1} + \cdots + a_1 x + a_0 \tag{7.2-2}$$

To evaluate $p_n(x)$ by computing all the powers of x and then multiplying by the coefficients and adding would require $2n - 2$ multiplications. A much better technique is to write

$$p_n(x) = x(x(\cdots (x(x + a_{n-1}) + a_{n-2}) + \cdots + a_2) + a_1) + a_0 \tag{7.2-3}$$

and then use Horner's rule

$$s_k = a_{k+1} + xs_{k+1} \qquad k = n-2, \ldots, 0, -1 \qquad s_{n-1} = 1 \tag{7.2-4}$$

from which it is easily verified that $s_{-1} = p_n(x)$. The number of multiplications required using (7.2-4) is $n - 1$, and the number of additions is n. When the polynomial is not going to be evaluated many times, the use of Eq. (7.2-4) is the recommended way to evaluate polynomials. But if $p_n(x)$ is part of a rational approximation to a function to be used in a computer subroutine or, in any case, is going to be evaluated very many times, more efficient techniques are worth searching for. We shall therefore develop an algorithm which generally results in a better computational procedure than the use of (7.2-4).

Our method will involve making a change of variable, whose purpose will become clear below, $y = x + \delta$, which converts the polynomial $p_n(x)$ to

$$q_n(y) = y^n + b_{n-1}y^{n-1} + b_{n-2}y^{n-2} + \cdots + b_1 y + b_0 \tag{7.2-5}$$

If the polynomial $q_n(y)$ is divided by $y^2 - \alpha_1$, the result is

$$q_n(y) = (y^2 - \alpha_1)(y^{n-2} + c_{n-3}y^{n-3} + c_{n-4}y^{n-4} + \cdots + c_1 y + c_0) + \gamma_1 y + \beta_1 \tag{7.2-6}$$

where

$$c_j = b_{j+2} + \alpha_1 c_{j+2} \qquad j = n-3, n-4, \ldots, 0, -1, -2$$
$$c_{n-2} = 1; \; c_{n-1} = 0 \tag{7.2-7}$$

with $c_{-1} = \gamma_1$, $c_{-2} = \beta_1$. Our interest here is in γ_1. Using (7.2-7) we write {1}

$$\begin{aligned}\gamma_1 = c_{-1} &= b_1 + \alpha_1 c_1 = b_1 + \alpha_1(b_3 + \alpha_1 c_3)\\ &= b_1 + \alpha_1 b_3 + \alpha_1^2(b_5 + \alpha_1 c_5)\\ &= b_1 + \alpha_1 b_3 + \alpha_1^2 b_5 + \cdots + \alpha_1^{r-1}b_{2r-1} + \alpha_1^r b_{2r+1}\end{aligned} \tag{7.2-8}$$

where $b_n = 1$ and $2r = n - 2$ if n is even and $n - 1$ if n is odd. Setting $\gamma_1 = 0$ in (7.2-8) gives us a polynomial equation for α_1. Suppose this equation has a real root. Let α_1 in (7.2-6) be this real root. Then

$$q_n(y) = (y^2 - \alpha_1)(y^{n-2} + c_{n-3}y^{n-3} + \cdots + c_1 y + c_0) + \beta_1 \tag{7.2-9}$$

Before proceeding further we return to (7.2-8) and write the auxiliary polynomial for α_1 in the general form

$$u_r(y) = b_1 + b_3 y + \cdots + b_{2r-1}y^{r-1} + b_{2r+1}y^r \tag{7.2-10}$$

If α_1 is a root of this polynomial, we can write

$$u_r(y) = (y - \alpha_1)(c'_{2r-1}y^{r-1} + c'_{2r-3}y^{r-2} + c'_{2r-5}y^{r-3} + \cdots + c'_3 y + c'_1) \tag{7.2-11}$$

where, using (7.2-10) and (7.2-11),

$$c'_{2j-1} = b_{2j+1} + \alpha_1 c'_{2j+1} \qquad j = r, \ldots, 1$$
$$c'_{2r+1} = 0 \tag{7.2-12}$$

Comparing this equation with (7.2-7), we see that $c'_{2j-1} = c_{2j-1}$ for all j. Now consider the polynomial of degree $n - 2$ in parentheses in (7.2-9). Dividing this polynomial by $y^2 - \alpha_2$ and proceeding as above, we get

$$q_n(y) = (y^2 - \alpha_1)[(y^2 - \alpha_2)(y^{n-4} + d_{n-5}y^{n-5} + d_{n-6}y^{n-6} + \cdots + d_0) + \beta_2] + \beta_1 \tag{7.2-13}$$

where the polynomial corresponding to that in (7.2-8) is

$$\gamma_2 = c_1 + \alpha_2 c_3 + \cdots + \alpha_2^{r-2}c_{2r-3} + \alpha_2^{r-1}c_{2r-1} \tag{7.2-14}$$

so that α_2 must be a root of the polynomial in parentheses in (7.2-11), which is to say that it, like α_1, must be a root of $u_r(y)$. Continuing in this way, if the polynomial corresponding to that in (7.2-8) has a real root at every stage, we get finally

$$q_n(y) = (\cdots\{[v(y^2 - \alpha_r) + \beta_r](y^2 - \alpha_{r-1}) + \beta_{r-1}\}\cdots)(y^2 - \alpha_1) + \beta_1 \tag{7.2-15}$$

where $$v = \begin{cases} y(y + \beta_{r+1}) + \alpha_{r+1} & n \text{ even} \\ y + \alpha_{r+1} & n \text{ odd} \end{cases}$$

results from the division of a quartic or cubic by $y^2 - \alpha_r$. The $\alpha_j, j = 1, \ldots, r$, are the roots of $u_r(y)$, which we have thus far assumed to be all real. Note also our implicit assumption that b_{2r+1} in (7.2-10) is nonzero [it cannot be zero if n is odd (why?)] for if not, $u_r(x)$ would not be a polynomial of degree r and therefore could not have r roots. If $b_{2r+1} = 0$, the process leading to (7.2-15) will fail at some point; i.e., no α_j will exist such that $\gamma_j = 0$.

It is an interesting fact, which we leave to a problem {2}, that the roots of $u_r(y)$ will all be real if at least $n - 1$ of the roots of $q_n(y)$ have real parts which are all nonnegative or all nonpositive. We have then the following procedure for determining the change of variable constant δ introduced above in order to assure that $q_n(y)$ satisfies the condition stated above. First determine the zeros of $p_n(x)$. Three cases need to be distinguished:

1. The zero with largest or smallest real part is $a + bi$, with $b \neq 0$. Then let $\delta = -a$. For if so, all zeros of $q_n(y)$ have nonnegative or all have nonpositive real parts. In particular $q_n(y)$ will have two zeros bi and $-bi$, so that with $\alpha_1 = -b^2$, $y^2 - \alpha_1$ is a factor of $q_n(y)$ and $\beta_1 = 0$ in (7.2-6).
2. The zero with largest or smallest real part is real, but the next pair are nonreal $a \pm bi$. Then we proceed just as in case 1 since only $n - 1$ of the zeros are required to all have nonpositive or all have nonnegative real parts.
3. The two zeros with largest or smallest real parts are real. We express them as $a + b$ and $a - b$ and let $\delta = -a$. Then $q_n(y)$ has zeros $+b$ and $-b$. With $\alpha_1 = b^2$ again we have $\beta_1 = 0$.

Note that almost always there are two choices of δ, one for the roots with positive real parts and one for the roots with negative real parts. In all three cases, therefore, we not only achieve a polynomial $u_r(y)$ with all real zeros but have the additional bonus that $\beta_1 = 0$. In case 2 above it could turn out that, with the choice of δ above and n even, $b_{2r+1} = b_{n-1} = 0$. But even this case can be finessed; we leave the details to a problem {3}.

Finally we note that the order of the α_j, $j = 2, \ldots, r$, is immaterial, so that any of the $(r - 1)!$ orderings can be chosen. This is important because the computational properties, i.e., the numerical accuracy, of (7.2-15) for the spectrum of necessary values of y may depend on the order in which the α_j, $j = 2, \ldots, r$ are chosen.

To summarize, our algorithm to evaluate $p_n(x)$ is as follows:

Input

$x, \delta, n \ (\geq 3)$

$\alpha_j, j = 1, \ldots, r+1$

$$\beta_j, j = 2, \ldots, \begin{cases} r+1 & n \text{ even} \\ r & n \text{ odd} \end{cases} \qquad (\beta_1 = 0)$$

Algorithm

$y \leftarrow x + \delta$

$z \leftarrow y^2$

if n even **then** $w \leftarrow y(y + \beta_{r+1}) + \alpha_{r+1}$

else $w \leftarrow y + \alpha_{r+1}$

endif

for $i = r, \ldots, 2$ **do** $w \leftarrow (z - \alpha_j)w + \beta_j$

$w \leftarrow (z - \alpha_1)w$

Output

$w \ (= p_n(x))$

The total number of operations in this algorithm is easily seen to be:

Multiplications: $\quad r + 2 = \dfrac{n}{2} + 1 \qquad n$ even

$\qquad r + 1 = \dfrac{n+1}{2} \qquad n$ odd

Additions: $\quad 2r + 2 = n \qquad n$ even

$\qquad 2r + 1 = n \qquad n$ odd

This can be summarized for any n as $\lfloor n/2 \rfloor + 1$ multiplications and n additions. Since it can be proved that any algorithm must require at least $\lfloor n/2 \rfloor$ multiplications and at least n additions, the algorithm above is very nearly optimal. Indeed, for $n \leq 7$ it is known that no algorithm can achieve both the $\lfloor n/2 \rfloor$ multiplication and n addition lower bounds.

The advantage of the quadratic-factor algorithm over Horner's rule as regards multiplication is illustrated in the following table.

Degree	Quadratic-factor algorithm	Horner's rule	Degree	Quadratic-factor algorithm	Horner's rule
3	2	2	7	4	6
4	3	3	8	5	7
5	3	4	9	5	8
6	4	5	10	6	9

The analytic effort required to express $p_n(x)$ in the form (7.2-15) becomes considerable as n increases. But it is clearly worthwhile when the polynomial in question is part of an approximation to a function which, once found, can be used forever.

This quadratic-factor algorithm by no means gives the best possible result. For cases when n is even, it is often (always when $n = 4$) possible by a simple trick to reduce the number of multiplications by 1 {4}. More generally, it is known, for example, that it is possible to evaluate any polynomial of sixth degree with three multiplications {6}. However, the quadratic-factor algorithm is nearly minimal for modest values of n and, for all but very small values of n, a distinct improvement over Horner's rule.

Example 7.1 Apply the above technique to the polynomial

$$p_5(x) = x^5 + 2x^4 - 9x^3 - 12x^2 + 38x - 20$$

The zeros of $p_5(x)$ are $+1, +1, +2, -3 + i, -3 - i$. Therefore, we choose $\delta = +3$ and make the change of variable $y = x + 3$ to obtain

$$q_5(y) = y^5 - 13y^4 + 57y^3 - 93y^2 + 56y - 80$$

and, since $r = (n - 1)/2 = 2$, we have

$$u_2(y) = 56 + 57y + y^2 = (y + 56)(y + 1)$$

from which we obtain

$$q_5(y) = [(y - 13)(y^2 + 56) + 648](y^2 + 1)$$

with β_1 in (7.2-15) equal to 0. Note that the relatively large value of $\beta_2 = 648$ should give rise to some disquiet concerning the roundoff error which may be incurred in evaluating $q_5(y)$.

In order to evaluate $R_{mk}(x)$, then, one approach would be to evaluate $P_m(x)$ and $Q_k(x)$ as described above and then to take their quotient. Another approach is to write $R_{mk}(x)$ in the form of a continued fraction. To convert a rational function to a continued fraction, we perform a series of divisions and reciprocations. For example,

$$\frac{2x^3 + x^2 + x + 3}{x^2 - x + 4} = 2x + 3 - \frac{4x + 9}{x^2 - x + 4}$$

$$= 2x + 3 - \cfrac{4}{\cfrac{x^2 - x + 4}{x + \frac{9}{4}}}$$

$$= 2x + 3 - \cfrac{4}{x - \frac{13}{4} + \cfrac{\frac{181}{16}}{x + \frac{9}{4}}}$$

An algorithm for performing this conversion in general is considered in {8}. Here we consider explicitly only the cases $m = k$ or $k + 1$ and leave the other cases to a problem {9}. For these cases, the continued-fraction form of $R_{mk}(x)$ is (except in certain degenerate cases {8})

$$R_{mk}(x) = C_0 x + D_0 + \cfrac{C_1}{x + D_1 + \cfrac{C_2}{x + D_2 + \cfrac{C_3}{x + D_3 + \ddots + \cfrac{C_k}{x + D_k}}}} \tag{7.2-16}$$

$$= C_0 x + D_0 + \frac{C_1|}{|x + D_1} + \frac{C_2|}{|x + D_2} + \cdots + \frac{C_k|}{|x + D_k}$$

where $C_0 = 0$ when $m = k$. To evaluate the continued fraction (7.2-16) we calculate

$$d_j = \frac{C_j}{x + D_j + d_{j+1}} \qquad \begin{array}{l} j = k, \ldots, 1 \\ d_{k+1} = 0 \end{array} \tag{7.2-17}$$

from which it follows that $R_{mk}(x) = C_0 x + D_0 + d_1$. The computation of (7.2-16) requires k divisions and, if $m = k + 1$, one multiplication. In the following table we compare the number of multiplications and divisions required to evaluate $R_{mk}(x)$ first by evaluating the numerator and denominator polynomials using the quadratic-factor algorithm discussed above and second by using continued fractions.†

† Note that in evaluating the two polynomials and then dividing, we can assume that one polynomial is in the form (7.2-2), but the other will in general have a coefficient multiplying the highest power of x which adds one multiplication to the total. Note also that the calculation of $z = y^2$ need be performed only once if the same δ is used in the numerator and denominator.

	Polynomial evaluations		Continued fraction	
(m, k)	Mult.	Div.	Mult.	Div.
(3, 3)	4	1	0	3
(4, 3)	5	1	1	3
(4, 4)	6	1	0	4
(5, 4)	6	1	1	4
(5, 5)	6	1	0	5
(6, 5)	7	1	1	5
(6, 6)	8	1	0	6

Note that when m or $k = 4$, the number of multiplications in the second column can be reduced by 1 since any monic quartic can be evaluated with two multiplications {4}.

Thus, we see that if division and multiplication are equally time-consuming, the continued-fraction approach is superior using this comparison, but if multiplication is somewhat faster than division, this may not be true. The number of additions and subtractions required by the two techniques in the same.

We considered the continued-fraction formulation with $m = k$ or $k + 1$ because these generally give the most accurate approximations among all rational approximations with the same index. In conclusion it is only fair to point out that the continued-fraction approach, like the quadratic-factor algorithm, may lead to certain computational difficulties, the main one being loss of significance due to the subtraction of nearly equal quantities. This difficulty can often be overcome quite easily {12, 13}, however.

7.3 PADÉ APPROXIMATIONS

Our first approach toward generating approximations of the form (7.2-1) will be, for a given m and k, to choose $P_m(x)$ and $Q_k(x)$ so that $f(x)$ and $P_m(x)/Q_k(x)$ are equal at $x = 0$ and have as many derivatives as possible equal at $x = 0$. In the case $k = 0$, the approximation is then just the Maclaurin expansion for $f(x)$. Implicit in what follows will be the assumption that the Maclaurin series for $f(x)$ exists in some neighborhood of $x = 0$. There are two reasons for the arbitrary choice of $x = 0$: (1) it makes the manipulations below substantially simpler than for any other x, and (2) the interval over which we wish to approximate most functions will contain 0, and when it does not, a simple change of variable can be used to make the interval contain 0. We also assume that $P_m(x)$ and $Q_k(x)$ have no common factors. Now let

$$P_m(x) = \sum_{j=0}^{m} a_j x^j \qquad Q_k(x) = \sum_{j=0}^{k} b_j x^j \qquad b_0 = 1 \tag{7.3-1}$$

It is permissible to let the constant term in $Q_k(x)$ equal 1 because (1) the constant term cannot be 0 if the approximation is to exist at $x = 0$ and (2) the value of $R_{mk}(x)$ is unchanged if numerator and denominator are divided by the same constant.

Now let $f(x)$ have a Maclaurin series

$$f(x) = \sum_{j=0}^{\infty} c_j x^j \tag{7.3-2}$$

Then we consider the difference

$$f(x) - \frac{P_m(x)}{Q_k(x)} = \frac{\left(\sum_{j=0}^{\infty} c_j x^j\right)\left(\sum_{j=0}^{k} b_j x^j\right) - \sum_{j=0}^{m} a_j x^j}{\sum_{j=0}^{k} b_j x^j} \tag{7.3-3}$$

Since we have $N + 1$ constants ($m + 1$ a_j's and k b_j's) at our disposal, we would hope to make $f(x) - R_{mk}(x)$ and its first N derivatives equal to 0 at $x = 0$. We shall achieve this if the numerator of the right-hand side of (7.3-3) is such that its leading power is of degree $N + 1$ (why?). Thus we write

$$\left(\sum_{j=0}^{\infty} c_j x^j\right)\left(\sum_{j=0}^{k} b_j x^j\right) - \sum_{j=0}^{m} a_j x^j = \sum_{j=N+1}^{\infty} d_j x^j \tag{7.3-4}$$

The vanishing of the coefficients of the first $N + 1$ powers of x on the left-hand side of (7.3-4) is equivalent to the equations {10}

$$\begin{aligned} &\sum_{j=0}^{k} c_{N-s-j} b_j = 0 && s = 0, 1, \ldots, N - m - 1 \\ & && c_i = 0 \text{ if } i < 0,\ b_0 = 1 \\ &a_r = \sum_{j=0}^{r} c_{r-j} b_j && r = 0, 1, \ldots, m \\ & && b_j = 0 \text{ if } j > k \end{aligned} \tag{7.3-5}$$

When this set of $N + 1$ linear equations in the $N + 1$ unknowns has a solution, it provides us with the desired approximation of the form (7.2-1) {10}. It should also be noted that Padé approximations for a given value of N can be computed via recurrence relations from Padé approximations for smaller values of N {11} and techniques also exist for computing all approximations for a particular N from $R_{N0}(x)$, also via recurrence relations.

One disadvantage of this derivation is that it does not provide us with an error term in closed form. Such an error term can be derived using other techniques {14 to 19}. But since our emphasis here is on finding the maximum error on an interval, error terms which contain a derivative of the function which we can bound or estimate but not evaluate are not of basic interest to us. Rather we must be able to find the error at any point in the interval, and this we shall do by actual evaluation of the approximation and comparison with the true function.

Estimates of the error are useful, however, in indicating how good a given approximation is likely to be in the minimum maximum error context. For the case of approximations as we have derived them in this section, it is often true that the coefficients d_i, $i = N + 1, \ldots,$ and b_i, $i = 1, 2, \ldots, k$, in (7.3-4) decrease very rapidly in magnitude. Thus a good estimate of the error in (7.2-1) may be given by the first term on the right-hand side of (7.3-4), which is $d_{N+1}x^{N+1}$, with d_{N+1} given by

$$d_{N+1} = \sum_{j=0}^{k} c_{N+1-j} b_j \tag{7.3-6}$$

This approximation to the error illustrates the fact that, in common with Maclaurin-series approximations, rational approximations of the kind we have developed have errors which are small near 0 and increase away from 0. This is just what we would expect because of our requirement that the approximation agree with $f(x)$ and its first N derivatives at $x = 0$. This means that in practice if $x = 0$ is not the center of the interval over which we are approximating, we should make a change of variable so that 0 becomes the center of the interval.

The development of approximations of the form (7.2-1) by the method of this section is due to the French mathematician Padé. The approximation $R_{mk}(x)$ is called the (m, k) entry in the Padé table of $f(x)$. We note the important empirical fact that, for most functions for which approximations are desired on computers, among all the entries in the Padé table, those for $m = k$ or $m = k + 1$ give the smallest minimum maximum error for a given N.

7.4 AN EXAMPLE

Let us consider the problem of approximating e^x on the interval $(-\infty, \infty)$.† The first thing we must consider is the infinite interval. We cannot expect to approximate a function over an infinite interval with good error behavior

† We shall use approximations to the exponential function for illustrative purposes throughout this chapter because they demonstrate the various approximation methods nicely and are relatively easy to derive. But we should note that the unique property of the exponential function that $e^{-x} = 1/e^x$ does in some cases lead to approximations which have this property and which are therefore atypical.

over the whole interval. Our first step then will be to convert the problem to that of approximating e^x on some finite interval.

We suppose that the approximation is going to be used on a digital computer which is binary internally.

For any x in $(-\infty, \infty)$ we write

$$x \log_2 e = X + F \tag{7.4-1}$$

where X is an integer and F a fraction such that $-1 < F < 1$.† From (7.4-1) we have

$$2^{x \log_2 e} = e^x = 2^X \times 2^F = 2^X \times e^{F \ln 2} \tag{7.4-2}$$

Since X is an integer, on a binary computer multiplication by 2^X represents a shift of a fixed-point number or a change in the exponent of a floating-point number, both of which are very simple operations. Thus we are left with the problem of approximating the exponential over the interval $(-\ln 2, \ln 2) \approx (-.69, .69)$.

Thus we have reduced our original problem to that of finding an approximation to e^x on the interval $(-.69, .69)$. For illustrative purposes, we consider the case $N = 4$. For $m = k = 2$ we shall now perform the calculations of Sec. 7.3 in detail. From the Maclaurin expansion for e^x, we have $c_0 = 1$, $c_1 = 1$, $c_2 = \frac{1}{2}$, $c_3 = \frac{1}{6}$, $c_4 = \frac{1}{24}$. With these values Eqs. (7.3-5) become

$$\frac{1}{24} + \frac{1}{6}b_1 + \frac{1}{2}b_2 = 0 \qquad \frac{1}{6} + \frac{1}{2}b_1 + b_2 = 0$$

$$a_0 = 1 \qquad a_1 = 1 + b_1 \qquad a_2 = \tfrac{1}{2} + b_1 + b_2 \tag{7.4-3}$$

Solving the first two equations for b_1 and b_2 and then using the last three to calculate a_0, a_1, and a_2, we find

$$b_1 = -\tfrac{1}{2} \qquad b_2 = \tfrac{1}{12} \qquad a_0 = 1 \qquad a_1 = \tfrac{1}{2} \qquad a_2 = \tfrac{1}{12}$$

so that

$$R_{2,2}(x) = \frac{12 + 6x + x^2}{12 - 6x + x^2} \tag{7.4-4}$$

Similarly we can calculate the other entries of the Padé table for $N = 4$ {20}:

$$R_{4,0}(x) = 1 + x + \tfrac{1}{2}x^2 + \tfrac{1}{6}x^3 + \tfrac{1}{24}x^4$$

$$R_{3,1}(x) = \frac{24 + 18x + 6x^2 + x^3}{24 - 6x}$$

$$R_{1,3}(x) = \frac{24 + 6x}{24 - 18x + 6x^2 - x^3}$$

$$R_{0,4}(x) = \frac{1}{1 - x + \frac{1}{2}x^2 - \frac{1}{6}x^3 + \frac{1}{24}x^4} \tag{7.4-5}$$

† Actually it would be easy to adjust X so that $-\frac{1}{2} < F < \frac{1}{2}$. We do not do so in order to keep the example in this section as consistent as possible with examples in subsequent sections of this chapter.

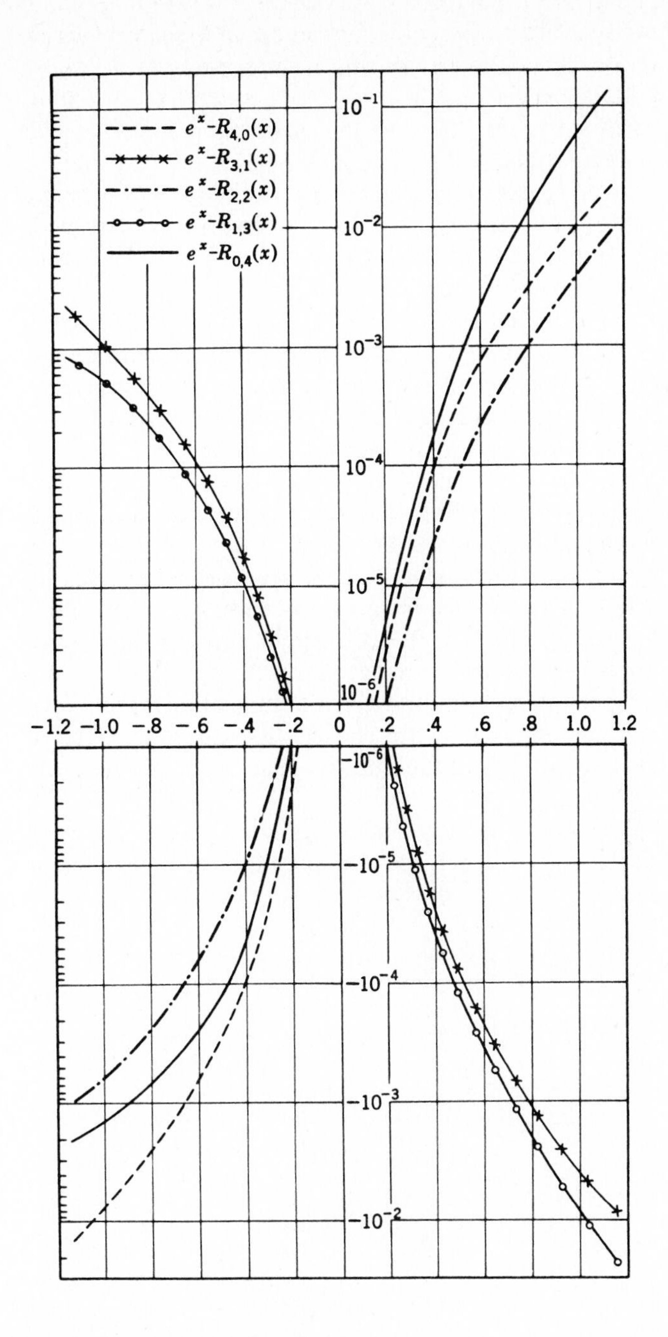

Figure 7.1 Errors in Padé approximations to e^x.

The approximation $R_{m,0}(x)$ is always just the truncated Maclaurin-series expansion of the function. Note also that in this example we have $R_{mk}(x) = 1/R_{km}(-x)$ (why would we expect this?).

Our interest is in the errors in these various approximations over the interval $[-.69, .69]$. In Fig. 7.1 we have plotted $e^x - R_{mk}(x)$, and in Table 7.1 we have listed some sample values of this error. We note first that over the range $[-.69, .69]$ $R_{2,2}(x)$ has the minimum maximum error, consistent with our previous assertion that approximations with $m = k$ or $k + 1$ would generally be best. However, over the larger interval $[-1.0, 1.0]$ shown in Table 7.1 $R_{3,1}(x)$ is the best approximation in a minimum maximum error sense. Note also that as the ends of the interval are approached, the errors grow very rapidly. This suggests that where possible we shall approximate functions over quite small intervals. Indeed our assertion about approximations with $m = k$ or $k + 1$ was made with small intervals in mind.

In general we have two alternatives when an approximation for a given N is not sufficiently good over the whole interval: (1) break up the interval into two or more subintervals and use an appropriate approximation over each subinterval or (2) use a larger value of N to get more accuracy. The disadvantage of the latter method is that the greater the value of N, the more multiplications and divisions required. But if different approximations are used on different subintervals, this is wasteful of memory and the time required to find the proper subinterval for a given argument may grow with the number of subintervals unless they are chosen very carefully. General rules for choosing the value of N and the number of subintervals are quite difficult to give.

With Padé approximations, the error increases rapidly away from the center of the interval, but, as we shall see, for minimum maximum error

Table 7.1 Errors $e^x - R_{mk}(x)$ in Padé approximations to e^x

	(m, k)				
x	(4, 0)	(3, 1)	(2, 2)	(1, 3)	(0, 4)
−1.00	−.00712	.00121	−.00054	.00053	−.00135
−.75	−.00175	.00033	−.00016	.00018	−.00050
−.69	−.00117	.00022	−.00011	.00013	−.00037
−.50	−.000240	.000049	−.000027	.000032	−.000104
−.25	−.0000078	.0000018	−.0000011	.0000014	−.0000052
.25	.0000086	−.0000023	.0000018	−.0000029	.0000129
.50	.000284	−.000088	.000073	−.000134	.000653
.69	.00147	−.00050	.00045	−.00088	.00463
.75	.00225	−.00079	.00072	−.00147	.00783
1.00	.00995	−.00394	.00400	−.00899	.05162

approximations, this is not the case. In the latter case, the argument for subintervals, while it still exists, is not nearly so cogent as for Padé approximations. Thus on digital computers most approximations are for the whole interval. In this case the accuracy desired, i.e., the maximum error that can be tolerated, determines N.

If we were using the approximation (7.4-4), we would, of course, not compute it in that form but in one of the forms of Sec. 7.2. We note that $R_{2,2}(x)$ requires two divisions to calculate as a continued fraction while all the other approximations with $N = 4$ require at least three operations i.e., multiplications and divisions. Because both numerator and denominator polynomials have a coefficient of the highest power of 1, $R_{2,2}(x)$ requires only two multiplications and one division if evaluated as the quotient of two polynomials. In this form the other approximations require at least three operations each, although $R_{4,0}(x)$ requires no division {22}.

7.5 CHEBYSHEV POLYNOMIALS

We would not expect Padé approximations to be best or even nearly best in a minimax sense. The basis for their derivation—equality of a function and its derivatives at a point—as we have seen, does not give good error behavior over a whole interval. In the remainder of this chapter, we shall develop methods which lead to approximations which are better in the minimax sense than Padé approximations. In particular in Sec. 7.7 we shall use Padé approximations as the starting point from which to develop better approximations. As a fundamental tool in much of the rest of this chapter, we shall use the Chebyshev polynomials, mentioned briefly in Chap. 4, and which we now discuss in some detail.

Before so doing, however, let us consider the motivation for the use of Chebyshev polynomials. The problem with using approximations based on Maclaurin series is that the error over an interval centered at 0 is extremely nonuniform—small near the center but growing very rapidly near the endpoints. It would seem more reasonable to use as approximating functions, instead of powers of x, polynomials whose behavior over an interval centered at 0 would be in some sense uniform. We would hope that rational functions formed from combinations of these polynomials would exhibit a more uniform error behavior. As we shall now show, the Chebyshev polynomials have ideal properties for these aims.

The Chebyshev polynomial of degree r is given by (see Prob. 26, Chap. 4)

$$T_r(x) = \cos (r \cos^{-1} x) \tag{7.5-1}$$

These polynomials satisfy the orthogonality relationship

$$\int_{-1}^{1} \frac{1}{(1-x^2)^{1/2}} T_r(x)T_s(x)\,dx = \begin{cases} 0 & r \neq s \\ \pi & r = s = 0 \\ \dfrac{\pi}{2} & r = s \neq 0 \end{cases} \tag{7.5-2}$$

and the recurrence relationship (see Prob. 26, Chap. 4)

$$T_{r+1}(x) = 2xT_r(x) - T_{r-1}(x) \qquad T_0(x) = 1 \qquad T_1(x) = x \tag{7.5-3}$$

In this and the next two sections we shall assume that the interval over which we wish to approximate a function is $[-1, 1]$. This will be convenient in the development here and involves no loss of generality.

From (7.5-1), it follows that in $[-1, 1]$ $T_r(x)$ has r zeros at

$$x = \cos\frac{(2j+1)\pi}{2r} \qquad j = 0, \ldots, r-1 \tag{7.5-4}$$

and $r + 1$ extrema of magnitude 1 at

$$x = \cos\frac{j\pi}{r} \qquad j = 0, \ldots, r \tag{7.5-5}$$

where (7.5-5) includes the $r - 1$ places where $T'_r(x) = 0$ as well as the two endpoints. For $T_6(x)$ this is illustrated in Fig. 7.2.

That property of the Chebyshev polynomials of particular interest to us here is expressed by the following theorem.

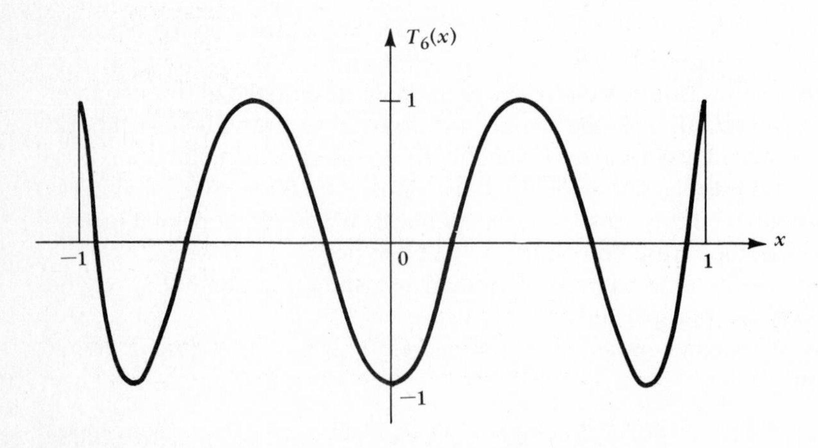

Figure 7.2 Graph of $T_6(x)$.

Theorem 7.1 (Chebyshev) Of all polynomials of degree r with coefficient of x^r equal to 1, the Chebyshev polynomial of degree r multiplied by $1/2^{r-1}$ oscillates with minimum maximum amplitude on the interval $[-1, 1]$.

PROOF From (7.5-3) it follows that the coefficient of x^r in $T_r(x)$ is 2^{r-1}. Thus when $r > 0$,

$$Q_r(x) = \frac{1}{2^{r-1}} T_r(x) \tag{7.5-6}$$

satisfies the conditions of the theorem. The requirement that the coefficient of x^r be 1 has the effect of normalizing all polynomials of degree r. Without this normalization the theorem would be meaningless.

The proof is by contradiction. Suppose there exists a polynomial $q_r(x)$ of degree r with leading coefficient 1 which has a smaller minimum maximum amplitude than $Q_r(x)$ on $[-1, 1]$. We consider

$$p_{r-1}(x) = Q_r(x) - q_r(x) \tag{7.5-7}$$

which is a polynomial of degree $r - 1$. The $r + 1$ extrema of the polynomial $Q_r(x)$, each of magnitude $1/2^{r-1}$, are given by (7.5-5). By our hypothesis $q_r(x)$ has a smaller magnitude than $Q_r(x)$ at each of these extrema, so that $p_{r-1}(x)$ has the same sign as $Q_r(x)$ at each of these extrema. From the definition of $T_r(x)$, it follows that the $r + 1$ extrema of $T_r(x)$ and thus of $Q_r(x)$ alternate in sign. It follows than that $p_{r-1}(x)$ alternates in sign from one extremum of $Q_r(x)$ to the next, which means that $p_{r-1}(x)$ has r zeros in $[-1, 1]$. But $p_{r-1}(x)$ is a polynomial of degree $r - 1$. Therefore, we have a contradiction. Now suppose there exists a polynomial $q_r(x)$ with minimum maximum amplitude equal to $Q_r(x)$. Unless $q_r(x) = Q_r(x)$ at an extremum of $Q_r(x)$, we get a contradiction as above. But, if $q_r(x) = Q_r(x)$ at such an extremum, then $p_{r-1}(x)$ has a double zero at this extremum, and, proceeding as above to count the zeros of $p_{r-1}(x)$, we again arrive at a contradiction, which completes the proof of the theorem

The Chebyshev polynomials are sometimes called *equal-ripple polynomials* because they oscillate between positive and negative extrema of the same magnitude. In the next two sections, we shall use the Chebyshev polynomials to derive approximations which are superior to Padé approximations in the minimax sense.

7.6 CHEBYSHEV EXPANSIONS

The expansion of $f(x)$ in a series of Chebyshev polynomials is given by

$$f(x) = \tfrac{1}{2}c_0 + \sum_{j=1}^{\infty} c_j T_j(x) \tag{7.6-1}$$

where, using the orthogonality property of the Chebyshev polynomials, we can write the coefficients in (7.6-1) as

$$c_j = \frac{2}{\pi}\int_{-1}^{1} \frac{f(x)T_j(x)}{(1-x^2)^{1/2}}\,dx \qquad j = 0, 1, \ldots \tag{7.6-2}$$

The series in (7.6-1) converges uniformly whenever $f(x)$ is continuous and of bounded variation in $[-1, 1]$. Our first object is to approximate $f(x)$ by truncating the expansion (7.6-1) after a finite number of terms. We shall call the approximation formed by truncating at $j = m$, $T_{m,0}(x)$. Having calculated $T_{m,0}(x)$ using (7.6-1), we shall then convert it to polynomial form by writing each $T_j(x)$ in its polynomial form.

Example 7.2 Let $f(x) = e^x$. Use (7.6-1) to calculate $T_{4,0}(x)$ and compare this with $R_{4,0}(x)$ in (7.4-5).

With $f(x) = e^x$, the integrals in (7.6-2) can be evaluated to give [see Watson (1962, p. 20, eq. 5)]

$$c_j = 2I_j(1) \tag{7.6-3}$$

where $I_j(x)$ is the modified Bessel function of the first kind. Using this result and truncating (7.6-1) after five terms, we have

$$\begin{aligned} T_{4,0}(x) = {} & 1.266066 + 1.130318T_1(x) + .271495T_2(x) \\ & + .044337T_3(x) + .005474T_4(x) \end{aligned} \tag{7.6-4}$$

which, converted to a polynomial, becomes

$$T_{4,0}(x) = 1.000044 + .997310x + .499200x^2 + .177344x^3 + .043792x^4 \tag{7.6-5}$$

However, it should be noted that for maximum accuracy $T_{4,0}(x)$ should be *evaluated* for particular values of x using (7.6-4) with the values of the Chebyshev polynomials calculated using the recurrence relation (7.5-3). Indeed, the evaluation of any $T_{mk}(x)$ should use the Chebyshev polynomial rather than powers-of-x form. In Fig. 7.3 and Table 7.2 the

Table 7.2 Errors in various polynomial approximations to e^x

x	$e^x - R_{4,0}(x)$	$e^x - T_{4,0}(x)$	$e^x - C_{4,0}(x)$
−1.00	−.00712	−.00050	.00069
−.75	−.00175	.00047	.00069
−.50	−.000240	−.00023	−.00024
−.25	−.0000078	−.00052	−.00050
0	.0000000	−.00004	.00000
.25	.0000085	.00051	.00050
.50	.000284	.00032	.00028
.75	.00225	−.00050	−.00019
1.00	.00994	.00059	.00214

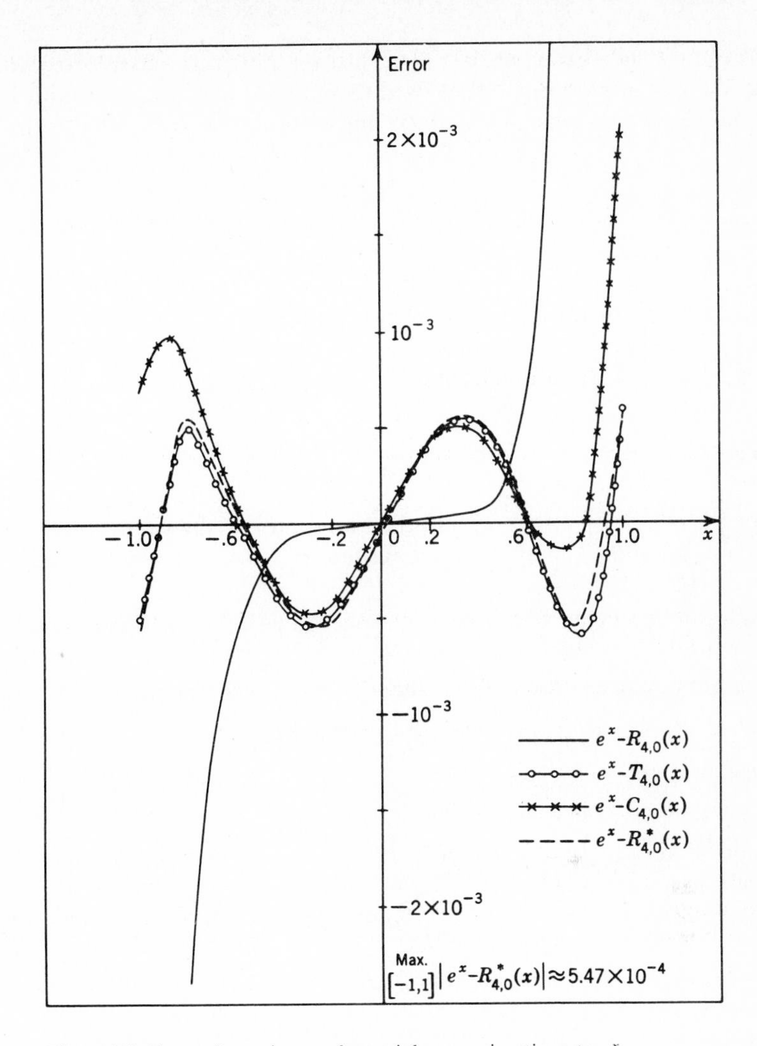

Figure 7.3 Errors in various polynomial approximations to e^x.

errors in this approximation and $R_{4,0}(x)$ are compared on the interval $[-1, 1]$.† From the figure and table the improved accuracy of $T_{4,0}(x)$ is clear. As we might expect, this is particularly notable near the endpoints, thus bearing out our hope that the smoothness of the Chebyshev polynomials would result in improved error behavior near the endpoints. For later reference (see Sec. 7.8) we note that the error has six maxima and minima, including the endpoints. Starting from -1, the values of these extrema are (multiplied by 10^4) -5.03, 5.08, -5.28, 5.56, -5.82, 5.92.

† We could compare over the interval $[-.69, .69]$ by changing variable from $x \to x/.69$ in (7.6-5).

We shall now use the expansion (7.6-1) to generate rational approximations in a manner analogous to that used to generate Padé approximations from Maclaurin expansions. We wish to find approximations of the form

$$T_{mk}(x) = \frac{\sum_{j=0}^{m} a_j T_j(x)}{\sum_{j=0}^{k} b_j T_j(x)} \tag{7.6-6}$$

where the a_j's and b_j's are to be determined so that in

$$f(x) - T_{mk}(x) = \frac{\left[\frac{1}{2}c_0 + \sum_{j=1}^{\infty} c_j T_j(x)\right]\left[\sum_{j=0}^{k} b_j T_j(x)\right] - \sum_{j=0}^{m} a_j T_j(x)}{\sum_{j=0}^{k} b_j T_j(x)} \tag{7.6-7}$$

the coefficients of $T_j(x), j = 0, \ldots, N$, in the numerator of the right-hand side vanish [cf. (7.3-3) and (7.3-4)]. Thus we write

$$\left[\frac{1}{2}c_0 + \sum_{j=1}^{\infty} c_j T_j(x)\right]\left[\sum_{j=0}^{k} b_j T_j(x)\right] - \sum_{j=0}^{m} a_j T_j(x) = \sum_{j=N+1}^{\infty} h_j T_j(x) \tag{7.6-8}$$

In order to get equations for the a_j's and b_j's, we use the identity {25}

$$T_{i+j}(x) + T_{|i-j|}(x) = 2T_i(x)T_j(x) \tag{7.6-9}$$

and with this rewrite (7.6-8) as

$$\frac{1}{2}c_0 \sum_{j=0}^{k} b_j T_j(x) + \frac{1}{2} \sum_{j=1}^{\infty} \sum_{i=0}^{k} b_i c_j [T_{i+j}(x) + T_{|i-j|}(x)] - \sum_{j=0}^{m} a_j T_j(x) = \sum_{j=N+1}^{\infty} h_j T_j(x) \tag{7.6-10}$$

From (7.6-10) we get the following set of equations for the a_j's and b_j's {25}:

$$a_0 = \frac{1}{2} \sum_{i=0}^{k} b_i c_i$$

$$a_r = \begin{cases} 1/2 \sum_{i=0}^{k} b_i (c_{|r-i|} + c_{r+i}) & r = 1, \ldots, N \\ 0 & r > m \end{cases} \tag{7.6-11}$$

These are $N + 1$ equations in the $N + 2$ constants a_j and b_j. This is what we would expect since one coefficient in (7.6-6) is arbitrary. We shall set $b_0 = 1$ in all cases where this leads to a system (7.6-11) which is solvable.

The first nonzero coefficient on the right-hand side of (7.6-8) is given by

$$h_{N+1} = \frac{1}{2} \sum_{i=0}^{k} b_i (c_{N+1-i} + c_{N+1+i}) \tag{7.6-12}$$

Table 7.3 Errors in various rational approximations to e^x

x	$e^x - R_{2,2}(x)$	$e^x - T_{2,2}(x)$	$e^x - C_{2,2}(x)$
-1.00	-.00054	-.000042	-.000032
-.75	-.00016	.000047	.000036
-.50	-.000027	-.000018	-.000027
-.25	-.0000011	-.000070	-.000063
0	.0000000	-.000020	.000000
.25	.0000018	.000088	.000105
.50	.000073	.000084	.000073
.75	.00072	-.000122	-.000163
1.00	.00400	.000189	.000239

In analogy with (7.3-6), this coefficient multiplied by $T_{N+1}(x)$ is often a good approximation of the error in $T_{mk}(x)$.

Example 7.3 Let $f(x) = e^x$. Find $T_{2,2}(x)$ and compare with $R_{2,2}(x)$ as given by (7.4-4). Equations (7.6-11) are

$$a_0 = \tfrac{1}{2}(c_0 + b_1 c_1 + b_2 c_2) \qquad (b_0 = 1)$$

$$a_1 = c_1 + \tfrac{1}{2}[b_1(c_0 + c_2) + b_2(c_1 + c_3)] \qquad a_2 = c_2 + \tfrac{1}{2}[b_1(c_1 + c_3) + b_2(c_0 + c_4)]$$

$$0 = c_3 + \tfrac{1}{2}[b_1(c_2 + c_4) + b_2(c_1 + c_5)] \qquad 0 = c_4 + \tfrac{1}{2}[b_1(c_3 + c_5) + b_2(c_2 + c_6)]$$

where for $j = 0, \ldots, 4$ the c_j's are given by (7.6-4) and $c_5 = .000543$, $c_6 = .000045$. Solving the last two equations for b_1 and b_2 and then solving the first three for a_0, a_1, and a_2, we get

$$a_0 = 1.0009875 \qquad a_1 = .4825306 \qquad a_2 = .0397096$$

$$b_1 = -.4783387 \qquad b_2 = .0387418$$

Converting $T_{2,2}(x)$ to a rational function, we get†

$$T_{2,2}(x) = \frac{1.0000205 + .5019781x + .0826200x^2}{1.0 - .4976173x + .0806064x^2} \tag{7.6-13}$$

In Fig. 7.4 and Table 7.3, the errors in this approximation are compared with those of $R_{2,2}(x)$ over $[-1, 1]$. Again we note that the Chebyshev approximation is substantially better than the Padé approximation, particularly near the ends of the interval. Note, however, that $T_{2,2}(x)$, while it also has six maxima and minima, including the endpoints, is not nearly so smooth as $T_{4,0}(x)$ over the whole interval. Since no condition at $x = 0$ is imposed in deriving $T_{mk}(x)$, it is not surprising that neither $T_{2,2}(x)$ nor $T_{4,0}(x)$ gives exact results at $x = 0$.

† Here and hereafter in this chapter, we shall normalize our approximations by setting $b_0 = 1$ in order to conform to the notation of (7.3-1). However, in actual computational practice, we would always normalize the approximation so that the coefficient of some, usually the highest, power of x in the denominator (or numerator) is equal to 1, for so doing always saves at least one multiplication (cf. Sec. 7.2).

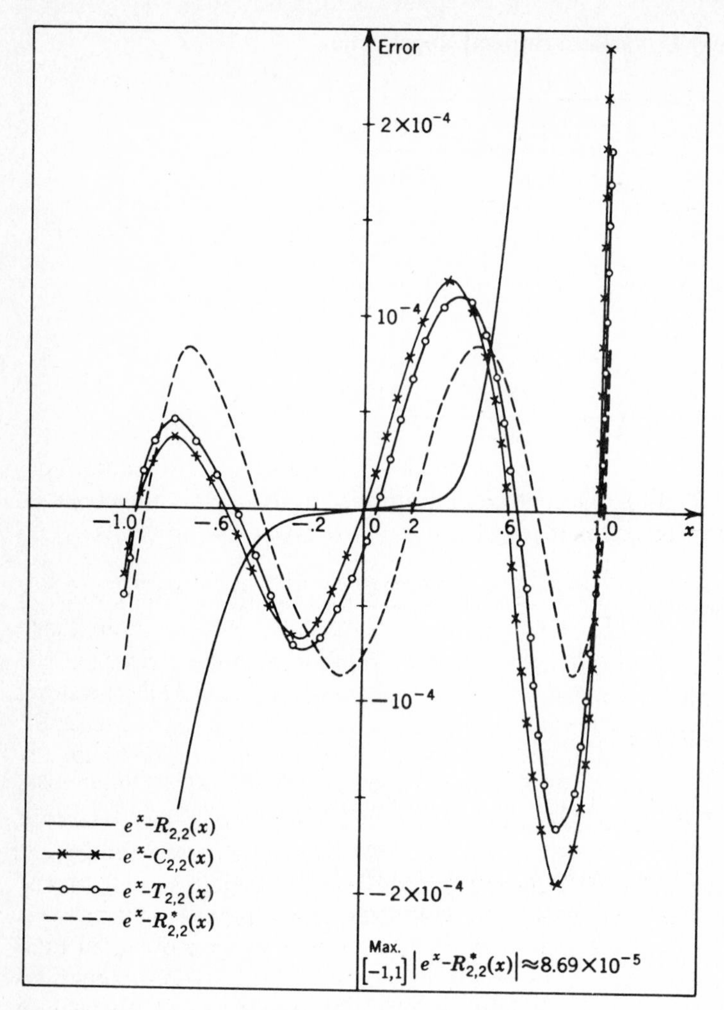

Figure 7.4 Errors in various rational approximations to e^x.

Our examples have illustrated the general empirical result that rational approximations derived using the Chebyshev expansion of a function give a smaller maximum error than Padé approximations. The chief drawback to the approach as we have presented it thus far is the necessity of evaluating the integrals (7.6-2), which cannot be done analytically in most cases. One way to avoid evaluation of the integrals (7.6-2) is to use trigonometric interpolation to approximate the first $L + 1$ coefficients in (7.6-1). We make the change of variable $x = \cos\theta$, and remembering that $T_j(\cos\theta) = \cos j\theta$, we see that (7.6-1) becomes

$$g(\theta) = f(\cos\theta) = \tfrac{1}{2}c_0 + \sum_{j=1}^{\infty} c_j \cos j\theta \tag{7.6-14}$$

and (7.6-2) becomes

$$c_j = \frac{2}{\pi}\int_0^\pi g(\theta)\cos j\theta \, d\theta \tag{7.6-15}$$

Equation (7.6-14) is just the Fourier-series expansion of $g(\theta)$. Therefore, the Fourier and Chebyshev expansions of a function are related to each other through the change of variable $x = \cos \theta$.

Our interest here is in approximating $g(\theta)$ by

$$y_L(\theta) = \tfrac{1}{2}\bar{c}_0 + \sum_{j=1}^{L} \bar{c}_j \cos j\theta \tag{7.6-16}$$

which is precisely equivalent to (6.6-36) since $b_j = 0$ for all j when the function being approximated is even. Therefore, applying the techniques of Sec. 6.6-2, we get approximations $\bar{c}_0, \bar{c}_1, \ldots, \bar{c}_L$ to the true coefficients $c_0, c_1, \ldots, c_L$.

Another approach which avoids evaluating the integrals (7.6-2), which is easy to mechanize, and which can achieve as much accuracy as desired is to use the Maclaurin expansion of $f(x)$ to calculate the coefficients in the Chebyshev expansion. This can be done by substituting the Maclaurin series into (7.6-2) and integrating term by term {27}. The resulting infinite series for c_j usually converges sufficiently rapidly to make the calculation of the c_j's quite feasible. In fact, this procedure can be mechanized on a digital computer so that the computer in effect generates the approximations to be used by the computer. This procedure does, however, involve a great deal of calculation even in the polynomial case ($k = 0$), and when we try to extend the method to rational functions, the amount of calculation becomes very great. In the next section we present another method of improving on Padé approximations using Chebyshev polynomials; this leads to approximations which are generally almost as good as those of this section and much easier to generate.

7.7 ECONOMIZATION OF RATIONAL FUNCTIONS

Our object here is to take the Padé approximations of Sec. 7.3 and perturb them so that the resulting approximation has a smaller minimum maximum error on the desired interval. Without losing generality, we shall assume that the interval is of the form $[-\alpha, \alpha]$ since a change of variable can always be used to bring any other interval into this form. We shall first consider the problem for $k = 0$, that is, for polynomial approximations, and then shall extend it to general rational functions.

7.7-1 Economization of Power Series

Given N, we begin with the Padé approximation with $m = N$, $k = 0$, which is just the first $N + 1$ terms of the Maclaurin expansion of $f(x)$. We write this as

$$R_{N,0}(x) = \sum_{j=0}^{N} d_j x^j \tag{7.7-1}$$

In order to get an improvement over this approximation, we shall use

$$R_{N+1,0}(x) = \sum_{j=0}^{N+1} d_j x^j \tag{7.7-2}$$

Then
$$C_{N,0}(x) = R_{N+1,0}(x) - d_{N+1}\frac{\alpha^{N+1}}{2^N} T_{N+1}\left(\frac{x}{\alpha}\right) \tag{7.7-3}$$

is a polynomial of degree N since the leading term of the Chebyshev polynomial $T_{N+1}(x/\alpha)$ is $x^{N+1}2^N/\alpha^{N+1}$. Moreover, since $|T_{N+1}(x)| \le 1$ for $x \in [-1, 1]$, the error in $C_{N,0}(x)$ is greater than that in $R_{N+1,0}(x)$ by no more than $(d_{N+1}\alpha^{N+1})/2^N$. Since α will be less than or equal to 1 in virtually all applications and d_{N+1} will be decreasing as N increases, this added error will be very small in general. What we have done then is to take the power-series approximation of degree $N + 1$ and from it derive an approximation of degree N whose maximum error is very little greater than that of the $(N + 1)$st degree approximation. Thus we have "economized" the power series in the sense of using fewer terms to achieve almost the same result.

Our object here, though, is to compare $C_{N,0}(x)$ with $R_{N,0}(x)$; that is, we wish to compare corresponding approximations of the same degree. We let $x = \alpha u$, so that the interval for u is $[-1, 1]$. We have from (7.7-1)

$$\lim_{\alpha \to 0} \frac{f(\alpha u) - R_{N,0}(\alpha u)}{\alpha^{N+1}} = d_{N+1} u^{N+1} \tag{7.7-4}$$

From (7.7-2) and (7.7-3)

$$\lim_{\alpha \to 0} \frac{f(\alpha u) - C_{N,0}(\alpha u)}{\alpha^{N+1}} = \lim_{\alpha \to 0} \frac{d_{N+1}(\alpha^{N+1}/2^N)T_{N+1}(u) + \sum_{j=N+2}^{\infty} d_j(\alpha u)^j}{\alpha^{N+1}}$$

$$= d_{N+1}\frac{T_{N+1}(u)}{2^N} \tag{7.7-5}$$

Comparing (7.7-4) and (7.7-5) and using Theorem 7.1, we conclude that, in the limit as $\alpha \to 0$, $C_{N,0}(x)$ has a smaller maximum error on $[-\alpha, \alpha]$ than $R_{N,0}(x)$. Therefore, for sufficiently small α, the economized approximation $C_{N,0}(x)$ is a better approximation than $R_{N,0}(x)$ in the minimax sense. In practice "sufficiently small α" includes almost all intervals of interest.

Example 7.4 Let $f(x) = e^x$. With $\alpha = 1$ find $C_{4,0}(x)$ and compare with $R_{4,0}(x)$ and $T_{4,0}(x)$.

We have

$$R_{5,0}(x) = 1 + x + \frac{x^2}{2} + \frac{x^3}{6} + \frac{x^4}{24} + \frac{x^5}{120} \qquad T_5(x) = 16x^5 - 20x^3 + 5x$$

so that (7.7-3) becomes

$$C_{4,0}(x) = R_{5,0}(x) - \tfrac{1}{120}(\tfrac{1}{16})(16x^5 - 20x^3 + 5x) = 1 + \tfrac{383}{384}x + \frac{x^2}{2} + \tfrac{17}{96}x^3 + \frac{x^4}{24}$$

In Fig. 7.3 and Table 7.2, the errors in this approximation are compared with those of $R_{4,0}(x)$ and $T_{4,0}(x)$. We note that (1) $C_{4,0}(x)$ is a much better approximation than $R_{4,0}(x)$ and (2) near $x = 0$, $C_{4,0}(x)$ and $T_{4,0}(x)$ are about equally good, but near the endpoints, $T_{4,0}(x)$ is substantially better. But as $\alpha \to 0$, the approximation using the Chebyshev expansion and that using the economized power series would be more nearly equivalent over the whole interval.

7.7-2 Generalization to Rational Functions

Corresponding to (7.7-4), we have for any Padé approximation $R_{mk}(x) = P_m(x)/Q_k(x)$

$$\lim_{\alpha \to 0} \frac{f(\alpha u) - R_{mk}(\alpha u)}{\alpha^{N+1}} = d_{N+1}^{(m,k)} u^{N+1} \tag{7.7-6}$$

The superscripts have been added to $d_{N+1}^{(m,k)}$ for later reference. Our object is, in analogy with (7.7-5), to find a rational approximation $C_{mk}(x)$ such that

$$\lim_{\alpha \to 0} \frac{f(\alpha u) - C_{mk}(\alpha u)}{\alpha^{N+1}} = d_{N+1}^{(m,k)} \frac{T_{N+1}(u)}{2^N} \tag{7.7-7}$$

because if we can do so, we shall again have a better approximation than $R_{mk}(x)$ for sufficiently small α.

To find such a $C_{mk}(x)$, we use a sequence of Padé approximations to $f(x)$ of the form

$$R_{i,j-i}^{(j)}(x) = \frac{P_i^{(j)}(x)}{Q_{j-i}^{(j)}(x)} \qquad j = 0, \ldots, N-1 \tag{7.7-8}$$

where the only restriction on i is that $0 \le i \le j$. Therefore, i is not uniquely determined except when $m = 0$, $k = 0$ and, for all m and k, when $j = 0$. Analogous to (7.7-6), we have

$$\lim_{\alpha \to 0} \frac{f(\alpha u) - R_{i,j-i}^{(j)}(\alpha u)}{\alpha^{j+1}} = d_{j+1}^{(i,j-i)} u^{j+1} \tag{7.7-9}$$

Now we define

$$C_{mk}(x) = \frac{P_m(x) + \sum_{j=0}^{N-1} \gamma_{j+1} P_i^{(j)}(x) + \gamma_0}{Q_k(x) + \sum_{j=0}^{N-1} \gamma_{j+1} Q_{j-i}^{(j)}(x)} \tag{7.7-10}$$

with

$$\gamma_{j+1} = \frac{d_{N+1}^{(m,k)}}{d_{j+1}^{(i,j-i)}} \frac{\alpha^{N-j}}{2^N} t_{j+1} \qquad j = 0, \ldots, N-1$$

$$\gamma_0 = -\frac{d_{N+1}^{(m,k)} \alpha^{N+1} t_0}{2^N} \tag{7.7-11}$$

where t_j is the coefficient of u^j in $T_{N+1}(u)$; then although we leave the algebra to a problem {31}, it can be shown that $C_{mk}(x)$ as defined in (7.7-10) satisfies (7.7-7).

From (7.7-11) it can be seen that this procedure will fail if $d_{j+1}^{(i,j-i)} = 0$ for any j. When this happens, however, it will often be possible to choose another member of the sequence (7.7-8). Note also that the foregoing derivation requires the use of only those γ_{j+1} for which t_{j+1} is nonzero. Thus, when $t_{j+1} = 0$ [as it is for every other coefficient of $T_{N+1}(u)$], the value of $d_{j+1}^{(i,j-i)}$ is immaterial {31}.

Example 7.5 Let $f(x) = e^x$. With $\alpha = 1$ find $C_{2,2}(x)$ and compare with $R_{2,2}(x)$ and $T_{2,2}(x)$.

For the sequence (7.7-8), we choose $R_{0,0}^{(0)}(x)$, $R_{1,0}^{(1)}(x)$, $R_{1,1}^{(2)}(x)$, and $R_{2,1}^{(3)}(x)$. Performing the calculations analogous to those in Sec. 7.4, we find

$$R_{0,0}^{(0)}(x) = 1 \qquad R_{1,0}^{(1)}(x) = 1 + x$$

$$R_{1,1}^{(2)}(x) = \frac{1 + \frac{1}{2}x}{1 - \frac{1}{2}x} \qquad R_{2,1}^{(3)}(x) = \frac{1 + \frac{2}{3}x + \frac{1}{6}x^2}{1 - \frac{1}{3}x}$$

and rewriting $R_{2,2}(x)$ so that $b_0 = 1$, we have

$$R_{2,2}(x) = \frac{1 + \frac{1}{2}x + \frac{1}{12}x^2}{1 - \frac{1}{2}x + \frac{1}{12}x^2}$$

Then using (7.3-6), we calculate

$$d_1^{(0,0)} = 1 \qquad d_2^{(1,0)} = \tfrac{1}{2} \qquad d_3^{(1,1)} = -\tfrac{1}{12} \qquad d_4^{(2,1)} = -\tfrac{1}{72} \qquad d_5^{(2,2)} = \tfrac{1}{720}$$

Further

$$T_5(x) = 16x^5 - 20x^3 + 5x$$

so from (7.7-11) with $N = 4$ and $\alpha = 1$

$$\gamma_0 = 0 \qquad \gamma_1 = \frac{\frac{1}{720}}{1}\frac{1}{16} \times 5 = \frac{1}{2304}$$

$$\gamma_2 = 0 \qquad \gamma_3 = \frac{\frac{1}{720}}{-\frac{1}{12}}\frac{1}{16} \times (-20) = \frac{1}{48} \qquad \gamma_4 = 0$$

Then from (7.7-10)

$$C_{2,2}(x) = \frac{(1 + \frac{1}{2}x + \frac{1}{12}x^2) + \frac{1}{48}(1 + \frac{1}{2}x) + \frac{1}{2304}}{(1 - \frac{1}{2}x + \frac{1}{12}x^2) + \frac{1}{48}(1 - \frac{1}{2}x) + \frac{1}{2304}}$$

$$= \frac{2353 + 1176x + 192x^2}{2353 - 1176x + 192x^2} = \frac{1.0 + .49978751x + .08159796x^2}{1.0 - .49978751x + .08159796x^2}$$

In Fig. 7.4 and Table 7.3 the errors in this approximation are compared with $R_{2,2}(x)$ and $T_{2,2}(x)$. As with the economized power series, the economized rational function is not quite so good as that derived using a Chebyshev expansion, but it is much superior to the Padé approximation.

This completes our discussion of the economization of rational functions. Although we have come close, we have not yet succeeded in deriving true minimax approximations. In the next section, we shall state the theorem which gives the characterization necessary for the true minimax approximation to a function over an interval.

7.8 CHEBYSHEV'S THEOREM ON MINIMAX APPROXIMATIONS

Let $f(x)$ be the continuous function we wish to approximate over the finite interval $[a, b]$ in the form (7.2-1). Let

$$r_{mk} = \max_{a \le x \le b} |f(x) - R_{mk}(x)| \tag{7.8-1}$$

for any rational function

$$R_{mk}(x) = \frac{P_m(x)}{Q_k(x)} = \frac{\sum_{j=0}^{m} a_j x^j}{\sum_{j=0}^{k} b_j x^j} \tag{7.8-2}$$

Then we can prove† the following theorem.

Theorem 7.2 (Chebyshev) There exists a unique rational function $R_{mk}(x)$ which minimizes r_{mk} (two rational functions are considered identical if they are equal when reduced to their lowest terms). Moreover, if we write this unique rational function as

$$R^*_{mk}(x) = \frac{\sum_{j=0}^{m-\nu} a_{j+\nu} x^j}{\sum_{j=0}^{k-\mu} b_{j+\mu} x^j} = \frac{P^*_m(x)}{Q^*_k(x)} \tag{7.8-3}$$

† Chebyshev proved only the characterization and uniqueness parts of the theorem. The existence portion was proved much later by Walsh.

where $$0 \le \mu \le k \qquad 0 \le \nu \le m \qquad a_m, b_k \ne 0 \tag{7.8-4}$$

and $P_m^*(x)/Q_k^*(x)$ is irreducible, then if $r_{mk}^* \ne 0$, the number of consecutive points of $[a, b]$ at which $f(x) - R_{mk}^*(x)$ takes on its maximum value of magnitude r_{mk}^* with alternate change of sign is not less than $L = m + k + 2 - d$, where $d = \min(\mu, \nu)$.

In particular, the theorem says that when $k = 0$, the number of points at which the error attains its maximum magnitude is at least $m + 2$. Note how nearly $T_{4,0}(x)$ as given by (7.6-5) meets this requirement.

We shall leave the existence and uniqueness parts of the proof to the problems {37, 38} and here prove only that $R_{mk}(x)$ has the characteristic form stated in the theorem. We note here that both this theorem and Theorem 7.3 below are true if we consider approximations to $f(x)$ of the form $s(x)R_{mk}(x)$, where $s(x)$ is a given continuous function which does not vanish in (a, b) {39}.

PROOF We suppose that L', the number of points at which $f(x) - R_{mk}^*(x)$ takes on its maximum value r_{mk}^* with alternating sign, is such that

$$L' < L$$

so that $$L' \le m + k + 1 - d \tag{7.8-5}$$

Note, for example, that if $f(x) - R_{mk}^*(x)$ has the form of Fig. 7.5, the number of points at which it takes on its extreme value *with alternating sign* is only 3 since the extrema labeled 2 and 3 have the same sign.

We can then subdivide the interval $[a, b]$ into L' subintervals

$$[a, x_1], [x_1, x_2], \ldots, [x_{L'-1}, b] \tag{7.8-6}$$

such that in alternate intervals the inequalities

$$-r_{mk}^* \le f(x) - R_{mk}^*(x) < r_{mk}^* - \alpha \tag{7.8-7}$$

and $$-r_{mk}^* + \alpha < f(x) - R_{mk}^*(x) \le r_{mk}^* \tag{7.8-8}$$

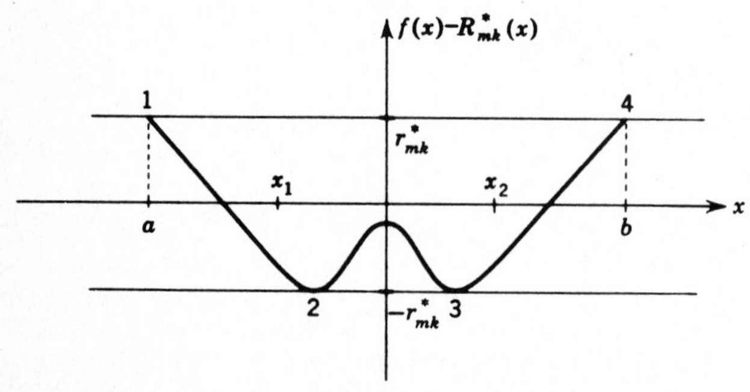

Figure 7.5 An example of alternating extrema of $f(x) - R_{mk}^*(x)$.

are satisfied for some positive number α.† In Fig. 7.5 the points a, x_1, x_2, b divide $[a, b]$ into subintervals in which (7.8-8) and (7.8-7) are alternately satisfied.

We consider the function

$$A(x) = (x - x_1)(x - x_2) \cdots (x - x_{L'-1}) \tag{7.8-9}$$

Since we have assumed that $P_m^*(x)/Q_k^*(x)$ is irreducible, we can write {36}

$$A(x) = Q_k^*(x)a(x) - P_m^*(x)b(x) \tag{7.8-10}$$

where $a(x)$ and $b(x)$ are, respectively, polynomials of degree less than or equal to m and k.

We shall now display a function $R_{mk}(x)$ which achieves a smaller minimum maximum error than $R_{mk}^*(x)$ on $[a, b]$ under the assumption that $L' < L$. This contradiction will prove the theorem. Let

$$R_{mk}(x) = \frac{P_m^*(x) - \beta a(x)}{Q_k^*(x) - \beta b(x)} \tag{7.8-11}$$

where β is a real number. The numerator is then a polynomial of degree $\leq m$, and the denominator is a polynomial of degree $\leq k$. Now we can write

$$f(x) - R_{mk}(x) = f(x) - R_{mk}^*(x) + \frac{\beta A(x)}{Q_k^*(x)[Q_k^*(x) - \beta b(x)]} \tag{7.8-12}$$

by making use of (7.8-10), (7.8-11), and (7.8-3). Since $Q_k^*(x)$ must be bounded away from zero in $[a, b]$ [for, if not, $R_{mk}^*(x)$ will certainly not be a minimax approximation], by choosing β sufficiently small in magnitude, the denominator of the last term on the right-hand side of (7.8-12) will also be bounded away from zero in $[a, b]$. The numerator changes sign at each point x_i because of (7.8-9) and is of constant sign in each subinterval (7.8-6). Thus we can choose β so that

$$\left| \frac{\beta A(x)}{Q_k^*(x)[Q_k^*(x) - \beta b(x)]} \right| < \alpha \qquad x \in [a, b] \tag{7.8-13}$$

and such that this term has sign opposite to the maximum error in $f(x) - R_{mk}^*(x)$ in each subinterval (7.8-6). With β chosen in this way (7.8-7), (7.8-8), and (7.8-13) imply that $R_{mk}(x)$ in (7.8-11) has a smaller minimum maximum error than $R_{mk}^*(x)$ on $[a, b]$, which is the desired contradiction. Therefore, we have proved that $L' \geq m + k + 2 - d$.

Theorem 7.2 gives us no way of constructing minimax approximations. Nor does it enable us to judge how close to the minimax approximation a

† That such a subdivision exists is not quite trivial; see Davis (1963, pp. 149–151).

given approximation is. The plot of $e^x - R^*_{4,0}(x)$ in Fig. 7.3 implies that $T_{4,0}(x)$ is very nearly a minimax approximation. Conversely the plot of $e^x - R^*_{2,2}(x)$ in Fig. 7.4 indicates that $T_{2,2}(x)$ deviates a good deal from the minimax approximation. We might surmise then that if L alternating extrema of the error in a given approximation are of nearly equal magnitude, this approximation achieves nearly as small a maximum error as the minimax approximation. The following theorem confirms this.

Theorem 7.3 Let

$$R_{mk}(x) = \frac{P_m(x)}{Q_k(x)} = \frac{\sum_{j=0}^{m-\nu} a_{j+\nu} x^j}{\sum_{j=0}^{k-\mu} b_{j+\mu} x^j} \tag{7.8-14}$$

be irreducible and let the difference

$$f(x) - R_{mk}(x)$$

remain finite in $[a, b]$. Further, let

$$x_1 < x_2 < \cdots < x_L$$

be in $[a, b]$ and let

$$f(x_i) - R_{mk}(x_i) = (-1)^i \lambda_i \qquad i = 1, \ldots, L \tag{7.8-15}$$

where $\lambda_i > 0$ for all i and $L = m + k + 2 - d$ with $d = \min(\mu, \nu)$. Let $S_{mk}(x) = p_m(x)/q_k(x)$ be any rational approximation to $f(x)$ with degrees of numerator and denominator less than or equal to, respectively, m and k and let

$$s_{mk} = \max_{[a, b]} |f(x) - S_{mk}(x)| \tag{7.8-16}$$

Then

$$s_{mk} \geq \min_i \lambda_i \tag{7.8-17}$$

We leave the proof of this theorem to a problem {38}.

By identifying $S_{mk}(x)$ with $R^*_{mk}(x)$, this theorem says that the error in the minimax approximation is greater than the magnitude of the smallest extremum in the error of $R_{mk}(x)$. Since the minimax error is certainly smaller than the magnitude of the largest extremum in the error of $R_{mk}(x)$, we thereby obtain an upper and lower bound on the minimax error from any approximation $R_{mk}(x)$ which satisfies the conditions of Theorem 7.3. For example, from Fig. 7.3 the six extrema of $T_{4,0}(x)$ vary in magnitude between 5.03×10^{-4} and 5.92×10^{-4} with the minimax error given by 5.47×10^{-4}. In Fig. 7.4 the extrema of $T_{2,2}(x)$ vary in magnitude between 4.20×10^{-5} and 18.88×10^{-5}, and the minimax error is equal to 8.69×10^{-5}.

It remains now only to consider some means for constructing the minimax approximation. This is the subject of the last section of this chapter.

7.9 CONSTRUCTING MINIMAX APPROXIMATIONS

A number of algorithms are known by which the true Chebyshev approximation to a function for a given m and k can be calculated. All of them proceed by generating a sequence of approximations which in the limit converges to the Chebyshev approximation. In this section, we shall present the two algorithms which are the most commonly used to generate minimax approximations on computers, namely the second algorithm of Remes and the differential correction algorithm.

7.9-1 The Second Algorithm of Remes

For simplicity, we shall assume in this section that the minimax approximation to $f(x)$ on $[a, b]$ of the form (7.8-3) has at least

$$N + 2 = m + k + 2$$

points at which the extreme value of the error is attained. That is, we assume that d in Theorem 7.2 is 0.† We also assume, without loss of generality, that the interval $[a, b]$ includes $x = 0$ so that we can set $b_0 = 1$. The problem then is:

Given: A continuous function $f(x)$ on an interval $[a, b]$ including zero and m and k

Find: $R^*_{mk}(x)$, the minimax approximation to $f(x)$ on $[a, b]$ of the form (7.8-3).

Let us suppose that we have somehow obtained (see comments after the algorithm) an approximation

$$R^{(0)}_{mk}(x) = \frac{\sum_{j=0}^{m} a_j x^j}{\sum_{j=0}^{k} b_j x^j} \qquad b_0 = 1 \tag{7.9-1}$$

such that $f(x) - R^{(0)}_{mk}(x)$ has $N + 2$ extrema which alternate in sign. Then the second algorithm of Remes is as follows:

1. Let $x_0^{(0)} < x_1^{(0)} < \cdots < x_{N+1}^{(0)}$ be $N + 2$ points at which $f(x) - R^{(0)}_{mk}(x)$ has local extrema which alternate in sign.

† Degeneracy, that is, $d \neq 0$, can occur when $f(x)$ is even or odd on $[a, b]$ {40}. Except in trivial cases like this, it occurs in only some quite pathological cases. But near degeneracy, in which $P^*_m(x)$ and $Q^*_k(x)$ have a nearly common factor, is more common and can cause severe computational difficulties.

2. Solve the system of $N + 2$ nonlinear equations

$$f(x_i^{(0)}) - \frac{\sum_{j=0}^{m} a_j (x_i^{(0)})^j}{\sum_{j=0}^{k} b_j (x_i^{(0)})^j} = (-1)^i E \qquad \begin{matrix} i = 0, \dots, N+1 \\ b_0 = 1 \end{matrix} \tag{7.9-2}$$

for the $N + 2$ unknowns $a_0, \dots, a_m, b_1, \dots, b_k$, and E. Call the solution $a_0^{(0)}, \dots, a_m^{(0)}, b_1^{(0)}, \dots, b_k^{(0)}, E^{(0)}$. Note that $|E^{(0)}|$ is the magnitude of the error in the approximation at each of the points $x_i^{(0)}$.

3. Define

$$h_0(x) = f(x) - \frac{\sum_{j=0}^{m} a_j^{(0)} x^j}{\sum_{j=0}^{k} b_j^{(0)} x^j} \qquad b_0^{(0)} = 1 \tag{7.9-3}$$

The function $h_0(x)$ then has magnitude $|E^{(0)}|$ with alternating sign at the points $x_i^{(0)}$, $i = 0, \dots, N + 1$. Therefore, it is not hard to show that in the neighborhood of each $x_i^{(0)}$ there is a point $x_i^{(1)}$ at which $h_0(x)$ has an extremum of the same sign as that of $f(x) - R_{mk}^{(0)}(x)$ at $x_i^{(0)}$. Replace each $x_i^{(0)}$ by the corresponding $x_i^{(1)}$. If $\bar{x}$, the point at which $h_0(x)$ has its maximum magnitude, is one of the points $x_i^{(1)}$, proceed to step 4. If not, replace one of the points $x_i^{(1)}$ by $\bar{x}$ in such a way that $h_0(x)$ still alternates in sign on the points $x_i^{(1)}$. This can always be done (why?).

4. Repeat steps 2 and 3 using the points $x_0^{(1)}, \dots, x_{N+1}^{(1)}$ in (7.9-2). This process then generates a sequence of rational approximations of the form (7.9-1) which converges uniformly to $R_{mx}^*(x)$ if the initial extrema $x_i^{(0)}$, $i = 0, \dots, N + 1$, are sufficiently close (see below) to the corresponding extrema of $R_{mk}^*(x)$.

We make the following comments on this algorithm:

1. If $k = 0$, the iteration will converge for an arbitrary choice of the $N + 2$ abscissas in Step 1. That is, an initial approximation of the form (7.9-1) is not necessary [Novodvorskii and Pinsker (1951)]. However, when $k \neq 0$, all that can be said is that there exists an $\epsilon > 0$ such that if each extremum of $f(x) - R_{mk}^{(0)}(x)$ lies within ϵ of the corresponding extremum of $f(x) - R_{mk}^*(x)$, the algorithm will converge. Thus what is really required is not an approximation $R_{mk}^{(0)}(x)$ but a set of extrema lying sufficiently close to the corresponding extrema of $f(x) - R_{mk}^*(x)$. For example we might use the $N + 2$ extrema of the Chebyshev polynomial $T_{N+1}(x)$, suitably related to $[a, b]$. But in many cases, in order to obtain a set of $N + 2$ $x_i^{(0)}$ for which the algorithm will converge, it is necessary first to derive an approximation $R_{mk}^{(0)}(x)$. When $f(x) - C_{mk}(x)$ has $N + 2$ extrema, $C_{mk}(x)$ can almost

always be used for $R_{mk}^{(0)}(x)$. In some particularly difficult cases a technique due to Werner (1962) can be used to generate an appropriate $R_{mk}^{(0)}(x)$.

2. When $k = 0$, the equations of step 2 of the algorithm are, in fact, linear (why?) and thus are easily solvable (see, for example, Sec. 9.3). When $k \neq 0$, the nonlinear system can be solved as follows:
   a. Write the system (7.9-2) as

$$\sum_{j=0}^{m} a_j (x_i^{(0)})^j - [f(x_i^{(0)}) - (-1)^i E_r] \sum_{j=1}^{k} b_j (x_i^{(0)})^j$$

$$= f(x_i^{(0)}) - (-1)^i E_{r+1} \qquad \begin{matrix} i = 0, \ldots, N+1 \\ r = 0, 1, \ldots \end{matrix} \qquad b_0 = 1 \tag{7.9-4}$$

   b. Starting from an assumed initial value of E_0, solve the *linear* system (7.9-4) in the unknowns $a_0, \ldots, a_m, b_1, \ldots, b_k$ and E_{r+1} for $r = 0, 1, \ldots$, until two successive values of E_r are in agreement. In the absence of other information (cf. Example 7.6 below), $E_0 = 0$ can be used. In practice, convergence of this method has seldom proved to be a problem.
3. The problem of finding the extrema of $h_0(x)$ is computationally intractable if we wish the exact solution. However, since the extrema of $h_0(x)$ will be close to those of $f(x) - R_{mk}^{(0)}(x)$ (and similarly at subsequent stages of the algorithm), it is sufficient in practice to search in the neighborhood of $x_i^{(0)}$ with a mesh of points until the approximate location of the extremum is found. Then $x_i^{(1)}$ is chosen, using a single stage of linear or quadratic inverse interpolation (cf. Sec. 3.6) as the point where the derivative of the error is zero.
4. This algorithm is also applicable to the case in which an approximation of the form $s(x)R_{mk}(x)$ is desired, where $s(x)$ is a given continuous function which does not vanish in (a, b) {39}. Choosing $s(x) = 1/f(x)$ and choosing the function to be approximated to be identically 1 enables us to compute the best *relative* minimax approximation to $f(x)$ on any interval on which $f(x)$ does not vanish since

$$1 - \frac{R_{mk}(x)}{f(x)} = \frac{1}{f(x)} [f(x) - R_{mk}(x)]$$

Example 7.6 Let $f(x) = e^x$. Starting with $C_{2,2}(x)$ as found in Example 7.5, use the above algorithm to find the minimax approximation to e^x on $[-1, 1]$ with $m = k = 2$.

Arbitrarily, we divide the interval $[-1, 1]$ into 201 points, equally spaced at an interval of .01. The six of these points at which $e^x - C_{2,2}(x)$ has its extreme values are (see Fig. 7.4)

$$-1.00, \ -.80, \ -.27, \ +.35, \ +.82, \ +1.00$$

Of these extrema, the one at $+1.00$ has the greatest magnitude, 2.39×10^{-4}. We expect the maximum error in the minimax approximation to be somewhat less than 2.39×10^{-4}. (In fact, we know from the previous section that $r_{22}^* < 1.89 \times 10^{-4}$.) Therefore, as E_0 in

(7.9-4), we choose -2.0×10^{-4}, the minus sign being used because the error at -1.00, that is, $i = 0$ in (7.9-2), is negative. Then solving (7.9-2) by the algorithm given above, we get

$$a_0 = 1.00007407 \qquad a_1 = .50802883 \qquad a_2 = .08549199$$

$$b_1 = -.49166131 \qquad b_2 = .07792980 \qquad E = -.745 \times 10^{-4}$$

The maximum error on $[-1, 1]$ is $.978 \times 10^{-4}$ at $x = .47333$. The six extrema of this approximation are at

$$-1.00, \qquad -.73773, \qquad -.13475, \qquad .47333, \qquad .86488, \qquad 1.00$$

The next stage of the algorithm gives

$$a_0 = 1.00007275 \qquad a_1 = .50864603 \qquad a_2 = .08583370$$

$$b_1 = -.49108231 \qquad b_2 = .07770411 \qquad E = -.867 \times 10^{-4}$$

Now the maximum error is $.870 \times 10^{-4}$ at $x = -.11898$. Note that this maximum error is very little greater than E, so that the process has almost converged. The extrema are now at

$$-1.00, \qquad -.72601, \qquad -.11898, \qquad .47363, \qquad .86571, \qquad 1.00$$

A third application of the algorithm gives

$$a_0 = 1.00007255 \qquad a_1 = .50863618 \qquad a_2 = .08582937$$

$$b_1 = -.49109193 \qquad b_2 = .07770847 \qquad E = -.8689990 \times 10^{-4}$$

The maximum error is now $.8689996 \times 10^{-4}$ at $x = .47357$. Thus, for all practical purposes, these coefficients give the Chebyshev approximation to e^x on $[-1, 1]$.

7.9-2 The Differential Correction Algorithm

In this algorithm we assume that the function $f(x)$ is defined not on an interval but on a discrete point set $X = \{x_1, x_2, \ldots, x_l\}$, and we attempt to find $R_{mk}^*(x)$ which is such that

$$\max_{1 \le i \le l} |f(x_i) - R_{mk}^*(x_i)| \le \max_{1 \le i \le l} |f(x_i) - R_{mk}(x_i)| \tag{7.9-5}$$

for all $R_{mk}(x)$. We note first that this problem is not significantly different from that of finding the best approximation on an interval $[a, b]$ because, as noted in the previous section, the implementation of the second algorithm of Remes results in searching for the extrema of $h_0(x)$ on a mesh of points in $[a, b]$. This mesh is essentially equivalent to the point set X. Or to put it another way, we expect that the best approximation on a rather dense point set in $[a, b]$ will normally be very close to the best approximation on the interval.

To describe the differential correction algorithm we first define

$$r_{mk}^{(s)} = \max_i \left| f(x_i) - \frac{P_m^{(s)}(x_i)}{Q_k^{(s)}(x_i)} \right| \tag{7.9-6}$$

where the $R_{mk}^{(s)}(x) = P_m^{(s)}(x)/Q_k^{(s)}(x)$ are a sequence of approximations to $R_{mk}^*(x)$. Starting from an initial approximation of the form (7.9-1), this algorithm considers at each step

$$w = \max_i \frac{|f(x_i)Q_k(x_i) - P_m(x_i)| - r_{mk}^{(s)} Q_k(x_i)}{Q_k^{(s)}(x_i)}$$
$$= \max_i \frac{(|\epsilon_{mk}(x_i)| - r_{mk}^{(s)})Q_k(x_i)}{Q_k^{(s)}(x_i)} \tag{7.9-7}$$

where $$\epsilon_{mk}(x) = f(x) - \frac{P_m(x)}{Q_k(x)} \tag{7.9-8}$$

At each step we seek $R_{mk}^{(s+1)}(x)$ which minimizes w in (7.9-7) subject to the additional condition that $Q_k(x_i) > 0$ for all i. The second form of w in (7.9-7) indicates that what we seek are coefficients a_j and b_j such that

$$r_{mk}^{(s+1)} = \max_i |\epsilon_{mk}(x_i)|$$

is as much less than $r_{mk}^{(s)}$ as possible, or, putting it another way, we wish (7.9-7) to be as negative as possible. That a negative value can be found at each step is certain, for otherwise $r_{mk}^{(s)}$ would equal r_{mk}^*.

To perform the computation which minimizes w we first note that (7.9-7) can be written

$$wQ_k^{(s)}(x_i) \geq |f(x_i)Q_k(x_i) - P_m(x_i)| - r_{mk}^{(s)} Q_k(x_i) \qquad i = 1, \ldots, l \tag{7.9-9}$$

which is equivalent to the two inequalities

$$wQ_k^{(s)}(x_i) + r_{mk}^{(s)} Q_k(x_i) - [f(x_i)Q_k(x_i) - P_m(x_i)] \geq 0$$
$$wQ_k^{(s)}(x_i) + r_{mk}^{(s)} Q_k(x_i) + [f(x_i)Q_k(x_i) - P_m(x_i)] \geq 0 \tag{7.9-10}$$

Our object is to minimize w subject to the constraints (7.9-10), the constraints $Q_k(x_i) > 0$, and the normalizing condition $b_0 = 1$ (or alternately, as is often done with this algorithm, $\max_j |b_j| = 1$). This is a *linear programming* problem and can be solved by any method for the solution of such problems such as the simplex method (see Sec. 9.10).

The major advantage of the differential correction algorithm over the second algorithm of Remes is that it converges for any starting approximation $R_{mk}^{(0)}(x)$ as long as $Q_k^{(0)}(x_i) > 0$ for all i. However, it should be noted that when the Remes algorithm does not converge, the number of points l will probably have to be large in order for the differential correction algorithm to lead to a solution close to that for the continuous interval. This results in a large linear programming problem and therefore a great deal of computation. Another, occasionally useful property of the differential correction algorithm is that it quite easily accommodates constraints on the approximation of the form $R_{mk}(x_i) = f(x_i)$ at selected points x_i. Both algorithms have their devotees, and both are in quite common use for calculating best rational approximations.

BIBLIOGRAPHIC NOTES

Section 7.1 The need to approximate functions on digital computers has given a great impetus to the well-established mathematical field of approximation theory. The books by Achieser (1956), Cheney (1966), Davis (1963), Fike (1968), and Rice (1964, 1969) all contain valuable material. Some of the earliest work on approximations for digital computers was done by Hastings (1955). The most up-to-date compilation of approximations for digital computers is contained in the handbook by Hart et al. (1968). Recent surveys of the field have been given by Kogbetliantz (1960), Stiefel (1959), Cheney and Southard (1963), and Cody (1970).

Section 7.2 The algorithm for evaluating polynomials is discussed in Knuth (1969). The standard text on continued fractions is that of Wall (1948). For techniques of converting rational functions to continued fractions, see Maehly (1960*a*).

Sections 7.3 and 7.4 The Padé table is extensively discussed by Wall (1948). Much of the material of these sections can be found in Kogbetliantz (1960). Lawson (1964) considers the problem of using different approximations on different subintervals.

Section 7.5 The classic reference on the Chebyshev polynomials and their many applications is Lanczos (1956). For more recent treatments see Snyder (1966) and Fox and Parker (1968).

Section 7.6 Much of the material in this section is due to Maehly and is discussed by Kogbetliantz (1960). Minnick (1957) discusses the use of the Maclaurin series to evaluate the coefficients in the Chebyshev expansion, and Spielberg (1961) uses this technique to mechanize the procedure on a digital computer. Powell (1967) shows that, for polynomial approximations, the Chebyshev expansion results in a minimum maximum error not much larger than the best approximation. Luke (1969, 1975) contains the coefficients for the Chebyshev expansion for many mathematical functions.

Section 7.7 The concept of economization of power series is due to Lanczos (1938) [see also Lanczos (1956)] and is discussed by Kogbetliantz (1960). The extension to rational functions was made by Maehly (1960*b*) and in the form presented here by Ralston (1963).

Section 7.8 The proof of Chebyshev's theorem here is from Achieser (1956).

Section 7.9 For rational functions the proofs of the convergence of the second algorithm of Remes and the related exchange algorithm of Remes are given by Ralston (1965). Werner (1962) has given a proof of the convergence of a modified form of the second algorithm. A proof of the convergence of the exchange algorithm for a class of functions including polynomials but not rational functions has been given by Novodvorskii and Pinsker (1951). The differential correction algorithm is discussed in detail by Barrodale, Powell, and Roberts (1972).

BIBLIOGRAPHY

Achieser, N. I. (1956): *Theory of Approximation* (trans. by C. J. Hyman), Frederick Ungar Publishing Co., New York.

Barrodale, I., M. J. D. Powell, and F. D. K. Roberts (1972): The Differential Correction Algorithm for Rational l_∞-Approximations, *SIAM J. Numer. Anal.*, vol. 9, pp. 493–504.

Cheney, E. W. (1966): *Introduction to Approximation Theory*, McGraw-Hill Book Company, New York.
——— and H. L. Loeb (1962): On Rational Chebyshev Approximation, *Numer. Math.*, vol. 4, pp. 124–127.
——— and T. H. Southard (1963): A Survey of Methods for Rational Approximation, with Particular Reference to a New Method Based on a Formula of Darboux, *SIAM Rev.*, vol. 5, pp. 219–231.
Cody, W. J. (1970): A Survey of Practical Rational and Polynomial Approximation of Functions, *SIAM Rev.*, vol. 12, pp. 400–423.
Davis, P. J. (1963): *Interpolation and Approximation*, Blaisdell Publishing Company, New York.
Fike, C. T. (1968): *Computer Evaluation of Mathematical Functions*, Prentice-Hall, Inc., Englewood Cliffs, N.J.
Fox, L., and I. P. Parker (1968): *Chebyshev Polynomials in Numerical Analysis*, Oxford University Press, London.
Fraser, W., and J. F. Hart (1962): On the Computation of Rational Approximations to Continuous Functions, *Commun. ACM*, vol. 5, pp. 401–403.
Hart, J. F., et al. (1968): *Handbook of Computer Approximations*, SIAM Series in Applied Mathematics, John Wiley and Sons, Inc., New York.
Hastings, C. (1955): *Approximations for Digital Computers*, Princeton University Press, Princeton, N.J.
Knuth, D. E. (1962): Evaluation of Polynomials by Computer, *Commun. ACM*, vol. 5, pp. 595–599.
——— (1969): *The Art of Computer Programming*, vol. 2, Addison-Wesley Publishing Company, Inc., Reading, Mass.
Kogbetliantz, E. G. (1960): Generation of Elementary Functions, pp. 7–35 in *Mathematical Methods for Digital Computers* (A. Ralston and H. S. Wilf, eds.), John Wiley & Sons, Inc., New York.
Lanczos, C. (1938): Trigonometric Interpolation of Empirical and Analytic Functions, *J. Math. and Phys.*, vol. 17, pp. 123–199.
——— (1956): *Applied Analysis*, Prentice-Hall, Inc., Englewood Cliffs, N.J.
Lawson, C. L. (1964): Characteristic Properties of the Segmented Rational Minimax Approximation Problem, *Numer. Math.*, vol. 6, pp. 293–301.
Luke, Y. L. (1969): *The Special Functions and Their Approximations*, vols. 1 and 2, Academic Press, New York.
——— (1975): *Mathematical Functions and Their Approximations*, Academic Press, New York.
Maehly, H. J. (1960*a*): Rational Approximation for Transcendental Functions, pp. 57–62 in *Information Processing, UNESCO, Paris.*
——— (1960*b*): Methods for Fitting Rational Approximations, I, *J. Ass. Comput. Mach.*, vol. 7, pp. 150–162.
——— (1963): Methods for Fitting Rational Approximations, II, III, *J. Ass. Comput. Mach.*, vol. 10, pp. 257–277.
Milne-Thomson, L. M. (1933): *Calculus of Finite Differences*, Macmillan & Co., Ltd., London.
Minnick, R. C. (1957): Tshebysheff Approximation for Power Series, *J. Ass. Comput. Mach.*, vol. 4, pp. 487–504.
Novodvorskii, E. N., and I. S. Pinsker (1951): On a Process of Equalization of Maxima, *Usp. Mat. Nauk.*, vol. 6, pp. 174–181 (trans. A. Shenitzer, available from New York University library).
Powell, M. J. D. (1967): On the Maximum Error of Polynomial Approximations Defined by Interpolation and by Least Squares Criteria, *Comput. J.*, vol. 9, pp. 404–407.
Ralston, A. (1963): On Economization of Rational Functions, *J. Ass. Comput. Mach.*, vol. 10, pp. 278–282.
——— (1965): Rational Chebyshev Approximation by Remes' Algorithms, *Numer. Math.* vol. 7, pp. 322–330.

——— (1965): Rational Chebyshev Approximation, pp. 264–284 in *Mathematical Methods for Digital Computers* (A. Ralston and H. S. Wilf, eds.) vol. II, John Wiley and Sons, Inc., New York.

Rice, J. R. (1964, 1969): *The Approximation of Functions*, vol. 1, *Linear Theory*, vol. 2, *Advanced Topics*, Addison-Wesley Publishing Company, Inc., Reading, Mass.

Snyder, M. A. (1966): *Chebyshev Methods in Numerical Approximation*, Prentice-Hall, Inc., Englewood Cliffs, N.J.

Spielberg, K. (1961): Representation of Power Series in Terms of Polynomials, Rational Approximations and Continued Fractions, *J. Ass. Comput. Mach.*, vol. 8, pp. 613–627.

Stiefel, E. (1959): Numerical Methods of Tchebysheff Approximation, pp. 217–232, in *On Numerical Approximation* (R. E. Langer, ed.), The University of Wisconsin Press, Madison, Wis.

Wall, H. S. (1948): *Analytic Theory of Continued Fractions*, D. Van Nostrand Company, Inc., Princeton, N.J.

Watson, G. N. (1962): *Theory of Bessel Functions*, Cambridge University Press, New York.

Werner, H. (1962): Die konstruktive Ermittlung der Tschebyscheff-Approximierenden im Bereich der rationalen Funktionen, *Arch. Ration. Mech. Anal.*, vol. 11, pp. 368–384.

PROBLEMS

Section 7.2

1 (*a*) Verify Eq. (7.2-7). (*b*) Use (7.2-7) to derive (7.2-8).

***2** With $q_n(z)$ as in (7.2-5) let

$$g_n(z) = z^n + b_{n-2} z^{n-2} + \cdots + b_{n(\text{mod } 2)} z^{n(\text{mod } 2)}$$

$$h_n(z) = b_{n-1} z^{n-1} + b_{n-3} z^{n-3} + \cdots + b_{(n-1)(\text{mod } 2)} z^{(n-1)(\text{mod } 2)}$$

Assume $q_n(z)$ has $n - 1$ zeros with nonnegative real parts and that $h_n(z)$ is not identically zero.

(*a*) Show by induction that if $q_n(0) = 0$, then $g_n(z)$ has at least $n - 2$ pure imaginary zeros and $h_n(z)$ has at least $n - 3$ pure imaginary zeros.

(*b*) Derive the same result if $q_n(\pm iy) = 0$ for some y.

(*c*) Now assume $q_n(z)$ has no roots with zero real part. By considering the path in the complex plane consisting of a semicircle in the left half plane of radius R for sufficiently large R and the diameter of this semicircle on the imaginary axis, show that the results of parts (*a*) and (*b*) also hold in this case.

(*d*) Deduce from this that the squares of the zeros of $g_n(z)$ and $h_n(z)$ are all real and, finally, from this deduce that if $u_r(y)$ as defined by (7.2-10) is not identically 0, it must have all real zeros.

3 (*a*) Show that if all zeros of $p_n(x)$ have the same real part, the quadratic factorization of $p_n(x)$ is easy.

(*b*) Show that cases 1 and 3 on p. 289 always lead to nonzero b_{n-1} when n is even unless all zeros of $p_n(x)$ have the same real part.

(*c*) Show that if case 2 on p. 289 leads to $b_{n-1} = 0$ when n is even, then unless $u_r(y)$ is identically 0, δ can be found for which $b_{n-1} \neq 0$.

4 (*a*) Show that the choice $\delta = (a_{n-1} - 1)/n$ leads to a polynomial $q_n(y)$ in which $b_{n-1} = 1$.

(*b*) If, for this choice of δ, $u_r(y)$ has all real roots, show that the number of multiplications for the quadratic factor algorithm is $n/2$ when n is even and $(n + 1)/2$ when n is odd.

(*c*) Show that if $u_r(y)$ does not have all real roots, the quadratic-factor algorithm may fail at some point but that when this happens, at most two applications of synthetic division will enable the quadratic-factor algorithm process to be continued.

5 (*a*) Apply the quadratic-factor algorithm as presented in Sec. 7.2 to the polynomial

$$(x + 7)(x^2 + 6x + 10)(x^2 + 4x + 5)(x + 1)$$

(*b*) Apply the algorithm of Prob. 4 to this polynomial and the polynomial of Example 7.1.

***6** (*a*) Given A, B, C, D, E show how the equations

$$\begin{aligned} 2p + 1 &= A \\ p(p + 1) + 2q + a &= B \\ p(2q + a) + q + r + s &= C \\ p(r + s) + r + q(q + a) &= D \\ ar + q(r + s) &= E \end{aligned}$$

can be solved for a, p, q, r, and s by finding a real root of a certain cubic equation in q.

(*b*) Then defining $b = q - ac$, $c = p - a$, $d = s - bc$, $e = r - bc$, $f = F - rs$, show that

$$P(x) = x^6 + Ax^5 + Bx^4 + Cx^3 + Dx^2 + Ex + F$$

can be evaluated with three multiplications by the algorithm

$$\begin{aligned} y &= x(x + a) \\ w &= (y + b)(x + c) \\ P(x) &= (w + y + d)(w + e) + f \end{aligned}$$

(*c*) Show how this algorithm can be used to evaluate $x^6 + 13x^5 + 49x^4 + 33x^3 - 61x^2 - 37x + 3$. [Ref. : Knuth (1962).]

***7** Consider the following algorithm for computing x^n (when no lower powers are required):

(*a*) Write n as a binary number, for example, $9 = 1001$.

(*b*) Cancel the high-order 1 (001).

(*c*) Replace each 0 by S (square) and each 1 by SX (square, multiply by x) ($001 \rightarrow SSSX$).

(*d*) Starting from the left with x compute by squaring or multiplying by x as specified in the "code word" $\{x^9 = [(x^2)^2]^2x\}$.

Prove that this algorithm is valid for any n. *Hint:* Use induction. [Ref.: Knuth (1962).]

***8** (*a*) Show that any rational function

$$R_{mk}(x) = \frac{\sum_{j=0}^{m} a_j x^j}{\sum_{j=0}^{k} b_j x^j} \qquad \begin{matrix} m \geq k \\ b_k \neq 0 \end{matrix}$$

can be written

$$R_{mk}(x) = x\left(\sum_{j=0}^{m-k-1} s_j x^j\right) + \frac{\sum_{j=0}^{k} p_j x^{k-j}}{\sum_{j=0}^{k} q_j x^{k-j}} \qquad q_0 = 1$$

Find a recurrence relation for the p_j's and q_j's in terms of the a_j's and b_j's.

(*b*) Except in certain degenerate cases, show that we can write

$$\frac{\sum_{j=0}^{k} p_j x^{k-j}}{\sum_{j=0}^{k} q_j x^{k-j}} = D_0 + \frac{C_1}{\left(x + \sum_{j=1}^{k} \bar{p}_j x^{k-j}\right) \Big/ \left(\sum_{j=1}^{k} \bar{q}_j x^{k-j}\right)} \qquad q_0 = \bar{q}_1 = 1$$

Find D_0 and C_1 in terms of p_0, p_1, and q_1. When do the degenerate cases occur?

(*c*) Derive recurrence relations for the $\bar{p}_j$ and $\bar{q}_j$ in terms of the p_j and q_j. Indicate how you would solve these relations.

(*d*) Thus deduce the correctness of (7.2-16) in nondegenerate cases.

9 (*a*) Use the previous problem to derive the relation analogous to (7.2-16) when $m > k + 1$.

(*b*) Do the same for $m < k$ by considering $1/R_{mk}(x)$.

(*c*) In both cases, how many multiplications and divisions are required to evaluate $R_{mk}(x)$?

(*d*) Use the recurrence relations of the previous problem to transform $(x^4 - 5x^3 + 12x^2 - 11x + 2)/(x^2 - 3x + 4)$ to the form of part (*a*).

Section 7.3

10 (*a*) Derive Eqs. (7.3-5) using (7.3-4).

(*b*) For $m = k = 1$, show that the Padé approximation to cos x does not exist.

11 Let the Padé approximation $R_{mk}(x)$ be written

$$R_{mk}(x) = \frac{P_m(x)}{Q_k(x)} = \frac{P_m^{(m,k)}}{Q_k^{(m,k)}}$$

Assuming the existence of all relevant Padé approximations, show that the following recurrence relations exist:

$$P_m^{(m,k)} = P_{m-1}^{(m-1,k)} - DxP_{m-1}^{(m-1,k-1)} \qquad Q_k^{(m,k)} = Q_k^{(m-1,k)} - DxQ_{k-1}^{(m-1,k-1)}$$

$$P_m^{(m,k)} = P_m^{(m,k-1)} - ExP_{m-1}^{(m-1,k-1)} \qquad Q_k^{(m,k)} = Q_{k-1}^{(m,k-1)} - ExQ_{k-1}^{(m-1,k-1)}$$

where D and E are constants which also depend upon k and m. Find expressions for D and E in terms of the Maclaurin coefficients and the constants in the Padé approximations. [Ref.: Wall (1948), p. 15 and chap. 20.]

***12** (*a*) Show that there exists a Padé approximation $R_{mk}(x)$ to cos x which contains only even powers of x for any m and k.

(*b*) Thus find $R_{6,6}(x) = R_{3,3}(z)$ $(z = x^2)$.

(*c*) In the continued-fraction form of $R_{3,3}(z)$ given by (7.2-16), calculate D_0. From the value of D_0, deduce that in calculating cos x using this approximation there will be a loss of significance of about two digits.

(*d*) For $R_{3,3}(z)$, calculate d_7 as given by (7.3-6).

***13** In the notation of Prob. 12, consider adding a term $-\xi b_3 z^4$ to the numerator of $R_{3,3}(z)$, where b_3 is the coefficient of z^3 in the denominator.

(*a*) Calculate b_1, b_2, b_3, and a_3 in terms of ξ by requiring, as in Prob. 12, that $d_j = 0, j = 0, \ldots, 6$, in (7.3-4).

(*b*) For this approximation, calculate C_0 and D_0 in (7.2-16) in terms of ξ.

(*c*) Calculate d_7 in terms of ξ.

(*d*) Calculate those values of ξ for which $D_0 = 0$. For each value of ξ calculate d_7. Which value of ξ would be the best one to use? [Ref.: Kogbetliantz (1960), pp. 14, 25.]

14 Reciprocal differences. Define the kth reciprocal difference of $f(x)$ as

$$\rho_k(x_0, x_1, \ldots, x_k) = \frac{x_0 - x_k}{\rho_{k-1}(x_0, \ldots, x_{k-1}) - \rho_{k-1}(x_1, \ldots, x_k)} + \rho_{k-2}(x_1, \ldots, x_{k-1}) \qquad k \geq 2$$

$$\rho_1(x_0, x_1) = \frac{x_0 - x_1}{f(x_0) - f(x_1)} \qquad \rho_0(x_0) = f(x_0)$$

(*a*) Show that if the two end arguments of ρ_k are interchanged or if any two interior arguments are interchanged, the value of the reciprocal difference is unchanged.

(*b*) Deduce then that if the value is unchanged if the first two arguments are interchanged, the value of a reciprocal difference is independent of the order of any of its arguments. [Reciprocal differences are in fact symmetric in all their arguments, but the proof of this result is quite difficult; see Milne-Thomson (1933) for one such proof.]

15 (*a*) Show how a reciprocal-difference table can be set up analogously to a finite-difference table.

(*b*) Given values of ln x at an interval of .1 from $x = .3$ to .9 (see Example 3.3), calculate the reciprocal-difference table.

16 (*a*) Use the definition of reciprocal differences with $x_0 = x$ to derive the identity

$$f(x) = f(x_1) + \frac{x - x_1|}{|\phi_1(x_1, x_2)} + \frac{x - x_2|}{|\phi_2(x_1, x_2, x_3)} + \cdots + \frac{x - x_n|}{|\rho_n(x, x_1, \ldots, x_n) - \rho_{n-2}(x_1, \ldots, x_{n-1})}$$

where $\qquad \phi_k(x_1, x_2, \ldots, x_{k+1}) = \rho_k(x_1, x_2, \ldots, x_{k+1}) - \rho_{k-2}(x_1, x_2, \ldots, x_{k-1})$

(*b*) Show that even if the last term on the right-hand side in this identity is deleted, the above is still an equality when $x = x_i$, $i = 1, \ldots, n$. This *interpolation formula* is called *Thiele's interpolation formula*.

(*c*) Use this interpolation formula and the table of Prob. 15*b* to estimate ln .54 (cf. Example 3.2). Do this by computing successive *convergents* of the continued fraction, i.e., using first one reciprocal difference, then two, etc.

***17** Reciprocal derivatives. Define the jth reciprocal derivative of $f(x)$ at a as

$$R_j f(a) = \lim_{x_1, \ldots, x_{j+1} \to a} \rho_j(x_1, \ldots, x_{j+1})$$

(*a*) Show that $R_1 f(a) = 1/f'(a)$.

(*b*) Using the definition of ϕ_k in Prob. 16, show that

$$\lim_{x_1, \ldots, x_{j+1} \to a} \phi_j(x_1, \ldots, x_{j+1}) = R_j f(a) - R_{j-2} f(a)$$

$$= \lim_{x_{j+1} \to a} \frac{x_{j+1} - a}{\rho_{j-1}(x_{j+1}, a, \ldots, a) - \rho_{j-1}(a, \ldots, a)} = \left. \frac{1}{(\partial/\partial x)\rho_{j-1}(x, \ldots, x)} \right|_{x=a}$$

if the various limits exist.

(*c*) Thus deduce that

$$R_j f(a) - R_{j-2} f(a) = j R_1 R_{j-1} f(a)$$

(*d*) Assuming that $f(x)$ has reciprocal derivatives of all orders, use Prob. 16 to show that as $n \to \infty$,

$$f(x) = f(a) + \frac{x-a|}{|R_1 f(a)} + \frac{x-a|}{|2R_1 R_1 f(a)} + \cdots + \frac{x-a|}{|kR_1 R_{k-1} f(a)} + \cdots$$

This result, known as *Thiele's theorem*, is the continued-fraction analog of Taylor's theorem.

***18** (*a*) Show that the reciprocal derivatives of $f(x)$ can conveniently be calculated using the recurrence relation

$$R_j(x) = R_{j-2}(x) + r_{j-1}(x) \qquad r_j(x) = \frac{j+1}{R_j'(x)} \qquad R_0(x) = f(x) \qquad R_{-1}(x) = 0$$

and that r_{j-1} therefore becomes the jth denominator in Thiele's theorem.

(*b*) Use Thiele's theorem to derive a continued-fraction expansion of $\ln(1+x)$ about $x = 0$. Derive a general form for $r_j(0)$. Use the first four convergents of this result to estimate ln .54, and compare with the result of Prob. 16*c*.

(*c*) Similarly, derive a continued-fraction expansion about $x = 0$ for e^x. Find the general form for $r_j(0)$. Use the first five convergents of this expansion to approximate e.

***19** (*a*) Write Thiele's interpolation formula as

$$f(x) = y(x) + E(x) = \frac{P(x)}{Q(x)} + E(x)$$

where $P(x)$ and $Q(x)$ are the polynomials that result when the continued fraction with terms through that in ϕ_{n-1} is converted to a rational function. By considering

$$F(z) = Q^2(z)[f(z) - y(z)] - Q^2(x)[f(x) - y(x)]\frac{p_n(z)}{p_n(x)}$$

with $p_n(x) = (x - x_1) \cdots (x - x_n)$ show that

$$E(x) = \frac{p_n(x)}{n!Q^2(x)} \frac{d^n}{dx^n}[Q^2(x)f(x)]_{x=\xi}$$

with ξ in the interval spanned by $x_1, \ldots, x_n$, and x.

(*b*) By taking the limit as $x_i \to a$, $i = 1, \ldots, n$ find the error incurred by truncating the continued-fraction expansion in Prob. 17*d* after the term in $(n-1)R_1 R_{n-2}(a)$.

(*c*) Deduce from the limiting form of the error term that the limiting form of the rational function in part (*a*) when $a = 0$ is identical to a Padé approximation with the same degree polynomials in numerator and denominator and thus find a general form for the error in a Padé approximation. [Ref.: Probs. 14 to 19, Milne-Thomson (1933).]

Section 7.4

20 (*a*) Derive each of the Padé approximations (7.4-5).

(*b*) Use (7.3-6) to estimate the errors in the approximations (7.4-4) and (7.4-5). In each case determine how good these estimates are by comparing them with the values of the errors in Table 7.1.

21 (*a*) Find all Padé approximations for (i) $f(x) = \sin x$ with $N = 5$; (ii) $f(x) = \cos x$ with $N = 4$.

(*b*) Display an algorithm for the calculation of $\sin x$ on $(-\infty, \infty)$ by first showing how any value of x outside of $[-(\pi/2), \pi/2]$ can be reduced to this interval. Then show how one of the approximations of part (*a*) can be applied with an argument less than $\pi/4$ in magnitude followed by an adjustment of the sign if necessary. Do the same for cos x on $(-\infty, \infty)$. What is the advantage of keeping the argument of the Padé approximation small?

(*c*) Draw a graph of the error on $[-(\pi/4), \pi/4]$ for each of the approximations of part (*a*).

22 (*a*) Use the technique of Probs. 8 and 9 to express each of the approximations (7.4-4) and (7.4-5) in the form of a polynomial plus a continued fraction.

(*b*) Use the results of Sec. 7.2 to determine how many multiplications and divisions are required to evaluate each of these five approximations (i) in continued-fraction form; (ii) in rational-function form.

(*c*) Which approximation can be computed most rapidly if the division time is (i) the same as the multiplication time; (ii) $1\frac{1}{2}$ times as great as the multiplication time; (iii) twice as great as the multiplication time?

Section 7.5

23 (*a*) By making an appropriate change of variable from $[-1, 1]$ to $[0, 1]$, derive the *shifted* Chebyshev polynomials $T_r^*(x)$ which are such that

$$T_r^*(x) = \cos r\theta \qquad x = \cos^2 \frac{\theta}{2}$$

(*b*) Find a recurrence relation for the shifted Chebyshev polynomials and use it to generate $T_r^*(x)$, $r = 1, \ldots, 6$.

(*c*) State and prove a theorem similar to Theorem 7.1 for the shifted Chebyshev polynomials.

***24** (*a*) With the Chebyshev polynomial of degree i written as

$$T_i(x) = \sum_{j=0}^{i} t_j^{(i)} x^j$$

show that the nonzero $t_j^{(i)}$ are given by

$$t_j^{(i)} = 2^{j-1}\left[2\binom{\frac{1}{2}(i+j)}{\frac{1}{2}(i-j)} - \binom{\frac{1}{2}(i+j)-1}{\frac{1}{2}(i-j)}\right](-1)^{(i-j)/2}$$

(*b*) Similarly, writing

$$T_i^*(x) = \sum_{j=0}^{i} u_j^{(i)} x^j$$

show that

$$u_j^{(i)} = 2^{2j-1}\left[2\binom{i+j}{i-j} - \binom{i+j-1}{i-j}\right](-1)^{i+j}$$

[Ref.: Lanczos (1956), pp. 454–457.]

Section 7.6

25 (*a*) Derive the identity (7.6-9). (*b*) Use this result to derive (7.6-10) and (7.6-11).

26 (*a*) With $f(x) = e^x$, use trigonometric interpolation to calculate $y_L(\theta)$ as given by (7.6-16) with $L = 4$. Compare this result with (7.6-4).

(*b*) Convert the result of part (*a*) to a polynomial analogous to (7.6-5). Plot the error in this approximation on $[-1, 1]$ and compare with the error in $T_{4,0}(x)$.

***27** (*a*) Let $f(x) = \sum_{j=0}^{\infty} a_j x^j$ be the Maclaurin series for $f(x)$ and let $T_i(x) = \sum_{j=0}^{i} t_j^{(i)} x^j$. By substituting these series in (7.6-2), show that

$$c_{2n} = \sum_{j=0}^{\infty} R_{2j}^{(2n)} a_{2j} \qquad c_{2n+1} = \sum_{j=0}^{\infty} R_{2j+1}^{(2n+1)} a_{2j+1}$$

where

$$R_{2j}^{(2n)} = \frac{(2j)!}{2^{2j-1}(j!)^2} \left[t_0^{(2n)} + t_2^{(2n)} \frac{2j+1}{2j+2} + \cdots + t_{2n}^{(2n)} \frac{(2j+1)(2j+3)\cdots(2j+2n-1)}{(2j+2)(2j+4)\cdots(2j+2n)} \right]$$

$$R_{2j+1}^{(2n+1)} = \frac{(2j)!}{2^{2j-1}(j!)^2} \left[t_1^{(2n+1)} \frac{2j+1}{2j+2} + t_3^{(2n+1)} \frac{(2j+1)(2j+3)}{(2j+2)(2j+4)} + \cdots + t_{2n+1}^{(2n+1)} \frac{(2j+1)(2j+3)\cdots(2j+2n+1)}{(2j+2)(2j+4)\cdots(2j+2n+2)} \right]$$

(*b*) Approximate the coefficients in (7.6-4) by truncating the series for c_{2n} and c_{2n+1} after the term in $j = 3$, and then convert the approximation to a power series.

(*c*) As in Prob. 26*b*, compare this approximation with $T_{4,0}(x)$. [Ref.: Minnick (1957).]

28 (*a*) Derive the Chebyshev expansions for sin x and cos x. By drawing a graph of the error on $[-(\pi/4), \pi/4]$ compare the approximations obtained by truncating these expansions with the corresponding results of Prob. 21 [see Watson (1962, p. 21) for the integrals (7.6-2)].

(*b*) Calculate $T_{3,2}(x)$ for $f(x) = \sin x$. Convert the result to rational-function form. Draw a graph of the error on $[-(\pi/4), \pi/4]$ and compare the result with $R_{3,2}(x)$ derived in Prob. 21.

(*c*) Calculate $T_{2,2}(x)$ for $f(x) = \cos x$. Convert the result to rational-function form, draw a graph of the error on $[-(\pi/4), \pi/4]$, and compare with $R_{2,2}(x)$ derived in Prob. 21.

29 For $f(x) = e^x$, calculate $T_{3,1}(x)$ and $T_{1,3}(x)$. Draw a graph of the error on $[-1, 1]$ and compare with the corresponding Padé approximations.

Section 7.7

30 (*a*) When the Maclaurin expansion of $f(x)$ contains only even powers or only odd powers, how should (7.7-3) be modified?

(*b*) Use this result to calculate $C_{5,0}(x)$ for $f(x) = \sin x$ with $\alpha = \pi/4$. Draw a graph of the error on $[-(\pi/4), \pi/4]$ and compare with the corresponding results of Probs. 21 and 28.

(*c*) Calculate $C_{4,0}(x)$ for $f(x) = \cos x$ with $\alpha = \pi/4$. Draw a graph of the error on $[-(\pi/4), \pi/4]$ and compare with the corresponding results of Probs. 21 and 28.

31 (*a*) Show that with $C_{mk}(x)$ defined as in (7.7-10) and (7.7-11), $f(x) - C_{mk}(x)$ satisfies (7.7-7).

(*b*) For $f(x) = e^x$ derive an approximation $C_{2,2}(x)$ with $\alpha = 1$ as in Example 7.5, using as the sequence (7.7-8) (i) $R_{0,0}^{(0)}(x)$, $R_{1,0}^{(1)}(x)$, $R_{2,0}^{(2)}(x)$, $R_{3,0}^{(3)}(x)$; (ii) $R_{0,0}^{(0)}(x)$, $R_{1,0}^{(1)}(x)$, $R_{0,2}^{(2)}(x)$, $R_{3,0}^{(3)}(x)$. Why doesn't the choice of the second and fourth rational functions make any difference? In each case compare the resulting approximation with $T_{2,2}(x)$, $R_{2,2}(x)$, and the approximation of Example 7.5.

32 (*a*) Calculate $C_{3,1}(x)$ and $C_{1,3}(x)$ for e^x with $\alpha = 1$. By drawing graphs of the errors on $[-1, 1]$, compare the errors with the corresponding Padé and Chebyshev approximations. [Use any convenient sequence (7.7-8).]

(*b*) Show that the derivation of $C_{1,1}(z)$ for $\sin(\sqrt{|z|})/\sqrt{|z|}$ on $[-(\pi^2/16), \pi^2/16]$ requires substantially less effort than the derivation of $C_{3,2}(x)$ for sin x on $[-(\pi/4), \pi/4]$. Is $xC_{1,1}(x^2)$ equal to $C_{3,2}(x)$?

(*c*) Actually derive $xC_{1,1}(x^2)$ with $\alpha = \pi/4$. By drawing the graph of the error on $[-(\pi/4), \pi/4]$, compare this approximation with corresponding Padé and Chebyshev approximations.

(*d*) Similarly derive $C_{1,1}(x^2)$ for cos x on $[-(\pi/4), \pi/4]$ and compare this approximation with corresponding Padé and Chebyshev approximations.

***33** A common iterative method to compute $\sqrt{a}$ for a in $(0, \infty)$ is (cf. Prob. 10, Chap. 2)

$$x_{n+1} = \tfrac{1}{2}\left(x_n + \frac{a}{x_n}\right)$$

This iteration converges very rapidly if the initial approximation x_1 is sufficiently close to $\sqrt{a}$ (see Sec. 8.4). To get a good initial approximation, it is convenient to use a rational approximation. We begin by writing $a = 10^{2m} \times b$, where $.01 < b \leq 1$.

(*a*) Derive a Padé approximation $R_{2,2}(x)$ to $\sqrt{x}$ on [.01, 1] by first making a change of variable so that the new interval is centered at the origin.

(*b*) Similarly, derive $C_{2,2}(x)$ with $\alpha = .495$ using $R^{(0)}_{0,0}(x)$, $R^{(1)}_{1,0}(x)$, $R^{(2)}_{1,1}(x)$, $R^{(3)}_{2,1}(x)$. By drawing a graph compare the errors in the approximations of parts (*a*) and (*b*). [Ref.: Kogbetliantz (1960), pp. 33–35.]

34 (*a*) How must the development in Sec. 7.7 be modified if the interval $[-\alpha, \alpha]$ is replaced by the interval $[0, \alpha]$?

(*b*) Use these modifications to derive an economized approximation $C^*_{2,2}(x)$ to e^x on [0, 1]. Draw a graph of the error.

***35** The τ method. Consider the differential equation

$$L(y) = p_n(x)\frac{d^n y}{dx^n} + p_{n-1}(x)\frac{d^{n-1}y}{dx^{n-1}} + \cdots + p_0(x)y + P(x) = 0$$

$$y(0) = y_0 \qquad y^{(i)}(0) = y_i \qquad i = 1, \ldots, n-1$$

where $p_j(x)$ is a polynomial of degree d_j and $P(x)$ is a polynomial of degree d.

(*a*) Assume a solution

$$y(x) = \sum_{j=0}^{m} a_j x^j \qquad m \geq n$$

and let
$$D = \max\,(d, d_0 + m, d_1 + m - 1, \ldots, d_n + m - n)$$

Show that substituting this solution into the differential equation leads in general to a system of $D + 1 + n$ equations in $m + 1$ unknowns if the initial conditions are to be satisfied.

(*b*) Show that, in general, the differential equation

$$L(y) = \sum_{i=1}^{D-m+n} \tau_i T^*_{m-n+i}(x)$$

does have a solution of the form of part (*a*), where the τ_i's are real numbers and T^*_{m-n+i} is the shifted Chebyshev polynomial of degree $m - n + i$ (see Prob. 23). Thus deduce that if the τ_i's are small, the solution of this differential equation is a good approximation to the solution of $L(y) = 0$.

(*c*) Use this method with $m = 4$ to approximate e^x. Draw a graph of the error on [0, 1] and compare this with the error in $R_{4,0}(x)$ on [0, 1]. [This method is of particular value when $f(x)$ does not have a convergent polynomial or continued-fraction expansion.] [Ref.: Lanczos (1956), pp. 464–469.]

Sections 7.8 and 7.9

36 Let $P_m^*(x)/Q_k^*(x)$ be irreducible where $P_m^*(x)$ and $Q_k^*(x)$ are polynomials of degree m and k, respectively. Prove that any polynomial $A(x)$ of degree $m + k$ or less can be written

$$A(x) = Q_k^*(x)a(x) - P_m^*(x)b(x)$$

where $a(x)$ and $b(x)$ are polynomials of degree less than or equal to m and k, respectively.

***37** (*a*) Let r be the greatest lower bound of all r_{mk} in (7.8-1) for given $f(x)$, $[a, b]$, m, and k. Why does an infinite sequence of rational approximations $R_{mk}^{(i)}(x) = P_m^{(i)}(x)/Q_k^{(i)}(x)$ exist such that $r_{mk}^{(i)} \to r$ as $i \to \infty$?

(*b*) Normalize $Q_k^{(i)}(x)$ so that $\sum_{j=0}^{k} (b_j^{(i)})^2 = 1$ and prove that for this normalization $\sum_{j=0}^{m} (a_j^{(i)})^2$ is bounded. Thus deduce that a subsequence $\{R_{mk}^{(i_n)}(x)\}$ exists such that

$$\lim_{n\to\infty} a_j^{(i_n)} = A_j \qquad \lim_{n\to\infty} b_j^{(i_n)} = B_j$$

for some constants A_j and B_j.

(*c*) Let

$$R(x) = \frac{\sum_{j=0}^{m} A_j x^j}{\sum_{j=0}^{k} B_j x^j} \qquad \text{and} \qquad \bar{r} = \max_{a \le x \le b} |f(x) - R(x)|$$

Prove that $\{R_{mk}^{(i)}(x)\}$ converges uniformly to $R(x)$.

(*d*) Use this result to show that $\bar{r} \le r$, and thus deduce the existence of the Chebyshev approximation.

***38** Suppose there exist two rational approximations $R_{mk}^{(1)}(x)$ and $R_{mk}^{(2)}(x)$ which minimize r_{mk} as given in Eq. (7.8-1). Let L_1 and L_2 be, respectively, the number of alternating extrema of the errors in $R_{mk}^{(1)}(x)$ and $R_{mk}^{(2)}(x)$. Assume $L_2 \ge L_1$ and let $\alpha_1, \ldots, \alpha_{L_2}$ be the abscissas of the error extrema for $R_{mk}^{(2)}(x)$.

(*a*) Define $\Delta(x) = R_{mk}^{(1)}(x) - R_{mk}^{(2)}(x)$. Show that

$$\operatorname{sgn}\, [f(\alpha_j) - R_{mk}^{(2)}(\alpha_j)] = \operatorname{sgn}\, \Delta(\alpha_j)$$

if $\Delta(\alpha_j) \ne 0$.

(*b*) Suppose $\Delta(\alpha_{i-1}) \ne 0$, $\Delta(\alpha_j) = 0$, $j = i, \ldots, i + k$, and $\Delta(\alpha_{i+k+1}) \ne 0$. Show that the number of zeros of $\Delta(x)$ (counting multiplicities) in $[\alpha_{i-1}, \alpha_{i+k+1}]$ is even if k is even and odd if k is odd. Thus deduce that the number of zeros (counting multiplicities) in this interval is at least $k + 2$.

(*c*) Thus deduce that $\Delta(x)$ has at least $L_2 - 1$ zeros in (a, b).

(*d*) From this deduce a contradiction and, therefore, the uniqueness of the Chebyshev approximation.

(*e*) Prove Theorem 7.3.

[Ref.: Probs. 37 and 38, Achieser (1956), pp. 52–57.]

39 (*a*) Show that the results of Sec. 7.8 and the previous three problems are unchanged if we consider $s(x)R_{mk}(x)$ as the approximation to $f(x)$ where $s(x)$ is continuous and does not vanish in (a, b).

(*b*) How do the algorithms of Sec. 7.9 need to be modified for approximations of the form $s(x)R_{mk}(x)$ where $s(x)$ is a continuous function which does not vanish in (a, b)?

40 Consider approximating $f(x) = x^2 - 1$ on $[-1, 1]$ by a rational function with $m = k = 1$. Show that the Chebyshev approximation of this type is a constant and thus deduce that in Theorem 7.2 $d = 1$.

41 Calculate the Chebyshev approximation to e^x on $[-1, 1]$ of the form

$$\frac{a_0 + a_1 x}{1 + b_1 x}$$

by (*a*) calculating the Padé approximation $R_{1,1}(x)$; (*b*) calculating the economized approximation $C_{1,1}(x)$; (*c*) calculating $R^*_{1,1}(x)$ using either of the algorithms of Sec. 7.9.

CHAPTER

EIGHT

THE SOLUTION OF NONLINEAR EQUATIONS

8.1 INTRODUCTION

The remaining three chapters of this book will be concerned with the solution of linear and nonlinear equations and systems of equations. In this chapter we shall be concerned with the solution of nonlinear algebraic and transcendental equations. It might seem illogical to discuss the solution of nonlinear before linear equations, but the relation between the two subjects is tenuous indeed. And since the solution of simultaneous linear equations provides the gateway to many of the advanced topics in numerical analysis not discussed in this book, it seems reasonable to leave linear equations to the last.

This chapter is divided naturally into two main sections which consider two not unrelated but mainly separate problems: (1) the search for *real* roots of the equation

$$f(x) = 0 \tag{8.1-1}$$

where x is a real variable and $f(x)$ is any reasonably well-behaved function, and (2) the search for *real and complex* roots of

$$P(x) = 0 \tag{8.1-2}$$

where $P(x)$ is a polynomial. This is not to say that we are never interested in complex roots of the general equation $f(x) = 0$, but this is a comparatively rare problem.

Except in Sec. 8.8, we shall restrict ourselves to single equations. One reason is that this case is more common than that of simultaneous nonlinear equations. An equally important reason, however, is that the solution of simultaneous nonlinear equations is a difficult problem about which much current research is centered.

In the case of the solution of simultaneous linear equations, the problem is not to find a closed form for the solution but rather to find an efficient algorithm for *computing* the known solution. For the single nonlinear equations of interest here, we assume no solution in closed form can be found, for if so, it would generally be easy to compute the solution, e.g., quadratic equations. Thus we must seek methods which lead to approximate solutions. Our technique throughout this chapter will be to develop iterative techniques for the solution of (8.1-1) and (8.1-2) (cf. Sec. 5.5-1). In considering these iterative techniques, we shall wish to answer two basic questions: (1) Does the iteration converge? (2) If so, how fast does it converge? Perhaps surprisingly to the reader, the first of these questions will occupy us much less than the second in our consideration of (8.1-1). This is because the convergence question is in one sense very easy to answer and in another sense too hard to answer. It is easy to answer because there is generally little difficulty in showing that if the initial approximation(s) to a root α of (8.1-1) are sufficiently close to the root, the iteration will converge to α. (In a very few cases the iteration will converge independent of the initial approximation.) But the phrase "sufficiently close" points the way to the hard part of the question. How close the initial approximation(s) have to be to α for convergence depends generally upon the value of a derivative of $f(x)$ at some unknown point on the interval spanned by the initial approximation(s) and α. This presents the same problems we have found previously in the estimation of error terms.

Moreover, in practical problems, enough a priori knowledge of the desired root of the equation is often known to ensure that convergence of the iterations is not a problem. When a priori knowledge is poor, it is often advisable to use a method which converges independent of the starting values (but, alas, usually slowly) until a good approximation is obtained and then to switch over to a more rapidly converging method. Thus we conclude that among those methods whose convergence depends upon the initial approximation, it is difficult to compare the convergence properties of various methods and often it is not significant.

However, it is both possible and important to compare the relative *rates* of convergence of various iteration methods. As we shall indicate, analogous to the case of the numerical solution of ordinary differential equations, the computational efficiency of a method depends upon the number of evaluations of $f(x)$ and its derivatives.

In the special case of polynomial equations, good a priori knowledge often is not available and, moreover, whereas often only a single root of

$f(x) = 0$ is desired, all the roots of $P(x) = 0$ may be desired. We shall therefore place rather more emphasis on methods whose convergence is assured independent of the starting value in the case of polynomial equations than in the general case.

It is worth noting that if an iterative method for the solution of (8.1-1) converges, the only limitation on the accuracy of the root is in the number of digits carried in the computation. That is, the roundoff error *in a single iteration* is the only inherent limitation on the accuracy. This seldom creates any problems in the solution of (8.1-1).

The methods we now discuss all yield a single real root of (8.1-1). In cases where we are interested in more than one root, once we have computed the first m roots, $\alpha_1, \ldots, \alpha_m$, we can compute the $(m + 1)$st root, α_{m+1}, by applying any of these methods to the function

$$f_m(x) = \frac{f(x)}{\prod_{i=1}^{m} (x - \alpha_i)}$$

In addition to this method of *implicit deflation* or *suppression*, it is possible to work with the original function $f(x)$ if there exists some knowledge about the location of the roots we are seeking. In this case, it may turn out that an iterative method will converge to the root closest to an initial approximation, although there is no guarantee for this. The only time we can guarantee convergence to a root different from previously computed roots, $\alpha_1, \ldots, \alpha_m$, is if we have an interval $[a, b]$ such that $f(a)f(b) < 0$ and $\alpha_i \notin [a, b]$, $i = 1, \ldots, m$. In this case, the methods of bisection (Sec. 2.2) and false position and its modifications (Sec. 8.3) will find some root α_{m+1} in $[a, b]$.

8.2 FUNCTIONAL ITERATION

Let $f(x)$ be a continuous real-valued function with as many derivatives as are required in what follows. Furthermore, we assume that in some neighborhood of the desired root α of $f(x) = 0$ the function $f(x)$ has an inverse; i.e., we assume α is a simple root. Our approach here is to derive iterative methods for the solution of $f(x) = 0$ by using inverse interpolation, a subject introduced briefly in Sec. 3.6.

Let $g(y)$ be the function inverse to $f(x)$. Given the points $y_j, j = 1, \ldots, n$, we can approximate $g(y)$ using a Lagrangian interpolation formula as

$$g(y) \approx h(y) = \sum_{j=1}^{n} l_j(y)g(y_j) = \sum_{j=1}^{n} l_j(y)x_{i+1-j} \tag{8.2-1}$$

using, for later convenience, the notation $x_{i+1-j} = g(y_j)$. Our object is to find a value α of x at which $y = 0$. Since $\alpha = g(0)$, we get from (8.2-1) an approximation x_{i+1} for α given by

$$x_{i+1} = h(0) = \sum_{j=1}^{n} l_j(0) x_{i+1-j} \tag{8.2-2}$$

where, using (3.2-3),

$$l_j(0) = \frac{(-1)^{n-1} y_1 y_2 y_3 \cdots y_{j-1} y_{j+1} \cdots y_n}{(y_j - y_1)(y_j - y_2) \cdots (y_j - y_{j-1})(y_j - y_{j+1}) \cdots (y_j - y_n)} \tag{8.2-3}$$

From (3.2-9)

$$\alpha - x_{i+1} = \frac{g^{(n)}(\eta)}{n!} (-1)^n y_1 y_2 \cdots y_n \tag{8.2-4}$$

with η in the interval spanned by $y_1, \ldots, y_n$ and 0. The derivative of the inverse function can be calculated in terms of derivatives of $f(x)$ {1}.

Equation (8.2-2) defines an iterative method for finding a root of (8.1-1) if the iteration converges. That is, using the points $x_i, \ldots, x_{i-n+1}$, we calculate x_{i+1}, and then replacing x_{i-n+1} by x_{i+1}, we calculate x_{i+2} using the remaining n points, and so on. This method is one example of an *n-point iteration function* whose most general form is

$$x_{i+1} = F_i(x_i, x_{i-1}, \ldots, x_{i-n+1}) \tag{8.2-5}$$

where the iteration function F_i will in general involve not only $x_i, \ldots, x_{i-n+1}$ but also values of $f(x)$ and some of its derivatives evaluated at one or more of the points x_k. For example, in (8.2-2) $l_j(0)$ involves values of $f(x)$ at the points $x_{i+1-j}, j = 1, \ldots, n$. We shall assume throughout this chapter that F_i has as many continuous derivatives as required in a neighborhood of α.

Methods for finding the roots of (8.1-1) based on (8.2-5) are called *functional iteration methods;* all the methods we shall consider for the solution of (8.1-1) are of this form. The subscript i on F_i is necessary only when the iteration function itself may change from one iteration to the next; generally the iteration function will be *stationary*, i.e., independent of i, in which case we shall write F instead of F_i. If the iteration using a stationary iteration function converges to a root α, we must have

$$\alpha = F(\alpha, \alpha, \ldots, \alpha) \tag{8.2-6}$$

A concept which will be basic to our discussion of iteration methods is that of the *order* of the method. First we define the error in the ith iterate to be

$$\epsilon_{i+1} = \alpha - x_{i+1} \tag{8.2-7}$$

where the starting values are taken to be $x_{2-j}, j = 1, \ldots, n$. Now assume that the iteration (8.2-5) converges so that $\lim_{i\to\infty} x_i = \alpha$. Then, if there exists a real number $p \geq 1$ such that

$$\lim_{i\to\infty} \frac{|x_{i+1} - \alpha|}{|x_i - \alpha|^p} = \lim_{i\to\infty} \frac{|\epsilon_{i+1}|}{|\epsilon_i|^p} = C \not\equiv 0 \tag{8.2-8}$$

we say that the method is of order p at α {2}. If $p = 1$, 2, or 3, convergence is said to be linear, quadratic, or cubic, respectively. The constant C is called the *asymptotic error constant;* it depends on $f(x)$. The requirement that $C \not\equiv 0$ is to be interpreted to mean that $C \neq 0$ for a general function $f(x)$. This assures the uniqueness of p. For a particular function $f(x)$, C may be 0, in which case, for this function, the iteration will converge more rapidly than usual. When $p = 1$, C must be less than or equal to 1 in order for the method to converge {2}, but for $p > 1$, C need not be less than 1 for convergence. If $p = 1$ and $C < 1$, we are assured of convergence if we start sufficiently close to α; this is not so for $C = 1$.

We assumed in the foregoing that α is a simple root. But we shall often apply iteration methods derived using this assumption to functions which have multiple roots. We shall see that the order of the iteration will depend on the multiplicity of the root.

8.2-1 Computational Efficiency

We could compare methods of functional iteration on the basis of how fast they converge, i.e., on their order, but it makes more sense to compare their *computational efficiency*, which is a measure of how much computation must be done to arrive at a given accuracy in a root. In order to arrive at a definition of computational efficiency, let us consider trying to find a root of (8.1-1) using two different iteration methods, both starting from the same initial approximation. Let the two methods have orders p_1 and p_2, respectively. For large enough i, that is, as we near convergence, the approximations

$$|\epsilon_{i+1}^{(1)}| = C_1 |\epsilon_i^{(1)}|^{p_1} \qquad |\epsilon_{i+1}^{(2)}| = C_2 |\epsilon_i^{(2)}|^{p_2} \tag{8.2-9}$$

are valid, where C_1 and C_2 are the asymptotic error constants of the two methods. Let

$$S_i = -\ln |\epsilon_i^{(1)}| \qquad T_i = -\ln |\epsilon_i^{(2)}| \tag{8.2-10}$$

Then $$S_{i+1} = -\ln C_1 + p_1 S_i \qquad T_{i+1} = -\ln C_2 + p_2 T_i \tag{8.2-11}$$

The solutions of these difference equations are {3}

$$S_i = S_1 p_1^i - \ln (C_1^{(p_1{}^i - 1)/(p_1 - 1)}) \qquad T_i = T_1 p_2^i - \ln (C_2^{(p_2{}^i - 1)/(p_2 - 1)}) \tag{8.2-12}$$

where the initial values S_1 and T_1 are equal. Let the number of steps required to get the desired degree of convergence be I and J, respectively, for methods 1 and 2. Hence $|\epsilon_I^{(1)}|$ and $|\epsilon_J^{(2)}|$, and therefore S_I and T_J, are essentially equal. Then using (8.2-12), we get

$$S_1(p_1^I - p_2^J) + \ln \frac{C_2^{(p_2{}^J - 1)/(p_2 - 1)}}{C_1^{(p_1{}^I - 1)/(p_1 - 1)}} = 0 \tag{8.2-13}$$

If θ and ϕ are, respectively, the costs per iteration, i.e., the amount of computation required per iteration, of the two methods, then the total costs of the computation are θI and ϕJ, respectively. The quantities θ and ϕ can be estimated from the iteration function, but (8.2-13) gives us no obvious way of relating I and J. It is often a good assumption, however, that the second term in (8.2-13) is small compared with the first (as will happen, for example, if C_1 and C_2 are both close to unity). In this case we get

$$\frac{I}{1/\ln p_1} \approx \frac{J}{1/\ln p_2} \tag{8.2-14}$$

Therefore, a measure of the cost of a method is the product of θ, the cost per iteration, and the reciprocal of the logarithm of the order p. The efficiency index, which is the reciprocal of the cost, may be defined as $(\ln p)/\theta$ or, to use the more usual definition,

$$EI = p^{1/\theta} \tag{8.2-15}$$

A particular case in which (8.2-13) can be used directly to estimate I/J is considered in {11}.

It is worth noting that since the order of a method is a property local to the neighborhood of a root, the efficiency index measures only how good a method is when it is near convergence. A determination of the efficiency of a method outside of the neighborhood of a root is generally extremely difficult.

The cost per iteration will, by analogy with our argument in Chap. 5, depend mainly on the number of evaluations of $f(x)$ and its derivatives required at each step and not on the arithmetic operations required to combine these quantities in the iteration function F_i of (8.2-5). We noted in Sec. 5.9-1 that, having computed $f(x)$, we can often compute $f'(x)$ quite cheaply on a digital computer. For example, if $f(x)$ is composed of elementary functions, the chief cost of evaluating $f(x)$ is in the evaluation of these elementary functions. Thus, since $f'(x)$ will also be some combination of these elementary functions, the evaluation of $f'(x)$ is simple {4}. Another example is the case where $f(x) = \int_0^x g(t)\,dt$; then $f'(x) = g(x)$, which will usually have been computed previously to get $f(x)$.

8.3 THE SECANT METHOD

This is one of the oldest methods known for the solution of $f(x) = 0$, but it has been surprisingly neglected until recently, when its important advantages, particularly for use on computers, have again been realized. For this reason and because it serves to illustrate various aspects of iteration methods, we shall discuss it before considering some more general aspects of functional iteration. In this section and the following two we assume that α is a simple root of (8.1-1).

Our approach in this section will be to use linear inverse interpolation, that is, (8.2-2) with $n = 2$, to derive methods of the form (8.2-5). One such method illustrated in Fig. 8.1, is the *method of false position* or *regula falsi.* Suppose we can find two points x_1 and x_2 such that $f(x_1)f(x_2) < 0$. The chord joining y_1 and y_2 intersects the x axis at a point x_3. Then choosing x_3 and x_i, $i = 1$ or 2, such that $f(x_3)f(x_i) < 0$ (in the case shown $i = 1$), we repeat the above procedure to obtain x_4 and so on. From a study of Fig. 8.1 it is clear that this method converges for any continuous function $f(x)$; we leave the details of the proof to a problem {5}. The method of bisection discussed in Sec. 2.2 also converges for all continuous functions.

The method of false position is a nonstationary iteration method in general. From Fig. 8.1 we have

$$
\begin{aligned}
x_3 &= \frac{y_2}{y_2 - y_1} x_1 + \frac{y_1}{y_1 - y_2} x_2 \\
x_4 &= \frac{y_3}{y_3 - y_1} x_1 + \frac{y_1}{y_1 - y_3} x_3 \qquad (8.3\text{-}1) \\
x_5 &= \frac{y_4}{y_4 - y_3} x_3 + \frac{y_3}{y_3 - y_4} x_4
\end{aligned}
$$

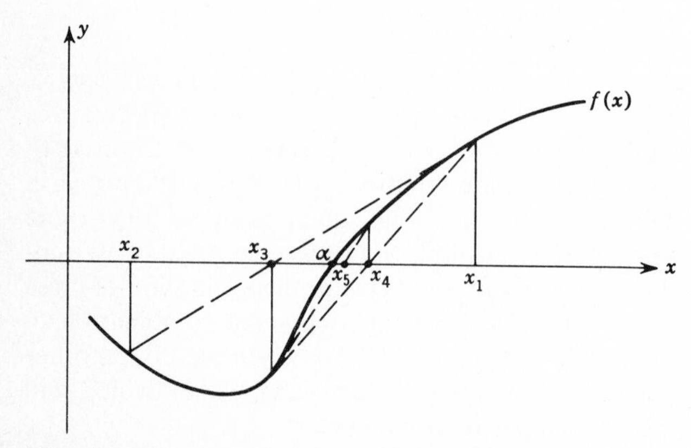

Figure 8.1 The method of false position.

Therefore,
$$F_i = \begin{cases} \dfrac{y_i}{y_i - y_{i-1}} x_{i-1} + \dfrac{y_{i-1}}{y_{i-1} - y_i} x_i & i = 2, 4 \\[2ex] \dfrac{y_i}{y_i - y_{i-2}} x_{i-2} + \dfrac{y_{i-2}}{y_{i-2} - y_i} x_i & i = 3 \end{cases} \tag{8.3-2}$$

For certain functions, however, the method of false position is a stationary method of functional iteration. For example, if $f(x)$ is convex between x_1 and x_2, the method is stationary. From Fig. 8.2 we see that the point x_1 is always one of the two points used to get the next iterate. Therefore, we have

$$x_{i+1} = \frac{y_i}{y_i - y_1} x_1 + \frac{y_1}{y_1 - y_i} x_i \tag{8.3-3}$$

for all i.

Since the method of false position is just a sequence of linear inverse interpolations, the error in this case can be written down, using (8.2-4), as

$$\epsilon_{i+1} = \alpha - x_{i+1} = \frac{g''(\eta)}{2} y_i y_1 = \frac{-f''(\xi)}{2[f'(\xi)]^3} y_i y_1 \tag{8.3-4}$$

since {1}
$$g''(y) = -\frac{f''(x)}{[f'(x)]^3} \tag{8.3-5}$$

Now since $f(\alpha) = 0$, we have, using the mean-value theorem,

$$y_1 = f(x_1) = f(x_1) - f(\alpha) = (x_1 - \alpha) f'(\xi_1) \tag{8.3-6}$$

and similarly
$$y_i = (x_i - \alpha) f'(\xi_i) \tag{8.3-7}$$

with ξ_1 and ξ_i in appropriate intervals. Using these two equations in (8.3-4), we have

$$\epsilon_{i+1} = \frac{-f''(\xi) f'(\xi_1) f'(\xi_i)}{2[f'(\xi)]^3} \epsilon_i \epsilon_1 \tag{8.3-8}$$

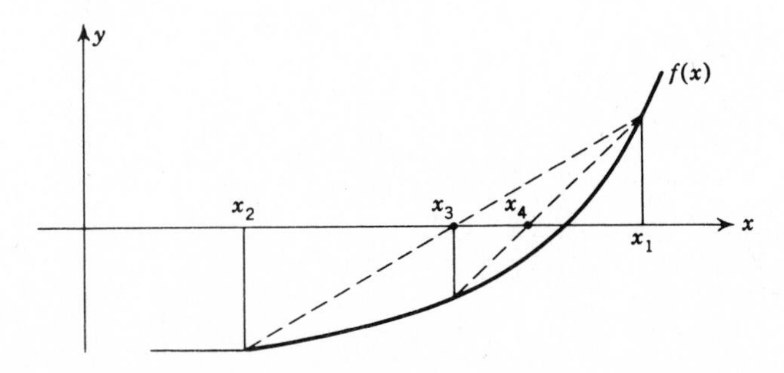

Figure 8.2 The method of false position for convex functions.

Therefore,
$$\lim_{i\to\infty}\frac{|\epsilon_{i+1}|}{|\epsilon_i|} = \left|\frac{-f''(\bar{\xi})f'(\xi_1)f'(\alpha)}{2[f'(\bar{\xi})]^3}\right||\epsilon_1| \tag{8.3-9}$$
since ξ_i approaches α and ξ approaches some limiting value $\bar{\xi}$ as $i \to \infty$. We have assumed that $f'(x)$ is bounded away from zero in a neighborhood of α. From this it follows that $p = 1$. Thus the method of false position has *linear* convergence for convex functions. A similar situation holds for concave functions. Since almost all functions are ultimately either concave or convex in the neighborhood of a root, we see that the method of false position has linear convergence for almost all functions.

The source of this poor convergence is that successive iterates lie on the same side of the root. By an appropriate modification of the method we can usually bring a new iterate over to the other side. Thus, suppose that x_{i-1}, x_i, and x_{i+1} are such that $y_{i-1}y_i < 0$ and $y_i y_{i+1} > 0$. In this situation, it is desirable to modify regula falsi by choosing a different method to compute x_{i+2} using the bracketing points x_{i-1} and x_{i+1}. The simplest approach is to use bisection (see Sec. 2.2), taking $x_{i+2} = \frac{1}{2}(x_{i-1} + x_{i+1})$. Then if x_{i+1} and x_{i+2} straddle the root, we proceed as in regula falsi. Otherwise we bisect again until we get a pair x_{i+j}, x_{i+j+1} which straddle the root.

A more efficient approach consists in applying the false-position formula to the points x_{i-1} and x_{i+1} but with y_{i-1} replaced by $\alpha y_{i-1} = \bar{y}_{i-1}$ for some α such that $0 < \alpha < 1$. The improvement such a procedure can bring about is obvious from Fig. 8.3. Our formula for x_{i+2} becomes

$$\begin{aligned} x_{i+2} &= \frac{\alpha y_{i-1}}{\alpha y_{i-1} - y_{i+1}} x_{i+1} + \frac{y_{i+1}}{y_{i+1} - \alpha y_{i-1}} x_{i-1} \\ &= \frac{\bar{y}_{i-1}}{\bar{y}_{i-1} - y_{i+1}} x_{i+1} + \frac{y_{i+1}}{y_{i+1} - \bar{y}_{i-1}} x_{i-1} \end{aligned} \tag{8.3-10}$$

If $y_{i+2}y_{i+1} < 0$, we continue as in regula falsi. Otherwise we do another modified step using a new α based on y_{i+2} and $\bar{y}_{i-1}$.

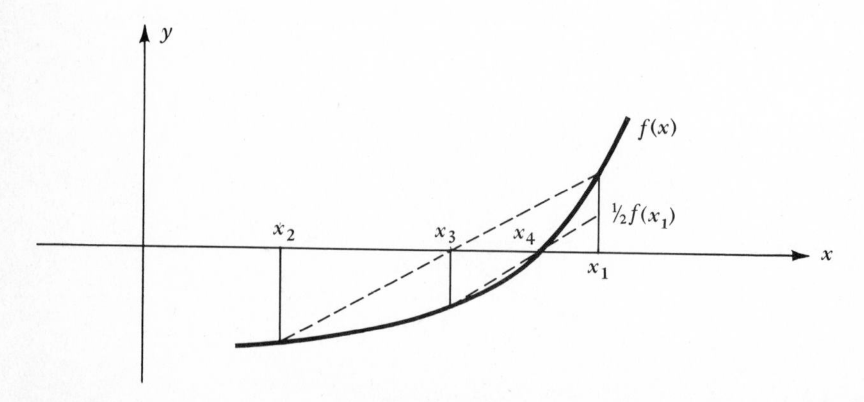

Figure 8.3 The Illinois method for convex functions.

The simplest choice for α is $\alpha = \frac{1}{2}$ (called the *Illinois method*). It can be shown that in this case, $\epsilon_{i+3} \approx k\epsilon_i^3$, so that the average order of the method is $3^{1/3} = 1.442$. If we take $\alpha = y_i/(y_i + y_{i+1})$ (called the *Pegasus method*), the average order increases to $(7.275)^{1/4} = 1.642$. A further refinement consists of taking

$$\alpha = \begin{cases} \beta & \beta > 0 \\ \frac{1}{2} & \beta \le 0 \end{cases} \qquad \beta = \frac{f[x_{i+1}, x_i]}{f[x_i, x_{i-1}]}$$

in which case the average order of the method is at least 1.68. Here, $f[x_k, x_j]$ is the divided difference $(y_k - y_j)/(x_k - x_j)$.

These modified versions of regula falsi are excellent methods when we have two points at which $f(x)$ has opposite signs. If we do not wish to expend the effort to find such points, we can use linear inverse interpolation through the last two computed points to generate the next point in the sequence. Thus, using x_i and x_{i-1} to generate x_{i+1}, we have

$$x_{i+1} = \frac{y_i}{y_i - y_{i-1}} x_{i-1} + \frac{y_{i-1}}{y_{i-1} - y_i} x_i \tag{8.3-11}$$

This stationary iteration method is called the *secant method.*

For (8.3-11) the analogous equation to (8.3-8) is

$$\epsilon_{i+1} = \frac{-f''(\xi)f'(\xi_i)f'(\xi_{i-1})}{2[f'(\xi)]^3} \epsilon_i \epsilon_{i-1} \tag{8.3-12}$$

If the initial approximations x_0 and x_1 are sufficiently close to α, then, since $f'(\xi)$ is bounded in some neighborhood of α, the iteration will converge to α. Assume now that the iteration does converge so that all iterates are contained in some interval I. On this interval let

$$0 < m_1 \le |f'(x)| \le M_1 \qquad |f''(x)| \le M_2 \tag{8.3-13}$$

Then from (8.3-12)

$$|\epsilon_{i+1}| \le K |\epsilon_i| \, |\epsilon_{i-1}| \tag{8.3-14}$$

where $K = M_2 M_1^2/2m_1^3$. Now let $K|\epsilon_i| = d_i$. Then we can write (8.3-14) as

$$d_{i+1} \le d_i d_{i-1} \tag{8.3-15}$$

Now let "sufficiently close" above mean that d_0 and d_1 are both less than or equal to $d < 1$. Then from (8.3-15)

$$d_2 \le d^2 \qquad d_3 \le d^3 \qquad d_4 \le d^5 \qquad d_5 \le d^8 \tag{8.3-16}$$

and in general

$$d_i \le d^{\gamma_i} \tag{8.3-17}$$

where

$$\gamma_{i+1} = \gamma_i + \gamma_{i-1} \qquad i = 1, 2, \ldots; \qquad \gamma_0 = \gamma_1 = 1 \tag{8.3-18}$$

Equation (8.3-18) is the difference equation for the Fibonacci numbers. The indicial equation for (8.3-18) is

$$\rho^2 = \rho + 1 \tag{8.3-19}$$

which has the solutions $(1 + \sqrt{5})/2$ and $(1 - \sqrt{5})/2$. Therefore, the solution of (8.3-18) is {6}

$$\gamma_i = \frac{1}{\sqrt{5}}\left[\left(\frac{1+\sqrt{5}}{2}\right)^{i+1} - \left(\frac{1-\sqrt{5}}{2}\right)^{i+1}\right] \tag{8.3-20}$$

Furthermore,

$$\lim_{i\to\infty}\frac{\gamma_{i+1}}{\gamma_i} = \frac{1+\sqrt{5}}{2} \approx 1.618 \tag{8.3-21}$$

Now from (8.3-17) and the definition of d_i we have

$$|\epsilon_i| \le \frac{1}{K}\, d^{\gamma_i} \tag{8.3-22}$$

This together with (8.3-21) suggests the result that {6}

$$\lim_{i\to\infty}\frac{|\epsilon_{i+1}|}{|\epsilon_i|^{(1+\sqrt{5})/2}} = \left|\frac{f''(\alpha)}{2f'(\alpha)}\right|^{(\sqrt{5}-1)/2} \tag{8.3-23}$$

since as x_i approaches α, the coefficient of $\epsilon_i\epsilon_{i-1}$ in (8.3-12) approaches $|f''(\alpha)/2f'(\alpha)|$. The reader will have noted that the foregoing by no means constitutes a proof of (8.3-23). A rigorous proof of (8.3-23) is beyond our scope here but will be found in Ostrowski (1973). From (8.3-23) it follows that the order of the secant method is $(1 + \sqrt{5})/2$ and that†

$$EI = \frac{1+\sqrt{5}}{2} \tag{8.3-24}$$

Not only is this substantially better than *regula falsi* ($EI = 1$), but, as we shall see, the efficiency of the secant method compares favorably with many seemingly more sophisticated methods.

The fact that the secant method does not require the evaluation of any derivatives can be a great advantage in certain problems. For example, a not uncommon problem is to find a root of

$$f(x; y_1, y_2, \ldots, y_k) = 0 \tag{8.3-25}$$

where

$$y_j = g_j(x) \qquad j = 1, \ldots, k \tag{8.3-26}$$

Evaluating derivatives of f with respect to x is often impractical in such a case.

A disadvantage of the secant method would seem to be that multiple-precision arithmetic would be required as the iteration nears convergence because (8.3-11) then involves the difference of two nearly equal quantities y_i and y_{i-1}. But let us rewrite (8.3-11) as

$$x_{i+1} = x_i - \frac{x_i - x_{i-1}}{y_i - y_{i-1}}\, y_i \tag{8.3-27}$$

† Here and hereafter in this chapter we shall estimate the cost of evaluating the iteration function F by considering only the evaluations of $f(x)$ and its derivatives.

The second term can be considered a correction term to x_i and as such requires only a very few significant digits as convergence is neared (why?). Therefore, although the quotient $(x_i - x_{i-1})/(y_i - y_{i-1})$ will have very few significant digits if multiple-precision arithmetic is not used, there will nevertheless generally be enough significant figures to compute α to nearly full single-precision accuracy.

Example 8.1 Find the positive root of $\sin x - x/2 = 0$ using the secant method, the method of false position, and some of its modifications.

Since the root lies between $\pi/2$ and π (see Fig. 8.4), we use $x_0 = \pi/2$, $x_1 = \pi$. Using (8.3-11), (8.3-3), and (8.3-10) with $\alpha = \frac{1}{2}$ and with $\alpha = y_i/(y_i + y_{i+1})$, we get (note that $f''(x) \leq 0$ in $[\pi/2, \pi]$ so that $f(x)$ is concave)

	Secant method	Method of false position	Illinois method	Pegasus method
x_2	1.75960	1.75960	1.75960	1.75960
x_3	1.93200	1.84420	1.91904	1.99169
x_4	1.89242	1.87701	1.89349	1.88772
x_5	1.89543	1.88895	1.89547	1.89509
x_6	1.89549	1.89320	1.89552	1.90226
x_7		1.89469	1.89549	1.89549
x_8		1.89521		
x_9		1.89540		
x_{10}		1.89546		
x_{11}		1.89548		
x_{12}		1.89549		

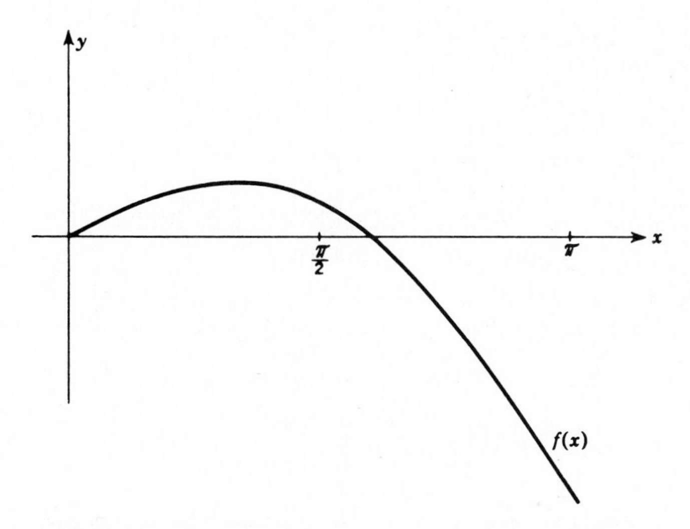

Figure 8.4 Graph of $f(x) = \sin x - x/2$.

As we would expect, the secant method converges substantially faster to the true value 1.89549 than the method of false position. On the other hand, both the Illinois and Pegasus methods perform almost as well as the secant method while retaining the bracketing property of the method of false position. Note also the slow convergence of the method of false position as the root is approached.

While the method of false position does not depend on the ordering of the initial values, all other methods illustrated above do. Thus, if we take $x_0 = \pi, x_1 = \pi/2$, we get the following results:

	Secant method	Illinois method	Pegasus method
x_2	1.75960	1.79560	1.79560
x_3	1.84420	1.84420	1.84420
x_4	1.90011	1.90820	1.95258
x_5	1.89535	1.89511	1.89381
x_6	1.89549	1.89549	1.89544
x_7			1.89708
x_8			1.89549

The switch improved the performance of the Illinois method while the convergence of the Pegasus method slowed down slightly.

8.4 ONE-POINT ITERATION FORMULAS

Equation (8.3-11) is a two-point iteration method; i.e., to compute x_{i+1} we use information at two previous values. Formulas of the class which use information at only one point are naturally called one-point iteration formulas. In this section we shall consider only stationary one-point iteration formulas which have the form

$$x_{i+1} = F(x_i) \tag{8.4-1}$$

with $\alpha = F(\alpha)$ if the method converges. We first prove a theorem:

Theorem 8.1 Assume that $F(x)$ has sufficiently many derivatives. The order of any one-point iteration function $F(x)$ is a positive integer. More specifically $F(x)$ has order p if and only if $F(\alpha) = \alpha$; $F^{(j)}(\alpha) = 0$, $1 \leq j < p$; $F^{(p)}(\alpha) \neq 0$.

PROOF We expand $F(x_i)$ in a Taylor series about α

$$F(x_i) = \alpha + (x_i - \alpha)F'(\alpha) + \cdots + \frac{(x_i - \alpha)^{p-1}}{(p-1)!} F^{(p-1)}(\alpha)$$

$$+ \frac{(x_i - \alpha)^p}{p!} F^{(p)}(\xi) = \alpha + \frac{(x_i - \alpha)^p}{p!} F^{(p)}(\xi) \tag{8.4-2}$$

where ξ lies between x_i and α. Since $F(x_i) = x_{i+1}$, we have

$$x_{i+1} - \alpha = \frac{(x_i - \alpha)^p}{p!} F^{(p)}(\xi) \tag{8.4-3}$$

Therefore,
$$\lim_{i \to \infty} \frac{|x_{i+1} - \alpha|}{|x_i - \alpha|^p} = \frac{1}{p!} |F^{(p)}(\alpha)| \neq 0 \tag{8.4-4}$$

if the iteration converges; this proves the "if" part of the theorem. On the other hand, it follows easily from (8.4-2) and (8.4-4) that if $F^{(j)}(\alpha) \neq 0$ for some j between 0 and p or if $F^{(p)}(\alpha) = 0$, then $F(x)$ cannot be of order p.

In this section we shall develop a particular class of one-point iteration functions which contains members of all integral orders. Although this class by no means exhausts the class of one-point iteration functions, there does exist a relationship between this class and all other one-point iteration functions which is, however, beyond our scope here.

We assume, as in Sec. 8.2, that in a neighborhood of a root α of (8.1-1), $f(x)$ has an inverse $g(y)$. The Taylor-series expansion of $g(y)$ about a point y_i is given by

$$\begin{aligned} x = g(y) &= \sum_{j=0}^{m+1} \frac{(y - y_i)^j}{j!} g^{(j)}(y_i) + \frac{(y - y_i)^{m+2}}{(m+2)!} g^{(m+2)}(\eta) \\ &= x_i + \sum_{j=1}^{m+1} \frac{(y - y_i)^j}{j!} g^{(j)}(y_i) + \frac{(y - y_i)^{m+2}}{(m+2)!} g^{(m+2)}(\eta) \end{aligned} \tag{8.4-5}$$

$$g^{(j)}(y_i) = \frac{d^j}{dy^j} g(y) \bigg|_{y = y_i}$$

where η is between y and y_i. Since $\alpha = g(0)$, we have

$$\begin{aligned} \alpha = x_i &+ \sum_{j=1}^{m+1} \frac{(-1)^j}{j!} y_i^j g^{(j)}(y_i) + \frac{(-1)^{m+2} y_i^{m+2}}{(m+2)!} g^{(m+2)}(\eta) \\ &= x_i + \sum_{j=1}^{m+1} \frac{(-1)^j}{j!} f_i^j g_i^{(j)} + \frac{(-1)^{m+2}}{(m+2)!} f_i^{m+2} g^{(m+2)}(\eta) \end{aligned} \tag{8.4-6}$$

where we have written $y_i = f(x_i) = f_i$ and $g^{(j)}(y_i) = g_i^{(j)}$.

Equation (8.4-6) suggests consideration of the iteration formula

$$x_{i+1} = x_i + \sum_{j=1}^{m+1} \frac{(-1)^j}{j!} f_i^j g_i^{(j)} \tag{8.4-7}$$

The value of (8.4-7) will depend partly on whether the $g_i^{(j)}$ can easily be calculated. Clearly

$$g_i^{(1)} = \frac{d}{dy} g(y) \bigg|_{y = y_i} = 1/f'_i$$

For $j > 1$, we can compute the $g_i^{(j)}$ in terms of derivatives of $f(x)$ at $x = x_i$ as follows. If we define the operator $D \equiv d/dy$, then

$$D = \frac{1}{f'}\frac{d}{dx}$$

Hence
$$D^j g = D^{j-1} g' = \left(\frac{1}{f'}\frac{d}{dx}\right)^{j-1} \frac{1}{f'}$$

so that we have expressed $g^{(j)}(y)$ in terms of derivatives of $f(x)$. The actual computation of $D^j g$ is given by the formula

$$D^j g = \frac{F_j}{(f')^{2j-1}} \qquad j = 1, 2, 3, \ldots \tag{8.4-8}$$

where the F_j are defined by the recursion

$$F_1 = 1 \qquad F_{j+1} = f'F_j' - (2j-1)f''F_j \tag{8.4-9}$$

which can be easily proved by induction {8}. Thus, the main problem in evaluating (8.4-7) is to evaluate f_i', f_i'', and F_j, $j = 2, \ldots, m+1$.

Subtracting (8.4-7) from (8.4-6), we have

$$\epsilon_{i+1} = \frac{(-1)^{m+2}}{(m+2)!} f_i^{m+2} g^{(m+2)}(\eta) \tag{8.4-10}$$

Since
$$f_i = f(x_i) = f(x_i) - f(\alpha) = (x_i - \alpha) f'(\xi_1) \tag{8.4-11}$$

with ξ_1 between x_i and α, we have

$$\epsilon_{i+1} = \frac{1}{(m+2)!}\{[f'(\xi_1)]^{m+2} g^{(m+2)}(\eta)\}\epsilon_i^{m+2} \tag{8.4-12}$$

If the root is simple, the term in braces in (8.4-12) is bounded in some neighborhood of α. Therefore, the order of (8.4-7) is $m + 2$, and if the initial approximation is sufficiently good, the iteration will converge. (Even for a bad initial approximation it may converge; see Sec. 8.7.)

The evaluation of (8.4-7) requires the evaluation of $f(x_i)$ and its first $m + 1$ derivatives (why?). If θ_j is the cost of evaluating $f^{(j)}(x_i)$ relative to the cost of evaluating $f(x)$, which as before we take to be 1, the efficiency index of (8.4-7) is given by

$$EI = (m+2)^{1/\psi} \qquad \text{where } \psi = 1 + \sum_{j=1}^{m+1} \theta_j \tag{8.4-13}$$

One important and familiar special case of (8.4-7) is that for $m = 0$. We have

$$x_{i+1} = x_i - u_i = x_i - \frac{f(x_i)}{f'(x_i)} \tag{8.4-14}$$

which is the familiar Newton-Raphson iteration. The error is given by (8.4-12) as

$$\epsilon_{i+1} = \tfrac{1}{2}[f'(\xi_1)]^2 g''(\eta)\epsilon_i^2 = -\frac{f''(\xi)[f'(\xi_1)]^2}{2[f'(\xi)]^3}\epsilon_i^2 \tag{8.4-15}$$

In fact it can be shown {10} that the terms in $f'(x)$ can be canceled as if $\xi = \xi_1 = x_i$, so that

$$\epsilon_{i+1} = -\frac{1}{2}\frac{f''(\xi)}{f'(x_i)}\epsilon_i^2 \tag{8.4-16}$$

From (8.4-13) the efficiency index of the Newton-Raphson method is $2^{1/(1+\theta_1)}$, where θ_1 is the cost of evaluating $f'(x)$. A straightforward calculation leads to the result that if $\theta_1 < .44$, the efficiency of the Newton-Raphson method is greater than that of the secant method (8.3-11). As we pointed out previously, the cost of evaluating the derivative is often much less than that of evaluating the function, a notable exception being polynomials. If a decision is to be made whether to use (8.3-11) or (8.4-14) to solve (8.1-1), then a perfectly reasonable basis for this decision is to estimate θ_1 and use (8.3-11) if it is greater than .44 and (8.4-14) otherwise.

Example 8.2 Repeat the calculation of Example 8.1 using the Newton-Raphson method starting with, first, $x_1 = \pi$ and, second, $x_1 = \pi/2$.

The results are

	$x_1 = \pi$	$x_1 = \pi/2$
x_2	2.09440	2.00000
x_3	1.91322	1.90100
x_4	1.89567	1.89551
x_5	1.89549	1.89549

The convergence in both cases requires 1 less iteration than the secant method. Since $f'(x) = \cos x - \frac{1}{2}$, the derivative is somewhat, but not a great deal, easier to compute than the function (because to compute the cosine from the sine a square root will have to be calculated). But if, for example, we had $f(x) = e^x - x$, the derivative would be much easier to calculate than the function.

8.5 MULTIPOINT ITERATION FORMULAS†

In this section we shall consider some examples of stationary multipoint iteration functions. Such iteration functions have the form (8.2-5) with $n > 1$. Of the various possible approaches to the derivation of multipoint iteration

† Traub (1964, chap. 8) uses "multipoint" in a different context. He calls the iteration functions of this section "one-point iteration functions with memory."

functions, we shall consider two in this section, the first because of its general theoretical interest and the second because it leads to some particularly interesting and useful formulas. Again we shall assume that α is a simple root.

8.5-1 Iteration Formulas Using General Inverse Interpolation

In Sec. 8.2 we used inverse Lagrangian interpolation to derive a class of methods of functional iteration. Here we shall generalize this technique by using a general polynomial interpolation formula which has the property that, at a point x_{i+1-j} the interpolation polynomial $y(x)$ and its first r_j derivatives agree with $f(x)$. By analogy with (3.2-9), using the points $x_{i+1-j}, j = 1, \ldots, n$, and assuming that f has as many continuous derivatives as we desire, we may write {17}

$$f(x) = y(x) + \frac{f^{(\beta+1)}(\xi)}{(\beta+1)!} \prod_{j=1}^{n} (x - x_{i+1-j})^{r_j+1} \qquad \beta = n - 1 + \sum_{j=1}^{n} r_j \quad (8.5\text{-}1)$$

where ξ lies in the interval spanned by $x_i, \ldots, x_{i-n+1}$ and x. For a direct application of this formula to functional iteration, see {19 to 21}. Our interest here, however, is in interpolating using the inverse function g. Assuming this function exists and has as many continuous derivatives as we desire, then corresponding to (8.2-1) we have

$$g(y) = q(y) + \frac{g^{(\beta+1)}(\eta)}{(\beta+1)!} \prod_{j=1}^{n} (y - y_j)^{r_j+1} \qquad (8.5\text{-}2)$$

where $y_j = f(x_{i+1-j})$ and η lies in the interval spanned by $y_1, \ldots, y_n$ and y. Since $\alpha = g(0)$, we get, by analogy with (8.3-2), an iteration formula given by $x_{i+1} = q(0)$, where $q(0)$ is a linear combination of $x_{i+1-j} = g(y_j)$, $j = 1, \ldots, n$, and derivatives of g evaluated at y_j, $j = 1, \ldots, n$. For example, if $r_j = 1$ for all j, then from the Hermite interpolation formula we get

$$x_{i+1} = \sum_{j=1}^{n} h_j(0) x_{i+1-j} + \sum_{j=1}^{n} \bar{h}_j(0) g'(y_j) \qquad (8.5\text{-}3)$$

This formulation of multipoint iteration formulas includes as a special case the one-point iteration formulas of the previous section. The Taylor-series expansion (8.4-5) is identical with the generalized interpolation formula when $n = 1$; that is, only one point is used. Note that the Newton-Raphson method is given by (8.5-3) with $n = 1$. Almost all the well-known methods of stationary functional iteration are derivable as special cases of this general approach. Besides the Newton-Raphson method another example of this that we have seen is the secant method, which is an application of linear inverse Lagrangian interpolation.

Of particular interest to us here is the order of an iteration formula derived from (8.5-2). Analogous to (8.2-4) we have

$$\alpha - x_{i+1} = (-1)^{\beta+1} \frac{g^{(\beta+1)}(\eta)}{(\beta+1)!} \prod_{j=1}^{n} y_j^{r_j+1} \tag{8.5-4}$$

Now since

$$\begin{aligned} y_j &= f(x_{i+1-j}) = f(x_{i+1-j}) - f(\alpha) \\ &= -f'(\xi_j)\epsilon_{i+1-j} = \frac{-\epsilon_{i+1-j}}{g'(\eta_j)} \qquad j = 1, \ldots, n \end{aligned} \tag{8.5-5}$$

we can rewrite (8.5-4) as

$$\epsilon_{i+1} = \frac{g^{(\beta+1)}(\eta)}{(\beta+1)!} \frac{\prod_{j=1}^{n} \epsilon_{i+1-j}^{r_j+1}}{\prod_{j=1}^{n} [g'(\eta_j)]^{r_j+1}} \tag{8.5-6}$$

We assume the iteration converges so that all iterates lie on some interval I. Let K be such that

$$\left| \frac{g^{(\beta+1)}(\eta)}{(\beta+1)!} \frac{1}{\prod_{j=1}^{n} [g'(\eta_j)]^{r_j+1}} \right| < K \tag{8.5-7}$$

for all η and η_j. Then

$$|\epsilon_{i+1}| \leq K \prod_{j=1}^{n} |\epsilon_{i+1-j}|^{r_j+1} \tag{8.5-8}$$

This equation is reminiscent of (8.3-14). Indeed we can show that

$$d_{i+1} \leq d^{\gamma_i} \qquad i = 1, 2, \ldots; \qquad d < 1 \tag{8.5-9}$$

where $d_i = K^{1/\beta}\epsilon_i$ and the γ_i's satisfy

$$\gamma_{i+1} = \sum_{j=1}^{n} (r_j + 1)\gamma_{i-j+1} \qquad \begin{array}{l} i = n-1, n, \ldots \\ \gamma_0 = \gamma_1 = \cdots = \gamma_{n-1} = 1 \end{array} \tag{8.5-10}$$

The indicial equation of (8.5-10) is

$$\rho^n = \sum_{j=1}^{n} (r_j + 1)\rho^{n-j} \tag{8.5-11}$$

so that γ_i is given by some linear combination of the zeros of the polynomial (8.5-11). From these zeros we might expect to be able to indicate the order of convergence of the iteration as we did in Sec. 8.3, and indeed this can be done. We shall content ourselves with stating the result that in the case $r_j = r$, for all j, the order of the iteration defined by $x_{i+1} = q(0)$, with $g(y)$ as in (8.5-2), is given by the only real root of (8.5-11) with magnitude greater than 1. This root is positive and lies in the interval $(r + 1, r + 2)$ for all n. The derivation of this result can be found in Traub (1964, pp. 62–67).

8.5-2 Derivative Estimated Iteration Formulas

Consider the Newton-Raphson method (8.4-14). Let us replace the derivative term by its approximation found by differentiating a two-point Lagrangian formula for $f(x)$, the two points being x_i and x_{i-1}. We have

$$f(x) \approx \frac{x - x_i}{x_{i-1} - x_i} f(x_{i-1}) + \frac{x - x_{i-1}}{x_i - x_{i-1}} f(x_i) \tag{8.5-12}$$

so that

$$f'(x_i) \approx \frac{f(x_{i-1}) - f(x_i)}{x_{i-1} - x_i} \tag{8.5-13}$$

If we substitute (8.5-13) into the Newton-Raphson method, we get precisely the secant method, which is not a very interesting result {22}.

Now consider the method (8.4-7) with $m = 1$, which is

$$x_{i+1} = x_i - \frac{f(x_i)}{f'(x_i)} - \frac{1}{2} \frac{[f(x_i)]^2 f''(x_i)}{[f'(x_i)]^3} \tag{8.5-14}$$

Let us replace the second derivative in (8.5-14) by its approximation found by differentiating a two-point Hermite interpolation formula for $f(x)$, the two points being x_i and x_{i-1}. We have {22}

$$\begin{aligned} f(x) \approx \frac{1}{(x_i - x_{i-1})^2} \Bigg[&\left(1 - 2\frac{x - x_i}{x_i - x_{i-1}}\right)(x - x_{i-1})^2 f(x_i) \\ &+ \left(1 - 2\frac{x - x_{i-1}}{x_{i-1} - x_i}\right)(x - x_i)^2 f(x_{i-1}) \\ &+ (x - x_i)(x - x_{i-1})^2 f'(x_i) + (x - x_{i-1})(x - x_i)^2 f'(x_{i-1}) \Bigg] \end{aligned} \tag{8.5-15}$$

so that

$$\begin{aligned} f''(x_i) \approx &\frac{-6}{(x_i - x_{i-1})^2} [f(x_i) - f(x_{i-1})] \\ &+ \frac{2}{x_i - x_{i-1}} [2f'(x_i) + f'(x_{i-1})] \end{aligned} \tag{8.5-16}$$

Substituting the right-hand side of (8.5-16) into (8.5-14), we obtain an iteration formula

$$x_{i+1} = x_i - \frac{f(x_i)}{f'(x_i)} - \frac{1}{2} \frac{[f(x_i)]^2}{[f'(x_i)]^3} \bar{f}''(x_i) \tag{8.5-17}$$

where

$$\bar{f}''(x_i) = -\frac{6}{h_i^2}[f(x_i) - f(x_{i-1})] + \frac{2}{h_i}[2f'(x_i) + f'(x_{i-1})]$$

$$h_i = x_i - x_{i-1} \qquad (8.5\text{-}18)$$

which, like the Newton-Raphson method, depends upon $f(x)$ and its first derivative. The order of (8.5-14) is 3. Of interest to us here is the order of the modified formula (8.5-17).

Let us call the right-hand side of (8.5-14) $F_1(x_i)$ and the right-hand side of (8.5-17) $F_2(x_i, x_{i-1})$. Then

$$F_1(x_i) - F_2(x_i, x_{i-1}) = \frac{f(x_i)^2}{2[f'(x_i)]^3}\left[\frac{(x_i - x_{i-1})^2}{12} f^{\text{iv}}(\xi)\right] \qquad (8.5\text{-}19)$$

where the term in brackets on the far right is the result of differentiating the error term of the Hermite interpolation formula twice and then evaluating the result at $x = x_i$. Now, using (8.4-12), we have

$$\alpha - F_1(x_i) = \tfrac{1}{6}\{[f'(\xi_1)]^3 g'''(\eta)\}\epsilon_i^3 \qquad \epsilon_i = \alpha - x_i \qquad (8.5\text{-}20)$$

Then from (8.5-19) and (8.5-20)

$$\epsilon_{i+1} = \alpha - F_2(x_i, x_{i-1}) = \frac{[f(x_i)]^2}{2[f'(x_i)]^3}\left[\frac{(x_i - x_{i-1})^2}{12} f^{\text{iv}}(\xi)\right]$$

$$+ \tfrac{1}{6}\{[f'(\xi_1)]^3 g'''(\eta)\}\epsilon_i^3 \qquad (8.5\text{-}21)$$

Now $\qquad (x_i - x_{i-1})^2 = (\alpha - x_{i-1} + x_i - \alpha)^2 = (\epsilon_i - \epsilon_{i-1})^2 \qquad (8.5\text{-}22)$

Substituting this into (8.5-21) and using (8.4-11), we have

$$\epsilon_{i+1} = \frac{f^{\text{iv}}(\xi)[f'(\xi_2)]^2}{2[f'(x_i)]^3}\epsilon_i^2(\epsilon_i - \epsilon_{i-1})^2 + \tfrac{1}{6}\{[f'(\xi_1)]^3 g'''(\eta)\}\epsilon_i^3 \qquad (8.5\text{-}23)$$

As we near convergence, the errors in (8.5-23) will be monotonically decreasing in magnitude. The dominant term in (8.5-23) is either ϵ_i^3 or $\epsilon_i^2\epsilon_{i-1}^2$ (why?). Our object is to show that the $\epsilon_i^2\epsilon_{i-1}^2$ term is in fact dominant. To do this, it is sufficient to consider the order of a hypothetical formula with an error†

$$\bar{\epsilon}_{i+1} = \frac{f^{\text{iv}}(\xi)[f'(\xi_2)]^2}{2[f'(x_i)]^3}\bar{\epsilon}_i^2\bar{\epsilon}_{i-1}^2 \qquad (8.5\text{-}24)$$

Proceeding as we have previously, we obtain

$$|\bar{\epsilon}_{i+1}| \le K|\bar{\epsilon}_i|^2|\bar{\epsilon}_{i-1}|^2 \qquad (8.5\text{-}25)$$

† The bars in (8.5-24) and (8.5-25) are to distinguish the errors in this hypothetical formula from the errors in (8.5-23).

where K is such that

$$\left| \frac{f^{\text{iv}}(\xi)[f'(\xi_2)]^2}{2[f'(x_i)]^3} \right| < K \tag{8.5-26}$$

on some interval including α. With $d_i = K^{1/3}|\bar{\epsilon}_i|$ we get, using (8.5-8) to (8.5-11), that

$$d_{i+1} \le d^{\gamma_i} \qquad i = 1, 2, \ldots; \qquad d < 1 \tag{8.5-27}$$

where

$$\gamma_i = \tfrac{1}{2}[(1 + \sqrt{3})^i + (1 - \sqrt{3})^i] \tag{8.5-28}$$

Therefore, by reasoning similar to that in Sec. 8.3, we conclude, but again do not prove, that

$$\lim_{i\to\infty} \frac{|\bar{\epsilon}_{i+1}|}{|\bar{\epsilon}_i|^{1+\sqrt{3}}} = \left| \frac{f^{\text{iv}}(\alpha)}{2f'(\alpha)} \right|^{1/\sqrt{3}} \tag{8.5-29}$$

Therefore, the order of (8.5-17) is $1 + \sqrt{3} \approx 2.732$, and the efficiency index is given by

$$EI = (1 + \sqrt{3})^{1/(1+\theta_1)} \tag{8.5-30}$$

which means that (8.5-17) is a distinct improvement over the Newton-Raphson method, its only relative disadvantage being the need for two starting values.†

The procedure described above can be generalized in a number of directions. One way would be to approximate the $(m + 1)$st-order derivative in the iteration (8.4-7) using an interpolation formula based on the first m derivatives. Another generalization would be to use more than two points in approximating $f''(x_i)$ in (8.5-16) {23}. All such methods of functional iteration are called *derivative estimated iteration formulas.*

Example 8.3 Repeat the calculations of Example 8.1 using the iteration formula (8.5-17) with $x_1 = \pi$, $x_2 = \pi/2$.

The calculations give

i	x_i	$\bar{f}''(x_i)$	$f''(x_i)$
2	$\pi/2$	−1.15847	−1.00000
3	1.78659	−.98064	−.97681
4	1.89414	−.94910	−.94818
5	1.89549	−.96230	−.94775

This method converges after three iterations, compared with four with the Newton-Raphson method in Example 8.2. This is as we would expect because the order of (8.5-17) is 2.732 while that of (8.4-14) is 2. The last two columns in the table give the approximation

† Note that we are assuming here that the additions and multiplications needed to evaluate (8.5-17) and (8.5-18) require a negligible amount of time compared with the evaluation of $f(x_i)$ and $f'(x_i)$. This assumption is not unreasonable in general.

to the second derivative given by (8.5-18) and the true value. Note that as the root is approached, the difference between the approximate and true values of the derivative first decreases and then increases. This is a phenomenon of numerical differentiation that we discussed in Chap. 4. When $x_i - x_{i-1}$ is comparatively large, truncation error dominates and decreases as $x_i - x_{i-1}$ decreases. But when $x_i - x_{i-1}$ gets very small, roundoff dominates and grows as $x_i - x_{i-1}$ decreases. Note, however, that as $x_i \to \alpha$, the coefficient of $\bar{f}''(x_i)$ gets very small; therefore, the loss of significance in $\bar{f}''(x_i)$ is not important.

8.6 FUNCTIONAL ITERATION AT A MULTIPLE ROOT

All the results of the past three sections depend upon the root α being simple. In particular, if α is a root of multiplicity $r > 1$, all our derivations based on inverse interpolation are invalid because the inverse function does not exist in any neighborhood of $x = \alpha$.

Nevertheless, we can still consider the behavior of such formulas in the neighborhood of a multiple root. We consider here the class of methods (8.4-7).

We show first, that, independent of m, (8.4-7) converges linearly when α is a root of multiplicity greater than 1. From (8.4-7) we have

$$\epsilon_{i+1} = \alpha - x_{i+1} = \alpha - x_i - \sum_{j=1}^{m+1} \frac{(-1)^j}{j!} f_i^j g_i^{(j)} = \epsilon_i + \sum_{j=0}^{m} Z_j(x_i) \qquad (8.6\text{-}1)$$

where we define

$$Z_j(x_i) = \frac{(-1)^{j+1}}{(j+1)!} f_i^{j+1} g_i^{(j+1)} \qquad (8.6\text{-}2)$$

Using the fact that $(g_i^{(j)})' = f_i' g_i^{(j+1)}$, where the prime denotes differentiation with respect to x, a simple calculation shows that

$$(j+1)Z_j = jZ_{j-1} - u_i Z_{j-1}' \qquad u_i = \frac{f_i}{f_i'} \qquad (8.6\text{-}3)$$

Now consider the Taylor-series expansion of $Z_j(x_i)$ about α. Since $Z_j(\alpha) = 0$ (why?), we have

$$Z_j(x_i) = \sum_{k=1}^{\infty} c_{jk} \epsilon_i^k \qquad (8.6\text{-}4)$$

From (8.6-1), (8.6-2), and (8.6-4) {24}

$$\epsilon_{i+1} = \epsilon_i \left(1 + \sum_{j=0}^{m} c_{j1}\right) + O(\epsilon_i^2) \qquad (8.6\text{-}5)$$

We suppose that α is a root of multiplicity r. Then {24}

$$u_i = -\frac{\epsilon_i}{r} + O(\epsilon_i^2) \qquad (8.6\text{-}6)$$

Substituting (8.6-4) into (8.6-3) and using (8.6-6), we get {24}

$$(j + 1)c_{j1} = jc_{j-1,1} - \frac{1}{r} c_{j-1,1} \tag{8.6-7}$$

or

$$c_{j1} = \frac{j - 1/r}{j + 1} c_{j-1,1} \qquad c_{01} = -\frac{1}{r} \tag{8.6-8}$$

Therefore,

$$c_{j1} = (-1)^{j+1} \binom{\frac{1}{r}}{j+1} \tag{8.6-9}$$

where $(1/r)_{j+1}$ is a binomial coefficient. Substituting (8.6-9) into (8.6-5), we get

$$\begin{aligned} \epsilon_{i+1} &= \epsilon_i \left[1 + \sum_{j=0}^{m} (-1)^{j+1} \binom{\frac{1}{r}}{j+1} \right] + O(\epsilon_i^2) \\ &= \epsilon_i \left[\sum_{j=0}^{m+1} (-1)^j \binom{\frac{1}{r}}{j} \right] + O(\epsilon_i^2) \\ &= (-1)^{m+1} \epsilon_i \binom{\frac{1}{r} - 1}{m+1} + O(\epsilon_i^2) \end{aligned} \tag{8.6-10}$$

since {24}

$$\sum_{j=0}^{m+1} (-1)^j \binom{\frac{1}{r}}{j} = (-1)^{m+1} \binom{\frac{1}{r} - 1}{m+1} \tag{8.6-11}$$

Finally we have from (8.6-10) that when the iteration converges,

$$\lim_{i \to \infty} \left| \frac{\epsilon_{i+1}}{\epsilon_i} \right| = (-1)^{m+1} \binom{\frac{1}{r} - 1}{m+1} \neq 0 \tag{8.6-12}$$

if $r \neq 1$, which proves that the order of (8.4-7) is 1 if α is not a simple root. But note that since $(1/r - 1)_{m+1}$ has magnitude less than 1, the methods of this class do converge in the neighborhood of a multiple root.

If the multiplicity of the root at α is known, the class of methods (8.4-7) can be modified so that they retain their order of convergence of $m + 2$. In particular, the Newton-Raphson method (8.4-14) can be modified so that it still has quadratic convergence for a root of multiplicity r if we write

$$x_{i+1} = x_i - r \frac{f(x_i)}{f'(x_i)} \tag{8.6-13}$$

We have

$$\alpha - x_{i+1} = \alpha - x_i + r \frac{f(x_i)}{f'(x_i)} \tag{8.6-14}$$

so that

$$(\alpha - x_{i+1}) f'(x_i) = G(x_i) \tag{8.6-15}$$

where we have defined

$$G(x) = (\alpha - x) f'(x) + r f(x) \tag{8.6-16}$$

Differentiating, we have

$$G^{(j)}(x) = rf^{(j)}(x) + (\alpha - x)f^{(j+1)}(x) - jf^{(j)}(x) \tag{8.6-17}$$

and, since α is a root of $f(x)$ of multiplicity r,

$$G^{(j)}(\alpha) = 0 \qquad j = 0, \ldots, r \qquad G^{(r+1)}(\alpha) \neq 0 \tag{8.6-18}$$

Therefore,

$$G(x) = \frac{(x - \alpha)^{r+1}}{(r+1)!} G^{(r+1)}(\xi_1) \tag{8.6-19}$$

Since

$$f'(x) = \frac{(x - \alpha)^{r-1}}{(r-1)!} f^{(r)}(\xi_2) \tag{8.6-20}$$

because α is a root of $f'(x)$ of multiplicity $r - 1$, we have, using (8.6-19) and (8.6-20) in (8.6-15),

$$\epsilon_{i+1} = \frac{1}{r(r+1)} \frac{G^{(r+1)}(\xi_1)}{f^{(r)}(\xi_2)} \epsilon_i^2 \tag{8.6-21}$$

Therefore, the order is 2 since $f^{(r)}(\xi_2)$ is bounded away from zero in a neighborhood of $x = \alpha$ and $G^{(r+1)}(\alpha) \neq 0$.

Generally, however, the multiplicity of the root is not known a priori. Thus it would be very desirable to have iteration methods whose order of convergence is independent of the multiplicity. Such methods can indeed be found; the key to finding them is to note that $u(x)$ has a zero of multiplicity 1 at $x = \alpha$ no matter what the multiplicity of the zero of $f(x)$ (why?). Therefore, if, instead of (8.1-1), we consider the equation

$$u(x) = 0 \qquad [u(x) = f(x)/f'(x)] \tag{8.6-22}$$

the roots of this equation are identical with those of (8.1-1) and they are all simple. We need then only replace $f(x)$ by $u(x)$ in any iteration formula we have developed thus far to get a formula whose order of convergence is independent of the multiplicity of the root. For example, the secant method (8.3-11) and the Newton-Raphson method (8.4-14) become

$$x_{i+1} = \frac{u(x_i)}{u(x_i) - u(x_{i-1})} x_{i-1} + \frac{u(x_{i-1})}{u(x_{i-1}) - u(x_i)} x_i \tag{8.6-23}$$

and

$$x_{i+1} = x_i - \frac{u(x_i)}{u'(x_i)} \tag{8.6-24}$$

The efficiency of each of these methods is, however, less than that of the secant and Newton-Raphson methods, respectively, because of the need to calculate one higher derivative in each case. Furthermore, $u(x)$ will have poles at those roots of $f'(x)$ which are not roots of $f(x)$ so that $u(x)$ may no longer be a continuous function.

Example 8.4 Find the positive root of $(\sin x - x/2)^2 = 0$ using (1) the Newton-Raphson method (8.4-14); (2) the modified Newton-Raphson method (8.6-13) with $r = 2$; (3) the modified Newton-Raphson method (8.6-24). This equation is, of course, identical with that of Example 8.1, but here we use the above form for illustrative purposes.

We have

$$f(x) = \left(\sin x - \frac{x}{2}\right)^2 \qquad f'(x) = 2\left(\sin x - \frac{x}{2}\right)(\cos x - \tfrac{1}{2})$$

$$f''(x) = 2\left[(\cos x - \tfrac{1}{2})^2 - \sin x\left(\sin x - \frac{x}{2}\right)\right]$$

$$u(x) = \frac{f(x)}{f'(x)} \qquad u'(x) = 1 - \frac{f(x)f''(x)}{[f'(x)]^2}$$

Using $x_1 = \pi/2$ in all cases, we calculate the following results:

	Method				
	1	2	3		Method 1
x_2	1.78540	2.00000	1.80175	x_{10}	1.89512
x_3	1.84456	1.90100	1.88963	x_{11}	1.89531
x_4	1.87083	1.89551	1.89547	x_{12}	1.89540
x_5	1.88335	1.89549	1.89549	x_{13}	1.89545
x_6	1.88946			x_{14}	1.89547
x_7	1.89249			x_{15}	1.89548
x_8	1.89399			x_{16}	1.89549
x_9	1.89475				

As we expect, methods 2 and 3 converge rapidly and much faster than method 1. Note that each successive iterate for method 1 has about one-half the error of the previous iterate {25}.

8.7 SOME COMPUTATIONAL ASPECTS OF FUNCTIONAL ITERATION

Each method of functional iteration we have discussed has the property that if the initial approximation is sufficiently close to the root α, the method will converge if α is a simple root. For a one-point iteration method, this is true even if the root is not simple although, as we have seen, the convergence will be slower in this case. In general, however, it is not possible to prove that a multipoint iteration function will always converge to a multiple root even if the initial approximations are arbitrarily close to the root. For the secant method it is easy to see that the iteration may not converge to a double root by considering the case in which x_1 and x_2 are on opposite sides of the root and are such that $f(x_1) = f(x_2)$ {27}.

If the initial error is not small enough to guarantee decreasing errors in every subsequent iteration, the iteration may (1) diverge or (2) converge anyhow because a small initial error is a sufficient but not a necessary condition for convergence. The two examples below illustrate these two types of behavior.

Example 8.5 Use the Newton-Raphson method to try to find the root of $xe^{-x} = 0$ starting with $x_1 = 2$.

The graph of the function is shown in Fig. 8.5. Since x_{i+1} in the Newton-Raphson method is the intersection of the x axis with the tangent to the curve at $f(x_i)$, the divergence of the method in this case is clear. On the other hand, if $0 < x_1 < 1$, the iteration would converge {28}.

Example 8.6 Use the Newton-Raphson method to find the positive root of $x^{20} - 1 = 0$ starting with $x_1 = \frac{1}{2}$.

For $f(x) = x^{20} - 1$ Eq. (8.4-14) becomes

$$x_{i+1} = x_i - \frac{x_i^{20} - 1}{20x_i^{19}}$$

so that

$$x_2 = \frac{1}{2} - \frac{(\frac{1}{2})^{20} - 1}{20/2^{19}} \approx \frac{1}{20} 2^{19} = 26{,}214.4$$

Thus, because $\frac{1}{2}$ was not close enough to the root $x = 1$, the first iterate leads to a far worse result. But

$$x_3 = x_2 - \frac{x_2^{20} - 1}{20x_2^{19}} \approx \frac{19}{20} x_2$$

and thus lies closer to $x = 1$ than x_2. In fact, it is not hard to see that successive iterates do in fact converge to 1, albeit very slowly {30}.

The slow convergence in Example 8.6 illustrates one of the computational problems associated with the use of functional iteration on a digital computer. It is important that the computer be programmed to recognize this slow convergence and take appropriate action. For example, if

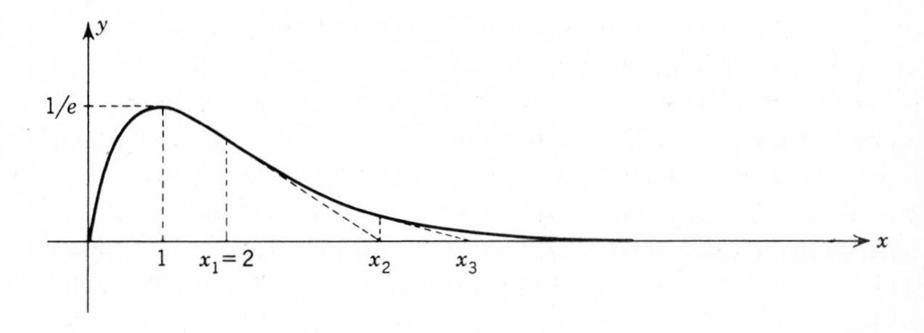

Figure 8.5 Graph of $f(x) = xe^{-x}$.

$|x_{i+1}/x_i|$ is greater than some specified constant k, then, instead of the computed x_{i+1}, the program would set $x_{i+1} = \pm K x_i$, where K is another specified constant, for example, $K = k/2$, and the sign agrees with that of x_{i+1}/x_i {30}.

It is clearly also important that divergence be recognized. When it is, a new starting value should be tried or an always-convergent method such as the modified method of false position or bisection should be used. Sometimes successive iterates may oscillate in such a way that it is not clear whether they are converging or not. In this case, it is often more efficient to err on the safe side and take action similar to that in the case of diveregence.

8.7-1 The δ^2 Process

When an iteration is converging linearly, it is possible to use a technique similar to Richardson extrapolation (see Sec. 4.2) in order to accelerate the convergence. If the iteration is converging, we have

$$\alpha - x_{i+1} = C_i(\alpha - x_i) \qquad |C_i| < 1 \tag{8.7-1}$$

where $|C_i| \to C$, the asymptotic error constant. Near convergence C_i will remain nearly constant, and we can write

$$\alpha - x_{i+1} \approx \bar{C}(\alpha - x_i) \qquad |\bar{C}| = C \tag{8.7-2}$$

Writing (8.7-2) with i replaced by $i + 1$ and eliminating $\bar{C}$, we have

$$\frac{\alpha - x_{i+2}}{\alpha - x_{i+1}} \approx \frac{\alpha - x_{i+1}}{\alpha - x_i} \tag{8.7-3}$$

Solving for α, we obtain

$$\alpha \approx \frac{x_i x_{i+2} - x_{i+1}^2}{x_{i+2} - 2x_{i+1} + x_i} = x_{i+2} - \frac{(\Delta x_{i+1})^2}{\Delta^2 x_i} \tag{8.7-4}$$

where Δ is the forward-difference operator of Chap. 3. This extrapolation procedure is associated with the name of Aitken. Because of the second difference in (8.7-4) (which could be expressed as a central difference), this procedure is called *Aitken's δ^2 process.*

As an example of the use of this technique, we can use the data in the second column of Example 8.1. Using x_5, x_6, and x_7 in (8.7-4), we obtain as the new approximation 1.89554, which is a substantial improvement over any of the values used. Because of the simplicity of this procedure, it should always be used to accelerate the convergence of linear iterations. Another application of this procedure is in the determination of the multiplicity of a multiple root {31}.

For iterations whose order of convergence is greater than 1, (8.7-4) should not be used. For such iterations it is sometimes possible to speed up the convergence using so-called *self-acceleration* procedures. A discussion of these is beyond our scope here; see Traub (1964, pp. 185–187).

8.8 SYSTEMS OF NONLINEAR EQUATIONS

The system of nonlinear equations

$$f_j(x^{(1)}, x^{(2)}, \ldots, x^{(n)}) = 0 \qquad j = 1, \ldots, n \tag{8.8-1}$$

can be rewritten in vector notation as†

$$\mathbf{f}(\mathbf{x}) = \mathbf{0} \tag{8.8-2}$$

where

$$\mathbf{f} = [f_1 \quad f_2 \quad \cdots \quad f_N] \qquad \mathbf{x} = [x^{(1)} \quad \cdots \quad x^{(n)}] \tag{8.8-3}$$

Using the form (8.8-2), we can then derive functional iteration methods as in Sec. 8.4. The general form of the functional iteration equation for stationary iterations is

$$\mathbf{x}_{i+1} = \mathbf{F}(\mathbf{x}_i, \mathbf{x}_{i-1}, \ldots, \mathbf{x}_{i-p}) \tag{8.8-4}$$

where

$$\mathbf{F} = [F_1 \quad F_2 \quad \cdots \quad F_n] \tag{8.8-5}$$

We suppose that in some neighborhood of the solution $\boldsymbol{\alpha} = (\alpha_1, \ldots, \alpha_N)$ of (8.8-2) the vector function $\mathbf{f}$ has an inverse

$$\mathbf{g} = [g_1 \quad g_2 \quad \cdots \quad g_N] \tag{8.8-6}$$

Then using the notation $\mathbf{y} = (y^{(1)}, \ldots, y^{(n)})$ for the point inverse to $\mathbf{x}$, we expand $\mathbf{g}(\mathbf{y})$ in a Taylor series about $\mathbf{y}_i$ [see Apostol (1957), pp. 123–124].

$$\mathbf{x} = \mathbf{g}(\mathbf{y}) = \mathbf{g}(\mathbf{y}_i) + \sum_{j=1}^{m+1} \frac{1}{j!} d^j\mathbf{g}(\mathbf{y}_i; \mathbf{y} - \mathbf{y}_i)$$
$$+ \frac{1}{(m+2)!} d^{m+2}\mathbf{g}(\boldsymbol{\xi}; \mathbf{y} - \mathbf{y}_i) \tag{8.8-7}$$

where $\boldsymbol{\xi}$ lies on the line segment joining $\mathbf{y}$ and $\mathbf{y}_i$ and the jth-order differential is defined by

$$d^j h(\mathbf{x}; \mathbf{s}) = \sum_{i_1=1}^{n} \sum_{i_2=1}^{n} \cdots \sum_{i_j=1}^{n} D_{i_1, i_2, \ldots, i_j} h(\mathbf{x}) s^{(i_1)} s^{(i_2)} \cdots s^{(i_j)} \tag{8.8-8}$$

where $D_{i_1, \ldots, i_j} h(\mathbf{x})$ is the partial derivative of h with respect to the variables

† For convenience and because it causes no confusion, we shall not distinguish between row and column vectors in this section.

$x_{i_1}, \ldots, x_{i_j}$ at the point $\mathbf{x}$ and $s^{(i_1)}, \ldots, s^{(i_j)}$ are components of $\mathbf{s}$. We assume, of course, that all needed partial derivatives exist. Then, since $\boldsymbol{\alpha} = \mathbf{g}(\mathbf{0})$, setting $\mathbf{y} = \mathbf{0}$ in (8.8-7) and dropping the remainder term, we get the equation analogous to (8.4-7). For example, with $m = 0$ we get the n-dimensional analog of the Newton-Raphson method

$$\mathbf{x}_{i+1} = \mathbf{x}_i + d\mathbf{g}(\mathbf{y}_i;\ -\mathbf{y}_i) = \mathbf{x}_i - \sum_{j=1}^{n} \frac{\partial}{\partial y^{(j)}} \mathbf{g}(\mathbf{y}_i) y_i^{(j)} \tag{8.8-9}$$

which can also be written as {32}

$$\mathbf{x}_{i+1} = \mathbf{x}_i - J(\mathbf{x}_i)^{-1}\mathbf{f}(\mathbf{x}_i) \equiv \mathbf{x}_i - J_i^{-1}\mathbf{f}_i \tag{8.8-10}$$

where the Jacobian

$$J_i = \left(\frac{\partial f_j}{\partial x^{(k)}} \right)_{\mathbf{x}=\mathbf{x}_i} \tag{8.8-11}$$

In particular, for $n = 2$, we use the explicit inverse of J_i to get

$$\mathbf{x}_{i+1} = \mathbf{x}_i - \left(\frac{1}{D} \begin{bmatrix} \dfrac{\partial f_2}{\partial x^{(2)}} & -\dfrac{\partial f_1}{\partial x^{(2)}} \\[2ex] -\dfrac{\partial f_2}{\partial x^{(1)}} & \dfrac{\partial f_1}{\partial x^{(1)}} \end{bmatrix} \begin{bmatrix} f_1 \\[2ex] f_2 \end{bmatrix} \right)_{\mathbf{x}=\mathbf{x}_i} \tag{8.8-12}$$

where

$$D = \begin{vmatrix} \dfrac{\partial f_1}{\partial x^{(1)}} & \dfrac{\partial f_1}{\partial x^{(2)}} \\[2ex] \dfrac{\partial f_2}{\partial x^{(1)}} & \dfrac{\partial f_2}{\partial x^{(2)}} \end{vmatrix} \tag{8.8-13}$$

As in the one-dimensional case, the order of convergence is 2 {33}. Here, however, we must note that the problems in the use of functional iteration methods with systems of equations are quite different from those for single equations. For single equations we have noted that good a priori information on the location of the root is often available; when it is not, we can use an always-convergent method to obtain a good approximation to the root. In this case we were therefore mainly interested in the efficiency of methods and comparatively little worried about whether or not a method converged. But with systems of equations convergence itself is such a serious problem that usually we shall be satisfied with any order of convergence if only the method will converge. Any reader who doubts this should try using the Newton-Raphson method to solve two simultaneous polynomial equations of degree 2 in two variables. Often, if the initial approximation is not quite close to the solution, the iteration will not converge {35}. The form of the error in (8.8-12), which we leave to a problem {33}, would make the reason for this clearer.

Before we tackle the problem of finding a good initial approximation, we shall indicate some modifications and generalizations of Newton's method. All these methods will be based on the iteration formula

$$\mathbf{x}_{i+1} = \mathbf{x}_i - t_i H_i \mathbf{f}_i \tag{8.8-14}$$

where H_i is an $n \times n$ matrix determined by the particular method used and t_i is a scaling factor, which is usually taken to be either unity or such that

$$\|\mathbf{f}_{i+1}\| < \|\mathbf{f}_i\| \tag{8.8-15}$$

for some convenient norm. In the latter case, t_i is usually determined by a one-dimensional search along the line $\mathbf{x}_i - tH_i\mathbf{f}_i$. In some methods, we try to find t_i such that $\|\mathbf{f}_{i+1}\| = \min_{t>0} \|\mathbf{f}(\mathbf{x}_i - tH_i\mathbf{f}_i)\|$ or approximately so. Since this may be too time-consuming, we may search only until we find a t such that (8.8-15) holds.

Newton's method (8.8-10) is of the form (8.8-14) with $H_i = J_i^{-1}$ and $t_i = 1$. With this H_i but with t_i chosen so that (8.8-15) holds, we obtain the so-called *damped Newton method*. In another variation of Newton's method, the *modified Newton method*, we do not recompute the value of the Jacobian $J(\mathbf{x})$ at each iteration point $\mathbf{x}_i$. Instead we use the same value of the Jacobian for a fixed number m of iterations, and if convergence is not reached within m iterations, we compute a new value.

Example 8.7 Solve the system of equations

$$f_1(x, y) = x^2 - y - 1 = 0$$

$$f_2(x, y) = (x - 2)^2 + (y - .5)^2 - 1 = 0$$

using (8.8-12).

This system has two roots

$$r_1 = (1.54634\,28833\,2,\ 1.39117\,63127\,9)$$

$$r_2 = (1.06734\,60858\,1,\ .139227\,66688\,7)$$

The Jacobian matrix of this system is

$$J(x, y) = \begin{bmatrix} 2x & -1 \\ 2x - 4 & 2y - 1 \end{bmatrix}$$

and (8.8-12) becomes

$$\begin{bmatrix} x_{i+1} \\ y_{i+1} \end{bmatrix} = \begin{bmatrix} x_i \\ y_i \end{bmatrix} - \frac{1}{4x_i y_i - 4} \begin{bmatrix} 2y_i - 1 & 1 \\ -2x_i + 4 & 2x_i \end{bmatrix} \begin{bmatrix} x_i^2 - y_i - 1 \\ (x_i - 2)^2 + (y_i - .5)^2 - 1 \end{bmatrix} \tag{8.8-16}$$

In the following table, we give various starting values, (x_1, y_1), the root to which (8.8-16) converged with that starting value and the number N of iterations required to achieve accuracy of 12 significant figures:

(x_1, y_1)	Root	N
(0, 0)	r_1	7
(.1, 2)	r_1	25
(1, 0)	r_1	4
(1.5, 1)	r_2	5
(2, 2)	r_2	5

We shall now show that the method of *steepest descent* can also be written in form (8.8-14). This method is used to solve a related minimum problem. Let

$$F(\mathbf{x}) = \mathbf{f}^T(\mathbf{x})\mathbf{f}(\mathbf{x}) = \sum_{i=1}^{n} \{f_i[x^{(1)}, \ldots, x^{(n)}]\}^2 \tag{8.8-17}$$

The function F takes on its absolute minimum, 0, at a solution of (8.8-1). Therefore, if we can find an absolute minimum for F, we shall have solved (8.8-1). If we define the gradient vector of a function $g(\mathbf{x})$ by

$$\nabla g(\mathbf{x}) = \left[\frac{\partial g}{\partial x_1} \quad \cdots \quad \frac{\partial g}{\partial x_n}\right] \tag{8.8-18}$$

then at the point $\mathbf{x}_i$, $g(\mathbf{x})$ decreases most rapidly in the direction $-\nabla g(\mathbf{x}_i)$. To minimize $g(\mathbf{x})$, the classical method of steepest descent searches along the direction $-\nabla g(\mathbf{x}_i)$ to find the point $\mathbf{x}_{i+1}$ which minimizes $g(\mathbf{x})$. However, to find the minimizing point $\mathbf{x}_{i+1}$ may require too much work as measured by the number of function evaluations. In many applications, it suffices to find a point $\mathbf{x}_{i+1}$ at which the magnitude of $g(\mathbf{x})$ is less than at $\mathbf{x}_i$. This is always possible if $\nabla g(\mathbf{x}_i) \neq 0$, that is, if $\mathbf{x}_i$ is not a stationary point of $g(\mathbf{x})$. In this case there is always an interval $I_i = (0, T_i)$ such that if $t \in I_i$, $|g(\mathbf{x}_i - t\,\nabla g(\mathbf{x}_i))| < |g(\mathbf{x}_i)|$. Here T_i depends on the function g as well as on the point $\mathbf{x}_i$ and is not usually known in advance. We also note that, by continuity, if $\mathbf{d}$ is a direction close to that of $\nabla g(\mathbf{x}_i)$, then there exists an interval $I(\mathbf{d})$ such that if $t \in I(\mathbf{d})$, $|g(\mathbf{x}_i - t\mathbf{d})| < |g(\mathbf{x}_i)|$, and as $\mathbf{d}$ approaches $\nabla g(\mathbf{x}_i)$, $I(\mathbf{d})$ approaches I_i.

If we now identify the function $F(\mathbf{x})$ with $g(\mathbf{x})$, we have that $\nabla F(\mathbf{x}) = 2J^T(\mathbf{x})\mathbf{f}(\mathbf{x})$ {36}. Hence, with the choice $H_i = J_i^T$, $0 < t_i < 2T_i$, (8.8-14) becomes the method of steepest descent and (8.8-15) is satisfied.

One problem with the method of steepest descent, which is always mentioned as a flaw of this method, is that we may converge to a local minimum $\mathbf{x}^*$ of $F(\mathbf{x})$, where $\mathbf{f}(\mathbf{x}^*) \neq 0$. In response to this, we point out that when this occurs, $J^T(\mathbf{x}^*)\mathbf{f}(\mathbf{x}^*) = 0$, and since $\mathbf{f}(\mathbf{x}^*) \neq 0$, this implies that $J(\mathbf{x}^*)$ is singular. But singularity of the Jacobian causes other methods, e.g., Newton's method, to break down too, so that this weakness is not special to steepest descent.

Whereas Newton's method converges quadratically in a neighborhood of the solution and steepest descent has only linear convergence, nevertheless the former method requires a good initial approximation to the solution if it is to converge at all, while the latter will converge from any initial approximation (if the Jacobian is nonsingular everywhere). Since, in practice, the convergence of steepest descent is very slow, it can be used initially, and when we get close to the solution, we can switch to Newton's method. The catch is, of course, to decide when we are sufficiently close to ensure that Newton's method will converge.

Another way to combine the methods of Newton and steepest descent is via the Levenberg-Marquardt algorithm, in which we set

$$H_i = (J_i^T J_i + \lambda_i I)^{-1} J_i^T \qquad t_i = 1.0$$

$$\lambda_i \geq 0 \tag{8.8-19}$$

For $\lambda_i = 0$, this yields Newton's method, whereas as λ_i increases, the step length decreases {37} and the direction tends to that of steepest descent. Since $H_i = (1/\lambda_i)(I + \lambda_i^{-1} J_i^T J_i)^{-1} J_i^T$, so that $H_i \to (1/\lambda_i) J_i^T$ as $\lambda_i \to \infty$, it follows from our remarks above that for λ_i sufficiently large, (8.8-15) will hold. The strategy in the use of this algorithm is thus to take λ_i initially large and to reduce it as the solution is approached. Note that for $\lambda_i > 0$ the inverse matrix always exists {37}. Thus, in effect, we use the method of steepest descent to get a good initial approximation to Newton's method.

An alternate way to get a good initial approximation to the solution of (8.8-1) is by *Davidenko's method*. Let $\mathbf{g}(\mathbf{x}, t)$ be a function of $\mathbf{x}$ and t such that $\mathbf{g}(\mathbf{x}, 0) \equiv \mathbf{f}(\mathbf{x})$ and $\mathbf{g}(\mathbf{x}, 1) = \mathbf{0}$ has a known solution. Then, starting with $t_0 = 1$, we choose a sequence t_j

$$t_0 = 1 > t_1 > t_2 > \cdots > t_n = 0$$

and solve in succession

$$\mathbf{g}(\mathbf{x}, t_i) = \mathbf{0} \tag{8.8-20}$$

for $\mathbf{x}_i$ using as initial approximation either $\mathbf{x}_{i-1}$ or an extrapolation from several previous values of $\mathbf{x}_j$. It follows then that the solution of $\mathbf{g}(\mathbf{x}, t_n) = \mathbf{0}$ is also a solution of $\mathbf{f}(\mathbf{x}) = \mathbf{0}$. If t_i is close to t_{i-1}, the initial approximation should be good enough to ensure that (8.8-20) can readily be solved by Newton's method or one of the variations discussed below. The problem of finding an appropriate function $\mathbf{g}(\mathbf{x}, t)$ is quite simple. In fact, if $\mathbf{x}_1$ is an initial estimate of the solution of (8.8-2), then two possibilities for $\mathbf{g}(\mathbf{x}, t)$ are

$$\mathbf{g}(\mathbf{x}, t) = \mathbf{f}(\mathbf{x}) - t\mathbf{f}(\mathbf{x}_1) \tag{8.8-21}$$

and

$$\mathbf{g}(\mathbf{x}, t) = (\mathbf{x} - \mathbf{x}_1)t + (1 - t)\mathbf{f}(\mathbf{x}) \tag{8.8-22}$$

Davidenko's method has connections with differential equations, which will be explored in a problem {38}.

Example 8.8 Solve the system of equations discussed in Example 8.7 by Davidenko's method using (8.8-21) with the sequence of $\{t_i\}$ given by $\{.95, .9, .8, .6, .3, 0\}$.

For each value of t_i, k iterations of Newton's method were performed with $k = 2, 3, 4$. For $t_i = .95$, the initial value $\mathbf{x}_1(t_i)$ for the Newton iterations was the corresponding initial value used in Example 8.7. For $t_i = .9$, the initial value was the result after k applications of the Newton iterations for $t_i = .95$ that is, $\mathbf{x}_1(.9) = \mathbf{x}_{k+1}(.95)$. For the remaining t_i, the initial values $\mathbf{x}_1(t_i)$ were computed by the extrapolation formula

$$\mathbf{x}_1(t_i) = \mathbf{x}_{k+1}(t_{i-1}) + [\mathbf{x}_{k+1}(t_{i-1}) - \mathbf{x}_{k+1}(t_{i-2})]\,\frac{t_i - t_{i-1}}{t_{i-1} - t_{i-2}} \tag{8.8-23}$$

where $\mathbf{x}_{k+1}(t_{i-1})$ is the solution for t_{i-1} after k iterations.

In the table below, we give the initial values (x_1, y_1), the root to which the final result appears to converge, and the number of correct significant figures in each component for each k.

(x_1, y_1)	Root	$k = 2$	$k = 3$	$k = 4$
(0, 0)	r_1	(8, 6)	(12, 12)	(12, 12)
(.1, 2)	r_1	(3, 1)	(4, 3)	(7, 6)
(1, 0)	r_1	(8, 6)	(12, 12)	(12, 12)
(1.5, 1)	r_2	(6, 5)	(10, 10)	(10, 10)
(2, 2)	r_2	(5, 5)	(9, 4)	(12, 12)

Newton's method and its modifications described above, the method of steepest descent and the Levenberg-Marquardt algorithm, all require the computation of the Jacobian J_i. However, even for moderate values of n, the differentiations and programming effort needed to evaluate J_i can become quite tedious, and for large values of n, even when a system for analytic differentiation is available, the computing time and storage requirements may become excessive. Thus, in practice, we usually approximate J_i by another matrix B_i, which is more easily computed. One way of defining B_i is by replacing the partial derivatives appearing in J_i by finite-difference approximations of the form

$$\left.\frac{\partial f_j}{\partial x_k}\right|_{\mathbf{x}=\mathbf{x}_i} \approx \frac{f_j(\mathbf{x}_i + \mathbf{e}_k h_i) - f_j(\mathbf{x}_i)}{h_i} \equiv b_{jk}^{(i)} \tag{8.8-24}$$

where $\mathbf{e}_k$ is the kth column of I_n, the identity matrix of order n, and h_i is small in absolute value. How small h_i should be is a problem since if h_i is too large, there will be a large truncation error, while if h_i is too small, a substantial roundoff error can occur. In any event, if we wish to retain the quadratic convergence of Newton's method, it has been shown that h_i must be chosen so that

$$|h_i| \le C\|\mathbf{f}_i\| \tag{8.8-25}$$

for some positive constant C.

While this discretization of the Jacobian may save on storage and computing time, it is still very expensive since it involves n^2 functional evaluations. One way of skimping on this is to retain the old Jacobian (or its approximation) over a series of iterations, updating it, for instance, after each m steps for some suitable m. This reduces the rate of convergence but should be more economical in the long run.

A better scheme is to generate an approximation B_{i+1} to J_{i+1} at step $i + 1$ by updating the current approximation B_i to the Jacobian J_i without further function evaluations. This possibility introduces a great economy in the computation. There are a variety of ways to do this, as we shall presently see. If we let $\mathbf{q}$ be an arbitrary vector, then

$$f_j(\mathbf{x}_i + \mathbf{q}) - f_j(\mathbf{x}_i) \approx \sum_{l=1}^{n} \frac{\partial f_j(\mathbf{x}_i)}{\partial x_l} q_l \qquad j = 1, \ldots, n \tag{8.8-26}$$

Hence, if we can find a matrix B such that

$$\mathbf{f}(\mathbf{x}_i + \mathbf{q}) - \mathbf{f}(\mathbf{x}_i) = B\mathbf{q} \tag{8.8-27}$$

then B will be a candidate for an approximation to the Jacobian J_i. Now, one can satisfy (8.8-27) simultaneously for a collection of vectors $\mathbf{q}_k$ provided they are linearly independent. For example, if $\mathbf{q}_k = \mathbf{e}_k h_i$, $k = 1, \ldots, n$, then $B = B_i = [b_{jk}^{(i)}]$, with $b_{jk}^{(i)}$ given by (8.8-24), satisfies (8.8-27).

Now, at step $i + 1$ we have already computed $\mathbf{x}_{i+1}$ and $\mathbf{f}_{i+1} = \mathbf{f}(\mathbf{x}_{i+1})$. Hence, a natural choice for $\mathbf{q}$ is $\mathbf{x}_{i+1} - \mathbf{x}_i$ since this does not require any further function evaluations. For this choice of $\mathbf{q}$, our previous approximation B_i will, in general, not satisfy (8.8-27). However, there exist many matrices which do satisfy (8.8-27), and from this class we select our new approximation B_{i+1}. Our first requirement, then, is that

$$B_{i+1}\mathbf{q}_i = \mathbf{v}_i \tag{8.8-28}$$

where

$$\mathbf{v}_i = \mathbf{f}_{i+1} - \mathbf{f}_i \qquad \mathbf{q}_i = \mathbf{x}_{i+1} - \mathbf{x}_i \tag{8.8-29}$$

Since the n equations (8.8-28) are not sufficient to specify the n^2 elements of B_{i+1}, various methods can be defined by imposing additional conditions. The generalized secant procedure {39} chooses B_{i+1} to satisfy the $n - 1$ additional sets of n equations

$$B_{i+1}\mathbf{q}_j = \mathbf{v}_j \qquad j = i - 1, i - 2, \ldots, i - n + 1 \tag{8.8-30}$$

but this can only be applied for $i \geq n$. A modified secant procedure requires that

$$B_{i+1}\mathbf{q}_j = \mathbf{v}_j \tag{8.8-31}$$

for $n - 1$ additional vectors $\mathbf{q}_j$ previously generated, $j < i$, where the $\mathbf{q}_j$ satisfy some criteria of linear independence.

The most sophisticated procedure is that of Broyden, which defines B_{i+1} by the equation

$$B_{i+1} = B_i - \frac{(B_i \mathbf{q}_i - \mathbf{v}_i)\mathbf{q}_i^T}{\mathbf{q}_i^T \mathbf{q}_i} \tag{8.8-32}$$

for which (8.8-28) is also satisfied. Since the rank of $B_{i+1} - B_i$ is 1, this is called a rank-one updating procedure. This formula has the advantage that if $A_i = B_i^{-1}$, then {42}

$$A_{i+1} = A_i - \frac{(A_i \mathbf{v}_i - \mathbf{q}_i)\mathbf{q}_i^T A_i}{\mathbf{q}_i^T A_i \mathbf{v}_i} \tag{8.8-33}$$

Thus, in those methods which require the inverse of the Jacobian, J_i^{-1}, we can avoid the computation of the inverse matrix involving $O(n^3)$ operations by working with the A_i instead. For example, corresponding to Newton's method, we have a *quasi-Newton method* given by the iteration

$$\mathbf{x}_{i+1} = \mathbf{x}_i - A_i \mathbf{f}(\mathbf{x}_i) \tag{8.8-34}$$

with $A_1 = J_1^{-1}$ and A_i given by the updating formulas (8.8-29) and (8.8-33).

Example 8.9 Solve the system of equations discussed in Example 8.7 using the matrix-update method with the same set of initial values.

The results corresponding to those in Example 8.7 for the number of iterations needed to achieve accuracy of 12 significant figures are:

(x_1, y_1)	N
(0, 0)	11
(.1, 2)	14
(1, 0)	7
(1.5, 1)	8
(2, 2)	9

As an example of the details of the calculation, we shall follow two steps in the computation starting with $x_1 = 0$, $y_1 = 0$. We have that

$$f_1(x_1, y_1) = -1 \qquad f_2(x_1, y_1) = 3.25 \qquad A_1 = J_1^{-1} = \begin{bmatrix} .25 & -.25 \\ -1 & 0 \end{bmatrix}$$

By Newton's method, $x_2 = 1.0625$, $y_2 = -1$, and $f_1(x_2, y_2) = 1.12890625$, $f_2(x_2, y_2) = 2.12890625$. From these values, we use (8.8-29) and (8.8-31) to compute

$$\mathbf{q}_2 = \begin{bmatrix} 1.0652 \\ -1 \end{bmatrix} \qquad \mathbf{v}_2 = \begin{bmatrix} 2.12890625 \\ -1.12109375 \end{bmatrix} \qquad \text{and}$$

$$A_2 = \begin{bmatrix} .3557441\cdots & -.2721932\cdots \\ -.5224991\cdots & -.1002162\cdots \end{bmatrix}$$

For comparison purposes, we note that the inverse of the true Jacobian is

$$J_2^{-1} = \begin{bmatrix} .3636363\cdots & .1212121\cdots \\ -.2272727\cdots & -.2575757\cdots \end{bmatrix}$$

Using the pure Newton method with J_2^{-1}, we find that $x_3 = .91003\,78787\ldots$, $y_3 = -.19507\,57575\cdots$ while for the update method, $x_3 = 1.24037\,2062\ldots$, $y_3 = -.19679\,74670\cdots$. We see that the difference between the values of x_3 is substantial while that between the y_3 is not.

In general, the results using the matrix-update method are worse than the results using Newton's method, except for the unusual case of $x_1 = .1$, $y_1 = 2$, in which case the matrix-update method required fewer iterations for convergence. For two equations the update algorithm is much more complicated than Newton's method, and, in general, matrix-update methods are of interest only for moderate and large values of n.

Many of the above ideas, together with some not mentioned here, have been incorporated into algorithms and programs for solving systems of nonlinear equations. These programs have been tested on a variety of suitably chosen test cases, and modifications, refinements, and improvements have been introduced in the light of the experience gained with these programs. While the current situation is far from ideal, the solution of a system of nonlinear equations is far from a hopeless task even when there is no good initial approximation to the solution.

8.9 THE ZEROS OF POLYNOMIALS: THE PROBLEM

In the remainder of this chapter, we shall be concerned with finding the complex and real roots of

$$f(z) \equiv a_n z^n + a_{n-1} z^{n-1} + \cdots + a_1 z + a_0 = 0 \tag{8.9-1}$$

where the coefficients a_i, $i = 0, \ldots, n$, are real numbers† and z is a complex variable. The methods of functional iteration discussed previously in this chapter can all be applied to finding the real roots of (8.9-1) and generally, with simple modifications, they can be applied to complex roots. However, the problem of finding the roots of (8.9-1) arises with such frequency that this alone justifies looking for methods particularly adapted to this problem. The particularly simple form of $f(z)$ in fact greatly aids us in finding such special methods. Also the need to find complex as well as real roots and often the need to find *all* the roots of (8.9-1) add another dimension to the problem which merits special attention.

The need to find all the roots of (8.9-1) commonly arises, as we have seen, for example, in Sec. 5.4 in the consideration of stability problems. Generally in such cases there is no good a priori information about the location of all or, sometimes, any of the roots. This implies the need to emphasize, more than we have done previously, methods which are always

† Many of the methods to be discussed are also applicable when the a_i's are allowed to be complex, but the case of real coefficients is by far the most important.

convergent, particularly for high-degree polynomials. We note that the methods of false position and bisection are not applicable to the case of complex roots. Our approach will be to consider the basic ideas of the classical methods, some of which may converge slowly but surely while others converge rapidly if the initial point is sufficiently close to a root. We shall then consider two modern algorithms which have been incorporated in current polynomial root-finding programs and which have yielded high-quality performance in speed and robustness.

But before considering directly methods for the solution of (8.9-1), it is important to note that there is a large literature on the *location* of the zeros of a polynomial as a function of its coefficients. The problem of solving (8.9-1) cannot be properly attacked without some knowledge of the theorems on the location of the zeros of polynomials. A general discussion of these theorems is beyond our scope here; see Marden (1966), Wall (1948), and Wilf (1962). One aspect of the location of real roots is so important both for the problems of this chapter and in other areas (see Sec. 10.3-2) that we shall discuss it in some detail here.

8.9-1 Sturm Sequences

Definition 8.1 Let

$$f_1(x), f_2(x), \ldots, f_m(x) \tag{8.9-2}$$

be a sequence of polynomials. Such a sequence is called a *Sturm sequence* on an interval (a, b) where either a or b may be infinite, if (1) $f_m(x)$ does not vanish in (a, b); (2) at any zero of $f_k(x)$, $k = 2, \ldots, m-1$, the two adjacent functions are nonzero and have opposite signs; that is,

$$f_{k-1}(x)f_{k+1}(x) < 0 \tag{8.9-3}$$

Definition 8.2 Let $\{f_i(x)\}$, $i = 1, \ldots, m$, be a Sturm sequence on (a, b), and let x_0 be a point of (a, b) at which $f_1(x) \neq 0$. We define $V(x_0)$ to be the number of changes of sign of $\{f_i(x_0)\}$, zero values being ignored. If a is finite, then $V(a)$ is defined as $V(a + \epsilon)$, where ϵ is such that no $f_i(x)$ vanishes in $(a, a + \epsilon)$ and similarly for b when b is finite. If $a = -\infty$, then $V(a)$ is defined to be the number of changes of sign of $\{\lim_{x \to -\infty} f_i(x)\}$ and similarly for $V(b)$ when $b = +\infty$.

Definition 8.3 Let $R(x)$ be any rational function. We define the *Cauchy index* of $R(x)$ on (a, b), denoted by $I_a^b R(x)$, to be the difference between the number of jumps of $R(x)$ from $-\infty$ to $+\infty$ and the number of jumps from $+\infty$ to $-\infty$ as x goes from a to b, excluding the endpoints. That is, at every real pole of $R(x)$ in (a, b) add 1 to the Cauchy index if $R(x) \to -\infty$ on the left of the pole and $R(x) \to +\infty$ on the right of the pole and subtract 1 if vice versa.

With these three definitions, we can prove the following theorem.

Theorem 8.2 (Sturm) If $f_i(x)$, $i = 1, \ldots, m$, is a Sturm sequence on an interval (a, b), then if neither $f_1(a)$ nor $f_1(b)$ equals 0,

$$I_a^b \frac{f_2(x)}{f_1(x)} = V(a) - V(b) \tag{8.9-4}$$

PROOF The value of $V(x)$ does not change when x passes through a zero of $f_k(x)$, $k = 2, \ldots, m$, because of (8.9-3). Thus $V(x)$ can change only when $f_1(x)$ goes through 0. If x_0 is a zero of $f_1(x)$, it is not a zero of $f_2(x)$ because of property 2 of Sturm sequences. Therefore, $f_2(x)$ has the same sign on both sides of x_0. If x_0 is a zero of $f_1(x)$ of even multiplicity, then $V(x)$ does not change as x increases through x_0 and there is no contribution to the Cauchy index. If the zero is of odd multiplicity, then $V(x)$ will increase by 1 if $f_1(x)$ and $f_2(x)$ have the same sign to the left of x_0 and will decrease by 1 if the signs to the left are different. Correspondingly for zeros of odd multiplicity, there is a -1 contribution to the Cauchy index if the signs of $f_1(x)$ and $f_2(x)$ are the same to the left of x_0 and a $+1$ contribution if they are different. This establishes the theorem.

Our chief interest here is in applying this theorem to find the real roots of (8.9-1) in an interval (a, b). Consider the sequence of functions $f_i(x)$, $i = 1, \ldots, m$, where

$$\begin{aligned} &f_1(x) = f(x) \qquad f_2(x) = f'(x) \\ &f_{j-1}(x) = q_{j-1}(x)f_j(x) - f_{j+1}(x) \qquad j = 2, \ldots, m-1 \\ &f_{m-1}(x) = q_{m-1}(x)f_m(x) \end{aligned} \tag{8.9-5}$$

where $q_{j-1}(x)$ is the quotient and $f_{j+1}(x)$ is the negative of the remainder when $f_{j-1}(x)$ is divided by $f_j(x)$. Thus $\{f_i(x)\}$ is a sequence of polynomials of decreasing degree which eventually must terminate in a polynomial $f_m(x)$, $m \le n + 1$, which divides $f_{m-1}(x)$ (why?). The polynomial $f_m(x)$ is the greatest common divisor of $f_1(x)$ and $f_2(x)$ and also of every other member of the sequence (8.9-5). Now suppose $f_m(x)$ does not vanish in (a, b) so that the first condition of Definition 8.1 is satisfied. But in this case, the second condition is also satisfied since, if $f_j(x) = 0$ for any j, $j = 2, \ldots, m-1$, then $f_{j-1}(x) = -f_{j+1}(x)$. Moreover, when $f_j(x) = 0$, $f_{j+1}(x) \neq 0$ since if it were 0, $f_m(x)$ would also be 0 (why?). Thus the sequence $\{f_i(x)\}$ is a Sturm sequence when $f_m(x)$ does not vanish in (a, b).

If $f_m(x)$ is not of constant sign in (a, b), then, in place of (8.9-5), we use the sequence $\{f_i(x)/f_m(x)\}$, $i = 1, \ldots, m$. Not only is this a Sturm sequence but also both sides of (8.9-4) are the same for this sequence and for the sequence (8.9-5) (why?). Therefore, we can use these two sequences interchangeably in applying Sturm's theorem.

Now for the sequence (8.9-5) we write

$$\frac{f_2(x)}{f_1(x)} = \frac{f'(x)}{f(x)} = \sum_{j=1}^{p} \frac{n_j}{x - a_j} + R_1(x) \tag{8.9-6}$$

where the a_j, $j = 1, \ldots, p$, are the distinct real zeros of $f(x)$, n_j is the multiplicity of the zero a_j, and $R_1(x)$ has no poles on the real axis {43}. Since the n_j are all positive, $I_a^b[f'(x)/f(x)]$ is equal to the number of distinct real zeros of $f(x)$ in the interval (a, b). Therefore, we have the following theorem.

Theorem 8.3 The number of distinct real zeros of the polynomial $f(x)$ in the interval $[a, b]$ is equal to $V(a) - V(b)$ if neither $f(a)$ nor $f(b)$ is equal to 0. Moreover, if $f(a)$ or $f(b)$ or both are equal to 0 and the root is simple, the result holds on $[a, b]$ if we define $V(x)$ to be the number of changes of sign in $f_2(x), \ldots, f_m(x)$ when $f_1(x) = 0$ {44}.

This result can be extremely useful in locating the roots of (8.9-1). If, for example, we are interested only in the real roots of (8.9-1), this theorem enables us to determine exactly how many such roots there are. In fact, by making use of $f_m(x)$, we can use this theorem to help find the multiplicity of these roots {44}.

Example 8.10 Apply Sturm's theorem to finding the number of real zeros of

$$f(x) = x^6 + 4x^5 + 4x^4 - x^2 - 4x - 4 = (x^2 + 1)(x^2 - 1)(x + 2)^2$$

Using (8.9-5), we calculate

$$f_1(x) = x^6 + 4x^5 + 4x^4 - x^2 - 4x - 4 \qquad f_2(x) = 6x^5 + 20x^4 + 16x^3 - 2x - 4$$

$$f_3(x) = 4x^4 + 8x^3 + 3x^2 + 14x + 16 \qquad f_4(x) = x^3 + 6x^2 + 12x + 8$$

$$f_5(x) = -17x^2 - 58x - 48 \qquad f_6(x) = -x - 2 = \frac{f_5(x)}{17x + 24}$$

where the coefficients have been made integers by multiplying by suitable positive constants. For some sample values of x the signs of the $f_i(x)$ are

	$-\infty$	∞	0	-1	$+1$	$-\frac{24}{17}$
$f_1(x)$	+	+	−	0	0	+
$f_2(x)$	−	+	−	−	+	−
$f_3(x)$	+	+	+	+	+	−
$f_4(x)$	−	+	+	+	+	+
$f_5(x)$	−	−	−	−	−	0
$f_6(x)$	+	−	−	−	−	−
Number of sign changes	4	1	2	2	1	3

Thus we have three distinct real zeros, two negative real zeros, and one positive real zero. Although -1 and $+1$ are zeros, the rule above shows that there are two distinct zeros in

$(-\infty, -1]$ and three in $(-\infty, +1]$. The point $-\frac{24}{17}$ illustrates the case when an $f_i(x) = 0$. For example, in $(-\infty, -\frac{24}{17}]$, there is one distinct zero. Since $f_6(x) = -x - 2$, the zero at -2 is a double zero {44}.

8.10 CLASSICAL METHODS

Before considering methods for finding zeros of a polynomial, we recall the algorithm given by Eq. (7.2-4) for evaluating a polynomial. To evaluate a polynomial $f(z)$ given by (8.9-1) at the point $z = z_j$, where z_j may be complex, we can use the recursion

$$b_{n-1} = a_n \qquad b_k = a_{k+1} + z_j b_{k+1} \qquad k = n-2, \ldots, -1 \tag{8.10-1}$$

with $f(z_j)$ given by b_{-1}. The coefficients b_k, $k = n-1, \ldots, 0$, are the coefficients resulting from the division of $f(z)$ by $z - z_j$, so that we have

$$f(z) = (z - z_j)(b_{n-1}z^{n-1} + b_{n-2}z^{n-2} + \cdots + b_0) + R_j \tag{8.10-2}$$

where $R_j = f(z_j) = b_{-1}$. This can readily be verified by equating the coefficients of the same power of z on both sides of (8.10-2) and comparing with (8.10-1). The equation (8.10-1) is said to yield a *synthetic-division algorithm* for dividing a polynomial by a linear factor.

The quotient in (8.10-2) can again be divided by $z - z_j$ to give

$$f(z) = (z - z_j)^2(c_{n-2}z^{n-2} + \cdots + c_0) + (z - z_j)R_j' + R_j \tag{8.10-3}$$

where the c_k are given by the recursion

$$c_{n-2} = b_{n-1} \qquad c_k = b_{k+1} + z_j c_{k+1} \qquad k = n-3, \ldots, -1 \tag{8.10-4}$$

and $R_j' = c_{-1}$. Clearly $R_j' = f'(z_j)$.

This can be repeated, up to n times, and at stage s we have

$$f(z) = (z - z_j)^s p_{n-s}(z) + \sum_{l=0}^{s-1} (z - z_j)^l R_j^{(l)} \tag{8.10-5}$$

where $p_{n-s}(z)$ is a polynomial of degree $n - s$ and the successive remainders $R_j^{(l)}$ are the reduced lth derivatives at $z = z_j$,

$$R_j^{(l)} = f^{(l)}(z_j)/l! \tag{8.10-6}$$

Thus, by $s + 1$ repetitions of the synthetic-division algorithm, we can compute the value of $f(z)$ and its first s reduced derivatives at the point $z = z_j$, using a total of

$$n + (n-1) + \cdots + (n-s) = \frac{(s+1)(2n-s)}{2}$$

multiplications and additions. However, we should note that there exists an algorithm {47} which requires only $2n - 1$ multiplications, s divisions, and

$(s+1)(2n-s)/2$ additions, which is a considerable saving if higher derivatives are required.

Now, if z_j is a zero of $f(z)$, $R_j = 0$ and the resulting quotient in (8.10-2) is the *deflated* polynomial of degree $n-1$, that is, the polynomial whose zeros are identical with the remaining $n-1$ zeros of $f(z)$. Thus, once we find a zero z_j of $f(z)$ by any method, synthetic division by $z - z_j$ yields the deflated polynomial of degree $n-1$. To find additional zeros of (8.9-1), the root-finding method can then be applied to this new polynomial. However, it is advisable to check first whether z_j is a multiple zero by applying algorithm (8.10-1) to the deflated polynomial.

Now, if $f(z)$ is a real polynomial and we have found a complex zero $z_j = x_j + iy_j$, we know that $\bar{z}_j = x_j - iy_j$ is also a zero, so that

$$(z - z_j)(z - \bar{z}_j) = z^2 - p_j z + q_j \qquad p_j = 2x_j \qquad q_j = x_j^2 + y_j^2 \tag{8.10-7}$$

is a real factor of $f(z)$. Hence, since we are interested in performing all our arithmetic in the real domain, it is of interest to develop an algorithm for synthetic division by a quadratic factor. We then have

$$f(z) = (z^2 + p_j z + q_j)(b_{n-2} z^{n-2} + \cdots + b_0) + b_{-1}(z + p_j) + b_{-2} \tag{8.10-8}$$

where the b_k are given by the recursion

$$b_{n-2} = a_n \qquad b_{n-3} = a_{n-1} - p_j b_{n-2}$$
$$b_k = a_{k+2} - p_j b_{k+1} - q_j b_{k+2} \qquad k = n-4, \ldots, 0, -1, -2 \tag{8.10-9}$$

The remainder can also be written in the form $R_j z + S_j$, with

$$R_j = a_1 - p_j b_0 - q_j b_1 \equiv b_{-1} \qquad S_j = a_0 - q_j b_0 \equiv b_{-2} + p_j b_{-1} \tag{8.10-10}$$

and if $z^2 + p_j z + q_j$ is a factor of $f(z)$, then $R_j = S_j = 0$. In this case, the quotient polynomial in (8.10-8) is the deflated polynomial of degree $n-2$.

We should point out that deflation can be an unstable process unless proper care is taken [see Wilkinson (1963), pp. 55–62]. The simplest way to avoid such instability is to evaluate the zeros of $f(z)$ in ascending order of magnitude, in which case deflation is quite stable. Alternatively, when the zeros are evaluated in descending order of magnitude, a different deflation algorithm exists, which is also stable {51}. In any case, it is advisable, once approximations to all zeros of $f(z)$ have been computed, to use these approximations as starting values of an iterative process such as Newton's method using the original polynomial. This is called *purifying* the zeros. Usually, one iteration will suffice to obtain the desired accuracy.

As we shall see in Sec. 8.12, Newton's method forms the basis of an efficient algorithm for finding all the zeros of a polynomial. However, not only Newton's method but all iterative methods discussed in the first half of this chapter, except those based on change of sign, can be used to find the complex zeros of polynomials. While the derivation of the order of a method in this case is beyond the scope of this book, nevertheless it can be shown, for example, that, as with real roots, Newton's method is of second order near a simple complex zero while the secant method is of order $(1 + \sqrt{5})/2$. The main disadvantage of these iterative methods is that they require a good initial approximation. When one is available, as in the situation described above, these methods are quite good.

Most of these methods for calculating complex zeros will not converge to complex roots unless we take complex values as initial approximations. However, Muller's method {21} has the desirable feature that it can generate a complex approximation starting with real ones. Thus, the search for the zeros is not influenced by the prejudices of the user.

If we have a real polynomial and a complex approximation z_j, then the algorithm for computing $f(z_j)$ can be expressed using real arithmetic only as follows. Let

$$\begin{aligned} z_j &= x_j + iy_j \\ b_k &= \gamma_k + i\delta_k \qquad i = \sqrt{-1} \\ c_k &= \epsilon_k + i\eta_k \end{aligned} \tag{8.10-11}$$

Then (8.10-1) and (8.10-4) become

$$\begin{aligned} &\gamma_k = a_{k+1} + x_j\gamma_{k+1} - y_j\delta_{k+1} && k = n-1, \ldots, 0, -1 \\ & && \gamma_n = \delta_n = 0 \\ &\delta_k = x_j\delta_{k+1} + y_j\gamma_{k+1} && k = n-2, \ldots, 0, -1 \\ &\epsilon_k = \gamma_{k+1} + x_j\epsilon_{k+1} - y_j\eta_{k+1} && \delta_{n-1} = \epsilon_{n-1} = \eta_{n-1} = 0 \\ &\eta_k = \delta_{k+1} + x_j\eta_{k+1} + y_j\epsilon_{k+1} && k = n-3, \ldots, 0, -1 \\ & && \eta_{n-2} = 0 \end{aligned} \tag{8.10-12}$$

$$\text{with} \quad R_j = b_{-1} = \gamma_{-1} + i\delta_{-1} \qquad R_j' = c_{-1} = \epsilon_{-1} + i\eta_{-1} \tag{8.10-13}$$

Thus Newton's method takes the form

$$x_{j+1} = x_j - \frac{\epsilon_{-1}\gamma_{-1} + \eta_{-1}\delta_{-1}}{\epsilon_{-1}^2 + \eta_{-1}^2} \qquad y_{j+1} = y_j - \frac{\delta_{-1}\epsilon_{-1} - \gamma_{-1}\eta_{-1}}{\epsilon_{-1}^2 + \eta_{-1}^2} \tag{8.10-14}$$

8.10-1 Bairstow's Method

In the case of a real polynomial, we know that the complex zeros occur as conjugate pairs. Hence, instead of looking for one zero at a time, we may look for pairs of zeros which generate a real quadratic factor. This is the basic idea of the iteration of Bairstow, which assumes a good initial approximation.

When p and q are used instead of p_j and q_j, Eqs. (8.10-10) can be written

$$R(p, q) = 0 \qquad S(p, q) = 0 \tag{8.10-15}$$

where b_1 and b_0 are also functions of p and q. In the Bairstow iteration, these two simultaneous equations for the two unknowns p and q are solved using the Newton-Raphson method for simultaneous equations (8.8-12). Let p_i and q_i and p_{i+1} and q_{i+1} denote, respectively, the results of the ith and $(i+1)$st steps in the iteration. Then from (8.8-12) we have

$$p_{i+1} = p_i - \frac{1}{D}\left[R\frac{\partial S}{\partial q} - S\frac{\partial R}{\partial q}\right]_{\substack{p=p_i\\q=q_i}}$$

$$q_{i+1} = q_i - \frac{1}{D}\left[S\frac{\partial R}{\partial p} - R\frac{\partial S}{\partial p}\right]_{\substack{p=p_i\\q=q_i}} \tag{8.10-16}$$

where

$$D = \begin{vmatrix} \dfrac{\partial R}{\partial p} & \dfrac{\partial S}{\partial p} \\ \dfrac{\partial R}{\partial q} & \dfrac{\partial S}{\partial q} \end{vmatrix}_{\substack{p=p_i\\q=q_i}} \tag{8.10-17}$$

Now using (8.10-10), we can write

$$\frac{\partial R}{\partial p} = -p\frac{\partial b_0}{\partial p} - q\frac{\partial b_1}{\partial p} - b_0 \qquad \frac{\partial R}{\partial q} = -p\frac{\partial b_0}{\partial q} - q\frac{\partial b_1}{\partial q} - b_1$$

$$\frac{\partial S}{\partial p} = -q\frac{\partial b_0}{\partial p} \qquad \frac{\partial S}{\partial q} = \frac{\partial b_{-2}}{\partial q} + p\frac{\partial b_{-1}}{\partial q} \tag{8.10-18}$$

From (8.10-9)

$$\frac{\partial b_k}{\partial p} = -b_{k+1} - p\frac{\partial b_{k+1}}{\partial p} - q\frac{\partial b_{k+2}}{\partial p} \qquad k = n-3, \ldots, 0, -1$$

$$\frac{\partial b_{n-2}}{\partial p} = \frac{\partial b_{n-1}}{\partial p} = 0 \tag{8.10-19}$$

$$\frac{\partial b_k}{\partial q} = -b_{k+2} - p\frac{\partial b_{k+1}}{\partial q} - q\frac{\partial b_{k+2}}{\partial q} \qquad k = n-4, \ldots, 0, -1, -2$$

$$\frac{\partial b_{n-3}}{\partial q} = \frac{\partial b_{n-2}}{\partial q} = 0 \tag{8.10-20}$$

If we define d_k by the recurrence relation

$$d_k = -b_{k+1} - pd_{k+1} - qd_{k+2} \qquad k = n-3, \ldots, 0, -1$$
$$d_{n-2} = d_{n-1} = 0 \tag{8.10-21}$$

then it follows from (8.10-19) and (8.10-20) that

$$\frac{\partial b_k}{\partial p} = d_k \qquad \frac{\partial b_{k-1}}{\partial q} = d_k \qquad k = n-3, \ldots, 0, -1 \tag{8.10-22}$$

and finally that

$$\frac{\partial R}{\partial p} = d_{-1} \qquad \frac{\partial R}{\partial q} = d_0 \qquad \frac{\partial S}{\partial p} = -qd_0 \qquad \frac{\partial S}{\partial q} = d_{-1} + pd_0 \tag{8.10-23}$$

Therefore, (8.10-16) and (8.10-17) become

$$p_{i+1} = p_i - \frac{1}{D}[b_{-1}(d_{-1} + p_i d_0) - (b_{-2} + p_i b_{-1})d_0]$$

$$q_{i+1} = q_i - \frac{1}{D}[(b_{-2} + p_i b_{-1})d_{-1} + d_0 b_{-1} q_i] \tag{8.10-24}$$

where
$$D = d_{-1}^2 + p_i d_0 d_{-1} + q_i d_0^2 \tag{8.10-25}$$

Example 8.11 Use the Bairstow iteration to find a quadratic factor of $z^3 - z - 1$ starting with $p_1 = q_1 = 1$.

We arrange the calculation in the form

	(p_i, q_i)	
a_n	b_{n-2}	d_{n-3}
a_{n-1}	$\vdots$	$\vdots$
$\vdots$	$\vdots$	d_0
$\vdots$	b_0	d_{-1}
$\vdots$	b_{-1}	
a_0	b_{-2}	

with (8.10-9) and (8.10-21) used to calculate the last two columns. For this problem we have, using (8.10-24) and (8.10-25),

	$(p_1, q_1) = (1, 1)$		$(p_2, q_2) = (\frac{4}{3}, \frac{2}{3})$	
1	1	−1	1	−1
0	−1	2	$-\frac{4}{3}$	$\frac{8}{3}$
−1	−1		$\frac{1}{9}$	
−1	1		$-\frac{7}{27}$	

and finally $p_3 = 1.3246$, $q_3 = .7544$, whereas the true values are $p = 1.3247$, $q = .7549$.

When it converges, Bairstow's method has the characteristic rapid convergence of the Newton-Raphson method. But, as is usually the case in solving simultaneous nonlinear equations, convergence requires quite a good initial approximation {53}.

8.10-2 Graeffe's Root-squaring Method

The essence of Graeffe's method is to replace (8.9-1) by an equation, still of degree n, whose roots are the squares of the roots of (8.9-1). By iterating this procedure, roots of (8.9-1) which are unequal in magnitude become more widely separated in magnitude. By separating the roots sufficiently we can, as we shall see, calculate the roots directly from the coefficients. When there are roots of equal magnitude, this process runs into difficulties, but they can be overcome.

Let the roots of (8.9-1) be α_i, $i = 1, \ldots, n$. We assume in the remainder of this section that $a_n = 1$. Then, writing $f_0(z)$ for $f(z)$, we have

$$f_0(z) = (z - \alpha_1)(z - \alpha_2) \cdots (z - \alpha_n) \tag{8.10-26}$$

Using this, we can write

$$f_1(w) = (-1)^n f_0(z) f_0(-z) = (w - \alpha_1^2)(w - \alpha_2^2) \cdots (w - \alpha_n^2) \qquad w = z^2 \tag{8.10-27}$$

so that the zeros of $f_1(w)$ are the squares of those of $f_0(z)$. Therefore, the sequence

$$f_{r+1}(w) = (-1)^n f_r(z) f_r(-z) \qquad r = 0, 1, \ldots \tag{8.10-28}$$

is such that the zeros of each polynomial are the squares of the zeros of the previous polynomial. If we denote the coefficients of $f_r(z)$ by $a_j^{(r)}$, $j = 0, \ldots, n$, we can show that {55}

$$a_j^{(r+1)} = (-1)^{n-j} \left[(a_j^{(r)})^2 + 2 \sum_{k=1}^{\min(n-j,\, j)} (-1)^k a_{j-k}^{(r)} a_{j+k}^{(r)} \right] \tag{8.10-29}$$

To use the sequence of polynomials $\{f_r(z)\}$, we need the well-known relationship between the coefficients of a polynomial and its zeros. This relationship is expressed by the equation

$$a_j^{(r)} = (-1)^{n-j} S_{n-j}(\alpha_1^{2^r}, \alpha_2^{2^r}, \ldots, \alpha_n^{2^r}) \qquad j = 0, \ldots, n-1 \tag{8.10-30}$$

where $S_k(x_1, \ldots, x_n)$ is the kth symmetric function of $x_1, \ldots, x_n$. This function is defined by the equation

$$S_k(x_1, \ldots, x_n) = \sum_{1}^{n}{}_c\, x_{r_1} x_{r_2} \cdots x_{r_k} \tag{8.10-31}$$

where the notation $\sum_{1}^{n}{}_c$ denotes that the sum is over all *combinations* of k out of the digits 1 to n in the subscripts. Thus, for example,

$$a_{n-1}^{(r)} = -S_1(\alpha_1^{2^r}, \ldots, \alpha_n^{2^r}) = -\sum_{k=1}^{n} \alpha_k^{2^r} \tag{8.10-32}$$

Let

$$\alpha_k = \rho_k e^{i\phi_k} \qquad k = 1, \ldots, n \tag{8.10-33}$$

Suppose first that all the roots are distinct in magnitude and ordered so that

$$\rho_1 > \rho_2 > \cdots > \rho_n \tag{8.10-34}$$

We write (8.10-32) as

$$a_{n-1}^{(r)} = -\alpha_1^{2^r}\left[1 + \sum_{k=2}^{n}\left(\frac{\alpha_k}{\alpha_1}\right)^{2^r}\right] \tag{8.10-35}$$

Then using (8.10-34), we have

$$\lim_{r\to\infty} |a_{n-1}^{(r)}|^{1/2^r} = |\alpha_1| \tag{8.10-36}$$

Therefore, for sufficiently large r

$$\rho_1 \approx |a_{n-1}^{(r)}|^{1/2^r} \tag{8.10-37}$$

Similarly, we have

$$a_{n-2}^{(r)} = \sum_{1}^{n}{}_c\, \alpha_{r_1}^{2^r}\alpha_{r_2}^{2^r} = \alpha_1^{2^r}\alpha_2^{2^r}\left[1 + \sum_{\substack{1 \\ (r_1, r_2)\neq(1, 2)}}^{n}{}_c \left(\frac{\alpha_{r_1}\alpha_{r_2}}{\alpha_1\alpha_2}\right)^{2^r}\right] \tag{8.10-38}$$

and, therefore, for sufficiently large r

$$\rho_2 \approx \frac{1}{\rho_1}|a_{n-2}^{(r)}|^{1/2^r} \approx \left|\frac{a_{n-2}^{(r)}}{a_{n-1}^{(r)}}\right|^{1/2^r} \tag{8.10-39}$$

Continuing in this way we have in general

$$\rho_k \approx \left|\frac{a_{n-k}^{(r)}}{a_{n-k+1}^{(r)}}\right|^{1/2^r} \qquad k = 3, \ldots, n \tag{8.10-40}$$

In practice "sufficiently large r" means only that we must continue the root-squaring process until the approximations to the magnitudes have stabilized to the number of decimal places we want.

When the roots are all separated, then once we have the magnitudes, determining the sign is easily accomplished by inserting the magnitude into (8.9-1).

Example 8.12 Use the root-squaring method to find the zeros of $z^3 - 5z^2 - 17z + 21$. Using (8.10-29), we calculate

r	$a_3^{(r)}$	$a_2^{(r)}$	$a_1^{(r)}$	$a_0^{(r)}$
1	1	-59	499	-441
2	1	$-2,483$	196,963	$-194,481$
3	1	$-5,771,363$	37,828,630,723	$-37,822,859,361$

and from these we can estimate at each stage

r	ρ_1	ρ_2	ρ_3
1	7.68	2.91	.94
2	7.06	2.98	.997
3	7.001	2.999	.99998

Insertion of these three magnitudes into the polynomial leads easily to the result that α_1 and α_3 are positive and α_2 is negative. The true roots are 7, -3, 1.

The difficulties in the use of the root-squaring procedure arise when some of the roots have equal magnitudes. These difficulties are of two kinds: (1) The relations (8.10-37), (8.10-39), and (8.10-40) no longer are correct in general. Therefore, determining the magnitudes of the roots is more difficult. (2) Since some roots may be complex, it is no longer simple to determine the root given the magnitude. While both these difficulties are not hard to overcome, we shall not go into the details here. Similarly, we shall not discuss the computational aspects of this method since we are interested principally in the ideas behind the method of this section, not in its implementation.

8.10-3 Bernoulli's Method

Consider the difference equation

$$a_n u_k + a_{n-1} u_{k-1} + \cdots + a_0 u_{k-n} = 0 \tag{8.10-41}$$

where the coefficients a_i, $i = 0, \ldots, n$, are those of (8.9-1). If the roots α_i of (8.9-1) are distinct, then the solution of this equation is given by {57}

$$u_k = \sum_{i=1}^{n} c_i \alpha_i^k \tag{8.10-42}$$

where the c_i's depend on the initial conditions used to solve (8.10-41). If the roots are ordered in magnitude as in (8.10-34), then by rewriting (8.10-42) as

$$u_k = c_1 \alpha_1^k \left[1 + \sum_{i=2}^{n} \frac{c_i}{c_1} \left(\frac{\alpha_i}{\alpha_1} \right)^k \right] \tag{8.10-43}$$

we have, if $c_1 \neq 0$,

$$\lim_{k \to \infty} \frac{u_k}{u_{k-1}} = \alpha_1 \tag{8.10-44}$$

The essence of Bernoulli's method is to use (8.10-41) to compute successive values of u_k and then to compute the ratio of successive values of u_k until these ratios converge to α_1.

For this method to work at all, it is necessary that $c_1 \neq 0$. The c_i's depend, as we said, on the n initial conditions required by (8.10-41). If we generate these initial values using the equation

$$a_n u_m + a_{n-1} u_{m-1} + \cdots + a_{n-m+1} u_1 + m a_{n-m} = 0 \qquad m = 1, \ldots, n \tag{8.10-45}$$

then it can be shown {56} that all c_i's are unity and thus

$$u_k = \sum_{i=1}^{n} \alpha_i^k \tag{8.10-46}$$

Therefore, (8.10-44) always holds for this choice of initial conditions.

The above was predicated on the assumption that α_1, the root of largest magnitude, is real and distinct. Nevertheless, (8.10-44) also holds if α_1 is multiple but real {57}. But when the root of largest magnitude is complex, or when there is some combination of real and complex roots of largest magnitude, (8.10-44) no longer holds. The number of possible special cases is therefore very large. Each such special case can be taken care of by a suitable modification of (8.10-44), but especially for automatic computation, it is extremely tedious to have to provide for all these cases. Moreover, if α_2 has nearly the same magnitude as α_1, the convergence of the process is extremely slow (why?). Thus, as a general-purpose method, Bernoulli's method has little in its favor. However, when the root of largest or smallest magnitude [by considering $f(1/z)$] is the only one that is desired and is distinct, Bernoulli's method can be very useful.

Example 8.13 Use Bernoulli's method to find the zero of greatest magnitude of the polynomial $z^3 - 5z^2 - 17z + 21$ of Example 8.12.

Using (8.10-45), we have

$$u_1 - 5 = 0 \to u_1 = 5$$

$$u_2 - 5u_1 - 34 = 0 \to u_2 = 59$$

$$u_3 - 5u_2 - 17u_1 + 63 = 0 \to u_3 = 317$$

Then using (8.10-41) in the form

$$u_k = 5u_{k-1} + 17u_{k-2} - 21u_{k-3}$$

we calculate

$$u_4 = 2483 \qquad \frac{u_4}{u_3} = 7.83 \qquad u_5 = 16{,}565 \qquad \frac{u_5}{u_4} = 6.67$$

$$u_6 = 118{,}379 \qquad \frac{u_6}{u_5} = 7.15 \qquad u_7 = 821{,}357 \qquad \frac{u_7}{u_6} = 6.94$$

$$u_8 = 5{,}771{,}363 \qquad \frac{u_8}{u_7} = 7.03 \qquad u_9 = 40{,}333{,}925 \qquad \frac{u_9}{u_8} = 6.989$$

This example illustrates two characteristic features of Bernoulli's method. First, it generally converges more slowly than the root-squaring method. However, because of the linear convergence of the method [note (8.10-44)], the δ^2 process can be used to accelerate the convergence. For example, when the first three approximations to the zero in the example above (7.83, 6.67, 7.15) are used, Eq. (8.7-4) gives 6.998 as the improved approximation. The second characteristic feature of Bernoulli's method is that, like the root-squaring method, the numbers involved grow very rapidly for roots of magnitude greater than 1. This latter problem can be avoided by working with the ratios u_{k-1}/u_k {58}.

8.10-4 Laguerre's Method

Suppose all the zeros of the polynomial $f(z)$ are real and ordered such that

$$\alpha_1 \leq \alpha_2 \leq \cdots \leq \alpha_n \qquad n > 2 \tag{8.10-47}$$

with $\alpha_1 < \alpha_n$. We define

$$I_i = [\alpha_i, \alpha_{i+1}] \qquad i = 0, \ldots, n \qquad \alpha_0 = -\infty \qquad \alpha_{n+1} = \infty \tag{8.10-48}$$

Let x be an approximation to a zero of $f(z)$. This approximation lies in some interval I_i. The essence of Laguerre's method is to construct a parabola with two real zeros in I_i at least one of which will be closer to a zero of $f(z)$ than x. It turns out that there are many such parabolas, depending on a real parameter λ, and we shall so choose λ that one of the zeros of the resulting parabola is as close as possible to a zero of $f(z)$.

For an arbitrary real λ, let

$$S(\lambda) = \sum_{i=1}^{n} \left(\frac{\lambda - \alpha_i}{x - \alpha_i} \right)^2 > 0 \tag{8.10-49}$$

The equation

$$\phi(y) = (x - y)^2 S(\lambda) - (\lambda - y)^2 = 0 \tag{8.10-50}$$

has two real roots, y_1, y_2, which are distinct if $\lambda \neq x$, which we shall henceforth assume. If $f(x) \neq 0$, it follows from (8.10-49) and (8.10-50) that

$\phi(x) < 0$ and that $\phi(\alpha_i) > 0, i = 0, 1, \ldots, n+1$. Therefore, if $x \in I_i, i = 0, \ldots, n$, the two roots y_1, y_2 both lie in I_i, one between α_i and x and one between x and α_{i+1}. Now, apparently, to know $\phi(y)$ as a function of λ requires a knowledge of the roots α_i. However, we shall show [cf. (8.10-54)] that this seeming dependence on the roots can be eliminated.

We now wish to choose λ so that one of the zeros of $\phi(y)$ is as close as possible to a zero of $f(z)$. This means that we wish to maximize $|x - y|$ as a function of λ or alternatively as a function of the real parameter $\mu = \lambda - x$ (why?). From (8.9-1), we can easily calculate {60}

$$\frac{f'(x)}{f(x)} = \sum_{i=1}^{n} \frac{1}{x - \alpha_i} \equiv S_1 \tag{8.10-51}$$

$$\frac{[f'(x)]^2 - f(x)f''(x)}{[f(x)]^2} = \sum_{i=1}^{n} \frac{1}{(x - \alpha_i)^2} \equiv S_2 \tag{8.10-52}$$

Also
$$\left(\frac{\lambda - \alpha_i}{x - \alpha_i}\right)^2 = \frac{\mu^2}{(x - \alpha_i)^2} + \frac{2\mu}{x - \alpha_i} + 1 \tag{8.10-53}$$

Using (8.10-49) and (8.10-51) to (8.10-53), we see that (8.10-50) becomes

$$\mu^2(\eta^2 S_2 - 1) + 2\mu\eta(\eta S_1 - 1) + (n-1)\eta^2 = 0 \tag{8.10-54}$$

where $\eta = x - y$. From this equation we see that, to solve the equation for the roots of $\phi(y) = 0$ knowledge of the α_i is not needed.

Now (8.10-54) is also a quadratic equation in μ whose roots are continuous functions of the parameter η. Since $\mu = \lambda - x$ takes on only real values in our discussion, our aim is to find the maximum value of $|\eta|$ such that μ is real. Since this implies that greater values of $|\eta|$ will result in complex values for μ, that is, complex roots of (8.10-54), the desired value of $|\eta|$ must be such that the discriminant D of the quadratic equation in μ is 0 (why?). This means that it is only necessary to determine the values of η which make $D = 0$. Thus, μ or, what amounts to the same thing, λ need not be determined at all.

From (8.10-54) we find that

$$D = \eta^2\{[S_1^2 - (n-1)S_2]\eta^2 - 2\eta S_1 + n\} \tag{8.10-55}$$

Solving $D = 0$ for η, we get

$$\eta = \frac{n}{S_1 \pm \sqrt{(n-1)(nS_2 - S_1^2)}} \tag{8.10-56}$$

which yields the equation

$$y = x - \frac{nf(x)}{f'(x) \pm \sqrt{H(x)}} \tag{8.10-57}$$

where
$$H(x) = (n-1)\{(n-1)[f'(x)]^2 - nf(x)f''(x)\} \tag{8.10-58}$$

When all the zeros are real, we can show that H is always nonnegative {60}. Equation (8.10-57) suggests the iteration

$$x_{i+1} = x_i - \frac{nf(x_i)}{f'(x_i) \pm \sqrt{H(x_i)}} \tag{8.10-59}$$

In order to determine which sign to use in the denominator of (8.10-59), let us use Theorem 8.1 to determine the order of this iteration. We have, with α any simple real zero of $f(z)$ {60}

$$\begin{aligned} F(x) &= x - \frac{nf(x)}{f'(x) \pm \sqrt{H(x)}} \\ F'(\alpha) &= 1 - \frac{nf'(\alpha)}{f'(\alpha) \pm \sqrt{H(\alpha)}} \\ &= 1 - \frac{nf'(\alpha)}{f'(\alpha) \pm (n-1)|f'(\alpha)|} \\ F''(\alpha) &= \frac{-nf''(\alpha)}{f'(\alpha) \pm (n-1)|f'(\alpha)|} \\ &\quad \times \left\{1 - \frac{2f'(\alpha)}{f'(\alpha) \pm (n-1)|f'(\alpha)|}\left[1 \pm \frac{n-2}{2}\frac{f'(\alpha)}{|f'(\alpha)|}\right]\right\} \end{aligned} \tag{8.10-60}$$

From (8.10-60) it is easy to see that if the sign is chosen to agree with the sign of $f'(\alpha)$, then both $F'(\alpha)$ and $F''(\alpha)$ are zero. Therefore, in practice we should choose the sign in (8.10-59) to agree with $f'(x_i)$. In particular we can then show that if the initial approximation $x_1 < \alpha_1$, then $x_i < x_{i+1} < \alpha_1$ and, similarly, if $\alpha_n < x_1$, then $\alpha_n < x_{i+1} < x_i$ {60}. Since it can be shown that $F'''(\alpha) \neq 0$, Laguerre's method is a third-order method for simple real zeros. This is obtained at the expense of calculating $f(x)$, $f'(x)$, and $f''(x)$ at every stage of the iteration.

The great advantage of Laguerre's method is that the method is sure to converge independent of the initial approximation x_1. From our construction, this is obvious if $x_1 \in [\alpha_1, \alpha_n]$, and from the comments above, it is also true if x_1 is outside this interval. This method is then a powerful, rapidly converging method for a polynomial all of whose zeros are real and simple. If all the zeros are real but some are not simple, the method still converges but is first order in the neighborhood of a multiple zero. For polynomials some of whose zeros are complex little is known about the overall convergence properties of the method. Note that a real initial approximation may nevertheless converge to a complex zero, since $H(x)$ can be negative in this case. It is known however, that when the method converges to a simple complex zero, the convergence is third order [Parlett (1964)]. Empirical evidence suggests that lack of convergence is extremely unusual. Laguerre's method is therefore a candidate for use as a general-purpose method for finding the zeros of polynomials. Its infrequent use for this purpose is due partially, at least, to an incomplete theoretical understanding of the method.

Example 8.14 Use Laguerre's method to find the largest positive zero of the polynomial of Example 8.12.

To illustrate the power of this method, we choose a very bad initial approximation $x_1 = 10^6$. We have

$$f(x) = x^3 - 5x^2 - 17x + 21$$

$$f'(x) = 3x^2 - 10x - 17$$

$$f''(x) = 6x - 10$$

Then using (8.10-58) and (8.10-59), we calculate

i	2	3	4
x_i	7.4785207	7.0001011	7.0000000

and the true zero is 7. We note that if the magnitude of x_1 is large, the arithmetic at the first iteration must be done to high precision or substantial loss of significance results.

8.11 THE JENKINS-TRAUB METHOD

Consider the monic polynomial

$$P(z) = z^n + a_{n-1}z^{n-1} + \cdots + a_1 z + a_0 = \prod_{j=1}^{p} (z - \alpha_j)^{m_j} \tag{8.11-1}$$

where the a_i are complex numbers with $a_0 \neq 0$ and where the zero α_j of $P(z)$ has multiplicity m_j, $j = 1, \ldots, p$, so that $\sum_{j=1}^{p} m_j = n$. We then have that

$$P'(z) = \sum_{j=1}^{p} m_j P_j(z) \tag{8.11-2}$$

where

$$P_j(z) = \frac{P(z)}{z - \alpha_j} \qquad j = 1, \ldots, p \tag{8.11-3}$$

We shall now generate a sequence of polynomials $H^{(\lambda)}(z)$ starting with $H^{(0)}(z) = P'(z)$, each of the form

$$H^{(\lambda)}(z) = \sum_{j=1}^{p} c_j^{(\lambda)} P_j(z) \tag{8.11-4}$$

with $c_j^{(0)} = m_j$, $j = 1, \ldots, p$, If we can choose such a sequence so that $H^{(\lambda)}(z) \to c_1^{(\lambda)} P_1(z)$, that is, so that the ratios

$$d_j^{(\lambda)} = \frac{c_j^{(\lambda)}}{c_1^{(\lambda)}} \to 0 \qquad j = 2, \ldots, p \tag{8.11-5}$$

then we can find a sequence $\{t_\lambda\}$ of approximations to α_1 by the formula

$$t_{\lambda+1} = s_\lambda - \frac{P(s_\lambda)}{\tilde{H}^{(\lambda+1)}(s_\lambda)} \tag{8.11-6}$$

where
$$\tilde{H}^{(\lambda)}(z) = \frac{H^{(\lambda)}(z)}{\sum_{j=1}^{p} c_j^{(\lambda)}} \tag{8.11-7}$$

is monic and $\{s_\lambda\}$ is an arbitrary sequence of complex numbers. This is so because, using (8.11-3) to (8.11-5),

$$t_{\lambda+1} = s_\lambda - \frac{P(s_\lambda)\sum_{j=1}^{p} c_j^{(\lambda+1)}}{\sum_{j=1}^{p} c_j^{(\lambda+1)}P_j(s_\lambda)}$$

$$= s_\lambda - \frac{P_1(s_\lambda)(s_\lambda - \alpha_1)c_1^{(\lambda+1)}\left(1 + \sum_{j=2}^{p} d_j^{(\lambda+1)}\right)}{P_1(s_\lambda)c_1^{(\lambda+1)}\left[1 + \sum_{j=2}^{p} d_j^{(\lambda+1)}P_j(s_\lambda)/P_1(s_\lambda)\right]}$$

which approaches $s_\lambda - (s_\lambda - \alpha_1) = \alpha_1$.

The $H^{(\lambda)}(z)$ are generated by the formula

$$H^{(\lambda+1)}(z) = \frac{1}{z - s_\lambda} Q^{(\lambda)}(z) \tag{8.11-8}$$

where
$$Q^{(\lambda)}(z) = H^{(\lambda)}(z) - \frac{H^{(\lambda)}(s_\lambda)}{P(s_\lambda)} P(z) \tag{8.11-9}$$

We can generate such a sequence so long as $P(s_\lambda) \neq 0$. Otherwise, of course, s_λ is already a zero of $P(z)$ and we can deflate $P(z)$ and start afresh. Now using (8.11-3), (8.11-4), and (8.11-9) in (8.11-8), we find that

$$H^{(\lambda+1)}(z) = \frac{P(z)}{z - s_\lambda}\left(\sum_{j=1}^{p} \frac{c_j^{(\lambda)}}{z - \alpha_j} - \sum_{j=1}^{p} \frac{c_j^{(\lambda)}}{s_\lambda - \alpha_j}\right) = \sum_{j=1}^{p} c_j^{(\lambda+1)}P_j(z) \tag{8.11-10}$$

where
$$c_j^{(\lambda+1)} = \frac{c_j^{(\lambda)}}{\alpha_j - s_\lambda} \qquad j = 1, \ldots, p \tag{8.11-11}$$

Hence

$$c_j^{(\lambda+1)} = \frac{c_j^{(\lambda)}}{\alpha_j - s_\lambda} = \frac{c_j^{(\lambda-1)}}{(\alpha_j - s_\lambda)(\alpha_j - s_{\lambda-1})} = \cdots = \frac{m_j}{\prod_{t=0}^{\lambda} (\alpha_j - s_t)} \qquad j = 1, \ldots, p \tag{8.11-12}$$

and if no s_t is a zero of $P(z)$, $c_j^{(\lambda)} \neq 0$ for all j.

We now discuss the choice of values of s_λ. Since, as we noted in Sec. 8.10, deflation is most stable when the zeros are generated in order of increasing magnitude, we are interested in converging to a zero of smallest magnitude. Hence, in the first stage, we choose $s_\lambda = 0$ for $\lambda = 0, \ldots, M - 1$, so that the coefficients of large zeros will be small while those corresponding to small zeros will be accentuated. This happens with $s_\lambda = 0$ because with such a choice

$$d_j^{(m)} = \frac{m_j}{m_1}\left(\frac{\alpha_1}{\alpha_j}\right)^m \qquad j = 2, \ldots, p \tag{8.11-13}$$

Now, were there a zero α_1 such that

$$|\alpha_1| < |\alpha_2| \leq |\alpha_3| \leq \cdots \leq |\alpha_p| \tag{8.11-14}$$

we could continue with $s_\lambda = 0$ and the sequence $\{t_\lambda\}$ would converge to α_1, with the rate of convergence determined by the size of the ratio $|\alpha_1/\alpha_2|$. However, (8.11-14) does not always hold, and even when it does hold, $|\alpha_1/\alpha_2|$ may be very close to unity; i.e., we may have a cluster of zeros. Therefore, we choose $s_\lambda = 0$ for a fixed number M of iterations and then subsequently handle the root-cluster problem. In practice, M is taken to be 5 on the basis of numerical experience.

In the second stage, we are interested in separating zeros of equal or almost equal magnitude. To this end we iterate with $s_\lambda = s$, a complex number which we hope will be closer to one zero than to any other. In the sequel, we shall call this zero α_1 even though it is not necessarily the zero smallest in magnitude. However, its magnitude will not differ much from the minimum, so that except in most unusual circumstances deflation will still be stable. With such a choice of s_λ we find that

$$d_j^{(L)} = \frac{m_j}{m_1}\left(\frac{\alpha_1}{\alpha_j}\right)^M \left(\frac{\alpha_1 - s}{\alpha_j - s}\right)^{L-M} \qquad j = 2, \ldots, p \tag{8.11-15}$$

where $\lambda = M, M + 1, \ldots, L - 1$. Hence, if $|\alpha_1 - s| < |\alpha_j - s|$, $d_j^{(L)} \to 0$ as $L \to \infty$, $j = 2, \ldots, p$, so that $t_\lambda \to \alpha_1$.

As an initial estimate for s we choose

$$s = e^{i\theta}\beta \tag{8.11-16}$$

where θ is an angle chosen at random and β is the unique positive zero of the polynomial

$$z^n + |a_{n-1}|z^{n-1} + \cdots + |a_1|z - |a_0| \tag{8.11-17}$$

By a theorem of Cauchy [Householder (1970), p. 73, example 5] β is a lower bound on the $|\alpha_j|$, $j = 1, \ldots, p$. θ is theoretically taken at random since we have no prior knowledge about the location of the zeros in the complex plane and with probability 1 it will yield a value of s closer to one zero than to any other zero. This way of choosing θ could be implemented in practice with a random-number generator. However, the standard implementation takes θ initially to be 49°, which gives a value close to the middle of the first quadrant of the complex plane. With this value of s, we could usually iterate to convergence. However, we would again have linear convergence with the rate of convergence determined by the value of the ratio $|(\alpha_1 - s)/(\alpha_2 - s)|$, where α_2 is the zero next closest to s (cf. Fig. 8.6). Hence we iterate only $L - M$ times, until the corresponding t_λ is sufficiently close to α_1. In theory, L depends on the distribution of the zeros of $P(z)$, as we shall see in Theorem 8.6. In practice, L is determined when

$$|t_\lambda - t_{\lambda-1}| \le \tfrac{1}{2}|t_{\lambda-1}| \qquad \text{and} \qquad |t_{\lambda-1} - t_{\lambda-2}| \le \tfrac{1}{2}|t_{\lambda-2}| \tag{8.11-18}$$

It may happen that by some misfortune s is equidistant or almost so from two or more zeros. In this case we switch to a new value of s if (8.11-18) does not hold after some maximum number of iterations; i.e., we choose a new value of θ in (8.11-16) either by generating a new random number or, in the standard implementation, by increasing the previous value of θ by 94°, which steps us through the four quadrants of the complex plane if necessary and then repeats the cycle at a point 16° away from the previous point in the same quadrant. We then restart with $\lambda = M$.

Once (8.11-18) is satisfied, we enter the third stage and use a variable shift, $s_\lambda = t_\lambda$, $\lambda = L, L + 1, \ldots$, since it is now assumed that t_λ is a reasonably good approximation to α_1. As we shall shortly see, this gives us better than quadratic convergence in the sense that the error coefficient also goes to zero. Before proceeding with the proof of convergence, we shall summarize

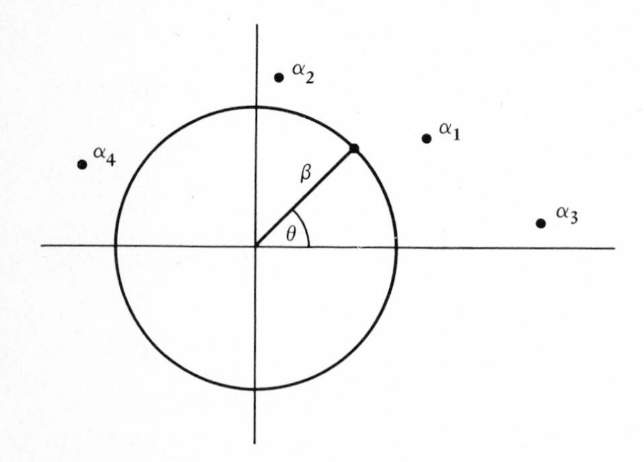

Figure 8.6 Ordering of the zeros in the Jenkins-Traub method.

the algorithm. Set

$$H^{(0)}(z) = P'(z)$$

$$s_\lambda = \begin{cases} 0 & \lambda = 0, \ldots, M-1 \\ s \text{ [from (8.11-16)]} & \lambda = M, \ldots, L-1 \\ t_\lambda \text{ [from (8.11-6)]} & \lambda = L, L+1, \ldots \end{cases} \tag{8.11-19}$$

$$H^{(\lambda+1)}(z) = \frac{1}{z - s_\lambda} Q^{(\lambda)}(z) \qquad Q^{(\lambda)}(z) \text{ from (8.11-8)}$$

where M is usually taken to be 5, L is determined by when (8.11-18) holds, and convergence to α_1 takes place when $|P(s_\lambda)| < \epsilon$, where ϵ is a measure of the roundoff error incurred in computing $P(s_\lambda)$.

We now state the theorem which ensures convergence if $s_L \equiv t_L$ is sufficiently close to α_1.

Theorem 8.4 If

1. $|s_L - \alpha_1| < \frac{1}{2}R$, where $R = \min_{2 \le j \le p} |\alpha_1 - \alpha_j|$

2. $D_L = \sum_{j=2}^{p} |d_j^{(L)}| < \frac{1}{3}$

then the s_λ in the third-stage iteration ($= t_\lambda$) converge to α_1. Furthermore, if we define

$$C(\lambda) = \frac{|s_{L+\lambda+1} - \alpha_1|}{|s_{L+\lambda} - \alpha_1|^2} \tag{8.11-20}$$

then

$$C(\lambda) \le \frac{2}{R} \tau_L^{\lambda(\lambda-1)/2} \tag{8.11-21}$$

where

$$\tau_\lambda = \frac{2D_L}{1 - D_L} < 1 \tag{8.11-22}$$

Thus, the process is second order with an error constant $C(\lambda)$ which approaches zero.

Proof Let us define

$$r_j^{(\lambda)} = \frac{s_\lambda - \alpha_1}{s_\lambda - \alpha_j} \qquad j = 1, \ldots, p \tag{8.11-23}$$

so that

$$d_j^{(\lambda+1)} = r_j^{(\lambda)} d_j^{(\lambda)} \tag{8.11-24}$$

Since $s_\lambda = t_\lambda$ in stage 3, we have from (8.11-6) that

$$s_{\lambda+1} = s_\lambda - \frac{P(s_\lambda)}{\tilde{H}^{(\lambda+1)}(s_\lambda)} \tag{8.11-25}$$

so that
$$\frac{s_{\lambda+1} - \alpha_1}{s_\lambda - \alpha_1} = 1 - \frac{P(s_\lambda)}{(s_\lambda - \alpha_1)\tilde{H}^{(\lambda+1)}(s_\lambda)} \tag{8.11-26}$$

Now using (8.11-3), (8.11-4), and (8.11-7)

$$\begin{aligned}(s_\lambda - \alpha_1)\tilde{H}^{(\lambda+1)}(s_\lambda) &= \frac{(s_\lambda - \alpha_1)\sum_{j=1}^{p} c_j^{(\lambda+1)}P_j(s_\lambda)}{\sum_{j=1}^{p} c_j^{(\lambda+1)}} \\ &= \frac{P(s_\lambda)\sum_{j=1}^{p} r_j^{(\lambda)}c_j^{(\lambda+1)}}{\sum_{j=1}^{p} c_j^{(\lambda+1)}} \\ &= \frac{c_1^{(\lambda+1)}P(s_\lambda)\sum_{j=1}^{p} r_j^{(\lambda)}d_j^{(\lambda+1)}}{c_1^{(\lambda+1)}\sum_{j=1}^{p} d_j^{(\lambda+1)}} \\ &= \frac{P(s_\lambda)\sum_{j=1}^{p} (r_j^{(\lambda)})^2 d_j^{(\lambda)}}{\sum_{j=1}^{p} r_j^{(\lambda)}d_j^{(\lambda)}}\end{aligned}$$

Substituting this into (8.11-26) and using the fact that $r_1^{(\lambda)} = d_1^{(\lambda)} = 1$, we obtain

$$\frac{s_{\lambda+1} - \alpha_1}{s_\lambda - \alpha_1} = \frac{\sum_{j=2}^{p} (r_j^{(\lambda)})^2 d_j^{(\lambda)} - \sum_{j=2}^{p} r_j^{(\lambda)}d_j^{(\lambda)}}{1 + \sum_{j=2}^{p} (r_j^{(\lambda)})^2 d_j^{(\lambda)}} \tag{8.11-27}$$

By the first hypothesis, $|s_L - \alpha_j| \geq |\alpha_1 - \alpha_j| - |s_L - \alpha_1| > R/2$, so that $|r_j^{(L)}| < 1, j = 2, \ldots, p$. If we now define

$$T_\lambda = \frac{|s_{\lambda+1} - \alpha_1|}{|s_\lambda - \alpha_1|} \tag{8.11-28}$$

it follows that

$$T_L \le \frac{\sum_{j=2}^{p} |d_j^{(L)}| + \sum_{j=2}^{p} |d_j^{(L)}|}{1 - \sum_{j=2}^{p} |d_j^{(L)}|} = \tau_L < 1 \tag{8.11-29}$$

Assume now that $T_t \le \tau_L$ for $t = L, \ldots, \lambda - 1$. Then since $|s_t - \alpha_1| = T_{t-1}|s_{t-1} - \alpha_1|$, it follows that

$$|s_t - \alpha_1| < \tau_L^{t-L}|s_L - \alpha_1| < \tfrac{1}{2}R \tag{8.11-30}$$

and
$$|s_t - \alpha_j| \ge |\alpha_1 - \alpha_j| - |s_t - \alpha_1| > \tfrac{1}{2}R \qquad j = 2, 3, \ldots, p \tag{8.11-31}$$

Hence $|r_j^{(t)}| < 1$, so that

$$\sum_{j=2}^{p} |d_j^{(\lambda)}| = \sum_{j=2}^{p} \left| d_j^{(L)} \prod_{t=L}^{\lambda-1} r_j^{(t)} \right| \le D_L$$

Therefore, by induction, $T_\lambda \le \tau_L < 1$ for $\lambda \ge L$, and since $s_\lambda - \alpha_1 = (s_L - \alpha_1) \prod_{t=L}^{\lambda-1} T_t$, this proves convergence. Now, using (8.11-23), (8.11-24), and (8.11-27) with λ replaced by $\lambda + L$, we have that

$$\frac{s_{L+\lambda+1} - \alpha_1}{(s_{L+\lambda} - \alpha_1)^2} = \frac{\sum_{j=2}^{p} \dfrac{r_j^{(L+\lambda)} d_j^{(L+\lambda)}}{s_{L+\lambda} - \alpha_j} - \sum_{j=2}^{p} \dfrac{d_j^{(L+\lambda)}}{s_{L+\lambda} - \alpha_j}}{1 + \sum_{j=2}^{p} (r_j^{(L+\lambda)})^2 d_j^{(L+\lambda)}} \tag{8.11-32}$$

From (8.11-30) and (8.11-31) we have for all λ and $j > 1$ {61}

$$|r_j^{(L+\lambda)}| \le \tau_L^\lambda \qquad \frac{1}{|s_{\lambda+L} - \alpha_j|} < \frac{2}{R} \tag{8.11-33}$$

From this and (8.11-24) it follows that

$$\sum_{j=2}^{p} |d_j^{(L+\lambda)}| = \sum_{j=2}^{p} |r_j^{(L+\lambda)}| \cdot |r_j^{(L+\lambda-1)}| \cdots |r_j^{(L)}|\,|d_j^{(L)}|$$

$$\le \tau_L^{\lambda(\lambda-1)/2} \sum_{j=2}^{p} |d_j^{(L)}| < \tfrac{1}{3}\tau_L^{\lambda(\lambda-1)/2} \tag{8.11-34}$$

Substituting these bounds into the numerator of (8.11-32) and using hypothesis 2 for the denominator establishes (8.11-21) {61}.

There is still one loose end in this proof, namely to show that the sequence $\{s_\lambda\}$ is always well defined for $\lambda \ge L$, that is, that $\tilde{H}^{(\lambda+1)}(s_\lambda) \ne 0$.

Now using (8.11-3) to (8.11-5), (8.11-7), and (8.11-11) gives {61}

$$\tilde{H}^{(\lambda+1)}(s_\lambda) = \frac{\sum_{j=1}^{p} c_j^{(\lambda)}(\alpha_j - s_\lambda)^{-1}P_j(s_\lambda)}{\sum_{j=1}^{p} c_j^{(\lambda)}(\alpha_j - s_\lambda)^{-1}}$$

$$= P_1(s_\lambda)\left[\frac{1 + \sum_{j=2}^{p} d_j^{(\lambda)}(r_j^{(\lambda)})^2}{1 + \sum_{j=2}^{p} d_j^{(\lambda)}r_j^{(\lambda)}}\right] \tag{8.11-35}$$

Since we assumed that s_λ is not a zero of $P(z)$, $P_1(s_\lambda) = P(s_\lambda)/(s_\lambda - \alpha_1) \neq 0$. Furthermore, as above,

$$\left|\sum_{j=2}^{p} d_j^{(\lambda)}(r_j^{(\lambda)})^2\right| \leq D_L < \tfrac{1}{3} \tag{8.11-36}$$

Hence $\tilde{H}^{(\lambda+1)}(s_\lambda) \neq 0$, and the third stage iteration is well defined so long as s_λ is not a zero of $P(z)$. This concludes the proof of Theorem 8.4.

It remains to show the existence of a theoretical L for which the hypotheses of Theorem 8.4 are satisfied.

Theorem 8.5 Let s in (8.11-16) be such that $|s - \alpha_1| < |s - \alpha_j|, j = 2, \ldots, p$. Then we can find an L such that hypotheses 1 and 2 of Theorem 8.4 hold.

PROOF From (8.11-15)

$$D_L = \sum_{j=2}^{p} |d_j^{(L)}| = \sum_{j=2}^{p} \frac{m_j}{m_1}\left|\frac{\alpha_1}{\alpha_j}\right|^M \left|\frac{\alpha_1 - s}{\alpha_j - s}\right|^{L-M} \tag{8.11-37}$$

If we fix M, then since the last term is less than 1, we can choose L sufficiently large to ensure that $D_L < \frac{1}{3}$ and that τ_L is sufficiently small for $|s - \alpha_1|\tau_L < \frac{1}{2}R$. As in (8.11-30),

$$|s_L - \alpha_1| < \tau_L|s - \alpha_1| < \tfrac{1}{2}R \tag{8.11-38}$$

which complete the proof.

There is an interesting connection between the Jenkins-Traub method and Newton's method, in that the formula (8.11-25) is identical with a single application of Newton's method to the function $W^{(\lambda)}(z) = P(z)/H^{(\lambda)}(z)$ {62}. Thus, we can interpret the third-stage iteration as the application of Newton's method to a sequence of rational functions $W^{(\lambda)}(z)$, which for λ sufficiently large are as close as desired to a linear polynomial with zero α_1.

The Jenkins-Traub method has been generalized to find quadratic factors of a real polynomial. This enables one to find the conjugate-complex zeros of a real polynomial using only real arithmetic. This makes for a faster algorithm at the cost of greater complexity.

Example 8.15 Find the zeros of the polynomial

$$\begin{aligned} P(z) &= z^5 - (13.999 + 5i)z^4 + (74.99 + 55.998i)z^3 \\ &\quad - (159.959 + 260.982i)z^2 + (1.95 + 463.934i)z \\ &\quad + (150 - 199.95i) \\ &= (z - 1 - i)^2(z - 4 + 3i)(z - 4 - 3i)(z - 3.999 - 3i) \end{aligned}$$

using the Jenkins-Traub method.

This polynomial has a zero of multiplicity 2 plus three almost equimodular zeros, two of which form a near-multiple pair. In the calculation of the zeros, M in stage 1 was set equal to 5. In the table we give the value of s used in stage 2, the number of steps in stage 2, $L - M$, the value of s_L used to start stage 3, and the values s_{L+j}, $j = 1, 2, \ldots$, until the stopping criterion was satisfied.

j		s_{L+j}
Zero(1): $s = .27281 + .30819i$	$L - M = 2$	$s_L = .99998 + 1.00003i$
1		.99999 99999 97 + .99999 99999 74i
Zero(2): $s = -.528885 + .41362i$	$L - M = 2$	$s_L = 1.00006 + 1.00034i$
1		.99999 99999 97 + 1.00000 00000 33i
2		1.00000 00000 04 + 1.00000 00000 26i
Zero(3): $s = -.85450 - 1.24126i$	$L - M = 6$	$s_L = 5.78433 - 1.78155i$
1		4.15089 03638 40 − 3.60756 77541 67i
2		3.98898 38448 95 − 3.00489 27129 73i
3		4.00000 02599 31 − 2.99999 95584 78i
4		4.00000 00000 00 − 3.00000 00000 00i
Zero(4): $s = 1.76985 - 1.06591i$	$L - M = 2$	$s_L = 3.99950 + 3.00000i$
1		3.99949 91925 97 + 3.00000 08282 43i
2		3.99959 49473 61 + 3.00016 17681 68i
3		4.00016 64468 29 + 3.00043 18443 90i
4		4.00007 84238 10 + 3.00000 41370 51i
5		4.00000 07928 07 + 2.99999 93289 69i
6		3.99999 99999 47 + 2.99999 99999 88i
Zero(5):		3.99900 00000 03 + 3.00000 00000 12i

8.12 A NEWTON-BASED METHOD

The aim of this method is to find the zero of smallest magnitude of a given polynomial so that it can deflate in a stable fashion and thus find all the zeros. Since it is based on Newton's method, the main problem is to find an approximation to this zero close enough to ensure convergence of Newton's method. This problem itself is solved with the help of the Newton formula in that this formula is used to provide a *direction* of search for the next iterate rather than its value. This direction is usually a descent direction for the real function of a complex argument $F(z) = |f(z)|$, so that we obtain a sequence of points giving decreasing function magnitudes. This is called stage 1. Since searching along any particular direction is expensive with respect to function evaluations, we would like to leave stage 1 as soon as possible. Hence, when we determine that we are close enough to a zero of $f(z)$ to ensure that Newton's method converges, we enter stage 2 and use the standard Newton formula. The details for stage 1 are as follows for the complex case. The modifications needed in the real case are only in the deflation and will be given later.

Given the polynomial $f(z)$ of (8.9-1), we wish to generate a sequence $\{z_k\}$ which converges to the zero of smallest magnitude or, at least, to one close to it in magnitude. Successive points are related by the formula

$$z_{k+1} = z_k + \beta dz_k \tag{8.12-1}$$

where dz_k is called the *tentative step* at iteration k and β is a scalar, possibly 0. We distinguish between a successful step, in which $z_{k+1} \neq z_k$, and an unsuccessful step, in which $z_{k+1} = z_k$. If the previous iteration was successful, the Newton correction

$$n_k = -\frac{f(z_k)}{f'(z_k)} \tag{8.12-2}$$

is computed and dz_k is taken as

$$dz_k = \begin{cases} n_k & \text{if } |n_k| \leq 3|z_k - z_{k-1}| \\ \dfrac{3|z_k - z_{k-1}|e^{i\theta}n_k}{|n_k|} & \text{otherwise} \end{cases} \tag{8.12-3}$$

where θ is chosen (arbitrarily) as $\tan^{-1} \frac{3}{4}$. If the previous step was unsuccessful, then

$$dz_k = -\tfrac{1}{2}e^{i\theta}dz_{k-1} \tag{8.12-4}$$

The motivation behind these choices of dz_k is as follows. After a successful iteration, we normally want to take a step in the Newton direction, since this is usually a descent direction for $F(z)$. However, in the neighborhood of a saddle point of $F(z)$ which occurs at a zero of $f'(z)$, the Newton direction is a worse direction than almost any other. Hence, if we see that $|n_k|$ is relatively large, we suspect that we are approaching such a point and therefore change the search direction by the amount θ, which seems to work well in practice. After an unsuccessful iteration, we want to change the search direction to one likely to be successful. Now, if $f(z_k) \neq 0$, there certainly exists a descent direction for $F(z)$ inasmuch as $F(z)$ has local minima (which are also global) only at the zeros of $f(z)$. Hence, since we search in a different direction each time, repeated use of (8.12-4) will certainly yield a descent direction. Furthermore, for reasons connected with the termination strategy, it is desirable to reduce the step size, as we shall see.

Once dz_k has been chosen, $f(z_k + dz_k)$ is computed and the inequality

$$F(z_k + dz_k) < F(z_k) \tag{8.12-5}$$

is tested. If it holds, the numbers $F(z_k + p\,dz_k)$, $p = 2, \ldots, n$, are computed until the sequence is no longer strictly decreasing. Since this search is designed to locate multiple zeros, we stop at $p = n$. Otherwise, we could end up searching indefinitely along a particular direction with only marginal improvement each time. If (8.12-5) does not hold, the numbers $F(z_k + \frac{1}{2}dz_k)$, $F(z_k + \frac{1}{4}dz_k)$, and $F(z_k + \frac{1}{4}e^{i\theta}dz_k)$ are computed until the sequence ceases to decrease. The reasoning here is that if the values of $F(z)$ decrease as we approach the point z_k, then this point may be near a saddlepoint and we hope to do better by switching to another direction. In either case, β is chosen as the last value for which $F(z_k + \beta dz_k)$ is strictly less than the previous one in the sequence. If $F(z_k + dz_k) \geq F(z_k)$ and $F(z_k + \frac{1}{2}dz_k) \geq F(z_k + dz_k)$, we take $\beta = 0$, that is, $z_{k+1} = z_k$, and we have an unsuccessful iteration. Note that if there is a true multiple zero of multiplicity m, then by (8.6-13), $z_k + mn_k$ is the proper Newton step, which we shall usually find if we are close enough to that zero. A similar situation will hold if we are at a fair distance from a cluster of m zeros but nearer to it than to any other zero.

Stage 1 is initialized by the following values:

$$\begin{aligned} z_0 &= 0 \\ dz_0 &= \begin{cases} \dfrac{-f(0)}{f'(0)} & \text{if } f'(0) \neq 0 \\ 1 & \text{otherwise} \end{cases} \\ z_1 &= \tfrac{1}{2}\min_{k>0}\left(\left|\frac{a_0}{a_k}\right|^{1/k}\right)\frac{dz_0}{|dz_0|} \end{aligned} \tag{8.12-6}$$

The choice of z_1 is such that its magnitude is less than that of any zero of $f(z)$ [cf. Householder (1970), p. 73, example 11], and it is in the direction of steepest descent of $|f(z)|$ from the origin {63}. It is therefore likely that we shall converge to a zero of near-minimal magnitude. We wish to switch to stage 2 (in which we shall be using straight Newton iteration) when we are reasonably sure of convergence of the Newton iteration, i.e., we are not converging to a saddle point, and when we are not converging to a multiple zero (for then the Newton iteration gives us no advantage). These conditions require that we switch to stage 2 only when $\beta = 1$. The test for the former condition is based on the Kantorovich theorem [Ostrowski (1973), p. 59], which states that if K_0 is the circle with center $z_k + n_k$ and radius n_k, the conditions

$$f(z_k)f'(z_k) \neq 0 \qquad 2|n_k| \max_{z \in K_0} |f''(z)| \leq |f'(z_k)| \tag{8.12-7}$$

ensure the convergence of Newton's method starting from z_k. In practice (8.12-7) is replaced by the test of the inequality

$$2|f(z_k)| \cdot |f'(z_{k-1}) - f'(z_k)| \leq |f'(z_k)|^2 |z_{k-1} - z_k| \tag{8.12-8}$$

which uses the crude approximation $|[f'(z_{k-1}) - f'(z_k)]/(z_{k-1} - z_k)|$ to $\max_{z \in K_0} |f''(z)|$. When (8.12-8) holds, we switch over to stage 2; in stage 2 (8.12-8) is checked at every iteration, and if it does not hold, we switch back to stage 1. Similarly, if (8.12-5) does not hold at any iteration, we switch back to stage 1 with a tentative step given by (8.12-3). These tests are necessary here since we used (8.12-8) rather than (8.12-7) to make our decision to enter stage 2. Had we used (8.12-7), we would be certain that, barring roundoff errors, (8.12-7) and (8.12-5) would hold at every stage 2 iteration. However, because of roundoff, (8.12-5) need not hold in practice near a zero, as we shall see, and even had we used (8.12-7), we would still have to check (8.12-5) at every iteration.

We can terminate the process with an approximation to a zero while in either stage 1 or stage 2. For termination, one of two criteria must be satisfied. Both involve ϵ, the largest number such that, to machine accuracy, $1 + \epsilon = 1$. These criteria are

$$\epsilon |z_{k+1}| > |z_{k+1} - z_k| > 0 \tag{8.12-9}$$

and

$$F(z_{k+1}) = F(z_k) \leq 16n|a_0|\epsilon \equiv \delta \tag{8.12-10}$$

The number δ is a generous overestimate of the roundoff error made in calculating $f(z)$ at the zero of smallest modulus; it is to be expected that such accuracy will be attainable. The normal convergence pattern is that $F(z_k)$ decreases until well below δ, and then roundoff errors cause a new iterate z_{k+1} to be equal to z_k, so that (8.12-10) holds. If such accuracy is unattainable and $F(z_k)$ does not decrease at each stage, we shall switch back to stage 1. Then the step will decrease steadily because of (8.12-4) until (8.12-9) is satisfied. With this combination of convergence criteria, we are almost certain to obtain the best possible solution, and usually at the cost of only one extra iteration.

If we have a real polynomial and have found a complex zero $\alpha_j = x_j + iy_j$, we must decide whether this is a true complex zero, in which case we deflate by the real quadratic factor $(z - \alpha_j)(z - \bar{\alpha}_j)$ using (8.10-9), or whether this is a simple real zero x_j. To this end, we compute $f(x_j)$ and check whether the difference between the computed values of $|f(x_j)|$ and $|f(\alpha_j)|$ is within the roundoff error in the computation of $f(x_j)$. If so, we deflate with the linear factor $z - x_j$ using (8.10-1).

A program based on this method has been compared with one based on the Jenkins-Traub method and was found to be from 2 to 4 times as fast and at least as accurate. An additional feature incorporated in this program is that it computes an error bound for each zero separately. This error bound is the radius of the circle with center at the calculated zero in which the true zero of $f(z)$ is almost certain to lie. The program and details are given in the reference cited in the Bibliographic Notes.

Example 8.16 Find the zeros of the polynomial given in Example 8.15 using the method of this section.

The zeros α_j were computed in the order given below. With each zero, we give m, the number of evaluations of $f(z)$, and n, the number of evaluations of $f'(z)$. We give the results to 12 figures. In this computation $\epsilon = 2.3 \times 10^{-16}$.

	α_j	m	n
1	.99999 99920 93 + 1.00000 00008 2i	25	7
2	1.00000 00079 1 + .99999 99991 77i	18	8
3	3.99899 99999 9 + 3.00000 00000 0i	48	14
4	4.00000 00000 0 − 3.00000 00000 0i	21	9
5	4.00000 00000 1 + 3.00000 00000 0i		

If we compare these numbers with the results in Example 8.15, we see that the accuracy of the Jenkins-Traub method was slightly better for the multiple zeros and marginally worse for the others. On the surface, it would appear that the Jenkins-Traub method was faster for this problem too, since the numbers corresponding to $m + n$ were respectively 12, 15, 29, and 27. However, there are more things involved in a computation than function evaluations, and, in fact, the measured computer time was almost half that using the Jenkins-Traub method. Furthermore, this ratio decreases as the degree of the polynomial increases, so that, for example, for the polynomial $x^{40} + 1$, the ratio was about $\frac{1}{3}$. Finally, the Jenkins-Traub program requires considerably more storage than this method.

8.13 THE EFFECT OF COEFFICIENT ERRORS ON THE ROOTS; ILL-CONDITIONED POLYNOMIALS

The coefficients of the polynomial whose zeros we actually compute are seldom the true coefficients of the polynomial whose zeros we desire. The coefficients we use may arise from empirical data, in which case we shall not know the true coefficients, or they may result from a lengthy computation that introduced many rounding errors, in which case we may have very conservative bounds on the roundoff errors. And even when we know the true coefficients, we must round them when inserting them into the computer. How then do these coefficient errors affect the accuracy of the calculated zeros?

Let the true polynomial be

$$F(z) = A_n z^n + A_{n-1} z^{n-1} + \cdots + A_1 z + A_0 \qquad (8.13\text{-}1)$$

and define

$$\delta_i = A_i - a_i \qquad i = 0, 1, \ldots, n \qquad (8.13\text{-}2)$$

If the computed zero is z_0 and the true zero is

$$Z_0 = z_0 + \epsilon \qquad (8.13\text{-}3)$$

where ϵ may be real or complex, we wish to find an estimate of the magnitude of ϵ. Assume for now that ϵ and the δ_i's are sufficiently small to permit all products of the errors to be neglected. Then substituting (8.13-2) and (8.13-3) into (8.13-1) and using the fact that z_0 is a root of (8.9-1), we get {66}

$$\sum_{i=0}^{n} \delta_i z_0^i + \epsilon f'(z_0) \approx 0 \qquad (8.13\text{-}4)$$

Therefore,

$$|\epsilon| \approx \frac{\left|\sum_{i=0}^{n} \delta_i z_0^i\right|}{|f'(z_0)|} \qquad (8.13\text{-}5)$$

One obvious limitation of this estimate occurs when $f'(z_0)$ is zero or small, in which case the previous assumption that products of errors could be neglected was unfounded {66}. We might expect, however, that when $f'(z_0)$ is not small and when the δ_i's are the result only of roundoff in entering the coefficients into the computer, $|\epsilon|$ would indeed be small and (8.13-5) would give a good estimate. The following example, due to Wilkinson (1959), indicates that this need not be so for polynomials of high degree.

Consider the polynomial

$$f(z) = (z + 1)(z + 2) \cdots (z + 20) \qquad (8.13\text{-}6)$$

with zeros $-1, -2, \ldots, -20$. If $z_0 = -20$, then $|f'(z_0)| = 19!$. Suppose that $\delta_i = 0$, $i = 0, 1, 2, \ldots, 17, 18, 20$, but that $\delta_{19} = 2^{-23} \approx 10^{-7}$. Then (8.13-5) becomes

$$|\epsilon| \approx \frac{10^{-7}(20)^{19}}{19!} \approx 4.4 \qquad (8.13\text{-}7)$$

which is not small. In fact, correct to nine decimal places, the zeros of $f(z) + 2^{-23}z^{19}$ are

$$\begin{array}{ll}
-1.00000\,0000 & -10.09526\,6145 \pm 0.64350\,0904i \\
-2.00000\,0000 & \\
-3.00000\,0000 & -11.79363\,3881 \pm 1.65232\,9728i \\
-4.00000\,0000 & \\
-4.99999\,9928 & -13.99235\,8137 \pm 2.51883\,0070i \\
-6.00000\,6944 & \\
-6.99969\,7234 & -16.73073\,7466 \pm 2.81262\,4894i \\
-8.00726\,7603 & \\
-8.91725\,0249 & -19.50243\,9400 \pm 1.94033\,0347i \\
-20.84690\,8101 &
\end{array}$$

For example, the zero corresponding to -19 in $f(z)$ has not only changed substantially but has become complex. It is therefore no surprise that (8.13-7) gives a poor result. The small changes in the zeros of small magnitude suggest that (8.13-5) would give accurate estimates for these errors, and in fact this is correct {66}.

A polynomial such as (8.13-6) in which a small change in a coefficient may cause a large change in one or more zeros is called ill-conditioned (cf. Sec. 1.7 and Sec. 9.5). If the coefficients are in fact known exactly so that the coefficient error is the result of roundoff in entering the coefficients into the computer, then by using multiple-precision arithmetic we can decrease this roundoff and increase the accuracy of the zeros. In fact, it is generally true that the solution of high-degree polynomial equations requires the use of multiple-precision floating-point arithmetic in order to achieve high accuracy.

BIBLIOGRAPHIC NOTES

Sections 8.1 to 8.7 The basis of much of these sections is the book by Traub (1964), which will not be referred to explicitly below. This book contains the best and most complete treatment available on functional iteration, including much that can be found nowhere else. It also contains an excellent bibliography. In some instances, our terminology differs somewhat from that of Traub. Another excellent book that we have used extensively is that by Ostrowski (1973). Other excellent general references for these sections and for the remainder of the chapter are Durand (1960–1961) and Householder (1970). See also Traub (1967).

Section 8.3 The secant method and the method of false position are discussed in detail and with insight by Ostrowski (1973). Most standard texts in numerical analysis consider one or both of these methods. The modifications to regula falsi are studied by Anderson and Björck (1973).

Section 8.4 Much of this section can be found in the papers by Traub (1961*a*,*b*). Ostrowski (1973) discusses the Newton-Raphson method in detail.

Section 8.5 A widely used multipoint iteration method which is not discussed in this section is that of Muller (1956) (see Prob. 21).

Section 8.7 Ostrowski (1973) has an interesting discussion of convergence. Wilkinson (1959) discusses the problems that arise in Example 8.6. The δ^2 process is due to Aitken (1926) and is discussed in a number of numerical analysis texts. For other acceleration procedures, see Ostrowski (1973).

Section 8.8 The most comprehensive work on the solution of systems of nonlinear algebraic equations is that of Ortega and Rheinboldt (1970). The third edition of Ostrowski (1973) contains much material not available elsewhere in book form. A concise survey of the field is given in the booklet by Rheinboldt (1974). Two recent books which contain proceedings of conferences devoted to the numerical solution of nonlinear algebraic equations are those edited by Rabinowitz (1970) and by Byrne and Hall (1973). Both books contain computer programs for the solution of systems of nonlinear equations; see also Rall (1969). Much of the material in this section is based on a paper by Broyden in the volume edited by Rabinowitz, where further details and references can be found. The methods discussed in Probs. 39 to 41 are due, respectively, to Ostrowski (1973), Wolfe (1959), and Kincaid (1961).

Section 8.9 Marden (1966) is the most complete reference on the location of the zeros of polynomials. Wall (1948) contains a number of results in this area; Wilf (1962) contains some selected results. The material on Sturm sequences is mainly from Gantmacher (1959). Wilf (1960) discusses the general problem of computing the zeros of polynomials, while Peters and Wilkinson (1971) discuss the practical aspects of the problem. The proceedings of a conference exclusively devoted to the problem of finding the zeros of a polynomial appear in the volume edited by Dejon and Henrici (1969).

Section 8.10 The convergence of the Newton-Raphson and secant methods in the complex case is given by Householder (1970). Hildebrand (1974) has a good discussion of Bairstow's method. A method very similar to Bairstow's is discussed by McAuley (1962). Graeffe's and Bernoulli's methods are discussed in many numerical analysis texts; see, for example, Hildebrand (1974) and Householder (1953). Many authors have discussed these two methods. Hildebrand (1974) contains a number of references to these papers. The mechanization of Graeffe's method for digital computation is due to Bareiss (1960, 1967). Durand (1960–1961) contains a good exposition of Laguerre's method.

Section 8.11 The Jenkins-Traub method is the culmination of a series of research works by Traub and Jenkins and appears in Jenkins and Traub (1970*a*). The computer implementation of this method for complex polynomials is given in Jenkins and Traub (1972). The corresponding references for the real case are Jenkins and Traub (1970*b*) and Jenkins (1975), respectively. A related method is given by Young and Gregory (1972).

Section 8.12 The original idea of this method can be found in Madsen (1973). This idea was further developed and incorporated in a computer program including error estimates by Madsen and Reid (1975). An algorithm in the same spirit has been proposed by Moore (1976).

Section 8.13 Wilkinson (1964) contains an excellent discussion of the computational problems that arise in the solution of high-degree polynomial equations when the polynomials are ill-conditioned. McCracken and Dorn (1964) give a discussion, similar to ours, of the estimation of errors.

BIBLIOGRAPHY

Aitken, A. C. (1926): On Bernoulli's Numerical Solution of Algebraic Equations, *Proc. Roy. Soc. Edinb.*, vol. 46, pp. 289–305.

Anderson, N., and Å. Björck (1973): A New High Order Method of Regula Falsi Type for Computing a Root of an Equation, *BIT*, vol. 13, pp. 253–264.

Apostol, T. M. (1957): *Mathematical Analysis*, Addison-Wesley Publishing Company, Inc., Reading, Mass.

Bareiss, E. H. (1960): Resultant Procedure and the Mechanization of the Graeffe Process, *J. Ass. Comput. Mach.*, vol. 7, pp. 346–386.

——— (1967): The Numerical Solution of Polynomial Equations and the Resultant Procedures, pp. 185–214 in *Mathematical Methods for Digital Computers*, vol. II, (A. Ralston and H. S. Wilf, eds.), John Wiley & Sons, Inc., New York.

Boggs, P. T. (1971): The Solution of Nonlinear Operator Equations by A-stable Integration Techniques, *SIAM J. Numer. Anal.*, vol. 8, pp. 767–785.

Byrne, G. D., and C. A. Hall (eds.) (1973): *Numerical Solution of Systems of Nonlinear Algebraic Equations*, Academic Press, Inc., New York.

Dejon, B., and P. Henrici (eds.) (1969): *Constructive Aspects of the Fundamental Theorem of Algebra*, Wiley-Interscience, New York.

Durand, E. (1960–1961): *Solutions numériques des équations algébriques*, vols. 1, 2, Masson et Cie, Paris.

Gantmacher, F. R. (1959): *Theory of Matrices*, vol. II, Chelsea Publishing Company, New York.

Hildebrand, F. B. (1974): *Introduction to Numerical Analysis*, 2d ed., McGraw-Hill Book Company, New York.

Householder, A. S. (1953): *Principles of Numerical Analysis*, McGraw-Hill Book Company, New York.

——— (1970): *The Numerical Treatment of a Single Nonlinear Equation*, McGraw-Hill Book Company, New York.

Jenkins, M. A. (1975): Algorithm 493: Zeros of a Real Polynomial, *ACM Trans. Math. Software*, vol. 1, pp. 178–189.

——— and J. F. Traub (1970*a*): A Three-Stage Variable-Shift Iteration for Polynomial Zeros and Its Relation to Generalized Rayleigh Iteration, *Numer. Math.*, vol. 14, pp. 252–263.

——— and ——— (1970*b*): A Three-Stage Algorithm for Real Polynomials Using Quadratic Iteration, *SIAM J. Numer. Anal.*, vol. 7, pp. 545–566.

——— and ——— (1972): Algorithm 419: Zeros of a Complex Polynomial, *Commun. ACM*, vol. 15, pp. 97–99.

Kincaid, W. M. (1961): A Two-Point Method for the Numerical Solution of Systems of Simultaneous Equations, *Q. Appl. Math.*, vol. 18, pp. 305–324.

McAuley, V. A. (1962): A Method for the Real and Complex Roots of a Polynomial, *J. Soc. Ind. Appl. Math.*, vol. 10, pp. 657–667.

McCracken, D. D., and W. S. Dorn (1964): *Numerical Methods and FORTRAN Programming*, John Wiley & Sons, Inc., New York.

Madsen, K. (1973): A Root Finding Algorithm Based on Newton's Method, *BIT*, vol. 13, pp. 71–75.

——— and J. K. Reid (1975): Fortran Subroutines for Finding Polynomial Zeros, *Harwell Rep.* AERE-R 7986, H.M. Stationery Office, London.

Marden, M. (1966): *Geometry of Polynomials*, American Mathematical Society, Providence, R.I.

Moore, J. B. (1976): A Consistently Rapid Algorithm for Solving Polynomial Equations, *J. Inst. Math. Appl.*, vol. 17, pp. 99–110.

Muller, D. E. (1956): A Method for Solving Algebraic Equations Using an Automatic Computer, *MTAC*, vol. 10, pp. 208–215.

Ortega, J., and W. Rheinboldt (1970): *Iterative Solution of Nonlinear Equations in Several Variables*, Academic Press, Inc., New York.

Ostrowski, A. M. (1973): *Solution of Equations in Euclidean and Banach Spaces*, Academic Press, Inc., New York.

Parlett, B. (1964): Laguerre's Method Applied to the Matrix Eigenvalue Problem, *Math. Comput.*, vol. 18, pp. 464–485.

Peters, G. and J. H. Wilkinson (1971): Practical Problems Arising in the Solution of Polynomial Equations, *J. Inst. Math. Appl.*, vol. 8, pp. 16–35.

Rabinowitz, P. (ed.) (1970): *Numerical Methods for Nonlinear Algebraic Equations*, Gordon and Breach, London.

Rall, L. B. (1969): *Computational Solution of Nonlinear Operator Equations*, John Wiley & Sons, Inc., New York.

Rheinboldt, W. (1974): Methods for Solving Systems of Nonlinear Equations, *CBMS Regional Conf. Ser. Appl. Math.*, vol. 14, Society for Industrial and Applied Mathematics, Philadelphia.

Shaw, M., and J. F. Traub (1974): On the Number of Multiplications for the Evaluation of a Polynomial and Some of Its Derivatives, *J. Ass. Comput. Mach.*, vol. 21, pp. 161–167.

Traub, J. F. (1961*a*): Comparison of Iterative Methods for the Calculation of *N*th Roots, *Commun. ACM*, vol. 4, pp. 143–145.

——— (1961*b*): On a Class of Iteration Formulas and Some Historical Notes, *Commun. ACM*, vol. 4, pp. 276–278.

——— (1964): *Iterative Methods for the Solution of Equations*, Prentice-Hall, Inc., Englewood Cliffs, N.J.

——— (1967): The Solution of Transcendental Equations, pp. 171–184 in *Mathematical Methods for Digital Computers*, vol. II (A. Ralston and H. S. Wilf, eds.), John Wiley & Sons, Inc., New York.

Wall, H. S. (1948): *Analytic Theory of Continued Fractions*, D. Van Nostrand Company, Inc., Princeton, N.J.

Wilf, H. S. (1960): The Numerical Solution of Polynomial Equations, pp. 233–241 in *Mathematical Methods for Digital Computers* (A. Ralston and H. S. Wilf, eds.), John Wiley & Sons, Inc., New York.

——— (1962): *Mathematiccs for the Physical Sciences*, John Wiley & Sons, Inc., New York.

Wilkinson, J. H. (1959): The Evaluation of Zeros of Ill-conditioned Polynomials, *Numer. Math.*, vol. 1, pp. 150–166, 167–180.

——— (1963): *Rounding Errors in Algebraic Processes*, Prentice-Hall, Inc., Englewood Cliffs, N.J.

Wolfe, P. (1959): The Secant Method for Simultaneous Non-linear Equations, *Commun. ACM*, vol. 2, pp. 12–13 (Dec.).

Young, D. M., and R. T. Gregory (1972): *A Survey of Numerical Mathematics*, vol. 1, Addison-Wesley Publishing Company, Inc., Reading, Mass.

PROBLEMS

Section 8.2

1 Let $x = g(y)$ be the function inverse to $y = f(x)$.

(*a*) By using induction, show that we can write

$$g^{(k)}(y) = \frac{X_k}{(y')^{2k-1}} \qquad k = 1, 2, \ldots$$

where X_k is a polynomial in $y', y'', \ldots, y^{(k)}$ which satisfies the recurrence relation

$$X_{n+1} = \frac{dX_n}{dx} y' - (2n-1)X_n y'' \qquad X_1 = 1 \qquad n = 1, 2, \ldots$$

(*b*) Use this result to find explicit expressions for $g^{(k)}(y)$ for $k = 1, 2, 3$. [Ref.: Ostrowski (1973), pp. 20–22.]

2 (*a*) Show that the convergence of a functional iteration method of order 1 implies that the asymptotic error constant is less than or equal to 1 but that for methods of order greater than 1 this need not be so. Why is the "or equal" part needed above?

(*b*) May functional iteration methods have order less than 1 and still converge? Why?

3 Solve the difference equations (8.2-11) and then use (8.2-12) to derive (8.2-13).

4 (*a*) On a digital computer, let the time required, i.e., the cost, to compute a multiplication or division be 1, a square root be 3, and any elementary function be 6. Ignoring the cost of additions and subtractions, what are the costs of evaluating $f(x)$ and $f'(x)$ when $f(x) = e^x \cos^2 x + \ln x \tan x$? Use this result to explain why it is generally quite inexpensive to compute $f'(x)$ once $f(x)$ has been computed.

(*b*) If $f(x)$ is a polynomial, what are the relative costs of computing $f(x)$ and $f'(x)$ if the algorithm (7.2-4) is used to evaluate a polynomial?

Section 8.3

5 (*a*) Prove that the sequence of iterates in the method of false position approaches a limit and that this limit is a solution of (8.1-1).

(*b*) Show that if the conditions (i) $f''(x) \neq 0$ in $[x_1, x_2]$; (ii) $f(x_1)f''(x_1) > 0$ are satisfied, then x_1 always remains one of the points in the false-position iteration. These conditions are called *Fourier's conditions.*

(*c*) Therefore, deduce that a sufficient condition for the method of false position to be a one-point iteration method is for $f(x)$ to be convex between $f(x_1)$ and $f(x_2)$.

6 (*a*) Solve the difference equation (8.3-18) to get the solution (8.3-20).

(*b*) Thus deduce the plausibility of (8.3-23).

7 (*a*) Calculate the smallest positive root of $\cos x - xe^x = 0$ using (i) the secant method, (ii) the method of false position, (iii) the Illinois method, (iv) the Pegasus method with $x_0 = 0$, $x_1 = 1$.

(*b*) Calculate the smallest positive root of $\tan x - \cos x = \frac{1}{2}$ using the secant method.

(*c*) Use the secant method to find the root of

$$e^z = z \qquad z = x + iy$$

with smallest positive imaginary part by eliminating x between the two equations for the real and imaginary parts of $e^z = z$.

Section 8.4

8 (*a*) Use the result of Prob. 1 to derive (8.4-9).

(*b*) Display the iteration formula (8.4-7) when $m = 1$ in terms of f and its derivatives.

(*c*) What is the efficiency index of this method? When would it tend to be a better method to use than the Newton-Raphson method?

9 Halley's method. Consider the iteration function

$$x_{i+1} = x_i - \frac{u_i}{Q(u_i)} \qquad u_i = \frac{f_i}{f'_i}$$

where Q is a polynomial.

(*a*) If Q is linear, derive a method of this form of order 3. This is called Halley's method.

(*b*) Use Halley's method and the method derived in Prob. 8*b* to compute $\sqrt{2}$ starting from $x_1 = 2$.

10 (*a*) Derive the Newton-Raphson method by expanding $f(x_{i+1})$ in a Taylor series about x_i and ignoring all but the first two terms.

(*b*) Use the Taylor-series expansion of $f(\alpha)$ about x_i to derive (8.4-16) directly.

11 (*a*) Use (8.3-23) and (8.4-16) to show that if the two iteration functions in (8.2-9) are the secant method and the Newton-Raphson method, respectively, then $C_1 = C_2^{(\sqrt{5}-1)/2}$.

(*b*) Use this result in (8.2-13) to show that the Newton-Raphson method has a lower computational efficiency than the secant method unless the cost of evaluating $f'(x)$ is less than .44 times the cost of evaluating $f(x)$. Is this the same result that would have been obtained using (8.2-15)?

12 (*a*) Find the smallest positive root of the equation of Prob. 7*a* using (i) Newton's method (ii) Halley's method (see Prob. 9). Use $x_1 = 0$; compare the results with those of Prob. 7*a* for speed of convergence.

(*b*) Repeat the calculations of Prob. 7*b* using Newton's and Halley's method with $x_1 = 0$.

13 (*a*) Use Newton's and Halley's method to compute $(10)^{1/5}$ starting with $x_1 = 10$.

(*b*) Some computers do not have the operation of division built into them. Newton's method can be used to find the reciprocal of a number without doing any divisions. Use this technique to calculate $\frac{1}{10}$ starting with $x_1 = .001$. Can Halley's method (Prob. 9) be used similarly to find a reciprocal? Why?

(*c*) Try to use the Newton-Raphson method to compute $\frac{1}{10}$ starting with $x_1 = 1.0$. Explain the behavior.

14 If $F(x_i)$ is an iteration function of order p, under what conditions on the function $U(x_i)$ will

$$G(x_i) = F(x_i) + U(x_i)u_i^r \qquad r \geq p$$

be an iteration function of order p?

15 Consider the iteration formula

$$x_{i+1} = x_i - cf(x_i)$$

where c may depend on f and x_i. Deduce that this method has linear convergence unless $c = 1/f'(x_i)$.

16 Consider the iteration

$$y_i = x_i - \frac{f(x_i)}{f'(x_i)} \qquad x_{i+1} = y_i - \frac{f(y_i)}{f'(x_i)}$$

This is the Newton-Raphson iteration with the derivative computed only every second step.

(*a*) Show that if the iteration converges,

$$\lim_{i\to\infty} \frac{x_{i+1} - \alpha}{(y_i - \alpha)(x_i - \alpha)} = \frac{f''(\alpha)}{f'(\alpha)}$$

(*b*) Thus deduce that

$$\lim_{i\to\infty} \frac{x_{i+1} - \alpha}{(x_i - \alpha)^3} = \frac{1}{2}\left[\frac{f''(\alpha)}{f'(\alpha)}\right]^2$$

(*c*) If the cost of computing $f(x)$ is 1 and $f'(x)$ is θ_1, for what values of θ_1 is this method more efficient than (i) the Newton-Raphson method; (ii) the secant method?

(*d*) Use this method to repeat the calculation of Prob. 7*a* with $x_1 = 0$.

Section 8.5

17 (*a*) Derive (8.5-1).

(*b*) Find the order of an iteration function derived from (8.5-2) with $n = 3$ and $r_j = 0$ for all j.

(*c*) Repeat part (*b*) with $n = 2$ and $r_1 = r_2 = 1$.

18 (*a*) Derive the form of the three-point iteration function whose order was calculated in Prob. 17*b*.

(*b*) Similarly, derive the two-point iteration function whose order was calculated in Prob. 17*c*.

(*c*) Use the methods of parts (*a*) and (*b*) to repeat the calculation of Prob. 7*a* using suitable starting values.

***19** Iteration functions by direct interpolation. Let $y(x)$ be the interpolation polynomial of (8.5-1). If the equation $y(x_{i+1}) = 0$ can be solved for x_{i+1}, then this defines an iteration function.

(*a*) Suppose all the points $x_{i-j}, j = 0, \ldots, n$, lie in an interval J which contains the root α and in which $f'(x) \neq 0$. If there is at least one x_{i-j} on each side of α, show that $y(x_{i+1}) = 0$ has a real solution in J. Need this solution be unique? (Even when all the points lie on one side of α, there is such a real solution in general.)

(*b*) Derive the following formula for the error

$$\epsilon_{i+1} = \frac{-f^{(\beta+1)}(\xi)}{(\beta+1)!\,y'(\eta)} \prod_{j=1}^{n} \epsilon_{i+1-j}^{r_j+1}$$

where η lies in the interval determined by x_{i+1} and α. [Ref.: Traub (1964), pp. 67–75.]

20 (*a*) Derive the secant method using direct interpolation.

(*b*) Derive the Newton-Raphson method using direct interpolation. Show that this derivation leads directly to (8.4-16) for the error.

***21** Muller's method. (*a*) Use a three-point Lagrangian interpolation formula and direct interpolation to derive the iteration formula

$$x_{i+1} = x_i + (x_i - x_{i-1}) \frac{-2f_i \delta_i}{c_i \pm [c_i^2 - 4f_i \delta_i \lambda_i (f_{i-2}\lambda_i - f_{i-1}\delta_i + f_i)]^{1/2}}$$

where $\lambda_i = (x_i - x_{i-1})/(x_{i-1} - x_{i-2})$, $\delta_i = 1 + \lambda_i$, $c_i = f_{i-2}\lambda_i^2 - f_{i-1}\delta_i^2 + f_i(\lambda_i + \delta_i)$. Why should the sign in the denominator be chosen to give the denominator the greatest magnitude?

(*b*) Use part (*b*) of Prob. 19 to deduce that $\epsilon_{i+1} = -\epsilon_i \epsilon_{i-1} \epsilon_{i-2}[f'''(\xi)/y'(\eta)]$.

(*c*) Proceed as in Sec. 8.3 to deduce the order of Muller's method.

(*d*) Use this method to repeat the calculation of Prob. 7*a*. [Ref.: Muller (1956).]

22 (*a*) Verify that substitution of (8.5-13) into the Newton-Raphson method gives the secant method.

(*b*) Verify Eqs. (8.5-15) and (8.5-16).

(*c*) Show that the order of (8.5-17) is given by the result of Prob. 17*c*.

***23** (*a*) Derive a new iteration method from the Newton-Raphson method by replacing $f'(x_i)$ by its approximation found by differentiating a three-point Lagrangian interpolation formula based on x_i, x_{i-1}, and x_{i-2}.

(*b*) Use reasoning similar to that in Sec. 8.5-2 to show that the order of this method is 1.84.

(*c*) Repeat the calculation of Prob. 7*a* using this method. What advantage does this method have over Muller's method (Prob. 21)? Any disadvantages?

Section 8.6

24 (*a*) Verify (8.6-5) and (8.6-6). (*b*) Then derive (8.6-7) and (8.6-9). (*c*) Derive (8.6-11) and from this deduce (8.6-10). (*d*) Finally deduce that the order of (8.4-7) is 1 if α is not a simple root.

25 (*a*) Approximately how fast will the error in the Newton-Raphson method decrease from one iterate to the next in the neighborhood of a root of multiplicity $r > 1$?

(*b*) Test this conclusion on the equation $(\cos x - xe^x)^3 = 0$ and compare with the results of Prob. 7*a*. Use $x_1 = 0$.

26 Apply the iterations of (*a*) Eq. (8.6-23) with $x_0 = 0$, $x_1 = 1$ and (*b*) Eq. (8.6-24) with $x_1 = 0$ to the equation in Prob. 25*b* and compare with the previous results.

Section 8.7

27 (*a*) Why can't the method of false position be used to find a root of even multiplicity?

(*b*) If α is a root of even multiplicity, show that the secant method may diverge no matter how close to α the initial approximations x_0 and x_1 are.

28 Prove that the iteration of Example 8.5 would have converged if $0 < x_1 < 1$.

29 (*a*) Why is it reasonable to test the convergence of an iteration by considering $|x_{i+1} - x_i|$?

(*b*) If the relative rather than the absolute error in the result is of interest, what would be a good quantity to use instead of $|x_{i+1} - x_i|$ to test for convergence?

30 (*a*) Show that the iteration in Example 8.6 eventually converges.

(*b*) Calculate the positive root of $x^{20} - 1 = 0$ using the Newton-Raphson method with $x_1 = \frac{1}{2}$ and the rule that if $|x_{i+1}/x_i| > 3$, then, in place of the computed x_{i+1}, use $x_{i+1} = \pm\frac{3}{2}x_i$, where the sign is chosen to agree with x_{i+1}/x_i.

31 Suppose the Newton-Raphson method is converging slowly, thereby indicating the presence of a multiple root.

(*a*) Show how (8.7-4) can be used in conjunction with (8.6-10) to get an estimate of the multiplicity. [Since the multiplicity is an integer, this estimate amounts to a determination of the multiplicity. When the multiplicity has been found, (8.6-13) can then be used to get rapid convergence.]

(*b*) Apply this technique to the equation of Prob. 25*b*.

Section 8.8

32 Let $J = (\partial f_i/\partial x^{(k)})$ and $K = (\partial g_j/\partial y^{(l)})$ be the Jacobian matrices of the vector function $\mathbf{y} = \mathbf{f}(\mathbf{x})$ and its inverse function $\mathbf{x} = \mathbf{g}(\mathbf{y})$, respectively. By differentiating the identity $\mathbf{x} = \mathbf{g}(\mathbf{f}(\mathbf{x}))$ with respect to each component $x^{(k)}$, $k = 1, \ldots, n$, show that $I = KJ$ and hence that (8.8-9) and (8.8-10) are equivalent.

***33** (*a*) Use the implicit-function theorem to derive (8.8-12).

(*b*) Derive (8.8-12) by expanding $f_1[x^{(1)}, x^{(2)}]$ and $f_2[x^{(1)}, x^{(2)}]$ in two Taylor series about the root $[\alpha^{(1)}, \alpha^{(2)}]$, setting $f_1[\alpha^{(1)}, \alpha^{(2)}] = f_2[\alpha^{(1)}, \alpha^{(2)}] = 0$, and dropping all derivative terms of order higher than 1.

(*c*) Derive the error term in (8.8-12) by using Eq. (8.8-7).

(*d*) By analogy with (8.2-8), how would you define order for iteration methods for simultaneous equations? Use this definition to show that (8.8-12) has order 2 at a simple root.

34 Find a root of the equation of Prob. 7*c* by solving the two equations for the real and imaginary parts using (8.8-12). Use $x_1^{(1)} = .2$, $x_1^{(2)} = 1.1$.

35 Consider the two simultaneous equations

$$42.25x^2 + 27.885x - .749y^2 - 2.54y - 2.466 = 0$$

$$-.052x - .0192 + .00359y^2 + .00356y = 0$$

(*a*) Attempt to solve these two equations using (8.8-12) with $x_1 = -.01$, $y_1 = .01$. Carry out 15 iterations.

(*b*) Now use $x_1 = .3$, $y_1 = 2.5$ in (8.8-12).

36 Verify that $\nabla F(\mathbf{x}) = 2J^T(\mathbf{x})\mathbf{f}(\mathbf{x})$ by differentiating (8.8-17) with respect to each component $x^{(j)}$ of $\mathbf{x}$, $j = 1, \ldots, n$.

37 A real symmetric matrix S is said to be *positive semidefinite* if for all real vectors $\mathbf{x}$, $(\mathbf{x}, S\mathbf{x}) \geq 0$. It is well known that for any such matrix S there exists a matrix R with $R^TR = I$, the identity matrix, such that $R^TSR = D = \text{diag}(d_1, \ldots, d_n)$, a diagonal matrix with diagonal elements $d_i \geq 0$.

(*a*) Show that for any real matrix A, A^TA is positive semidefinite.

(*b*) Show that if $\lambda > 0$, then the inverse of $B(\lambda) = A^TA + \lambda I$ exists and is given by $R[D(\lambda)]^{-1}R^T$, where R is such that $R^T(A^TA)R = D$ and $D(\lambda) = \text{diag}\,(d_1 + \lambda, \ldots, d_n + \lambda)$.

(*c*) Show that $(H(\lambda)\mathbf{f}, H(\lambda)\mathbf{f})$ is a decreasing function of λ, where $H(\lambda) = B(\lambda)^{-1}A^T$, and consequently that in the Levenberg-Marquardt algorithm, the step length $\|\mathbf{x}_{i+1} - \mathbf{x}_i\|_2$ decreases as λ_i increases.

38 Consider the system of nonlinear equations (8.8-2) and an initial approximation $\mathbf{x} = \mathbf{x}_0$. Define

$$\mathbf{g}(\mathbf{x}, t) = \mathbf{f}(\mathbf{x}) - e^{-t}\mathbf{f}(\mathbf{x}_0) \qquad 0 \leq t < \infty$$

so that $\mathbf{g}(\mathbf{x}_0, 0) = \mathbf{0}$ and $\mathbf{g}(\mathbf{x}, t) \to \mathbf{f}(\mathbf{x})$ as $t \to \infty$.

(*a*) Show that if we define $\mathbf{x}(t)$ to be the solution of $\mathbf{g}(\mathbf{x}(t), t) = \mathbf{0}$, then $\mathbf{x}(t)$ satisfies the differential equation

$$J(\mathbf{x})\mathbf{x}'(t) = -\mathbf{f}(\mathbf{x}) \qquad \mathbf{x}(0) = \mathbf{x}_0 \qquad 0 \leq t < \infty$$

where $J(\mathbf{x})$ is the Jacobian matrix of the vector function $\mathbf{f}(\mathbf{x})$.

(*b*) Show that the application of the Newton-Raphson iteration (8.8-10) to (8.8-2) is equivalent to the solution of the above system of ordinary differential equations by Euler's method (5.5-13) with $h = 1$. [Ref.: Boggs (1971).]

***39** (*a*) Given three approximations (x_i, y_i), (x_{i-1}, y_{i-1}), (x_{i-2}, y_{i-2}) to the solution of $f_1(x, y) = 0$, $f_2(x, y) = 0$, find the equations of two planes $z = L_1(x, y)$ and $z = L_2(x, y)$ such that $L_j(x_{i-k}, y_{i-k}) = f_j(x_{i-k}, y_{i-k})$, $j = 1, 2$; $i = 0, 1, 2$.

(*b*) Calculate the next approximation (x_{i+1}, y_{i+1}) to be the intersection of $z = L_1(x, y)$, $z = L_2(x, y)$ and $z = 0$. When will this procedure fail?

(*c*) Use this method to solve the equation of Prob. 7*c* by considering the two equations for the real and imaginary parts. Use as initial points (.4, 1.4), (.2, 1.4), and (.3, 1.1) and do three iterations. Will this method always converge? Is there a two-dimensional analog of the method of false position which will always converge? [Ref.: Ostrowski (1973), pp. 294–295.]

***40** Let $\mathbf{x}_i, \mathbf{x}_{i-1}, \ldots, \mathbf{x}_{i-n}$ be $n + 1$ approximations to the solution of (8.8-2) and let $\pi_0, \ldots, \pi_n$ be such that

$$\sum_{j=0}^{n} \pi_j = 1 \qquad \text{and} \qquad \sum_{j=0}^{n} \pi_j f_k(\mathbf{x}_{i-j}) = 0 \qquad k = 1, \ldots, n$$

Define

$$\mathbf{x}_{i+1} = \sum_{j=0}^{n} \pi_j \mathbf{x}_{i-j}$$

(*a*) Show that when $n = 1$, this iteration method is the secant method.

(*b*) Let $\boldsymbol{\alpha}$ be the solution of (8.8-2) and $G_k(\mathbf{x})$ and $2Q_k(\mathbf{x})$, respectively, the vector of first partial derivatives of $f_k(\mathbf{x})$ and the matrix of second partial derivatives of $f_k(\mathbf{x})$. Use the first two nonzero terms of the Taylor-series expansion of $f_k(\mathbf{x}_{i+1})$ about $\boldsymbol{\alpha}$ to get the approximation

$$f_k(\mathbf{x}_{i+1}) \approx \sum_{j=0}^{n} \pi_j(G_k(\boldsymbol{\alpha}), \mathbf{x}_{i-j} - \boldsymbol{\alpha}) + (\mathbf{x}_{i+1} - \boldsymbol{\alpha})^T Q_k(\boldsymbol{\alpha})(\mathbf{x}_{i+1} - \boldsymbol{\alpha})$$

(*c*) Use the first two nonzero terms of the Taylor-series expansion of $f_k(\mathbf{x}_{i-j})$ to eliminate $G_k(\boldsymbol{\alpha})$ in the above approximation and thereby derive

$$f_k(\mathbf{x}_{i+1}) \approx -\sum_{j=0}^{n} \pi_j(\mathbf{x}_{i+1} - \mathbf{x}_{i-j})^T Q_k(\boldsymbol{\alpha})(\mathbf{x}_{i+1} - \mathbf{x}_{i-j})$$

Can you infer from this that this method has quadratic convergence? Why?

(*d*) Use this method to repeat the calculation of Prob. 7*c*. Use the starting values of Prob. 39. At every step replace that $\mathbf{x}_{i-j}$ with $\mathbf{x}_{i+1}$ for which

$$\sum_{k=1}^{n} |f_k(\mathbf{x}_{i-j})|^2$$

is maximal. Why is this latter a reasonable rule? [Ref.: Wolfe (1959).]

***41** Consider the same two equations as in Prob. 39. Let $f_3(x, y) = -f_1(x, y) - f_2(x, y)$.

(*a*) Let R_1, S_1, and T_1 be three points in the xy plane which are the initial approximations to the solution of the pair of equations. Let

$$S_1' = R_1 f_1 S_1$$

where $R_1 f_1 S_1$ denotes the intersection with the xy plane of the line joining $f_1(R_1)$ and $f_1(S_1)$. Similarly, let

$$T_1' = R_1 f_1 T_1 \qquad T_2 = S_1' f_2 T_1' \qquad R_1' = S_1' f_2 R_1 \qquad R_2 = T_2 f_3 R_1' \qquad S_2 = T_2 f_3 S_1'$$

Show that if this process leads to a solution of the system, then the points R_i, S_i, T_i form a sequence of nearly similar triangles of decreasing size.

(*b*) Use this method to repeat the computation of Prob. 7*c*. Use the starting values of Prob. 39. [Ref.: Kincaid (1961).]

42 Verify that if A^{-1} exists and that if $B = A - (A\mathbf{q} - \mathbf{v})\mathbf{q}^T/\mathbf{q}^T\mathbf{q}$, then

$$B^{-1} = A^{-1} - \frac{(A^{-1}\mathbf{v} - \mathbf{q})\mathbf{q}^T A^{-1}}{\mathbf{q}^T A^{-1}\mathbf{v}}$$

provided only that $\mathbf{q}$ and $\mathbf{q}^T A^{-1}\mathbf{v} \neq 0$. *Hint:* For any scalar s and matrices C and D, $C(sD) = s(CD)$.

Section 8.9

43 Derive (8.9-6).

44 (*a*) If a zero of $f(x)$ in $[a, b]$ has been found, how can the Sturm sequence (8.9-5) be used to find the multiplicity of the zero?

(*b*) Prove that part of Theorem 8.3 relating to the case where $f(a)$ or $f(b)$ or both is zero. What happens when the zero is not simple?

45 If $\{f_i(x)\}$, $i = 1, \ldots, m$, is a Sturm sequence, a *generalized Sturm sequence* is any sequence $\{p(x)f_i(z)\}$, where $p(x)$ is an arbitrary polynomial.

(*a*) Does Theorem 8.2 also hold for generalized Sturm sequences?

(*b*) If $f(x)$ and $g(x)$ are two polynomials, use the Euclidean algorithm in a fashion analogous to (8.9-5) to generate a sequence of polynomials using $f_1(x) = f(x)$ and $f_2(x) = g(x)$. Show that this sequence is either a Sturm sequence or a generalized Sturm sequence. Thus deduce a method for finding the Cauchy index of any rational function.

46 (*a*) Determine the number of positive and negative real zeros of $z^4 - 3z^3 - 54z^2 - 150z - 100$.

(*b*) Determine the number of real zeros greater than 1 of $z^4 - 10z^3 + 34z^2 - 50z + 25$.

Section 8.10

47 (*a*) Verify that the following algorithm computes the value of the polynomial $P(z) = \sum_{i=0}^{n} a_{n-i} z^i$ and its first p reduced derivatives at the point $z = x$, $p \le n$, using $(p+1) \times (2n - p)/2$ additions, $2n - 1$ multiplications, and p divisions.

$$T_i^{-1} = a_{i+1} x^{n-i-1} \qquad i = 0, 1, \ldots, n-1$$

$$T_j^j = a_0 x^n \qquad j = 0, 1, \ldots, p$$

$$T_i^j = T_{i-1}^{j-1} + T_{i-1}^j \qquad \begin{array}{l} j = 0, 1, \ldots, p \\ i = j+1, \ldots, n \end{array}$$

$$\frac{P^{(j)}(x)}{j!} = x^{-j} T_n^j \qquad j = 0, 1, \ldots, p$$

(*b*) How many storage locations are required for this algorithm and how many for $p + 1$ applications of synthetic division? [Ref.: Shaw and Traub (1974)].

48 (*a*) Derive a synthetic-division algorithm analogous to those in Sec. 8.10 for the division of a polynomial of degree n by a polynomial of degree $j < n$.

(*b*) Use this algorithm to determine whether $z^5 - z^4 - 2z^3 + 2z^2 + z - 1$ has a triple zero at (i) $z = 1$; (ii) $z = -1$. Find all the zeros of this polynomial.

49 The polynomial $P_3(x) = x^3 + 9813.18x^2 + 8571.08x + .781736$ has the zeros -9812.31, $-.873412$, and $-.00009\,12157$.

(*a*) Assume that -9812.30 has been accepted as a zero of $P_3(x)$. Deflate $P_3(x)$ using this value in (8.10-1) and then use the quadratic formula to determine the remaining zeros of the deflated polynomial.

(*b*) Do the same with the values $-.873413$ and $-.00009\,12156$

50 Backward deflation. (*a*) Show that if $f(z)$ given by (8.9-1) is written in the form $a_0 + a_1 z + \cdots + a_n z^n$, it can be deflated by a zero α by dividing through by $-\alpha + z$ to yield the deflated polynomial $q_0 + q_1 z + \cdots + q_{n-1} z^{n-1}$, where the q_i are given by the recursion

$$a_0 = -\alpha q_0 \qquad a_r = -\alpha q_r + q_{r-1} \qquad r = 1, \ldots, n-1$$

(*b*) Apply backward deflation to the polynomial of the previous problem using each of the three accepted zeros given there and determine in each case the remaining zeros of the deflated polynomial. [Ref.: Peters and Wilkinson (1971)].

51 (*a*) Show that if the zeros of $f(z)$ given by (8.9-1) with $a_0 \ne 0$ are $\alpha_1, \ldots, \alpha_n$, then the zeros of the polynomial $g(z) = a_0 z^n + a_1 z^{n-1} + \cdots + a_{n-1} z + a_n$ are $\alpha_1^{-1}, \ldots, \alpha_n^{-1}$.

(*b*) Let the polynomial resulting from the deflation of $g(z)$ by a zero α^{-1} be $d_0 z^{n-1} + \cdots + d_{n-2} z + d_{n-1}$, where the d_k are given as in (8.10-1) by the recursion

$$d_0 = a_0 \qquad d_k = a_k + \alpha^{-1} d_{k-1} \qquad k = 1, \ldots, n-1$$

Show that these are the same coefficients that would result if we performed backward deflation by dividing through by $1 - x/\alpha$ instead of by $-\alpha + x$. Hence, conclude that the reason backward deflation is stable if we deflate with the zero of largest magnitude is that we are essentially performing deflation of $g(z)$ with the zero of smallest magnitude.

52 Derive the equations analogous to (8.10-11) to (8.10-14) for the secant method.

53 Do the calculation of Example 8.11 with an initial approximation $p_1 = -1$, $q_1 = -1$. Would you expect the convergence of the Bairstow iteration to be very sensitive to the initial approximation? Why?

54 (*a*) Use (8.10-11) to (8.10-14) to find the zero of maximum magnitude of $z^4 + 2z^3 + 3z^2 + 4z + 5$. Use $.3 + 1.4i$ as the initial approximation.

(*b*) Repeat this calculation using the equations derived in Prob. 52.

55 (*a*) Verify (8.10-29).

(*b*) Use (8.10-30), (8.10-31), and (8.10-33) to show that

$$a_{n-j}^{(r)} = (-1)^j a_n^{(r)} \sum_c^n{}_1 (\rho_{k_1}\rho_{k_2} \cdots \rho_{k_j})^{2^r} \exp\,[i2^r(\phi_{k_1} + \cdots + \phi_{k_j})]$$

(*c*) Use this to deduce (8.10-40) when (8.10-34) holds.

56 Let the polynomial $f(z)$ of degree n have zeros z_i, $i = 1, \ldots, n$.

(*a*) Show that if z is sufficiently large, then

$$(z - z_i)^{-1} = z^{-1} + z_i z^{-2} + z_i^2 z^{-3} + \cdots$$

and, thus, $\displaystyle\sum_{i=1}^n (z - z_i)^{-1} = nz^{-1} + s_1 z^{-2} + s_2 z^{-3} + \cdots$ where $s_j = \displaystyle\sum_{i=1}^n z_i^j$.

(*b*) Show that

$$f(z) \sum_{i=1}^n (z - z_i)^{-1} = f'(z)$$

(*c*) Thus deduce *Newton's identities*

$$a_n s_m + a_{n-1} s_{m-1} + \cdots + a_{n-m+1} s_1 + m a_{n-m} = 0 \qquad m = 1, \ldots, n$$

$$a_n s_{n+j} + a_{n-1} s_{n+j-1} + \cdots + a_0 s_j = 0 \qquad j = 1, 2, \ldots$$

(*d*) From this deduce that, if (8.10-45) is used to generate the starting values in Bernoulli's method, then all the c_i's are unity.

57 (*a*) How must (8.10-42) be modified when some of the roots of (8.9-1) are multiple?

(*b*) Nevertheless, show that if α_1 is multiple but real, then (8.10-44) still holds.

58 In Bernoulli's method, define $\lambda_k = u_{k-1}/u_k$ so that $\lim_{k\to\infty} (1/\lambda_k) = \alpha_1$. Show, by rewriting (8.10-41) that

$$-\frac{1}{\lambda_k} = b_{n-1} + b_{n-2}\lambda_{k-1} + b_{n-3}\lambda_{k-2}\lambda_{k-1} + \cdots + b_0(\lambda_{k-n+1} \cdots \lambda_{k-1})$$

(*b*) Show that for $k \le n$, (8.10-51) can be expressed in terms of the λ_i's.

(*c*) Show that for $k > n$, λ_k can be computed using the recurrence relation

$$\gamma_{k0} = b_0$$

$$\gamma_{kr} = \gamma_{k,r-1}\lambda_{k-n+r} + b_r \qquad r = 1, \ldots, n-1$$

$$\lambda_k = -\frac{1}{\gamma_{k,n-1}}$$

59 Use Bernoulli's method to find the zero of maximum magnitude of the polynomial of Prob. 54. Use the result of Prob. 54 to explain the slow convergence (you will need to calculate at least to u_{70} in order to stabilize the first decimal of β_1). Then calculate the remaining zeros of the polynomial.

***60** (*a*) Verify (8.10-51) to (8.10-54).

(*b*) Show that, if all the zeros of $f(z)$ are real, then H is always nonnegative.

(*c*) Verify (8.10-60).

(*d*) Show that if the sign is chosen in (8.10-59) to agree with $f'(x_i)$, then if $x_1 < \alpha_1$, $x_i < x_{i+1} < \alpha_1$ and if $\alpha_n < x_1$, $\alpha_n < x_{i+1} < x_i$.

(*e*) Use Laguerre's method with $x_1 = 10^6$ to find a zero of (i) $z^4 + 8.1z^3 - 19.8z^2 - 5.9z + 21$; (ii) $z^4 + 2z^3 + 3z^2 + 4z + 5$.

Section 8.11

61 (*a*) Verify (8.11-33) to (8.11-36).

(*b*) Use (8.11-33) and (8.11-34) to verify (8.11-21).

62 Let $V^{(\lambda)}(z) = H^{(\lambda)}(z)/P(z)$, where $P(z)$ is given by (8.11-1) and $H^{(\lambda)}(z)$ by (8.11-8) and (8.11-9).

(*a*) Show that $V^{(\lambda+1)}(z) = [V^{(\lambda)}(z) - V^{(\lambda)}(s_\lambda)]/(z - s_\lambda)$, where s_λ is the variable shift given by t_λ of (8.11-6).

(*b*) Show that $V^{(\lambda+1)}(s_\lambda) = [V^{(\lambda)}(s_\lambda)]'$ and that $h_0^{(\lambda+1)} = -V^{(\lambda)}(s_\lambda)$, where $h_0^{(\lambda)}$ is the coefficient of z^{n-1} in $H^{(\lambda)}(z)$.

(*c*) Show that $t_{\lambda+1} = s_\lambda + V^{(\lambda)}(s_\lambda)/[V^{(\lambda)}(s_\lambda)]' = s_\lambda - W^{(\lambda)}(s_\lambda)/[W^{(\lambda)}s_\lambda)]'$, where $W^{(\lambda)}(z) = 1/V^{(\lambda)}(z) = P(z)/H^{(\lambda)}(z)$, and consequently, that formula (8.11-6) is precisely one Newton iteration step performed on the rational function $P(z)/H^{(\lambda)}(z)$. Note that for λ sufficiently large, this function is close to a linear polynomial whose zero is α_1.

Section 8.12

63 Consider the function $f(z) = g(x, y) + ih(x, y)$, where $z = x + iy$.

(*a*) Show that $f'(z) = g_x(x, y) + h_y(x, y) + i(h_x(x, y) - g_y(x, y))$.

(*b*) Define $f(z)/f'(z) = G(x, y) + iH(x, y)$ and $|f(z)| = \sqrt{g^2(x, y) + h^2(x, y)} = F(x, y)$. Use the Cauchy-Riemann equations

$$g_x(x, y) = h_y(x, y) \qquad g_y(x, y) = -h_x(x, y)$$

to show that $-H(x, y)/G(x, y) = -F_y(x, y)/F_x(x, y)$, which is the direction of steepest descent of $|f(z)|$ at the point $z = (x, y)$.

64 Use any method or combination of methods to find the roots of (*a*) $z^3 - 2z - 5 = 0$; (*b*) $z^3 - 16z^2 + 3 = 0$; (*c*) $z^4 - 3z^2 + 2z - 1 = 0$; (*d*) $z^4 + 4z^2 - 3z - 1 = 0$; (*e*) $z^6 + 5z^3 + 7z^2 + 1 = 0$.

65 Use any method to find the root of largest magnitude of (*a*) $z^3 - 20z^2 - 3z + 18 = 0$; (*b*) $z^4 - 3z^3 - 60z^2 + 150z + 300 = 0$; (*c*) $10z^3 - 21z^2 - 40z + 84 = 0$.

Section 8.13

66 (*a*) Derive (8.13-4) under the assumption that products of errors can be neglected.

(*b*) Why is the assumption that products of errors can be neglected generally unfounded when $f'(z_0)$ is small?

(*c*) Apply (8.13-5) to $f(z)$ of (8.13-6) with $\delta_i = 0, i \neq 19$ and $\delta_{19} = 2^{-23}$ for $z_0 = -1, -5, -8$, and -15.

CHAPTER

NINE

THE SOLUTION OF SIMULTANEOUS LINEAR EQUATIONS

9.1 THE BASIC THEOREM AND THE PROBLEM

Our concern in this chapter is with the solution of n simultaneous linear equations in n unknowns

$$\sum_{j=1}^{n} a_{ij}x_j = b_i \qquad i = 1, \ldots, n \tag{9.1-1}$$

Equation (9.1-1) is conveniently written in the matrix form

$$A\mathbf{x} = \mathbf{b} \tag{9.1-2}$$

where $A = [a_{ij}]$ is the $n \times n$ matrix of coefficients, $\mathbf{x}^T = (x_1, \ldots, x_n)$ and $\mathbf{b}^T = (b_1, \ldots, b_n)$ with T denoting the transpose.† We shall use matrix algebra and matrix notation extensively but not exclusively in this chapter. We assume everywhere in this chapter that A and $\mathbf{b}$ are real.

We denote by A_b the $n \times (n + 1)$ matrix which has the column vector $\mathbf{b}$ appended as an $(n + 1)$st column to A. We denote the rank of any matrix A by $r(A)$. The basic theorem on the existence of solutions of (9.1-2) is as follows.

† In this and the next chapter column vectors will be in boldface and row vectors will be in boldface with a superscript T.

Theorem 9.1

1. The system of equations (9.1-2) has a solution if and only if

$$r(A) = r(A_b)$$

2. If $r(A) = r(A_b) = k < n$, then if $x_{i_1}, x_{i_2}, \ldots, x_{i_k}$ are k variables whose corresponding columns are linearly independent in A, the remaining $n - k$ variables can be arbitrarily assigned; i.e., since there must be some set of k linearly independent columns of A (why?), there is an $(n - k)$-parameter family of solutions.
3. If $r(A) = r(A_b) = n$, there is a unique solution.

Corollary 9.1 From 2 and 3 it follows that in the homogeneous case ($\mathbf{b} = \mathbf{0}$), there is a nontrivial solution if and only if $r(A) < n$.

This familiar theorem and its corollary can be proved using Gaussian elimination, which will be discussed in Sec. 9.3-1; the proofs themselves we leave to a problem {1}.

In contrast to the equations of Chaps. 5 and 8, there is no problem in finding an analytic solution of (9.1-1). Cramer's rule gives us such a solution. Instead the problem is in *computing* the solution. Even for quite low-order systems, the large amount of computation required to evaluate determinants makes the use of Cramer's rule impractical. Therefore, our major aim in this chapter will be to develop more efficient computational algorithms for the solution of (9.1-1).

The efficiency of an algorithm will be judged by two main criteria: (1) How fast is it; i.e., how many operations are involved? (2) How accurate is the computed solution?

These two criteria are aimed at the evaluation of algorithms for the solution of high-order systems (up to 500 equations or more) on a digital computer. Because of the formidable amount of computation required to solve (9.1-1) for large systems, the need to answer the first question is clear. The need to answer the second question arises because small roundoff errors may cause errors in the computed solution out of all proportion to their size. Furthermore, because of the large number of operations involved in solving a high-order system, the potential accumulated roundoff error is nontrivial. In Sec. 6.3-1 we had a glimpse of how such roundoff errors could cause substantial loss of accuracy.

Before getting into the details of solving (9.1-1), in the next section we shall consider the problem in rather general terms in order to get an intuitive feeling for the difficulties that will be encountered.

9.2 GENERAL REMARKS

Sources and types of problems The matrices of coefficients that occur in practice generally fall into one of two categories:

1. Filled but not large. By filled we mean that there are few zero elements and by not large we mean matrices of order, say, less than 100. Such matrices occur in a wide variety of problems in statistics, mathematical physics, engineering, etc.
2. Sparse and perhaps very large. In contrast to the above a sparse matrix has few nonzero elements. In most cases, these elements lie on or near the main diagonal. Very large may mean of order 1000 or more. Such matrices arise commonly in the numerical solution of partial differential equations {2}.

It should not be surprising that different approaches are commonly used for matrices in these two categories. One cause of this is that the size of the matrices in the second category often makes memory space a problem even on the largest computers. But, basically, it is the different characters—sparse and filled—of the matrices that make the direct methods to be described in Secs. 9.3 to 9.5 generally superior for the first category, while the iterative methods of Secs. 9.6 and 9.7 are most often used for problems in the second category. It should be emphasized that there are no hard-and-fast rules in this, however, and indeed there is some controversy. Almost no one would recommend iterative methods for filled, low-order matrices, but there is substantial opinion in favor of direct methods for medium-sized sparse matrices.

Ill condition Assuming that A is nonsingular, as we shall throughout this chapter, the solution of (9.1-2) can be written

$$\mathbf{x} = A^{-1}\mathbf{b} \tag{9.2-1}$$

Suppose that the elements of A have been normalized so that the largest in magnitude has order of magnitude unity. Suppose also that $B = A^{-1}$ has some very large elements, one of which is

$$b_{ji} = \frac{A_{ij}}{|A|} \tag{9.2-2}$$

where $|A|$ denotes the determinant of A and A_{ij} is the cofactor of a_{ij} and is therefore unaffected by a change in a_{ij}. The assumption that b_{ji} is large means that A_{ij} must be large relative to $|A|$. Since one of the terms in the expansion of $|A|$ about the ith row or jth column is $a_{ij}A_{ij}$, a small error in a_{ij} (relative to unity, i.e., relative to the normalization of A) may cause a large relative error in $|A|$ and therefore a large relative error in b_{ji}. This in turn can cause a large relative error in $\mathbf{x}$. Similarly, a small change in an element of $\mathbf{b}$ could cause a large change in $\mathbf{x}$. This effect can also be produced by roundoff errors in the course of the computation (cf. Sec. 6.3-1) because a roundoff error introduced during the computation is equivalent in effect to an initial error in the elements of A (why?—see Sec. 9.4).

An alternative way of looking at this problem is to consider the residual vector

$$\mathbf{r} = \mathbf{b} - A\mathbf{x}_c \tag{9.2-3}$$

where $\mathbf{x}_c$ is the computed solution. If A^{-1} has some large elements, then $\mathbf{r}$ may be very small even if $\mathbf{x}_c$ is substantially different from the true solution. For let $\mathbf{x}_t$ be the true solution of (9.1-2), so that $A\mathbf{x}_t = \mathbf{b}$. Then (9.2-3) can be written

$$\mathbf{r} = A(\mathbf{x}_t - \mathbf{x}_c) \tag{9.2-4}$$

or

$$\mathbf{x}_t - \mathbf{x}_c = A^{-1}\mathbf{r} \tag{9.2-5}$$

Therefore, if some elements of A^{-1} are large, a small component of $\mathbf{r}$ can still mean a large difference between $\mathbf{x}_t$ and $\mathbf{x}_c$, or conversely, $\mathbf{x}_c$ may be far from $\mathbf{x}_t$ but $\mathbf{r}$ can nevertheless still be small. This implies that we cannot test the correctness of a computed solution of (9.1-2) merely by substituting the result into the equations and calculating the residuals. Or to put it another way, an accurate solution, i.e., a small difference between $\mathbf{x}_c$ and $\mathbf{x}_t$ (see Sec. 9.4), will always produce small residuals if the matrix A is normalized, but small residuals do not guarantee an accurate solution.

If the matrix A is normalized as described above and is such that A^{-1} contains some very large elements, then we say the matrix and therefore the system of equations is *ill-conditioned* [see Sec. 1.7]. (Conversely, if the largest element in magnitude of A^{-1} has order of magnitude unity, the matrix may be said to be well-conditioned.) The folowing simple example will illustrate the dangers inherent in solving ill-conditioned systems. Consider the system [see Prob. 27, Chap. 1]

$$2x + 6y = 8 \qquad 2x + 6.00001y = 8.00001 \tag{9.2-6}$$

which has the solution $x = 1$, $y = 1$, and the system

$$2x + 6y = 8 \qquad 2x + 5.99999y = 8.00002 \tag{9.2-7}$$

which has the solution $x = 10$, $y = -2$. Here a change of .00002 in a_{22} and .00001 in b_2 has caused a gross change in the solution. The inverse of the matrix of coefficients in (9.2-6) has elements whose order of magnitude is 10^5, which indicates the ill condition of A. The necessity in the preceding discussion of the requirement that A be normalized can be seen by considering the systems (9.2-6) and (9.2-7) both multiplied by 10^5. Now a small *relative* change in a_{22} and b_2 causes the same gross change in the solution as above although the elements of A^{-1} are all of order magnitude unity.

The coefficients in (9.2-7) might, for example, be empirical values of those in (9.2-6). If empirical values of the coefficients accurate to more than five decimal places cannot be obtained, then the solution of (9.2-7), *no matter how accurately it is calculated*, may be grossly in error. How to calculate solutions of (9.1-2), *as accurate as the data warrant*, when the system is ill-conditioned is probably the single most difficult problem encountered in the solution of simultaneous linear equations. In Sec. 9.5 we shall consider the solution of ill-conditioned systems in some detail.

Sources of error There are three sources of error in the solution of systems of linear equations, two of which were mentioned above. The first is caused by errors in the coefficients and the elements of **b**. When such errors occur because these quantities are empirical, we must live with them. If a bound on the empirical errors is known, we can do no more than use this to get bounds on the errors in the solution (see Sec. 9.4). When the coefficients and the vector **b** are known exactly (as they are, for example, when a partial differential equation is approximated by differences) but must be rounded when they are inserted into the computer, we can control this source of error by using double-precision arithmetic, if necessary.

The second source of error is the roundoff error introduced in calculating the solution. The third source is truncation error. In direct methods, e.g., Cramer's rule or Gaussian elimination, which would lead to an exact solution in the absence of roundoff, there is no truncation error. But the iterative methods to be discussed in Secs. 9.6 and 9.7 generally converge only as the number of iterations goes to infinity. They are therefore subject to truncation error. One of the determining factors in deciding to use an iterative instead of a direct method is whether the truncation error can be made extremely small with an amount of computation comparable with, or less than, that required for a direct method. Truncation error is therefore almost always a minor source of error in the computed solution of (9.1-1).

9.3 DIRECT METHODS

As we indicated above, a direct method for the solution of (9.1-1) is one which if all computations were carried out without roundoff would lead to the true solution of the *given* system. Most direct methods involve some variation of the elimination procedure associated with the name of Gauss, which we shall now consider in detail.

9.3-1 Gaussian Elimination

We write out the system (9.1-1) in the form

$$
\begin{aligned}
a_{11}x_1 + a_{12}x_2 + \cdots + a_{1n}x_n &= a_{1,\,n+1} \\
a_{21}x_1 + a_{22}x_2 + \cdots + a_{2n}x_n &= a_{2,\,n+1} \\
\cdots\cdots\cdots\cdots\cdots\cdots\cdots\cdots\cdots & \\
a_{n1}x_1 + a_{n2}x_2 + \cdots + a_{nn}x_n &= a_{n,\,n+1}
\end{aligned}
\tag{9.3-1}
$$

where for notational simplicity we have written $b_j = a_{j,\,n+1}$. We assume, of course, that the matrix of coefficients is nonsingular. Suppose $a_{11} \neq 0$. We subtract the multiple a_{i1}/a_{11} of the first equation from the ith equation, $i = 2, \ldots, n$, to get the *first derived system*

$$
\begin{aligned}
a_{11}x_1 + a_{12}x_2 + \cdots + a_{1n}x_n &= a_{1,\,n+1} \\
a_{22}^{(1)}x_2 + \cdots + a_{2n}^{(1)}x_n &= a_{2,\,n+1}^{(1)} \\
\cdots\cdots\cdots\cdots\cdots\cdots\cdots & \\
a_{n2}^{(1)}x_2 + \cdots + a_{nn}^{(1)}x_n &= a_{n,\,n+1}^{(1)}
\end{aligned}
\tag{9.3-2}
$$

The new coefficients $a_{ij}^{(1)}$ are given by

$$
a_{ij}^{(1)} = a_{ij} - m_{i1}a_{1j} \qquad \begin{aligned} i &= 2, \ldots, n \\ j &= 2, \ldots, n+1 \end{aligned}
\tag{9.3-3}
$$

where $m_{i1} = a_{i1}/a_{11}$, $i = 2, \ldots, n$. If $a_{11} = 0$, then, because A is nonsingular, by interchanging two rows of (9.3-1) we can get a nonzero element in the upper left-hand corner (why?). We can also interchange two columns of A to achieve the same effect, but in this case the order in which the elements of the solution are computed will not be the natural order, $x_1, \ldots, x_n$, but some permutation of it.

Now, if $a_{22}^{(1)}$ in (9.3-2) is nonzero, we subtract $m_{i2} = a_{i2}^{(1)}/a_{22}^{(1)}$ times the second equation from the ith equation in (9.3-2), $i = 3, \ldots, n$, and get the second derived system

$$
\begin{aligned}
a_{11}x_1 + a_{12}x_2 + \cdots\cdots\cdots\cdots + a_{1n}x_n &= a_{1,\,n+1} \\
a_{22}^{(1)}x_2 + \cdots\cdots\cdots\cdots + a_{2n}^{(1)}x_n &= a_{2,\,n+1}^{(1)} \\
a_{33}^{(2)}x_3 + \cdots + a_{3n}^{(2)}x_n &= a_{3,\,n+1}^{(2)} \\
\cdots\cdots\cdots\cdots\cdots\cdots\cdots & \\
a_{n3}^{(2)}x_3 + \cdots + a_{nn}^{(2)}x_n &= a_{n,\,n+1}^{(2)}
\end{aligned}
\tag{9.3-4}
$$

where

$$
a_{ij}^{(2)} = a_{ij}^{(1)} - m_{i2}a_{2j}^{(1)} \qquad \begin{aligned} i &= 3, \ldots, n \\ j &= 3, \ldots, n+1 \end{aligned}
\tag{9.3-5}
$$

Again, if $a_{22}^{(1)} = 0$, we can interchange two rows or columns to get a nonzero element in the (2, 2) position. Continuing with this process through $n - 1$ steps, we arrive at the final system

$$
\begin{aligned}
a_{11}x_1 + a_{12}x_2 + \cdots\cdots\cdots + a_{1n}x_n &= a_{1,n+1} \\
a_{22}^{(1)}x_2 + \cdots\cdots\cdots + a_{2n}^{(1)}x_n &= a_{2,n+1}^{(1)} \\
a_{33}^{(2)}x_3 + \cdots + a_{3n}^{(2)}x_n &= a_{3,n+1}^{(2)} \\
\ddots \qquad \vdots \qquad &\quad \vdots \\
a_{nn}^{(n-1)}x_n &= a_{n,n+1}^{(n-1)}
\end{aligned}
\tag{9.3-6}
$$

with the diagonal elements all nonzero and where

$$
\begin{aligned}
a_{ij}^{(k)} = a_{ij}^{(k-1)} - m_{ik}a_{kj}^{(k-1)} \qquad & k = 1, \ldots, n-1 \\
& j = k+1, \ldots, n+1 \\
& i = k+1, \ldots, n \\
a_{ij}^{(0)} = a_{ij} &
\end{aligned}
\tag{9.3-7}
$$

with $m_{ik} = a_{ik}^{(k-1)}/a_{kk}^{(k-1)}$. Given (9.3-6), the solution is easily calculated as†

$$
x_i = \frac{1}{a_{ii}^{(i-1)}}\left[a_{i,n+1}^{(i-1)} - \sum_{j=i+1}^{n} a_{ij}^{(i-1)}x_j\right] \qquad i = n, \ldots, 1 \tag{9.3-8}
$$

The process leading to (9.3-6) is called *Gaussian elimination;* the calculation of the solution by (9.3-8) is called the *back substitution.* Using Gaussian elimination, it is easy to prove Theorem 9.1 {1}.

A variant of the above process, which it is convenient to consider here, is the *Gauss-Jordan reduction.* In this technique we proceed as before to get (9.3-2), but in place of (9.3-4) we derive the system

$$
\begin{aligned}
a_{11}x_1 + \qquad\qquad a_{13}^{(2)}x_3 + \cdots + a_{1n}^{(2)}x_n &= a_{1,n+1}^{(2)} \\
a_{22}^{(1)}x_2 + a_{23}^{(1)}x_3 + \cdots + a_{2n}^{(1)}x_n &= a_{2,n+1}^{(2)} \\
a_{33}^{(2)}x_3 + \cdots + a_{3n}^{(2)}x_n &= a_{3,n+1}^{(2)} \\
\cdots\cdots\cdots\cdots\cdots\cdots\cdots\cdots & \\
a_{n3}^{(2)}x_3 + \cdots + a_{nn}^{(2)}x_n &= a_{n,n+1}^{(2)}
\end{aligned}
\tag{9.3-9}
$$

in which the element in the first row and second column has also been reduced to zero, the remaining elements in the first row are given by (9.3-5) with $i = 1$, and $a_{2,n+1}^{(2)} = a_{2,n+1}^{(1)}$. Continuing in this way, so that at each

† Here we use the convention that $\sum_{j=n+1}^{n} = 0$.

stage all the elements in a column except the diagonal element are reduced to zero, we get finally

$$\begin{aligned} a_{11}x_1 \qquad\qquad\qquad &= a_{1,n+1}^{(n-1)} \\ a_{22}^{(1)}x_2 \qquad\quad &= a_{2,n+1}^{(n-1)} \\ \ddots \qquad &\quad \vdots \\ a_{nn}^{(n-1)}x_n &= a_{n,n+1}^{(n-1)} \end{aligned} \tag{9.3-10}$$

with the $a_{ii}^{(i-1)}$ given by (9.3-7) and with

$$a_{i,n+1}^{(k)} = \begin{cases} a_{i,n+1}^{(k-1)} - m_{ik}a_{k,n+1}^{(k-1)} & i = 1, \ldots, k-1, k+1, \ldots, n \\ a_{i,n+1}^{(k-1)} & i = k \end{cases}$$

$$k = 1, \ldots, n-1 \tag{9.3-11}$$

The solution of (9.3-10) is then simply given by

$$x_i = \frac{a_{i,n+1}^{(n-1)}}{a_{ii}^{(i-1)}} \qquad i = 1, \ldots, n \tag{9.3-12}$$

At first glance it might seem that the Gauss-Jordan reduction is to be preferred to Gaussian elimination, but we shall show now that in fact Gaussian elimination is the more efficient of the two.

As usual in determining the number of operations required, we shall consider only multiplications and divisions. To estimate the number of operations in Gaussian elimination, we use (9.3-7). For each k we need $n - k$ divisions† $(a_{ik}^{(k-1)}/a_{kk}^{(k-1)})$ and $(n - k)(n - k + 1)$ multiplications. The total number of multiplications and divisions is then {4}

$$M = \sum_{k=1}^{n-1} [(n-k)(n-k+1) + (n-k)] = \tfrac{1}{3}n^3 + O(n^2)‡ \tag{9.3-13}$$

We single out the n^3 term since it is only for large n that we are interested in M. The back substitution adds to M a term $n(n + 1)/2$ and does not affect the n^3 term.

For the Gauss-Jordan reduction, we again use (9.3-7), but this time with i running from 1 to $k - 1$ as well as from $k + 1$ to n. The calculation corresponding to (9.3-13) results in {4}

$$M = \tfrac{1}{2}n^3 + O(n^2) \tag{9.3-14}$$

† Or one division $(1/a_{kk}^{(k-1)})$ and $n - k$ multiplications $(1/a_{kk}^{(k-1)})a_{ik}^{(k-1)}$.

‡ In this chapter, in contrast to Chap. 5, the notation $O(n^k)$ will always refer to the situation as $n \to \infty$.

For large n, therefore, the Gauss-Jordan reduction requires about 50 percent more operations than Gaussian elimination. For this reason, in the remainder of this section we shall consider only Gaussian elimination.

9.3-2 Compact Forms of Gaussian Elimination

During the days of hand computation, the classical Gaussian elimination method involved many recordings of intermediate results. This was tiresome in itself and was also a source of copying errors. To eliminate the need for these intermediate recordings, compact forms of Gaussian elimination were developed. Although the recording of intermediate results on a digital computer is neither tiresome nor error-prone, compact forms have still been found useful in digital computation. The reason is that they reduce roundoff error by accumulating inner products and performing other arithmetic operations in double precision. The reader may ask: Why not perform the entire computation in double precision to cut down on roundoff errors? The answer is that this would involve doubling the storage requirements of the problem whereas doing selected portions of the computation in double precision requires only minimal storage increase. The increase in computation time is about the same whether all or just some of the computation, as discussed above, is done in double precision; this increase may be quite small on sophisticated computers. Thus, throughout this chapter, we shall point out situations where double-precision computation is possible at almost no extra cost in storage and computation time. One such situation is the back substitution given by (9.3-8). The quantity in brackets can be computed quite economically using double-precision arithmetic and, after division by $a_{ii}^{(i-1)}$, rounded to a single-precision number and stored as such.

Matrix formulation Before deriving a compact form, we shall show how to express the Gaussian elimination process in matrix form. To this end, we introduce two types of special matrices which are modifications of the identity matrix. The first type is denoted by P_{ij} and has the form

$$P_{ij} = \begin{bmatrix} 1 & & & & & & & \\ & 1 & & & & & \bigcirc & \\ & & \ddots & & & & & \\ & & & 0 & 1 & & & \\ & & & 1 & 0 & & & \\ & & & & & \ddots & & \\ & \bigcirc & & & & & 1 & \\ & & & & & & & \ddots \\ & & & & & & & & 1 \end{bmatrix} \begin{matrix} \\ \\ \\ \text{row } i \\ \text{row } j \\ \\ \\ \\ \\ \end{matrix} \qquad (9.3\text{-}15)$$

(col i and col j label the columns containing the interchanged entries.)

Premultiplying a matrix A by P_{ij} interchanges rows i and j, while postmultiplication by P_{ij} interchanges the corresponding columns. Clearly P_{ij} is symmetric and $P_{ij}^2 = I$, so that $P_{ij}^{-1} = P_{ij}$. By convention, $P_{ii} = I$ {5}.

The second class of matrices we require are unit lower triangular matrices L_i of the form

$$L_i = \begin{bmatrix} 1 & & & & \bigcirc \\ & \ddots & & & \\ & & 1 & & \\ & & -m_{i+1,\,i} & & \\ \bigcirc & & \vdots & \ddots & \\ & & -m_{n,\,i} & & 1 \end{bmatrix} \qquad \text{(column } i\text{)} \tag{9.3-16}$$

where the m_{ji} are defined following (9.3-7). An interesting property of the L_i, which can be verified by multiplication {6}, is that $L_i^{-1} = \hat{L}_i$ has a form identical to that of L_i with the signs of the m_{ji} reversed.

If we now denote by A_1 the matrix of coefficients in (9.3-2), then

$$A_1 = L_1 P_{1,r_1} A \qquad r_1 \geq 1 \tag{9.3-17}$$

where we have assumed the interchange of rows 1 and r_1 before the elimination step. (If no interchange was needed, $r_1 = 1$ and $P_{1,r_1} = I$.) In general, if we denote the matrix of coefficients in the ith derived system by A_i, then

$$A_i = L_i P_{i,r_i} A_{i-1} \qquad i = 1, \ldots, n-1;\ r_i \geq i \tag{9.3-18}$$

where $A_0 = A$ and where we interchange rows i and r_i before the ith elimination step. Finally, denoting the upper triangular matrix of coefficients A_{n-1} in the final system (9.3-6) by $U = [u_{ij}]$, we have

$$U = L_{n-1} P_{n-1,r_{n-1}} L_{n-2} P_{n-2,r_{n-2}} L_{n-3} \cdots L_1 P_{1,r_1} A \tag{9.3-19}$$

If we had known the matrices P_{i,r_i} in advance, we would have applied them to A initially to yield the matrix

$$\tilde{A} = P_{n-1,r_{n-1}} P_{n-2,r_{n-2}} \cdots P_{1,r_1} A = PA \tag{9.3-20}$$

where P is a permutation matrix, $P = P_{n-1,r_{n-1}} \cdots P_{1,r_1}$, and we would have been able to apply Gaussian elimination without interchanges. This implies that we can rewrite (9.3-19) as

$$U = L_{n-1} \cdots L_1 (P_{n-1,r_{n-1}} \cdots P_{1,r_1}) A = \bar{L}\tilde{A} \tag{9.3-21}$$

where $\bar{L}$ is unit lower triangular since the product of two unit lower triangular matrices is again unit lower triangular {7}. The inverse of $\bar{L}$ is

$L = L_1^{-1} \cdots L_{n-1}^{-1}$, which is also unit lower triangular and indeed of the form {7}

$$L = \begin{bmatrix} 1 & & & \bigcirc \\ m_{21} & 1 & & \\ \cdots & \cdots & \cdots & \cdots \\ m_{n1} & \cdots & m_{n,n-1} & 1 \end{bmatrix} \tag{9.3-22}$$

We thus have that

$$\tilde{A} = PA = LU \tag{9.3-23}$$

This shows that if A is a nonsingular matrix, there exists a permutation matrix P such that $\tilde{A}$ can be decomposed into the product of a unit lower triangular matrix L and an upper triangular matrix U. This product is called the triangular or LU decomposition of A. Furthermore, this decomposition is unique. For assume also that $\tilde{A} = L_1 U_1$, where L_1 is unit lower triangular and U_1 upper triangular, so that $LU = L_1 U_1$. Then since L and U_1 are nonsingular, we have $UU_1^{-1} = L^{-1}L_1$. Since the inverse of a (unit) triangular matrix is of the same form and the product of two (unit) triangular matrices is also of the same form, we have that an upper triangular matrix is equal to a unit lower triangular matrix. This is possible only if both matrices are equal to the identity matrix, i.e., that $U_1 = U$ and $L_1 = L$, proving uniqueness.

If we denote by $\mathbf{b}_f$ the right-hand side of (9.3-6), then we have similarly that

$$\mathbf{b}_f = \bar{L}P\mathbf{b} \tag{9.3-24}$$

so that

$$\tilde{\mathbf{b}} = P\mathbf{b} = L\mathbf{b}_f \tag{9.3-25}$$

The details of the triangular decomposition are usually stored in the computer as follows. Initially, we define a vector $\mathbf{v} = (1, 2, \ldots, n)^T$. At stage j, we interchange the indices v_j and v_{r_j}. Then we compute the m_{ij}, $i > j$, and store them in place of those elements of A_j which are set to 0, that is, eliminated at that stage. At the end of stage $n - 1$, the initial matrix A has been replaced by U in the upper triangle and by L, without the unit diagonal, below the diagonal. From the final vector $\mathbf{v}$, we can reconstruct the permutation matrix P explicitly to compute $\tilde{\mathbf{b}}$, or, as is more usually the case, we use $\mathbf{v}$ directly, to compute $\tilde{\mathbf{b}}$.

From this matrix formulation, we see that it is not necessary to carry $\mathbf{b}$ along during the Gaussian elimination process. In fact, once we have found L, U, and P such that (9.3-23) holds, we can solve (9.1-2) for any right-hand side as follows:

$$LU\mathbf{x} = PA\mathbf{x} = P\mathbf{b} \tag{9.3-26}$$

is equivalent to (9.1-2) since P is nonsingular. Setting $\mathbf{y} = U\mathbf{x}$, we first solve the equation

$$L\mathbf{y} = P\mathbf{b} = \tilde{\mathbf{b}} \tag{9.3-27}$$

for $\mathbf{y}$ by the process of forward substitution using a formula similar to (9.3-8),

$$y_i = \tilde{b}_i - \sum_{j=1}^{i-1} l_{ij} y_j \qquad i = 1, \ldots, n \tag{9.3-28}$$

We then solve the equation $U\mathbf{x} = \mathbf{y}$ by back substitution. As in the case of (9.3-8), we can do the arithmetic in (9.3-28) in double precision at little extra cost.

From the above, we see that we can divide the Gaussian elimination algorithm for solving (9.1-2) into two distinct processes. The first is the triangular decomposition of A (9.3-23), which is independent of $\mathbf{b}$. The second is the combination of forward and back substitution applied to $P\mathbf{b}$ to get the solution $\mathbf{x}$. This means that once we have found the LU decomposition of A, we can solve (9.1-2) for *any* right-hand side $\mathbf{b}$. In particular, we need not know the value of $\mathbf{b}$ at the time we perform the LU decomposition. This is in contrast to the Gaussian elimination algorithm where we carry $\mathbf{b}$ along during the elimination and do not save the m_{ij}. The importance of this feature will become clear later on in our development.

9.3-3 The Doolittle, Crout, and Cholesky Algorithms

We shall initially assume that we are working with a matrix A in which no interchanges are necessary; subsequently we shall discuss how to introduce interchanges into the algorithms. The Doolittle algorithm is essentially another way to arrive at an LU decomposition of A while allowing for double-precision calculation of inner products. Since we assume that A has an LU decomposition, let us try to find L and U by equating corresponding elements of A and LU. We thus have

$$a_{ij} = \sum_{k=1}^{n} l_{ik} u_{kj} = \sum_{k=1}^{\min(i,j)} l_{ik} u_{kj} \qquad i, j = 1, \ldots, n \tag{9.3-29}$$

There are n^2 equations in n^2 unknowns, the $n(n+1)/2$ elements of U and the $n(n-1)/2$ subdiagonal elements of L. These equations can be solved recursively by using the properties of triangular matrices in several ways. In view of the fact that we wish to allow the possibility of row interchanges, we shall organize the computation as follows, computing in succession one row of U followed by the corresponding column of L.

Setting $i = 1$ in (9.3-29), we immediately have that

$$u_{1j} = a_{1j} \qquad j = 1, \ldots, n \tag{9.3-30}$$

since $l_{11} = 1$. Then setting $j = 1$ and $i > j$ in (9.3-29), we have

$$l_{i1} = \frac{a_{i1}}{u_{11}} \qquad i = 2, \ldots, n \tag{9.3-31}$$

Setting now $i = 2$ and $j \geq i$ in (9.3-29), we find

$$u_{2j} = a_{2j} - l_{21}u_{1j} \qquad j = 2, \ldots, n \tag{9.3-32}$$

Then, setting $j = 2$ and $i > j$, we have

$$l_{i2} = \frac{a_{i2} - l_{i1}u_{12}}{u_{22}} \qquad i = 3, \ldots, n \tag{9.3-33}$$

In general, we compute row r, $r = 1, \ldots, n$, of U by

$$u_{rj} = a_{rj} - \sum_{k=1}^{r-1} l_{rk}u_{kj} \qquad j = r, \ldots, n \tag{9.3-34}$$

followed by column r, $r = 1, \ldots, n - 1$, of L, using

$$l_{ir} = \frac{a_{ir} - \sum_{k=1}^{r-1} l_{ik}u_{kr}}{u_{rr}} \qquad i = r + 1, \ldots, n \tag{9.3-35}$$

At each stage, all values entering into (9.3-34) and (9.3-35) have been computed at previous stages. Our assumption that no interchanges are necessary ensures that all $u_{jj} \neq 0$, so that the algorithm is fully defined. As is evident from the equations, we can compute the right-hand sides of (9.3-34) and (9.3-35) in double precision, rounding only at the end to yield single-precision values u_{rj} and l_{ir}. It is worth noting that if we compute (9.3-34) and (9.3-35) using single-precision arithmetic, the results are identical with the values of L and U arising in the LU decomposition based on Gaussian elimination {9}.

The Crout algorithm differs from the Doolittle one in that it generates a unit upper triangular matrix $\hat{U}$ and a general lower triangular matrix $\hat{L}$ such that

$$\hat{L}\hat{U} = A \tag{9.3-36}$$

Following the same method as before, we first generate column r of $\hat{L}$ using the equations

$$\hat{l}_{ir} = a_{ir} - \sum_{k=1}^{r-1} \hat{l}_{ik}\hat{u}_{kr} \qquad i = r, \ldots, n \tag{9.3-37}$$

followed by row r of $\hat{U}$, given by

$$\hat{u}_{rj} = \frac{a_{rj} - \sum_{k=1}^{r-1} \hat{l}_{rk}\hat{u}_{kj}}{\hat{l}_{rr}} \qquad j = r + 1, \ldots, n \tag{9.3-38}$$

The Crout algorithm has a slight advantage over the Doolittle algorithm when we wish to do row interchanges, as we shall see in the next section.

The two algorithms are closely related. In fact, the diagonal of U is identical with that of $\hat{L}$ {10}. If we write $D = \text{diag}\,(u_{ii})$, then

$$A = LU = (LD)(D^{-1}U) = \hat{L}\hat{U} = LD\hat{U} \tag{9.3-39}$$

Thus, both the Doolittle and Crout factorizations are special cases of the LDU factorization of A into a unit lower triangular matrix, a diagonal matrix, and a unit upper triangular matrix.

We shall return to the question of row interchanges after we discuss the subject of pivoting in the next section. First, however, we shall give an algorithm for the triangular decomposition of a positive definite symmetric matrix A. Such a matrix can be decomposed without row interchanges in the form

$$A = LL^T \tag{9.3-40}$$

where L is a real nonsingular lower triangular matrix. The proof of this fact is instructive in that it gives an algorithm for decomposing matrices formed by adjoining a row and a column to a matrix A_{n-1} of similar form using the result of the decomposition of A_{n-1}. This feature is useful in solving least-square problems using normal equations when the desired order of the approximation is not known in advance [see Sec. 6.3-2].

Assume now that we have a decomposition of $A_{n-1} = L_{n-1}L_{n-1}^T$, where

$$A_n = \begin{bmatrix} A_{n-1} & \mathbf{b}_{n-1} \\ \mathbf{b}_{n-1}^T & a_{nn} \end{bmatrix} \tag{9.3-41}$$

We can assume this induction hypothesis since A_{n-1} is also positive definite. This follows from the fact that if a matrix is positive definite, all its principal minors are positive definite {14}. Now define $c_{n-1} = L_{n-1}^{-1}b_{n-1}$, $x = (a_{nn} - \mathbf{c}_{n-1}^T\mathbf{c}_{n-1})^{1/2}$ and

$$L_n = \begin{bmatrix} L_{n-1} & 0 \\ \mathbf{c}_{n-1}^T & x \end{bmatrix} \tag{9.3-42}$$

Then it is easy to see that {11}

$$L_n L_n^T = A_n \tag{9.3-43}$$

The only thing remaining is to show that x is real. To this end, we take determinants of both sides of (9.3-43) and use (9.3-42) to get

$$\det\,(L_{n-1})^2 x^2 = \det\,(A_n) \tag{9.3-44}$$

Since $\det\,(A_n) > 0$ because A_n is positive definite, it follows that x is real.

The elements of the rth row of L can be computed directly from (9.3-40) using the equations

$$\sum_{j=1}^{i-1} l_{rj} l_{ij} + l_{ri} l_{ii} = a_{ri} \qquad i = 1, \ldots, r-1 \tag{9.3-45}$$

$$\sum_{j=1}^{r-1} l_{rj}^2 + l_{rr}^2 = a_{rr} \tag{9.3-46}$$

so that this decomposition, called the *Cholesky factorization* of A, requires n square roots and about $n^3/6$ multiplications. From (9.3-46) we see that

$$\sum_{j=1}^{r} l_{rj}^2 = a_{rr}$$

which implies that all elements of L are bounded by max $a_{ii}^{1/2}$.

Example 9.1 Apply the Cholesky algorithm to the positive definite symmetric matrix A below using 5-digit floating-point decimal arithmetic.

$$A = \begin{bmatrix} 136.01 & 90.860 & 0 & 0 \\ 90.860 & 98.810 & -67.590 & 0 \\ 0 & -67.590 & 132.01 & 46.260 \\ 0 & 0 & 46.260 & 177.17 \end{bmatrix}$$

For $r = 1$: $\qquad l_{11}^2 = a_{11} \qquad l_{11} = 11.662$

For $r = 2$: $\qquad l_{21} l_{11} = a_{21} \qquad l_{21} = 7.7911$

$$l_{21}^2 + l_{22}^2 = a_{22} \qquad l_{22} = 6.1733$$

For $r = 3$: $\qquad l_{31} l_{11} = a_{31} \qquad l_{31} = 0$

$$l_{31} l_{21} + l_{32} l_{22} = a_{32} \qquad l_{32} = -10.949$$

$$l_{31}^2 + l_{32}^2 + l_{33}^2 = a_{33} \qquad l_{33} = 3.4827$$

For $r = 4$: $\qquad l_{41} l_{11} = a_{41} \qquad l_{41} = 0$

$$l_{41} l_{21} + l_{42} l_{22} = a_{42} \qquad l_{42} = 0$$

$$l_{41} l_{31} + l_{42} l_{32} + l_{43} l_{33} = a_{43} \qquad l_{43} = 13.283$$

$$l_{41}^2 + l_{42}^2 + l_{43}^2 + l_{44}^2 = a_{44} \qquad l_{44} = .85552$$

Thus

$$L = \begin{bmatrix} 11.662 & 0 & 0 & 0 \\ 7.7911 & 6.1733 & 0 & 0 \\ 0 & -10.949 & 3.4827 & 0 \\ 0 & 0 & 13.283 & .85552 \end{bmatrix}$$

As a check we compute LL^T and find that

$$LL^T = \begin{bmatrix} 136.00 & 90.860 & 0 & 0 \\ 90.860 & 98.811 & -67.591 & 0 \\ 0 & -67.591 & 132.01 & 46.261 \\ 0 & 0 & 46.261 & 177.17 \end{bmatrix}$$

which agrees with A to within rounding error.

Although A is tridiagonal, we did not use this fact in setting up the equations for the elements of L.

If we actually need the factorization of A in the form LL^T, then we must compute the square roots implied by (9.3-46). However, if we are only solving a system of equations, we can save the computation of the square roots as follows.

Let $D = \text{diag.}\,(l_{ii})$; then

$$A = (LD^{-1})D^2(D^{-1}L^T) = \bar{L}\bar{D}\bar{L}^T \tag{9.3-47}$$

where $\bar{L}$ is unit lower triangular and $\bar{D}$ is a positive diagonal matrix. The rth row of $\bar{L}$ and rth element of $\bar{D}$ are given by the equations {12}

$$\bar{d}_i\,\bar{l}_{ri} = a_{ri} - \sum_{j=1}^{i-1} \hat{a}_{rj}\,\bar{l}_{ij} \equiv \hat{a}_{ri} \qquad i = 1, \ldots, r-1 \tag{9.3-48}$$

$$\bar{d}_r = a_{rr} - \sum_{j=1}^{r-1} \hat{a}_{rj}\,l_{rj} \tag{9.3-49}$$

Note that in both of the decompositions, we can compute in double precision.

9.3-4 Pivoting and Equilibration

Our discussion of Gaussian elimination makes it clear that when the element $a_{ii}^{(i-1)} = 0$, we must interchange two rows in order to be able to continue the process. Now it is a general rule in numerical analysis that when a certain process cannot be carried out in a particular situation, it probably cannot be carried out successfully in a neighboring situation. Thus, if $a_{ii}^{(i-1)}$ is not exactly 0 but is close to 0, we can expect that the elimination process may suffer from numerical instability. Therefore to minimize the possibility of such instability, we seek at each stage the number farthest from 0; that is, we find the row $r_i \geq i$ for which $|a_{ri}^{(i-1)}|$ is maximal, $r = i, \ldots, n$, and interchange it with row i. Thus, the current value of $a_{ii}^{(i-1)}$ will be the furthest from 0. The element $a_{r_i,i}^{(i-1)}$ is called the *pivot element* and the row r_i, the *pivot row*. The process of so choosing $a_{r_i,i}^{(i-1)}$ is called *partial pivoting*, in contrast to *complete pivoting*, where we interchange both rows and columns to find the row r_i and column c_i for which $|a_{rc}^{(i-1)}|$ is maximal, $r = i, \ldots, n$, $c = i, \ldots, n$. While this latter process has some theoretical advantage over partial pivoting, it is almost never used in practice since it is more time-consuming and does not give better results in realistic situations.

Example 9.2 Solve the following system of linear equations by Gaussian elimination using 10-digit decimal floating-point arithmetic.

$$\begin{aligned} .3 \times 10^{-11} x_1 + x_2 &= .7 \\ x_1 + x_2 &= .9 \end{aligned} \tag{9.3-50}$$

Without pivoting, we compute first $m_{21} = .33333\,33333 \times 10^{12}$ and end up with the triangular system

$$\begin{aligned} .3 \times 10^{-11} x_1 + x_2 &= .7 \\ -.33333\,33333 \times 10^{12} x_2 &= .23333\,33333 \times 10^{12} \end{aligned} \tag{9.3-51}$$

which yields by back substitution the values

$$x_2 = .70000\,00000 \qquad x_1 = .00000\,00000$$

in contrast to the true solution correct to 10 digits,

$$x_1 = .20000\,00000 \qquad x_2 = .70000\,00000$$

With pivoting, we interchange rows 1 and 2 and then find that $m_{21} = .3 \times 10^{-11}$, giving the triangular system

$$\begin{aligned} x_1 + x_2 &= .9 \\ x_2 &= .7 \end{aligned} \tag{9.3-52}$$

which gives the correct solution after back substitution.

Note that the error in the first case is not due to any accumulation of roundoff error but only to a single large rounding error which arises from the large value of m_{21}. This results in $a_{22}^{(1)}$ being essentially independent of the value of $a_{22}^{(0)}$, which can range over a large set of values without changing (9.3-51).

Now, let us multiply the first row of (9.3-50) by 10^{12} and solve the resulting system using partial pivoting. Since this time the coefficient of x_1 in the first equation is larger than that in the second, we do not interchange rows and we find that our results are identical with the results from the solution of (9.3-50) without pivoting. For pivoting to be effective, all the rows of A should be of the same order of magnitude, so that in searching for the pivot element we are comparing numbers which originated in vectors of about the same size. The simplest way to accomplish this is to require that the L_∞ norm of each row of A be unity. This is called *equilibration by rows*. It is not always satisfactory, but it generally works {19}. In practice we can do either an explicit equilibration by dividing each element in a row by the element of maximum magnitude in that row or an implicit equilibration, as below. The main disadvantage of explicit equilibration is that it introduces an additional rounding error in the elements of A.

We now show how to implement the Doolittle and Crout algorithms with partial pivoting and implicit equilibration. Initially we compute the values

$$d_i = \max_{1 \le j \le n} |a_{ij}| \qquad i = 1, \ldots, n \tag{9.3-53}$$

For the Doolittle algorithm assume that we have reached the rth stage and have computed u_{ij}, $i = 1, \ldots, r - 1, j \ge i$, and l_{ij}, $j = 1, \ldots, r - 1$, $i > j$, and that these have overwritten the corresponding a_{ij}. We now compute quantities s_i in double precision defined by

$$s_i = a_{ir} - \sum_{k=1}^{r-1} l_{ik} u_{kr} \qquad i = r, \ldots, n \tag{9.3-54}$$

and store s_i at the end of row i. The s_i's are the quantities which would have resulted in rows r to n of column r after the first $r - 1$ stages of Gaussian elimination {9}. Now let p be such that

$$\left| \frac{s_p}{d_p} \right| = \max_{r \le i \le n} \left| \frac{s_i}{d_i} \right| \tag{9.3-55}$$

Then this p is the index of that row from rows r to n which would have had the element of largest magnitude in column r after $r - 1$ elimination stages if the original matrix had been explicitly equilibrated. The calculation of p by (9.3-55) therefore represents implicit equilibration and determination of the pivot element. Using this p, we interchange rows r and p of the complete array augmented by the s_i and also interchange the elements v_r and v_p of the vector $\mathbf{v}$.† If we still refer to the new element in the (i, j) position as l_{ij} or a_{ij}, whichever is relevant, we have [cf. (9.3-34) and (9.3-35)]

$$\begin{aligned} u_{rr} &= s_r \qquad u_{rj} = a_{rj} - \sum_{k=1}^{r-1} l_{rk} u_{kj} \qquad && j = r + 1, \ldots, n \\ l_{ir} &= \frac{s_i}{u_{rr}} && i = r + 1, \ldots, n \end{aligned} \tag{9.3-56}$$

Note that the Doolittle method with partial pivoting requires the storage of a double-precision vector $\mathbf{s}$.

In the Crout algorithm, on the other hand, we proceed as follows at the rth step. Compute $\hat{l}_{ir}$ by the formula [cf. (9.3-37)]

$$\hat{l}_{ir} = a_{ir} - \sum_{k=1}^{r-1} \hat{l}_{ik} \hat{u}_{kr} \qquad i = r, \ldots, n \tag{9.3-57}$$

and overwrite a_{ir} by $\hat{l}_{ir}$. The $\hat{l}_{ir}$ here are also computed in double precision, as the s_i were in the Doolittle algorithm. However, since we do not do any

† We do not actually have to interchange the two rows explicitly. It can be done implicitly by replacing the row index, say r, throughout by the corresponding index v_r.

further computation with these values, we store them as properly rounded single-precision numbers in place of the a_{ir}. If

$$\left|\frac{\hat{l}_{pr}}{d_p}\right| = \max_{r \le i \le n} \left|\frac{\hat{l}_{ir}}{d_i}\right| \tag{9.3-58}$$

interchange rows r and p of the complete array and also v_r and v_p, as before. Then compute $\hat{u}_{rj}$ by means of [cf. (9.3-38)]

$$\hat{u}_{rj} = \frac{a_{rj} - \sum_{k=1}^{r-1} \hat{l}_{rk}\hat{u}_{kj}}{\hat{l}_{rr}} \qquad j = r+1, \ldots, n \tag{9.3-59}$$

As in the Doolittle method, (9.3-58) determines the pivot element using implicit equilibration.

With partial pivoting and no equilibration or explicit equilibration we have that in the Doolittle method $|l_{ij}| \le 1$ and in the Crout method $|\hat{l}_{ij}| \le |\hat{l}_{jj}|$. However, this is not so when implicit equilibration is used. Then we only have that $|l_{ij}/d_i| \le 1$ and $|\hat{l}_{ij}/d_i| \le |\hat{l}_{jj}/d_j|$, respectively.

Since the Crout algorithm is a little more economical in storage, it should be used, since its stability properties are essentially the same as those of the Doolittle algorithm. We could modify the Doolittle algorithm so that it required the same amount of storage as the Crout algorithm by rounding the s_i to single precision and overwriting a_{ir} by s_i, $i = r, \ldots, n$. However, in this case, we would lose accuracy, so that again the Crout algorithm is preferable. Of course, in using the Crout algorithm, Eqs. (9.3-28) and (9.3-8) for forward and back substitution must be modified accordingly {20}.

Example 9.3 Apply the Doolittle algorithm with partial pivoting and equilibration to the matrix A below using 5-digit floating-point decimal arithmetic.

$$A = \begin{bmatrix} 136.01 & 90.860 & 0 & 0 \\ 90.860 & 98.810 & -67.590 & 0 \\ 0 & -67.590 & 132.01 & 46.260 \\ 0 & 0 & 46.260 & 177.17 \end{bmatrix}$$

We first initialize the vector $\mathbf{v}$ to be $(1, 2, 3, 4)^T$. The values d_i, $i = 1, \ldots, 4$, are

$$d_1 = 136.01 \qquad d_2 = 98.810 \qquad d_3 = 132.01 \qquad d_4 = 177.17$$

For $r = 1$, we have that $s_i = a_{i1}$, $i = 1, \ldots, 4$.

Since $|s_1/d_1| = \max_{1 \le i \le 4} |s_i/d_i|$, we do not interchange rows. We now set

$$u_{11} = s_1 \qquad u_{1j} = a_{1j} \qquad j = 2, \ldots, 4$$

$$l_{i1} = \frac{s_i}{u_{11}} \qquad i = 2, \ldots, 4$$

The new matrix has the form

$$\begin{bmatrix} 136.01 & 90.860 & 0 & 0 \\ .66804 & 98.810 & -67.590 & 0 \\ 0 & -67.590 & 132.01 & 46.260 \\ 0 & 0 & 46.260 & 177.17 \end{bmatrix}$$

and $\mathbf{v} = (1, 2, 3, 4)^T$ is unchanged. For $r = 2$

$$s_i = a_{i2} - l_{i1}u_{12} \qquad i = 2, \ldots, 4$$

$$s_2 = 38.11188\,560 \qquad s_3 = -67.590 \qquad s_4 = 0$$

Since $|s_3/d_3| = \max_{2 \le i \le 4} |s_i/d_i|$, we interchange rows 2 and 3 in the matrix as well as s_2 and s_3. After this interchange

$$u_{22} = s_2 = -67.590 \qquad u_{2j} = a_{2j} - l_{21}u_{1j} \qquad j = 3, 4$$

$$l_{i2} = \frac{s_i}{u_{22}} \qquad i = 3, 4$$

The new matrix is

$$\begin{bmatrix} 136.01 & 90.860 & 0 & 0 \\ 0 & -67.590 & 132.01 & 46.260 \\ .66804 & -.56387 & -67.690 & 0 \\ 0 & 0 & 46.260 & 177.17 \end{bmatrix}$$

and $\mathbf{v}$ is now $(1, 3, 2, 4)^T$.

For $r = 3$,

$$s_i = a_{i3} - l_{i1}u_{13} - l_{i2}u_{23} \qquad i = 3, 4$$

$$s_3 = 6.84647\,8700 \qquad s_4 = 46.260$$

Since $|s_4/d_4| > |s_3/d_3|$, we interchange rows 3 and 4 in the matrix as well as s_3 and s_4. We then have

$$u_{33} = s_3 = 46.260 \qquad u_{34} = a_{34} - l_{31}u_{14} - l_{32}u_{24} \qquad l_{43} = \frac{s_4}{u_{33}}$$

The new matrix is

$$\begin{bmatrix} 136.01 & 90.860 & 0 & 0 \\ 0 & -67.590 & 132.01 & 46.260 \\ 0 & 0 & 46.260 & 177.17 \\ .66804 & -.56387 & .14800 & 0 \end{bmatrix}$$

and $\mathbf{v} = (1, 3, 4, 2)^T$.

Finally, for $r = 4$, $s_4 = a_{44} - l_{41}u_{14} - l_{42}u_{24} - l_{43}u_{34} = -.1365338$, and $u_{44} = -.13653$.

We now have the factorization $PA = LU$, where

$$L = \begin{bmatrix} 1 & 0 & 0 & 0 \\ 0 & 1 & 0 & 0 \\ 0 & 0 & 1 & 0 \\ .66804 & -.56387 & .14800 & 1 \end{bmatrix}$$

$$U = \begin{bmatrix} 136.01 & 90.860 & 0 & 0 \\ 0 & -67.590 & 132.01 & 46.260 \\ 0 & 0 & 46.260 & 177.17 \\ 0 & 0 & 0 & -.13653 \end{bmatrix}$$

and
$$P = \begin{bmatrix} 1 & 0 & 0 & 0 \\ 0 & 0 & 1 & 0 \\ 0 & 0 & 0 & 1 \\ 0 & 1 & 0 & 0 \end{bmatrix}$$

Note that although A is symmetric and tridiagonal, we have not used these properties of A in this decomposition.

9.4 ERROR ANALYSIS

Were we to carry out the Gaussian elimination algorithm in any of its equivalent forms using exact arithmetic operating on a true matrix A and a true right-hand side $\mathbf{b}$, we would obtain a true answer $\mathbf{x}_t = A^{-1}\mathbf{b}$. Indeed, there now exist programs which compute an exact rational solution of (9.1-2) when the elements of A and $\mathbf{b}$ are rational numbers in a reasonable amount of time for moderate values of n, say up to 100 or so. However, since most matrices arising in practice are either based on experimental data or result from a lengthy computation, the values of the elements of these matrices are not exact. Thus, it makes no sense to invest the considerable effort needed to compute an exact solution to an inexact problem, inasmuch as the time required for an exact solution is greater by several orders of magnitude than that required for an approximate solution using Gaussian elimination and floating-point arithmetic.

In the course of this section we shall use backward error analysis [see Sec. 1.6-1] to show that the computed solution $\mathbf{x}_c$ of the problem (9.1-2) is the exact solution of a neighboring problem

$$(A + E)\mathbf{x} = \mathbf{b} \tag{9.4-1}$$

where E is small and we can determine bounds on the elements of E. Note that the right-hand side of (9.4-1) is the same as that of (9.1-2). This does not mean that $\mathbf{b}$ has no effect on the matrix E. Indeed, E depends on $\mathbf{b}$. However, as we shall see in Sec. 9.4-1, we can find bounds on the elements of E independent of $\mathbf{b}$. Hence, it is important to know how this perturbation of A affects the solution. This problem is of interest in itself aside from considerations of roundoff error. For if A arises from experimental data, the elements of A are determined up to experimental error, so that the value of the (i, j) element of the true matrix may be anywhere in the interval $[a_{ij} - \epsilon_{ij}, a_{ij} + \epsilon_{ij}]$, where $\epsilon_{ij} > 0$ is the experimental error in the accepted value a_{ij}. Hence it is important to see how the exact solution of (9.1-2) is affected by changes in the elements of A, and similarly in those of $\mathbf{b}$, even though in this latter case, the results do not affect the bounds obtained by roundoff error analysis.

We, therefore, consider first *exact* solutions of related problems. The simplest case is that where only $\mathbf{b}$ changes, that is,

$$A\mathbf{x} = \mathbf{b} + \delta\mathbf{b} \tag{9.4-2}$$

If we denote the solution of (9.4-2) by $\mathbf{x}_t + \boldsymbol{\delta}\mathbf{x}$, we find that $\boldsymbol{\delta}\mathbf{x}$ satisfies the equation

$$A\boldsymbol{\delta}\mathbf{x} = \boldsymbol{\delta}\mathbf{b} \tag{9.4-3}$$

so that

$$\boldsymbol{\delta}\mathbf{x} = A^{-1}\,\boldsymbol{\delta}\mathbf{b} \tag{9.4-4}$$

Taking norms (see Sec. 1.3-3), we find that

$$\|\boldsymbol{\delta}\mathbf{x}\| \leq \|A^{-1}\| \cdot \|\boldsymbol{\delta}\mathbf{b}\| \tag{9.4-5}$$

Since we are almost always interested in the error relative to the solution $\mathbf{x}_t$, we should divide by $\|\mathbf{x}_t\|$. Now clearly

$$\|\mathbf{b}\| \leq \|A\| \cdot \|\mathbf{x}_t\| \tag{9.4-6}$$

so that

$$\|\mathbf{x}_t\| \geq \frac{\|\mathbf{b}\|}{\|A\|} \tag{9.4-7}$$

Dividing inequality (9.4-5) by inequality (9.4-7) yields the result

$$\frac{\|\boldsymbol{\delta}\mathbf{x}\|}{\|\mathbf{x}_t\|} \leq \|A\| \cdot \|A^{-1}\| \frac{\|\boldsymbol{\delta}\mathbf{b}\|}{\|\mathbf{b}\|} \tag{9.4-8}$$

The quantity $\|A\| \cdot \|A^{-1}\|$ will appear frequently in the sequel and will be denoted $K(A)$. These numbers, defined for the various matrix norms we have been using, give a measure of the *condition* of A and are always greater than or equal to 1 {21}. In particular

$$K_2(A) = \|A\|_2\,\|A^{-1}\|_2 \tag{9.4-9}$$

is called the *spectral condition number* of A.

A lower bound on $\|\boldsymbol{\delta}\mathbf{x}\|/\|\mathbf{x}_t\|$ can be similarly derived by reversing the roles of $\mathbf{x}$ and $\mathbf{b}$ and is given by {22}

$$\frac{\|\boldsymbol{\delta}\mathbf{x}\|}{\|\mathbf{x}_t\|} \geq \frac{1}{K(A)} \frac{\|\boldsymbol{\delta}\mathbf{b}\|}{\|\mathbf{b}\|} \tag{9.4-10}$$

From (9.4-8) we see that the larger $K(A)$ is, the greater can be the influence of an error in $\mathbf{b}$ on the accuracy of the solution of (9.1-2). Thus, if $K(A)$ is close to unity, we say that the matrix A is well-conditioned and if it is large, we say that A is ill-conditioned.

If we were just interested in the effects of perturbations in the data on the results of exact computations, we could compute $K(A)$ by computing A^{-1} and using a computable norm such as the L_1 or L_∞ norm. However, our main purpose in introducing this perturbation analysis is to study the effect of roundoff error on the computed solution. Generally we cannot compute A^{-1} exactly, and our computed value of $K(A)$ may be very inaccurate.

Fortunately, there are some situations in which we can determine a lower bound on $K(A)$ based on the following theorem.

Theorem 9.2 Let A be a nonsingular matrix and B any singular matrix. Then

$$\frac{1}{\|A - B\|} \leq \|A^{-1}\| \tag{9.4-11}$$

PROOF Since B is singular, there exists a vector $\mathbf{x} \neq \mathbf{0}$ such that $B\mathbf{x} = \mathbf{0}$. Hence for this $\mathbf{x}$

$$\|A\mathbf{x}\| = \|A\mathbf{x} - B\mathbf{x}\| = \|(A - B)\mathbf{x}\| \leq \|A - B\| \cdot \|\mathbf{x}\| \tag{9.4-12}$$

But

$$\|\mathbf{x}\| = \|A^{-1}A\mathbf{x}\| \leq \|A^{-1}\| \cdot \|A\mathbf{x}\| \tag{9.4-13}$$

from which (9.4-11) follows.

The meaning of this theorem is that, for normalized matrices A such that $\|A\| = 1$, the condition of A depends on how close A is to a singular matrix. Theorem 9.2 has the following interesting corollaries.

Corollary 9.2 If A is nonsingular and B is a matrix such that $1/\|A - B\| > \|A^{-1}\|$, then B is nonsingular.

Corollary 9.3 If C is a matrix such that $\|I - C\| < 1$ for some norm, then C is nonsingular. Similarly, if $\|D\| < 1$, then $I - D$ is nonsingular.

PROOF In Corollary 9.2, take $A = I$ so that $\|A^{-1}\| = 1$ and $B = C$. The second part follows from the first by setting $D = I - C$.

From Corollary 9.3, it follows {23} that diagonally dominant matrices are nonsingular, where a matrix A is said to be *diagonally dominant* if

$$|a_{ii}| > \sum_{\substack{j=1 \\ j \neq i}}^{n} |a_{ij}| \qquad i = 1, \ldots, n \tag{9.4-14}$$

Since all the principal minors of a diagonally dominant matrix are themselves diagonally dominant (why?), it follows that pivoting is not necessary in theory in the LU decomposition of such matrices {23}.

We now turn to the case where there is an error in A and see how it affects the accuracy of the solution. To this end we look at the equation

$$(A + \delta A)(\mathbf{x}_t + \delta\mathbf{x}) = \mathbf{b} \tag{9.4-15}$$

This situation is more complicated than that in the previous case in that $A + \delta A$ may be singular. Therefore we seek a condition on δA which assures us that $A + \delta A$ is nonsingular. Since

$$A + \delta A = A(I + A^{-1}\,\delta A) \tag{9.4-16}$$

it follows from Corollary 9.3 that a sufficient condition for this to hold is that $\|A^{-1}\,\delta A\| < 1$, which we shall henceforth assume. In fact, we shall assume a stronger condition, namely that $\|A^{-1}\| \cdot \|\delta A\| < 1$. With this assumption, we have that

$$\mathbf{x}_t + \boldsymbol{\delta}\mathbf{x} = (I + A^{-1}\,\delta A)^{-1}A^{-1}\mathbf{b} \tag{9.4-17}$$

so that

$$\boldsymbol{\delta}\mathbf{x} = [(I + A^{-1}\,\delta A)^{-1} - I]\mathbf{x}_t \tag{9.4-18}$$

Now, if for some matrix C such that $\|C\| < 1$ we define

$$D = (I + C)^{-1} - I \tag{9.4-19}$$

then

$$(I + C)D = -C \tag{9.4-20}$$

$$\|D\| - \|C\| \cdot \|D\| \le \|D\| - \|CD\| \le \|C\| \tag{9.4-21}$$

and

$$\|D\| \le \frac{\|C\|}{1 - \|C\|} \tag{9.4-22}$$

Hence, we have that

$$\|\boldsymbol{\delta}\mathbf{x}\| \le \frac{\|A^{-1}\,\delta A\| \cdot \|\mathbf{x}_t\|}{1 - \|A^{-1}\,\delta A\|} \le \frac{\|A^{-1}\| \cdot \|\delta A\| \cdot \|\mathbf{x}_t\|}{1 - \|A^{-1}\| \cdot \|\delta A\|} \tag{9.4-23}$$

From this it follows that

$$\frac{\|\boldsymbol{\delta}\mathbf{x}\|}{\|\mathbf{x}_t\|} \le \frac{K(A)\|\delta A\|/\|A\|}{1 - K(A)\|\delta A\|/\|A\|} \tag{9.4-24}$$

giving an upper bound for the relative error in $\mathbf{x}$ in terms of the relative error in A. Here again the condition number $K(A)$ figures prominently.

A simpler bound on the relative error of $\boldsymbol{\delta}\mathbf{x}$ can be derived with the error relative to the perturbed solution $\mathbf{x}_t + \boldsymbol{\delta}\mathbf{x}$. From (9.4-15) we have that

$$A\,\boldsymbol{\delta}\mathbf{x} + \delta A(\mathbf{x}_t + \boldsymbol{\delta}\mathbf{x}) = 0 \tag{9.4-25}$$

so that

$$\boldsymbol{\delta}\mathbf{x} = -A^{-1}\,\delta A(\mathbf{x}_t + \boldsymbol{\delta}\mathbf{x}) \tag{9.4-26}$$

and

$$\frac{\|\boldsymbol{\delta}\mathbf{x}\|}{\|\mathbf{x}_t + \boldsymbol{\delta}\mathbf{x}\|} \le \|A^{-1}\| \cdot \|\delta A\| = K(A)\frac{\|\delta A\|}{\|A\|} \tag{9.4-27}$$

Returning to the problem of roundoff error, we see that if we can show that the computed solution $\mathbf{x}_c$ satisfies exactly the equation

$$(A + E)\mathbf{x}_c = \mathbf{b} \tag{9.4-28}$$

then using (9.4-15) and (9.4-24) gives

$$\frac{\|\mathbf{x}_t - \mathbf{x}_c\|}{\|\mathbf{x}_t\|} \le \frac{K(A)\|E\|/\|A\|}{1 - K(A)\|E\|/\|A\|} \tag{9.4-29}$$

Hence, two factors determine how good the computed solution is, the size of E and the condition of A. There is nothing we can do to change $K(A)$. However, as we shall see, the size of E depends on the precision of the arithmetic used in the computation. Hence, if the bound given by (9.4-29) is too large, we can remedy the situation by going to higher precision, provided, of course, that A is known exactly. This higher precision will also help in reducing the initial roundoff error which occurs when an exact A is entered into the computer. If, on the other hand, A is uncertain to within the matrix δA, the best we can do is to reduce E to the same order of magnitude as δA. Further reduction of E will not contribute to any significant improvement in the error bound.

9.4-1 Roundoff Error Analysis

We shall now show that the computed solution $\mathbf{x}_c$ of (9.1-2) using the Doolittle and Crout algorithms satisfies exactly the equation

$$(A + E)\mathbf{x}_c = \mathbf{b} \tag{9.4-30}$$

where E is a matrix, depending on both A and $\mathbf{b}$, with elements which are usually small relative to A. We assume that A has been explicitly row-equilibrated and that the rows of A have been permuted in such a way that no interchanges are required throughout the algorithms. Note that the latter assumption results in no loss of generality. We further assume that all the arithmetic operations, namely the accumulation of inner products and their addition to single-precision numbers and subsequent division by single-precision numbers, have been carried out in double precision. As we have noted above, this does not require much extra storage and usually not too much additional computation time and hence is the recommended mode of operation. We shall first study the Doolittle algorithm since it is easier to analyze, in that it is more closely connected to Gaussian elimination. We shall then make the necessary modifications to accommodate the Crout algorithm.

Recalling (9.3-34) and (9.3-35), we have that the *computed* value u_{rj} satisfies exactly the equation [cf. (1.5-29) and Prob. 22 of Chap. 1]

$$u_{rj} = \left(a_{rj} - \sum_{k=1}^{r-1} l_{rk} u_{kj}\right)(1 + \epsilon_{rj}) \qquad j = r, \ldots, n$$

$$|\epsilon_{rj}| \le 2^{-p} \tag{9.4-31}$$

where the l_{rk} and u_{kj} are previously computed values and where we are working with p-digit binary floating-point arithmetic. Here and in the remainder of this section, we shall be neglecting errors of the order of 2^{-2p}. Therefore, the only roundoff error of order 2^{-p} is that committed when the double-precision right-hand side of (9.3-34) is rounded to single precision. Similarly the computed value of l_{ir} exactly satisfies the equation

$$l_{ir} = \frac{a_{ir} - \sum_{k=1}^{r-1} l_{ik} u_{kr}}{u_{rr}} (1 + \epsilon_{ir}) \qquad i = r+1, \ldots, n$$

$$|\epsilon_{ir}| \le 2^{-p} \tag{9.4-32}$$

Hence, in view of the fact that $l_{rr} = 1$ and that $1/(1 + \epsilon_{ij}) = 1 - \epsilon_{ij} + O(2^{-2p})$, we have from (9.4-31) and (9.4-32), respectively, that

$$a_{rj} + \epsilon_{rj} u_{rj} = \sum_{k=1}^{r} l_{rk} u_{kj} \qquad j = r, \ldots, n \tag{9.4-33}$$

and

$$a_{ir} + \epsilon_{ir} l_{ir} u_{rr} = \sum_{k=1}^{r} l_{ik} u_{kr} \qquad i = r+1, \ldots, n \tag{9.4-34}$$

Combining these two equations for $r = 2, \ldots, n$, and recalling that $u_{1j} = a_{1j}$, $j = 1, \ldots, n$, we have that

$$A + F = LU \tag{9.4-35}$$

where the elements f_{ij} of F satisfy

$$f_{ij} = \begin{cases} 0 & & i = 1 \\ \epsilon_{ij} u_{ij} & j \ge i & \\ \epsilon_{ij} l_{ij} u_{jj} & j < i & i > 1 \end{cases} \qquad j = 1, \ldots, n \tag{9.4-36}$$

Now by the assumptions of no required interchanges and explicit equilibration, $|l_{ij}| \le 1, j < i$, so that

$$|f_{ij}| \le 2^{-p} g \tag{9.4-37}$$

where $g = \max_{i,j} |u_{ij}|$ is called the *growth factor*. This growth factor can be determined a posteriori and is usually of the order of unity. A precise a priori bound for g is 2^{n-1}, as we shall see below, but in practice it is extremely rare for g to exceed 4. Furthermore, for various special types of matrices such as symmetric positive definite, diagonally dominant, and band (Sec. 9.11) matrices, much smaller theoretical bounds than 2^{n-1} exist for g {24, 68}. Thus, in most practical situations, we see that under our above assumptions, the elements of F are of the order of a single roundoff error.

The bound 2^{n-1} on g is determined as follows. The element u_{ij} can be identified with the element $a_{ij}^{(i-1)}$ in Gaussian elimination. We shall prove by induction that $|a_{kl}^{(i-1)}| \leq 2^{i-1}$, $k, l = i, \ldots, n$. From this it will follow that $|u_{ij}| \leq 2^{i-1} \leq 2^{n-1}$. By assumption $a_{kl}^{(0)} \leq 1$. Assume now that $|a_{kl}^{(i-1)}| \leq 2^{i-1}$. We shall show that $|a_{kl}^{(i)}| \leq 2^i$, $k, l = i+1, \ldots, n$. From (9.3-7)

$$a_{kl}^{(i)} = a_{kl}^{(i-1)} - m_{ki} a_{il}^{(i-1)} \tag{9.4-38}$$

Since $|m_{ki}| \leq 1$ by assumption, we have that

$$|a_{kl}^{(i)}| \leq |a_{kl}^{(i-1)}| + |a_{il}^{(i-1)}| \leq 2^i \tag{9.4-39}$$

as claimed.

In the Crout decomposition, we have, by similar arguments {25} that

$$A + \hat{F} = \hat{L}\hat{U} \tag{9.4-40}$$

where
$$\hat{f}_{ij} = \begin{cases} 0 & & j = 1 \\ \hat{\epsilon}_{ij}\hat{l}_{ij} & j \leq i & j > 1 \\ \hat{\epsilon}_{ij}\hat{l}_{ii}\hat{u}_{ij} & j > i & \end{cases} \qquad i = 1, \ldots, n \tag{9.4-41}$$

Since neither $|\hat{l}_{ij}|$ nor $|\hat{u}_{ij}|$ is bounded by unity, it would appear that the situation here is worse than in the Doolittle case. However, with exact computation, we have that

$$\hat{l}_{ii} = u_{ii} \qquad \hat{u}_{ij} = \frac{u_{ij}}{u_{ii}} \qquad \hat{l}_{ij} = l_{ij}u_{ii} \tag{9.4-42}$$

Hence, except for the first row and column, the elements of $\hat{f}_{ij}$ are essentially the same as those of f_{ij}, and we have the same bound as in (9.4-37).

We now investigate the processes of forward and back substitution. In the Doolittle case, we have from (9.3-28) that

$$y_i = \left(b_i - \sum_{j=1}^{i-1} l_{ij} y_j\right)(1 + \epsilon_i) \qquad i = 1, \ldots, n$$

$$|\epsilon_i| \leq 2^{-p} \tag{9.4-43}$$

from which we easily calculate

$$b_i = \sum_{j=1}^{i} l_{ij} y_j + y_i \epsilon_i \qquad i = 1, \ldots, n \tag{9.4-44}$$

or
$$(L + \delta L)\mathbf{y} = \mathbf{b} \qquad \delta L = \text{diag}\,(\epsilon_i) \tag{9.4-45}$$

The back substitution then gives in a similar fashion {25}

$$(U + \delta U)\mathbf{x} = \mathbf{y} \tag{9.4-46}$$

where
$$\delta U = \text{diag}\,(\eta_i u_{ii}) \qquad |\eta_i| \leq 2^{-p} \tag{9.4-47}$$

Hence $\mathbf{x}_c$ is the exact solution of

$$(L + \delta L)(U + \delta U)\mathbf{x} = \mathbf{b} \tag{9.4-48}$$

which, using (9.4-35), is equivalent to

$$(A + F + \delta L\, U + L\, \delta U + \delta L\, \delta U)\mathbf{x} = \mathbf{b} \tag{9.4-49}$$

If we set

$$E = F + \delta L\, U + L\, \delta U + \delta L\, \delta U \tag{9.4-50}$$

and ignore terms of order 2^{-2p}, we see that (9.4-30) holds with

$$e_{ij} = f_{ij} + \epsilon_i u_{ij} + l_{ij} u_{jj} \eta_j \tag{9.4-51}$$

Note that, through (9.4-43) and a similar equation from back substitution, the values of e_{ij} depend on **b** but (9.4-45) and (9.4-47) show that we have *a bound on them independent of* **b**. A similar situation holds for the modified forward and back substitution formulas connected with the Crout algorithm {25}.

If we now use (9.4-37) and the fact that L and U are triangular to bound the elements e_{ij}, we find that

$$|e_{ij}| \leq (2 + \delta_{ij} - \delta_{i1})2^{-p}g \tag{9.4-52}$$

where δ_{ij} is the Kronecker symbol; similarly for the $\hat{e}_{ij}$ in the Crout method, we have {25}

$$|\hat{e}_{ij}| \leq (2 + \delta_{ij} - \delta_{1j})2^{-p}g \tag{9.4-53}$$

Hence we have that

$$\|E\|,\ \|\hat{E}\| \leq (2n + 1)g2^{-p} \tag{9.4-54}$$

for all standard norms.

If the entire computation is carried out in single precision, we have the markedly weaker bound [Wilkinson (1967), p. 82]

$$\|E\|_\infty,\ \|\hat{E}\|_\infty \leq (1.06)(3n^2 + n^3)g2^{-p} \tag{9.4-55}$$

However, it should be noted that these are extreme upper bounds and take no account of the statistical distribution of roundoff errors.

9.5 ITERATIVE REFINEMENT

Once we have computed a solution $\mathbf{x}_c$ to (9.1-2), we can calculate the *residual* vector

$$\mathbf{r} = \mathbf{b} - A\mathbf{x}_c \tag{9.5-1}$$

Now, in some situations, we are not really interested in a solution which is close to the true solution of (9.1-2), $\mathbf{x}_t = A^{-1}\mathbf{b}$, but only in a vector $\mathbf{x}$ for which the residual $\mathbf{r}$ is small. In such a case, if $\|\mathbf{r}\|$ or, better, $\|\mathbf{r}\|/\|\mathbf{b}\|$ is less than some preassigned tolerance, we are through. Later in this section, we give an algorithm for computing a solution with a smaller residual when $\|\mathbf{r}\|$ or $\|\mathbf{r}\|/\|\mathbf{b}\|$ for $\mathbf{x}_c$ is not sufficiently small. This algorithm is also applicable to the more usual situation when we want a solution $\mathbf{x}_c$ such that the error vector

$$\mathbf{e} = \mathbf{x}_t - \mathbf{x}_c \tag{9.5-2}$$

is small relative to $\mathbf{x}_t$. In this case, the size of $\mathbf{r}$ may not give any indication as to the size of $\mathbf{e}$ even though

$$A\mathbf{e} = \mathbf{r} \tag{9.5-3}$$

as can be verified by premultiplying (9.5-2) by A. Proceeding as in Sec. 9.4 [cf. (9.4-3), (9.4-8), and (9.4-10)], we find that {26}

$$\frac{1}{K(A)}\frac{\|\mathbf{r}\|}{\|\mathbf{b}\|} \le \frac{\|\mathbf{e}\|}{\|\mathbf{x}_t\|} \le K(A)\frac{\|\mathbf{r}\|}{\|\mathbf{b}\|} \tag{9.5-4}$$

Thus, if the condition number of A is close to unity, then small relative errors in $\mathbf{r}$ and $\mathbf{e}$ go together. However, for ill-conditioned matrices, small relative errors in $\mathbf{r}$ can be associated with large relative errors in $\mathbf{e}$ and vice versa. Of one thing we can be certain. If A is normalized such that $\|A\| = 1$ and $\|\mathbf{e}\|$ is small, then $\mathbf{r}$ is small. This follows from (9.5-3), which implies that

$$\|\mathbf{r}\| \le \|A\| \cdot \|\mathbf{e}\| = \|\mathbf{e}\| \tag{9.5-5}$$

But notice that this does not say anything about the *relative* error in $\mathbf{r}$. Of this we know only that

$$\frac{\|\mathbf{r}\|}{\|\mathbf{b}\|} \le K(A)\frac{\|\mathbf{e}\|}{\|\mathbf{x}_t\|} = \|A^{-1}\| \frac{\|\mathbf{e}\|}{\|\mathbf{x}_t\|} \tag{9.5-6}$$

The perceptive reader may notice that since $\mathbf{r}$ can be calculated for each computed solution $\mathbf{x}_c$, we can solve (9.5-3) for $\mathbf{e}$, which when added to $\mathbf{x}_c$ will yield the true solution $\mathbf{x}_t$. This would indeed be true if we could solve (9.5-3) exactly. However, roundoff errors enter here also, so that we are only able to compute an approximation to $\mathbf{e}$, $\mathbf{e}_c$. Will this help us? The answer is that it depends on the condition of A and the precision of the arithmetic. But before entering into a discussion of this question, let us extend and formalize the above process, called *iterative refinement*.

The first requirement, and a crucial one, is that the residual $\mathbf{r}$ be computed using double-precision arithmetic throughout, i.e., not only in the computation of $A\mathbf{x}_c$ but also in the subtraction of $A\mathbf{x}_c$ from $\mathbf{b}$. The reason is that $\mathbf{r}$ will usually be of the order of magnitude of the roundoff error, and hence unless it is computed to higher accuracy, the relative error in the computed **r** could be almost 100 percent, so that the solution of (9.5-3) with such an **r** would bear no relation at all to the error **e**.

We now set up an iterative scheme as follows. Start with the initial computed solution $\mathbf{x}^{(1)}$. Now define the residual at stage m by

$$\mathbf{r}^{(m)} = \mathbf{b} - A\mathbf{x}^{(m)} \qquad m = 1, 2, \ldots \tag{9.5-7}$$

Define the (computed) error $\mathbf{e}^{(m)}$ as the computed solution of

$$A\mathbf{e} = \mathbf{r}^{(m)} \tag{9.5-8}$$

and the next approximate solution as

$$\mathbf{x}^{(m+1)} = \mathbf{x}^{(m)} + \mathbf{e}^{(m)} \tag{9.5-9}$$

If $\mathbf{x}^{(m+1)}$ is not satisfactory, we proceed to stage $m + 1$. We shall accept $\mathbf{x}^{(m+1)}$ as our solution if $\|\mathbf{e}^{(m)}\|/\|\mathbf{x}^{(m)}\|$ is less than a prescribed tolerance, i.e., if the computed error at stage m does not change the significant digits of interest in $\mathbf{x}^{(m)}$. This stopping criterion is not foolproof and counterexamples have been constructed {27}. However, in practice this criterion works quite well.

We must discuss the conditions for the convergence of this process, for it may happen that $\|\mathbf{e}^{(m)}\|$ will never decrease. First, however, we shall mention some practical aspects of the computation. The first is that we must save the original A and **b** in order to compute the residuals. It is important to point this out since in the algorithms we have given, we have assumed that A was overwritten, and in practice **b** is also usually destroyed. The second point is that (9.5-8) can be solved in the $O(n^2)$ operations of forward and back substitution if we have a triangular decomposition of A. Here is the point alluded to previously about the importance of being able to solve (9.1-2) for a right-hand side unknown at the time of the LU decomposition of A. Each stage in the iterative refinement process is thus seen to be quite fast relative to the time required for the solution of the original problem. Finally, we must remember to compute (9.5-7) to double-precision accuracy.

We now return to the question of when the iterative refinement process will converge. The answer is usually if $\|\mathbf{e}^{(1)}\|_\infty/\|\mathbf{x}^{(1)}\|_\infty$ is less than $\frac{1}{2}$, that is, if $\mathbf{x}^{(1)}$ has at least one correct binary digit. Otherwise, the matrix is too ill-conditioned, and if a solution is desired, the LU decomposition must be carried out with greater precision. To indicate why this is so, let us assume that the residual $\mathbf{r}^{(m)}$ is computed exactly and that $\mathbf{e}^{(m)}$ is added to $\mathbf{x}^{(m)}$ exactly. Assume now that

$$\|E^{(m)}\| \cdot \|A^{-1}\| = 2^{-p} < \tfrac{1}{2} \tag{9.5-10}$$

where $\mathbf{e}^{(m)}$ is the exact solution of

$$(A + E^{(m)})\mathbf{e}^{(m)} = \mathbf{r}^{(m)} \tag{9.5-11}$$

$E^{(m)}$ is a function of $\mathbf{r}^{(m)}$, but, as we have seen above, we have a uniform bound for all $E^{(m)}$. Now

$$\mathbf{e}^{(m)} = (A + E^{(m)})^{-1}(\mathbf{b} - A\mathbf{x}^{(m)}) = (A + E^{(m)})^{-1}A(\mathbf{x}_t - \mathbf{x}^{(m)}) \quad (9.5\text{-}12)$$

and

$$\mathbf{x}^{(m+1)} = \mathbf{x}^{(m)} + (A + E^{(m)})^{-1}A(\mathbf{x}_t - \mathbf{x}^{(m)}) \quad (9.5\text{-}13)$$

so that

$$\mathbf{x}_t - \mathbf{x}^{(m+1)} = [I - (A + E^{(m)})^{-1}A](\mathbf{x}_t - \mathbf{x}^{(m)}) \quad (9.5\text{-}14)$$

Hence, as in Sec. 9.4 and using (9.5-10), we have

$$\begin{aligned} \|\mathbf{x}_t - \mathbf{x}^{(m+1)}\| &\le \frac{\|A^{-1}\| \cdot \|E^{(m)}\| \cdot \|\mathbf{x}_t - \mathbf{x}^{(m)}\|}{1 - \|A^{-1}\| \cdot \|E^{(m)}\|} \\ &\le \frac{2^{-p}}{1 - 2^{-p}} \|\mathbf{x}_t - \mathbf{x}^{(m)}\| \end{aligned} \quad (9.5\text{-}15)$$

Since $2^{-p}/(1 - 2^{-p}) < 1$, we have that $\mathbf{x}^{(m)} \to \mathbf{x}_t$, so that the process converges. We do not know the value of $\|E^{(m)}\| \cdot \|A^{-1}\|$, but we do know, as in (9.4-27), that

$$\frac{\|\mathbf{e}^{(1)}\|}{\|\mathbf{x}^{(1)}\|} \le \|E^{(1)}\| \cdot \|A^{-1}\| \quad (9.5\text{-}16)$$

Hence, we use the criterion that $\|\mathbf{e}^{(1)}\|/\|\mathbf{x}^{(1)}\| < \frac{1}{2}$ to start the iteration. It still may not converge, but chances are slight. Nevertheless, we must make provision for such a likelihood in our algorithm by stopping after a maximum number of iterations. Since each iteration should give at least one additional significant digit, we stop after p iterations.

As can be seen, $\|E^{(1)}\| \cdot \|A^{-1}\|$ can be quite large even though $\|\mathbf{e}^{(1)}\|/\|\mathbf{x}^{(1)}\| < \frac{1}{2}$, and that is why counterexamples exist. However, a program based on the above considerations and those given in the previous sections will almost invariably give the correct solution to (9.1-2) when the condition of the matrix warrants it and will indicate the need for higher precision otherwise.

9.6 MATRIX ITERATIVE METHODS

The methods we shall consider in the next two sections are analogous to the methods of functional iteration of the last chapter. Starting with an initial vector $\mathbf{x}_1$, we shall generate a sequence of vectors

$$\mathbf{x}_{i+1} = F_i[\mathbf{x}_i, \quad \mathbf{x}_{i-1}, \quad \ldots, \quad \mathbf{x}_{i-k}] \quad (9.6\text{-}1)$$

where the i subscript on F denotes that the iteration function itself may change from one iteration to the next. Analogously to Chap. 8, if the iteration function is not dependent on i, we say that the iteration is *stationary*.

For most matrices A iterative methods require more computation to achieve a desired degree of convergence than the direct methods we have discussed. Why consider them then? We noted the basic answer to this question in Sec. 9.2. This is that for sparse matrices—which arise commonly in solving partial differential equations by difference techniques—iterative techniques may indeed compare favorably with direct methods in terms of total amount of computation. Furthermore, because they are economical in their use of the computer memory, iterative methods are particularly advantageous for the very large matrices that often occur in the numerical solution of partial differential equations.†

In some situations, the matrix A need not be stored at all. Instead, each element a_{ij} is computed every time it is needed in the calculation. Usually the computation is trivial, such as setting a_{ij} to some constant, but in more involved problems, the computation may be quite time-consuming. However, this may still be the most efficient way to solve the problem since it may not be possible to store even the sparse matrix A in the high-speed memory, and so even a time-consuming computation of a_{ij} may still be faster than the retrieval of a_{ij} from auxiliary storage.

We shall restrict ourselves to linear iterative processes, i.e., iterations in which F_i is a linear function of $\mathbf{x}_i, \mathbf{x}_{i-1}, \ldots, \mathbf{x}_{i-k}$. The primary reason for this is not that we are dealing with linear equations but rather that nonlinear iterations are much more difficult to analyze in general. Moreover, they tend to be computationally inefficient because of the number of matrix-vector products which must be calculated. As in Chap. 8, our interest will be focused on one-point iteration functions. A *linear one-point matrix iteration* has the form

$$\mathbf{x}_{i+1} = B_i \mathbf{x}_i + \mathbf{c}_i \tag{9.6-2}$$

where the matrix B_i and vector $\mathbf{c}_i$ are independent of i in a stationary iteration.

To motivate considering iterations of the type (9.6-2) to solve

$$A\mathbf{x} = \mathbf{b} \tag{9.6-3}$$

let us write (9.6-3) in the form

$$(I + A)\mathbf{x} = \mathbf{x} + \mathbf{b} \tag{9.6-4}$$

or

$$\mathbf{x} = (I + A)\mathbf{x} - \mathbf{b} \tag{9.6-5}$$

Equation (9.6-5) suggests the iteration

$$\mathbf{x}_{i+1} = (I + A)\mathbf{x}_i - \mathbf{b} \tag{9.6-6}$$

Equation (9.6-2) is then just a generalization of (9.6-6).

† Computer programs to handle systems of up to order 108,000 have been written!

As in Chap. 8, we require that the true solution $\mathbf{x}_t$ of (9.6-3) be a fixed point of (9.6-2). Therefore, analogously to (9.6-5) we must have

$$\mathbf{x}_t = B_i \mathbf{x}_t + \mathbf{c}_i \tag{9.6-7}$$

for all i. Since $\mathbf{x}_t = A^{-1}\mathbf{b}$, we have

$$A^{-1}\mathbf{b} = B_i A^{-1}\mathbf{b} + \mathbf{c}_i \tag{9.6-8}$$

or

$$\mathbf{c}_i = (I - B_i)A^{-1}\mathbf{b} = C_i \mathbf{b} \tag{9.6-9}$$

We assume that B_i and C_i are independent of $\mathbf{b}$. Therefore, we must have

$$(I - B_i)A^{-1} = C_i \tag{9.6-10}$$

or

$$B_i + C_i A = I \tag{9.6-11}$$

This is called the *condition of consistency* for B_i and C_i.

Because of (9.6-9) we can rewrite (9.6-2) in its more usual form

$$\mathbf{x}_{i+1} = B_i \mathbf{x}_i + C_i \mathbf{b} \tag{9.6-12}$$

In order to consider the convergence of (9.6-12), we define

$$\boldsymbol{\epsilon}_i = \mathbf{x}_i - \mathbf{x}_t \tag{9.6-13}$$

Using (9.6-7), (9.6-9), and (9.6-12), we have

$$\boldsymbol{\epsilon}_{i+1} = B_i \mathbf{x}_i + C_i \mathbf{b} - \mathbf{x}_t = B_i \mathbf{x}_i - B_i \mathbf{x}_t = B_i \boldsymbol{\epsilon}_i \tag{9.6-14}$$

If $\mathbf{x}_1$ is the initial approximation to the solution of (9.6-3), then

$$\boldsymbol{\epsilon}_{i+1} = K_i \boldsymbol{\epsilon}_1 \tag{9.6-15}$$

where

$$K_i = B_i B_{i-1} \cdots B_1 \tag{9.6-16}$$

Therefore, a necessary and sufficient condition for the convergence of the sequence $\{\mathbf{x}_i\}$ to $\mathbf{x}_t$ *for arbitrary* $\mathbf{x}_1$ is that

$$\lim_{i \to \infty} K_i \mathbf{y} = 0 \qquad \text{for all } \mathbf{y} \tag{9.6-17}$$

Another necessary and sufficient condition for convergence is that

$$\lim_{i \to \infty} \|\boldsymbol{\epsilon}_i\| = 0 \tag{9.6-18}$$

and since

$$\max_{\epsilon_1} \frac{\|\boldsymbol{\epsilon}_{i+1}\|}{\|\boldsymbol{\epsilon}_1\|} = \max_{\epsilon_1} \frac{\|K_i \boldsymbol{\epsilon}_1\|}{\|\boldsymbol{\epsilon}_1\|} = \|K_i\| \tag{9.6-19}$$

$$\lim_{i \to \infty} \|K_i\| = 0 \tag{9.6-20}$$

is also such a condition. For stationary processes still another necessary and sufficient condition for convergence is

$$\lim_{i \to \infty} \rho(K_i) = 0 \tag{9.6-21}$$

9.7 STATIONARY ITERATIVE PROCESSES AND RELATED MATTERS

Because they are easier to analyze and computationally more desirable, most matrix iterative methods are stationary. Our main concern in this section will be with stationary iterations. In Sec. 9.7-4, however, we shall consider briefly the use of nonstationary iterations to accelerate the convergence of stationary iterative processes.

When an iteration is stationary, then $B_i = B$, $C_i = C$, and from (9.6-16)

$$K_i = B^i \tag{9.7-1}$$

The eigenvalues of B^i are the ith powers of the eigenvalues of B. Therefore, the condition (9.6-21) is equivalent to requiring that all eigenvalues of B lie within the unit circle (why?).

If the iteration converges, the rate at which it does so is of interest. From (9.6-15)

$$\|\epsilon_{i+1}\| \leq \|K_i\| \cdot \|\epsilon_i\| \leq \|B\|^i \|\epsilon_i\| \tag{9.7-2}$$

Therefore, the smaller some norm of B is, the more rapidly we expect the iteration to converge. If B is symmetric, the spectral norm and spectral radius are equal and the latter can then be used as a measure of the rate of convergence. If, as is more usual, B is not symmetric, then although we know that $\rho(B) \leq \|B\|$, it is not obvious that the spectral radius can be used to measure the rate of convergence. However, it can be shown, although it is beyond our scope here [see Varga (1962), pp. 61–68], that even in the nonsymmetric case the spectral radius does measure the rate of convergence. Although it is usually impractical actually to calculate the spectral radius, the effect of the spectral radius on the rate of convergence has important ramifications; see, for example, Sec. 9.7-4.

9.7-1 The Jacobi Iteration

We write the matrix A in the form

$$A = D + L + U \tag{9.7-3}$$

where D is a diagonal matrix and L and U are, respectively, lower and upper triangular matrices with zeros on the diagonal. Then (9.6-3) can be written

$$D\mathbf{x} = -(L + U)\mathbf{x} + \mathbf{b} \tag{9.7-4}$$

which suggests the iteration

$$\mathbf{x}_{i+1} = -D^{-1}(L + U)\mathbf{x}_i + D^{-1}\mathbf{b} \tag{9.7-5}$$

We assume here naturally that the diagonal of A contains no zero elements. If it does have zero terms but A is nonsingular, then by permuting rows and columns it is always possible to get a nonsingular matrix D. It is in fact desirable to have the diagonal elements as large as possible in relation to the off-diagonal elements. For suppose we set $\mathbf{x}_1 = \mathbf{0}$. Then $\mathbf{x}_2 = D^{-1}\mathbf{b}$, and if the diagonal terms are dominant, this is already a good approximation to the solution.

The iteration (9.7-5) is known by many names, but the most usual are the *Jacobi iteration* or the method of simultaneous displacements. The latter name follows from the fact that every element of the solution vector is changed before any of the new elements are used in the iteration (cf. Sec. 9.7-2). For this method the matrix B is

$$B = -D^{-1}(L + U) \tag{9.7-6}$$

It may be quite difficult to determine whether B is such that the iteration converges. Of course, if $\|D^{-1}(L + U)\| < 1$ for some easily computable matrix norm, then we are certain that the iteration does converge.

The Jacobi method as we have presented it here is seldom used in practice for the solution of (9.6-3). This is largely because the Gauss-Seidel method to be considered below almost always converges when the Jacobi method does, may converge when the Jacobi method does not, and generally converges faster than the Jacobi method. Furthermore, the implementation of the Gauss-Seidel method on a computer is more efficient than that of the Jacobi method.

9.7-2 The Gauss-Seidel Method

The difference between the Jacobi and Gauss-Seidel methods is that in the latter, as each component of $\mathbf{x}_{i+1}$ is computed, we use it immediately in the iteration. For this reason the Gauss-Seidel method is sometimes called the method of successive displacements.

Denote the jth component of $\mathbf{x}_i$ by $x_i^{(j)}$. Consider the system of equations written in the form (9.1-1). To compute $x_{i+1}^{(1)}$ we use the first equation in the form

$$x_{i+1}^{(1)} = -\frac{1}{a_{11}}\left(\sum_{j=2}^{n} a_{1j}x_i^{(j)} - b_1\right) \tag{9.7-7}$$

(We assume here as before that D is nonsingular.) To get $x_{i+1}^{(2)}$ we use the second equation, but with $x_i^{(1)}$ replaced by the result of (9.7-7)

$$x_{i+1}^{(2)} = -\frac{1}{a_{22}}\left(a_{21}x_{i+1}^{(1)} + \sum_{j=3}^{n} a_{2j}x_i^{(j)} - b_2\right) \tag{9.7-8}$$

In general

$$x_{i+1}^{(r)} = -\frac{1}{a_{rr}}\left(\sum_{j=1}^{r-1} a_{rj}x_{i+1}^{(j)} + \sum_{j=r+1}^{n} a_{rj}x_i^{(j)} - b_r\right) \qquad r = 1, \ldots, n \tag{9.7-9}$$

With A written in the form (9.7-3), this method becomes {31}

$$\mathbf{x}_{i+1} = -D^{-1}(L\mathbf{x}_{i+1} + U\mathbf{x}_i) + D^{-1}\mathbf{b} \tag{9.7-10}$$

which can be solved for $\mathbf{x}_{i+1}$ to give

$$\mathbf{x}_{i+1} = -(D + L)^{-1}U\mathbf{x}_i + (D + L)^{-1}\mathbf{b} \tag{9.7-11}$$

Since $B = -(D + L)^{-1}U$, a sufficient condition for convergence is then that $\|(D + L)^{-1}U\| < 1$, but this is not very useful in practice (why?). In the important case, however, in which A is positive definite, we can prove that the Gauss-Seidel method converges.

Theorem 9.3 If the matrix A is positive definite, the Gauss-Seidel iteration (9.7-11) converges independently of the initial vector.

PROOF We write A as

$$A = L + D + L^T \tag{9.7-12}$$

since A is symmetric. The matrix B is then

$$B = -(D + L)^{-1}L^T \tag{9.7-13}$$

Let $-\lambda$ and $\mathbf{v}$ be, respectively, an eigenvalue and eigenvector of B. Then

$$(D + L)^{-1}L^T\mathbf{v} = \lambda\mathbf{v} \tag{9.7-14}$$

or

$$L^T\mathbf{v} = \lambda(D + L)\mathbf{v} \tag{9.7-15}$$

Even though A is positive definite, the eigenvalues of B may still be complex. We have†

$$\mathbf{v}^*L^T\mathbf{v} = \mathbf{v}^*\lambda(D + L)\mathbf{v} \tag{9.7-16}$$

Adding $\mathbf{v}^*(D + L)\mathbf{v}$ to both sides of (9.7-16), we get

$$\mathbf{v}^*A\mathbf{v} = (1 + \lambda)\mathbf{v}^*(D + L)\mathbf{v} \tag{9.7-17}$$

Since A is real and symmetric, the conjugate transpose of the left-hand side of (9.7-17) leaves this quantity unchanged. Therefore,

$$(1 + \bar{\lambda})\mathbf{v}^*(D + L)^T\mathbf{v} = (1 + \lambda)\mathbf{v}^*(D + L)\mathbf{v} = (1 + \lambda)(\mathbf{v}^*D\mathbf{v} + \mathbf{v}^*L\mathbf{v}) = (1 + \lambda)[\mathbf{v}^*D\mathbf{v} + \bar{\lambda}\mathbf{v}^*(D + L)^T\mathbf{v}] \tag{9.7-18}$$

the last line following from use of the conjugate transpose of (9.7-15). Rearranging the terms, we have

$$(1 - |\lambda|^2)\mathbf{v}^*(D + L)^T\mathbf{v} = (1 + \lambda)\mathbf{v}^*D\mathbf{v} \tag{9.7-19}$$

† $\mathbf{v}^*$ denotes the conjugate transpose of $\mathbf{v}$.

Multiplying both sides of (9.7-19) by $1 + \bar{\lambda}$ and then using the conjugate transpose of (9.7-17), we get

$$(1 - |\lambda|^2)\mathbf{v}^*A\mathbf{v} = |1 + \lambda|^2\mathbf{v}^*D\mathbf{v} \tag{9.7-20}$$

Since A is positive definite so is D; moreover, no eigenvalue $-\lambda$ of B can equal 1 (why?). Therefore, we must have $1 - |\lambda|^2 > 0$, which means that the eigenvalues of B lie within the unit circle. This completes the proof. A partial converse of this theorem will be found in a problem {32}.

We have noted that much of the application of iterative methods such as the Jacobi and Gauss-Seidel methods is in the numerical solution of partial differential equations. The coefficient matrices that arise in the numerical solution of partial differential equations are often such that the iteration matrix B in (9.7-1) is nonnegative; i.e., all the elements of B are nonnegative. The Perron-Froebenius theory of nonnegative matrices, a subject which is beyond the scope of this book [see Varga (1962), pp. 26–33], provides the basis for analysis of iterative methods when B is nonnegative. We content ourselves here with stating the Stein-Rosenberg theorem, which follows from the Perron-Frobenius theory. Suppose that by dividing each equation by its diagonal element we make D in (9.7-3) the identity matrix. Suppose further that $B_J = -(L + U)$ in (9.7-6) is nonnegative. Then it follows {33} that $B_G = -(I + L)^{-1}U$ in (9.7-11) is also nonnegative. Let ρ_J and ρ_G be the spectral radii, respectively, of B_J and B_G. Then this theorem states that one of the following conditions holds:

Condition 1: $\rho_J = \rho_G = 0$

Condition 2: $\rho_J = \rho_G = 1$

Condition 3: $0 < \rho_G < \rho_J < 1$

Condition 4: $1 < \rho_J < \rho_G$

Thus the Jacobi and Gauss-Seidel iterations both converge or both diverge, and when they both converge, the Gauss-Seidel method converges faster (except for the trivial case, condition 1). This theorem is the basis for our previous statement that the Jacobi iteration is seldom used in preference to the Gauss-Seidel iteration.

Equation (9.6-14) implies that matrix iterative methods converge linearly. They are therefore candidates for the δ^2 process (see Sec. 8.7-1). Analogous to Eq. (8.7-4) we can write

$$x_i^{(r)} \approx x_{i+2}^{(r)} - \frac{(\Delta x_{i+1}^{(r)})^2}{\Delta^2 x_i^{(r)}} \qquad r = 1, \ldots, n \tag{9.7-21}$$

where the superscripts denote components of the relevant vectors.

Example 9.4 For the system

$$\begin{aligned} 4x^{(1)} - x^{(2)} \qquad\quad &= 2 \\ -x^{(1)} + 4x^{(2)} - x^{(3)} &= 6 \\ - x^{(2)} + 4x^{(3)} &= 2 \end{aligned}$$

do four iterations of the Jacobi and Gauss-Seidel methods and then use (9.7-21) to get an improved solution. (Although of too low an order to arise in a practical problem the form of the coefficient matrix is typical of those which arise in the numerical solution of partial differential equations.) Use $\mathbf{x}_1 = \mathbf{0}$ for both methods.

For the Jacobi iteration

$$B = -D^{-1}(L+U)$$

$$= -\begin{bmatrix} \frac{1}{4} & 0 & 0 \\ 0 & \frac{1}{4} & 0 \\ 0 & 0 & \frac{1}{4} \end{bmatrix} \begin{bmatrix} 0 & -1 & 0 \\ -1 & 0 & -1 \\ 0 & -1 & 0 \end{bmatrix} = \begin{bmatrix} 0 & \frac{1}{4} & 0 \\ \frac{1}{4} & 0 & \frac{1}{4} \\ 0 & \frac{1}{4} & 0 \end{bmatrix}$$

From (9.7-5) we calculate

$$\mathbf{x}_2^T = (\tfrac{1}{2}, \tfrac{3}{2}, \tfrac{1}{2})$$

$$\mathbf{x}_3^T = (\tfrac{7}{8}, \tfrac{7}{4}, \tfrac{7}{8})$$

$$\mathbf{x}_4^T = (\tfrac{15}{16}, \tfrac{31}{16}, \tfrac{15}{16})$$

$$\mathbf{x}_5^T = (\tfrac{63}{64}, \tfrac{63}{32}, \tfrac{63}{64})$$

For the Gauss-Seidel iteration, using (9.7-9) we calculate

$$\mathbf{x}_2^T = (\tfrac{1}{2}, \tfrac{13}{8}, \tfrac{29}{32})$$

$$\mathbf{x}_3^T = (\tfrac{29}{32}, \tfrac{125}{64}, \tfrac{253}{256})$$

$$\mathbf{x}_4^T = (\tfrac{253}{256}, \tfrac{1021}{512}, \tfrac{2045}{2048})$$

$$\mathbf{x}_5^T = ({}^{2045}/_{2048}, {}^{8189}/_{4096}, {}^{16,381}/_{16,384})$$

and, since the true solution is $\mathbf{x}_t^T = (1, 2, 1)$, it is clear that both iterations are converging and that the Gauss-Seidel is converging more rapidly. Calculation of the spectral radii of the B matrices for the two iterations would indicate just how fast each is converging {34}.

If we apply (9.7-21) with $i = 2$ to the Gauss-Seidel results, we have

$$\Delta x_3^{(1)} = \tfrac{21}{256} \qquad \Delta x_3^{(2)} = \tfrac{21}{512} \qquad \Delta x_3^{(3)} = \tfrac{21}{2048}$$

$$\Delta^2 x_2^{(1)} = -\tfrac{85}{256} \qquad \Delta^2 x_2^{(2)} = -\tfrac{147}{512} \qquad \Delta^2 x_2^{(3)} = -\tfrac{147}{2048}$$

This gives as the new approximation

$$(\tfrac{335}{332}, 2, 1)$$

which is an improved result in all components and a perfect result in two of them. However, it is important that the δ^2 process not be used too early in the computation, in which case it may give poorer results than the last iteration {34}.

9.7-3 Roundoff Error in Iterative Methods

The roundoff error incurred in an iterative method is only that which is incurred in the last iteration. This is because we always use the original coefficient matrix and, so to speak, start from scratch at each iteration using the last iterate as the initial vector. It is perhaps natural to expect then that roundoff error is a much less severe problem in iterative methods than it is in direct methods. However, not only can roundoff be a serious source of error in iterative methods, but also it may be very nearly as serious as in direct methods. In this section we shall consider roundoff error in the Gauss-Seidel iteration. In analogy with the error analysis of Sec. 9.4, we ask: What are the perturbations of the matrix A and vector $\mathbf{b}$ such that, starting from $\mathbf{x}_i$ as computed, the result of using these perturbations in (9.7-11) would have been the computed value $\mathbf{x}_{i+1}$ if no roundoff error were incurred?

Before studying the effect of roundoff errors, let us rewrite (9.7-9) to take into account only nonzero elements a_{rj}. We then have

$$x_{i+1}^{(r)} = -\frac{1}{a_{rr}}\left(\sum_{j=k_{1r}}^{k_{pr}} a_{rj}x_{i+1}^{(j)} + \sum_{j=k_{p+1,r}}^{k_{qr}} a_{rj}x_i^{(j)} - b_r\right) \qquad r = 1, \ldots n \tag{9.7-22}$$

where

$$1 \le k_{1r} < \cdots < k_{pr} \le r-1 < r+1 \le k_{p+1,r} < \cdots < k_{qr} \le n \tag{9.7-23}$$

and $a_{rj} \ne 0$ if and only if $j = k_{lr}$, $l = 1, \ldots, q$.

If we now calculate $x_{i+1}^{(r)}$ using p-digit binary floating-point operations, we find that the $x_{i+1}^{(r)}$ satisfy exactly the equation

$$\begin{aligned} x_{i+1}^{(r)} = -\frac{1}{a_{rr}}\Bigg\{\Bigg[&\sum_{j=k_{1r}}^{k_{pr}} a_{rj}x_{i+1}^{(j)}(1+\epsilon_j)\prod_{s=m(j)}^{p}(1+\eta_s) \\ &+ \sum_{j=k_{p+1,r}}^{k_{qr}} a_{rj}x_i^{(j)}(1+\epsilon_j)\prod_{s=m'(j)}^{q-p}(1+\eta_s)\Bigg](1+\delta_1) - b_r\Bigg\}(1+\delta_2)(1+\delta_3) \end{aligned} \tag{9.7-24}$$

where $m(j) = 2 \quad j = k_{1r} \qquad m'(j) = 2 \quad j = k_{p+1,r}$

$m(j) = 1 \quad j = k_{lr};\ l > 1 \qquad m'(j) = l - p \quad j = k_{lr};\ l > p+1$

and ϵ_j = relative error in the multiplication of a_{rj} by $x_i^{(j)}$ or $x_{i+1}^{(j)}$

$\Pi(1+\eta_s)$ = relative error in the accumulation of the sum with η_s corresponding to ϵ_s in (1.5-26)

δ_1 = relative error in the addition of the two sums

δ_2 = relative error in the addition of b_r

δ_3 = relative error in the division by a_{rr}

Multiplying (9.7-24) by $a_{rr}/(1+\delta_2)(1+\delta_3)$ and expressing the result in vector-matrix notation, we have

$$\hat{D}\mathbf{x}_{i+1} = -\hat{L}\mathbf{x}_{i+1} - \hat{U}\mathbf{x}_i - \mathbf{b} \tag{9.7-25}$$

where $\hat{l}_{rj} = \begin{cases} a_{rj}(1+\epsilon_j)\prod(1+\eta_s)(1+\delta_1) & r > j;\ j = k_{jr} \\ 0 & \text{otherwise} \end{cases}$

$$\hat{u}_{rj} = \begin{cases} a_{rj}(1+\epsilon_j)\prod(1+\eta_s)(1+\delta_1) & r<j;\, j=k_{jr} \\ 0 & \text{otherwise} \end{cases} \tag{9.7-26}$$

$$\hat{d}_{rr} = \frac{a_{rr}}{(1+\delta_2)(1+\delta_3)}$$

and $|\epsilon_j|,\ |\eta_s|,\ |\delta_l| \le 2^{-p}$.

If, however, the computation of (9.7-9) is carried out in double-precision arithmetic operating on the single-precision numbers a_{rj}, $x_i^{(j)}$, $x_{i+1}^{(j)}$, b_r with only a final roundoff to obtain a single precision value for $x_{i+1}^{(r)}$, then (9.7-9) takes the form

$$x_{i+1}^{(r)} = -\frac{1}{a_{rr}}\left(\sum_{j=1}^{r-1} a_{rj}x_{i+1}^{(j)} + \sum_{j=r+1}^{n} a_{rj}x_i^{(j)} - b_r\right)(1+\delta_3) \tag{9.7-27}$$

in which case we have

$$\tilde{D}\mathbf{x}_{i+1} = -L\mathbf{x}_{i+1} - U\mathbf{x}_i - \mathbf{b} \tag{9.7-28}$$

with $\tilde{d}_{rr} = a_{rr}/(1+\delta_3)$.

Thus, when the computations are carried out in double precision, the computed vector $\mathbf{x}_{i+1}$ is the *exact* result of operating on $\mathbf{x}_i$ with a matrix $\tilde{A}$ which differs from A by at most one unit on the diagonal. In the single-precision case, $\hat{A}$ is not that close to A. However, for sparse matrices, $\hat{A}$ will differ from A by at most of the order of Q units in the last place, where Q is the maximum number of nonvanishing elements in a row of A. Note that the $\mathbf{b}$ which appears in all the equations above is the true right-hand side.

In general, this amount of roundoff error will not be serious, especially since in an iterative process the truncation error incurred by terminating the iteration will usually be of a greater order of magnitude. However, it can be a serious problem when the system is ill-conditioned. In {35} a simple example illustrating this point is considered.

9.7-4 Acceleration of Stationary Iterative Processes

Since the rate of convergence of a stationary iterative process depends on the spectral radius of B, any modification of the matrix B that will reduce the spectral radius will increase the rate of convergence. We consider briefly here a method of accelerating the convergence of an iterative process called the method of *successive overrelaxation.* First we rearrange the system (9.6-3) so that each diagonal element of A is 1. This can be done by arranging the system so that no diagonal term is zero and then dividing each equation by its diagonal element. For positive definite matrices, we preserve symmetry by pre- and postmultiplying A by $D^{-1/2}$. When we have done this, Eq. (9.7-9) for the Gauss-Seidel iteration becomes

$$\begin{aligned} x_{i+1}^{(r)} &= -\sum_{j=1}^{r-1} a_{rj} x_{i+1}^{(j)} - \sum_{j=r+1}^{n} a_{rj} x_i^{(j)} + b_r \\ &= x_i^{(r)} - \sum_{j=1}^{r-1} a_{rj} x_{i+1}^{(j)} - \sum_{j=r}^{n} a_{rj} x_i^{(j)} + b_r \end{aligned} \tag{9.7-29}$$

since $a_{rr} = 1$. Consider replacing (9.7-29) by

$$x_{i+1}^{(r)} = x_i^{(r)} - \omega\left(\sum_{j=1}^{r-1} a_{rj} x_{i+1}^{(j)} + \sum_{j=r}^{n} a_{rj} x_i^{(j)} - b_r \right) \tag{9.7-30}$$

where ω is the overrelaxation factor. Using (9.7-30), we can write this iteration in matrix form as {36}

$$\mathbf{x}_{i+1} = B_\omega \mathbf{x}_i + C_\omega \mathbf{b} \tag{9.7-31}$$

where $\quad B_\omega = (I + \omega L)^{-1}[(1 - \omega)I - \omega U] \qquad C_\omega = \omega(I + \omega L)^{-1} \quad$ (9.7-32)

with L and U as in (9.7-3). The crux of the matter now, of course, is to choose ω so as to minimize $\rho(B_\omega)$. The reader will probably not be surprised to find that this is a problem of some difficulty. It is in fact the subject of a large literature. We content ourselves here with stating that for a large class of matrices that arise in the numerical solution of partial differential equations, there is a simple relationship between the optimum value of ω and $\rho(L + U)$. This does not get rid of the problem, but finding the eigenvalues or at least estimates of the eigenvalues of $L + U$ may be reasonably simple. Moreover, the insight gained from the relationship between ω and $\rho(L + U)$ is useful, for example, in indicating that it is preferable to overestimate ω than underestimate it.

Successive overrelaxation is only one of a large class of methods that have been developed to accelerate the convergence of iterative processes. Some of these accelerations lead to nonstationary iterations; for example,

$$\mathbf{x}_{i+1} = [(1 + \omega_i)B - \omega_i I]\mathbf{x}_i + (1 + \omega_i)C\mathbf{b} \tag{9.7-33}$$

where the relaxation factor ω_i changes from step to step. The proper use of such acceleration procedures lies at the heart of much of the utilization of high-speed computing equipment for the solution of partial differential equations.

9.8 MATRIX INVERSION

The solution of simultaneous linear equations can certainly be accomplished by inverting the matrix A. In fact, if the system (9.1-1) is to be solved for many different right-hand sides, one may first invert A and then calculate

$A^{-1}\mathbf{b}$ for each right-hand side. However, it is more efficient to compute the LU decomposition of A and then solve for each right-hand side using forward and back substitution. The only time it is necessary to compute the inverse of a matrix is when the elements of A^{-1} are explicitly needed, as in certain statistical calculations or in the matrix-updating procedures for solving systems of nonlinear equations (Sec. 8.8). One very efficient way to invert A in general is a simple extension of one of the algorithms of Sec. 9.3-3.

We suppose that the matrix A which we wish to invert has been decomposed into a product LU by the techniques of Sec. 9.3-3, where L has the form (9.3-22) and U the form

$$U = \begin{bmatrix} u_{11} & \cdots\cdots & u_{1n} \\ & u_{22} & \\ \bigcirc & \ddots & \vdots \\ & & u_{nn} \end{bmatrix} \tag{9.8-1}$$

Then since

$$A^{-1} = U^{-1}L^{-1} \tag{9.8-2}$$

our problem is to invert the triangular matrices L and U. We leave to a problem {40} the derivation of the result that the inverse of a lower (upper) triangular matrix is lower (upper) triangular.

Let $L^{-1} = \{r_{ij}\}$ with $r_{ij} = 0$, $i < j$, and let L_i and L_j^{-1} be, respectively, the ith column of L and the jth row of L^{-1}. Then for $1 \le k \le n$

$$\begin{aligned} L_k^{-1}L_k &= 1 = r_{kk} \\ L_k^{-1}L_j &= 0 = \sum_{i=j}^{k} r_{ki}l_{ij} \qquad j = k-1, \ldots, 1 \end{aligned} \tag{9.8-3}$$

from which we can calculate r_{ki}, $i = k, k-1, \ldots, 1$, for any k, thereby obtaining all the elements of L^{-1}.

Similarly we can calculate U^{-1}, the only difference being the fact that the diagonal elements of U are not 1s {41}. The calculation of L^{-1} and U^{-1} followed by the matrix multiplication $U^{-1}L^{-1}$ requires $\frac{2}{3}n^3 + O(n^2)$ multiplications and divisions {41}, so that $n^3 + O(n^2)$ are required for the complete inversion. Some variants of the procedure of this section and other methods of matrix inversion are considered in the problems {42 to 47}.

9.9 OVERDETERMINED SYSTEMS OF LINEAR EQUATIONS

In this section we consider an alternative approach to the least-squares problem considered in Chap. 6, an approach which has application not only to the case of fitting a function to experimental data but also to such problems as, for example, the approximation of a function given on a finite set of points. The problem stated in Eq. (6.2-1), namely approximating observed values $\bar{f}_i$ by a linear combination of a set of functions, can be reformulated as an attempt to solve the system of linear equations

$$A\mathbf{y} = \mathbf{b} \tag{9.9-1}$$

where A is a matrix of n rows and m columns. [In the notation of Chap. 6, $b_i = \bar{f}_i$ and $a_{ij} = \phi_j(x_i)$, and we have replaced m by $m - 1$.] In the usual case, $n > m$, so that the system (9.9-1) is *overdetermined* and cannot be solved. Instead we attempt to minimize the norm $\|\mathbf{r}\|$ of the residual vector

$$\mathbf{r} = \mathbf{b} - A\mathbf{y} \tag{9.9-2}$$

where the norms of interest are the L_1, L_2, and L_∞ norms. We shall first discuss the case of the L_2 norm and return to the other two later.

When we try to minimize the L_2, or least-squares, norm, we obtain the normal equations of Chap. 6, which, in our formulation, have the form [cf. Eq. (6.2-6)]

$$A^T A\mathbf{y} = A^T\mathbf{b} \tag{9.9-3}$$

It can be shown that the matrix $A^T A$ is positive definite and hence nonsingular if the columns of A are linearly independent {51}. Nevertheless, as discussed in Sec. 6.3-1, the matrix $A^T A$ may be ill-conditioned and for moderate values of m, the solution $\mathbf{y}$ will not be accurate. Hence, an alternative method must be used. One such method for the case where the coefficients $a_{ij} = x_i^{j-1}$ was given in Sec. 6.4, namely the use of orthogonal polynomials. Here we present another method applicable to the general case (9.9-1). Before we can give the details of this method, however, we need some preliminary definitions and results on special kinds of matrices.

Definition 9.1 A matrix Q is said to be *orthogonal* if $Q^T Q = I$.

Since this says that $Q^T = Q^{-1}$, it follows that $QQ^T = I$. Orthogonal matrices have the property that for any vector $\mathbf{x}$ the Euclidean length of $Q\mathbf{x}$, $\|Q\mathbf{x}\|_2 = \|\mathbf{x}\|_2$. This follows immediately from the definition since

$$\|Q\mathbf{x}\|_2^2 = (Q\mathbf{x})^T Q\mathbf{x} = \mathbf{x}^T Q^T Q\mathbf{x} = \mathbf{x}^T\mathbf{x} = \|\mathbf{x}\|_2^2 \tag{9.9-4}$$

Definition 9.2 An *elementary reflector* P is a matrix of the form

$$P = I - 2\mathbf{v}\mathbf{v}^T \qquad \text{where } \mathbf{v}^T\mathbf{v} = 1 \tag{9.9-5}$$

These matrices are also known as elementary Hermitian matrices or Householder transformations (cf. Sec. 10.3-3). They reflect the space in the hyperplane through the origin orthogonal to $\mathbf{v}$. The matrices P are symmetric and orthogonal {52} and have the important property that, given any two vectors of equal length, $\mathbf{x}$ and $\mathbf{y}$, we can find a matrix P such that $P\mathbf{x} = \mathbf{y}$. To this end we take $\mathbf{v} = (\mathbf{x} - \mathbf{y})/\|\mathbf{x} - \mathbf{y}\|_2$. We prove this as follows:

$$\left[I - \frac{2(\mathbf{x} - \mathbf{y})(\mathbf{x} - \mathbf{y})^T}{\|\mathbf{x} - \mathbf{y}\|_2^2}\right]\mathbf{x} = \mathbf{x} - \frac{2(\mathbf{x}^T\mathbf{x} - \mathbf{y}^T\mathbf{x})}{\mathbf{x}^T\mathbf{x} - \mathbf{y}^T\mathbf{x} - \mathbf{x}^T\mathbf{y} + \mathbf{y}^T\mathbf{y}}(\mathbf{x} - \mathbf{y}) \quad (9.9\text{-}6)$$

Since $\mathbf{x}^T\mathbf{x} = \mathbf{y}^T\mathbf{y}$ by hypothesis and $\mathbf{y}^T\mathbf{x} = \mathbf{x}^T\mathbf{y}$ since this is a scalar quantity which is invariant under transposition, it follows that the denominator in (9.9-6) reduces to $2(\mathbf{x}^T\mathbf{x} - \mathbf{y}^T\mathbf{x})$ and we have that $P\mathbf{x} = \mathbf{y}$.

With the aid of this result, we can find a sequence of Householder transformations $P_1, P_2, \ldots, P_m$ which when applied to A yields a matrix of the form $(R, 0)^T$, where R is an upper triangular matrix. Let us assume for the moment that we have found a matrix $P = P_m P_{m-1} \cdots P_1$ such that

$$PA = \begin{bmatrix} R \\ 0 \end{bmatrix} \qquad P\mathbf{b} = \begin{bmatrix} \mathbf{c} \\ \mathbf{d} \end{bmatrix} \quad (9.9\text{-}7)$$

Then

$$P\mathbf{r} = \begin{bmatrix} \mathbf{c} \\ \mathbf{d} \end{bmatrix} - PA\mathbf{y} = \begin{bmatrix} \mathbf{c} \\ \mathbf{d} \end{bmatrix} - \begin{bmatrix} R \\ 0 \end{bmatrix}\mathbf{y} = \begin{bmatrix} \mathbf{c} - R\mathbf{y} \\ \mathbf{d} \end{bmatrix} \quad (9.9\text{-}8)$$

Since P is orthogonal,

$$\|\mathbf{r}\|_2^2 = \|P\mathbf{r}\|_2^2 = \|\mathbf{c} - R\mathbf{y}\|_2^2 + \|\mathbf{d}\|_2^2 \quad (9.9\text{-}9)$$

and $\|\mathbf{r}\|_2^2$ is minimized if $\mathbf{c} - R\mathbf{y} = 0$. Thus, our least-squares solution is

$$\mathbf{y} = R^{-1}\mathbf{c} \quad (9.9\text{-}10)$$

and the residual vector $\mathbf{r}$ is given by $\mathbf{r} = P^T(\mathbf{0}, \mathbf{d})^T$.

If the computations are carried out in the proper sequence, it can be shown [Lawson and Hanson (1974)] that there is a matrix E and a vector $\mathbf{e}$ such that

$$P(A + E) = \begin{bmatrix} \hat{R} \\ 0 \end{bmatrix} \qquad P(\mathbf{b} + \mathbf{e}) = \begin{bmatrix} \hat{\mathbf{c}} \\ \hat{\mathbf{d}} \end{bmatrix}$$

where $\hat{R}$, $\hat{\mathbf{c}}$, and $\hat{\mathbf{d}}$ are the computed values and where the norms of E and $\mathbf{e}$ are small relative to those of A and $\mathbf{b}$, respectively. In fact, if inner products are computed using double-precision arithmetic, the elements of E and $\mathbf{e}$ are of the order of the rounding error. In addition, the computed solution $\hat{\mathbf{y}}$ satisfies $(\hat{R} + F)\hat{\mathbf{y}} = \hat{\mathbf{c}}$, where F is also small relative to $\hat{R}$. If we now set

$$G = E + P^T \begin{bmatrix} F \\ 0 \end{bmatrix}$$

then, because P is orthogonal, G is also small and we have

$$P(A + G) = \begin{bmatrix} \hat{R} + F \\ 0 \end{bmatrix}$$

Hence, $\hat{\mathbf{y}}$ is the true solution to the least-squares problem of minimizing $\|(A + G)\mathbf{y} - (\mathbf{b} + \mathbf{e})\|_2$. We thus have a stable algorithm. Furthermore, the system $R\mathbf{y} = \mathbf{c}$ is better-conditioned than (9.9-3). In fact, if we use the spectral condition number $K_2(B) = \|B\|_2 \|B^{-1}\|_2$, it is easy to see that $K_2(A^T A) = K_2^2(R)$ {53}. Hence, the above method is recommended for the solution of the general least-squares problem.

We note in passing that when A is an $n \times n$ matrix, i.e., when we have the usual system of linear equations $A\mathbf{x} = \mathbf{b}$, we have that

$$PA = R \qquad \mathbf{c} = P\mathbf{b} \qquad R\mathbf{x} = \mathbf{c} \qquad \mathbf{x} = R^{-1}\mathbf{c}$$

and we have a solution that is very stable and does not require pivoting. However, this process is more costly than LU decomposition with equilibration and partial pivoting {54} and since the latter scheme is stable for all practical purposes, it is preferable.

It remains to show how to choose the P_i. It is easy to find P_1 so that the first column of $A_2 = P_1 A_1$ has zeros below the diagonal where we have set $A_1 = A$. Using the result above, we set $\mathbf{x}$ equal to the first column $\mathbf{a}_1$ of A_1, and $\mathbf{y} = (\pm \|\mathbf{a}_1\|_2, 0, \ldots, 0)^T$. With $\mathbf{v}_1 = (\mathbf{x} - \mathbf{y})/\|\mathbf{x} - \mathbf{y}\|_2$, $P_1 = I - 2\mathbf{v}_1\mathbf{v}_1^T$ does the trick, since the first column of $P_1 A_1$ is then $P_1\mathbf{a}_1 = P_1\mathbf{x} = \mathbf{y}$. For computational purposes we choose the sign to be equal to $-\operatorname{sgn}(a_{11})$ so as to reduce the roundoff error.

The problem arises with P_2. We must choose P_2 so that the subdiagonal elements in the second column of $A_3 = P_2 A_2$ vanish while at the same time ensuring that the first column remains unchanged. To this end, we notice that if we take $\mathbf{v}$ to be of the form $\mathbf{v} = (0, v_2, \ldots, v_n)^T$, then the corresponding P is of the form

$$\begin{bmatrix} 1 & 0 & 0 & \cdots & 0 \\ 0 & x & x & \cdots & x \\ \cdots & \cdots & \cdots & \cdots & \cdots \\ 0 & x & x & \cdots & x \end{bmatrix}$$

and PA_2 will have the same first row and column as A_2. Hence, to get P_2 we take $\mathbf{x} = \mathbf{a}_2$, $\mathbf{y} = (a_{12}, \pm(\|\mathbf{a}_2\|_2^2 - a_{12}^2)^{1/2}, 0, \ldots, 0]^T$ and with $\mathbf{v}_2 = (\mathbf{x} - \mathbf{y})/\|\mathbf{x} - \mathbf{y}\|_2$, $P_2 = I - 2\mathbf{v}_2\mathbf{v}_2^T$.

In the kth step, we take $\mathbf{x} = \mathbf{a}_k$,

$$\mathbf{y} = \left[a_{1k}, \quad \ldots, \quad a_{k-1,k}, \quad \pm\left(\|\mathbf{a}_k\|_2^2 - \sum_{i=1}^{k-1} a_{ik}^2\right)^{1/2} \quad 0, \quad \ldots, \quad 0\right]^T$$

$\mathbf{v}_k = (\mathbf{x} - \mathbf{y})/\|\mathbf{x} - \mathbf{y}\|_2$, and $P_k = I - 2\mathbf{v}_k\mathbf{v}_k^T$. In each case $\mathbf{a}_k$ refers to the kth column of the matrix A_k with the sign chosen as $-\operatorname{sgn}(a_{kk})$.

After m steps, we have that $A_{m+1} = P_m \cdots P_1 A_1 = (R, 0)^T$.

Example 9.5 Solve the problem of Sec. 6.5 using the method of Householder transformations.

In this problem we are given the empirical data

x_i	.1	.2	.3	.4	.5	.6	.7	.8	.9
$\bar{f}_i$	5.1234	5.3057	5.5687	5.9378	6.4370	7.0978	7.9493	9.0253	10.3627

and we wish to find the least-squares approximation by a polynomial of degree 4. To this end, we set up the 9 × 5 matrix A with elements defined by $a_{ij} = x_i^{j-1}$ and the vector $\mathbf{b}$ with $b_i = \bar{f}_i$. After applying a sequence of five Householder transformations to (9.9-1), we find that

$$R = \begin{bmatrix} -3.00000 & -1.50000 & -.95000 & -.67500 & -.51110 \\ 0 & .77460 & .77460 & .67235 & .57010 \\ 0 & 0 & .17550 & .26325 & .29208 \\ 0 & 0 & 0 & -.03776 & -.07551 \\ 0 & 0 & 0 & 0 & -.00767 \end{bmatrix}$$

$$\mathbf{c} = [-20.93590 \quad 4.91058 \quad 1.43284 \quad -.18832 \quad -.00766]^T$$

and $\qquad \mathbf{d} \times 10^3 = [.4407 \quad .3020 \quad -.3129 \quad -.3233]^T$

Solving the system $R\mathbf{y} = \mathbf{c}$ by back substitution, we find the following values for the coefficients of the polynomial $y_4(x) = \sum_{j=0}^{4} y_j x^j$:

$$y_0 = 5.00097 \qquad y_1 = .99199 \qquad y_2 = 2.01720 \qquad y_3 = 2.98987 \qquad y_4 = .9988$$

which is almost identical with those given by (6.5-5).

If we apply the transformation P^T to $(\mathbf{0}, \mathbf{d})^T$, we get the vector $\mathbf{r}$ of residuals given by

$$\mathbf{r} \times 10^3 = [-.0326 \quad .1256 \quad -.2328 \quad .3600 \quad -.4254 \quad .1843 \quad .1652 \quad -.2037 \quad .0594]^T$$

We now proceed to the L_1 and L_∞ cases. To do this, we must first discuss the linear programming (LP) problem. A quite general formulation of this problem is as follows.

Find a vector $\mathbf{x} = (x_1, x_2, \ldots, x_t)$ which minimizes the linear function

$$\mathbf{c}^T\mathbf{x} = c_1 x_1 + c_2 x_2 + \cdots + c_t x_t \tag{9.9-11}$$

subject to the constraints†

$$\begin{aligned} \mathbf{a}_i^T\mathbf{x} &= b_i & i &= 1, \ldots, s_1 \\ \mathbf{a}_i^T\mathbf{x} &\le b_i & i &= s_1 + 1, \ldots, s \\ x_j &\ge 0 & j &= t_1, t_1 + 1, \ldots, t \end{aligned} \tag{9.9-12}$$

where the $\mathbf{a}_i$ are vectors of dimension t. In the next section, on the simplex

† If we have a constraint of the form $\mathbf{a}_i^T\mathbf{x} \ge b_i$, we multiply through by -1 to get the inequality $-\mathbf{a}_i^T\mathbf{x} \le -b_i$, which fits into the framework of (9.9-12).

method, we shall show how to solve this problem. Here we are concerned with the question of how to convert the L_1 and L_∞ minimization problems to LP problems.

Consider first the L_∞ case. We are seeking a vector $\mathbf{y} = (y_1, \ldots, y_m)^T$ such that $\|\mathbf{r}\|_\infty = \|\mathbf{b} - A\mathbf{y}\|_\infty = \max_{1 \le j \le n} |b_j - A_j\mathbf{y}| = \text{minimum}$, where A_j denotes the jth row of A. To this end, we introduce a new variable y_{m+1} which will be equal to $\max_j |b_j - A_j\mathbf{y}|$. Then, for every j we have that

$$-y_{m+1} \le b_j - A_j\mathbf{y} \le y_{m+1} \qquad j = 1, \ldots, n$$

Since we wish to minimize this maximum value, our problem reduces to the LP problem: find a vector $\bar{\mathbf{y}} = (y_1, \ldots, y_{m+1})^T$ which minimizes the linear function

$$\mathbf{c}^T\bar{\mathbf{y}} = y_{m+1}$$

subject to the constraints

$$\begin{aligned} A_j\mathbf{y} - y_{m+1} &\le b_j \\ -A_j\mathbf{y} - y_{m+1} &\le -b_j \qquad j = 1, \ldots, m \\ y_{m+1} &\ge 0 \end{aligned} \tag{9.9-13}$$

which is of the form (9.9-11) and (9.9-12) with $\mathbf{c} = (0, \ldots, 0, 1)^T$, $s_1 = 0$, $t_1 = t = m + 1$, $s = 2m$ and

$$\begin{aligned} \mathbf{a}_i &= (A_i, -1) \qquad & i &= 1, \ldots, m \\ \mathbf{a}_i &= (-A_{i-m}, -1) \qquad & i &= m+1, \ldots, s \end{aligned}$$

The L_1 case can be treated in a similar fashion. We are seeking a vector $\mathbf{y} = (y_1, \ldots, y_m)^T$ such that

$$\|\mathbf{r}\|_1 = \|\mathbf{b} - A\mathbf{y}\|_1 = \sum_{j=1}^{n} |b_j - A_j\mathbf{y}| = \text{minimum}$$

Now the residual $r_j = b_j - A_j\mathbf{y}$ can be expressed as the difference, $p - q$, of two nonnegative numbers p and q in many ways. If we consider the sum $p + q$, we see that $p + q = |r_j|$ if and only if $p = r_j, q = 0$ or $p = 0, q = -r_j$. Furthermore, the minimum value of $p + q$ subject to the constraints $p \ge 0$, $q \ge 0$ occurs when either $p = 0$ or $q = 0$. Therefore, for each j we introduce two new nonnegative variables y_{m+j} and y_{m+n+j} such that

$$b_j - A_j\mathbf{y} = y_{m+j} - y_{m+n+j} \qquad j = 1, \ldots, n$$

Our problem then is to find a vector $\bar{\mathbf{y}} = (y_1, \ldots, y_{m+2n})^T$ which minimizes the linear function

$$\sum_{k=m+1}^{m+2n} y_k \tag{9.9-14}$$

subject to the constraints

$$A_j\mathbf{y} + y_{m+j} - y_{m+n+j} = b_j \qquad j = 1, \ldots, n$$
$$y_k \geq 0 \qquad k = m+1, \ldots, m+2n \tag{9.9-15}$$

This problem is clearly of the form (9.9-11) and (9.9-12) {57}. The minimum is attained with either y_{m+j} or y_{m+n+j} (or both) equal to zero for each j. In this case, (9.9-14) is equal to $\sum_{j=1}^{n} |b_j - A_j\mathbf{y}|$, which is minimal.

9.10 THE SIMPLEX METHOD FOR SOLVING LINEAR PROGRAMMING PROBLEMS

As we have seen in the previous section, the problem of finding the solution of an overdetermined system of linear equations which minimizes the L_1 or L_∞ norm of the error vector can be formulated as an LP problem. This is only one example of many in which LP can be applied to numerical analysis [Rabinowitz (1968)]. However, the principal applications of LP have been in industry, where it is used to determine the optimum allocation of resources subject to various constraints imposed by a given situation. Because of the economic importance of LP, much effort has been invested in building sophisticated systems for solving large LP problems† on computers, and many algorithms have been developed to this end. Among these algorithms, the simplex algorithm, developed by Dantzig and his coworkers in the late forties, is still the one in most prevalent use, in its original form or in one of its many modifications. While there are deficiencies in the simplex algorithm, such as its possible numerical instability, nevertheless, for all practical purposes, it has proved itself over the years and is perfectly adequate for the solution of LP problems arising in numerical analysis. We shall discuss here only the basic simplex method and refer the reader to the literature for further details on solving LP problems.

The simplex method assumes that the LP problem has been formulated in the following *standard* form: find $\mathbf{x} = (x_1, x_2, \ldots, x_n)^T$ which minimizes the objective or cost function

$$f(\mathbf{x}) \equiv \mathbf{c}^T\mathbf{x} = c_1x_1 + c_2x_2 + \cdots + c_nx_n \tag{9.10-1}$$

subject to the constraints

$$A\mathbf{x} = \mathbf{b} \geq 0 \tag{9.10-2}$$

$$\mathbf{x} \geq 0 \tag{9.10-3}$$

where A is an $m \times n$ matrix, $m < n$, and $\mathbf{b}$ is a nonnegative vector of dimension m.

† Problems involving tens of thousands of variables and thousands of constraints have been successfully solved with some of these systems.

The more general formulation of the previous section can be converted to standard form by the following devices. For any inequality

$$a_{j1}x_1 + \cdots + a_{jn}x_n \le b_n \tag{9.10-4}$$

introduce a *slack* variable $\tilde{x}_j \ge 0$ which converts (9.10-4) to an equation

$$a_{j1}x_1 + \cdots + a_{jn}x_n + \tilde{x}_j = b_n$$

To this variable assign a cost coefficient $\tilde{c}_j = 0$. Then replace a variable x_k which is unrestricted in sign by the difference between two nonnegative variables

$$x_k = x_{k1} - x_{k2} \qquad x_{k1}, x_{k2} \ge 0$$

Finally, multiply any equation in which the right-hand side is negative by -1 to yield a nonnegative $\mathbf{b}$.

Our standard problem may thus have many more variables than the original problem but the same number of rows. Fortunately, it is the number of rows which measures the amount of work required in the solution.

In some problems, one may wish to maximize the objective function rather than minimize it. This is readily accomplished since

$$\max \mathbf{c}^T\mathbf{x} = -\min\,(-\mathbf{c}^T\mathbf{x})$$

and the solution which minimizes $-\mathbf{c}^T\mathbf{x}$ will maximize $\mathbf{c}^T\mathbf{x}$.

Returning now to our standard form, we notice that (9.10-2) is an *underdetermined* system of linear equations. If the rank of A is m, which we shall henceforth assume to be the case, then (9.10-2) has an infinite number of solutions in which $n - m$ of the variables may vary freely. Suppose, for example, the first m columns of A are linearly independent; then partitioning A into a square matrix A_{mm} formed from the first m columns of A and a matrix $A_{m,\,n-m}$, we premultiply both sides of (9.10-2) by A_{mm}^{-1} to obtain

$$\begin{aligned} x_1 &= \bar{b}_1 - \bar{a}_{1,\,m+1}x_{m+1} - \cdots - \bar{a}_{1,\,n}x_n \\ &\cdots\cdots\cdots\cdots\cdots\cdots\cdots\cdots\cdots\cdots \\ x_m &= \bar{b}_m - \bar{a}_{m,\,m+1}x_{m+1} - \cdots - \bar{a}_{m,\,n}x_n \end{aligned} \tag{9.10-5}$$

where the overbars indicate the result of premultiplying $A_{m,\,n-m}$ and $\mathbf{b}$ by A_{mm}^{-1}. (As we shall see below, we do not have to calculate A_{mm}^{-1} explicitly.)

Thus, for any choice of values $x_{m+1}^*, \ldots, x_n^*$, the vector $\mathbf{x}^* = (x_1^*, \ldots, x_n^*)$ is a solution of (9.10-2), where $x_1^*, \ldots, x_m^*$ are calculated by substituting $x_{m+1}^*, \ldots, x_n^*$ into (9.10-5). In particular the choice $x_{m+1} = x_{m+2} = \cdots = x_n = 0$ yields a *basic* solution $\mathbf{x} = (\bar{b}_1, \ldots, \bar{b}_m, 0, \ldots, 0)$. For every subset of m linearly independent columns of A, we can similarly find a basic solution.

We now introduce the constraints (9.10-3) and call a vector $\mathbf{x}$ *feasible* if $\mathbf{x}$ satisfies (9.10-2) and (9.10-3). A basic solution of (9.10-2) which satisfies (9.10-3) is naturally called a basic feasible solution. A basic feasible solution in which all m nonzero components are positive is called a nondegenerate basic feasible solution. We shall assume for the present that all basic feasible solutions are nondegenerate.

Let us assume that we have found a general solution (9.10-5) in which all the $\bar{b}_j > 0$. Then for any nonnegative values of $x_{m+1}, \ldots, x_n$ for which $x_1, \ldots, x_m$ remain nonnegative, we have a feasible solution $\mathbf{x} = (x_1, x_2, \ldots, x_n)^T$. Furthermore, for $x_{m+1} = x_{m+2} = \cdots = x_n = 0$ we have a nondegenerate basic feasible solution. Substituting (9.10-5) into the objective function (9.10-1) eliminates the variables $x_1, \ldots, x_m$ from (9.10-1). We obtain

$$f(\mathbf{x}) = z_0 + \bar{c}_{m+1}x_{m+1} + \cdots + \bar{c}_n x_n \tag{9.10-6}$$

where

$$z_0 = c_1\bar{b}_1 + c_2\bar{b}_2 + \cdots + c_m\bar{b}_m \tag{9.10-7}$$

and where the $\bar{c}_j$, $j = m + 1, \ldots, n$, are called the *reduced cost* coefficients. For the basic feasible solution $\mathbf{x}^* = (\bar{b}_1, \ldots, \bar{b}_m, 0, \ldots, 0)$ the value of the objective function $f(\mathbf{x}^*) = z_0$. Furthermore, if all the reduced cost coefficients are nonnegative, any change in the values of the independent variables $x_{m+1}, \ldots, x_n$ which yields a feasible solution cannot decrease $f(\mathbf{x})$. Hence $f(\mathbf{x})$ has a local minimum at $\mathbf{x}^*$, which can be shown to be a global minimum over the set of all feasible solutions {59}. Conversely, if some $\bar{c}_k < 0$, then, as we shall see, we can increase x_k to yield a feasible solution $\hat{\mathbf{x}}$, in fact a basic feasible solution, with $f(\hat{\mathbf{x}}) < f(\mathbf{x}^*)$ so that $\mathbf{x}^*$ is not an optimal solution. Thus, we have that a necessary and sufficient condition for optimality of a basic feasible solution is that the reduced cost coefficients $\bar{c}_k$ be nonnegative. Furthermore, if an optimal solution exists, there exists a basic feasible solution which is also optimal {60}.

Assume now that in (9.10-6) some $\bar{c}_k < 0$. Then we can increase x_k and so decrease $f(\mathbf{x})$, and the more we increase x_k the greater the decrease in $f(\mathbf{x})$. What limits us in increasing x_k? From (9.10-5) we see that a change in x_k will cause $x_1, \ldots, x_m$ to change. If we change x_k so much that one of the variables x_i, $i = 1, \ldots, m$, changes sign, we shall no longer have a feasible solution. Thus, we increase x_k by the maximum amount consistent with feasibility. If for a particular variable x_p, $1 \le p \le m$, $\bar{a}_{pk} \le 0$, then any increase in x_k will not decrease x_p, so that x_p will always remain positive. For $\bar{a}_{pk} > 0$, x_p will remain nonnegative provided that $x_k \le \bar{b}_p/\bar{a}_{pk}$. Hence all x_i, $1 \le i \le m$, will remain nonnegative (and x_i, $m + 1 \le i \le n$, $i \ne k$, will remain 0) as long as

$$x_k \le \min_{\bar{a}_{pk} > 0} \frac{\bar{b}_p}{\bar{a}_{pk}} = \frac{\bar{b}_q}{\bar{a}_{qk}} \tag{9.10-8}$$

If we set $x_k = \bar{b}_q/\bar{a}_{qk} > 0$, we see that $x_q = 0$ and we have a new basic feasible solution

$$[\bar{b}'_1 \quad \cdots \quad \bar{b}'_{q-1} \quad 0 \quad \bar{b}'_{q+1} \quad \cdots \quad \bar{b}'_m \quad 0 \quad \cdots \quad 0 \quad x_k \quad 0 \quad \cdots \quad 0]^T$$

Since we have assumed that all basic feasible solutions are nondegenerate, there is a unique q such that $x_k = \bar{b}_q/\bar{a}_{qk}$; otherwise one of the $\bar{b}'_i$ would vanish. Before we show how to express this new solution in form (9.10-5), let us see what happens if all the coefficients $\bar{a}_{pk}$ are nonpositive. In this case, we can increase x_k without bound, and the vector

$$\mathbf{x}' = [\bar{b}'_1 \quad \cdots \quad \bar{b}'_m \quad 0 \quad \cdots \quad 0 \quad x_k \quad 0 \quad \cdots \quad 0)^T$$

always remains a feasible solution. At the same time, the objective function $f(\mathbf{x}')$ decreases to $-\infty$; that is, we have an *unbounded* solution. This usually indicates that the LP problem has not been formulated properly and that probably some constraint has been omitted. At any rate, the combination $\bar{c}_k < 0$ and $\bar{a}_{pk} \le 0$, $p = 1, \ldots, m$, terminates the algorithm with an indication that the solution is unbounded.

We now show how to write the new solution in form (9.10-5), which will then allow us to start a new iteration. With q defined by (9.10-8) we rewrite row q of (9.10-5) as

$$x_k = \frac{1}{\bar{a}_{qk}}(\bar{b}_q - \bar{a}_{q,\,m+1}x_{m+1} - \cdots - \bar{a}_{q,\,k-1}x_{k-1} - \bar{a}_{q,\,k+1}x_{k+1} - \cdots - \bar{a}_{qn}x_n - x_q) \tag{9.10-9}$$

and substitute this value of x_k into the other equations of (9.10-5) and into (9.10-6). This yields the new system

$$\begin{aligned} x_i = \bar{b}'_i - \bar{a}'_{i,\,m+1}x_{m+1} - \cdots - \bar{a}'_{i,\,k-1}x_{k-1} - \bar{a}'_{i,\,k+1}x_{k+1} - \cdots \\ - \bar{a}'_{i,\,n}x_n - \bar{a}'_{i,\,q}x_q \qquad i = 1, \ldots, q-1, q+1, \ldots, m, k \end{aligned} \tag{9.10-10}$$

and the new objective function

$$f(\mathbf{x}) = z'_0 + \bar{c}'_{m+1}x_{m+1} + \cdots + \bar{c}'_{k-1}x_{k-1} + \bar{c}'_{k+1}x_{k+1} + \cdots + \bar{c}'_n x_n + \bar{c}'_q x_q$$

where

$$\bar{a}'_{ij} = \bar{a}_{ij} - \frac{\bar{a}_{ik}\bar{a}_{qj}}{\bar{a}_{qk}} \qquad \bar{b}'_i = \bar{b}_i - \frac{\bar{a}_{ik}\bar{b}_q}{\bar{a}_{qk}} \qquad \begin{aligned} i &= 1, \ldots, m; \quad i \ne q \\ j &= m+1, \ldots, n; j \ne k \end{aligned} \tag{9.10-11}$$

$$\bar{c}'_j = \bar{c}_j - \frac{\bar{c}_k\bar{a}_{qj}}{\bar{a}_{qk}} \qquad \bar{a}'_{iq} = -\frac{\bar{a}_{ik}}{\bar{a}_{qk}} \qquad \bar{a}'_{kj} = \frac{\bar{a}_{qj}}{\bar{a}_{qk}} \qquad \bar{a}'_{kq} = \frac{1}{\bar{a}_{qk}}$$

$$\bar{c}'_q = -\frac{\bar{c}_k}{\bar{a}_{qk}} \qquad \bar{b}'_k = \frac{\bar{b}_q}{\bar{a}_{qk}} \qquad z'_0 = z_0 + \frac{\bar{c}_k\bar{b}_q}{\bar{a}_{qk}}$$

Since $\bar{c}_k < 0$, $\bar{b}_q > 0$, and $\bar{a}_{qk} > 0$, we see that $z'_0 < z_0$, so that each iteration decreases the value of the objective function.

We shall now give the algorithm for a general step, at the same time simplifying the notation.

Let $N = \{1, 2, \ldots, n\}$ be partitioned into two sets, $N = D \cup I$, where $D = \{P_1, \ldots, P_m\}$ is the set of indices of the dependent or basic variables and $I = \{P_{m+1}, \ldots, P_n\}$ is the set of indices of the independent variables. In the following, any index i used refers to P_i, $i = 1, \ldots, n$. Thus x_1 refers to x_{P_1}, a basic variable, etc., and we can describe the initial state of each iteration by (9.10-5) and (9.10-6). We now simplify the notation a little by writing $\bar{a}_{i0}$ in place of $\bar{b}_i$, $\bar{a}_{0j}$ in place of $\bar{c}_j$, $\bar{a}_{00}$ in place of $-z_0$, and x_0 in place of $-f(x)$. Thus (9.10-5) and (9.10-6) are combined into the system

$$x_i = \bar{a}_{i0} - \sum_{j=m+1}^{n} \bar{a}_{ij} x_j \qquad i = 0, \ldots, m \tag{9.10-12}$$

The matrix $\bar{A} = [\bar{a}_{ij}]$, $i = 0, \ldots, m$, $j = m + 1, \ldots, n$, together with the vector $\bar{\mathbf{a}}_0$ and the permutation vectors D and I form what is called the tableau (see Fig. 9.1). The simplex methods proceeds as follows:

1. Compute $\min_{m+1 \le j \le n} \bar{a}_{0j} = \bar{a}_{0k}$ (to yield the most negative $\bar{c}_k$). In case of ties, choose the first index k. If $\bar{a}_{0k} \ge 0$, terminate with the basic solution $\bar{\mathbf{x}}$ with positive components $x_{P_i} = \bar{a}_{i0}$, $i = 1, \ldots, m$, and zero components $x_{P_{m+1}}, \ldots, x_{P_n}$. (Recall our assumption of nondegeneracy.) The optimal value of the objective function is $-\bar{a}_{00}$.
2. If $\bar{a}_{0k} < 0$, compute $\min_{\bar{a}_{ik} > 0} (\bar{a}_{i0}/\bar{a}_{ik}) = \bar{a}_{q0}/\bar{a}_{qk}$ [cf. (9.10-8)]. If all $\bar{a}_{ik} \le 0$, terminate with an indication that the solution is unbounded.
3. Introduce P_k into D in place of P_q to yield D' and replace P_k by P_q in I to yield I'. Corresponding to (9.10-11), compute

$$\begin{aligned} &\bar{a}'_{ij} = \bar{a}_{ij} - \frac{\bar{a}_{ik}\bar{a}_{qj}}{\bar{a}_{qk}} \qquad \begin{array}{l} i = 0, \ldots, m;\ i \ne q \\ j = 0,\ m + 1, \ldots, n;\ j \ne k \end{array} \\ &\bar{a}'_{ik} = -\frac{\bar{a}_{ik}}{\bar{a}_{qk}} \qquad \bar{a}'_{qj} = \frac{\bar{a}_{qj}}{\bar{a}_{qk}} \qquad \bar{a}'_{qk} = \frac{1}{\bar{a}_{qk}} \end{aligned} \tag{9.10-13}$$

   and return to step 1 with an updated matrix $\bar{A}'$, vector $\bar{\mathbf{a}}'_0$ and permutation vectors D' and I'.

This algorithm is guaranteed to terminate in either step 1 or 2, since each iteration reduces the value of $-\bar{a}_{00}$ and hence any particular partition (I, D) can appear only once while the number of distinct partitions of N of

		P_{m+1}		P_n
	$\bar{a}_{00} = -z_0$	$\bar{a}_{0,m+1} = \bar{c}_{m+1}$	$\cdots$	$\bar{a}_{0n} = \bar{c}_n$
P_1	$\bar{a}_{10} = \bar{b}_1$	$\bar{a}_{1,m+1}$	$\cdots$	$\bar{a}_{1n}$
$\cdots$	$\cdots$	$\cdots$	$\cdots$	$\cdots$
P_m	$\bar{a}_{m0} = \bar{b}_m$	$\bar{a}_{m,m+1}$	$\cdots$	$\bar{a}_{mn}$

Figure 9.1 The tableau of a linear programming problem.

the form $N = I \cup D$ equals $\binom{n-m}{m}$. In practice the number of iterations required to terminate is of the order of $2m$.

There remains the problem how to start the process, namely how to get a representation in form (9.10-5). That this is not always possible can be seen from the following trivial example.

$$\text{Minimize } x_1 + x_2 \qquad \text{subject to } -x_1 - x_2 = 1;\ x_1, x_2 \geq 0 \tag{9.10-14}$$

This is a case where there is no feasible solution to the problem. Hence, we must find an algorithm to find a basic feasible solution if one exists and to terminate with an indication of infeasibility if there are no feasible solutions. Fortunately, we can use the same simplex algorithm applied to a related problem. By doing so we shall find a basic feasible solution for our problem and, at the same time, formulate the problem in the form (9.10-5), without having actually computed A_{mm}^{-1} and premultiplied A by it.

Let us introduce a vector $\mathbf{y} = (y_1, \ldots, y_m)^T$ of so-called artificial variables, and consider the LP problem

$$\text{Minimize } \sum_{i=1}^{m} y_i \tag{9.10-15}$$

subject to the constraints

$$A\mathbf{x} + I_m\mathbf{y} = \mathbf{b} \tag{9.10-16}$$

$$\mathbf{x}, \mathbf{y} \geq \mathbf{0} \tag{9.10-17}$$

where I_m is the unit $m \times m$ matrix. This problem has a basic feasible solution $\mathbf{y} = \mathbf{b}, \mathbf{x} = \mathbf{0}$, so that we can apply the simplex algorithm. Since the objective function is bounded below by 0, there cannot be an unbounded solution, so that the simplex algorithm must terminate with an optimal solution. Now if in this solution any component of $\mathbf{y}$ is positive, it means that there is no feasible solution to the problem (9.10-1) to (9.10-3). For if a feasible solution $\mathbf{x}$ existed, then $(\mathbf{x}, \mathbf{y})$ with $\mathbf{y} = \mathbf{0}$ would be a feasible solution of (9.10-15) to (9.10-17) with the value of the objective function (9.10-15) equal of 0. Since the algorithm terminated with some $y_j > 0$, so that the value of (9.10-15) > 0, this means that a solution of the form $(\mathbf{x}, \mathbf{y})$ with $\mathbf{y} = \mathbf{0}$ to (9.10-15) to (9.10-17) does not exist; i.e., no feasible solution exists for the problem (9.10-1) to (9.10-3). If every component of $\mathbf{y}$ vanishes, we have an initial basic feasible solution consisting of the m positive components of $\mathbf{x}$ in the optimal solution. (Recall our assumption of nondegeneracy.) With this solution we can compute the reduced cost coefficients and thus set up our tableau and apply the simplex algorithm until termination. Moreover, the solution of (9.10-15) to (9.10-17) will automatically result in a tableau with Eqs. (9.10-2) in the form (9.10-5). (See also Example 9.6, below.) Thus we have two phases in our problem, phase I, in which we find a feasible solution if one exists, and phase II, in which we find an optimal solution or determine unboundedness.

We conclude with a word about degeneracy. There are two aspects to degeneracy, a theoretical side and a practical side. In theory, the simplex method need not terminate in the presence of degeneracy and may cycle. However, by the use of a perturbation argument [see Dantzig (1963)] it can be shown in step 2 of the simplex method, when there may exist several rows such that $\bar{a}_{i0}/\bar{a}_{ik}$ are equal to the minimum, thus leading to degeneracy, that a choice procedure exists for choosing the proper row to avoid cycling. The simplex method so modified will terminate even in the presence of degeneracy. In practice, degeneracy is just ignored, and in step 2 an arbitrary choice is made in the case of a tie. There have been almost no cases of cycling reported in practical problems, even though degeneracy is very common.

Example 9.6 Solve the following LP problem using the simplex method.

$$\text{Minimize } f(\mathbf{x}) = x_1 + 6x_2 - 7x_3 + x_4 + 5x_5 \tag{9.10-18}$$

subject to the constraints

$$\begin{aligned} 5x_1 - 4x_2 + 13x_3 - 2x_4 + x_5 &= 20 \\ x_1 - x_2 + 5x_3 - x_4 + x_5 &= 8 \\ \text{all } x_i &\geq 0 \end{aligned} \tag{9.10-19}$$

Since there is no obvious initial basic feasible solution, we shall apply phase I to find one. To this end we set up the following problem which we solve first:

$$\text{minimize } g(\mathbf{x}) = x_6 + x_7 \tag{9.10-20}$$

subject to the constraints

$$\begin{aligned} 5x_1 - 4x_2 + 13x_3 - 2x_4 + x_5 + x_6 &= 20 \\ x_1 - x_2 + 5x_3 - x_4 + x_5 + x_7 &= 8 \\ \text{all } x_i &\geq 0 \end{aligned} \tag{9.10-21}$$

The initial basic feasible solution to this problem is $x_6 = 20$, $x_7 = 8$. The corresponding tableau is

		1	2	3	4	5
	−28	−6	5	−18	3	−2
6	20	5	−4	(13)	−2	1
7	8	1	−1	5	−1	1

where we shall circle the pivot element in each tableau. This element is determined by the rules of the simplex method. Applying the transformations (9.10-13), we get the following two tableaus:

		1	2	6	4	5
	$-\frac{4}{13}$	$\frac{12}{13}$	$-\frac{7}{13}$	$\frac{18}{13}$	$\frac{3}{13}$	$-\frac{8}{13}$
3	$\frac{20}{13}$	$\frac{5}{13}$	$-\frac{4}{13}$	$\frac{1}{13}$	$-\frac{2}{13}$	$\frac{1}{13}$
7	$\frac{4}{13}$	$-\frac{12}{13}$	$\frac{7}{13}$	$-\frac{5}{13}$	$-\frac{3}{13}$	$\textcircled{\frac{8}{13}}$

		1	2	6	4	7
	0	0	0	1	0	1
3	$\frac{3}{2}$	$\frac{1}{2}$	$-\frac{3}{8}$	$\frac{1}{8}$	$\frac{1}{8}$	$-\frac{1}{8}$
5	$\frac{1}{2}$	$-\frac{3}{2}$	$\frac{7}{8}$	$-\frac{5}{8}$	$-\frac{3}{8}$	$\frac{13}{8}$

Since all elements $a_{0i} \geq 0$, $i \geq 1$, we have reached an optimal solution of phase I. This solution, $x_3 = \frac{3}{2}$, $x_5 = \frac{1}{2}$ is a basic feasible solution of (9.10-19), as we readily verify. Note that the final reduced cost coefficients are precisely the original coefficients of (9.10-15) because this last tableau corresponds to a premultiplication of (9.10-16) by A_{mm}^{-1}, which must leave (9.10-15) unchanged since the solution variables have 0 cost coefficients in the objective function.

We now are ready to start phase II and return to (9.10-18) and (9.10-19) to set up the following tableau:

		1	2	4
	8	12	-1	2
3	$\frac{3}{2}$	$\frac{1}{2}$	$-\frac{3}{8}$	$\frac{1}{8}$
5	$\frac{1}{2}$	$-\frac{3}{2}$	$\textcircled{\frac{7}{8}}$	$-\frac{3}{8}$

All the rows but the first arise from the final tableau of phase I by deleting the columns corresponding to the artificial variables x_6 and x_7. To get the first row, we insert the values of x_3 and x_5 into (9.10-18), where, from the tableau,

$$\begin{aligned} x_3 &= \tfrac{3}{2} - \tfrac{1}{2}x_1 + \tfrac{3}{8}x_2 - \tfrac{1}{8}x_4 \\ x_5 &= \tfrac{1}{2} + \tfrac{3}{2}x_1 - \tfrac{7}{8}x_2 + \tfrac{3}{8}x_4 \end{aligned} \tag{9.10-22}$$

so that (9.10-18) becomes

$$f(\mathbf{x}) = -8 + 12x_1 - x_2 + 2x_4 \tag{9.10-23}$$

The next (and final) tableau follows by applying the simplex method as before

		1	5	4
	$\frac{60}{7}$	$\frac{72}{7}$	$\frac{8}{7}$	$\frac{11}{7}$
3	$\frac{12}{7}$	$-\frac{1}{7}$	$\frac{3}{7}$	$-\frac{2}{7}$
2	$\frac{4}{7}$	$-\frac{12}{7}$	$\frac{8}{7}$	$-\frac{3}{7}$

From this, we read off the solution $x_2 = \frac{4}{7}$, $x_3 = \frac{12}{7}$, $x_j = 0$, $j = 1, 4, 5$. For this solution, the value of $f(\mathbf{x})$ is $-\frac{60}{7}$.

9.11 MISCELLANEOUS TOPICS

Complex matrices The most efficient way to solve the system of linear equations with complex elements

$$C\mathbf{z} = \mathbf{w} \tag{9.11-1}$$

with

$$C = A + iB \qquad \mathbf{z} = \mathbf{x} + i\mathbf{y} \qquad \mathbf{w} = \mathbf{u} + i\mathbf{v} \tag{9.11-2}$$

is to follow the real algorithm, replacing all real operations by complex ones. This can readily be accomplished in a programming language which allows for the declaration of complex variables. In the absence of such a facility, it can become tiresome to convert every real operation into a series of such operations. In this case, one may be willing to pay the cost and convert the problem to one involving real numbers. Inserting (9.11-2) into (9.11-1) and equating real and imaginary parts of both sides, we get the real system {62}

$$A\mathbf{x} - B\mathbf{y} = \mathbf{u} \qquad B\mathbf{x} + A\mathbf{y} = \mathbf{v} \tag{9.11-3}$$

This is a $2n \times 2n$ system requiring $O(4n^2)$ storage locations, as against $O(2n^2)$ for the complex system. Similarly, the amount of computation is $O(8n^3/3)$ multiplications as against $O(4n^3/3)$ real multiplications for the complex case. Thus, if complex systems arise infrequently, it may not pay to invest the programming effort in writing a special program for complex matrices. However, if they appear frequently and the language in use does not accommodate complex variables, the superiority of the complex formulation in both storage requirements and computation time calls for writing a separate program for complex systems.

Determinants Whereas determinants are usually thought of as an aid in solving systems of linear equations via Cramer's rule, there are other situations in which the value of the determinant of a matrix A is required, e.g., in some methods for solving eigenvalue and generalized eigenvalue problems. As another example, the determinant of the Jacobian of a transformation is needed when a change of variable is made in a multiple integral. The most efficient way to compute the value of det (A) is via the LU decomposition of A. From (9.3-23) we have that

$$A = P^{-1}LU$$

from which

$$\det(A) = \det(P^{-1})\det(L)\det(U) = \det(P^{-1})\prod_{i=1}^{n} u_{ii} \tag{9.11-4}$$

Recalling that

$$P = P_{n-1,r_{n-1}} \cdots P_{1,r_1} \tag{9.11-5}$$

and that

$$\det(P_{k,r_k}) = -1 \qquad \text{if } k \neq r_k \tag{9.11-6}$$

we have that det $(P) = \det(P^{-1}) = N$, the number of row interchanges. Thus finally,

$$\det(A) = (-1)^N \prod_{i=1}^{n} u_{ii} \qquad (9.11\text{-}7)$$

Band matrices A *band* matrix of *bandwidth* $[w_1, w_2]$ is a matrix for which $a_{ij} = 0$ if $i - j > w_1$ or $j - i > w_2$. Thus, upper and lower triangular matrices are band matrices of bandwidth $[0, n-1]$ and $[n-1, 0]$, respectively. Other important examples of band matrices are the tridiagonal matrices with bandwidth [1, 1] and the upper and lower Hessenberg matrices, which are defined to be matrices of bandwidths $[1, n-1]$ and $[n-1, 1]$, respectively.

The importance of band matrices lies in the fact that in the LU decomposition (9.3-23) of a band matrix A, the bandwidth of L is always $[w_1, 0]$ {64}. Furthermore, in the absence of pivoting (as in the case of symmetric positive definite matrices or diagonally dominant matrices) the bandwidth of U is $[0, w_2]$ {64}, so that in this case, if the elements of L and U are overwritten on those of A, no additional storage space is required. And even in the case where pivoting is required, it is easy to see {64} that the bandwidth of U is at most $[0, \min(w_1 + w_2, n-1)]$. Thus, for upper Hessenberg matrices, the same situation holds as above, while in the general case, additional storage is required but usually much less than for a full matrix. (The lower Hessenberg matrices are an exception to this rule; however, see {69}.) In addition, the computation time is much smaller for certain band matrices. In the important case of upper Hessenberg matrices, we need $O(n^2)$ multiplications for LU decomposition while for general band matrices for which $w_1 + w_2 \ll n$ we need only $O(n)$ multiplications. A similar reduction in the number of operations also occurs in the forward and back substitutions {67}.

Of particular interest is the triangular decomposition of a tridiagonal matrix without row interchanges, which arises in many situations in numerical analysis, e.g., in the computation of cubic splines (cf. Sec. 3.8). The algorithm for the complete solution of a tridiagonal system is very simple and elegant in this case and proceeds as follows.

The matrix A can be represented as three vectors **a**, **b**, and **c** consisting of the subdiagonal, diagonal, and superdiagonal elements, respectively. [For uniformity of notation $\mathbf{a} = (0, a_2, \ldots, a_n)^T$ and $\mathbf{c} = (c_1, \ldots, c_{n-1}, 0)^T$.]

In the LU decomposition of A, L has bandwidth [1, 0] and U, [0, 1]. Furthermore it turns out, as can be verified by direct multiplication, that the superdiagonal of U is equal to that of A. It thus remains to compute $\boldsymbol{\alpha}$ the subdiagonal of L and $\boldsymbol{\beta}$, the diagonal of U. The formulas for this are given by {65}

$$\beta_1 = b_1 \qquad \alpha_k = \frac{a_k}{\beta_{k-1}} \qquad \beta_k = b_k - \alpha_k c_{k-1} \qquad k = 2, 3, \ldots, n \qquad (9.11\text{-}8)$$

Once we have the triangular decomposition, we can solve the system $A\mathbf{x} = \mathbf{d}$ by forward and back substitution as follows:

$$\delta_1 = d_1 \qquad \delta_k = d_k - \alpha_k \delta_{k-1} \qquad k = 2, 3, \ldots, n \tag{9.11-9}$$

$$x_n = \frac{\delta_n}{\beta_n} \qquad x_k = \frac{\delta_k - c_k x_{k+1}}{\beta_k} \qquad k = n-1, \ldots, 1 \tag{9.11-10}$$

Throughout this process, the vectors $\boldsymbol{\alpha}$, $\boldsymbol{\beta}$, $\boldsymbol{\delta}$ can overwrite **a**, **b**, **d** and the solution **x** can in turn overwrite $\boldsymbol{\delta}$. The total number of operations is seen to be $3n - 3$ additions, $3n - 3$ multiplications, and $2n - 1$ divisions, which is a very satisfactory result.

When pivoting is required, the algorithm is not as elegant and requires an additional vector for storage. The number of operations also increases to at most $4n - 5$ additions, $5n - 6$ multiplications, and $2n - 1$ divisions, which is still quite satisfactory {66}.

Example 9.7 Solve the following tridiagonal system using 5-digit floating-point decimal arithmetic.

$$\begin{aligned}
136.01x_1 + 90.860x_2 \qquad\qquad\qquad\qquad\qquad &= -33.254 \\
90.860x_1 + 98.810x_2 - 67.590x_3 \qquad\qquad\quad &= 49.790 \\
-67.590x_2 + 132.01x_3 + 46.260x_4 &= 28.067 \\
46.260x_3 + 177.17x_4 &= -7.3244
\end{aligned}$$

We have that

$$\begin{aligned}
\mathbf{a} &= [0 \quad 90.860 \quad -67.590 \quad 46.260]^T \\
\mathbf{b} &= [136.01 \quad 98.810 \quad 132.01 \quad 177.17]^T \\
\mathbf{c} &= [90.860 \quad -67.590 \quad 46.260 \quad 0]^T \\
\mathbf{d} &= [-33.254 \quad 49.790 \quad 28.067 \quad -7.3244]^T
\end{aligned}$$

Using (9.11-8), we compute $\boldsymbol{\alpha}$ and $\boldsymbol{\beta}$ as follows:

$$\beta_1 = 136.01 \qquad \alpha_2 = \frac{90.860}{\beta_1} = .66804 \qquad \beta_2 = 98.810 - 90.860\alpha_2 = 38.112$$

$$\alpha_3 = -\frac{67.590}{\beta_2} = -1.7735 \qquad \beta_3 = 132.01 + 67.590\alpha_3 = 12.139$$

$$\alpha_4 = \frac{46.260}{\beta_3} = 3.8109 \qquad \beta_4 = 177.17 - 46.260\alpha_4 = .87777$$

We then compute $\boldsymbol{\delta}$ as follows, using (9.11-9),

$$\delta_1 = -33.254 \qquad \delta_2 = 49.790 - \alpha_2\delta_1 = 72.005$$

$$\delta_3 = 28.067 - \alpha_3\delta_2 = 155.77 \qquad \delta_4 = -7.3244 - \alpha_4\delta_3 = -600.95$$

Finally, we compute the solution $\mathbf{x}$, using (9.11-10),

$$x_4 = \frac{\delta_4}{\beta_4} = -684.63 \qquad x_3 = \frac{\delta_3 - 46.260x_4}{\beta_3} = 2621.9$$

$$x_2 = \frac{\delta_2 - (-67.590x_3)}{\beta_2} = 4651.7 \qquad x_1 = \frac{\delta_1 - 90.860x_2}{\beta_1} = -3105.7$$

In Example 9.3, we computed the LU decomposition of the matrix of this system. If we now solve the system

$$LU\hat{\mathbf{x}} = P\mathbf{d}$$

by forward and back substitution, using the matrices L, U, and P of Example 9.3, we find that

$$\hat{\mathbf{x}} = [-2953.3 \quad 4420.5 \quad 2491.5 \quad -650.59]^T.$$

This solution differs considerably from that computed above. However, if we solve the tridiagonal system using 10-digit arithmetic, the solution $\tilde{\mathbf{x}}$ rounded to 5 digits is

$$\tilde{\mathbf{x}} = [-2957.4 \quad 4426.6 \quad 2495.0 \quad -651.49]^T$$

which is much closer to $\hat{\mathbf{x}}$ than to $\mathbf{x}$.

The reason for this discrepancy is that the matrix A is ill-conditioned. If the inverse matrix A^{-1} is computed, we have that

$$A^{-1} = \begin{bmatrix} 22.229 & -33.264 & -18.746 & 4.8948 \\ -33.264 & 49.793 & 28.062 & -7.3271 \\ -18.746 & 28.062 & 15.823 & -4.1315 \\ 4.8948 & -7.3271 & -4.1315 & 1.0844 \end{bmatrix}$$

so that $\|A^{-1}\|_\infty \approx 118$. Since $\|A\|_\infty \approx 257$, we have that $K_\infty(A) > 30{,}000$. Thus if we assume that $\mathbf{x}$ is the exact solution of $(A + \delta A)\mathbf{x} = \mathbf{d}$, where $\|\delta A\|_\infty \simeq \frac{1}{2} \times 10^{-5}\|A\|_\infty$, which is one unit of roundoff error, we have from (9.4-23) that

$$\frac{\|\mathbf{x}_t - \mathbf{x}_c\|_\infty}{\|\mathbf{x}_t\|_\infty} \le \frac{\|A^{-1}\|_\infty \|\delta A\|_\infty}{1 - \|A^{-1}\|_\infty \|\delta A\|_\infty} > .15$$

and indeed

$$\frac{\|\tilde{\mathbf{x}} - \mathbf{x}\|_\infty}{\|\mathbf{x}\|_\infty} = \frac{225.1}{4426.6} \approx .05$$

BIBLIOGRAPHIC NOTES

The literature on the solution of simultaneous linear equations is vast. A classification of this literature together with an exhaustive bibliography up to 1952 is given in Forsythe (1953*b*). More recent extensive bibliographies can be found in Faddeev and Faddeeva (1963) and Householder (1964); see also Westlake (1968) and Stewart (1973). Bibliographies on iterative methods appear in Varga (1962) and Young (1971).

For the material of this chapter on direct methods, we have drawn heavily on the work of Wilkinson (1960, 1961, 1964, 1967). For the use of iterative techniques in the solution of partial differential equations, a good source is Varga (1962). Other good general sources for the material of this chapter are Bodewig (1959), Faddeeva (1959), and Fox (1965). Householder (1964) presents an extensive survey of the theoretical aspects of the solution of linear systems.

A comprehensive survey of the various methods for solving systems of linear equations is given in Westlake (1968). An interesting attempt to combine the theory, applications, and numerical aspects of linear algebra in one framework appears in Noble (1969). A collection of algorithms implementing many of the methods treated in this chapter appears in Wilkinson and Reinsch (1971).

Section 9.1 The material in this section is classical; see, for example, Birkhoff and MacLane (1953) or Faddeev and Faddeeva (1963).

Section 9.2 Discussion of some of the matters considered here can be found in Bodewig (1959), Newman (1962), Forsythe (1953*a*), and also in *Modern Computing Methods* (1961).

Sections 9.3 to 9.5 The material in these sections is based principally on Wilkinson (1967), which is the source of Example 9.0. Other good references are Forsythe and Moler (1967) and Stewart (1973).

Section 9.3 Gaussian elimination and the Gauss-Jordan reduction are discussed in any source that considers direct methods; see, for example, Bodewig (1959) or Hildebrand (1974). The Doolittle algorithm is discussed in *Modern Computing Methods* (1961) and by Bodewig (1959), that of Crout (1941) in Hildebrand (1974), and that of Cholesky in Schwarz et al. (1973). Bodewig (1959) discusses the number of operations required for various methods.

Section 9.4 Fraenkel and Lowenthal (1971) discuss a method for finding the exact solution of a system of linear equations with rational coefficients and give a Fortran program which implements this method.

The approach to error analysis based on bounding the roundoff accumulation from stage to stage is exhaustively discussed in the now classical papers of Von Neumann and Goldstine (1947) and Goldstine and Von Neumann (1951). Our approach here is due to Wilkinson (1964), which is a mine of information, and Wilkinson (1960). Wilkinson (1961) discusses the errors in a number of direct methods in terms of both fixed- and floating-point arithmetic. Bodewig (1959) discusses errors at some length.

Section 9.5 Our approach to finding improved solutions of ill-conditioned equations will be found in *Modern Computing Methods* (1961); Bodewig (1959) also discusses ill-conditioned equations. The pitfalls of iterative refinement are vividly illustrated by Kahan (1966).

Section 9.6 The most complete coverage of iterative methods is that in Young (1971), which also contains an extensive bibliography. Iterative methods are discussed by many other authors; see Varga (1962), Bodewig (1959), Faddeeva (1959), Newman (1962), and Sheldon (1960). The paper by Martin and Tee (1961) contains a useful survey of iterative methods.

Section 9.7 The Jacobi and Gauss-Seidel iterations have been widely analyzed. Varga (1962) considers both; the Gauss-Seidel method is discussed by Van Norton (1960). An excellent source on the acceleration of stationary iterative processes and their use in partial differential equations is Varga (1962); see also Sheldon (1960).

Section 9.8 Thorough discussions of matrix inversion are given by Bodewig (1959). For other techniques than we have presented here, see Wilf (1960), Householder (1953), and Westlake (1968).

Section 9.9 An extensive treatment of the problem of solving overdetermined systems of linear equations in the least-squares sense is given in Lawson and Hanson (1974), which also contains a collection of Fortran programs implementing the theory; see also Stewart (1973) and the contribution by Businger and Golub in Wilkinson and Reinsch (1971). Another approach to the problem via the singular-value decomposition of a matrix is also discussed in Lawson and Hanson and in the contribution by Golub and Reinsch in Wilkinson and Reinsch (1971).

The application of linear programming to the solution of overdetermined systems of linear equations in the L_1 and L_∞ norms is discussed by Rabinowitz (1968), who describes many applications of linear programming in numerical analysis.

Section 9.10 The simplex method was developed by Dantzig and his coworkers in the late forties and is treated in great detail in Dantzig (1963). Our treatment is based on that of Luenberger (1973), which is also the source of Example 9.6. An algorithm implementing a more stable variation of the simplex method is given in the contribution of Bartels, Stoer, and Zenger in Wilkinson and Reinsch (1971).

Section 9.11 Algorithms for the solution of complex systems of linear equations using the Crout factorization are given in the contribution by Bowdler, Martin, Peters, and Wilkinson in Wilkinson and Reinsch (1971). The accurate evaluation of determinants is discussed in Forsythe and Moler (1967). Algorithms for the solution of symmetric and unsymmetric band equations are given in two contributions by Martin and Wilkinson in Wilkinson and Reinsch (1971). The algorithm for the solution of a tridiagonal system appears in Dahlquist and Björck (1974) and in many other references.

BIBLIOGRAPHY

Birkhoff, G., and S. MacLane (1953): *A Survey of Modern Algebra*, rev. ed., The Macmillan Company, New York.

Bodewig, E. (1959): *Matrix Calculus*, 2d ed., Interscience Publishers, Inc., New York.

Crout, P. D. (1941): A Short Method for Evaluating Determinants and Solving Systems of Linear Equations with Real or Complex Coefficients, *Trans. AIEE*, vol. 60, pp. 1235–1240.

Dahlquist, G., and Å. Björck (1974): *Numerical Methods* (trans. N. Anderson), Prentice-Hall, Inc., Englewood Cliffs, N.J.

Dantzig, G. B. (1963): *Linear Programming and Extensions*, Princeton University Press, Princeton, N.J.

Faddeev, D. K., and V. N. Faddeeva (1963): *Computational Methods of Linear Algebra*, (trans. R. C. Williams), W. H. Freeman and Company, San Francisco.

Faddeeva, V. N. (1959): *Computational Methods of Linear Algebra* (trans. C. D. Benster), Dover Publications, Inc., New York.

Forsythe, G. E. (1953*a*): Solving Linear Algebraic Equations Can Be Interesting, *Bull. Am. Math. Soc.*, vol. 59, pp. 299–329.

——— (1953*b*): Tentative Classification of Methods and Bibliography on Solving Systems of Linear Equations, pp. 1–28 in *Simultaneous Linear Equations and the Determination of Eigenvalues* (L. J. Paige and O. Taussky, eds.), Natl. Bur. Std. Appl. Math. Ser. 29.

——— and C. B. Moler (1967): *Computer Solution of Linear Algebraic Systems*, Prentice-Hall, Inc., Englewood Cliffs, N.J.

Fox, L. (1965): *An Introduction to Numerical Linear Algebra*, Oxford University Press, New York.

Fraenkel, A. S., and D. Loewenthal (1971): Exact Solutions of Linear Equations with Rational Coefficients, *J. Res. Natl. Bur. Std.*, vol. 75B, pp. 67–75.

Goldstine, H. H., and J. Von Neumann (1951): Numerical Inverting of Matrices of High Order, II, *Proc. Am. Math. Soc.*, vol. 2, pp. 188–202.

Hildebrand, F. B. (1974): *Introduction to Numerical Analysis*, 2d ed., McGraw-Hill Book Company, New York.

Householder, A. S. (1953): *Principles of Numerical Analysis*, McGraw-Hill Book Company, New York.

——— (1958): Unitary Triangularization of a Nonsymmetric Matrix, *J. Ass. Comput. Mach.*, vol. 5, pp. 339–342.

——— (1964): *The Theory of Matrices in Numerical Analysis*, Blaisdell Publishing Company, New York.

Kahan, W. (1966): Numerical Linear Algebra, *Can. Math. Bull.*, vol. 9, 757–801.

Lawson, C. L., and R. J. Hanson (1974): *Solving Least Squares Problems*, Prentice-Hall, Inc., Englewood Cliffs, N.J.

Luenberger, D. G. (1973): *Introduction to Linear and Nonlinear Programming*, Addison-Wesley Publishing Company, Inc., Reading, Mass.

Martin, D. W., and G. J. Tee (1961): Iterative Methods for Linear Equations with Symmetric Positive Definite Matrix, *Comput. J.*, vol. 4, pp. 242–254.

Miles, E. P. (1960): Generalized Fibonacci Numbers and Associated Matrices, *Am. Math. Monthly*, vol. 67, pp. 745–752.

Modern Computing Methods (1961): Philosophical Library, New York.

Newman, M. (1962): Matrix Computations, pp. 225–254 in *Survey of Numerical Analysis* (J. Todd, ed.), McGraw-Hill Book Company, New York.

Noble, B. (1969): *Applied Linear Algebra*, Prentice-Hall, Inc., Englewood Cliffs, N.J.

Rabinowitz, P. (1968): Applications of Linear Programming to Numerical Analysis, *SIAM Rev.*, vol. 10, pp. 121–159.

Schwarz, H. R., H. Ruthishauser, and E. Stiefel (1973): *Numerical Analysis of Symmetric Matrices* (trans. P. Hertelendy), Prentice-Hall, Inc., Englewood Cliffs, N.J.

Sheldon, J. W. (1960): Iterative Methods for the Solution of Partial Differential Equations, pp. 144–156 in *Mathematical Methods for Digital Computers* (A. Ralston and H. S. Wilf, eds.), John Wiley & Sons, Inc., New York.

Stewart, G. W. (1973): *Introduction to Matrix Computations*, Academic Press, Inc., New York.

Van Norton, R. (1960): The Solution of Linear Equations by the Gauss-Seidel Method, pp. 56–61 in *Mathematical Methods for Digital Computers* (A. Ralston and H. S. Wilf, eds.), John Wiley & Sons, Inc., New York.

Varga, R. (1962): *Matrix Iterative Analysis*, Prentice-Hall, Inc., Englewood Cliffs, N.J.

Von Neumann, J., and H. H. Goldstine (1947): Numerical Inverting of Matrices of High Order, *Bull. Am. Math. Soc.*, vol. 53, pp. 1021–1099.

Westlake, J. R. (1968): *A Handbook of Numerical Matrix Inversion and Solution of Linear Equations*, John Wiley & Sons, Inc., New York.

Wilf, H. S. (1960): Matrix Inversion by the Method of Rank Annihilation, pp. 73–77 in *Mathematical Methods for Digital Computers* (A. Ralston and H. S. Wilf, eds.), John Wiley & Sons, Inc., New York.

Wilkinson, J. H. (1960): Rounding Errors in Algebraic Processes, pp. 44–53 in *Information Processing*, UNESCO, Paris.

——— (1961): Error Analysis of Direct Methods of Matrix Inversion, *J. Ass. Comput. Mach.*, vol. 8, pp. 281–330.

——— (1964): *Rounding Errors in Algebraic Processes*, Prentice-Hall, Inc., Englewood Cliffs, N.J.

——— (1967): The Solution of Ill-conditioned Linear Equations, pp. 65–93 in *Mathematical Methods for Digital Computers* (A. Ralston and H. S. Wilf, eds.), vol. II, John Wiley & Sons, Inc., New York.

——— and C. Reinsch (1971): *Linear Algebra: Handbook for Automatic Computation*, vol. II, Springer-Verlag, New York.

Young, D. (1971): *Iterative Solution of Large Linear Systems*, Academic Press, Inc., New York.

PROBLEMS

Section 9.1

1 Use Gaussian elimination to prove Theorem 9.1 and its corollary.

Section 9.2

2 (*a*) Is the matrix of the coefficients in the normal equations of polynomial least-squares approximations usually filled or sparse? Explain your answer.

(*b*) By considering simple difference approximations to derivatives, explain why the numerical solution of partial differential equations often results in systems of equations with sparse coefficient matrices.

3 The Fibonacci sequence is generated using the difference equation

$$f_j = f_{j-1} + f_{j-2} \qquad j > 1 \qquad f_0 = 0 \qquad f_1 = 1$$

(*a*) Show that $f_n f_{n+2} - f_{n+1}^2 = (-1)^{n+1}$, $n = 0, 1, \ldots$

(*b*) Thus find the unique solution of

$$f_n x_1 + f_{n+1} x_2 = f_{n+2} \qquad f_{n+1} x_1 + f_{n+2} x_2 = f_{n+3}$$

(*c*) How do you know that the system in part (*b*) becomes increasingly ill-conditioned as n increases?

(*d*) In particular let $n = 10$ in part (*b*) and replace f_{n+2} in the second equation by $f_{n+2} + \epsilon$. Calculate the solution for $\epsilon = .018$ and $\epsilon = .02$. For what value of ϵ does the solution not exist? [Ref.: Miles (1960).]

Section 9.3

4 Verify Eqs. (9.3-13) and (9.3-14).

5 Let A be an $m \times n$ matrix.

(*a*) Show that if P_{ij} is an $m \times m$ matrix defined by (9.3-15), then $P_{ij}A$ is the matrix A with rows i and j interchanged.

(*b*) Show that if P_{ij} is an $n \times n$ matrix defined by (9.3-15), then AP_{ij} is the matrix A with columns i and j interchanged.

(*c*) Use part (*a*) to show that $P_{ij}^2 = I$.

6 Show that if L_i is defined as in (9.3-16), then with $L_i^{-1} = \hat{L}$ we have $\hat{l}_{kj} = \delta_{kj}$, $j \neq i$, $\hat{l}_{ki} = \delta_{ki}$, $k \leq i$, $\hat{l}_{ki} = m_{ki}$, $k > i$.

7 (*a*) Prove that the product of two (unit) triangular matrices is a (unit) triangular matrix of the same form.

(*b*) Show that if L is a unit lower triangular matrix such that $l_{kj} = 0$ for $j \leq i, j \neq k$, and L_i is defined as in (9.3-16), then $L_i L = \bar{L}$ where $\bar{l}_{kj} = l_{kj}$, $j \neq i$, $\bar{l}_{ki} = l_{ki}$, $k \leq i$, $\bar{l}_{ki} = -m_{ki}$, $k > i$.

(*c*) Hence verify (9.3-22).

8 (*a*) Show that if $r_k \geq k > i$ and L_i is defined as in (9.3-16), then for any matrix B, $P_{k,r_k} L_i B = (P_{k,r_k} L_i P_{k,r_k}) P_{k,r_k} B$ where $P_{k,r_k} L_i P_{k,r_k}$ has the same form as L_i.

(*b*) Hence, show directly that (9.3-21) follows from (9.3-19).

9 (*a*) Use (9.3-7) to derive

$$a_{ij}^{(k-1)} = a_{ij} - m_{i1} a_{1j}^{(0)} - m_{i2} a_{2j}^{(1)} - \cdots - m_{i,k-1} a_{k-1,j}^{(k-2)} \qquad i, j \geq k$$

$$a_{1j}^{(0)} \equiv a_{1j}$$

(*b*) By making the identifications $m_{ir} = l_{ir}$, $i > r$ and $a_{rj}^{(r-1)} = u_{rj}$, $j \geq r$, show that the equations in part (*a*) for $i, j = k = r$ are identical with (9.3-34) and (9.3-35).

(*c*) Hence, conclude that if (9.3-34) and (9.3-35) are computed using single-precision arithmetic, the results are identical with those arising from Gaussian elimination.

10 Show that if we define $\hat{l}_{ir} = u_{rr} l_{ir}$, $i = r, \ldots, n$ (recall that $l_{rr} = 1$), and $\hat{u}_{rj} = u_{rj}/u_{rr}$, $j = r, \ldots, n$, then (9.3-34) and (9.3-35) go over to (9.3-38) and (9.3-37), respectively, and hence that $\hat{l}_{rr} = u_{rr}$, $r = 1, \ldots, n$.

11 (*a*) Verify (9.3-43).

(*b*) Compute the exact number of multiplications and divisions in the Cholesky factorization of a positive definite matrix.

(*c*) Repeat part (*b*) for the factorization (9.3-47).

12 (*a*) Use (9.3-47) to show that (9.3-48) and (9.3-49) follow from (9.3-45) and (9.3-46), respectively.

(*b*) Assume that we have a decomposition of A_{n-1} in the form $\bar{L}_{n-1}\bar{D}_{n-1}\bar{L}_{n-1}^T$, where $\bar{L}_{n-1}$ is unit lower triangular and $\bar{D}_{n-1}$ is a diagonal matrix with positive elements. Show that a decomposition of A_n exists of the form

$$A_n = \bar{L}_n \bar{D}_n \bar{L}_n^T \qquad \text{where } \bar{L}_n = \begin{bmatrix} \bar{L}_{n-1} & 0 \\ \bar{\mathbf{c}}_{n-1}^T & 1 \end{bmatrix} \qquad \bar{D} = \begin{bmatrix} \bar{D}_{n-1} & 0 \\ 0 & \bar{d}_n \end{bmatrix} \qquad \bar{d}_n > 0$$

and exhibit the values of $\bar{\mathbf{c}}_{n-1}$ and $\bar{d}_n$.

13 (*a*) Show that if we apply Gaussian elimination without pivoting to a symmetric matrix A, then $m_{i1} = a_{1i}/a_{11}$.

(*b*) Deduce from this that $A^{(1)}$ is symmetric and consequently that with $m_{ik} = a_{ki}^{(k-1)}/a_{kk}^{(k-1)}$ all $A^{(k)}$ are symmetric.

(*c*) Show that the computation required is almost halved compared with the nonsymmetric case.

(*d*) Use this simplification to solve the system

$$\begin{aligned} .6428x_1 + .3475x_2 - .8468x_3 &= .4127 \\ .3475x_1 + 1.8423x_2 + .4759x_3 &= 1.7321 \\ -.8468x_1 + .4759x_2 + 1.2147x_3 &= -.8621 \end{aligned}$$

14 Let A be a symmetric positive definite matrix of order n and M a nonvoid subset of $N = \{1, 2, \ldots, n\}$. By considering the set of vectors $\mathbf{x} = (x_1, x_2, \ldots, x_n)^T$ such that $x_i = 0$, $i \in N - M$, show that the submatrix $A_M = (a_{ij} \mid i, j \in M)$ is positive definite; i.e., show that every principal minor of A is positive definite.

15 Consider the system

$$\begin{aligned} .4096x_1 + .1234x_2 + .3678x_3 + .2943x_4 &= .4043 \\ .2246x_1 + .3872x_2 + .4015x_3 + .1129x_4 &= .1550 \\ .3645x_1 + .1920x_2 + .3781x_3 + .0643x_4 &= .4240 \\ .1784x_1 + .4002x_2 + .2786x_3 + .3927x_4 &= -.2557 \end{aligned}$$

(*a*) Solve this system without pivoting by Gaussian elimination, carrying four decimal places throughout.

(*b*) Solve the system of part (*a*) using partial pivoting and compare the results with those of part (*a*).

***16** Let A_n be the matrix of order n

$$a_{ii} = 1 \quad i \neq n \qquad a_{ij} = \begin{cases} (-1)^{i+j-1} & i > j \quad \text{or} \quad j = n \\ 0 & i < j < n \end{cases}$$

(*a*) Show that the elements of the inverse of this matrix are given by

$$a_{ii}^{-1} = 2^{-1} \qquad i \neq n \qquad a_{nj}^{-1} = (-1)^{n+j-1}2^{-j} \qquad j \neq n$$

$$a_{ij}^{-1} \begin{cases} = 0 & i > j;\ i \neq n \\ = (-1)^{i+j-1}2^{-(j-i+1)} & i < j;\ j \neq n \end{cases} \qquad \begin{aligned} a_{in}^{-1} &= (-1)^{n+i-1}2^{-n+i} \quad i \neq n \\ a_{nn}^{-1} &= -2^{-n+1} \end{aligned}$$

(*b*) Write out A_6 and A_6^{-1}.

(*c*) Why is A_n very well-conditioned?

(*d*) Consider A_{31} and let B_{31} be the matrix formed by replacing $a_{31,31}$ by $-\frac{1}{2}$. Show that $B_{31}^{-1} = A_{31}^{-1} - \frac{1}{2}$ (last column of A_{31}^{-1}) (last row of A_{31}^{-1})/$(1 - 2^{-31})$, and thus deduce that some of the elements of A_{31}^{-1} differ from those of B_{31}^{-1} in the first decimal.

(*e*) Show that, if partial pivoting is used, Gaussian elimination applied to A_{31} and B_{31} is identical until the final step; i.e., show that the thirty-first row never has the element with *greatest* magnitude in a column.

(*f*) Show that the final element in the triangle should be -2^{30} for A_{31} and $-2^{30} + \frac{1}{2}$ for B_{31}. Thus deduce that if fewer than 30 binary digits are used in the computation, Gaussian elimination applied to A_{31} and B_{31} leads to identical L and U matrices, and thus to identical inverses [or solutions of (9.1-1)] in spite of the fact that the true inverses differ substantially.

(*g*) Why would complete pivoting avoid this difficulty? [Ref.: Wilkinson (1961), pp. 327–328.]

***17** (*a*) Let A in (9.1-1) be symmetric and positive definite and let B be the matrix of coefficients in (9.3-2) excluding the first row. Show that

$$\sum_{i=1}^{n}\sum_{j=1}^{n} a_{ij}x_ix_j - a_{11}\left(x_1 + \sum_{i=2}^{n}\frac{a_{i1}}{a_{11}}x_i\right)^2 = \sum_{i=2}^{n}\sum_{j=2}^{n} a_{ij}^{(1)}x_ix_j$$

(*b*) Deduce from this that B is positive definite and therefore that the matrices of all the derived systems are positive definite.

(*c*) Further, show that

$$a_{ii}^{(1)} \leq a_{ii} \qquad i = 2, \ldots, n$$

and thus deduce that

$$\max_{2 \leq i,\, j \leq n} |a_{ij}^{(1)}| \leq \max_{2 \leq i,\, j \leq n} |a_{ij}|$$

(*d*) From parts (*b*) and (*c*) deduce that if $|a_{ij}| < 1$, then $|a_{ij}^{(k)}| < 1$ for all k.

(*e*) By giving a 2×2 example, show that with no pivoting some of the multipliers m_{ij} may be very large. [Ref.: Wilkinson (1961), pp. 285–286.]

18 (*a*) Solve the system

$$\begin{aligned} .2641x_1 + .1735x_2 + .8642x_3 &= -.7521 \\ .9411x_1 - .0175x_2 + .1463x_3 &= .6310 \\ -.8641x_1 - .4243x_2 + .0711x_3 &= .2501 \end{aligned}$$

using Gaussian elimination with (*i*) no pivoting, (*ii*) partial pivoting, (*iii*) complete pivoting.

(*b*) Repeat part (*a*) for the symmetric system

$$\begin{aligned} .4721x_1 + .2352x_2 - .2613x_3 + .8421x_4 &= -.2317 \\ .2352x_1 + .7411x_2 - .0463x_3 + .1569x_4 &= .3219 \\ -.2613x_1 - .0463x_2 + .8955x_3 + .1748x_4 &= .6217 \\ .8421x_1 + .1569x_2 + .1748x_3 + .9841x_4 &= .9835 \end{aligned}$$

19 (*a*) Consider the equilibrated system $A\mathbf{x} = \mathbf{b}$, where

$$A = \begin{bmatrix} \epsilon & -1 & 1 \\ -1 & 1 & 1 \\ 1 & 1 & 1 \end{bmatrix} \qquad |\epsilon| \ll 1$$

Verify that

$$A^{-1} = \begin{bmatrix} 0 & -2 & 2 \\ -2 & 1-\epsilon & 1+\epsilon \\ 2 & 1+\epsilon & 1-\epsilon \end{bmatrix}$$

(*b*) Use (1.3-16) to show that $K_\infty(A) = 12$ so that A is a well-conditioned matrix [See page 431 for the definition of $K(A)$].

(*c*) Make the substitutions $x_2' = x_2/\epsilon$, $x_3' = x_3/\epsilon$ into the system $A\mathbf{x} = \mathbf{b}$ and equilibrate the resulting system to yield $A'\mathbf{x}' = \mathbf{b}'$. Show that

$$A' = \begin{bmatrix} 1 & -1 & 1 \\ -1 & \epsilon & \epsilon \\ 1 & \epsilon & \epsilon \end{bmatrix}$$

(*d*) Verify that

$$A'^{-1} = \begin{bmatrix} 0 & -\dfrac{1}{2} & \dfrac{1}{2} \\[2ex] -\dfrac{1}{2} & \dfrac{1-\epsilon}{4\epsilon} & \dfrac{1+\epsilon}{4\epsilon} \\[2ex] \dfrac{1}{2} & \dfrac{1+\epsilon}{4\epsilon} & \dfrac{1-\epsilon}{4\epsilon} \end{bmatrix}$$

(*e*) Use (1.3-16) to show that $K'_\infty(A') = \frac{3}{2} + 3/2\epsilon$, so that A' is an ill-conditioned matrix even though A' is equilibrated.

(*f*) Show that the result of solving $A'\mathbf{x}' = \mathbf{b}'$ with partial pivoting or (if $\epsilon < 0$) with complete pivoting is equivalent to solving $A\mathbf{x} = \mathbf{b}$ without pivoting.

(*g*) For $\epsilon = -10^{-3}$ and $\mathbf{b} = (1, 1, 3)^T$ and using 4-digit floating point decimal arithmetic with rounding, solve the system $A\mathbf{x} = \mathbf{b}$ without pivoting.

(*h*) Repeat part (*g*) with partial pivoting.

(*i*) Repeat part (*g*) for the system $A'\mathbf{x}' = \mathbf{b}'$ and compare the result transformed by the equations $x_1 = x_1'$, $x_2 = \epsilon x_2'$, $x_3 = \epsilon x_3'$ with the results of parts (*g*) and (*h*). [Ref.: Dahlquist and Björck (1974), p. 182, and Fox (1965), p. 169.]

20 (*a*) Give formulas corresponsing to (9.3-28) and (9.3-8) to solve the system $\hat{L}\hat{U}\mathbf{x} = \mathbf{b}$, where $\hat{L}$ is a nonsingular lower triangular matrix and $\hat{U}$ is unit upper triangular (the Crout case).

(*b*) Repeat part (*a*) for the system $LL^T\mathbf{x} = \mathbf{b}$ (the Cholesky case).

(*c*) Repeat part (*a*) for the system $LDU\mathbf{x} = \mathbf{b}$, where L and U are unit triangular and D is diagonal and nonsingular.

(*d*) Specialize part (*c*) to the system $\bar{L}\bar{D}\bar{L}^T\mathbf{x} = \mathbf{b}$ arising from (9.3-47).

Section 9.4

21 (*a*) Show that for any matrix norm $\|I\| \geq 1$, where I is the identity matrix.

(*b*) Show that for any matrix norm $\|A\| \cdot \|A^{-1}\| \geq 1$.

22 Verify (9.4-10).

23 Diagonally dominant matrices. (*a*) If A is diagonally dominant, that is, $\sum_{j=1, j\neq i}^{n} |a_{ij}| < |a_{ii}|$, $i = 1, \ldots, n$, show that $a_{ii} \neq 0$, $i = 1, \ldots, n$.

(*b*) Let $A = D^{-1}B$, where $D = \text{diag}(a_{ii})$. Show that $B = I - C$, where $\|C\|_\infty < 1$, so that by Corollary 9.3, B and hence A are nonsingular.

(*c*) Show that if we apply Gaussian elimination to a diagonally dominant matrix A, all the elements $a_{kk}^{(k-1)} \neq 0$, so that in theory, pivoting is never necessary.

24 Let A be a matrix such that A^T is diagonally dominant, i.e. such that $\sum_{i=1, i\neq j}^{n} |a_{ij}| < |a_{jj}|$, $j = 1, \ldots, n$ and apply Gaussian elimination without pivoting.

(*a*) Show that $\sum_{i=2}^{n} |m_{i1}| < 1$.

(*b*) Show that $\sum_{i=2}^{n} |a_{ij}^{(1)}| < \sum_{i=1}^{n} |a_{ij}^{(0)}|$, $a_{ij}^{(0)} = a_{ij}$, $j = 2, \ldots, n$.

(*c*) Show that $|a_{jj}^{(1)}| > \sum_{i=2, i\neq j}^{n} |a_{ij}^{(1)}|$, $j = 2, \ldots, n$, so that $A^{(1)T}$ is also diagonally dominant.

(*d*) Repeat the arguments above to show that $\sum_{i=r+1}^{n} |a_{ij}^{(r)}| < \sum_{i=r}^{n} |a_{ij}^{(r-1)}|$, $r = 2, \ldots, n-1$, and hence that $|a_{ij}^{(r)}| < \sum_{i=1}^{n} |a_{ij}^{(0)}| < 2|a_{jj}|$, $r = 1, \ldots, n$.

(*e*) Hence conclude that if A is column diagonally dominant, i.e., if A^T is (row) diagonally dominant and we apply Gaussian elimination without pivoting, the growth factor $g < 2$. [Ref.: Wilkinson (1961).]

25 (*a*) Verify (9.4-40) and (9.4-41).

(*b*) Derive from (9.3-8), a formula similar to (9.4-43) and use it to verify (9.4-46) and (9.4-47).

(*c*) Derive formulas similar to (9.4-43) and that of part (*b*) for the case of the forward and back substitution formulas connected with the Crout algorithm given in Prob. 20.

(*d*) Derive formulas analogous to (9.4-44) to (9.4-51) for the Crout case.

(*e*) Verify (9.4-53).

(*f*) Verify (9.4-54).

Section 9.5

26 Verify (9.5-4).

27 (*a*) Using 4-digit floating-point decimal arithmetic with chopping (not rounding), obtain a solution correct to three significant figures to the system

$$.8647x_1 + .5766x_2 = .2885$$

$$.4322x_1 + .2882x_2 = .1442$$

by using iterative refinement.

(*b*) Repeat part (*a*) using 9-digit arithmetic, either chopped or rounded.

(*c*) Compute the exact residual vectors for the solutions obtained in parts (*a*) and (*b*) and thus verify that the solution obtained in part (*b*) is the exact solution.

(*d*) Account for the failure of the iterative refinement process in part (*a*). [Ref.: Kahan (1966).]

28 Consider the system

$$.05x_1 + .07x_2 + .06x_3 + .05x_4 = .23$$

$$.07x_1 + .10x_2 + .08x_3 + .07x_4 = .32$$

$$.06x_1 + .08x_2 + .10x_3 + .09x_4 = .33$$

$$.05x_1 + .07x_2 + .09x_3 + .10x_4 = .31$$

(*a*) What is the true solution of this system?

(*b*) Solve this system using the Doolittle method with partial pivoting.

(*c*) Repeat part (*b*) using the Crout method with partial pivoting.

(*d*) Repeat part (*b*) using the LDL^T decomposition.

(*e*) Apply iterative refinement for each of the above methods until the solution is correct to four significant figures.

Section 9.6

29 (*a*) Show that the matrix iteration $B_{i+1} = B_i(2I - AB_i)$, B_1 arbitrary, to find A^{-1} is the matrix analog of the Newton-Raphson method for finding the reciprocal of a number.

(*b*) Defining $C_i = I - AB_i$, show that $C_i = C_1^{2i-1}$.

(*c*) Thus deduce that a sufficient condition for the convergence of the iteration of part (*a*) to A^{-1} is that all the eigenvalues of C_1 lie within the unit circle. (This condition is also necessary.)

(*d*) How could this iteration be used to solve systems of linear equations? Do two iterations on the system of Example 9.4 with $B_1 = \frac{1}{4}I$. What is bad about this method from the computational point of view? [Ref.: Newman (1962), pp. 223–226.]

30 (*a*) Let $\mathbf{x}_t$ be a solution of $B\mathbf{x} = 0$, where $\det(B) = 0$, and let $x_t^{(i)}$ be the component of $\mathbf{x}_t$ of largest magnitude. Show that

$$|b_{ii}| \leq \sum_{\substack{k=1 \\ k \neq i}}^{n} |b_{ik}|$$

(*b*) Use this result to prove that

$$|\lambda_{\max}| \leq \max_i \sum_{k=1}^{n} |a_{ik}| \qquad |\lambda_{\min}| \geq \min_i \left(|a_{ii}| - \sum_{\substack{k=1 \\ k \neq i}}^{n} |a_{ik}| \right)$$

where $\lambda_{\max}$ and $\lambda_{\min}$ are the eigenvalues of $A = [a_{ij}]$ of maximum and minimum magnitude, respectively.

(*c*) Thus deduce that the Jacobi iteration for a matrix A with diagonal terms equal to 1 converges if

$$\sum_{\substack{k=1 \\ k \neq i}}^{n} |a_{ik}| < 1 \qquad i = 1, \ldots, n \qquad \text{or if} \qquad \sum_{\substack{k=1 \\ k \neq i}}^{n} |a_{ki}| < 1 \qquad i = 1, \ldots, n$$

Section 9.7

31 Use (9.7-9) to derive Eqs. (9.7-10) and (9.7-11).

***32** In the Gauss-Seidel method let $x_{i+1}^{(k)}$ be the kth component of $\mathbf{x}_{i+1}$ and let

$$r_{i+1}^{(k)} = b_k - \sum_{j=1}^{k-1} a_{kj} x_{i+1}^{(j)} - \sum_{j=k}^{n} a_{kj} x_i^{(j)}$$

(*a*) Show that

$$x_{i+1}^{(k)} = x_i^{(k)} + \frac{r_{i+1}^{(k)}}{a_{kk}}$$

(*b*) If $\boldsymbol{\epsilon}_i = \mathbf{x}_t - \mathbf{x}_i$, show that

$$\epsilon_{i+1}^{(k)} = \epsilon_i^{(k)} - \frac{r_{i+1}^{(k)}}{a_{kk}} \qquad \text{and} \qquad r_{i+1}^{(k)} = \sum_{j=1}^{k-1} a_{kj} \epsilon_{i+1}^{(j)} + \sum_{j=k}^{n} a_{kj} \epsilon_i^{(j)}$$

(*c*) Let A be symmetric and consider the quadratic form $Q(\epsilon_i) = \epsilon_i^T A \epsilon_i$. Show that

$$Q(\epsilon_{i+1}) - Q(\epsilon_i) = -\sum_{j=1}^{n} \frac{(r_{i+1}^{(j)})^2}{a_{jj}}$$

(*d*) From this deduce that if A is a nonsingular symmetric matrix with positive diagonal elements, and if the Gauss-Seidel method converges for any $\mathbf{x}_1$, then A must be positive definite. [Ref.: Van Norton (1960).]

33 (*a*) Let L and U be, respectively, lower and upper triangular nonnegative matrices with zeros on the main diagonal. Prove that $(I - L)^{-1}U$ is nonnegative.

(*b*) Thus deduce that if A is a matrix with 1s on the main diagonal and nonpositive entries otherwise, both B_J and B_G, the Jacobi and Gauss-Seidel iteration matrices, are nonnegative. [Ref.: Varga (1962), p. 69.]

34 (*a*) By calculating the spectral radius of the B matrices for the Jacobi and Gauss-Seidel iterations in Example 9.4, show that the Gauss-Seidel method would be expected to converge substantially faster than the Jacobi method.

(*b*) Apply the δ^2 process to the results of the Jacobi iteration in Example 9.4 with $i = 2$. How do you explain this result?

(*c*) Apply the δ^2 process to the Gauss-Seidel results with $i = 3$. How do you explain this result?

35 (*a*) Determine the spectral radius of the B matrix for the Gauss-Seidel iteration for the system

$$.96326x^{(1)} + .81321x^{(2)} = .88824$$

$$.81321x^{(1)} + .68654x^{(2)} = .74988$$

(*b*) Compute the exact solution of this system to five decimal places using Cramer's rule.

(*c*) Attempt to use the Gauss-Seidel method to find a solution of this system starting with $x^{(1)} = .33116$, $x^{(2)} = .70000$. How do you explain your result? [Ref.: Wilkinson (1961), pp. 328–329.]

36 Use (9.7-30) to derive (9.7-31) and (9.7-32).

***37** Consider the nonstationary iteration

$$\mathbf{x}_{i+1} = [(1 + \omega_i)B - \omega_i I]\mathbf{x}_i + (1 + \omega_i)C\mathbf{b}$$

where B and C satisfy the consistency condition (9.6-11).

(*a*) Show that this iteration also satisfies the consistency condition and that K_i defined by (9.6-16) is

$$K_i(B) = \prod_{j=1}^{i} [(1 + \omega_j)B - \omega_j I]$$

(*b*) By letting $\omega_i = \gamma_i/(1 - \gamma_i)$ show that

$$K_i(B) = \prod_{j=1}^{i} \frac{B - \gamma_j I}{1 - \gamma_j}$$

(*c*) If B is symmetric, deduce that $\rho[K_i(B)] = \max_j |K_i(u_j)|$, where $u_j, j = 1, \ldots, n$, are the eigenvalues of B.

(*d*) Suppose we know that $-1 < x_0 \le \mu_j \le x_1 < 1$ for all μ_j. Show then that in order to minimize the magnitude of $\rho[K_i(B)]$ we should like to have

$$K_i(x) = \frac{T_i(ax + b)}{T_i(a + b)}$$

where $T_i(x)$ is the Chebyshev polynomial of degree i and

$$a = \frac{2}{x_1 - x_0} \qquad b = -\frac{x_1 + x_0}{x_1 - x_0}$$

(*e*) Thus deduce that the γ_j's should be chosen as the zeros of $T_i(ax + b)$ with the ω_i's then given by part (*b*).

(*f*) Suppose these ω_i's have been used to compute $\mathbf{x}_{i+1}$. Besides requiring some knowledge of the eigenvalues of B, what disadvantage does this technique have if, having computed $\mathbf{x}_{i+1}$, we now desire $\mathbf{x}_{i+2}$? [Ref.: Sheldon (1960), pp. 146–147.]

38 Consider the system

$$\begin{aligned} x^{(1)} \qquad\qquad\quad - \tfrac{1}{4}x^{(3)} - \tfrac{1}{4}x^{(4)} &= \tfrac{1}{2} \\ x^{(2)} - \tfrac{1}{4}x^{(3)} - \tfrac{1}{4}x^{(4)} &= \tfrac{1}{2} \\ -\tfrac{1}{4}x^{(1)} - \tfrac{1}{4}x^{(2)} + x^{(3)} \qquad\quad &= \tfrac{1}{2} \\ -\tfrac{1}{4}x^{(1)} - \tfrac{1}{4}x^{(2)} \qquad\quad + x^{(4)} &= \tfrac{1}{2} \end{aligned}$$

(*a*) Starting with $\mathbf{x}_1 = \mathbf{0}$, do four iterations of the Jacobi method.

(*b*) Using the same starting vector do four iterations of the Gauss-Seidel method.

(*c*) What is the true solution of the system?

39 (*a*) Calculate the B matrix for the Jacobi method for the system of the previous problem and find its spectral radius.

(*b*) Similarly, find the spectral radius of B for the Gauss-Seidel iteration.

(*c*) Are the results of the previous problem in accord with these results?

Section 9.8

40 Prove that the inverse of a triangular matrix is a triangular matrix of the same form.

41 (*a*) Derive the algorithm for inverting the upper triangular matrix U which is analogous to (9.8-3).

(*b*) Calculate that the inversion of L and U and the computation of $U^{-1}L^{-1}$ requires $\frac{2}{3}n^3 + O(n^2)$ multiplications and divisions and thus deduce that the complete inversion of A by this method requires $n^3 + O(n^2)$ operations.

42 (*a*) Show that the inverse of a matrix A can be found by solving the n systems

$$A\mathbf{x} = \mathbf{e}_i \qquad i = 1, \ldots, n$$

where $\mathbf{e}_i$ is the vector with a 1 in the ith position and zeros elsewhere.

*(*b*) How many operations are required to find the inverse in this manner?

(*c*) Show that the Gauss-Jordan reduction of A to I is equivalent to the premultiplication of A by a sequence of elementary matrices and thus deduce that A^{-1} can be found by applying to I the same operations required to reduce A to I by the Gauss-Jordan reduction. What is the connection of this method with that of part (*a*)?

43 (*a*) If $A = LU$, where L and U are as in Sec. 9.8, show that $A^{-1}L = U^{-1}$

(*b*) If U^{-1} has been calculated as in Prob. 41, derive an algorithm for computing A^{-1} directly one column at a time.

(*c*) How many operations are required to calculate A^{-1} in this manner? [Ref.: Bodewig (1959), pp. 214–215.]

44 (*a*) Consider the two equations

$$UA^{-1} = L^{-1} \qquad A^{-1}L = U^{-1}$$

Show that these are $2n^2$ equations for the elements of A^{-1}, L^{-1}, and U^{-1} but that $n^2 - n$ of the equations involve only zero elements of L^{-1} and U^{-1} and n equations involve elements of L^{-1} known to be 1.

(*b*) Thus derive n^2 equations for the n^2 elements of A^{-1} which involve only known elements of U^{-1} or L^{-1}.

(*c*) Derive an algorithm for solving these equations. [Ref.: Bodewig (1959), pp. 215–216.]

45 Indicate how the matrix inversion scheme of Sec. 9.8 can be simplified if A is symmetric. How many operations are required for the inversion in this case?

46 (*a*) Let $\mathbf{u}$ and $\mathbf{v}$ be column vectors and A a nonsingular square matrix. Verify that

$$(A + \mathbf{u}\mathbf{v}^T)^{-1} = A^{-1} - \frac{(A^{-1}\mathbf{u})(\mathbf{v}^T A^{-1})}{1 + \mathbf{v}^T A^{-1}\mathbf{u}}$$

(*b*) Let

$$B = D + \sum_{i=1}^{m} \mathbf{u}_i \mathbf{v}_i^T$$

where D is a nonsingular diagonal matrix. Define

$$C_k = \left(\sum_{i=1}^{k} \mathbf{u}_i \mathbf{v}_i^T + D \right)^{-1}$$

Use the result of part (*a*) to prove that

$$C_{k+1} = C_k - \frac{(C_k \mathbf{u}_{k+1})(\mathbf{v}_{k+1}^T C_k)}{1 + \mathbf{v}_{k+1}^T C_k \mathbf{u}_{k+1}}$$

(*c*) Use the result of part (*b*) to deduce an algorithm for calculating B^{-1} with B as given in part (*b*). [Ref.: Wilf (1960), p. 73.]

47 Let B be an $n \times n$ matrix such that $b_{11} \neq 1$ and let $A = B - I$. Define a sequence of matrices $A^{(k)} = [a_{ij}^{(k)}]$ such that

$$A^{(1)} = A$$

$$a_{ij}^{(k+1)} = a_{ij}^{(k)} - \frac{a_{ik}^{(k)} a_{kj}^{(k)}}{a_{kk}^{(k)}} \qquad k = 1, \ldots, n$$

(*a*) Prove that $a_{ij}^{(k)} = 0$ if $i < k$ or $j < k$ for $k = 1, \ldots, n$, and thus deduce that $A^{(n+1)} = 0$.

(*b*) Thus show that B may be expanded as in the previous problem with $D = I$, $u_i^{(j)} = a_{ji}^{(i)}/a_{ii}^{(i)}$, $v_i^{(j)} = a_{ij}^{(i)}$.

(*c*) Specialize the algorithm of part (*c*) of the previous problem for this expansion. [Ref.: Wilf (1960), p. 74.]

48 Invert the matrix of coefficients in Prob. 18*a* using (*a*) the method of Sec. 9.8; (*b*) the method of Prob. 42; (*c*) the method of Prob. 43; (*d*) the method of Prob. 44; (*e*) the method of Probs. 46 and 47.

49 Repeat the calculations of the previous problem using the matrix of Prob. 18*b*.

Section 9.9

50 Show that if the matrix A is positive (semi-) definite, then all of its eigenvalues are positive (nonnegative).

51 (*a*) Show that $\mathbf{x}^T A^T A \mathbf{x} = \|A\mathbf{x}\|_2^2$ and hence that $A^T A$ is positive semidefinite.

(*b*) Show that if the columns of A are linearly independent, then $A^T A$ is positive definite.

52 (*a*) Show that if P is defined by (9.9-5), then $P^2 = I$.

(*b*) Verify that P is symmetric and hence, using part (*a*), orthogonal.

53 (*a*) Show that for any nonsingular matrix B, the spectral condition number of $B^T B$ is $K_2(B^T B) = [K_2(B)]^2$.

(*b*) Show that if $PA = (R, 0)^T$, where P is orthogonal, then $A^T A = R^T R$.

(*c*) Hence show that $K_2(A^T A) = [K_2(R)]^2$ if R is nonsingular.

54 Determine the number of multiplications and divisions required to solve the system $A\mathbf{x} = \mathbf{b}$ by the method of Householder transformations.

55 Solve parts (*a*) and (*b*) of Prob. 19 of Chap. 6 using the method of Householder transformations.

56 Denote by a_{ij} and a'_{ij}, $i = 1, \ldots, n, j = 1, \ldots, m$, the elements of A_k and $A_{k+1} = P_k A_k$, respectively. Show that a'_{ij} can be computed as follows:

$$a'_{ij} = a_{ij} \qquad i = 1, \ldots, n; j = 1, \ldots, k-1$$

$$i = 1, \ldots, k-1; j = k, \ldots, m$$

$$a'_{i,k} = 0 \qquad i = k+1, \ldots, n$$

$$u_i = a_{i,k} \quad i = k+1, \ldots, n \qquad \alpha = \sum_{i=k+1}^{n} u_i^2 \qquad \delta = \operatorname{sgn}(a_{kk})(a_{kk}^2 + \alpha)^{1/2}$$

$$u_k = a_{k,k} + \delta \qquad \beta = \frac{2}{\alpha + u_k^2} \qquad a'_{kk} = \delta$$

$$\gamma_j = \beta \sum_{i=k}^{n} u_i a_{ij} \qquad a'_{ij} = a_{ij} - \gamma_j u_i \qquad i = k, \ldots, n \qquad j = k+1, \ldots, m$$

57 What are the values of $\mathbf{c}$, t, t_1, s_1, s, and $\mathbf{a}_i$ of the linear programming problem defined by (9.9-11) and (9.9-12) for the problem defined by (9.9-14) and (9.9-15)?

58 (*a*) Formulate the problem of finding the best L_∞ approximation to the data of Example 9.5 by a polynomial of degree 4 as a linear programming problem.

(*b*) Repeat part (*a*) for the best L_1 approximation.

Section 9.10

59 (*a*) Show that if $\mathbf{x}_1$ and $\mathbf{x}_2$ are feasible solutions, then so is $\alpha\mathbf{x}_1 + (1-\alpha)\mathbf{x}_2$ for any α such that $0 \le \alpha \le 1$.

(*b*) Show that if $f(\mathbf{x})$ is a linear function of $\mathbf{x}$, then

$$f(\alpha\mathbf{x}_1 + (1-\alpha)\mathbf{x}_2) = \alpha f(\mathbf{x}_1) + (1-\alpha) f(\mathbf{x}_2).$$

(*c*) Use the results of parts (*a*) and (*b*) to show that any local minimum of $f(\mathbf{x})$ over the set of feasible vectors is also a global minimum.

60 (*a*) Let $\mathbf{x} = (x_1, \ldots, x_n)^T$ be feasible solution of (9.10-2) and (9.10-3) such that the columns of A corresponding to nonzero components of $\mathbf{x}$ are linearly dependent. Show that we can find another feasible solution $x^* = (x_1^*, \ldots, x_n^*)^T$ with at least one less nonzero component.

(*b*) Use part (*a*) to show that if there is a feasible solution to (9.10-2) and (9.10-3), there exists a basic feasible solution.

(*c*) Let $\hat{\mathbf{x}} = (\hat{x}_1, \ldots, \hat{x}_n)^T$ be an optimal solution of (9.10-1) to (9.10-3) such that the columns of A corresponding to nonzero components of $\hat{\mathbf{x}}$ are linearly dependent; i.e., there exists a vector $\mathbf{y} = (y_1, \ldots, y_n)^T$ such that $\sum_{i=1}^{n} y_i \mathbf{a}_i = \mathbf{0}$ and $\hat{x}_k y_k \neq 0$ for some k. Show that $\mathbf{c}^T\mathbf{y} = 0$.

(*d*) Use parts (*b*) and (*c*) to show that if there is an optimal solution to (9.10-1) to (9.10-3), then there exists a basic optimal solution.

61 Solve the following linear programming problem by the simplex method:

$$\text{Minimize } x_1 + 6x_2 - 7x_3 + x_4 + 5x_5$$

subject to the constraints

$$5x_1 - 4x_2 + 13x_3 - 2x_4 + x_5 = 20$$

$$x_1 - x_2 + 5x_3 - x_4 + x_5 = 8$$

$$x_i \ge 0 \qquad i = 1, \ldots, 5$$

[Ref.: Dantzig (1963), pp. 108–110.]

Section 9.11

62 Verify (9.11-3).

63 Evaluate the determinants of the matrices of Probs. 13, 15, 18, 27, and 28.

64 (*a*) Show that if A has bandwidth $[w_1, w_2]$, then, even with partial pivoting, $m_{i1} = 0$, $i < w_1 + 1$.

(*b*) Let $A^{(k)}$ be the matrix of the kth derived system in Gaussian elimination. Show that in the absence of pivoting, $A^{(1)}$ and consequently all $A^{(k)}$, $k = 1, \ldots, n-1$, have bandwidth at most $[w_1, w_2]$.

(*c*) Show that if $w_1 + w_2 \le n - 1$, then the maximum bandwidth of $A^{(1)}$ is $[w_1, w_2 + w_1]$ if we allow partial pivoting. Show further that the maximum bandwidth of all subsequent $A^{(k)}$ is $[w_1, w_2 + w_1]$, so that $U = A^{(n-1)}$ has bandwidth $[0, w_2 + w_1]$.

(*d*) Show that in both cases, $m_{ik} = 0$, $i > w_1 + k$, so that the matrix $L = [m_{ij}]$ has bandwidth $[w_1, 0]$.

65 Determine the LU decomposition of a tridiagonal matrix A using (9.3-34) and (9.3-35) and verify (9.11-8) and that the superdiagonal of U is equal to that of A.

66 Give an algorithm to solve a tridiagonal system by Gaussian elimination, partial pivoting, and implicit row equilibration which requires five vectors for storage if we allow overwriting. Determine the maximum number of additions, multiplications, and divisions required.

67 Show that if we do forward and back substitution with band matrices L and U of bandwidth $[w_1, 0]$ and $[0, w_2]$, respectively, where L is unit triangular, then the total number of operations is

$$n(w_1 + w_2 - 2) - \frac{w_1(w_1 - 1)}{2} - \frac{w_2(w_2 - 1)}{2}$$

additions and multiplications and n divisions.

68 (*a*) Let A be an upper Hessenberg equilibrated matrix. Prove by induction that if we apply Gaussian elimination with partial pivoting, then $|a_{ij}^{(r-1)}| \le r$, so that the growth factor $g \le n$.

(*b*) Let A be a tridiagonal matrix and apply Gaussian elimination with pivoting. Show that $g \le 2$.

(*c*) Let A be a band matrix of bandwidth $[k, k]$ and apply Gaussian elimination without pivoting. Show that $g \le 2^k$.

69 Assume that we carry out the process of Gaussian elimination with partial pivoting from bottom to top and from right to left rather than the usual way of top to bottom and left to right.

(*a*) Show that we arrive at the decomposition $PA = UL$, where U is a unit upper triangular matrix and L is a lower triangular matrix.

(*b*) Give the formulas for the elements of U and L in the case where P is the unit matrix, i.e., no pivoting.

(*c*) Show that with no pivoting, if A is a band matrix of bandwidth $[w_1, w_2]$, then U and L have bandwidths $[0, w_2]$ and $[w_1, 0]$, respectively.

(*d*) Show that with partial pivoting U still has bandwidth $[0, w_2]$ while L has bandwidth $[0, \min(w_1 + w_2, n-1)]$. Hence, for lower Hessenberg matrices, we have a decomposition requiring only $O(n^2)$ multiplications.

CHAPTER
TEN

THE CALCULATION OF EIGENVALUES AND EIGENVECTORS OF MATRICES

10.1 BASIC RELATIONSHIPS

Let A be a square matrix of order n. Its eigenvalues $\lambda_1, \ldots, \lambda_n$ are the solutions of the determinantal equation, called the characteristic equation†

$$|A - \lambda I| = 0 \tag{10.1-1}$$

Since $|A^T - \lambda I| = |A - \lambda I|$, it follows that the set of eigenvalues of A^T is identical to that of A. Corresponding to each distinct eigenvalue λ_i there exists at least one nontrivial solution (determined to within a multiplicative constant) of the system of linear equations

$$A\mathbf{x} = \lambda_i \mathbf{x} \tag{10.1-2}$$

This solution $\mathbf{x}_i^T = (x_i^{(1)}, x_i^{(2)}, \ldots, x_i^{(n)})$ is a *right eigenvector* of A. (In what follows the term *eigenvector* will refer exclusively to right eigenvectors.) A *left eigenvector* corresponding to λ_i is a nontrivial solution of

$$\mathbf{y}^T A = \lambda_i \mathbf{y}^T \tag{10.1-3}$$

† Since the great majority of computational problems involve real matrices, we shall assume for simplicity throughout this chapter that A is real. As will generally be clear from the context, many of the theorems we shall consider are also true when A is complex or, if the real matrix is assumed symmetric, when A is Hermitian. Of course, even for real matrices, eigenvalues and eigenvectors may be complex. However, the eigenvectors corresponding to real eigenvalues are real.

Therefore, a left eigenvector of A is an eigenvector of A^T. It is easily shown {1} that if $\mathbf{y}_k$ and $\mathbf{x}_j$ are, respectively, left and right eigenvectors corresponding to distinct eigenvalues, then $\mathbf{y}_k$ and $\mathbf{x}_j$ are orthogonal. In this chapter we shall be interested in methods for the calculation of some or all of the eigenvalues and some or all of the eigenvectors of A. First, however, we shall review some of the basic theorems and relationships which are necessary to understand the development that follows.

10.1-1 Basic Theorems

In this section we list four basic theorems concerning the eigenvalues and eigenvectors of a matrix with which the reader should be familiar. Proofs of three of them are left to a problem {2}.

Theorem 10.1 If $\lambda_1, \lambda_2, \ldots, \lambda_n$ are the eigenvalues of A, then the eigenvalues of A^k are $\lambda_1^k, \lambda_2^k, \ldots, \lambda_n^k$. More generally, if $p(x)$ is a polynomial, the eigenvalues of $p(A)$ are $p(\lambda_1), \ldots, p(\lambda_n)$.

Theorem 10.2 If A is real and symmetric, all eigenvalues and eigenvectors are real. Moreover, eigenvectors corresponding to distinct eigenvalues are orthogonal, and the left eigenvector corresponding to the eigenvector $\mathbf{x}_i$ is $\mathbf{x}_i^T$.

Theorem 10.3 Any similarity transformation PAP^{-1} applied to A leaves the eigenvalues of the matrix unchanged.

PROOF Let λ be an eigenvalue of A and $\mathbf{x}$ the associated eigenvector. Then

$$A\mathbf{x} = \lambda\mathbf{x}$$

so that

$$PA\mathbf{x} = \lambda P\mathbf{x} \tag{10.1-4}$$

Let $\mathbf{y} = P\mathbf{x}$, so that $\mathbf{x} = P^{-1}\mathbf{y}$. Substituting in (10.1-4), we get

$$PAP^{-1}\mathbf{y} = \lambda\mathbf{y} \tag{10.1-5}$$

Thus λ is an eigenvalue of PAP^{-1}, and $\mathbf{y}$ is the associated eigenvector.

Theorem 10.4 (Cayley-Hamilton) Let

$$f(\lambda) = |A - \lambda I| = 0 \tag{10.1-6}$$

be the characteristic equation of A. Then $f(A) = 0$.

10.1-2 The Characteristic Equation

One method of determining the eigenvalues of A is to find the roots of the polynomial equation (10.1-1). For this purpose the methods of Chap. 8 can be used. But before we can use most of these methods to find the roots of (10.1-1) we must determine the coefficients of the characteristic equation itself. Direct calculation of these coefficients from the definition of the determinant is never recommended except for very low-order matrices, since it involves an astronomical amount of calculation. Here we present two methods for finding the coefficients of the powers of λ in the characteristic equation without direct evaluation of the determinant.

Krylov's method We write the characteristic equation as

$$f(\lambda) = \lambda^n + \sum_{i=0}^{n-1} b_i \lambda^i = 0 \tag{10.1-7}$$

From the Cayley-Hamilton theorem we have

$$A^n + \sum_{i=0}^{n-1} b_i A^i = 0 \tag{10.1-8}$$

Then, for any vector $\mathbf{y}$,

$$A^n\mathbf{y} + \sum_{i=0}^{n-1} b_i A^i\mathbf{y} = 0 \tag{10.1-9}$$

Equation (10.1-9) is a system of n linear equations in the n unknowns $b_0, \ldots, b_{n-1}$ which can be solved by any of the methods of the previous chapter. Note that the calculation of $A^i\mathbf{y} = A(A^{i-1}\mathbf{y})$ requires n^2 multiplications so that about n^3 multiplications are required to establish (10.1-9) followed by about $\frac{1}{3}n^3$ to solve (10.1-9) by one of the methods of Sec. 9.3.

LeVerrier's method Using the property that the sum of the eigenvalues of any matrix is equal to the trace of the matrix, and using Theorem 10.1, we can write

$$\sum_{i=1}^{n} \lambda_i^k = t_k \qquad k = 1, \ldots, n \tag{10.1-10}$$

where t_k is the trace of A^k. In Sec. 8.10-3 we noted that the sums of the powers of the roots of a polynomial could easily be computed using the coefficients of the polynomial. This process is easily reversed; the coefficients of the polynomial are easily computed if the sums of the powers of its roots are known {3}. Thus, using (10.1-10), we can compute the coefficients of the characteristic equation. But LeVerrier's method is much inferior to Krylov's because of the necessity of actually computing A^k, $k = 1, \ldots, n$.

In a problem {4} we consider another technique of computing the coefficients of the characteristic polynomial. In general, direct computation of these coefficients is less efficient than the methods which will be discussed later in this chapter. Moreover, it may convert a well-conditioned problem into an ill-conditioned one, since, as we have seen in Sec. 8.13, small errors in the coefficients of a polynomial can induce large errors in some of its roots. Thus, direct computation of the coefficients of the characteristic polynomial is not recommended as a method for computing eigenvalues and is only of historical interest.

10.1-3 The Location of, and Bounds on, the Eigenvalues

In this section we consider some of the more important among the many theorems which deal with the location of the eigenvalues of a matrix, i.e., the location of the zeros of the characteristic polynomial. These theorems can be used, for example, to estimate the magnitude of the largest and smallest eigenvalues in magnitude and thus to estimate $\rho(A)$ and the condition of A. Such estimates can also be used to generate initial approximations to be used in iterative methods for determining eigenvalues (see Sec. 10.2).

Gerschgorin's theorem Let $A = [a_{ij}]$ and let C_i, $i = 1, \ldots, n$, be the circles with centers a_{ii} and radii

$$r_i = \sum_{\substack{k=1 \\ k \neq i}}^{n} |a_{ik}| \qquad i = 1, \ldots, n \tag{10.1-11}$$

Further let

$$D = \bigcup_{i=1}^{n} C_i \tag{10.1-12}$$

Then we can state the following theorem.

Theorem 10.5 (Gerschgorin)† All the eigenvalues of A lie within the domain D.

PROOF We use the result of Prob. 30 of Chap. 9 that if the component of largest magnitude of the solution of $B\mathbf{x} = 0$ is the ith component, then with $B = [b_{ij}]$

$$|b_{ii}| \leq \sum_{\substack{k=1 \\ k \neq i}}^{n} |b_{ik}| \tag{10.1-13}$$

† This theorem is also valid when the elements of A are complex.

Letting $B = A - \lambda I$, where λ is an eigenvalue of A, we have

$$|\lambda - a_{ii}| \leq \sum_{\substack{k=1 \\ k \neq i}}^{n} |a_{ik}| \tag{10.1-14}$$

Equation (10.1-14) holds for any eigenvalue (although i may vary from one eigenvalue to another), and this is sufficient to prove the theorem.

As a consequence of Theorem 10.5 we get one of the results of Prob. 30b of Chap. 9.

Corollary 10.1 The spectral radius of A^{-1} is such that

$$\frac{1}{\rho(A^{-1})} \geq \min_i \left(|a_{ii}| - \sum_{\substack{k=1 \\ k \neq i}}^{n} |a_{ik}| \right) \tag{10.1-15}$$

By making use of the result of Theorem 10.3, it may be possible to improve upon the bound for the eigenvalues given by Gerschgorin's theorem by first applying a similarity transformation to A {5}.

Other theorems A number of theorems give information about the eigenvalues of a matrix by using related norms and analogous quantities. The following are theorems of this type.

Theorem 10.6 Let A be symmetric and positive definite. Then

$$\rho(A) = \max_{\mathbf{x}} \frac{\mathbf{x}^* A \mathbf{x}}{\mathbf{x}^* \mathbf{x}} \tag{10.1-16}$$

$$\frac{1}{\rho(A^{-1})} = \min_{\mathbf{x}} \frac{\mathbf{x}^* A \mathbf{x}}{\mathbf{x}^* \mathbf{x}} \tag{10.1-17}$$

where $\mathbf{x}$ is an arbitrary real or complex nonzero vector, and, as it will throughout this chapter, the superscript * denotes the conjugate transpose.

Proof Since A is symmetric and positive definite, all its eigenvalues are real and positive and can be ordered as follows:

$$\lambda_1 \geq \lambda_2 \geq \cdots \geq \lambda_n > 0 \tag{10.1-18}$$

so that $\rho(A) = \lambda_1$. Now for any nonzero vector $\mathbf{x}$,

$$\lambda_1 - \frac{\mathbf{x}^* A \mathbf{x}}{\mathbf{x}^* \mathbf{x}} = \frac{\mathbf{x}^*(\lambda_1 I - A)\mathbf{x}}{\mathbf{x}^* \mathbf{x}} \geq 0 \tag{10.1-19}$$

since all the eigenvalues of $\lambda_1 I - A$ are nonnegative, so that $\lambda_1 I - A$ is positive semidefinite. On the other hand,

$$\lambda_1 = \frac{\mathbf{x}_1^* A \mathbf{x}_1}{\mathbf{x}_1^* \mathbf{x}_1} \tag{10.1-20}$$

where $\mathbf{x}_1$ is the eigenvector corresponding to λ_1, proving (10.1-16).

Similarly, we can show that

$$\lambda_n = \min_{\mathbf{x}} \frac{\mathbf{x}^* A \mathbf{x}}{\mathbf{x}^* \mathbf{x}} \tag{10.1-21}$$

Since the eigenvalues of A^{-1} are $\lambda_1^{-1}, \ldots, \lambda_n^{-1}$, so that $\rho(A^{-1}) = \lambda_n^{-1}$, (10.1-17) holds.

From this proof, it is readily seen that if A is symmetric but not positive definite, the theorem is still true if $\rho(A)$ is replaced by $\lambda_{\max}$ and $1/\rho(A^{-1})$ is replaced by $\lambda_{\min}$, where $\lambda_{\min} \le \lambda_i \le \lambda_{\max}$ for any eigenvalue λ_i.

Theorem 10.7 For an arbitrary nonsingular matrix A

$$\frac{1}{\rho[(A^T A)^{-1}]} \le |\lambda_i|^2 \le \rho(A^T A) \tag{10.1-22}$$

where λ_i is any eigenvalue of A.

PROOF Let $\mathbf{x}_i$ be the eigenvector corresponding to λ_i. Then

$$A\mathbf{x}_i = \lambda_i \mathbf{x}_i \tag{10.1-23}$$

and

$$\mathbf{x}_i^* A^T = \bar{\lambda}_i \mathbf{x}_i^* \tag{10.1-24}$$

Therefore,

$$\mathbf{x}_i^* A^T A \mathbf{x}_i = |\lambda_i|^2 \mathbf{x}_i^* \mathbf{x}_i \tag{10.1-25}$$

Using Theorem 10.6 we have, since $A^T A$ is positive definite,

$$\rho(A^T A) \ge \frac{\mathbf{x}_i^* A^T A \mathbf{x}_i}{\mathbf{x}_i^* \mathbf{x}_i} \ge \frac{1}{\rho[(A^T A)^{-1}]} \tag{10.1-26}$$

which with (10.1-25) proves the theorem.

If A is singular, the theorem is true if the left-hand side of the inequality (10.1-22) is replaced by 0.

10.1-4 Canonical Forms

One of the most important techniques in the calculation of eigenvalues and eigenvectors of matrices is to transform the given matrix into some canonical form. Here we consider some of the theorems dealing with these canonical forms. In particular we shall indicate the basic differences that occur in considering symmetric and nonsymmetric matrices since they have important implications on what follows in this chapter.

Theorem 10.8 Eigenvectors of an arbitrary matrix A corresponding to distinct eigenvalues are linearly independent.

PROOF By induction. Let $\mathbf{x}_1$ and $\mathbf{x}_2$ be eigenvectors corresponding to λ_1 and λ_2, $\lambda_1 \neq \lambda_2$. Then if $a_1\mathbf{x}_1 + a_2\mathbf{x}_2 = \mathbf{0}$, we also have

$$a_1\lambda_1\mathbf{x}_1 + a_2\lambda_2\mathbf{x}_2 = \mathbf{0}$$

which implies $a_1 = a_2 = 0$ (why?). Now suppose $\mathbf{x}_1, \ldots, \mathbf{x}_k$ corresponding to distinct eigenvalues $\lambda_1, \ldots, \lambda_k$ are independent. Then if $\mathbf{x}_{k+1}$ corresponds to λ_{k+1}, consider

$$a_1\mathbf{x}_1 + \cdots + a_{k+1}\mathbf{x}_{k+1} = \mathbf{0} \qquad (10.1\text{-}27)$$

Premultiplying by A, we have

$$a_1\lambda_1\mathbf{x}_1 + \cdots + a_{k+1}\lambda_{k+1}\mathbf{x}_{k+1} = \mathbf{0} \qquad (10.1\text{-}28)$$

If $\lambda_{k+1} = 0$, it follows immediately from (10.1-28) and the induction hypothesis that $a_{k+1} = 0$. If $\lambda_{k+1} \neq 0$, divide (10.1-28) by λ_{k+1} and subtract from (10.1-27) to obtain

$$a_1\left(1 - \frac{\lambda_1}{\lambda_{k+1}}\right)\mathbf{x}_1 + \cdots + a_k\left(1 - \frac{\lambda_k}{\lambda_{k+1}}\right)\mathbf{x}_k = \mathbf{0} \qquad (10.1\text{-}29)$$

Since $\lambda_1, \ldots, \lambda_{k+1}$ are all distinct, the induction hypothesis again gives $a_1 = a_2 = \cdots = a_k = 0$, and (10.1-27) then gives $a_{k+1} = 0$. This proves the theorem.

Theorem 10.9 Let A be an arbitrary matrix whose eigenvalues are all distinct. Then there exists a similarity transformation such that

$$P^{-1}AP = D$$

where D is a diagonal matrix whose diagonal elements are the eigenvalues of A.

PROOF Let P be the matrix whose columns are the (right) eigenvectors of A. Since the eigenvalues are distinct, P exists (but may be complex). Then

$$AP = PD \qquad (10.1\text{-}30)$$

where D is a diagonal matrix with the eigenvalues on the diagonal. Since the eigenvectors are linearly independent by Theorem 10.8, P^{-1} exists and

$$P^{-1}AP = D \qquad (10.1\text{-}31)$$

which proves the theorem. Note that the rows of P^{-1} are the left eigenvectors of A.

The computational disadvantage of Theorem 10.9 is the need to find a matrix P and its inverse. It would be more convenient if we could diagonalize a matrix using an orthogonal or unitary transformation. The most general theorem we can prove in this case is the following.

Theorem 10.10 For an arbitrary matrix A there exists a unitary transformation Q such that

$$Q^*AQ = T \tag{10.1-32}$$

where T is triangular (but may have complex elements).

Note that this does not require that A have distinct eigenvalues, as in Theorem 10.9.

PROOF The proof is by induction on the order n of the matrix. If $n = 1$, the theorem is true since a matrix of order 1 is triangular. Suppose the theorem is true for matrices of order $n - 1$ and let A be a matrix of order n. Let $\mathbf{u}_1$ be an eigenvector of A of magnitude 1 corresponding to any eigenvalue, say λ_1. Let

$$\mathbf{u}_1, \mathbf{v}_1, \ldots, \mathbf{v}_{n-1} \tag{10.1-33}$$

be an orthonormal set of vectors.† If Q_1 is the matrix whose columns are $\mathbf{u}_1, \mathbf{v}_1, \ldots, \mathbf{v}_{n-1}$, we have

$$A_1 = Q_1^* A Q_1 = \begin{bmatrix} \lambda_1 & \mathbf{w}^T \\ \mathbf{0} & B \end{bmatrix} \tag{10.1-34}$$

By the induction hypothesis there exists a unitary matrix P of order $n - 1$ such that

$$P^*BP = T_{n-1} \tag{10.1-35}$$

Now let

$$Q_2 = \begin{bmatrix} 1 & \mathbf{0}^T \\ \mathbf{0} & P \end{bmatrix} \tag{10.1-36}$$

so that Q_2 is unitary of order n. Then from (10.1-34) to (10.1-36)

$$Q_2^* Q_1^* A Q_1 Q_2 = Q_2^* \begin{bmatrix} \lambda_1 & \mathbf{w}^T \\ \mathbf{0} & B \end{bmatrix} Q_2 = \begin{bmatrix} \lambda_1 & \mathbf{w}^T P \\ \mathbf{0} & T_{n-1} \end{bmatrix} = T_n \tag{10.1-37}$$

where T_n is triangular of order n. Setting $Q = Q_1 Q_2$ proves the theorem.

The proof of Theorem 10.10 implies that, if all the eigenvalues of A are real, then Q is a real orthogonal matrix (why?). In particular, since the eigenvalues of a symmetric matrix are real and since the orthogonal transformation of a symmetric matrix is symmetric, we have the following corollary.

† That is, $\mathbf{u}_1^* \mathbf{v}_i = 0$, $i = 1, \ldots, n - 1$, and $\mathbf{v}_i^* \mathbf{v}_j = \delta_{ij}$, $i, j = 1, \ldots, n - 1$.

Corollary 10.2 If A is symmetric, then there exists an orthogonal matrix Q such that

$$Q^T A Q = D \tag{10.1-38}$$

with D as in Theorem 10.9.

For computational purposes this corollary is an improvement over Theorem 10.9 not only in that the similarity transformation is replaced by an orthogonal one but also in that there is no requirement that the eigenvalues be distinct. Since it follows from (10.1-38) and the orthogonality of Q that

$$AQ = QD \tag{10.1-39}$$

we have the result that the columns of Q are the eigenvectors of A.

Since a triangular matrix T yields its eigenvalues just as easily as a diagonal matrix, the reader may think that Theorem 10.10, except for the necessity of using complex arithmetic, will be as useful in practice as Corollary 10.2. But in practice it is much easier to diagonalize symmetric matrices than it is to triangularize nonsymmetric matrices, as we shall see in Secs. 10.3 and 10.4.

Theorem 10.9 and Corollary 10.2 take care of the diagonalization of all but nonsymmetric matrices with multiple eigenvalues. The theorem covering this case is

Theorem 10.11 Given an arbitrary matrix A, there exists a nonsingular matrix P, whose elements may be complex, such that

$$P^{-1}AP = \begin{bmatrix} J_1 & & & \bigcirc \\ & J_2 & & \\ & & \ddots & \\ \bigcirc & & & J_K \end{bmatrix} \tag{10.1-40}$$

where J_k, $k = 1, \ldots, K \le n$, is a matrix with an eigenvalue λ_i of A on its main diagonal and 1s on the diagonal above the main diagonal.

Thus

$$J_k = \begin{bmatrix} \lambda_i & 1 & & & \bigcirc \\ & \lambda_i & 1 & & \\ & & \ddots & \ddots & \\ & & & \ddots & 1 \\ \bigcirc & & & & \lambda_i \end{bmatrix} \tag{10.1-41}$$

Note that a given eigenvalue may appear as the diagonal element of more than one J_k. The matrix in (10.1-40) is called the *Jordan canonical form* of A. The determinants

$$\det(J_k - \lambda I) = (\lambda_i - \lambda)^{v_k} \tag{10.1-42}$$

where v_k is the order of J_k, are called the *elementary divisors* of A. If $v_k = 1$, we say the elementary divisor is linear. The proof of Theorem 10.11 is beyond the scope of this book [see, for example, Bodewig (1959), pp. 82–88].

It follows from Theorem 10.11 that, if the eigenvalues are distinct, all the elementary divisors are linear, which gives us another proof of Theorem 10.9. Theorem 10.11 together with Corollary 10.2 implies that for a symmetric matrix the elementary divisors are linear whether or not the eigenvalues are distinct. More important is the result that if the elementary divisors are linear, then corresponding to an eigenvalue of multiplicity k there are k linearly independent eigenvectors. Therefore, Theorem 10.9 can be generalized to include all matrices with linear elementary divisors. If, however, there are nonlinear elementary divisors, then Theorem 10.11 implies that there are eigenvalues whose multiplicity is greater than the number of independent eigenvectors corresponding to them. The simplicity of the result of Corollary 10.2 in comparison with the results of Theorems 10.10 and 10.11 is an indication that the calculation of the eigenvalues and eigenvectors of symmetric matrices will cause fewer difficulties than that of nonsymmetric matrices. The remainder of this chapter will bear out this indication.

With the theorems of this section as background, we are now ready to consider methods for the calculation of eigenvalues and eigenvectors. We begin with methods for the calculation of the largest eigenvalue in magnitude.

10.2 THE LARGEST EIGENVALUE IN MAGNITUDE BY THE POWER METHOD

The basis of our techniques for determining the largest eigenvalue in magnitude is very similar to Bernoulli's method (Sec. 8.10-3) for determining the zeros of polynomials. And this, of course, should not be surprising because of the similarity between the problems of finding the eigenvalues of a matrix and the zeros of a polynomial.

The basic assumption of this section is that all the elementary divisors of A are linear. However, the method of this section is often applicable even when A has nonlinear elementary divisors {9}. This assumption implies, as noted in the previous section, that there are n linearly independent eigenvectors of A and thus that the eigenvectors span n space. Therefore, any vector $\mathbf{v}_0$ can be expressed as a linear combination

$$\mathbf{v}_0 = \sum_{i=1}^{n} \alpha_i \mathbf{x}_i \tag{10.2-1}$$

where $\mathbf{x}_i$, $i = 1, \ldots, n$, are the eigenvectors of A. If λ_i is the eigenvalue corresponding to $\mathbf{x}_i$, then

$$A\mathbf{v}_0 = \sum_{i=1}^{n} \alpha_i \lambda_i \mathbf{x}_i \tag{10.2-2}$$

and in general the mth *iterated vector* is given by

$$\mathbf{v}_m = A^m \mathbf{v}_0 = \sum_{i=1}^{n} \alpha_i \lambda_i^m \mathbf{x}_i \tag{10.2-3}$$

Now we order the eigenvalues so that

$$|\lambda_1| \geq |\lambda_2| \geq |\lambda_3| \geq \cdots \geq |\lambda_n| \tag{10.2-4}$$

If then, in particular, λ_1 is dominant, that is, $|\lambda_1| > |\lambda_2|$, and therefore real, we have the following theorem.

Theorem 10.12 (Von Mises) If the matrix A has n linearly independent eigenvectors, and if the largest eigenvalue in magnitude λ_1 is dominant, then if $\mathbf{v}_0$ has a component in the direction of $\mathbf{x}_1$,

$$\lim_{m \to \infty} \frac{1}{\lambda_1^m} A^m \mathbf{v}_0 = \alpha_1 \mathbf{x}_1 \tag{10.2-5}$$

The proof follows immediately from (10.2-3). The requirement that $\mathbf{v}_0$ have a component in the direction of $\mathbf{x}_1$ is assurance that $\alpha_1 \neq 0$.

As a consequence of this theorem we see that if $\mathbf{y}$ is any vector not orthogonal to $\mathbf{x}_1$, then, using (10.2-5),

$$\lambda_1 = \lim_{m \to \infty} \frac{\mathbf{y}^T \mathbf{v}_{m+1}}{\mathbf{y}^T \mathbf{v}_m} \tag{10.2-6}$$

The numbers $\mathbf{y}^T \mathbf{v}_{m+1} = \mathbf{y}^T A \mathbf{v}_m$ are called *Schwarz constants.* A convenient choice of $\mathbf{y}$ in practice is the vector with a component of 1 in the position corresponding to the maximum component of $\mathbf{v}_m$ and 0 elsewhere. This minimizes the computation in (10.2-6). Of course, early in the computation the largest component of $\mathbf{v}_m$ may vary, but ultimately it will be the largest component of the eigenvector. In practice, then, we compute successive approximations to λ_1 as the ratio of the largest component of successive $\mathbf{v}_m$'s.

Example 10.1 Find the dominant eigenvalue of the matrix

$$A = \begin{bmatrix} 1.0 & 1.0 & .5 \\ 1.0 & 1.0 & .25 \\ .5 & .25 & 2.0 \end{bmatrix}$$

using $\mathbf{v}_0^T = (1, 1, 1)$.

In the table below each $\mathbf{v}_m$ was calculated using (10.2-3) and then "normalized" to make its largest component 1 before doing the next iteration. The quantity λ is the normalizing factor and thus represents the ratio of the largest components of $\mathbf{v}_m$ and $\mathbf{v}_{m-1}$; for example, the first entry in the λ column is the ratio of the third component of the unnormalized $\mathbf{v}_1$ to the third component of $\mathbf{v}_0$.

m	$\mathbf{v}_m^T$ (normalized)	λ	m	$\mathbf{v}_m^T$ (normalized)	λ
1	(.9091, .8182, 1.0000)	2.7500000	15	(.7483, .6497, 1.0000)	2.5366256
5	(.7651, .6674, 1.0000)	2.5587918	16	(.7483, .6497, 1.0000)	2.5365840
10	(.7494, .6508, 1.0000)	2.5380029	17	(.7482, .6497, 1.0000)	2.5365598
11	(.7489, .6504, 1.0000)	2.5373873	18	(.7482, .6497, 1.0000)	2.5365456
12	(.7486, .6501, 1.0000)	2.5370284	19	(.7482, .6497, 1.0000)	2.5365374
13	(.7484, .6499, 1.0000)	2.5368188	20	(.7482, .6497, 1.0000)	2.5365323
14	(.7484, .6498, 1.0000)	2.5366969			

The calculations were performed on a digital computer using floating-point arithmetic with an 8-digit fractional part. The values of $\mathbf{v}_m$ are rounded values. To eight figures the true value of λ is 2.5365258 and that of $\mathbf{x}_1$ is (.74822116, .64966116, 1.00000000). These values were reached in 28 iterations. Although a symmetric matrix was used in this example, remember that the power method requires only that the elementary divisors be linear.

From (10.2-3)

$$\mathbf{v}_m = \lambda_1^m \left[\alpha_1 \mathbf{x}_1 + \sum_{i=2}^{n} \left(\frac{\lambda_i}{\lambda_1} \right)^m \alpha_i \mathbf{x}_i \right] \tag{10.2-7}$$

Therefore, the *rate* of convergence of the power method depends upon how fast the ratios $(\lambda_i/\lambda_1)^m$ go to zero; in particular this rate depends upon the ratio $|\lambda_2|/|\lambda_1|$. The number of iterations required to get a desired degree of convergence depends upon both the rate of convergence and on how large α_1 is compared with the other α_i, the latter depending in turn on the choice of $\mathbf{v}_0$. In Example 10.1 the convergence was quite slow. In later examples we shall see how these two factors—$|\lambda_2/\lambda_1|$ and α_1—affect the convergence. Later in this section we shall also consider means of speeding up the convergence.

If $\mathbf{v}_0$ has no component in the $\mathbf{x}_1$ direction and $|\lambda_2| > |\lambda_3|$, then the iteration should converge to λ_2 and $\mathbf{x}_2$ (why?). In practical computation, however, roundoff error will generally introduce a component in the $\mathbf{x}_1$ direction so that eventually the iteration will converge to λ_1 and $\mathbf{x}_1$. Nevertheless a good approximation to λ_2 and $\mathbf{x}_2$ can be obtained before the term in λ_1 dominates.

Example 10.2 Apply the power method to the matrix of Example 10.1 using $\mathbf{v}_0^T = (-.64966116, .7482216, 0)$ which is orthogonal to $\mathbf{x}_1$. Therefore, since A is symmetric, $\mathbf{v}_0$ has no component in the $\mathbf{x}_1$ direction.

Corresponding to the table in Example 10.1 we have

m	$\mathbf{v}_m^T$ (normalized)	λ	m	$\mathbf{v}_m^T$ (normalized)	λ
1	(−.7154, −.7154, 1.0000)	−.1377753	10	(−.6369, −.8058, 1.0000)	1.4801240
2	(−.6360, −.8068, 1.0000)	1.4634741	15	(−.6368, −.8057, 1.0000)	1.4801606
3	(−.6369, −.8058, 1.0000)	1.4803108	20	(−.6356, −.8044, 1.0000)	1.4807007
4	(−.6369, −.8058, 1.0000)	1.4801194	30	(−.4008, −.5576, 1.0000)	1.5931941
5	(−.6369, −.8058, 1.0000)	1.4801216	40	(.7180, .6179, 1.0000)	2.4976711
6	(−.6369, −.8058, 1.0000)	1.4801217	50	(.7481, .6495, 1.0000)	2.5363412
7	(−.6369, −.8058, 1.0000)	1.4801219	61	(.7482, .6497, 1.0000)	2.5365253

Initially the iteration converges to a value very nearly equal to $\lambda_2 = 1.4801215$, but although $\mathbf{v}_0$ is orthogonal to $\mathbf{x}_1$, roundoff introduces a component in the $\mathbf{x}_1$ direction and slowly, but very slowly, the term in $\mathbf{x}_1$ dominates, as in Example 10.1. The reason the convergence to λ_2 is so good and so rapid is that, as we shall see in a later example, the ratio of $|\lambda_2/\lambda_3|$ is very large.

The drawbacks to this method are similar to those of Bernoulli's method, in that special cases require special treatment. However, when the dominant eigenvalue is multiple but real, the method does converge. For if λ_1 has multiplicity k,

$$\mathbf{v}_m = A^m \mathbf{v}_0 = \lambda_1^m \sum_{i=1}^{k} \alpha_i \mathbf{x}_i + \sum_{i=k+1}^{n} \lambda_i^m \alpha_i \mathbf{x}_i \tag{10.2-8}$$

and again the term in λ_1 dominates. Further, since $\sum_{i=1}^{k} \alpha_i \mathbf{x}_i$ is an eigenvector the process converges to an eigenvector as before. A procedure to get the other eigenvectors corresponding to λ_1 is considered in a problem {10}.

10.2-1 Acceleration of Convergence

Because the power method may converge slowly, means of accelerating the convergence are clearly desirable. In this section we consider four such means.

The δ^2 process This application of the δ^2 process is similar to our previous applications. We assume that both λ_1 and λ_2 are real and that neither $-\lambda_1$ nor $-\lambda_2$ is an eigenvalue. Let $\mathbf{e}_i$ be the ith column of the identity matrix I. Then from (10.2-3) we calculate the Schwarz constant

$$\mathbf{e}_i^T \mathbf{v}_m = \sum_{j=1}^{n} a_j \lambda_j^m \tag{10.2-9}$$

where a_j depends on α_j and the ith component of $\mathbf{x}_j$. Then

$$\frac{\mathbf{e}_i^T \mathbf{v}_{m+1}}{\mathbf{e}_i^T \mathbf{v}_m} = \frac{a_1 \lambda_1^{m+1} + \sum_{j=2}^{n} a_j \lambda_j^{m+1}}{a_1 \lambda_1^m + \sum_{j=2}^{n} a_j \lambda_j^m} \tag{10.2-10}$$

Dividing numerator and denominator by $a_1 \lambda_1^m$ and expanding the denominator in a power series, we get

$$\frac{\mathbf{e}_i^T \mathbf{v}_{m+1}}{\mathbf{e}_i^T \mathbf{v}_m} = \lambda_1 + \beta_i \left(\frac{\lambda_2}{\lambda_1}\right)^m + \left[\text{terms in } \left(\frac{\lambda_3}{\lambda_1}\right)^m, \ldots, \left(\frac{\lambda_n}{\lambda_1}\right)^m \text{ and higher powers}\right] \tag{10.2-11}$$

If we are near convergence, the terms in brackets are small and we have

$$\lambda_1 \approx R_m - \beta_i r^m \tag{10.2-12}$$

with $R_m = \mathbf{e}_i^T \mathbf{v}_{m+1} / \mathbf{e}_i^T \mathbf{v}_m$ and $r = \lambda_2 / \lambda_1$. Then proceeding as in Sec. 8.7-1, we get as a better approximation to λ_1 {15}

$$\lambda_1 \approx R_{m+2} - \frac{(\Delta R_{m+1})^2}{\Delta^2 R_m} \tag{10.2-13}$$

If we apply (10.2-13) with $m = 10$ and $i = 3$ in (10.2-11) to the results in Example 10.1, we get as our new approximation $\lambda_1 = 2.5365266$, which is better than that achieved after 20 iterations.

Wilkinson's method Suppose that all the eigenvalues are real but that the power method is converging slowly because there are two eigenvalues nearly equal in magnitude. Consider the matrix $A - pI$, which has eigenvalues $\lambda_i - p$. By a judicious choice of p, it may be possible to speed up the convergence markedly. The optimal choice of p if we wish to converge to λ_1 is that value which minimizes

$$\max_{2 \leq i \leq n} \left| \frac{\lambda_i - p}{\lambda_1 - p} \right|$$

Hence, some knowledge of the eigenvalues of A is needed to apply this technique. For example, if all the eigenvalues are positive, the optimal value of p is $(\lambda_2 + \lambda_n)/2$ {16}, where λ_n is the smallest eigenvalue of A. The choice $p = \lambda_n$ will also yield an improved rate of convergence.

This method is an excellent example of one which, when used by an experienced numerical analyst, can be very powerful; but if used haphazardly, i.e., if the value of p is not chosen judiciously, it will be little better than the power method. As an example of this method, if we use $p = .75$ in Example 10.1, then, in place of the table in that example, we get

m	$\mathbf{v}_m^T$ (normalized)	λ
5	(.7516, .6522, 1.0000)	1.7914011
6	(.7491, .6511, 1.0000)	1.7888443
7	(.7488, .6501, 1.0000)	1.7873300
8	(.7484, .6499, 1.0000)	1.7869152
9	(.7483, .6497, 1.0000)	1.7866587
10	(.7482, .6497, 1.0000)	1.7865914

which is a better result than after 15 iterations in Example 10.1. Convergence to $\lambda_1 = 1.7865258$ was achieved in 19 iterations. The reason why $p = .75$ was a good choice will become apparent when we calculate all the eigenvalues of the matrix A in Example 10.4.

The Rayleigh quotient If A is symmetric, then the eigenvectors are orthogonal, and if we consider them to be orthonormal,

$$\mathbf{v}_m^T A \mathbf{v}_m = \mathbf{v}_m^T \mathbf{v}_{m+1} = \sum_{i=1}^{n} \alpha_i^2 \lambda_i^{2m+1} \tag{10.2-14}$$

and

$$\mathbf{v}_m^T \mathbf{v}_m = \sum_{i=1}^{n} \alpha_i^2 \lambda_i^{2m} \tag{10.2-15}$$

We can write

$$\frac{\mathbf{v}_m^T A \mathbf{v}_m}{\mathbf{v}_m^T \mathbf{v}_m} = \frac{\mathbf{v}_m^T \mathbf{v}_{m+1}}{\mathbf{v}_m^T \mathbf{v}_m} = \lambda_1 + O\left[\left(\frac{\lambda_i}{\lambda_1}\right)^{2m}\right] \tag{10.2-16}$$

By comparison for an arbitrary vector $\mathbf{y}$

$$\frac{\mathbf{y}^T \mathbf{v}_{m+1}}{\mathbf{y}^T \mathbf{v}_m} = \lambda_1 + O\left[\left(\frac{\lambda_i}{\lambda_1}\right)^{m}\right] \tag{10.2-17}$$

Since the higher-order terms in (10.2-16) will usually be smaller than those in (10.2-17), the former will generally give a better approximation to λ_1 than the power method itself. For example, consider Example 10.1 with $m = 11$. The unrounded values of $\mathbf{v}_{11}^T$ and $\mathbf{v}_{12}^T$ are, respectively, (.74888011, .65035358, 1.0) and (.74860561, .65006512, 1.0). For these vectors the left-hand side of (10.2-16) is 2.5365256, which is a slightly better result than that achieved using the δ^2 process, which also used data through the twelfth iteration. It is generally true that this technique will give better results than the δ^2 process {17}. Therefore, for symmetric matrices, it is to be preferred. The quotient in (10.2-16) is the *Rayleigh quotient* and bears a close resemblance to the quotient in Theorem 10.6.

Matrix powers If two eigenvalues are nearly equal in magnitude, then in order to separate the eigenvalues we can compute A^2, A^4, A^8, ... This technique is directly analogous to the root-squaring procedure of Sec. 8.10-2 but is, of course, very inefficient because the computation of each power of A requires n^3 operations. It is therefore not recommended.

10.2-2 The Inverse Power Method

If A is a nonsingular matrix with eigenvalues $\lambda_1, \ldots, \lambda_n$, then the inverse matrix A^{-1} has the eigenvalues $\lambda_1^{-1}, \ldots, \lambda_n^{-1}$. Hence, if $A - pI$ is nonsingular, the eigenvalues of $(A - pI)^{-1}$ are $(\lambda_1 - p)^{-1}, \ldots, (\lambda_n - p)^{-1}$. If we now apply the power method with $(A - pI)^{-1}$, we find that, starting as before with $\mathbf{v}_0$ satisfying (10.2-1), the mth iterated vector $\mathbf{v}_m$ is given by

$$\mathbf{v}_m = (A - pI)^{-m}\mathbf{v}_0 = \sum_{i=1}^{n} \alpha_i(\lambda_i - p)^{-m}\mathbf{x}_i \tag{10.2-18}$$

If now p is close to one of the distinct eigenvalues λ_j, which may be a multiple eigenvalue but which for simplicity we assume to be of multiplicity 1, and if λ_j is separated from all the other eigenvalues so that $|p - \lambda_j| < |p - \lambda_i|$, $i \neq j$, then, provided that $\alpha_j \neq 0$, the term $\alpha_j(\lambda_j - p)^{-m}\mathbf{x}_j$ will dominate the sum on the right-hand side of (10.2-18) for sufficiently large m. The vector $\mathbf{v}_m$ will then be a very good approximation to the eigenvector corresponding to λ_j. This dominance shows up quite early if p is very close to λ_j; in this case, even $\mathbf{v}_2$ will be effectively an eigenvector. This is so even if $\mathbf{v}_0$ has a relatively small component in the $\mathbf{x}_j$ direction and λ_j is not particularly well separated from the other eigenvalues. For example, if $\lambda_j - p = 10^{-10}$, $|\lambda_k - p| > 10^{-3}$, $k \neq j$, and $\alpha_j = 10^{-4}$, then {20}

$$10^{-16}\mathbf{v}_2 = \mathbf{x}_j + \mathbf{y} \tag{10.2-19}$$

where $|\mathbf{y}| < 10^{-10}$.

The inverse power method is applied in two different contexts. The first is in an iterative method to find eigenvalues and eigenvectors using a generalized version of the Rayleigh iteration given in Sec. 10.2-1. We give here the application of this method to symmetric matrices and treat the general case in a problem {21}. Starting with an initial approximation $\mathbf{v}_0$ to an eigenvector, we compute the Rayleigh quotient

$$\mu_1 = \frac{\mathbf{v}_0^T A \mathbf{v}_0}{\mathbf{v}_0^T \mathbf{v}_0} \tag{10.2-20}$$

followed by

$$\mathbf{v}_1 = k_1(A - \mu_1 I)^{-1}\mathbf{v}_0 \tag{10.2-21}$$

where k_1 is a normalizing factor chosen so that $\|\mathbf{v}_1\| = 1$ for one of the standard norms and μ_1 plays the role of p in (10.2-18). In general, we compute

$$\mu_{i+1} = \frac{\mathbf{v}_i^T A \mathbf{v}_i}{\mathbf{v}_i^T \mathbf{v}_i} \tag{10.2-22}$$

$$\mathbf{v}_{i+1} = k_{i+1}(A - \mu_{i+1} I)^{-1} \mathbf{v}_i \tag{10.2-23}$$

Under appropriate hypotheses, it can be shown that the sequence $\{\mu_i, \mathbf{v}_i\}$ converges cubically to an eigenvalue-eigenvector pair $\{\lambda, \mathbf{v}\}$ which depends on the initial approximation $\mathbf{v}_0$.

In practice, the inverse $(A - \mu_{i+1} I)^{-1}$ is not computed explicitly. Instead, the set of linear equations

$$(A - \mu_{i+1} I)\mathbf{z}_{i+1} = \mathbf{v}_i \tag{10.2-24}$$

is solved, and then we use $\mathbf{v}_{i+1} = k_{i+1}\mathbf{z}_{i+1}$. Even though the coefficient matrix $A - \mu_i I$ approaches singularity as μ_i approaches an eigenvalue, this does not detract from the accuracy of the calculation provided that partial pivoting is used to solve (10.2-24) (see Sec. 9.3). We mention at this point that the generalized Rayleigh iteration as applied to nonsymmetric matrices is intimately related to the Jenkins-Traub method for finding the zeros of a polynomial (see Sec. 8.11).

The second context in which the inverse power method is used is to find eigenvectors corresponding to computed eigenvalues by the method of *inverse iteration*, which uses (10.2-18) with a fixed value of p corresponding to an accurate approximation to an eigenvalue. Of course, if the eigenvalue has been computed by the power method, we have the corresponding eigenvector as well. However, there are many methods studied later in this chapter which compute only eigenvalues and not eigenvectors. Combining the inverse power method with one of the better methods for computing eigenvalues given in succeeding sections yields one of the most efficient ways to compute the eigensystem of a matrix. It is especially efficient if A has been reduced to tridiagonal or Hessenberg form (see Secs. 10.3 and 10.4), for then the amount of work per eigenvector is reduced from $O(n^3)$ operations to $O(n)$ or $O(n^2)$ operations, respectively (see Sec. 9.11). (The same reduction in work per iteration for these special matrices occurs also in the Rayleigh quotient method.)

In inverse iteration, we assume that we have computed an accurate approximation $\hat{\lambda}$ to an eigenvalue λ and wish to compute the corresponding eigenvector $\mathbf{x}$. Even if $\hat{\lambda}$ differs from λ only by rounding error, the matrix $A - \hat{\lambda} I$ is still nonsingular, so that, as we have seen in Sec. 9.3, [cf. (9.3-23)], we can compute a triangular decompositon of some permutation of $A - \hat{\lambda} I$

$$P(A - \hat{\lambda} I) = LU \tag{10.2-25}$$

in $O(n)$, $O(n^2)$, or $O(n^3)$ operations, depending on whether A is tridiagonal, Hessenberg, or general, respectively. We now wish to solve

$$(A - \hat{\lambda}I)\mathbf{z}_1 = P^{-1}LU\mathbf{z}_1 = \mathbf{v}_0 \tag{10.2-26}$$

for some initial vector $\mathbf{v}_0$ which does not have an excessively small component in the direction $\mathbf{x}$. In the absence of any additional information about $\mathbf{x}$, this may be quite difficult to ensure. One way which has been very successful in practice, at the same time saving some computing effort, is to choose $\mathbf{v}_0$ so that

$$L^{-1}P\mathbf{v}_0 = \mathbf{u} \equiv [1, \quad 1, \quad \ldots, \quad 1]^T \tag{10.2-27}$$

Hence, we start our iteration by solving $U\mathbf{z}_1 = \mathbf{u}$. After normalizing $\mathbf{z}_1$, as above, we apply one more iteration involving a forward and back substitution and accept $\mathbf{z}_2$ as our eigenvector $\mathbf{x}$, which we can normalize in any way desired. Except in the case of pathologically close roots, $\mathbf{z}_2$ is a very accurate approximation to $\mathbf{x}$.

One apparent drawback to this method is that if λ is a multiple eigenvalue, we shall be able to compute only one corresponding eigenvector. However, we can usually compute a different eigenvector if we start with a perturbation of $\hat{\lambda}$ by as little as three units in the last significant digit. This may affect the speed of convergence in that more than two iterations may be necessary, but the accuracy of the computed eigenvector will not be affected. Alternatively, we may choose $\mathbf{v}_0$ to be a vector whose components are (pseudo-) random numbers between -1 and 1. Then, if λ is a multiple eigenvalue, different choices of $\mathbf{v}_0$ will generally yield linearly independent eigenvectors.

Example 10.3 Apply inverse iteration to find the eigenvector $\mathbf{x}$ corresponding to the computed eigenvalue $\hat{\lambda} = -256.01$ of the matrix

$$A = \begin{bmatrix} -120.0 & 90.86 & 0 & 0 \\ 90.86 & -157.2 & -67.59 & 0 \\ 0 & -67.59 & -124.0 & 46.26 \\ 0 & 0 & 46.26 & -78.84 \end{bmatrix}$$

using 5-digit floating-point decimal arithmetic.

Since $A - \hat{\lambda}I$ is the matrix of Example 9.3, the triangular factorization of $A - \hat{\lambda}I$ is given in that example [cf. (10.2-25)] with

$$L = \begin{bmatrix} 1 & 0 & 0 & 0 \\ 0 & 1 & 0 & 0 \\ 0 & 0 & 1 & 0 \\ .66804 & -.56387 & .14800 & 1 \end{bmatrix}$$

$$U = \begin{bmatrix} 136.01 & 90.860 & 0 & 0 \\ 0 & -67.590 & 132.01 & 46.260 \\ 0 & 0 & 46.260 & 177.17 \\ 0 & 0 & 0 & -.13653 \end{bmatrix}$$

$$P = \begin{bmatrix} 1 & 0 & 0 & 0 \\ 0 & 0 & 1 & 0 \\ 0 & 0 & 0 & 1 \\ 0 & 1 & 0 & 0 \end{bmatrix}$$

Setting $\mathbf{u} = (1, 1, 1, 1)^T$, we solve $U\mathbf{z}_1 = \mathbf{u}$ by back substitution and find

$$\mathbf{z}_1 = [-33.254 \quad 49.790 \quad 28.067 \quad -7.3244]^T$$

We then solve the system $LU\mathbf{z}_2 = P\mathbf{z}_1$ as in Example 9.7 to get the final solution

$$\mathbf{z}_2 = [-2953.3 \quad 4420.5 \quad 2491.5 \quad -650.59]^T$$

The value $\hat{\lambda}$ is an approximation to the true eigenvalue of A, $\lambda = 256.00$. The eigenvector corresponding to λ is

$$\mathbf{x} = [-.66810 \quad 1.0000 \quad .56363 \quad -.14718]^T$$

A similar normalization of $\mathbf{z}_2$ yields the vector

$$\hat{\mathbf{x}} = [-.66809 \quad 1.0000 \quad .56362 \quad -.14718]^T$$

which agrees with $\mathbf{x}$ to within rounding error.

10.3 THE EIGENVALUES AND EIGENVECTORS OF SYMMETRIC MATRICES

Our object here is to develop methods which will enable us to compute all the eigenvalues and eigenvectors when the matrix A is symmetric, as we shall assume it is throughout this section. The three methods we shall consider all use as their basic tool orthogonal transformations of A. In particular the first of these three methods has its theoretical basis in Corollary 10.2 of Sec. 10.1-4.

10.3-1 The Jacobi Method

Corollary 10.2 assures us that there exists an orthogonal matrix Q such that

$$Q^TAQ = D \tag{10.3-1}$$

where D is a diagonal matrix with the eigenvalues of A on the diagonal. Our technique will be to find a sequence $\{S_k\}$ of orthogonal matrices with the property that

$$\lim_{k\to\infty} S_1S_2 \cdots S_k = Q \tag{10.3-2}$$

We shall use the notation

$$T_k = S_k^TS_{k-1}^T \cdots S_1^TAS_1S_2 \cdots S_k \qquad T_0 = A \tag{10.3-3}$$

We denote the elements of T_k by $t_{ij}^{(k)}$ and of S_k by $s_{ij}^{(k)}$. We define

$$v_k = \sum_{\substack{i=1 \\ i\neq j}}^{n} \sum_{j=1}^{n} (t_{ij}^{(k)})^2 \qquad k = 0, 1, \ldots \tag{10.3-4}$$

and

$$w_k = \sum_{i=1}^{n} \sum_{j=1}^{n} (t_{ij}^{(k)})^2 \qquad k = 0, 1, \ldots \tag{10.3-5}$$

Thus w_k is the square of the Euclidean norm of T_k and v_k is the sum of the squares of the off-diagonal elements of T_k. Our object is to choose the sequence $\{S_k\}$ so that

$$w_{k+1} = w_k \qquad v_{k+1} < v_k \qquad \text{for all } k \tag{10.3-6}$$

and

$$\lim_{k\to\infty} v_k = 0 \tag{10.3-7}$$

in which case

$$\lim_{k\to\infty} T_k = D \tag{10.3-8}$$

Let $t_{pq}^{(k-1)}$ be a nonzero off-diagonal element of T_{k-1}. We wish to choose S_k so that $t_{pq}^{(k)} = 0$. In doing so we shall show that (10.3-6) is satisfied. Let

$$\begin{aligned} s_{pp}^{(k)} &= s_{qq}^{(k)} = \cos\theta_k \qquad & s_{ii}^{(k)} &= 1 \qquad & i &\neq p \text{ or } q \\ s_{pq}^{(k)} &= -s_{qp}^{(k)} = \sin\theta_k \qquad & s_{ij}^{(k)} &= 0 \qquad & &\text{otherwise} \end{aligned} \tag{10.3-9}$$

where θ_k will be chosen below so that the pq element is annihilated; that is, $t_{pq}^{(k)} = 0$. The orthogonal matrix defined by (10.3-9) is called a *plane rotation matrix* because the linear transformation defined by S_k consists of a rotation of the axes of the pth and qth coordinates through an angle θ_k. From (10.3-3)

$$T_k = S_k^T T_{k-1} S_k \tag{10.3-10}$$

so that we have, using (10.3-9),

$$\begin{aligned} t_{pj}^{(k)} &= t_{pj}^{(k-1)} \cos\theta_k - t_{qj}^{(k-1)} \sin\theta_k \\ t_{qj}^{(k)} &= t_{pj}^{(k-1)} \sin\theta_k + t_{qj}^{(k-1)} \cos\theta_k \end{aligned} \qquad j \neq p \text{ or } q \tag{10.3-11}$$

$$\begin{aligned} t_{ip}^{(k)} &= t_{ip}^{(k-1)} \cos\theta_k - t_{iq}^{(k-1)} \sin\theta_k \\ t_{iq}^{(k)} &= t_{ip}^{(k-1)} \sin\theta_k + t_{iq}^{(k-1)} \cos\theta_k \end{aligned} \qquad i \neq p \text{ or } q \tag{10.3-12}$$

$$\begin{aligned} t_{pp}^{(k)} &= t_{pp}^{(k-1)} \cos^2\theta_k + t_{qq}^{(k-1)} \sin^2\theta_k - 2t_{pq}^{(k-1)} \sin\theta_k \cos\theta_k \\ t_{qq}^{(k)} &= t_{pp}^{(k-1)} \sin^2\theta_k + t_{qq}^{(k-1)} \cos^2\theta_k + 2t_{pq}^{(k-1)} \sin\theta_k \cos\theta_k \\ t_{pq}^{(k)} &= t_{qp}^{(k)} = \tfrac{1}{2}(t_{pp}^{(k-1)} - t_{qq}^{(k-1)}) \sin 2\theta_k + t_{pq}^{(k-1)} \cos 2\theta_k \end{aligned} \tag{10.3-13}$$

$$t_{ij}^{(k)} = t_{ij}^{(k-1)} \qquad i \neq p;\, j \neq q \tag{10.3-14}$$

Now we shall choose θ_k to make $t_{pq}^{(k)}$ vanish. From the last equation of (10.3-13) we have

$$\alpha = \cot 2\theta_k = \frac{t_{qq}^{(k-1)} - t_{pp}^{(k-1)}}{2t_{pq}^{(k-1)}} \tag{10.3-15}$$

so that a θ_k always exists. In practice we do not calculate θ_k itself but only $\sin\theta_k$ and $\cos\theta_k$ since they are all that are required by (10.3-11) to (10.3-13). A convenient way to calculate $\sin\theta_k$ and $\cos\theta_k$ is as follows. Let $T = \tan\theta_k$.

Then using the identity $\tan^2 \theta + 2 \tan \theta \cot 2\theta - 1 = 0$, choose T to be the smaller root in magnitude of the equation

$$t^2 + 2At - 1 = 0 \tag{10.3-16}$$

Thus using the form of the quadratic equation formula with the discriminant in the denominator, we have

$$T = \frac{1}{|A| + \sqrt{1 + A^2}} \operatorname{sign}(A) \tag{10.3-17}$$

Then†

$$C = \cos \theta_k = \frac{1}{\sqrt{1 + T^2}} \qquad S = \sin \theta_k = TC \tag{10.3-18}$$

Equations (10.3-11) to (10.3-13) can be reformulated {22} to yield the following set of equations which give more accurate results in computation by expressing all elements of T_k as perturbations of elements of T_{k-1}:

$$\begin{aligned} &t_{pp}^{(k)} = t_{pp}^{(k-1)} - h \qquad t_{qq}^{(k)} = t_{qq}^{(k-1)} + h \qquad t_{pq}^{(k)} = 0 \\ &t_{jp}^{(k)} = t_{pj}^{(k)} = t_{pj}^{(k-1)} - S(t_{qj}^{(k-1)} + \tau t_{pj}^{(k-1)}) \\ &t_{jq}^{(k)} = t_{qj}^{(k)} = t_{qj}^{(k-1)} + S(t_{pj}^{(k-1)} - \tau t_{qj}^{(k-1)}) \end{aligned} \qquad j \neq p \text{ or } q \tag{10.3-19}$$

where

$$\tau = \tan\left(\tfrac{1}{2}\theta_k\right) = \frac{S}{1 + C} \qquad \text{and} \qquad h = T t_{pq}^{(k-1)} \tag{10.3-20}$$

Using (10.3-11) to (10.3-13), we can easily calculate that, independent of θ_k,

$$\begin{aligned} (t_{pj}^{(k)})^2 + (t_{qj}^{(k)})^2 &= (t_{pj}^{(k-1)})^2 + (t_{qj}^{(k-1)})^2 \qquad j \neq p \text{ or } q \\ (t_{ip}^{(k)})^2 + (t_{iq}^{(k)})^2 &= (t_{ip}^{(k-1)})^2 + (t_{iq}^{(k-1)})^2 \qquad i \neq p \text{ or } q \end{aligned} \tag{10.3-21}$$

and

$$\begin{aligned} &(t_{pp}^{(k)})^2 + (t_{qq}^{(k)})^2 + (t_{pq}^{(k)})^2 + (t_{qp}^{(k)})^2 \\ &\qquad = (t_{pp}^{(k-1)})^2 + (t_{qq}^{(k-1)})^2 + (t_{pq}^{(k-1)})^2 + (t_{qp}^{(k-1)})^2 \end{aligned} \tag{10.3-22}$$

But with θ_k chosen as in (10.3-15), $t_{pq}^{(k)} = t_{qp}^{(k)} = 0$. Therefore, (10.3-21) and (10.3-22) together with (10.3-14) show that $w_{k-1} = w_k$ and (10.3-22) implies that $v_k < v_{k-1}$. In fact, (10.3-22) implies that since $t_{pq}^{(k-1)} = t_{qp}^{(k-1)}$, because all the T_k are symmetric, the off-diagonal sum of squares is reduced by $2(t_{pq}^{(k-1)})^2$ and the sum of squares of the diagonal elements is therefore increased by a like amount. At each stage of the Jacobi iteration we (1) choose a *nonzero* off-diagonal element; (2) calculate $\sin \theta_k$ and $\cos \theta_k$ from (10.3-18); (3) calculate those elements in T_k which differ from those in T_{k-1} using (10.3-19).

† Note that (10.3-17) and (10.3-18) always result in a rotation angle such that $|\theta_k| \leq \pi/4$ (why?).

Note that an off-diagonal element made zero at one stage will generally become nonzero at some later stage (why?).

This process can be continued for as long as is desired. The convergence of the method can be proved for a variety of techniques, including those considered below, for choosing the nonzero off-diagonal element at each stage. For some techniques the proof is quite difficult, but in the case where the off-diagonal element of greatest magnitude is annihilated at each stage, the proof is quite straightforward {23}. The most common convergence criterion used in practice is to require that the off-diagonal norm $\sqrt{v_k}$ be less than some preset tolerance (cf. {24}).

The eigenvectors of A are also readily calculated using this iteration. We noted in Sec. 10.1-4 [cf. (10.1-38)] that the columns of the orthogonal matrix used to reduce A to diagonal form are the eigenvectors of A. To calculate the eigenvectors we must then calculate the product of the S_k matrices in (10.3-2). To see how this can be done simply we write (10.3-3) as

$$T_k = R_k^T A R_k \tag{10.3-23}$$

with

$$R_k = S_1 S_2 \cdots S_k \tag{10.3-24}$$

Thus

$$R_{k+1} = R_k S_{k+1} \tag{10.3-25}$$

and from (10.3-9), we can then calculate the elements of $R_{k+1} = [r_{ij}^{(k+1)}]$ in terms of those of $R_k = [r_{ij}^{(k)}]$:

$$\begin{aligned} r_{ip}^{(k+1)} &= r_{ip}^{(k)} \cos\theta_k - r_{iq}^{(k)} \sin\theta_k \\ r_{iq}^{(k+1)} &= r_{ip}^{(k)} \sin\theta_k + r_{iq}^{(k)} \cos\theta_k \\ r_{ij}^{(k+1)} &= r_{ij}^{(k)} \qquad j \neq p \text{ or } q \end{aligned} \tag{10.3-26}$$

starting with $R_1 = S_1$.

The one problem remaining in the use of the Jacobi method is how to choose the off-diagonal element to be annihilated at each stage. We should like to choose that element of greatest magnitude since this would result in the greatest reduction of the off-diagonal norm $\sqrt{v_k}$ at every stage. For hand computation this is easily done, but on a digital computer this requires a search through all the off-diagonal elements (actually just through half of them since the matrix is symmetric at every stage). Thus, it is more convenient to annihilate all the subdiagonal elements in serial order, starting with the (2, 1) element, proceeding down the first column, continuing with the (3, 2) element, etc. This is called the *serial Jacobi method* and can be shown to converge provided the rotation angle θ_k satisfies $|\theta_k| \leq \pi/4$, as indeed it does in our formulation. However, since the annihilation of a small subdiagonal element may not be worth the computational effort invested, a procedure called the *threshold Jacobi method* may be used. The essence of this procedure is to search through the off-diagonal elements until an element of magnitude greater than some threshold value is found. This element is then annihilated. This method assures some minimum reduction of the off-diagonal norm at each stage. The details of this method, in particular how to choose the threshold at every stage, are considered in a problem {24}.

Example 10.4 Apply the Jacobi method to the matrix A of Example 10.1.

The largest off-diagonal element is 1.0 in the first row and second column ($p = 1$, $q = 2$). Using (10.3-17) and (10.3-18), we have

$$\sin \theta_1 = -\frac{1}{\sqrt{2}} \qquad \text{and} \qquad \cos \theta_1 = \frac{1}{\sqrt{2}}$$

Using (10.3-19) we compute

$$T_1 = S_1^T A S_1 = \begin{bmatrix} 2 & 0 & \frac{3}{4\sqrt{2}} \\ 0 & 0 & \frac{-1}{4\sqrt{2}} \\ \frac{3}{4\sqrt{2}} & \frac{-1}{4\sqrt{2}} & 2 \end{bmatrix}$$

By continuing this process we obtain

$$D = \begin{bmatrix} 2.5365258 & 0 & 0 \\ 0 & -.0166473 & 0 \\ 0 & 0 & 1.4801215 \end{bmatrix}$$

and, using (10.3-26),

$$Q = \begin{bmatrix} .53148338 & -.72120712 & -.44428106 \\ .46147338 & .68634928 & -.56210938 \\ .71032933 & .09372796 & .69760117 \end{bmatrix}$$

As a check on the results we can compute QDQ^T, which should equal A.

The accuracy of the results of the Jacobi method depend upon how accurately the square roots leading to $\sin \theta_k$ and $\cos \theta_k$ given by (10.3-18) are calculated and on how the roundoff error accumulates. The error analysis of the Jacobi method is quite involved and beyond us here [see Goldstine, Murray, and Von Neumann (1959) and Wilkinson (1962)], but if the square roots are calculated with appropriate accuracy, the method is completely stable with respect to roundoff error; i.e., no significant growth of error occurs because of roundoff.

The Jacobi method has been widely applied on digital computers to find the eigenvalues of symmetric matrices, some of very high order. Using a relatively simple, compact program, it yields all the eigenvalues and, if desired, all the eigenvectors. Furthermore, the computed eigenvectors are always orthogonal, up to roundoff error, even when there are multiple eigenvalues. Its major disadvantage is that it is an infinite iterative method, and when high accuracy is desired in the eigenvalues or the diagonal elements of A are not large compared with the off-diagonal elements, it may lead to a lengthy computation. In the next two sections, we consider two methods, both of which are finite iterative processes, which lead not to the matrix D but instead to a new matrix whose eigenvalues are much easier to compute than those of a general symmetric matrix A.

10.3-2 Givens' Method

The basis of this method is to use orthogonal matrices not to diagonalize A but rather to tridiagonalize A, that is, reduce A to a form in which the only nonzero elements are on the main diagonal and the two diagonals directly above and below it, as shown in Fig. 10.1. We shall then consider how to find the eigenvalues of such a matrix.

The orthogonal matrices S_k used in the Jacobi method had the property that only the pth and qth rows and columns of T_{k-1} were changed in calculating $T_k = S_k^T T_{k-1} S_k$. If then we arrange the order of the calculations carefully, we should not be surprised to find that—up to a point—we can annihilate off-diagonal elements while at the same time keeping previously annihilated elements zero. First we note that using S_k as defined by (10.3-9) (with θ_k unspecified as yet) we can annihilate, instead of $t_{pq}^{(k-1)}$, $t_{rq}^{(k-1)}$ with $r \neq p$ or q (and by symmetry $t_{qr}^{(k-1)}$ is annihilated also). This follows from (10.3-12) by writing

$$0 = t_{rq}^{(k)} = t_{rp}^{(k-1)} \sin \theta_k + t_{rq}^{(k-1)} \cos \theta_k \tag{10.3-27}$$

which will be satisfied if

$$\sin \theta_k = -\alpha t_{rq}^{(k-1)} \qquad \cos \theta_k = \alpha t_{rp}^{(k-1)} \tag{10.3-28}$$

with

$$\alpha = 1/[(t_{rp}^{(k-1)})^2 + (t_{rq}^{(k-1)})^2]^{1/2} \tag{10.3-29}$$

Let us denote the matrix whose elements are given by (10.3-9), with $\sin \theta_k$ and $\cos \theta_k$ given by (10.3-28) and (10.3-29), by the triplet (p, q, r), these three *distinct* integers denoting the row (and column) indices of significance in the transformation. In particular consider the sequence of transformations

$$(p, q, r) = (2, i, 1) \qquad i = 3, \ldots, n \tag{10.3-30}$$

applied in succession to the original matrix A. That is, each triplet defines a transformation which we apply by premultiplying the current matrix by the transpose of the matrix defined by (10.3-9) and postmultiplying by the

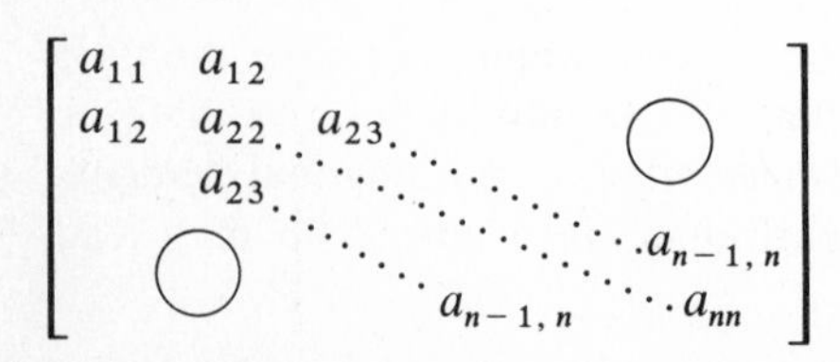

Figure 10.1 A symmetric tridiagonal matrix.

matrix itself. Denoting this matrix now by S_{pqr}, we have after the first step

$$S_{231}^T A S_{231} \tag{10.3-31}$$

Thus the transformation (2, 3, 1) annihilates the element in the first row (column) and third column (row) with p and q in (10.3-9) being 2 and 3, respectively. In general the (2, i, 1) transformation annihilates the element in the first row (column) and ith column (row). Of equal significance, however, is the fact that *each annihilated element remains zero.* This is because the (2, i, 1) transformation changes only the second row (column) and the ith column (row) and therefore does not affect previously annihilated elements. Therefore, after the sequence of transformations (10.3-30), all elements in the first row (column) except the first two (a_{11} and a_{12}) are zero.

Next we consider the sequence

$$(p, q, r) = (3, i, 2) \qquad i = 4, \ldots, n \tag{10.3-32}$$

which annihilates the elements of the second row (column) from a_{24} to a_{2n}. By reasoning similar to that in the previous paragraph and use of (10.3-32) we can show that {26} the sequence of transformations (10.3-32) leaves *all* previously annihilated elements in the *first* and *second* rows (columns) zero. The general algorithm is now clear. We apply the sequence of transformations

$$(p, q, r) = (j, i, j-1) \qquad \begin{matrix} j = 2, \ldots, n-1 \\ i = j+1, \ldots, n \end{matrix} \tag{10.3-33}$$

and in so doing get a new symmetric matrix B with the form

$$B = \begin{bmatrix} b_1 & c_1 & & & \bigcirc \\ c_1 & b_2 & c_2 & & \\ & c_2 & \ddots & \ddots & \\ & & \ddots & \ddots & c_{n-1} \\ \bigcirc & & & c_{n-1} & b_n \end{bmatrix} \tag{10.3-34}$$

From (10.3-33) the number of transformations required is $(n-2)(n-1)/2$, and we may show that the total number of operations required is of the order of $\frac{4}{3}n^3$ plus, of course, the $(n-2)(n-1)/2$ square roots required by (10.3-29) {31}.

In Sec. 10.5-1 we shall consider a method particularly suited to finding eigenvalues of tridiagonal matrices, symmetric or otherwise, having real eigenvalues. Here, however, we shall show that B has a form which permits another effective way of computing its eigenvalues. We consider the matrix

$$\lambda I - B = \begin{bmatrix} -b_1 + \lambda & -c_1 & & \bigcirc \\ -c_1 & -b_2 + \lambda & \ddots & \\ & \ddots & \ddots & -c_{n-1} \\ \bigcirc & & -c_{n-1} & -b_n + \lambda \end{bmatrix} \tag{10.3-35}$$

If we denote the principal minor of order i in the above matrix by $f_{n-i}(\lambda)$, we can easily show that {27}

$$f_{n-(i+1)}(\lambda) = (\lambda - b_{i+1})f_{n-i}(\lambda) - c_i^2 f_{n-(i-1)}(\lambda) \qquad i = 1, \ldots, n-1 \tag{10.3-36}$$

with $f_n(\lambda) = 1$ and $f_{n-1}(\lambda) = -b_1 + \lambda$. The characteristic equation is

$$f_0(\lambda) = 0 \tag{10.3-37}$$

We can, of course, solve (10.3-37) by the methods of Chap. 8 [since the matrix is symmetric, all roots of each $f_{n-i}(\lambda)$ are real]. But here we shall show how our results about Sturm sequences in Sec. 8.9-1 can be applied to this problem.

We assume that no c_i in (10.3-34) is zero. For if any $c_i = 0$, the determinant of the matrix in (10.3-35) can be written as the product of two determinants of smaller tridiagonal matrices {27} and the results below will apply to each. Under this assumption we can show that the sequence $f_0(\lambda), \ldots, f_n(\lambda)$ forms a Sturm sequence as defined in Definition 8.1; we leave the proof of this to a problem {27}. This suggests trying to get some idea about the roots of (10.3-37) by applying Theorem 8.3, i.e., by calculating $V(a)$ and $V(b)$ for various a and b and thereby determining the number of roots in various intervals $[a, b]$. However, Theorem 8.3 requires the Sturm sequence in question to have as its second member the derivative of the first, but $f_1(\lambda)$ is not $f'_0(\lambda)$. However, if $f_1(\lambda)$ has the same sign as $f'_0(\lambda)$ at each zero of $f_0(\lambda)$, then Theorem 8.3 is also valid if the sequence used to compute $V(a) - V(b)$ is $\{f_i(\lambda)\}$ (why?). Thus we wish to prove the following theorem.

Theorem 10.13 At a zero of $f_0(\lambda)$, $f_1(\lambda)$ has the same sign as $f'_0(\lambda)$.

To prove this theorem we first prove a lemma.

Lemma 10.1 The i zeros s_j, $j = 1, \ldots, i$, of $f_{n-i}(\lambda)$ separate the $i-1$ zeros r_j, $j = 1, \ldots, i-1$, of $f_{n-(i-1)}(\lambda)$ for $i = 2, \ldots, n$. That is,

$$-\infty < s_1 < r_1 < s_2 < r_2 \cdots < r_{i-1} < s_i < \infty \tag{10.3-38}$$

PROOF OF LEMMA 10.1 By induction. For $i = 2$ this can be proved directly using (10.3-36) {27}. Suppose (10.3-38) holds for some i. Then, using property 2 of a Sturm sequence at each s_j, the function $f_{n-(i+1)}(\lambda)$ has sign opposite to $f_{n-(i-1)}(\lambda)$. But (10.3-38) implies that the sign of $f_{n-(i-1)}(\lambda)$ alternates from one zero of $f_{n-i}(\lambda)$ to the next. Thus between s_1 and s_i, $f_{n-(i+1)}(\lambda)$ has $i-1$ zeros lying between the zeros of $f_{n-i}(\lambda)$. At s_1, $f_{n-(i+1)}(\lambda)$ has sign opposite to $f_{n-(i-1)}(\lambda)$, but as $\lambda \to -\infty$, both $f_{n-(i-1)}(\lambda)$ and $f_{n-(i+1)}(\lambda)$ have the same sign since they differ by degree 2. Therefore, $f_{n-(i+1)}(\lambda)$ has a zero between $-\infty$ and s_1 and by a similar argument it has a zero between s_i and $+\infty$, which proves the lemma.

PROOF OF THEOREM 10.13 Since the zeros of $f_0(\lambda)$ and $f_1(\lambda)$ separate each other, the zeros of $f_0(\lambda)$ are distinct. [Note that even though B may have multiple eigenvalues, $f_0(\lambda)$ has only simple zeros. This is so since some $c_i = 0$ for there to be multiple eigenvalues, and we have assumed above that we have decomposed B into submatrices in each of which no c_i is 0]. Therefore, between each two zeros of $f_0(\lambda)$, $f'_0(\lambda)$ also has a zero. Moreover, for sufficiently large λ, $f'_0(\lambda)$ and $f_1(\lambda)$ have the same sign since both are polynomials with leading term λ^{n-1}. Therefore, they both have the same sign at the largest zero of $f_0(\lambda)$. The argument above then guarantees that they have the same sign at every zero of $f_0(\lambda)$, which proves the theorem.

To compute the value of $V(x)$ requires the evaluation of each $f_{n-i}(\lambda)$ at $\lambda = x$, but this is easily accomplished using (10.3-36). However, there is a danger that when there are clusters of eigenvalues, the sequence $f_{n-i}(\lambda)$ may have some very small numbers for values of λ close to such eigenvalues. Now, if underflow occurs in the computation of these numbers and they are replaced by 0, the sign determination may be incorrect. To avoid such a situation, we work with the quotients

$$q_{n-i}(\lambda) = \frac{f_{n-i-1}(\lambda)}{f_{n-i}(\lambda)} \tag{10.3-39}$$

which satisfy the recurrence formulas

$$q_n(\lambda) = -b_1 + \lambda \qquad q_{n-i}(\lambda) = (\lambda - b_{i+1}) - \frac{c_i^2}{q_{n-i+1}(\lambda)} \qquad i = 1, \ldots, n-1 \tag{10.3-40}$$

Then $V(\lambda)$ equals the number of negative $q_{n-i}(\lambda)$. Now (10.3-40) in itself does not give a process free of problems. However, by prescaling our matrix and modifying the recurrence appropriately, we get an algorithm which will almost always work.

Let m and M be respectively the smallest and largest floating-point numbers which can be represented in the computer and define

$$\tau = m^{1/4}M^{-1/2}$$

We now find σ which is the largest power of the base of our number system such that $\sigma|b_i|$, $\sigma|c_i| \le \tau M$, $i = 1, \ldots, n$. We work with the matrix with elements σb_i, σc_i, which we rename b_i, c_i. Note that the multiplication by σ does not introduce any roundoff error. The eigenvalues of this new matrix will be σ times the original eigenvalues, so that we must remember to "unscale" at the end. As a result of this prescaling we have that $\|B\|_\infty \le 3\tau M$, from which it follows that $\rho(B) \le \|B\|_\infty \le 3\tau M$ so that $\lambda \le 3\tau M$ for any λ for which we compute $V(\lambda)$. We are now ready to start the recurrence (10.3-40), modified as follows. If, for any i, $i = 1, \ldots, n$, $|q_{n-i+1}(\lambda)| \le \sqrt{m}$, set $q_{n-i+1}(\lambda) = -\sqrt{m}$. This modification can be interpreted to mean that we have perturbed our matrix slightly by working with $b_i + \epsilon_i$ with $|\epsilon_i| \le 2\sqrt{m}$. Note that $|\epsilon_i|$ is usually much smaller than roundoff error. With this modification, we cannot have overflow since $c_i^2/\sqrt{m} \le \tau^2 M^2/\sqrt{m} = M$ and the addition of $\lambda - b_{i+1}$ leaves the result less than or equal to M (why?) and clearly underflow does not cause any problems either. Finally, a vanishing c_i^2 does not cause any trouble since we can look at $c_i^2/q_{n-i+1} = 0$ as resulting from an underflow with $c_i \ne 0$, so that we need not worry about decomposing our matrix into submatrices but can work all the time with the scaled original matrix.

Therefore, after the reduction of the matrix to tridiagonal form, the calculation of the eigenvalues of the tridiagonal matrix can be accomplished without too much difficulty using the properties of Sturm sequences in conjunction with one of the methods of Chap. 8. In particular, the method of bisection (see Sec. 2.2), although somewhat slow, is a very convenient method for calculating the eigenvalues. We proceed as follows. Evaluate $V(\lambda)$ for the set of functions (10.3-40) for a sequence of values of λ. Whenever two values, λ_1 and λ_2, are found such that $V(\lambda_1) \ne V(\lambda_2)$, then there is an eigenvalue between λ_1 and λ_2. By evaluating $V[(\lambda_1 + \lambda_2)/2]$ the interval in which λ lies can be halved, and by continuing in this way the interval can be made as small as desired.

Since the tridiagonalization is a finite iterative process, the error in generating B can be strictly controlled, and then the eigenvalues of B can be determined with arbitrary accuracy. Thus Givens' method is generally preferable to the Jacobi method. Only for matrices where the diagonal terms dominate would we expect the Jacobi method to be competitive with Givens' method (why?).

The eigenvectors of A bear the same relation to those of B as in the Jacobi method. That is, by keeping track of the successive orthogonal transformations as in (10.3-24) to (10.3-26) the eigenvectors of A can be calculated from those of B {28}. Note that the eigenvectors of B can be computed using the method of inverse iteration given in Sec. 10.2-2, which is very efficient for tridiagonal matrices.

Example 10.5 Apply Givens' method to the matrix A of Example 10.1.

The only off-tridiagonal element is a_{13}, so that there is only one transformation to perform. From (10.3-28) and (10.3-29) with $p = 2$, $q = 3$, $r = 1$

$$\alpha = \frac{2}{\sqrt{5}} \qquad \sin\theta = -\frac{1}{\sqrt{5}} \qquad \cos\theta = \frac{2}{\sqrt{5}}$$

Using these to get the matrix S of (10.3-9), we then calculate

$$B = S^T A S = \begin{bmatrix} 1 & \frac{\sqrt{5}}{2} & 0 \\ \frac{\sqrt{5}}{2} & 1.40 & .55 \\ 0 & .55 & 1.60 \end{bmatrix}$$

Then from (10.3-36) we get

$$f_3(\lambda) = 1 \qquad f_2(\lambda) = \lambda - 1$$

$$f_1(\lambda) = \lambda^2 - 2.4\lambda + .15 \qquad f_0(\lambda) = \lambda^3 - 4\lambda^2 + 3.6875\lambda + .0625$$

with $f_0(\lambda)$ the characteristic equation of A.

10.3-3 Householder's Method

This method, a variation of Givens' method, enables us to reduce A to tridiagonal form with about half as much computation as Givens' method requires. The technique is to reduce a whole row and column (except for the tridiagonal elements) to zero at a time.

Let $\mathbf{v}$ be a vector such that

$$\mathbf{v}^T\mathbf{v} = 1 \tag{10.3-41}$$

Then it is easy to show {29} that the matrix

$$P = I - 2\mathbf{v}\mathbf{v}^T \tag{10.3-42}$$

is orthogonal and symmetric. In particular we choose $\mathbf{v}_k$ to be a vector whose first $k - 1$ components are zero, so that

$$\mathbf{v}_k^T = [0 \quad 0 \quad \cdots \quad 0 \quad v_k^{(k)} \quad v_k^{(k+1)} \quad \cdots \quad v_k^{(n)}] \tag{10.3-43}$$

Then with

$$P_k = I - 2\mathbf{v}_k\mathbf{v}_k^T \tag{10.3-44}$$

we define

$$A_k = P_k^T A_{k-1} P_k \qquad k = 2, \ldots, n-1 \qquad A_1 = A \tag{10.3-45}$$

Now suppose that the symmetric matrix $A_{k-1} = [g_{ij}]$ has zeros in its first $k - 2$ rows and columns except for the tridiagonal elements:

$$A_{k-1} = \begin{bmatrix} g_{11} & g_{12} & 0 & \cdots & \cdots & \cdots & \cdots & 0 \\ g_{12} & g_{22} & g_{23} & \ddots & & & & \vdots \\ 0 & g_{23} & \ddots & \ddots & & & & \vdots \\ \vdots & \ddots & \ddots & \ddots & \ddots & & & \vdots \\ \vdots & & \ddots & g_{k-2,k-2} & g_{k-2,k-1} & 0 & \cdots & 0 \\ \vdots & & & g_{k-2,k-1} & g_{k-1,k-1} & \cdots & \cdots & g_{k-1,n} \\ \vdots & & & 0 & \vdots & \ddots & & \vdots \\ \vdots & & & \vdots & \vdots & & \ddots & \vdots \\ 0 & \cdots & \cdots & 0 & g_{k-1,n} & \cdots & \cdots & g_{nn} \end{bmatrix} \begin{matrix} \\ \\ \\ \\ \\ \text{row } k-1 \\ \\ \\ \\ \end{matrix} \tag{10.3-46}$$

The matrix P_k has the form

$$I - 2\mathbf{v}_k\mathbf{v}_k^T = \begin{bmatrix} 1 & 0 & \cdots & & & \cdots & 0 \\ 0 & \ddots & & & & & \vdots \\ \vdots & & 1 & 0 & \cdots & \cdots & 0 \\ \vdots & & 0 & 1 - 2(v_k^{(k)})^2 & \cdots & \cdots & -2v_k^{(n)}v_k^{(k)} \\ \vdots & & \vdots & \vdots & \ddots & & \vdots \\ 0 & \cdots & 0 & -2v_k^{(n)}v_k^{(k)} & \cdots & \cdots & 1 - 2(v_k^{(n)})^2 \end{bmatrix} \qquad \text{row } k \tag{10.3-47}$$

Using (10.3-45) to (10.3-47) we can verify that A_k has zeros in the positions shown as zero for A_{k-1} in (10.3-46). Our object is to choose the $n - k + 1$ numbers $v_k^{(k)}, \ldots, v_k^{(n)}$ to satisfy (10.3-41) and so that the $n - k$ off-tridiagonal elements in row (column) $k - 1$ of A_k are zero.

We define

$$S = \sum_{j=k}^{n} g_{k-1,j}^2 \tag{10.3-48}$$

and then let

$$(v_k^{(k)})^2 = \tfrac{1}{2}\left(1 \pm \frac{g_{k-1,k}}{\sqrt{S}}\right) \tag{10.3-49}$$

and

$$v_k^{(j)} = \pm \frac{g_{k-1,j}}{2v_k^{(k)}\sqrt{S}} \qquad j = k+1, \ldots, n \tag{10.3-50}$$

where the plus or minus sign will be chosen below. The motivation for (10.3-49) and (10.3-50) can be found in the algebra leading to the proof that the desired $n - k$ elements in the $(k - 1)$st row (column) of A_k are zero and that (10.3-41) is satisfied. We leave this algebra to a problem {29}. Proceeding as above at each step, we arrive at a tridiagonal matrix A_{n-1}.

The accuracy of this method depends naturally on the accuracy of the matrices P_k, and these in turn depend upon the accuracy of the components of (10.3-43). The key to making this accuracy as great as possible is to make $v_k^{(k)}$ as accurate as possible, which we do by choosing the sign in (10.3-49) to be that of $g_{k-1,k}$, thus avoiding a possible loss of significant figures resulting from the subtraction of almost equal quantities. We then use the same sign in (10.3-50). We leave to a problem {31} the result that the total number of operations is of the order of $\frac{2}{3}n^3$, compared with $\frac{4}{3}n^3$ for Givens' method. At each stage it would appear that two square roots are required, one for $\sqrt{S}$ and one for $[(v_k^{(k)})^2]^{1/2}$. However, by arranging the calculations properly, the latter of these two need not be calculated {31}. Therefore, Householder's method requires $n - 2$ square roots compared with $(n - 2)(n - 1)/2$ for Givens' method. For large matrices, then, Householder's method is a more efficient way than Givens' method to reduce a symmetric matrix to tridiagonal form. The discussion in Sec. 10.3-2 on finding the eigenvalues and eigenvectors of tridiagonal matrices also applies here. Calculation of the eigenvectors of A from the eigenvectors of a tridiagonal matrix found using Householder's method is considered in a problem {30}.

Example 10.6 Apply Householder's method to the matrix of Example 10.1. As in Example 10.5, there is only one step to perform. We have

$$S = 1^2 + (\tfrac{1}{2})^2 = \tfrac{5}{4} \qquad \sqrt{S} \approx 1.11803$$

Since $a_{12} = 1$, we choose the $+$ sign in (10.3-49) and get

$$v_2^{(2)} \approx \left[\frac{1}{2}\left(1 + \frac{1}{1.11803}\right)\right]^{1/2} \approx .97325 \qquad v_2^{(3)} \approx \frac{1}{4 \times (1.11803) \times .97325} \approx .22975$$

(In fact the calculation of the square root required to get $v_2^{(2)}$ can be avoided, as mentioned above {31}.) The best way to proceed with the computation is to note that

$$A_{k-1}P_k = A_{k-1} - 2\mathbf{w}_k \mathbf{v}_k^T \qquad \text{with } \mathbf{w}_k = A_{k-1}\mathbf{v}_k$$

and then to use the result {30}

$$A_k = P_k^T A_{k-1} P_k = A_{k-1} - 2\mathbf{v}_k \mathbf{q}_k^T - 2\mathbf{q}_k \mathbf{v}_k^T \qquad \text{with } \mathbf{q}_k = \mathbf{w}_k - (\mathbf{v}_k^T \mathbf{w}_k)\mathbf{v}_k$$

In this example then we compute

$$\mathbf{w}_2^T = \mathbf{v}_2^T A = [1.08813 \quad 1.03069 \quad .70281]$$

Then $\qquad \mathbf{v}_2^T \mathbf{w}_2 = 1.16459 \qquad$ and $\qquad \mathbf{q}_2^T = [1.08813 \quad -.10275 \quad .43525]$

Finally

$$A_2 = \begin{bmatrix} 1 & -1.11803 & 0 \\ -1.11803 & 1.40000 & -.55000 \\ 0 & -.55000 & 1.60000 \end{bmatrix}$$

which except for roundoff and some sign changes is the same matrix as B in Example 10.5. Because only a single orthogonal transformation is needed in this case, we expect Givens' and Householder's methods to lead to essentially the same tridiagonal matrix. But for higher-order matrices, this will not be the case. Note the desirability of computing double-precision scalar products in this method in order to minimize roundoff.

10.4 METHODS FOR NONSYMMETRIC MATRICES

Our approach to nonsymmetric matrices will be similar to that of Givens' and Householder's methods in the sense that we shall perform a series of transformations on the matrix A—similarity more often than orthogonal—in order to reduce A to a matrix B with the same eigenvalues as A but whose eigenvalues are more easily calculable. In the method to be considered in Sec. 10.4-1 the matrix B will be tridiagonal, as in the case of Givens' and Householder's methods. In Sec. 10.4-2 we shall consider techniques for reducing A to a matrix B which has the form shown in Fig. 10.2. Such a matrix is said to be in supertriangular or, more commonly, in lower *Hessenberg* form (whereas the transpose of such a matrix is said to be in upper Hessenberg form).

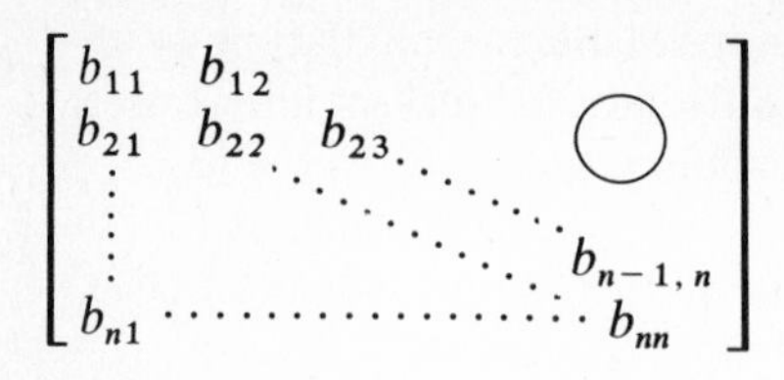

Figure 10.2 A matrix in lower Hessenberg form.

A vital aspect of the calculation of the eigenvalues of nonsymmetric matrices is the stability of the calculation with respect to the growth of roundoff errors. The details of the roundoff-error analysis of the methods we are about to consider are beyond the scope of this book. We shall content ourselves with making some general comments on this matter in what follows; for more details the reader is referred to the references mentioned in the Bibliographic Notes.

10.4-1 Lanczos' Method

Our object here is to construct a matrix S such that the result of the similarity transformation

$$T = S^{-1}AS \tag{10.4-1}$$

is a tridiagonal matrix T. Our approach will be to construct a sequence of vectors $\mathbf{x}_1, \ldots, \mathbf{x}_n$ which, as the columns of S, achieve the desired result. In the construction of this set of vectors, we shall also construct another set, $\mathbf{y}_1, \ldots, \mathbf{y}_n$, such that the sequences $\{\mathbf{x}_i\}$ and $\{\mathbf{y}_i\}$ are *biorthogonal;* that is,

$$\mathbf{y}_j^T\mathbf{x}_i = 0 \qquad \text{if } i \neq j \tag{10.4-2}$$

The two sequences of vectors are generated by the following recursion

$$\begin{aligned} \mathbf{x}_{k+1} &= A\mathbf{x}_k - b_k\mathbf{x}_k - c_{k-1}\mathbf{x}_{k-1} \\ \mathbf{y}_{k+1} &= A^T\mathbf{y}_k - b_k\mathbf{y}_k - c_{k-1}\mathbf{y}_{k-1} \end{aligned} \qquad k = 1, \ldots, n-1 \tag{10.4-3}$$

with $\mathbf{x}_0 = \mathbf{y}_0 = \mathbf{0}$, $\mathbf{x}_1$ and $\mathbf{y}_1$ arbitrary vectors and with the b_k's and c_k's defined by

$$\begin{aligned} b_k &= \frac{\mathbf{y}_k^T A\mathbf{x}_k}{\mathbf{y}_k^T\mathbf{x}_k} \\ c_{k-1} &= \frac{\mathbf{y}_{k-1}^T A\mathbf{x}_k}{\mathbf{y}_{k-1}^T\mathbf{x}_{k-1}} = \frac{\mathbf{y}_k^T\mathbf{x}_k}{\mathbf{y}_{k-1}^T\mathbf{x}_{k-1}} \qquad c_0 = 0 \end{aligned} \qquad k = 1, \ldots, n-1 \tag{10.4-4}$$

Now clearly this recursion requires that $\mathbf{y}_j^T\mathbf{x}_j \neq 0$, $j = 1, \ldots, n-1$. For the present let us assume this is true. We can then prove the following theorem.

Theorem 10.14 For the system of vectors defined by (10.4-3) and (10.4-4), Eq. (10.4-2) holds for $i, j = 1, \ldots, n$.

PROOF By induction. We have

$$\mathbf{x}_2 = A\mathbf{x}_1 - b_1\mathbf{x}_1 = A\mathbf{x}_1 - \frac{\mathbf{y}_1^T A\mathbf{x}_1}{\mathbf{y}_1^T\mathbf{x}_1}\mathbf{x}_1$$

from which we have immediately $\mathbf{y}_1^T\mathbf{x}_2 = 0$. Similarly we get $\mathbf{x}_1^T\mathbf{y}_2 = 0$. Now suppose (10.4-2) holds for all i and j less than or equal to k. We have from (10.4-3) and (10.4-4)

$$\mathbf{y}_j^T\mathbf{x}_{k+1} = \mathbf{y}_j^T A\mathbf{x}_k - \frac{\mathbf{y}_k^T A\mathbf{x}_k}{\mathbf{y}_k^T\mathbf{x}_k}\mathbf{y}_j^T\mathbf{x}_k - \frac{\mathbf{y}_{k-1}^T A\mathbf{x}_k}{\mathbf{y}_{k-1}^T\mathbf{x}_{k-1}}\mathbf{y}_j^T\mathbf{x}_{k-1} \tag{10.4-5}$$

For $j = k$ $(k - 1)$ the first term on the right-hand side cancels the second (third) and the other term is zero by the induction hypothesis. For $j < k - 1$ the last two terms are zero by the induction hypothesis. The remaining term is $\mathbf{y}_j^T A\mathbf{x}_k = \mathbf{x}_k^T A^T\mathbf{y}_j$. For $j < k - 1$ this term can be shown to be 0 by multiplying the second equation of (10.4-3) with $k = j$ by $\mathbf{x}_k^T$ and using the induction hypothesis. Since we can similarly show that $\mathbf{x}_j^T\mathbf{y}_{k+1} = 0$ for $j = 1, \ldots, k$, the theorem is proved.

The second form of c_{k-1} in (10.4-4) follows using the second equation of (10.4-3) and the biorthogonality. From this theorem we obtain a corollary.

Corollary 10.3 Under the assumption that $\mathbf{y}_j^T\mathbf{x}_j \neq 0, j = 1, \ldots, n$,

1. The vectors $\mathbf{x}_i$, $i = 1, \ldots, n$, are linearly independent.
2. If (10.4-3) is used with $k = n$ to generate $\mathbf{x}_{n+1}$, then $\mathbf{x}_{n+1} = \mathbf{0}$.

PROOF If, for some $j \leq n$, $\mathbf{x}_j$ were a linear combination of the $\mathbf{x}_k$, $k < j$, we would have

$$\mathbf{x}_j = \sum_{k=1}^{j-1} \alpha_k\mathbf{x}_k \tag{10.4-6}$$

But this would mean that $\mathbf{y}_j^T\mathbf{x}_j = 0$ because of the biorthogonality. Since this contradicts the hypothesis, we have proved the first part of the corollary. To prove the second part we note first that the proof of Theorem 10.14 implies that if (10.4-3) is used with $k = n$ and $\mathbf{y}_n^T\mathbf{x}_n \neq 0$, then Eq. (10.4-2) still holds if the vectors $\mathbf{x}_{n+1}$ and $\mathbf{y}_{n+1}$ are included in the sequences. But then the first part of the corollary would also hold for the vectors $\mathbf{x}_1, \ldots, \mathbf{x}_{n+1}$. Since $n + 1$ vectors cannot be linearly independent in n space, the only way out of this impasse is to have $\mathbf{x}_{n+1} = \mathbf{0}$.

Using this corollary we rewrite the first equation of (10.4-3) as

$$\begin{aligned} A\mathbf{x}_1 &= \mathbf{x}_2 + b_1\mathbf{x}_1 \\ A\mathbf{x}_k &= \mathbf{x}_{k+1} + b_k\mathbf{x}_k + c_{k-1}\mathbf{x}_{k-1} \qquad k = 2, \ldots, n-1 \qquad (10.4\text{-}7) \\ A\mathbf{x}_n &= b_n\mathbf{x}_n + c_{n-1}\mathbf{x}_{n-1} \end{aligned}$$

Thus, if S is the nonsingular matrix with columns $\mathbf{x}_1, \ldots, \mathbf{x}_n$, it follows from (10.4-7) that

$$AS = S\begin{bmatrix} b_1 & c_1 & & & \bigcirc \\ 1 & b_2 & \ddots & & \\ & \ddots & \ddots & \ddots & \\ & & \ddots & \ddots & c_{n-1} \\ \bigcirc & & & 1 & b_n \end{bmatrix} = ST \qquad (10.4\text{-}8)$$

Therefore, the tridiagonal matrix T of (10.4-8) is the desired matrix of (10.4-1). To find a tridiagonal matrix with the same eigenvalues as a general matrix A, we need then only calculate the b_k's and c_k's given by (10.4-4) starting from arbitrary $\mathbf{x}_1$ and $\mathbf{y}_1$.

Since the method of Lanczos, which is also called the method of minimized iterations {37}, is not used in practice because of its numerical instability, we shall not pursue this method further.

Example 10.7 Use Lanczos' method to convert to tridiagonal form

$$A = \begin{bmatrix} 2 & -2 & 3 \\ 1 & 1 & 1 \\ 1 & 3 & -1 \end{bmatrix}$$

With $\mathbf{x}_1^T = \mathbf{y}_1^T = (0, 0, 1)$ we calculate

$$\begin{array}{ll} (A\mathbf{x}_1)^T = [3 \quad 1 \quad -1] & (A^T\mathbf{y}_1)^T = [1 \quad 3 \quad -1] \\ b_1 = -\frac{1}{1} = -1 & \\ \mathbf{x}_2^T = [3 \quad 1 \quad 0] & \mathbf{y}_2^T = [1 \quad 3 \quad 0] \\ (A\mathbf{x}_2)^T = [4 \quad 4 \quad 6] & (A^T\mathbf{y}_2)^T = [5 \quad 1 \quad 6] \\ b_2 = \frac{8}{3} & c_1 = 6 \\ \mathbf{x}_3^T = [-4 \quad \frac{4}{3} \quad 0] & \mathbf{y}_3^T = [\frac{7}{3} \quad -7 \quad 0] \\ (A\mathbf{x}_3)^T = [-\frac{32}{3} \quad -\frac{8}{3} \quad 0] & (A^T\mathbf{y}_3)^T = [-\frac{7}{3} \quad -\frac{35}{3} \quad 0] \\ b_3 = \frac{1}{3} & c_2 = -\frac{28}{9} \end{array}$$

so that

$$T = \begin{bmatrix} -1 & 6 & 0 \\ 1 & \frac{8}{3} & -\frac{28}{9} \\ 0 & 1 & \frac{1}{3} \end{bmatrix}$$

10.4-2 Supertriangularization

If Givens' or Householder's method is applied to a nonsymmetric matrix A, the result is a Hessenberg matrix, as shown in Fig. 10.2 (why?). However, it is possible to reduce a general matrix A to Hessenberg form with considerably less computation than is required by the methods mentioned above. This can be achieved by using Gaussian elimination instead of orthogonal transformations. Moreover, by using partial pivoting, this process can be made extremely stable with respect to roundoff error.

As in Givens' and Householder's methods, we proceed row by row to make the off-tridiagonal elements (in the upper triangle) zero. Suppose we have done this for rows $1, \ldots, k-1$. For convenience let the elements in the matrix at this stage still be denoted by a_{ij}. Then for row k we

1. Select the largest element a_{kl} in magnitude among $a_{k,k+1}, \ldots, a_{kn}$ and interchange columns $k+1$ and l.
2. Calculate

$$m_{kj} = -\frac{a_{kj}}{a_{k,\,k+1}} \qquad j = k+2, \ldots, n \tag{10.4-9}$$

   where, because of step 1, $|m_{kj}| \leq 1$.
3. Add m_{kj} times column $k+1$ to column j, $j = k+2, \ldots, n$

Step 1 and each part of step 3 are equivalent to postmultiplying the matrix by elementary column matrices {38}. Therefore, to complete the similarity transformation for row k, it is only necessary to premultiply by the inverses of these (nonsingular) elementary matrices {38}. It is quite easy to see that the zero elements in rows $1, \ldots, k-1$ remain zero. Therefore, performing this algorithm for $k = 1, \ldots, n-2$ results in a matrix $B = [b_{ij}]$ in Hessenberg form. The stability of this method with respect to roundoff results from the fact that the m_{kj} in (10.4-9) are all no greater than 1 in magnitude.

The number of multiplications and divisions required is $\frac{2}{3}n^3 + O(n^2)$ while Householder's method for nonsymmetric matrices requires $\frac{4}{3}n^3 + O(n^2)$ operations plus $n-2$ square roots {38}.

In Sec. 10.5-2 we shall consider a powerful method for calculating all the eigenvalues of a matrix in Hessenberg form. In the remainder of this section, however, we shall consider how we can calculate an eigenvalue and eigenvector of such a matrix. We assume throughout the remainder of this section that no $b_{i,\,i+1}$ is 0, for if so we can then consider a reduced matrix {39}.

The system of equations $(B - \lambda I)\mathbf{x} = 0$ can be written

$$\begin{aligned}
&(b_{11} - \lambda)x_1 + b_{12}x_2 = 0 \\
&b_{21}x_1 + (b_{22} - \lambda)x_2 + b_{23}x_3 = 0 \\
&\cdots\cdots\cdots\cdots\cdots\cdots\cdots\cdots \\
&b_{n1}x_1 + b_{n2}x_2 + \cdots + (b_{nn} - \lambda)x_n = 0
\end{aligned} \tag{10.4-10}$$

Now let us assume a value of λ and set $x_1 = 1$ in the first equation. The first $n-1$ equations can then be solved recursively for $x_2, \ldots, x_n$. But the last equation will be satisfied only if λ is an eigenvalue (why?). Denote the value of the left-hand side of the last equation by $F(\lambda)$. We shall now show that $F(\lambda)$ is a multiple of the characteristic equation.

The matrix $B - \lambda I$ has the form

$$B - \lambda I = \begin{bmatrix} b_{11} - \lambda & b_{12} & & & \bigcirc \\ b_{21} & b_{22} - \lambda & b_{23} & & \\ \vdots & & \ddots & \ddots & \\ \vdots & & & \ddots & b_{n-1,n} \\ b_{n1} & \cdots & \cdots & \cdots & b_{nn} - \lambda \end{bmatrix} \tag{10.4-11}$$

For $i = 2, \ldots, n$ we multiply the ith column by x_i as found above and add this to the first column, thereby obtaining the matrix

$$\begin{bmatrix} 0 & b_{12} & & & \bigcirc \\ \vdots & b_{22} - \lambda & b_{23} & & \\ \vdots & \vdots & \ddots & \ddots & \\ 0 & \vdots & & \ddots & b_{n-1,n} \\ F(\lambda) & b_{n2} & \cdots & \cdots & b_{nn} - \lambda \end{bmatrix} \tag{10.4-12}$$

We have then

$$|B - \lambda I| = F(\lambda) \prod_{i=1}^{n-1} b_{i,\,i+1} = cF(\lambda) \tag{10.4-13}$$

where c is nonzero since we have assumed no $b_{i,\,i+1} = 0$. Since our object is to find a zero of $F(\lambda)$, we can use one of the techniques of Chap. 8 to find this zero. Since the eigenvalues of B may be complex, an appropriate choice is Muller's method (Chap. 8, Prob. 21), which can find a complex root starting with real approximations and which has good convergence properties. Alternatively, we could compute one or two derivatives of $F(\lambda)$ and use the Newton-based method of Sec. 8.12 or Laguerre's method (Sec. 8.10-4).

To compute $F'(\lambda)$, we differentiate (10.4-10) with respect to λ, taking into account that the x_i, $i = 1, \ldots, n$, are functions of λ. We get the system

$$\begin{aligned} &(b_{11} - \lambda)x_1' - \lambda x_1 + b_{12}x_2' = 0 \\ &b_{21}x_1' + (b_{22} - \lambda)x_2' - \lambda x_2 + b_{23}x_3' = 0 \\ &\cdots\cdots\cdots\cdots\cdots\cdots\cdots\cdots \\ &b_{n1}x_1' + b_{n2}x_2' + \cdots + (b_{nn} - \lambda)x_n' - \lambda x_n = 0 \end{aligned} \tag{10.4-14}$$

Setting $x_1 = 1$, $x_1' = 0$, we solve (10.4-14) successively for $x_2', \ldots, x_n'$ using the previously computed values $x_2, \ldots, x_n$. The value of the left-hand side of the last equation is then $F'(\lambda)$. We proceed similarly for $F''(\lambda)$.

It might seem that the accuracy of this procedure would be severely curtailed if any of the $b_{i,i+1}$ are very small in magnitude because of the necessity of dividing by $b_{i,i+1}$ in the recursion used to solve the first $n-1$ equations of (10.4-10). But, in fact, it can be shown that there is little correlation between the accuracy of the method and the magnitude of the $b_{i,i+1}$, $i = 1, \ldots, n-1$ [see Wilkinson (1959*a*)].

Having computed an eigenvalue of B by the procedure above, we compute the corresponding eigenvector by inverse iteration (Sec. 10.2-2), which is quite efficient for Hessenberg matrices.

Once we have computed one eigenvalue λ_1, we apply an implicit deflation and find a zero of $F_1(\lambda) \equiv F(\lambda)/(\lambda - \lambda_1)$. More generally, once we have computed $\lambda_1, \ldots, \lambda_p$, we seek a zero of

$$F_p(\lambda) \equiv \frac{F(\lambda)}{\prod_{i=1}^{p} (\lambda - \lambda_i)} \tag{10.4-15}$$

evaluating $F(\lambda)$ by (10.4-10). The derivatives needed for the Newton-based method and Laguerre's method can be computed using the formulas {40}

$$\frac{F_p'(\lambda)}{F_p(\lambda)} = \frac{F'(\lambda)}{F(\lambda)} - \sum_{i=1}^{p} (\lambda - \lambda_i)^{-1} \tag{10.4-16}$$

$$\left[\frac{F_p'(\lambda)}{F_p(\lambda)}\right]^2 - \frac{F_p''(\lambda)}{F_p(\lambda)} = \left[\frac{F'(\lambda)}{F(\lambda)}\right]^2 - \frac{F''(\lambda)}{F(\lambda)} - \sum_{i=1}^{p} (\lambda - \lambda_i)^{-2} \tag{10.4-17}$$

Example 10.8 Apply the Gaussian elimination and deflation methods of this section to the matrix of Example 10.7.

Interchanging the second and third columns of the matrix and then eliminating the element in the (1, 3) position, we obtain the matrix

$$\begin{bmatrix} 2 & 3 & 0 \\ 1 & 1 & \frac{5}{3} \\ 1 & -1 & \frac{7}{3} \end{bmatrix} \tag{10.4-18}$$

Then premultiplying by the inverses of the elementary matrices used to derive (10.4-18), we obtain {38}

$$B = \begin{bmatrix} 2 & 3 & 0 \\ \frac{1}{3} & -\frac{5}{3} & \frac{11}{9} \\ 1 & 1 & \frac{5}{3} \end{bmatrix}$$

To calculate an eigenvalue of B let us take as initial approximations $\lambda = 0$, $\lambda = -\frac{1}{2}$, and $\lambda = \frac{1}{2}$. Then using Eqs. (10.4-10) we calculate $F(0) = -\frac{18}{11}$, $F(-\frac{1}{2}) = -\frac{189}{88}$, and $F(\frac{1}{2}) = -\frac{75}{88}$. Then using Muller's method, we obtain as the next approximation $\lambda = .912537$. Convergence to $\lambda_1 = 1$ is very rapid (five iterations).

Now suppose we have found $\lambda_1 = 1$. Then, to calculate the next eigenvalue of B, we use the same initial approximations, and, using (10.4-15), we calculate $F_1(0) = \frac{18}{11}$, $F_1(-\frac{1}{2}) = \frac{63}{44}$, and $F_1(\frac{1}{2}) = \frac{75}{44}$. Using Muller's method again, we converge to $\lambda_2 = -2$ in one iteration. Similarly, we converge to $\lambda_3 = 3$ in one iteration.

10.4-3 Jacobi-Type Methods

The Jacobi method for symmetric matrices was based on the result that any symmetric matrix can be diagonalized by an orthogonal transformation. Theorem 10.10 assures us that any nonsymmetric matrix can be triangularized by a unitary transformation. However, it has not been proved that this triangularization always can be accomplished by unitary matrices analogous to the plane rotation matrices of Sec. 10.3-1. In fact, for certain procedures directly analogous to those in Sec. 10.3-1, examples can be given for which the process will not converge. Nevertheless, by successively annihilating the largest off-diagonal element in one triangle of the matrix or by using the threshold technique it has been found possible in many cases to triangularize A and thus to find its eigenvalues and eigenvectors {41}.

A more sophisticated and theoretically sound Jacobi-type method is the norm-reducing method of Eberlein. To explain the basic idea of the method, we need the concept of a *normal* matrix, which is a matrix unitarily similar to a real or complex diagonal matrix. It follows that a matrix A is normal if and only if $AA^* = A^*A$ {42}. Thus, the class of normal matrices includes all unitary and Hermitian matrices. A unitary extension of the Jacobi method exists which, when applied to a normal matrix, converges to a diagonal matrix.

Now, there is a theorem which states that for *any* matrix A, there exists a nonsingular matrix P such that $P^{-1}AP$ is arbitrarily close to a normal matrix and

$$\inf \|P^{-1}AP\|_E^2 = \sum_{i=1}^{n} |\lambda_i|^2 \tag{10.4-19}$$

Eberlein's method in the complex form, which is more efficient than the real form when there are complex eigenvalues, constructs a sequence of matrices C_i, where

$$C_{i+1} = W_i^{-1} C_i W_i \tag{10.4-20}$$

and the W_i are two-dimensional transformations, depending on a pair of indices, (k, l), of the form RS. R is a complex rotation defined by

$$r_{kk} = r_{ll} = \cos x \qquad r_{kl} = -r_{lk}^* = e^{i\alpha} \sin x \qquad r_{ij} = \delta_{ij} \text{ otherwise} \tag{10.4-21}$$

and S is a complex shear defined by

$$s_{kk} = s_{ll} = \cosh y \qquad s_{kl} = s_{lk}^* = -ie^{i\beta} \sinh y \qquad s_{ij} = \delta_{ij} \text{ otherwise} \tag{10.4-22}$$

The index pairs are usually chosen in serial order. The parameters x, α of the matrix R are chosen so that if C_i is Hermitian, they reduce to the ordinary unitary Jacobi parameters. The parameters y, β of S are chosen to reduce the

Euclidean norm of C_{i+1} in case the (k, l) element of C_i deviates from normality, i.e., if $\gamma_{kl} \equiv (C_i^* C_i - C_i C_i^*)_{kl} \neq 0$. In this case, we have that

$$\|C_i\|_E^2 - \|C_{i+1}\|_E^2 \geq \frac{\frac{1}{3}|\gamma_{kl}|^2}{\|C_i\|_E^2} \tag{10.4-23}$$

If $\gamma_{kl} = 0$, then $\beta = y = 0$, $S = I$, and $\|C_{i+1}\|_E = \|C_i\|_E$ and W_i reduces to a complex rotation. Thus, as long as the matrix C_i is not normal, some succeeding matrix C_{i+p} is reduced in norm, and as soon as a C_i becomes normal to machine accuracy, the above-mentioned extension of the Jacobi method can be applied. The end result is a complex matrix $W = W_1 W_2 \cdots W_N$ such that the off-diagonal elements of

$$\hat{C} = W^{-1} C W \tag{10.4-24}$$

are arbitrarily small. The diagonal elements of $\hat{C}$ are approximations to the eigenvalues of C, and the columns of W are approximations to the corresponding eigenvectors. The details will be found in the references given in the Bibliographic Notes.

10.5 THE *LR* AND *QR* ALGORITHMS

The basis of these methods is the successive factorization of a sequence of matrices $\{A_k\}$, all of which have the same form as the original matrix $A_1 \equiv A$; for example, if A is tridiagonal, so is every A_k. The key to these methods is the observation that if A_1 is factored into the product $F_1 G_1$, where F_1 is nonsingular, then if we multiply F_1 and G_1 in reverse order, the matrix $A_2 = G_1 F_1$ has the same eigenvalues as A_1. This is true because

$$A_2 = G_1 F_1 = F_1^{-1} A_1 F_1 \tag{10.5-1}$$

so that A_1 and A_2 are similar. Now A_2 itself can likewise be decomposed into $A_2 = F_2 G_2$, and in this way we define a sequence of matrices

$$A_k = F_k G_k = G_{k-1} F_{k-1} \qquad k = 2, 3, \ldots \qquad A_1 = A = F_1 G_1 \tag{10.5-2}$$

The following properties of the matrices A_k, F_k, and G_k are of interest to us:

1. All the matrices A_k are similar and therefore have the same eigenvalues. This follows from (10.5-1).
2. Let $E_k = F_1 \cdots F_k$, so that E_k is nonsingular. Since, as in (10.5-1), $A_{k+1} = F_k^{-1} A_k F_k$, it follows inductively that

$$A_{k+1} = E_k^{-1} A_1 E_k \tag{10.5-3}$$

3. Let $H_k = G_k \cdots G_1$. Since $F_j G_j = G_{j-1} F_{j-1}$,

$$\begin{aligned} E_k H_k &= F_1 \cdots F_{k-1} F_k G_k G_{k-1} \cdots G_1 \\ &= F_1 \cdots F_{k-1} G_{k-1} F_{k-1} G_{k-1} \cdots G_1 = E_{k-1} A_k H_{k-1}. \end{aligned}$$

Hence, using (10.5-3) with k replaced by $k-1$, $E_k H_k = A_1 E_{k-1} H_{k-1}$. By repetition of this process we arrive at

$$E_k H_k = A_1^k \tag{10.5-4}$$

The LR and QR algorithms come from the following two factorizations of A.

LR If we assume that there exists a unique triangular decomposition for each A_k, $A_k = L_k U_k$† with L_k unit lower triangular and hence nonsingular, then $F_k = L_k$, $G_k = U_k$ defines the LR transformation. Such a decomposition may not necessarily exist even if A_k is nonsingular. In this case, all we know is that for some permutation matrix P, $PA_k = L_k U_k$. Hence, the above assumption is nontrivial.

QR If we assume that it is possible to decompose an arbitrary real matrix A into a product QR, where Q is orthogonal and R‡ is upper triangular with nonnegative diagonal elements, and that when A is nonsingular, this decomposition is unique, then $F_k = Q_k$, $G_k = R_k$ defines the QR transformation. In contrast to the LR case, we now show that a QR decomposition always exists by constructing it.

To obtain this decomposition, we apply to A a sequence $\{P_k\}$ of Householder transformations, as in Sec. 9.9, making sure at each stage that the diagonal element becomes nonnegative. This is always possible since such transformations are determined up to a sign. Thus, we find that

$$P_{n-1} P_{n-2} \cdots P_1 A = R \tag{10.5-5}$$

Since each P_j is orthogonal, we have that $A = QR$, where

$$Q = [P_{n-1} \quad \cdots \quad P_1]^T \tag{10.5-6}$$

Since each P_j is uniquely determined by the nonnegativity condition if the diagonal element $a_{jj}^{(j)}$ is not 0 when P_j is applied, it follows that when A is nonsingular, the decomposition is unique.

Since the product of triangular matrices is triangular and that of orthogonal matrices is orthogonal, we have that E_k is the lower triangular or

† Its discoverer, Rutishauser (1955), used the mnemonic left-right (LR) whereas, in keeping with the notation of Chap. 9, we use lower-upper.

‡ Francis (1961), the originator of this method, used R as a mnemonic for right triangular.

orthogonal factor of A_1^k. Hence the convergence of either the *LR* or *QR* process is determined by the behavior of the sequence $\{E_k\}$, since $A_{k+1} = E_k^{-1} A_1 E_k$.

Theorem 10.15 If $\{E_k\}$ converges to a nonsingular matrix E_∞ as $k \to \infty$, and if each G_k is an upper triangular matrix, then $\lim_{k\to\infty} A_k$ exists and is an upper triangular matrix.

PROOF Since $\{E_k\}$ converges, the following limits also exist:

$$\lim_{k\to\infty} F_k = \lim_{k\to\infty} E_{k-1}^{-1} E_k = I \tag{10.5-7}$$

$$\begin{aligned} G_\infty &= \lim_{k-\infty} G_k = \lim_{k\to\infty} A_{k+1} F_k^{-1} = \lim_{k\to\infty} E_k^{-1} A_1 E_{k-1} \\ &= E_\infty^{-1} A_1 E_\infty \end{aligned} \tag{10.5-8}$$

Furthermore G_∞ is upper triangular since each G_k is. Therefore,

$$A_\infty = \lim_{k\to\infty} A_k = \lim_{k\to\infty} F_k G_k = G_\infty \tag{10.5-9}$$

exists and is upper triangular, which proves the theorem.

An investigation of the convergence of the E_k in general is beyond the scope of this book. However, we shall prove a quite general theorem for the *QR* case, since this is the case of practical importance, and quote some results for cases not covered by the theorem. First we make the following definition.

If the sequence $\{A_k\}$ produced by either algorithm tends to *upper triangular form* even though elements *above* the main diagonal may not converge, we say that the algorithm *converges essentially*. If the sequence $\{A_k\}$ converges, we say that the algorithm converges.

Theorem 10.16 Let the real $n \times n$ matrix $A_1 = X \Lambda X^{-1}$, where $\Lambda = \operatorname{diag}\{\lambda_1, \ldots, \lambda_n\}$. If $|\lambda_1| > |\lambda_2| > \cdots > |\lambda_N| > 0$, and if $X^{-1} = Y$ has an *LU* factorization $L_y U_y$, then the *QR* algorithm converges essentially.

PROOF $$A_1^k = X \Lambda^k X^{-1} = X \Lambda^k L_y U_y = X(\Lambda^k L_y \Lambda^{-k})(\Lambda^k U_y) \tag{10.5-10}$$

Now $\Lambda^k L_y \Lambda^{-k} = I + B_k$ {44}, where

$$(B_k)_{ij} = \begin{cases} l_{ij}\left(\dfrac{\lambda_i}{\lambda_j}\right)^k & i > j;\ l_{ij} \in L_y \\ 0 & i \le j \end{cases} \tag{10.5-11}$$

We thus have that

$$A_1^k = X(I + B_k)(\Lambda^k U_y) \tag{10.5-12}$$

where, since $|\lambda_i/\lambda_j| < 1$, $i > j$, $B_k \to 0$ as $k \to \infty$.

Now, by our previous construction, X can be decomposed into $Q_x R_x$, where R_x has positive diagonal elements. Therefore,

$$A_1^k = Q_x R_x(I + B_k)(\Lambda^k U_y) = Q_x(I + R_x B_k R_x^{-1})(R_x \Lambda^k U_y) \tag{10.5-13}$$

Since $B_k \to 0$, $I + R_x B_k R_x^{-1}$ will eventually become nonsingular and hence will have a unique factorization $\tilde{Q}_k \tilde{R}_k$, where $\tilde{Q}_k \to I$, $\tilde{R}_k \to I$ as $k \to \infty$. Thus

$$A_1^k = (Q_x \tilde{Q}_k)(\tilde{R}_k R_x \Lambda^k U_y) \tag{10.5-14}$$

This need not be the QR factorization of A_1^k since the diagonal of the second factor need not be positive because of Λ^k and U_y. We therefore define diagonal orthogonal matrices D_1 and D_2 such that $D_1 \Lambda$ and $D_2 U_y$ have positive diagonals. Then the matrix $D_1^k D_2 \tilde{R}_k R_x \Lambda^k U_y$ also has a positive diagonal. Therefore, $Q_x Q_k D_2^{-1} D_1^{-k}$ is the orthogonal factor of A_1^k. Hence, with $Q_x \tilde{Q}_k D_2^{-1} D_1^{-k}$ playing the role of E_k,

$$\begin{aligned} A_{k+1} &= D_1^k D_2 \tilde{Q}_k^T Q_x^T A_1 Q_x \tilde{Q}_k D_2^{-1} D_1^{-k} \to D_1^k D_2 Q_x^T A_1 Q_x D_2^{-1} D_1^{-k} \\ &= D_1^k (D_2 R_x \Lambda R_x^{-1} D_2^{-1}) D_1^{-k} \end{aligned} \tag{10.5-15}$$

as $k \to \infty$ since $\tilde{Q}_k \to I$ and $A_1 = Q_x R_x \Lambda R_x^{-1} Q_x^T$. If D_1^k converges, so does A_k. But $D_1 = \text{diag}(\pm 1, \ldots, \pm 1)$, and so D_1^k will not converge if any diagonal element is negative. However, this has no effect on the diagonal elements of $R_x \Lambda R_x^{-1}$ (nor on the magnitudes of the elements in the upper triangle). Hence, we have essential convergence. Furthermore, since the diagonal of $R_x \Lambda R_x^{-1} = \Lambda$, we see that the eigenvalues appear on the diagonal in decreasing order of magnitude.

Further results on convergence which we state without proof are the following.

If, in addition to the hypotheses of Theorem 10.16, X has an LU factorization, then the LR algorithm converges. If Y does not have an LU factorization, then since Y is nonsingular, there exists a permutation matrix P such that PY does have such a factorization. It is then not difficult to show that the QR algorithm still converges essentially. However, in contrast to the case of Theorem 10.16, the eigenvalues do not appear on the diagonal in decreasing order of magnitude. In fact, the sequence $\{A_k\}$ always converges to block triangular form, each diagonal block having roots of equal magnitude. In general, this does not cause problems. Only when there are *distinct* eigenvalues of equal modulus, do we have convergence to a form which is troublesome. Thus, even with real multiple eigenvalues, the convergence is to triangular form, while for multiple complex-conjugate pairs of eigenvalues, the limiting form of A_k will yield these roots in a string of 2×2 submatrices along the diagonal. Hence, except for rare cases, the limiting form of A_k is some modification of the form

$$\begin{bmatrix} \lambda_1 & x & \cdots & x & x & x & \cdots & x & x \\ 0 & \lambda_2 & \cdots & x & x & x & \cdots & x & x \\ \cdots & \cdots & \cdots & \cdots & \cdots & \cdots & \cdots & \cdots & \cdots \\ 0 & 0 & \cdots & \lambda_m & x & x & \cdots & x & x \\ 0 & 0 & \cdots & 0 & \multicolumn{2}{c}{\multirow{2}{*}{B_1}} & \cdots & x & x \\ 0 & 0 & \cdots & 0 & & & \cdots & x & x \\ \cdots & \cdots & \cdots & \cdots & \cdots & \cdots & \cdots & \cdots & \cdots \\ 0 & 0 & \cdots & 0 & 0 & 0 & \cdots & \multicolumn{2}{c}{\multirow{2}{*}{B_l}} \\ 0 & 0 & \cdots & 0 & 0 & 0 & \cdots & & \end{bmatrix} \tag{10.5-16}$$

where $m + 2l = n$, each B_j is a 2×2 real submatrix with complex conjugate eigenvalues which are eigenvalues of A_1, and the real eigenvalues λ_i and the B_j may appear in any order along the diagonal. Cases in which this limiting form is not achieved almost never arise in practice since, as we shall see below, the QR transformations are modified by shifts, which greatly reduce the possibility of distinct eigenvalues of equal modulus.

Returning to Theorem 10.16, we see that $a_{ii}^{(k)} \to \lambda_i$ as $k \to \infty$. In fact, it can be shown [see Parlett (1965)] that

$$a_{ii}^{(k)} = \lambda_i + O(r_i^k) \qquad a_{i+1,i}^{(k)} = O(r_i^k) \tag{10.5-17}$$

where $$r_i = \max\left(\left|\frac{\lambda_i}{\lambda_{i-1}}\right|, \left|\frac{\lambda_{i+1}}{\lambda_i}\right|\right) \qquad \lambda_0 = \infty \qquad \lambda_{n+1} = 0 \tag{10.5-18}$$

which is linear convergence. In particular, $r_n = |\lambda_n/\lambda_{n-1}|$, and our strategy is to try to modify the algorithm to make r_n very small so that $a_{nn} \to \lambda_n$ and $a_{n,n-1} \to 0$ very rapidly. Looking ahead a little, if the A_k were upper Hessenberg or tridiagonal, then once $a_{n,n-1}$ converged to 0 to within machine accuracy, we would have computed $\lambda_n = a_{nn}$. Then, we could deflate the matrix and work only with the matrix of order $n - 1$ consisting of the first $n - 1$ rows and columns of A_k. Thus successive deflations would result in less and less work to compute succeeding QR transformations.

Recalling our discussion of the power method (Sec. 10.2), we know that the eigenvalues of $A_k - pI$ are $\lambda_i - p$, $i = 1, \ldots, n$, so that if we can choose an appropriate value of p, the ratio $|(\lambda_n - p)/(\lambda_{n-1} - p)|$ will converge to 0 very rapidly. A good estimate of λ_n would serve this purpose very well. In order to allow us to choose the best estimate at each iteration, we must work with $A_k - p_k I$ involving the variable shift p_k. This gives us the modified algorithm

$$\begin{aligned} &\text{Factor } A_k - p_k I \text{ into } F_k G_k \\ &\text{Form } \; A_{k+1} = G_k F_k + p_k I \end{aligned} \qquad k = 1, 2, \ldots \tag{10.5-19}$$

Then, as before {45},

$$A_{k+1} = F_k^{-1} \cdots F_1^{-1} A_1 F_1 \cdots F_k = E_k^{-1} A_1 E_k \tag{10.5-20}$$

and

$$E_k H_k = \prod_{j=1}^{k} (A_1 - p_j I) \tag{10.5-21}$$

If we define $\phi_k(\lambda) = \prod_{j=1}^{k} (\lambda - p_j)$, the proof of Theorem 10.16 will carry over to the modified algorithm when λ_i^k is replaced by $\phi_k(\lambda_i)$. Thus, provided the p_j are chosen so that $|\phi_k(\lambda_i)| \neq |\phi_k(\lambda_s)|$, $i \neq s$, for $k > K$ and so that $\phi_k(\lambda_i) \neq 0$, then the modified algorithm will converge. We defer until later our discussion of the choice of p_j.

10.5-1 The Simple *QR* Algorithm

As indicated above, we shall restrict our attention to the *QR* algorithm. We do this because the *LR* algorithm is numerically unstable. This is clearly true for the general situation since the *LU* decomposition of a matrix without pivoting may lead to disaster. While it is possible to modify this algorithm to include pivoting, the theoretical basis of the convergence proof is lost, and, in fact, simple examples can be constructed for which convergence does not take place. Even when the *LU* decomposition without pivoting can be justified, as in the case of positive definite symmetric matrices, in practice, there is a loss of accuracy. Hence, even though the *QR* algorithm is more expensive than the *LR* algorithm, its superior numerical properties more than make up for the extra labor.

If we count the number of operations involved in a single *FG* transformation applied to a full matrix, we find that it is of the order of n^3. Since many iterations may be necessary before convergence, this could be quite an expensive task. Fortunately, the workload decreases substantially if we work with Hessenberg† or tridiagonal matrices, which we can do inasmuch as the *QR* transformation leaves these forms invariant {46}. The work involved in a single transformation is of the order of n^2 operations for Hessenberg matrices and of n operations for tridiagonal matrices. We shall henceforth restrict our discussion to such matrices, where we further assume that all elements on the subdiagonal are nonzero, since otherwise we could decompose our problem into smaller ones {46}.

We now distinguish between two cases. In this section, we shall deal with the case where we know that all the eigenvalues are real. In Sec. 10.5-2 we deal with the case where some of the eigenvalues may be complex. In the first case there is a very efficient scheme for implementing the *QR* transformation.

Since A_k is now assumed to be of Hessenberg or tridiagonal form, we

† In this section, Hessenberg will always mean upper Hessenberg.

need not use Householder transformations to transform A_k to upper triangular form but can use plane rotations. Recalling the notation of Sec. 10.3-2, we apply the sequence of transformations $(i, i+1, i)$, $i = 1, \ldots, n-1$ to $A_k - p_k I$, where premultiplication by the corresponding matrix $S^T_{i,\,i+1,\,i}$ annihilates the element in the ith column and $(i+1)$st row. We thus have

$$S^T_{n-1,\,n,\,n-1} \cdots S^T_{2,\,3,\,2} S^T_{1,\,2,\,1}(A_k - p_k I) = R_k \tag{10.5-22}$$

so that $Q_k = S_{1,\,2,\,1} \cdots S_{n-1,\,n,\,n-1}$, and

$$A_{k+1} = R_k S_{1,\,2,\,1} \cdots S_{n-1,\,n,\,n-1} + p_k I \tag{10.5-23}$$

Each such transformation takes about $4n^2$ multiplications and $n-1$ square roots {46}. The shift p_k is determined from the eigenvalues μ_k and ν_k of T_k, the bottom 2×2 submatrix of A_k. If both are real, we take p_k to be μ_k or ν_k according as $|\mu_k - a^{(k)}_{nn}|$ or $|\nu_k - a^{(k)}_{nn}|$ is smaller. Otherwise we set $p_k = \text{Re}\,\mu_k$.

If our matrix A_1 is symmetric and tridiagonal, then since the QR transformation preserves symmetry {43}, all subsequent matrices A_k will be symmetric and hence tridiagonal (why?). In this case, the QR algorithm with shifts is very efficient. Thus the combined algorithm of first reducing a symmetric matrix to tridiagonal form by Householder transformations and then applying the QR algorithm is probably the most effective way to evaluate all the eigenvalues of a symmetric matrix.

Example 10.9 Apply the simple QR algorithm to find all the (real) eigenvalues of the symmetric tridiagonal matrix A, where

$$A = A_1 = \begin{bmatrix} 120.0 & -90.86 & 0 & 0 \\ -90.86 & 157.2 & 67.59 & 0 \\ 0 & 67.59 & 124.0 & -46.26 \\ 0 & 0 & -46.26 & 78.84 \end{bmatrix}$$

The computations below were carried out to about 14 significant figures. We give the results rounded to five figures. We give the transformation from A_1 to A_2 in full detail. After that, we shall only give the shifts, p_k, and the matrices A_k.

$$p_1 = 49.943 \qquad S^T_{121} = \begin{bmatrix} .61061 & -.79193 & 0 & 0 \\ .79193 & .61061 & 0 & 0 \\ 0 & 0 & 1 & 0 \\ 0 & 0 & 0 & 1 \end{bmatrix}$$

$$S^T_{121}(A_1 - p_1 I) = \begin{bmatrix} 114.73 & -140.42 & -53.527 & 0 \\ 0 & -6.4629 & 41.271 & 0 \\ 0 & 67.590 & 74.057 & -46.260 \\ 0 & 0 & -46.260 & 28.897 \end{bmatrix}$$

$$S^T_{232} = \begin{bmatrix} 1 & 0 & 0 & 0 \\ 0 & -.095185 & .99546 & 0 \\ 0 & -.99546 & -.095185 & 0 \\ 0 & 0 & 0 & 1 \end{bmatrix}$$

$$S_{232}^T S_{121}^T (A_1 - p_1 I) = \begin{bmatrix} 114.73 & -140.42 & -53.527 & 0 \\ 0 & 67.898 & 69.792 & -46.050 \\ 0 & 0 & -48.133 & 4.4032 \\ 0 & 0 & -46.260 & 28.897 \end{bmatrix}$$

$$S_{343}^T = \begin{bmatrix} 1 & 0 & 0 & 0 \\ 0 & 1 & 0 & 0 \\ 0 & 0 & -.72099 & -.69294 \\ 0 & 0 & .69294 & -.72099 \end{bmatrix}$$

$$R_1 = S_{343}^T S_{232}^T S_{121}^T (A_1 - p_1 I) = \begin{bmatrix} 114.73 & -140.42 & -53.527 & 0 \\ 0 & 67.898 & 69.792 & -46.050 \\ 0 & 0 & 66.759 & -23.198 \\ 0 & 0 & 0 & -17.783 \end{bmatrix}$$

$$R_1 S_{121} = \begin{bmatrix} 181.26 & 5.1182 & -53.527 & 0 \\ -53.771 & 41.459 & 69.792 & -46.050 \\ 0 & 0 & 66.759 & -23.198 \\ 0 & 0 & 0 & -17.783 \end{bmatrix}$$

$$R_1 S_{121} S_{232} = \begin{bmatrix} 181.26 & -53.771 & 0 & 0 \\ -53.771 & 65.529 & -47.914 & -46.050 \\ 0 & 66.456 & -6.3544 & -23.198 \\ 0 & 0 & 0 & -17.783 \end{bmatrix}$$

$$A_2 - p_1 I = R_1 S_{121} S_{232} S_{343} = \begin{bmatrix} 181.26 & -53.771 & 0 & 0 \\ -53.771 & 65.529 & 66.456 & 0 \\ 0 & 66.456 & 20.657 & 12.323 \\ 0 & 0 & 12.323 & 12.822 \end{bmatrix}$$

$$A_2 = \begin{bmatrix} 231.20 & -53.771 & 0 & 0 \\ -53.771 & 115.47 & 66.456 & 0 \\ 0 & 66.456 & 70.600 & 12.323 \\ 0 & 0 & 12.323 & 62.765 \end{bmatrix}$$

$$p_2 = 53.752 \qquad A_3 = \begin{bmatrix} 251.32 & -23.029 & 0 & 0 \\ -23.029 & 136.29 & 38.237 & 0 \\ 0 & 38.237 & 28.568 & -2.8312 \\ 0 & 0 & -2.8312 & 63.861 \end{bmatrix}$$

$$p_3 = 64.087 \qquad A_4 = \begin{bmatrix} 255.18 & -9.6151 & 0 & 0 \\ -9.6151 & 140.163 & 24.060 & 0 \\ 0 & 24.060 & 20.690 & -.0047934 \\ 0 & 0 & -.0047934 & 64.003 \end{bmatrix}$$

$$p_4 = 64.003 \qquad A_5 = \begin{bmatrix} 255.86 & -3.9842 & 0 & 0 \\ -3.9842 & 142.45 & 14.730 & 0 \\ 0 & 14.730 & 17.730 & <10^{-10} \\ 0 & 0 & <10^{-10} & 64.003 \end{bmatrix}$$

We now deflate and continue on $\bar{A}_5$, the 3×3 submatrix of A_5.

$$p_5 = 16.014 \qquad \bar{A}_6 = \begin{bmatrix} 255.96 & -2.1128 & 0 \\ -2.1128 & 144.07 & -.00010338 \\ 0 & -.00010338 & 16.013 \end{bmatrix}$$

$$p_6 = 16.013 \qquad \bar{A}_7 = \begin{bmatrix} 255.99 & -1.1273 & 0 \\ -1.1273 & 144.04 & < 10^{-10} \\ 0 & < 10^{-10} & 16.013 \end{bmatrix}$$

We now deflate again and continue on $\tilde{A}_7$ the 2×2 submatrix of $\bar{A}_7$.

$$p_7 = 144.03 \qquad \tilde{A}_8 = \begin{bmatrix} 256.00 & < 10^{-10} \\ < 10^{-10} & 144.03 \end{bmatrix}$$

We are now through, and the eigenvalues, read off the diagonal of the full matrix A_8, are

$$256.00 \quad 144.03 \quad 16.013 \quad 64.003$$

Note that they are not in decreasing order of magnitude.

We see that each premultiplication by $S_{i,i+1,i}^T$ affects only rows i and $i + 1$ and, similarly, each postmultiplication by $S_{j,j+1,j}$ affects only columns j and $j + 1$. From this we see that if A is tridiagonal, then R is a band matrix of bandwidth [0, 2]. This implies that the storage requirements in this case are $O(n)$ rather than $O(n^2)$ and similarly for the number of operations per iteration. We also notice that each A_k is tridiagonal and symmetric, as expected.

We also notice that at the same time that the element in the lower right-hand corner of the matrix is converging to an eigenvalue, the other diagonal elements are also converging, although at a slower rate. Similarly, the other off-diagonal elements are tending to 0 although not as fast as $a_{n,n-1}$. Thus, subsequent iterations on the later deflated matrices converge in fewer iterations than are required for the earlier ones.

We point out one final item connected with the organization of the computation. We have applied formulas (10.5-22) and (10.5-23) in a straightforward fashion here. This requires saving the matrices $S_{i,i+1,i}$, $i = 1, \ldots, n - 1$, or more accurately, the two values, $\sin \theta_i$ and $\cos \theta_i$ which determine $S_{i,i+1,i}$. However, we can reorganize the computation to avoid saving these values by noticing that, for example, after computing $S_{232}^T S_{121}^T(A_1 - p_1 I)$, the first two columns remain unchanged and no information contained in them is needed for the rest of the computation of R_1. Hence, we can compute $S_{232}^T S_{121}^T(A_1 - p_1 I)S_{121}$ before premultiplying by S_{343}^T and therefore we need not save S_{121} any more. In general, we postmultiply by $S_{i-1,i,i-1}$ after the premultiplication by $S_{i,i+1,i}^T$, $i = 2, \ldots, n - 1$, and finally, postmultiply by $S_{n-1,n,n-1}$. This saves about $2n$ storage locations.

Further savings in computation and storage can be obtained in the symmetric tridiagonal case if we take into account the fact that each A_k is of the same form. We leave the details as a problem {47}.

10.5-2 The Double *QR* Algorithm

We now discuss the more general situation in which the matrix A, which we assume to be Hessenberg, may have complex roots which occur in pairs as conjugate-complex numbers. In this case it is important for rapid convergence of the QR algorithm that shifts be made with complex numbers. For example, if $\lambda_n = \bar{\lambda}_{n-1} = a + ib$ and $\lambda_{n-2} = \bar{\lambda}_{n-3} = a + i(b + \epsilon)$, $\epsilon > 0$, then the real number p which minimizes

$$r^2 = \left| \frac{\lambda_n - p}{\lambda_{n+2} - p} \right|^2$$

is $p = a$, and the corresponding value of r which gives the best rate of convergence is $|b/(b + \epsilon)|$, which is close to 1 for small ϵ. This shows that the convergence of the QR algorithm may be very slow if we are restricted to real shifts. Thus, assuming that $\lambda_n = \bar{\lambda}_{n-1}$ is the eigenvalue of smallest magnitude (we shall see below that the procedure to be described works even if λ_n is real), we wish to drive the element $a_{n-1,n-2}$ in position $(n-1, n-2)$ to zero leaving a 2×2 matrix at the bottom of A_k whose eigenvalues are λ_n and $\bar{\lambda}_n$. Since $a_{n-1,n-2}^{(k)}$ goes to 0 as $|\lambda_n/\lambda_{n-2}|^k$, a shift by an approximation to λ_n is called for. However, we would prefer not to leave the real field, if possible. This can be accomplished if we follow a shift with a complex p by a shift with $\bar{p}$. Thus if we perform the following pair of QR transformations

$$\begin{aligned} A_1 - pI &= Q_1 R_1 \qquad & R_1 Q_1 + pI &= A_2 \\ A_2 - \bar{p}I &= Q_2 R_2 \qquad & R_2 Q_2 + \bar{p}I &= A_3 \end{aligned} \tag{10.5-24}$$

it can be shown {50} that

$$A_3 = (Q_1 Q_2)^* A (Q_1 Q_2) = Q^* A Q \tag{10.5-25}$$

and $$(Q_1 Q_2)(R_2 R_1) = (A_1 - pI)(A_1 - \bar{p}I) \equiv B(p) \tag{10.5-26}$$

The matrices Q_1, Q_2, R_1, R_2 will be complex, but Q and $R = R_2 R_1$ are real since they correspond to the factorization of the real matrix $B(p)$, so that A_3 is also real.

The *double QR algorithm* is a method for determining Q and A_3 without computing Q_1, R_1, Q_2, R_2, or A_2. It is based on the following theorem about Hessenberg matrices.

Theorem 10.17 Let $H = Q^T A Q$, where A is any matrix, Q is orthogonal, and H is a Hessenberg matrix which is known to have nonvanishing subdiagonal elements. Then given A and the first column of Q, H and the remaining columns of Q are determined.

PROOF We shall proceed in an inductive manner to generate H and the remaining columns of Q. Denote by $\mathbf{h}_j = (h_{1j}, h_{2j}, \ldots, h_{j+1,j}, 0, \ldots, 0)^T$ and by $\mathbf{q}_j$, the jth column of H and Q, respectively. Equating the first columns of QH and AQ, we have that

$$h_{11}\mathbf{q}_1 + h_{21}\mathbf{q}_2 = A\mathbf{q}_1 \tag{10.5-27}$$

Hence, since Q is orthogonal and $\mathbf{q}_1$ is known, we can determine

$$h_{11} = \mathbf{q}_1^T A \mathbf{q}_1 \tag{10.5-28}$$

Once h_{11} is known, we have

$$h_{21}\mathbf{q}_2 = A\mathbf{q}_1 - h_{11}\mathbf{q}_1 \equiv \tilde{\mathbf{q}}_2 \tag{10.5-29}$$

Since Q is orthogonal, $|\mathbf{q}_2| = 1$. Hence we have

$$h_{21} = |\tilde{\mathbf{q}}_2| \qquad \mathbf{q}_2 = \frac{\tilde{\mathbf{q}}_2}{h_{21}} \tag{10.5-30}$$

Thus we have determined $\mathbf{h}_1$ and $\mathbf{q}_2$.

Assume now that we have determined $\mathbf{q}_1, \ldots, \mathbf{q}_j, j < n$, and $\mathbf{h}_1, \ldots, \mathbf{h}_{j-1}$. We determine $\mathbf{h}_j$ and $\mathbf{q}_{j+1}$ as follows. Since

$$A\mathbf{q}_j = Q\mathbf{h}_j = \sum_{i=1}^{j+1} h_{ij}\mathbf{q}_i \tag{10.5-31}$$

we find as before that

$$h_{ij} = \mathbf{q}_i^T A\mathbf{q}_j \qquad i = 1, \ldots, j \tag{10.5-32}$$

Furthermore $$h_{j+1,j}\mathbf{q}_{j+1} = A\mathbf{q}_j - \sum_{i=1}^{j} h_{ij}\mathbf{q}_i \equiv \tilde{\mathbf{q}}_{j+1} \tag{10.5-33}$$

from which $$h_{j+1,j} = |\tilde{\mathbf{q}}_{j+1}| \qquad \mathbf{q}_{j+1} = \frac{\tilde{\mathbf{q}}_{j+1}}{h_{j+1,j}} \tag{10.5-34}$$

Once we have determined $\mathbf{q}_n$, we compute h_{in}, $i = 1, \ldots, n$, using (10.5-32). The only point at which this procedure can break down is if some $h_{j+1,j}$ vanishes. However, our hypothesis on H precludes this, and the proof is complete.

This theorem implies that if we can find the first column of Q in (10.5-25), then we can find Q and H such that

$$A_1 Q = QH \tag{10.5-35}$$

with $H = A_3$ as long as no subdiagonal elements of A_3 vanish. Now, if p is not an eigenvalue of A_1, it can be shown that if A_1 has nonvanishing subdiagonal elements, so does A_3. On the other hand, if p happens to be an eigenvalue of A_1, then since, as we shall see below, we do not compute A_3 using the algorithm indicated by the proof of Theorem 10.17 but a more efficient algorithm based on matrix transformations, we shall find that the last two rows of A_3 will be in block triangular form, thus revealing the eigenvalues.

The double QR transformation derives a Hessenberg matrix H from A_1, which guarantees that $H = A_3$. From (10.5-26) the correct Q is that orthogonal matrix for which

$$QR = B(p) \qquad \text{or} \qquad R = Q^T B(p) \tag{10.5-36}$$

We are interested in the first column of Q or the first row of Q^T. Q^T is a matrix which converts $B(p)$ to upper triangular form, i.e., to R. Hence it can be derived as a product $P_{n-1} \cdots P_1$ of Householder transformations as shown in Sec. 9.9. Now, the first row of $Q^T = P_{n-1} \cdots P_1$, that is, the first *column* of Q, is the first row of P_1 {51}. Hence our problem reduces to finding the first row of P_1. But P_1 is the matrix which introduces zeros into the first column when triangularizing the real matrix $B(p)$ and it is determined solely by the first column of $B(p)$.

Because $B(p)$ is the product of two upper Hessenberg matrices, only the first three elements of its first column are nonzero. They are given by

$$\begin{aligned} b_{11} &= (a_{11} - p)(a_{11} - \bar{p}) + a_{12}a_{21} = a_{11}^2 - a_{11}(p + \bar{p}) + p\bar{p} + a_{12}a_{21} \\ b_{21} &= a_{21}[a_{11} + a_{22} - (p + \bar{p})] \qquad (10.5\text{-}37) \\ b_{31} &= a_{32}a_{21} \end{aligned}$$

Having computed b_{11}, b_{21}, and b_{31} from the values of A_1 and the shift p, we compute P_1 as in Sec. 9.9 by

$$P_1 = I - \frac{2\mathbf{u}\mathbf{u}^T}{\|\mathbf{u}\|_2^2} \qquad (10.5\text{-}38)$$

where

$$\mathbf{u} = [b_{11} + \text{sgn}\,(b_{11})(b_{11}^2 + b_{21}^2 + b_{31}^2)^{1/2} \quad b_{21} \quad b_{31} \quad 0 \quad \cdots \quad 0]^T$$

$$(10.5\text{-}39)$$

Consider now the matrix $P_1 A_1 P_1$ (recall that P_1 is orthogonal and symmetric). It can be reduced to Hessenberg form by a series of orthogonal similarity transformations using Householder transformations as in Sec. 10.4. Thus we have that

$$TP_1 A_1 P_1 T = H \qquad (10.5\text{-}40)$$

where T is the product of Householder matrices and H is Hessenberg. Now the first row of TP_1 is the first row of P_1 {51}, which is itself the first row of Q^T. Therefore, $P_1 T$ is the Q of Theorem 10.17. Thus, by (10.5-25), H is the matrix A_3 obtained by the two steps of the QR algorithm (10-5.24).

Now, while the double QR algorithm was originally derived to do two QR transformations with conjugate-complex shifts, it works just as well with two real shifts p_1 and p_2, where we replace $p + \bar{p}$ by $p_1 + p_2$ and $p\bar{p}$ by $p_1 p_2$ in (10.5-37). We shall therefore refer to a pair of shifts (p_1, p_2), where p_1 and p_2 are either both real or complex conjugates. One recommended strategy for choosing them is as follows. At each stage μ_k, ν_k are eigenvalues of T_k, the lower 2×2 matrix of A_k. Initially $p_1^{(1)} = p_2^{(1)} = 0$. If $|\mu_k - \mu_{k-2}| > \frac{1}{2}|\mu_k|$ and $|\nu_k - \nu_{k-2}| > \frac{1}{2}|\nu_k|$, then $p_1^{(k)} = p_1^{(k-2)}$, $p_2^{(k)} = p_2^{(k-2)}$. If $|\mu_k - \mu_{k-2}| \le \frac{1}{2}|\mu_k|$ and $|\nu_k - \nu_{k-2}| \le \frac{1}{2}|\nu_k|$, then $p_1^{(k)} = \mu_k$, $p_2^{(k)} = \nu_k$.

Otherwise, if $|\mu_k - \mu_{k-2}| \le \frac{1}{2}|\mu_k|$, $p_1^{(k)} = p_2^{(k)} = \text{Re } \mu_k$, while if $|\nu_k - \nu_{k-2}| \le \frac{1}{2}|\nu_k|$, $p_1^{(k)} = p_2^{(k)} = \text{Re } \nu_k$. The reasoning behind this strategy is as follows. If both eigenvalues of T_k have settled down to the extent that the magnitude of the ratio of the difference between an eigenvalue of T_k and the corresponding eigenvalue of T_{k-2} to the eigenvalue of T_k is less than $\frac{1}{2}$, then we use them as shifts. If neither eigenvalue has settled down, we use the previous shifts. If only one of the eigenvalues has settled down, we assume that it indicates the presence of a real eigenvalue and use a real shift.

To sum up, a complete step of the double QR algorithm is as follows:

1. Compute $p_1^{(k)} + p_2^{(k)}$, $p_1^{(k)} p_2^{(k)}$ based on the eigenvalues of T_k.
2. Compute b_{11}, b_{21}, and b_{31} using (10.5-37) with $p_1^{(k)}$, $p_2^{(k)}$ in place of p, $\bar{p}$, determine the matrix P_1 using (10.5-38) and (10.5-39) and compute $P_1 A_k P_1$.
3. Reduce $P_1 A_k P_1$ to Hessenberg form using Householder similarity transformations yielding A_{k+1}.

As soon as $a_{n,n-1}^{(k)}$ or $a_{n-1,n-2}^{(k)}$ is 0 to within machine accuracy, we deflate to a matrix of order $n-1$ or $n-2$, respectively and accept a_{nn} as a single eigenvalue or the pair of eigenvalues of T_k as a pair of eigenvalues of A. If, at any stage, one of the elements on the subdiagonal, say $a_{j,j-1}^{(k)}$, becomes effectively 0, we decompose A_k and continue working on the lower matrix of order $n-j$. One conservative criterion for accepting an off-diagonal element $a_{j,j-1}^{(k)}$, where j may also be n or $n-1$, as 0 is that

$$|a_{j,j-1}^{(k)}| \le \epsilon \min\left(|a_{jj}^{(k)}|, |a_{j-1,j-1}^{(k)}|\right) \tag{10.5-41}$$

where ϵ is our relative accuracy tolerance. Alternatively, we may use the less stringent test

$$|a_{j,j-1}^{(k)}| \le \epsilon\left(|a_{jj}^{(k)}| + |a_{j-1,j-1}^{(k)}|\right) \tag{10.5-42}$$

Since the matrix $B(p)$ is not a full matrix but is of a special form (in particular, its first column has only three nonzero elements), the matrix P_1 will also be of the following form

$$P_1 = \begin{bmatrix} x & x & x & \\ x & x & x & \bigcirc \\ x & x & x & \\ & \bigcirc & & I \end{bmatrix} \tag{10.5-43}$$

and $P_1 A_k P_1$ will be nearly Hessenberg. Hence, the transformation of $P_1 A_k P_1$ to Hessenberg form will be very efficient, and a double QR step will take only of the order of $5n^2$ multiplications, which is a very satisfactory result. If we add to this the fact that we deflate the matrix each time we find a single root or a pair of roots, and that convergence is quite rapid so that an average of fewer than two steps per eigenvalue is required, we can understand why the double QR algorithm has become the accepted method for finding the eigenvalues of Hessenberg matrices and, therefore, in view of Sec. 10.4, of general real nonsymmetric matrices.

Finally, we mention one apparent problem with the QR algorithm. If there are different roots of the same magnitude, we need not converge to a matrix of the form (10.5-16). In general, this does not cause any difficulties since the introduction of shifts will eliminate this problem. However, it is possible to find matrices for which the shift strategy given above will not work. Thus, the matrix

$$\begin{bmatrix} 0 & 0 & 0 & 0 & 1 \\ 1 & 0 & 0 & 0 & 0 \\ 0 & 1 & 0 & 0 & 0 \\ 0 & 0 & 1 & 0 & 0 \\ 0 & 0 & 0 & 1 & 0 \end{bmatrix} \tag{10.5-44}$$

which has the eigenvalues $\exp \frac{1}{2} i\pi r$, $r = 0, \ldots, 4$, is left unchanged by the double QR algorithm as given above. Therefore, any program implementing the QR algorithm should introduce a random shift if there is no convergence after a fixed number of iterations. This will usually suffice to bring about convergence.

Example 10.10 Apply the double QR algorithm to find all the eigenvalues of the following Hessenberg matrix $A = A_1$.

$$A_1 = \begin{bmatrix} 5.0 & -2.0 & -5.0 & -1.0 \\ 1.0 & 0.0 & -3.0 & 2.0 \\ & 2.0 & 2.0 & -3.0 \\ & & 1.0 & -2.0 \end{bmatrix}$$

The computations below were carried out to about 14 significant figures using the listed values of $p_1^{(k)}$ and $p_2^{(k)}$. We give the results rounded to 5 decimal places. As in Example 10.9, we give the transformation from A_1 to A_3 in detail. Subsequently, we give only the eigenvalues μ_k, ν_k of T_k, the pairs $(p_1^{(k)}, p_2^{(k)})$, and the matrices A_{k+2}, $k = 3, \ldots, 11$.

$$T_1 = \begin{bmatrix} 2.0 & -3.0 \\ 1.0 & -2.0 \end{bmatrix}$$

so that $\mu_1 = 1.0$, $\nu_1 = -1.0$. The initial values for $p_1^{(1)}$ and $p_2^{(1)}$ are $p_1^{(1)} = p_2^{(1)} = 0$.

Using (10.5-37), we compute, for these values of p_1 and p_2,

$$b_{11} = a_{11}^2 + a_{12}a_{21} = 23.0$$

$$b_{21} = a_{21}(a_{11} + a_{22}) = 5.0$$

$$b_{31} = a_{32}a_{21} = 2.0$$

From (10.5-39) we have

$$\mathbf{u} = [46.62202 \quad 5.0 \quad 2.0 \quad 0]^T$$

and from (10.5-38)

$$P_1 = \begin{bmatrix} .97368 & -.21167 & -.08467 & 0 \\ -.21167 & .97730 & -.00908 & 0 \\ -.08467 & -.00908 & .99637 & 0 \\ 0 & 0 & 0 & 1.0 \end{bmatrix}$$

so that
$$P_1 A_1 P_1 = \begin{bmatrix} 4.11828 & 2.76448 & 5.72860 & .80433 \\ .15331 & .43031 & -1.88167 & 2.19351 \\ -.24348 & 2.18233 & 2.45142 & -2.92260 \\ -.08467 & -.00908 & .99637 & -2.0 \end{bmatrix}$$

In order to reduce $P_1 A_1 P_1$ to Hessenberg form, we generate Householder matrices as in Sec. 10.3-3. We first compute

$$P_2 = \begin{bmatrix} 1.0 & 0 & 0 & 0 \\ 0 & -.51115 & .81181 & .15165 \\ 0 & .81181 & .56389 & -.15165 \\ 0 & .28230 & -.15165 & .94726 \end{bmatrix}$$

and then
$$P_2 P_1 A_1 P_1 P_2 = \begin{bmatrix} 4.11828 & 3.46451 & 5.35253 & .67356 \\ -.29992 & .68720 & 3.69620 & -3.89738 \\ 0 & -.92577 & 1.05049 & .90434 \\ 0 & -.09024 & -.02781 & -.85597 \end{bmatrix}$$

Next, we compute

$$P_3 = \begin{bmatrix} 1.0 & 0 & 0 & 0 \\ 0 & 1.0 & 0 & 0 \\ 0 & 0 & -.99528 & -.09702 \\ 0 & 0 & -.09702 & .99528 \end{bmatrix}$$

and finally

$$P_3 P_2 P_1 A_1 P_1 P_2 P_3 = A_3 = \begin{bmatrix} 4.11828 & 3.46451 & -5.39263 & 1.51084 \\ -0.29992 & .68721 & -3.30064 & -4.23760 \\ & .93016 & 1.11718 & -.71200 \\ & & .22015 & -.92266 \end{bmatrix}$$

$$\mu_3 = 1.03720 \qquad \nu_3 = -.84268 \qquad p_1^{(3)} = \mu_3 \qquad p_2^{(3)} = \nu_3$$

$$A_5 = \begin{bmatrix} 3.88102 & 5.50657 & -2.83249 & -.98166 \\ .08301 & 2.29698 & -3.41894 & 3.17136 \\ & 1.65698 & -.18116 & 2.41387 \\ & & .01060 & -.99684 \end{bmatrix}$$

$$\mu_5 = -.15091 \qquad \nu_5 = -1.02709 \qquad p_1^{(5)} = p_2^{(5)} = \nu_5$$

$$A_7 = \begin{bmatrix} 4.01900 & .73616 & -6.17208 & .85558 \\ -.02762 & -.19762 & -3.26500 & -3.83089 \\ & 1.68124 & 2.17861 & 1.23705 \\ & & -\epsilon & -1.00000 \end{bmatrix}$$

$$\mu_7 = 2.17861 \qquad \nu_7 = -1.00000 \qquad p_1^{(7)} = p_2^{(7)} = \nu_7$$

$$A_9 = \begin{bmatrix} 4.01231 & 5.45684 & -3.02039 & -.83340 \\ -.00903 & 2.11330 & -3.33301 & 3.23455 \\ & 1.57244 & .12561 & 2.40441 \\ & & 0 & -1.00000 \end{bmatrix}$$

$$\mu_9 = 1.00191 + 1.99905i \qquad \nu_9 = \bar{\mu}_9 \qquad p_1^{(9)} = 0 \qquad p_2^{(9)} = 0$$

$$A_{11} = \left[\begin{array}{ccc:c} 3.99618 & 4.15400 & 4.64849 & * \\ .99182 & .37161 & 1.17200 & * \\ & -3.74868 & 1.63220 & * \\ \hdashline & & & -1.0 \end{array}\right]$$

$$\mu_{11} = 1.11946 + 2.06234i \qquad \nu_{11} = \bar{\mu}_{11} \qquad p_1^{(11)} = \mu_{11} \qquad p_2^{(11)} = \nu_{11}$$

$$A_{13} = \left[\begin{array}{ccc:c} 4.00000 & 5.04835 & -3.65643 & * \\ 0 & 1.87894 & -3.59100 & * \\ & 1.32902 & .12106 & * \\ \hdashline & & & -1.0 \end{array}\right]$$

A_{13} is in block diagonal form, and we compute from it the eigenvalues λ_i of A:

$$\lambda_1 = 4.0 \qquad \lambda_2 = 1.0 + 2.0i \qquad \lambda_3 = 1.0 - 2.0i \qquad \lambda_4 = -1.0$$

Note that since we deflated A_9, we chose $p_1^{(9)} = p_2^{(9)} = 0$.

10.6 ERRORS IN COMPUTED EIGENVALUES AND EIGENVECTORS

In most of the methods considered above, we do not compute the eigenvalues directly from the given matrix A but from a similar matrix B, to which we have transformed A. In the course of the computation of B from A, roundoff errors enter, so that B is not strictly similar to A. However, by backward error analysis, it can be shown that for all the methods discussed in this chapter B is similar to $A + \delta A$, where the elements of δA are small. Thus, in the reduction of A to Hessenberg or tridiagonal form by Householder transformations using t-digit floating-point binary arithmetic, it can be shown that

$$\|\delta A\|_E \le \gamma n^2 2^{-t} \|A\|_E \tag{10.6-1}$$

where γ is a constant of order 1. If inner products are accumulated to double precision, the bound (10.6-1) becomes

$$\|\delta A\|_E \le \gamma n 2^{-t} \|A\|_E \tag{10.6-2}$$

The important question to consider is the effect of perturbations of the elements of A on its eigenvalues (cf. Sec. 8.13).

Let us first consider the symmetric case, where both A and B, and hence also δA, are symmetric. If the eigenvalues of these matrices are α_i, β_i, and δ_i, respectively, arranged in nonincreasing order, it can be shown that

$$\alpha_i + \delta_n \le \beta_i \le \alpha_i + \delta_1 \tag{10.6-3}$$

If the elements of δA are all less than ϵ in magnitude, we have further that

$$-n\epsilon \leq \delta_n \leq \delta_1 \leq n\epsilon \tag{10.6-4}$$

so that

$$-n\epsilon \leq \beta_i - \alpha_i \leq n\epsilon \tag{10.6-5}$$

These results hold even when there are multiple eigenvalues, and so it follows that the eigenvalue problem for a symmetric matrix is always well-conditioned.

In the nonsymmetric case, we may have ill conditioning. Let α_i be a simple eigenvalue of A and $\mathbf{x}_i$ and $\mathbf{y}_i$ the corresponding right and left eigenvectors normalized so that $\|\mathbf{x}_i\|_2 = \|\mathbf{y}_i\|_2 = 1$. Then as δA tends to the null matrix, $A + \delta A$ has an eigenvalue $\alpha_i + \delta\alpha_i$ such that

$$\delta\alpha_i \sim \frac{\mathbf{y}_i^T \delta A \mathbf{x}_i}{\mathbf{y}_i^T \mathbf{x}_i} \tag{10.6-6}$$

Thus, for δA small enough,

$$|\delta\alpha_i| \leq \frac{\|\delta A\|_2}{|\mathbf{y}_i^T \mathbf{x}_i|} \tag{10.6-7}$$

where $\mathbf{y}_i^T \mathbf{x}_i$ is the cosine of the angle θ_i between $\mathbf{y}_i$ and $\mathbf{x}_i$. Matrices exist with simple eigenvalues for which the $\cos\theta_i$ are arbitrarily small, and any such eigenvalue is very sensitive to perturbations in the elements. Still, if $\|\delta A\|_2 \leq \epsilon$, we have that $|\delta\alpha_i| \leq \epsilon/|\cos\theta_i|$ and the right-hand side is linear in ϵ. If α_i is not a simple root, the situation may be worse. For example, the matrix

$$\begin{bmatrix} a & 0 \\ 1 & a \end{bmatrix} \tag{10.6-8}$$

has the double eigenvalue $\alpha_1 = \alpha_2 = a$, while the perturbed matrix

$$\begin{bmatrix} a & \epsilon \\ 1 & a \end{bmatrix} \tag{10.6-9}$$

has the eigenvalues $a + \epsilon^{1/2}$.

We see that for a particular matrix A, some eigenvalues may be sensitive while others are not. For simple eigenvalues, it depends on the angle between the corresponding right and left eigenvectors. In case $\cos\theta_i$ is not small, α_i is well-determined and a good algorithm should give accurate results.

The sensitivity of the eigenvectors of a matrix with respect to perturbations in its elements is a more complicated problem. If A has distinct eigenvalues, the eigenvector $\mathbf{x}_i'$ of $A + \delta A$ is such that as $\|\delta A\| \to 0$

$$\mathbf{x}_i' - \mathbf{x}_i \sim \sum_{k \neq i} (\mathbf{y}_k^T \delta A \mathbf{x}_i)\mathbf{x}_k/(\alpha_i - \alpha_k)\cos\theta_k \tag{10.6-10}$$

In the symmetric case, all the $\cos \theta_k = 1$, so that the sensitivity of the eigenvector $\mathbf{x}_i$ is dependent only on the closeness of α_i to other eigenvalues. In the nonsymmetric case, the factors $\cos \theta_k$ make things more complicated. However, if none of the $\cos \theta_i$ is small, the behavior is much the same as for symmetric matrices.

BIBLIOGRAPHIC NOTES

The literature on the calculation of eigenvalues and eigenvectors is almost as vast as that on simultaneous linear equations. The best general source for the material of this chapter is the book by Wilkinson (1965). Some good classical sources are Bodewig (1959), Fadeeva (1959), and Householder (1953). More recent sources are Faddeev and Faddeva (1963), Fox (1965), Stewart (1973) and Gourlay and Watson (1973). A compendium of fully documented Algol programs which implement almost all the methods discussed in this chapter, in addition to some not treated here, is given in Wilkinson and Reinsch (1971). The corresponding Fortran programs appear in Smith et al. (1974). A good summary comparing various methods is given by Wilkinson (1966). An extensive survey of the theoretical aspects of the calculation of eigenvalues and eigenvectors as well as an excellent bibliography is contained in Householder (1964).

Section 10.1 The material in this section is all classical. Good sources are Bodewig (1959) and Householder (1953), both of which contain extensive bibliographies.

Section 10.2 Our major source for the material of this section is Bodewig (1959). Much of the material can also be found in Faddeeva (1959). Wilkinson's technique can be found in Wilkinson (1955). The Rayleigh quotient appears in a number of areas of numerical analysis; see, for example, Kopal (1961). The generalized Rayleigh iteration is studied in great detail in a series of papers by Ostrowski (1958–1959). Computational aspects of inverse iteration are discussed in Ortega (1967).

Section 10.3 The various methods for computing eigensystems of symmetric matrices are discussed at length by Schwarz et al. (1973). There is a large literature on the Jacobi method. For a discussion emphasizing computational aspects, see Greenstadt (1960). The reformulation (10.3-19) of the Jacobi equations (10.3-11) to (10.3-13) is given in the contribution by Rutishauser in Wilkinson and Reinsch (1971). Pope and Tompkins (1957) discuss the threshold Jacobi method {24}. An error analysis is given by Goldstine, Murray, and Von Neumann (1959). Wilkinson (1962) analyzes the errors in the Jacobi method as well as those in other methods based on orthogonal transformations. For original papers on the other methods of this section, see Givens (1954) and Householder and Bauer (1959). The use of the method of bisection is considered by Wilkinson (1959*a*) and Ortega (1967). The analysis of the recurrence (10.3-40) is due to Kahan (1966). Householder's method, especially in its computational aspects, is carefully explained by Wilkinson (1960). Wilkinson (1958b) illustrates the problems involved in computing the eigenvectors of tridiagonal matrices.

Section 10.4 The biorthogonalization technique is due to Lanczos (1950). Wilkinson (1958*a*) discusses some computational aspects of the method. The use of Gaussian elimination to reduce a matrix to Hessenberg form is due to Wilkinson (1959*b*). The technique to find the eigenvalues of such a matrix is due to Hyman (1957). Wilkinson (1959*a*) discusses some computational aspects of Hyman's and other methods. See Greenstadt (1955) for a discussion of the Jacobi method for nonsymmetric matrices. White (1958) considers all these methods and contains an extensive bibliography. Goldstine and Horwitz (1959) have generalized Jacobi's method to normal matrices. The norm-reducing method appears in Eberlein (1962). Algorithms implementing this method are given in the contributions by Eberlein and Boothroyd and by Eberlein in Wilkinson and Reinsch (1971).

Section 10.5 The *LR* transformation method is due to Rutishauser (1955). The best paper on this method in English is by Rutishauser (1958). The *QR* transformation is due to Francis (1961). Parlett (1964*a*) gives an excellent exposition of both transformations. The computer implementation of the double *QR* algorithm is discussed in Parlett (1967), which is the source of Theorem 10.15 and Example 10.8. Stewart (1973) contains a good treatment of both the single and double *QR* algorithms, including explicit algorithms for both methods. For the application of the *QR* algorithm to the computation of eigenvectors, see Stewart (1973) and the contribution of Peters and Wilkinson in Wilkinson and Reinsch (1971).

Section 10.6 Chapter 2 of Wilkinson (1965) discusses in detail the perturbation theory of eigenvalues and eigenvectors for both symmetric and nonsymmetric matrices. The bounds (10.6-1) and (10.6-2) are from Stewart (1973). For an extensive and excellent discussion of errors in eigenvalue computations see Wilkinson (1964).

BIBLIOGRAPHY

Birkhoff, G., and S. MacLane (1953): *A Survey of Modern Algebra*, rev. ed., The Macmillan Company, New York.

Bodewig, E. (1959): *Matrix Calculus*, 2d ed., Interscience Publishers, Inc., New York.

Eberlein, P. J. (1962): A Jacobi-like Method for the Automatic Computation of Eigenvalues and Eigenvectors of an Arbitrary Matrix, *J. Soc. Ind. Appl. Math.*, vol. 10, pp. 74–88.

Faddeev, D. K., and V. N. Faddeeva (1963): *Computational Methods of Linear Algebra*, (trans. R. C. Williams), W. H. Freeman and Company, San Francisco.

Faddeeva, V. N. (1959): *Computational Methods of Linear Algebra* (trans. C. D. Benster), Dover Publications, Inc., New York.

Fox, L. (1965): *An Introduction to Numerical Linear Algebra*, Oxford University Press, New York.

Francis, J. G. F. (1961): The QR Transformation: A Unitary Analogue to the LR Transformation, *Comput J.*, vol. 4, pp. 265–271, 332–345.

Givens, W. (1954): Numerical Computation of the Characteristic Values of a Real Symmetric Matrix, Oak Ridge Nat. Lab. Rep. ORNL 1574.

——— (1958): Computation of Plane Unitary Rotations Transforming a General Matrix to Triangular Form, *J. Soc. Ind. Appl. Math.*, vol. 6, pp. 26–50.

Goldstine, H. H., and L. P. Horwitz (1959): A Procedure for the Diagonalization of Normal Matrices, *J. Ass. Comput. Mach.*, vol. 6, pp. 176–195.

———, F. J. Murray and J. Von Neumann (1959): The Jacobi Method for Real Symmetric Matrices, *J. Ass. Comput. Mach.*, vol. 6, 59–96.

Gourlay, A. R., and G. A. Watson (1973): *Computational Methods for Matrix Eigenproblems*, John Wiley & Sons, London.

Greenstadt, J. (1955): A Method for Finding Roots of Arbitrary Matrices, *MTAC*, vol. 9, pp. 47–52.

——— (1960): The Determination of the Characteristic Roots of a Matrix by the Jacobi Method, pp. 84–91 in *Mathematical Methods for Digital Computers* (A. Ralston and H. S. Wilf, eds.), John Wiley & Sons, Inc., New York.

Householder, A. S. (1953): *Principles of Numerical Analysis*, McGraw-Hill Book Company, New York.

——— (1964): *The Theory of Matrices in Numerical Analysis*, Blaisdell Publishing Company, New York.

——— and F. L. Bauer (1959): On Certain Methods for Expanding the Characteristic Polynomial, *Numer. Math.*, vol. 1, pp. 29–37.

Hyman, M. (1957): Eigenvalues and Eigenvectors of General Matrices, presented at *12th Ann. Meet. Ass. Comput. Mach., June 1957, Houston, Tex.*

Kahan, W. (1966): Accurate Eigenvalues of a Symmetric Tridiagonal Matrix, *Comput. Sci. Dept. Stanford Univ. Tech. Rep.* CS41.

Kopal, Z. (1961): *Numerical Analysis*, 2d ed., John Wiley & Sons, Inc., New York.

Lanczos, C. (1950): An Iteration Method for the Solution of the Eigenvalue Problem of Linear Differential and Integral Operators, *J. Res. Natl. Bur. Std.*, vol. 45, pp. 255–282.

Ortega, J. M. (1967): The Givens-Householder Method for Symmetric Matrices, pp. 94–115 in *Mathematical Methods for Digital Computers* (A. Ralston and H. S. Wilf, eds.), vol. II, John Wiley & Sons, Inc., New York.

——— and H. F. Kaiser (1963): The LL^T and *QR* Methods for Symmetric Tridiagonal Matrices, *Comput. J.*, vol. 6, pp. 99–101.

Ostrowski, A. M. (1958–1959): On the Convergence of the Rayleigh Quotient Iteration for the Computation of Characteristic Roots and Vectors, *Arch. Ration. Mech. Anal.*, vol. 1, pp. 233–241, vol. 2, pp. 423–428, vol. 3, pp. 325–340, 341–347, 372–481, vol. 4, pp. 153–165.

Parlett, B. (1964*a*): The Development and Use of Methods of *LR* Type, *SIAM Rev.*, vol. 6, pp. 275–295.

——— (1964*b*): Laguerre's Method Applied to the Matrix Eigenvalue Problem, *Math. Comput.*, vol. 18, pp. 464–485.

——— (1965): Convergence of the *QR* Algorithm. *Numer. Math.*, vol. 7, pp. 187–193.

——— (1967): The *LU* and *QR* Algorithms, pp. 116–130 in *Mathematical Methods for Digital Computers* (A. Ralston and H. S. Wilf, eds.), vol. II, John Wiley & Sons, Inc., New York.

Pope, D. A., and C. Tompkins (1957): Maximizing Functions of Rotations, *J. Ass. Comput. Mach.*, vol. 4, pp. 459–466.

Rutishauser, H. (1955): Une méthode pour la détermination des valeurs propres d'une matrice, *CR Acad. Sci. Paris*, vol. 240, pp. 34–36.

——— (1958): Solution of Eigenvalue Problems with the *LR* Transformation in *Further Contributions to the Solution of Simultaneous Linear Equations and the Determination of Eigenvalues*, vol. 49, pp. 47–81, National Bureau of Standards Applied Mathematics Series.

Schwarz, H. R., H. Ruthishauser, and E. Stiefel (1973): *Numerical Analysis of Symmetric Matrices* (trans. P. Hertelendy), Prentice-Hall, Inc., Englewood Cliffs, N.J.

Smith, B. T., J. M. Boyle, B. S. Garbow, Y. Ikebe, V. C. Klema, and C. B. Moler (1974): Matrix Eigensystem Routines: EISPACK Guide, *Lect. Notes Comput. Sci.*, vol. 6, Springer-Verlag, New York.

Stewart, G. W. (1973): *Introduction to Matrix Computations*, Academic Press, Inc., New York.

White, P. A. (1958): The Computation of Eigenvalues and Eigenvectors of a Matrix, *J. Soc. Ind. Appl. Math.*, vol. 6, pp. 393–437.

Wilkinson, J. H. (1955): The Use of Iterative Methods of Finding the Latent Roots and Vectors of Matrices, *MTAC*, vol. 9, pp. 184–191.

——— (1958*a*). The Calculation of Eigenvectors by the Method of Lanczos, *Comput. J.*, vol. 1, pp. 148–152.

——— (1958*b*): The Calculation of the Eigenvectors of Codiagonal Matrices, *Comput. J.*, vol. 1, pp. 90–96.

——— (1959*a*): Error Analysis of Floating-Point Computation. *Numer. Math.*, vol. 2, pp. 319–340.

——— (1959*b*): Stability of the Reduction of a Matrix to Almost Triangular and Triangular Forms by Elementary Similarity Transformations, *J. Ass. Comput. Mach.*, vol. 6, pp. 336–359.

——— (1959*c*): The Evaluation of Zeros of Ill-conditioned Polynomials, *Numer. Math.*, vol. 1, pp. 150–166, 167–180.

——— (1960): Householder's Method for the Solution of the Algebraic Eigenproblem, *Comput. J.*, vol. 3, pp. 23–27.
——— (1962): Error Analysis of Eigenvalue Techniques Based on Orthogonal Transformations, *J. Soc. Ind. Appl. Math.*, vol. 10, pp. 162–195.
——— (1964): *Rounding Errors in Algebraic Processes*, Prentice-Hall, Inc., Englewood Cliffs, N.J.
——— (1965): *The Algebraic Eigenvalue Problem*, Oxford University Press, Fair Lawn, N.J.
——— (1966): Calculation of Eigensystems of Matrics, pp. 27–61 in *Numerical Analysis: An Introduction* (J. Walsh, ed.), Academic Press, Inc., London.
——— and C. Reinsch (1971): *Linear Algebra: Handbook for Automatic Computation*, vol. II, Springer-Verlag, Berlin.

PROBLEMS

Section 10.1

1 (*a*) Let $\mathbf{x}_i$ and $\mathbf{y}_j$ be, respectively, a right and left eigenvector of a matrix A. Prove that $\mathbf{x}_i$ and $\mathbf{y}_j$ are orthogonal if they correspond to distinct eigenvalues.

(*b*) Therefore, prove that the (right) eigenvectors of a symmetric matrix corresponding to distinct eigenvalues are orthogonal.

2 (*a*) Prove Theorems 10.1 and 10.2.

(*b*) Use Theorem 10.1 to prove Theorem 10.4 in the case where A is symmetric.

(*c*) Prove Theorem 10.4 when A has a number of independent eigenvectors equal to its order. [Ref.: Birkhoff and MacLane (1953), pp. 306–307, and Bodewig (1959), pp. 59–60.]

3 Derive an algorithm for computing the coefficients of a polynomial of degree n given the sums of the first n powers of its zeros.

***4** Danilevsky's method. Let $A = [a_{ij}]$ be a matrix of order n.

(*a*) Suppose $a_{n,n-1} \neq 0$. Let M_{n-1} be the identity matrix of order n with its $(n-1)$st row replaced by $-a_{n1}/a_{n,n-1}, -a_{n2}/a_{n,n-1}, \ldots, 1/a_{n,n-1}, -a_{nn}/a_{n,n-1}$. Show that M_{n-1}^{-1} is the identity with its $(n-1)$st row replaced by the nth row of A.

(*b*) Show that $M_{n-1}^{-1}AM_{n-1}$ has zeros in the nth row except in the $(n-1)$st column, where there is a 1.

(*c*) Show that, by applying this technique $n-1$ times, if the element to the left of the diagonal term is not 0 at any stage we can reduce A to a similar matrix B of the form

$$B = \begin{bmatrix} p_1 & p_2 & \cdots & \cdots & p_n \\ 1 & 0 & \cdots & \cdots & 0 \\ & 1 & \ddots & & \vdots \\ & \bigcirc & \ddots & \ddots & \vdots \\ & & & 1 & 0 \end{bmatrix}$$

(*d*) Show that the characteristic equation of B is

$$P(\lambda) = (-1)^n(\lambda^n - p_1\lambda^{n-1} - p_2\lambda^{n-2} - \cdots - p_n) = 0$$

(*e*) Show that the number of multiplications and divisions required to calculate the characteristic equation is $(n-1)(n^2+n)$.

(*f*) Suppose the process has proceeded to the stage where the $k+1$ to n rows have been reduced to the desired form. Suppose also that the element in the $(k, k-1)$ position is 0 but that for some $j < k-1$ the term in the (k, j) position is not 0. How can the process be continued?

(*g*) Suppose now that the elements in the (k, j) positions for all $j < k$ are all 0. What can be done in this case? [Ref.: Fadeeva (1959), pp. 166–176.]

5 Let A be the matrix (cf. Example 10.7)

$$\begin{bmatrix} 2 & -2 & 3 \\ 1 & 1 & 1 \\ 1 & 3 & -1 \end{bmatrix}$$

(*a*) Use Gerschgorin's theorem to find a domain in which the eigenvalues must lie.

(*b*) Repeat part (*a*) using A^T in place of A.

(*c*) Let S be a diagonal matrix of order 3 with elements 1, 2, 2 on the diagonal. Calculate $B = SAS^{-1}$ and apply Gerschgorin's theorem to B^T.

(*d*) Generalize part (*c*) by indicating what happens to A when a diagonal matrix S with α in its last m positions and ones in the remaining positions is used to effect a similarity transformation on A.

(*e*) With $m = 2$, what value of α minimizes the total length of the domain on the real axis in which Gerschgorin's theorem says the eigenvalues of A must lie?

6 Prove that two similar matrices have the same trace.

7 (*a*) Let A and B be two matrices such that AB is defined. Let the ranks of A and B be $r(A)$ and $r(B)$. Prove that $r(AB) \leq \min\,[r(A), r(B)]$.

(*b*) Thus deduce that if A is nonsingular, then $r(AB) = r(B)$.

(*c*) Finally deduce that two similar matrices have the same rank.

***8** Generalized eigenvectors.

(*a*) Show that the number of eigenvectors that a matrix A has corresponding to an eigenvalue λ_i is equal to the number of different elementary divisors (10.1-42) in which λ_i appears.

(*b*) Suppose A has an elementary divisor of order ν corresponding to λ_i. Show that there exists a solution $\mathbf{x}_\nu$ of the system

$$(A - \lambda_i I)^\nu \mathbf{x} = \mathbf{0}$$

such that

$$(A - \lambda_i I)^{\nu-1} \mathbf{x}_\nu \neq \mathbf{0}$$

(*c*) If we define

$$(A - \lambda_i I)^j \mathbf{x}_\nu = \mathbf{x}_{\nu-j} \qquad j = 1, \ldots, \nu - 1$$

show that

$$(A - \lambda_i I)^j \mathbf{x}_j = \mathbf{0} \qquad j = 1, \ldots, \nu - 1$$

(The vectors $\mathbf{x}_j$, $j = 1, \ldots, \nu$, are called *generalized eigenvectors* of rank j corresponding to λ_i. The vector $\mathbf{x}_1$ is an eigenvector.)

(*d*) Prove that the vectors $\mathbf{x}_j$, $j = 1, \ldots, \nu$, are linearly independent.

Section 10.2

9 (*a*) Where does the derivation of the power method fail if A has nonlinear elementary divisors?

(*b*) Show that the matrix

$$A = \begin{bmatrix} 6 & 2 & 2 \\ -2 & 2 & 0 \\ 0 & 0 & 2 \end{bmatrix}$$

has nonlinear elementary divisors.

(*c*) Do 10 iterations of the power method on this matrix using (1, 1, 1) as the starting vector. Does the method seem to be converging?

10 (*a*) Suppose the dominant eigenvalue λ_1 of A is multiple and that the power method has been used to compute λ_1 and a corresponding eigenvector $\mathbf{x}_1$. In order to compute another eigenvector corresponding to λ_1, is it sufficient to choose an initial vector $\mathbf{v}_0$ which is orthogonal to $\mathbf{x}_1$? Why? Is it necessary that $\mathbf{v}_0$ be orthogonal to $\mathbf{x}_1$? Why?

(*b*) From this deduce a technique to calculate all the eigenvectors corresponding to a given eigenvalue.

(*c*) If it is not known a priori what the multiplicity of the eigenvalue is, what will happen when all the eigenvectors for a given eigenvalue have been found?

(*d*) Apply this technique to find all the eigenvectors corresponding to the dominant eigenvalue of

$$A = \begin{bmatrix} 7\frac{1}{8} & 2\frac{5}{8} & -1\frac{3}{4} \\ \frac{7}{8} & 5\frac{3}{8} & 1\frac{3}{4} \\ -1\frac{3}{4} & 5\frac{1}{4} & 4\frac{1}{2} \end{bmatrix}$$

11 The matrix B in part (*c*) of Prob. 4 is called the *companion matrix* of the polynomial $P(x)$ in part (*d*) of that problem.

(*a*) Show why the power method applied to the companion matrix is precisely equivalent to Bernoulli's method applied to $P(x)$.

(*b*) What operation applied to the companion matrix corresponds to Graeffe's method?

12 (*a*) Let $f(x)$ be a function of x with a Maclaurin expansion $F(x)$. If A is a square matrix, define $f(A) = F(A)$. What are the eigenvalues of $f(A)$? (Cf. Theorem 10.1.)

(*b*) Consider the iteration

$$\mathbf{v}_{m+1} = f(A)\mathbf{v}_m \qquad m = 1, 2, \ldots$$

where $\mathbf{v}_1$ is arbitrary. Generalize Theorem 10.12 for this iteration.

(*c*) If γ is a good approximation to an eigenvalue λ_1 of A, why should $f(x) = 1/(x - \gamma)$ be a good function to use in part (*b*)? But what happens if this function is actually used?

(*d*) But deduce from part (*c*) that for computational purposes using

$$f(A) = A^k + \gamma A^{k-1} + \cdots + \gamma^{k-1}A + \gamma^k I \qquad k \geq 1$$

should produce more rapid convergence than using A if λ_1 is the dominant eigenvalue. [Ref.: Bodewig (1959), pp. 322–323.]

13 (*a*) Use the technique of the previous problem with $k = 2$ and $\gamma = 2$ to find the dominant eigenvalue of the matrix of Example 10.1.

(*b*) Show the connection between the technique of the previous problem and Wilkinson's method.

14 (*a*) What does the iteration of part (*b*) of Prob. 12 converge to if $f(x) = e^x$? If $f(x) = e^{-x}$?

(*b*) Show how $f(x) = e^{-x}$ might be used to determine if a matrix is positive definite.

15 (*a*) Derive (10.2-13) from (10.2-12).

(*b*) Apply the δ^2 process to the calculation of Example 10.2 with $m = 5$. Using the results of Example 10.4, explain whether or not this result is just fortuitous.

16 (*a*) If the eigenvalues of A are all positive, show that the optimal value of p in Wilkinson's method is $(\lambda_2 + \lambda_n)/2$.

(*b*) What is the optimal choice for p in Example 10.1 given the results of Example 10.4?

17 Explain why you would expect the Rayleigh quotient to generally give better results than the δ^2 process.

18 Use the power method to calculate the dominant eigenvalue and corresponding eigenvector of

$$(a)\quad A_1 = \begin{bmatrix} 7 & 3 & -2 \\ 3 & 4 & -1 \\ -2 & -1 & 3 \end{bmatrix} \qquad (b)\quad A_2 = \begin{bmatrix} 3 & -4 & 3 \\ -4 & 6 & 3 \\ 3 & 3 & 1 \end{bmatrix}$$

Stop each iteration when three decimal places of the eigenvalue have stabilized.

19 (*a*) Use the δ^2 process, where applicable, to get an improved value of the eigenvalue for each part of the previous problem.

(*b*) Use Wilkinson's method with $p = 2$ to speed up the convergence of the power method for the matrix A_1 of the previous problem.

(*c*) Use the Rayleigh quotient to get an improved value of the eigenvalue for A_1 and A_2 in the previous problem.

20 Show that (10.2-19) holds for the inverse power method given the assumptions preceeding this equation.

21 (*a*) Let $\mathbf{v}_i = \mathbf{x}_j + \sum_{k=1}^{n}{}' \epsilon_{jk}\mathbf{x}_k$, where the eigenvectors of the symmetric matrix A are $\mathbf{x}_1$, ..., $\mathbf{x}_n$ normalized so that $\|\mathbf{x}_k\|_2 = 1$, $k = 1, \ldots, n$, and where $\sum'$ denotes that the index $k = j$ is omitted in the summation. Show that $\mathbf{v}_{i+1}$ given by (10.2-22) and (10.2-23) satisfies, apart from a normalizing factor, the relation

$$\mathbf{v}_{i+1} = \mathbf{x}_j + \frac{\sum'(\lambda_k - \lambda_j)\epsilon_{jk}^2}{1 + \sum'\epsilon_{jk}^2} \frac{\sum'\epsilon_{jk}\mathbf{x}_k}{\lambda_k - \mu_{i+1}}$$

so that all coefficients other than that of $\mathbf{x}_j$ are cubic in the ϵ_{jk}.

(*b*) Let A be a general matrix with n distinct eigenvalues $\lambda_1, \ldots, \lambda_n$ and corresponding right and left eigenvectors $\mathbf{x}_1, \ldots, \mathbf{x}_n$ and $\mathbf{y}_1, \ldots, \mathbf{y}_n$, respectively. Consider the following iteration starting with $\mathbf{v}_0$ and $\mathbf{w}_0$:

$$\mu_{i+1} = \frac{\mathbf{w}_i^T A \mathbf{v}_i}{\mathbf{w}_i^T \mathbf{v}_i} \qquad \mathbf{v}_{i+1} = k_{i+1}(A - \mu_{i+1}I)^{-1}\mathbf{v}_i \qquad \mathbf{w}_{i+1} = \hat{k}_{i+1}(A^T - \mu_{i+1}I)^{-1}\mathbf{w}_i$$

where k_{i+1} and $\hat{k}_{i+1}$ are normalizing factors chosen so that $\|\mathbf{v}_{i+1}\| = \|\mathbf{w}_{i+1}\| = 1$. Show that if $\mathbf{v}_i = \mathbf{x}_j + \sum'\epsilon_{jk}\mathbf{x}_k$ and $\mathbf{w}_i = \mathbf{y}_j + \sum'\eta_{jk}\mathbf{y}_k$, then

$$\mu_{i+1} = \frac{\lambda_j \mathbf{y}_j^T\mathbf{x}_j + \sum'\lambda_k\epsilon_{jk}\eta_{jk}\mathbf{y}_k^T\mathbf{x}_k}{\mathbf{y}_j^T\mathbf{x}_j + \sum'\epsilon_{jk}\eta_{jk}\mathbf{y}_k^T\mathbf{x}_k}$$

(*c*) Show that, apart from normalizing factors

$$\mathbf{v}_{i+1} = \mathbf{x}_j + (\lambda_j - \mu_{i+1})\textstyle\sum'\epsilon_{jk}\mathbf{x}_k/(\lambda_k - \mu_{i+1}) \qquad \mathbf{w}_{i+1} = \mathbf{y}_j + (\lambda_j - \mu_{i+1})\sum'\eta_{jk}\mathbf{y}_k/(\lambda_k - \mu_{i+1})$$

so that all coefficients other than those of $\mathbf{x}_j$ and $\mathbf{y}_j$ are cubic in the ϵ_{jk} and η_{jk}.

Section 10.3

22 Show that (10.3-19) is equivalent to (10.3-11) to (10.3-13).

23 Give a formal proof of the convergence of the Jacobi method to a diagonal matrix similar to A in the case where the off-diagonal element of greatest magnitude is annihilated at each stage. Why isn't this proof sufficient to show that the method converges for any choice of a nonzero off-diagonal element at each stage?

***24** The threshold Jacobi method.

(*a*) With v_0 given by (10.3-4) define $\gamma_1 = \sqrt{v_0}/\sigma$, where σ is a positive number called the *threshold constant*. If $\sigma \geq n$, show that there is at least one off-diagonal element in the original matrix A greater than or equal to γ_1 in magnitude.

(*b*) If all off-diagonal elements of magnitude greater than or equal to γ_1 are annihilated, show that the remaining sum of squares of the off-diagonal elements is no greater than $(1 - 2/\sigma^2)v_0$.

(*c*) Define $\gamma_{i+1} = \gamma_i/\sigma$, $i = 1, 2, \ldots$. Then at the ith stage of the method, all elements of magnitude greater than or equal to γ_i will be annihilated. Let $\{\gamma_{i_j}\}$ be the subsequence of $\{\gamma_i\}$

such that at the i_j stage at least one element is annihilated. Deduce that, after i_m stages, the sum of squares of the off-diagonal elements is no greater than $(1 - 2/\sigma^2)^m v_0$.

(*d*) Suppose we set an accuracy requirement that the final sum of squares of the off-diagonal elements should be less than $\rho^2 v_0$ where ρ is some constant. Show that this requirement will be satisfied if the final threshold γ_F is such that $\gamma_F \le (\rho/n)\sqrt{v_0}$. [Ref.: Greenstadt (1960).]

25 Do three stages of the threshold Jacobi method, in which at least one element is annihilated, for the matrix of Example 10.1 using $\sigma = 3$.

26 (*a*) Show that the sequence of orthogonal transformations defined by (10.3-32) annihilates the elements $a_{24}, a_{25}, \ldots, a_{2n}$ and does not affect the zeros in the first row obtained using (10.3-30).

(*b*) Thus deduce that the sequence of transformations (10.3-33) does indeed reduce A to the form B in (10.3-34).

27 (*a*) Derive (10.3-36).

(*b*) Show that if any c_i in (10.3-34) is 0, then the determinant of B can be written as the product of the determinants of two smaller tridiagonal matrices.

(*c*) Prove that the sequence defined by (10.3-36) is a Sturm sequence if no $c_i = 0$.

(*d*) Verify (10.3-38) when $i = 2$.

28 Derive the relationship between the eigenvectors of B in (10.3-34) and those of A.

***29** (*a*) Show that the matrix P defined by (10.3-42) is symmetric and orthogonal.

(*b*) Use (10.3-45) to (10.3-47) to show that A_k has zeros in the same positions in its first $k-2$ rows and columns as A_{k-1}.

(*c*) Show that with the elements of $\mathbf{v}$ chosen as in (10.3-49) and (10.3-50), A_k has zeros in the desired positions in the $(k-1)$st row and column and that (10.3-41) is satisfied.

30 Let $\mathbf{y}$ be an eigenvector of A_{n-1} given by (10.3-45).

(*a*) Show that the corresponding eigenvector $\mathbf{x}$ of A is given by

$$\mathbf{x} = P_2 P_3 \cdots P_{n-1}\mathbf{y}$$

How would you calculate $\mathbf{x}$ given $\mathbf{y}$?

(*b*) Compare this result with that of Prob. 28 to show that about half as much computation is required to compute the eigenvectors of A in Householder's method as in Givens' method. Assume that both $\mathbf{v}_k$ and $2\mathbf{v}_k$ are available from the computation of the eigenvalues (cf. Example 10.6).

(*c*) Derive the equation for A_k given in Example 10.6.

***31** (*a*) Verify that the total number of operations required in Givens' method to reduce the matrix to tridiagonal form is of the order of $\frac{4}{3}n^3$ and that $(n-2)(n-1)/2$ square roots must be calculated.

(*b*) Verify that the corresponding figures for Householder's method are $\frac{2}{3}n^3$ operations and $2n - 4$ square roots if the scheme of Example 10.6 is used.

(*c*) Show, however, that the $v_k^{(k)}$ need not be calculated explicitly and that therefore only $n - 2$ square roots are required. [Ref.: Wilkinson (1960).]

32 Carry through the computation of Example 10.6 using the minus instead of the plus sign in (10.3-49). Keep five decimals in all computations. Compare the results with those of Example 10.6 and explain the differences.

33 (*a*) Use the Jacobi method as described in Sec. 10.3-1 to find the eigenvalues of A_1 and A_2 of Prob. 18. Carry through five rotations. Why are the results for A_1 better than those for A_2?

(*b*) Use Givens' method to tridiagonalize A_1 and A_2 and then calculate the characteristic equation.

(*c*) Use Householder's method to tridiagonalize A_1 and A_2 and then calculate the characteristic equation.

34 (*a*) Show that if P_k and A_k are defined by (10.3-44) and (10.3-45), respectively, where now A_{k-1} is not necessarily symmetric, A_k can be computed by the following series of operations:

$$\mathbf{p}_k^T = \mathbf{v}_k^T A_{k-1} \qquad B_k = A_{k-1} - 2\mathbf{v}_k \mathbf{p}_k^T$$

$$\mathbf{q}_k = B_k \mathbf{v}_k \qquad A_k = B_k - 2\mathbf{q}_k \mathbf{v}_k^T$$

(*b*) Determine the number of operations needed to reduce a general matrix A to upper Hessenberg form using this formulation with the improvement given in Prob. 31*c*. [Ref.: Wilkinson (1966), p. 42.]

Section 10.4

35 (*a*) Use Lanczos' method to tridiagonalize the matrix of Example 10.1. Then calculate the characteristic equation and find its roots.

(*b*) How many operations are required in the application of Lanczos' method to a symmetric matrix of order n to reduce the matrix to tridiagonal form?

(*c*) Apply Lanczos' method to the matrix

$$A = \begin{bmatrix} 5 & 1 & -1 \\ -5 & 0 & 1 \\ 1 & 0 & 1 \end{bmatrix}$$

using $\mathbf{x}_1^T = (.6, -1.4, .3)$ and $\mathbf{y}_1^T = (.6, .3, -.1)$.

(*d*) Repeat the calculations of part (*c*) using $\mathbf{y}_1^T = \mathbf{x}_1^T = (.6, -1.4, .3)$. Then calculate the characteristic equation. [Ref.: Wilkinson (1958*a*).]

36 (*a*) Calculate the eigenvectors of the matrix T in Example 10.7.

(*b*) Use the results of part (*a*) to calculate the eigenvectors of A.

***37** Let A be an arbitrary matrix and let $\mathbf{x}_1$ and $\mathbf{y}_1$ be arbitrary vectors. Define two sequences of vectors

$$\mathbf{x}_{i+1} = A\mathbf{x}_i - \sum_{j=1}^{i} c_{ij}\mathbf{x}_j \qquad \mathbf{y}_{i+1} = A^T\mathbf{y}_i - \sum_{j=1}^{i} c_{ij}\mathbf{y}_j \qquad i = 1, 2, \ldots$$

(*a*) Show that

$$\mathbf{x}_{i+1} = P_i(A)\mathbf{x}_1 \qquad \text{and} \qquad \mathbf{y}_{i+1} = P_i(A^T)\mathbf{y}_1$$

where $P_i(A)$ is a polynomial of degree i in A.

(*b*) Thus deduce that $\mathbf{y}_{i+1}^T\mathbf{x}_j = \mathbf{y}_j^T\mathbf{x}_{i+1}$.

(*c*) Let the c_{ij} be chosen so that $\mathbf{y}_{i+1}^T\mathbf{x}_{i+1}$ is a minimum. Show that this requires that $0 = -\mathbf{y}_j^T\mathbf{x}_{i+1} - \mathbf{y}_{i+1}^T\mathbf{x}_j, j = 1, \ldots, i$.

(*d*) From parts (*b*) and (*c*) deduce that this minimum requirement means that the vectors $\mathbf{x}_i$ and $\mathbf{y}_i$ must form a biorthogonal sequence.

(*e*) Deduce then that those c_{ij} which give the minimum are given by $c_{ij} = \mathbf{y}_j^T A\mathbf{x}_i/\mathbf{y}_j^T\mathbf{x}_j$.

(*f*) Then deduce that $c_{ij} = 0, j < i - 1$.

(*g*) Finally deduce that these sequences of biorthogonal vectors are precisely those of Lanczos' method. Because of the requirement of part (*c*), Lanczos' method is therefore often called the method of minimized iterations. [Ref.: Lanczos (1950).]

38 (*a*) Display the elementary column matrices used in the reduction of a matrix A to Hessenberg form by Gaussian elimination with partial pivoting.

(*b*) Display the inverses of the matrices of part (*a*).

(*c*) Verify that the number of multiplications and divisions required for this method is $\frac{2}{3}n^3 + O(n^2)$ as opposed to $\frac{4}{3}n^3 + O(n^2)$ for Householder's method.

(*d*) Use this Gaussian elimination technique to derive the matrix B in Example 10.8. [Ref.: Wilkinson (1959*b*).]

39 Suppose an element $b_{i,i+1}$ above the principal diagonal in Fig. 10.2 is zero.

(*a*) Show that B can be written

$$B = \begin{bmatrix} B_1 & 0 \\ B_3 & B_2 \end{bmatrix} \}\, i \text{ rows}$$

where B_1 and B_2 are both in Hessenberg form.

(*b*) Show how the eigenvectors of B can be found from those of B_1 and B_2.

40 By taking the logarithmic derivation of $F_p(\lambda)$, that is, computing $d[\log F_p(\lambda)]/d\lambda$, show that (10.4-16), and consequently (10.4-17), hold.

41 (*a*) If all the eigenvalues of a matrix are real, does it follow from Corollary 10.2 of Sec. 10.1 that the matrix can be triangularized using a sequence of plane rotation matrices of the type (10.3-9)? Why? If the eigenvalues are real, is it always possible to annihilate any given off-diagonal element by an orthogonal transformation using plane rotation matrices?

(*b*) What modifications must be made in (10.3-11) to (10.3-15) for nonsymmetric matrices?

(*c*) Use plane rotation matrices to annihilate successively the largest element in the upper triangle of the matrix of Example 10.7. Do six iterations. [Ref.: Greenstadt (1955).]

42 (*a*) Show that if A is normal, then $AA^* = A^*A$.

(*b*) Show that if $AA^* = A^*A$, then the triangular matrix T in Theorem 10.10 is diagonal, so that A is normal.

Section 10.5

43 (*a*) Let $A = [a_{ij}]$ be a band matrix; that is, $a_{ij} = 0$ for $|i - j| > m$ for some $m < n - 1$. Prove that the LR transformation applied to A results in a sequence of matrices A_k all of which are band matrices with the same value of m.

(*b*) Prove that the QR transformation applied to a symmetric band matrix results in a sequence of symmetric band matrices with the same value of m.

(*c*) Show that if A is in upper Hessenberg form, the QR transformation results in a sequence of matrices A_k each of which is in upper Hessenberg form. [Ref.: Rutishauser (1958), p. 71, and Francis (1961).]

44 Show that if $\Lambda = \text{diag}(\lambda_1, \ldots, \lambda_n)$ and L is a unit lower triangular matrix with elements l_{ij}, $i > j$, then $\Lambda^k L \Lambda^{-k} = I + B_k$, where B_k is defined by (10.5-11).

45 Verify (10.5-20) and (10.5-21).

46 (*a*) Show that one cycle of the simple QR algorithm applied to a Hessenberg matrix results in a Hessenberg matrix and takes about $4n^2$ multiplications and $n - 1$ square roots.

(*b*) Show that in either the Hessenberg or tridiagonal cases a zero subdiagonal element allows decomposition of the problem into smaller ones.

47 Let A_k be a symmetric tridiagonal matrix. Show that R_k given by (10.5-22) has nonzero elements only on the main diagonal and on the two diagonals above the main diagonal. Derive algorithms for the computation of the elements of R_k and the elements of A_{k+1}. (These algorithms can be combined to form one compact algorithm for the QR transformation for symmetric tridiagonal matrices.) [Ref.: Ortega and Kaiser (1963).]

48 Apply the QR transformation to the two tridiagonal matrices obtained in Prob. 33*b*.

49 Apply the LR transformation directly to the matrices of Prob. 18, and compare these results with those of the previous problem.

50 Verify that (10.5-25) and (10.5-26) hold.

51 (*a*) What is the structure of the Householder matrix P_j which causes the subdiagonal elements of column j of a matrix to vanish?

(*b*) Show that pre- or postmultiplication of any matrix A by $P_j, j > 1$, leaves the first row and column of A unchanged.

(*c*) Thus show that the matrix T such that TBT, for any matrix B, is in Hessenberg form has the structure of P_2.

Section 10.6

52 (*a*) Assume that A has n distinct eigenvalues α_j and corresponding right and left eigenvectors $\mathbf{x}_j$ and $\mathbf{y}_j$, respectively, $j = 1, \ldots, n$. Show how (10.6-6) follows from the fact that the right and left eigenvectors of $A + \delta A$ corresponding to $\alpha_i + \delta\alpha_i$ can be written as

$$\mathbf{x}_i + \sum_{\substack{j=1 \\ j \neq i}}^{n} \epsilon_{ij} \mathbf{x}_j \qquad \text{and} \qquad \mathbf{y}_i + \sum_{\substack{j=1 \\ j \neq i}}^{n} \eta_{ij} \mathbf{y}_j$$

respectively, where the ϵ_{ij} and $\eta_{ij} \to 0$ as $\delta A \to 0$.

(*b*) Prove that for any vectors $\mathbf{x}$, $\mathbf{y}$ and any matrix A

$$|\mathbf{y}^T A \mathbf{x}| \leq \|\mathbf{y}\|_2 \|A\|_2 \|\mathbf{x}\|_2$$

53 (*a*) Let λ and $\mathbf{x}$ be approximations to an eigenvalue and eigenvector of a symmetric matrix A. If $\mathbf{r} = A\mathbf{x} - \lambda\mathbf{x}$, show that there exists an eigenvalue λ_i of A such that

$$|\lambda_i - \lambda|^2 \leq \mathbf{r}^T \mathbf{r} = \epsilon^2$$

(*b*) By considering

$$A = \begin{bmatrix} a & \epsilon \\ \epsilon & a \end{bmatrix} \qquad \lambda = a \qquad \text{and } \mathbf{x}^T = [1 \quad 0]$$

show that no useful bound on the error in the eigenvector can be obtained. [Ref.: Wilkinson (1959*c*, 1964).]

INDEX

Adams-Bashforth formula, 170, 187, 200
Adams-Moulton formula, 170, 188, 195, 200
Adams-type method, 170, 181
Adaptive integration, 126–130
Aitken's δ^2-process (*see* δ^2-process)
Algorithms, 39–42
Aliasing, (Prob. 37) 284
Approximating functions, classes of, 33
Approximation, 31–39
  exponential, (Prob. 34) 87
  of function by least squares techniques, 222–274
  interpolating, 34
  least squares (*see* Least squares approximation)
  linear, 33
  minimax (*see* Minimax approximation)
  polynomial, 34–39
  rational, 34, 287–320
    piecewise, 34
    (*See also* Padé approximations)
  types of, 32
Arithmetic:
  fixed-point, 12
  floating-point, 15
Asymptotic error constant, 336

Back substitution, 416
Backward difference, 58
Backward error analysis, 21
Bairstow's method, 374-376
Band-limited function, 39, (Prob. 38) 284
Band matrix, 466
Bandwidth, 466
Bernoulli numbers, 136
Bernoulli polynomials, 136
Bernoulli's method, 378–380
Bernstein polynomial, 36, 38–39
Bessel's interpolation formula, 63
Biorthogonal vector, 514
Bisection, method of, 40–42
Broyden's method, 366

Canonical form, 488
  Jordan, 492
Cauchy index, 368
Cauchy-Schwarz inequality, (Prob. 6) 25
Cayley-Hamilton theorem, 485
Central difference, 58
Characteristic equation, 485, 508
Chebyshev approximation, 286
Chebyshev expansion, 301–307
  and Fourier series, 307
  rational, 304
Chebyshev nodes, 66
Chebyshev norm, 286
Chebyshev polynomials, 109, 299–301, (Prob. 37) 478
  orthogonality property, 300, 302
  oscillation property, 301
  of second kind, 111
  shifted, (Prob. 23) 327
  zeros and extrema of, 300
Chebyshev quadrature, 109
Chebyshev's theorem on minimax approximations, 311–314
Cholesky factorization, 424
Christoffel-Darboux identity, 104
Companion matrix, (Prob. 11) 543
Complete set of functions, 35
Complex matrices, 465
Composite quadrature formula, 116
Composite rule, 116
Compound rule, 116

Computational efficiency, 336
Condition:
  of a matrix, 431
  of a problem, 22, 23
Consistency condition, 442
Consistent method, 169, 177
Continued fraction, (Prob. 5) 47, 292
Convergence:
  essential, 523
  of numerical integration methods, 176, 181
    rate of, 333
Convergence factor, 199
Correctors, 186–189
  Adams-Moulton, 188
Cost of computation, 337
Cost coefficient, reduced, 459
Cotes numbers, 119
Cramer's rule, 395
Crout algorithm, 422
Cubic spline, natural, 74, 77

Danilevsky's method, (Prob. 4) 541
Davidenko's method 363
Deflated polynomial, 372
Deflation, implicit, 334
Degeneracy:
  in rational approximations, 315
  in simplex method, 463
δ^2-process, 358–359, 446, 495
Derivative-estimated iteration formulas,
  350–352
Determinants, 465
Diagonally dominant matrix, 432
Difference equation, 174
Difference operators, (Prob. 9) 82
Differences, 57–61
  backward, 58
  central, 58
  divided, (Prob. 22) 85
  forward, 58
  table of, 59
Differential correction algorithm, 318–320
Differential equations:
  numerical solution of, 164–232
  (*See also* Numerical integration methods; Runge-Kutta methods)
Digit reversal, 266
Discrete Fourier transform, 263
Divided differences, (Prob. 22) 85
Doolittle algorithm, 421
Double precision arithmetic, 19

Eberlein's method, 520
Economization:
  of power series, 308–309
Economization:
  of rational functions, 309–311
Efficiency index, 337
Eigenvalues:
  of matrices: basic theorems on, 484
    bounds on, 486
    calculation of, 483–538
      errors in, 536–538
      power method for, 492–501
    location of, 486
  of nonsymmetric matrices: calculation of,
      513–521
    Jacobi-type methods, 520–521
    Lanczos' method, 514–516
    supertriangularization, 516–519
  of symmetric matrices: calculation of,
      501–513
    Givens' method, 506–511
    Householder's method, 511–513
    Jacobi method, 501–506
Eigenvectors, 483
  calculation of, 483–538
  errors in, 536–538
  generalized, (Prob. 8) 542
  of symmetric matrices, 501–513
Elementary divisor, 473
Elementary reflector, 452
Equal ripple polynomial, 301
Equilibration, 425–430
  for Crout algorithm, 427
  for Doolittle algorithm, 427
Error:
  absolute, 4
  accumulated, 178, 179
  definitions of, 4
  in eigenvalues, 536–538
  in functional evaluation, 5
  probable, 11
  propagated, bounds and estimates for,
    182–183, 215
  relative, 4, 438
  roundoff, 4, 9–12
    (*See also* Roundoff error)
  sources of, 2
  truncation, 3
    (*See also* Truncation error)
  types of, 3
Error analysis, 17, 20–22, 43–44
  backward, 21
  forward, 20
  roundoff, 9–12, 216, 434
  of solution of linear systems, 430–437
Error bound, propagated, 182
Error propagation, 166
Euclidean matrix norm, 8

Euclidean norm, 8
Euler-Maclaurin sum formula, 136
second, (Prob. 62) 161
Euler transformation, 143–145
Euler's constant, (Prob. 64) 161
Euler's method, 178, 188
Everett's interpolation formula, (Prob. 14) 83
Explicit integration formula, 167
Exponential approximation, (Prob. 34) 87
Extrapolation, 78
active, 227
passive, 227
Extrapolation method, 226
of numerical integration, 165

Factorial functions, 141
False position, method of, 338
Fast Fourier transform, 263–270
algorithm for, 265
digit reversal, 266
Sande-Tukey form, (Prob. 27) 281
Fibonacci sequence, 341, (Prob. 3) 472
Finite difference interpolation formulas, 61–63
Finite differences, 57–63
(*See also* Differences)
Fixed-point arithmetic, 12
Floating-point arithmetic, 15
Floating-point numbers, 13
Forward difference, 58
Forward integration formula, 167
Fourier functions, 33
Fourier transform, discrete, 263
Fourier's conditions, (Prob. 5) 401
Fraser diagram, 59
Frobenius norm, 8
Functional iteration, 334–337
computational aspects, 336–337
at a multiple root, 353–356
Functional iteration methods, 335
asymptotic error constant, 336
computational efficiency, 336
order of, 335
stationary, 335
Functionals, 42–46

Gauss' backward formula, 62
Gauss-Chebyshev quadrature, 109
Gauss' formula, (Prob. 63) 161
Gauss' forward formula, 62
Gauss-Hermite quadrature, 107
Gauss-Jacobi quadrature, 108
Gauss-Jordan reduction, 416
Gauss-Laguerre quadrature, 106
Gauss-Legendre quadrature, 101, 115
Gauss-Seidel method, 444–447
convergence of, 446
Gaussian elimination, 415–418, 517
compact forms for, 418-421
error analysis for, 430–437
matrix formulation, 418–421
Gaussian quadrature, 98–101
applied to singular integrals, 111–113
in composite formulas, 113–117
over infinite intervals, 105–108
and orthogonal polynomials, 104–105
weight functions in, 102–103
Gaussian quadrature formulas, 101, 133, 134
summary of, 114
Gear's method, 230–231
Generalized eigenvector, (Prob. 8) 542
Generating function, (Prob. 16) 49
Gerschgorin's theorem, 486
Givens' method, 506–511
Graeffe's method, 376–378
Gram polynomials, 259
Gram-Schmidt process, 256
Gregory's formula, 140

Halley's method, (Prob. 9) 401
Hamming's method, 194, 202–208
Hardy's rule, (Prob. 44) 158
Harmonic number, (Prob. 16) 50
Hermite interpolation, 70–73
Hermite interpolation formula, 72, 98, 188, 224
modified, 72
Hermite polynomials, 107
Hermitian matrix, 452
Hessenberg matrix, 466, 513
lower, 466
upper, 466
Hilbert matrix, 252
Horner's rule, 287, 291
Householder transformation, 452, 453, 527, 533
Householder's method, 511–513
Hurwitz-Routh criterion, (Prob. 16) 239

Ill-condition, 412–413
Ill-conditioned matrix, 252
Ill-conditioned polynomials, 395
Ill-conditioned problem, 22
Illinois method, 341
Implicit deflation, 334

Implicit integration formula, 168
Index of a rational function, 287
Induced norm, 7
Influence function, 43, 172
Inherent error, 21
Inner product, 18
Instability, 22
Interpolant, 66
Interpolation, 52–79
  at equal intervals, 56–63
  finite difference, 61–63
  Hermite, 70–73
  inverse, 68, 334, 348–350
    near a singularity, (Prob. 27) 86
  iterated, 66–68
  Lagrangian (*see* Lagrangian interpolation)
  near a singularity, (Prob. 25) 86
  polynomial, 52–79
  rational, (Prob. 37) 88
  spline, 73–78
  trigonometric, 271–273
  using reciprocal differences, (Prob. 16) 325
Interpolation formulas:
  finite difference, 61–63
  use of, 63–66
Interpolation series, (Prob. 19) 84
Interpolatory approximation, 34
Interval, changing the, 197
Inverse interpolation, 68, 334, 348–350
  near a singularity, (Prob. 27) 86
Inverse iteration, 499
Inverse power method, 498–501
Iterated interpolation, 66–68
Iterated vector, 493
Iteration:
  convergence of, 184
  functional (*see* Functional iteration)
  inverse, 499
Iteration formulas:
  computational efficiency of, 346
  derivative-estimated, 350–352
  by direct interpolation, (Prob. 19) 403
  multipoint, 347–353
  one-point, 344–347
  order of, 344
  using general inverse interpolation, 348–350
Iteration function:
  n-point, 335
  (*See also* Functional iteration)
Iterative formula, 168
Iterative process:
  acceleration of, 449–450
Iterative process:
  roundoff error in, 447–449
  stationary, 443–450
Iterative refinement, 437–440

Jacobi iteration, 443–444
Jacobi method, 501–506
  for nonsymmetric matrices, 520–521
  serial, 504
  threshold, 504
Jacobi polynomials, 108
Jacobian, 231–232
Jenkins-Traub method, 383–391, 499
Jordan canonical form, 492

Knot, spline, 73
Kronecker delta, 54
Krylov's method, 485
Kummer's method, 142

Lagrangian interpolation, 53–57
  at equal intervals, 56–57
  error in, 55
  formula for, 55, 118, 197
  used to generate iteration formulas, 334
Lagrangian interpolation polynomial, 55, 249, 274
Laguerre polynomials, 106
Laguerre's method, 380–383
Lanczos' method, 514–516
Least-squares, principle of, 248
Least squares approximation, 34, 247–274
  orthogonal polynomial, 254–260
  polynomial, 251–254
  using Fourier functions, 271–274
Least squares problem, 454
Left eigenvector, 483
Legendre polynomial, 99
Levenberg-Marquadt algorithm, 363
LeVerrier's method, 485
Linear approximation, 33
Linear equations:
  overdetermined systems of, 451–457
  solution of simultaneous, 410–468
    basic theorem on, 411
    with complex coefficients, 465
    direct methods for, 414–430
      (*See also* Gaussian elimination)
    error analysis of, 430–437
    iterative refinement, 437–440

Linear equations:
  solution of simultaneous: matrix iterative methods for, 440–442
    roundoff error analysis, 434
    sources of error in, 414
  underdetermined systems of, 458
Linear one-point matrix iteration, 441
Linear programming, 319
  simplex method for, 457–464
    basic solution for, 458
    degeneracy in, 463
    feasible solution for, 458
    standard form, 457
    tableau, 461
    unbounded solution of, 460
Lozenge diagram, 59
Lp norm, 7
LR algorithm, 521–522

Matrix:
  band, 466
  companion, (Prob. 11) 543
  complex, 465
  diagonally dominant, 432
  Hermitian, 452
  Hessenberg, 466, 513
    (*See also* Hessenberg matrix)
  normal, 520
  orthogonal, 452
  plane rotation, 502
  triangular, 419
  tridiagonal, 466, 506
Matrix inversion, 450–451
Matrix iterative methods for solution of linear systems, 440–442
  linear one-point, 441
Matrix norm, 8
Matrix powers, 498
Maximally stable formula, (Prob. 28) 241
Mehler quadrature, 108
Midpoint method, 188
  modified, 226
Midpoint rule, 120
Milne-type estimate, 189
Milne's method, 186
Minimax approximation, 34, 250, 286
  constructing, 315–320
  relative, 317
Minimum maximum error techniques, 285–320
  (*See also* Minimax approximation)
Muller's method, 373, (Prob. 21) 403, 518
Multipoint iteration formulas, 347–353
Multistep method, 165

Newton-based method for polynomials, 392–395
Newton-Cotes formulas, 118–126, 131, 132, 134, 170, 184, 187
  closed, 118
  composite, 121
  open, 120
  tables of, 120–121
Newton interpolation formula, 56
Newton-Raphson method, 230, 347, 355, 360, (Prob. 29) 477
  for multiple roots, 354
Newton's backward formula, 62
Newton's divided-difference formula, (Prob. 22) 85
Newton's forward formula, 61
Newton's identities, (Prob. 56) 408
Newton's method:
  damped, 361
  modified, 361
  quasi, 366
  (*See also* Newton-Raphson method)
Node, spline, 73
Nonlinear equations:
  solution of, 333–397
  solution of systems of, 359–367
Norm, 6–9, 431
  Euclidean, 8
  induced, 7
  matrix, 8
  spectral, 9
  subordinate, 7
  uniform, 7
  vector, 7
  weighted, 7
Normal equations, 250
  solution of, 251–253
Normal matrix, 520
Normalized numbers, 14
Null hypothesis, 254
Numerical differentiation, 89–95
  of data, 89–92
  of functions, 93–95
Numerical integration of data, 97–98
Numerical integration methods, 166–183
  based on higher derivatives, 224–226
  consistency of, 169
  convergence of, 176–178
  error estimation for, 189–191
  extrapolation-based, 226–228
  order of, 168
  self-starting, 195
  stability of, 173–182, 191–195
  truncation error in, 171–173

Numerical quadrature, 96–145
(*See also* Gaussian quadrature; Newton-Cotes formulas; Quadrature formulas)
Nystrom's method, 188
Nystrom's predictor, (Prob. 23) 240

Obrechkoff's method, (Prob. 3) 242
One-point iteration formulas, 344–347
Operator:
difference, (Prob. 9) 82
differentiation, (Prob. 3) 148
shifting, (Prob. 9) 82
Order:
of accuracy, 55, 96
of iteration, 338
Ordinary differential equations, 164–232
(*See also* Differential equations, numerical solution of)
Orthogonal matrix, 452
Orthogonal polynomials, 99, 104–105, 254–260
approximation with, 254–260
and Gaussian quadrature, 104–105
recurrence relations for, (Prob. 16) 151, 256–260
(*See also* Chebyshev polynomials; Gram polynomials; Hermite polynomials; Jacobi polynomials; Laguerre polynomials; Legendre polynomial)
Overdetermined systems of linear equations, 451–457
Overflow, 13, 19

Padé approximations, 293–299
error in, 295
use in economization of rational functions, 309–311
Parabolic rule, 1, 122, 127
Parasitic solution, 175
Parseval's theorem, (Prob. 24) 281
Peano kernel, 43, 172
Peano's theorem, 43
Pegasus method, 341
Perron-Frobenius theory, 446
Picard's method, 196
Pivot element, 425
Pivoting, 425–430
complete, 425
partial, 425
Plane rotation, 527
Plane rotation matrix, 502
Point-slope method, 188
Polynomial approximation, 34, 251
Polynomial equations (*see* Zeros of polynomials)
Power method, 492–501
acceleration of convergence of, 495–498
inverse, 498–501
Predictor-corrector methods, 183–195
compared with Runge-Kutta methods, 223–224
convergence of iterations in, 184–185
error estimation, 189–191
stability of, 191–195
use of, 198–200, 202–208
Predictors, 185–187, 224
Adams-Bashforth, 187
Hermite, 188
Probability density, 11, (Prob. 10) 26
Probability distribution, (Prob. 10) 26
Probable error, 11
Prony's method, (Prob. 34) 87, (Prob. 35) 88
Propagation of error, 166
in Runge-Kutta methods, 215
Purifying zeros, 372

QR algorithm, 521–536
double, 529–536
simple, 526–529
Quadratic factor algorithm, 287–291
Quadrature:
Gaussian (*see* Gaussian quadrature)
numerical, 96–145
(*See also* Gaussian quadrature; Newton-Cotes formulas; Quadrature formulas)
Quadrature formulas:
choosing, 130–135
composite, 113–117
Gaussian, 98–117
Newton-Cotes, 118–126
Quasi-Newton method, 366

Rational approximations, 34, 286–320
(*See also* Minimax approximation; Padé approximations)
Rational functions:
index of, 287
summation of, 141–143
Rational interpolation, (Prob. 37) 88
Rayleigh quotient, 497
Reciprocal derivative, (Prob. 17) 325
Reciprocal difference, (Prob. 14) 325
Recurrence relation, (Prob. 16) 151, 256–260
Regula falsi, 338

Relative error, 4, 179, 438
Relative stability, 180
Remes' second algorithm, 315
Residual, 248, 413
Richardson extrapolation, 94, 123, 227
Riemann sum, 117, 252
Right eigenvector, 483
Robustness, 130
Rodrigues formula, (Prob. 13) 151
Romberg integration, 123–126, 132, 133
Root-squaring, 376–378
Roundoff error, 4, 5, 9–12
  in iterative methods, 447–449
  probabilistic approach, 9–12
  in Runge-Kutta methods, 216
  in solution of linear systems, 433–437
  in solution of nonlinear equations, 334
Runge effect, 66
Runge-Kutta methods, 209–224
  compared to predictor-corrector methods, 223
  error estimation, 219
  fourth-order, 217
  higher-order, 218
  propagated errors in, 215
  roundoff error in, 216
  second-order, 216
  stability, 221
  third-order, 216
  truncation error in, 213

Sampling theorem, (Prob. 38) 284
Sande-Tukey algorithm, (Prob. 27) 281
Scaling, 13
Schur's theorem, (Prob. 17) 239
Schwarz constant, 493
Schwarz inequality, 25
Secant method, 338–344
  efficiency index for, 342
Second Euler-Maclaurin sum formula, (Prob. 62) 161
Self-starting method, 195
Serial Jacobi method, 504
Shifting operator, (Prob. 9) 82
Significant digit, 5
Simplex method, 457–464
  (*See also* Linear programming, simplex method for)
Simpson's rule, 120, 127, 186
Simultaneous linear equations (*see* Linear equations, solution of simultaneous)
Single step method, 165
Singular integrals, 111–113
Slack variable, 458
Spectral condition number, 431
Spectral norm, 9
Spectral radius, 8, 487
Spline, 73
  knot of, 73
  node of, 73
Spline interpolation, 73–78
Square root calculation, 31, (Prob. 10) 48, (Prob. 33) 329
Stability, 22
  of numerical integration methods, 173, 176
  of predictor-corrector methods, 191
  relative, 180
  of Runge-Kutta methods, 221
Stationary iteration, 335, 440
Stationary iterative process, 443–450
Steady-state solution, 229
Steepest descent, method of, 362
Steffensen's interpolation formula, (Prob. 14) 83
Stein-Rosenberg theorem, 446
Stiff equations, 228–232
Stiffness, 229
Stiffness ratio, 230
Stirling numbers of the second kind, (Prob. 67) 162
Stirling's interpolation formula, 63
Sturm sequence, 368–371, 508
  generalized, (Prob. 45) 406
Subordinate norm, 7
Subtractive cancellation, 18
Successive approximations, method of, 196
Successive overrelaxation, 449
Summation, 136–145
  by parts, (Prob. 11) 278
  of rational functions, 141–143
Supertriangular matrix (*see* Hessenberg matrix)
Supertriangularization, 516–519
Suppression, method of, 334
Synthetic division algorithm, 371

t-method, (Prob. 35) 329
Tabular point, 53
Taylor-series methods, 166, 195
Thiele's interpolation formula, (Prob. 16) 325, (Prob. 19) 326
Thiele's theorem, (Prob. 17) 326
Threshold constant, (Prob. 24) 544
Threshold Jacobi method, 504, (Prob. 24) 544
Toeplitz convergence, 125
Transient solution, 229
Trapezoidal rule, 1, 120, 124, 230

Triangular matrix, 419
Tridiagonal matrix, 75, 466, 506
Trigonometric interpolation, 271
Truncation error, 3
  local, 166
  in numerical integration, 171–173
  in Runge-Kutta methods, 213–215

Underdetermined system of linear equations, 458
Underflow, 19
Undetermined coefficients, method of, 44–46, 168
Uniform norm, 7, 286

Variable-order method, 188
Variable-order, variable-step methods, 201–202

Von Mises' theorem, 493

Weddle's rule, (Prob. 44) 157
Weierstrass theorem, 36–38, 134
Weight functions, 102–103, 248
Wilkinson's method, 496

Zeros of polynomials:
  computation of, 367–397
    classical methods, 371–383
    effect of coefficient errors, 395–397
  location of, 368
  purifying, 372
  (*See also* Bairstow's method; Bernoulli's method; Jenkins-Traub method; Laguerre's method; Newton-based method for polynomials; Root-squaring)